Roberta

SCIENCEPOWER™ 8

SCIENCE • TECHNOLOGY • SOCIETY • ENVIRONMENT

McGraw-Hill Ryerson SCIENCEPOWER™ Program

SCIENCEPOWER™ 7

SCIENCEPOWER™ 8

SCIENCEPOWER™ 9

SCIENCEPOWER™ 10

Chenelière/McGraw-Hill OMNISCIENCES Program

OMNISCIENCES 7

OMNISCIENCES 8

OMNISCIENCES 9

OMNISCIENCES 10

This program is available directly from Chenelière/McGraw-Hill.

Teacher Support for Each Grade Level

Teacher's Resource Binder
Blackline Masters
Computerized Assessment Bank
Web site: *http://www.mcgrawhill.ca*
Videotape series

Our cover Water droplets cling to a leaf. Why does this make the veins of the leaf appear larger? Try to answer this question now. After you have studied Units 1 and 3, see if you still agree with your answer.

SCIENCEPOWER™ 8

SCIENCE • TECHNOLOGY • SOCIETY • ENVIRONMENT

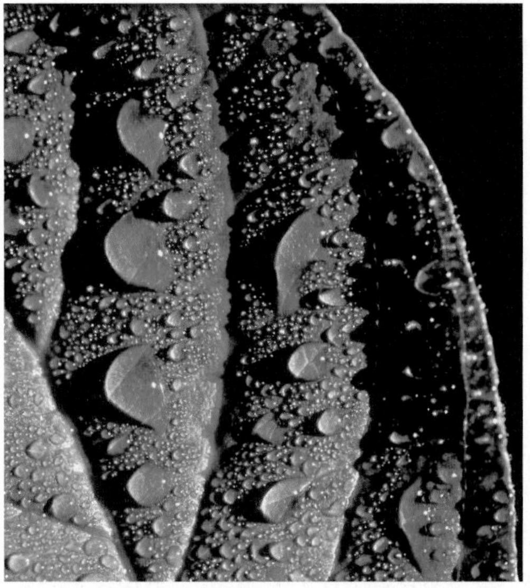

Author Team

Christina Clancy
Loyola Catholic Secondary School
Mississauga, Ontario

Lois Edwards
Professional Writer
Cochrane, Alberta

Eric Grace
Professional Writer
Victoria, British Columbia

Eric Brown
Formerly Assistant Head of Science at
Applewood Heights Secondary School
Mississauga, Ontario

Les Jolliffe
Formerly Assistant Head of Science at
Sir Wilfrid Laurier Secondary School
Gloucester, Ontario

Contributing Author

Sandy Wohl
Hugh Boyd Secondary School
Richmond, British Columbia

Senior Program Consultants

Malisa Mezenberg
Loyola Catholic Secondary School
Mississauga, Ontario

Douglas A. Roberts
University of Calgary
Calgary, Alberta

Toronto Montréal New York Burr Ridge Bangkok Bogotá Caracas
Lisbon London Madrid Mexico City Milan New Delhi
Seoul Singapore Sydney Taipei

COPIES OF THIS BOOK MAY BE OBTAINED BY CONTACTING:

McGraw-Hill Ryerson Ltd.

WEBSITE:
http://www.mcgrawhill.ca

E-MAIL:
Orders@mcgrawhill.ca

TOLL FREE FAX:
1-800-463-5885

TOLL FREE CALL:
1-800-565-5758

OR BY MAILING YOUR ORDER TO:
McGraw-Hill Ryerson
Order Department,
300 Water Street
Whitby, ON L1N 9B6

Please quote the ISBN and title when placing your order.

SCIENCEPOWER™ 8
Science • Technology • Society • Environment

0-07-560359-4

http//www.mcgrawhill.ca

4 5 6 7 8 9 0 GTC 00 09 08 07 06 05 04 03 02 01

Printed and bound in Canada

Canadian Cataloguing in Publication Data

Main entry under title:

Sciencepower 8: science, technology, society, environment

Includes index.

ISBN 0-07-560359-4

1. Science – Juvenile literature. I. Clancy, Christina. II Title: Sciencepower eight.

Q161.2.S385 1999 500 C99-931205-7

The SCIENCEPOWER™ Development Team

SCIENCE PUBLISHER: Trudy Rising
SENIOR DEVELOPMENTAL EDITOR: Jane McNulty
DEVELOPMENTAL EDITORS: Gerry De Iuliis, Tom Gamblin, Dan Kozlovic, Lauri Seidlitz
SENIOR SUPERVISING EDITOR: Nancy Christoffer
PROJECT CO-ORDINATORS: Kelli Legros, Crystal Shortt
ASSISTANT PROJECT CO-ORDINATOR: Janie Reeson
EDITORIAL ASSISTANT: Joanne Murray
FIELD-TEST CO-ORDINATOR: Jill Bryant
SPECIAL FEATURES: Jill Bryant, Gerry De Iuliis, Trudee Romanek, Elma Schemenauer
COPY EDITOR: Grace D'Alfonso
PERMISSIONS EDITOR: Jane Affleck
SENIOR PRODUCTION CO-ORDINATORS: Yolanda Pigden, Nicla Dattolico
COVER AND INTERIOR DESIGN: Pronk&Associates
ELECTRONIC PAGE MAKE-UP: Pronk&Associates
SET-UP PHOTOGRAPHY: Ian Crysler, Dave Starrett
SET-UP PHOTOGRAPHY CO-ORDINATOR: Jane Affleck
TECHNICAL ILLUSTRATIONS: Imagineering Scientific and Technical Artworks Inc./Pronk&Associates
ILLUSTRATIONS: Steve Attoe, Deborah Crowle, Tina Holdcroft, Jun Park, Dusan Petricic, Dino Pulera, Theresa Sakno, Margo Stahl, Bart Vallecoccia, Dave Whamond
COVER IMAGE: Mark Tomalty/Masterfile

Acknowledgements

Our ability to offer you this high-quality teaching resource is only possible thanks to the honest and frank feedback from the individual consultants and reviewers listed below. In particular, Erminia Pedretti's special interest in STSE (science, technology, society, and the environment), Sylvia Constancio's ESL (English as a second language) specialty and her interest in providing a resource for non-specialist as well as specialist teachers, and Dan Forbes' understanding of students' comprehension at different levels were all critical to the development of ***SCIENCEPOWER™ 8***. Likewise, each and every pedagogical reviewer across the country, and our highly experienced safety reviewer, brought different perspectives based on their special interests, their regions, and their own teaching expertise. The authors, editors, and publisher sincerely thank them all. We extend sincere thanks to Exclusive Educational Games of Excellence Inc. and Boreal Laboratories Ltd. for supplying us with equipment gratis for use in our photographs of students conducting various investigations.

Consultants

Sylvia Constancio
York University, Assessment of Science and Technology Achievement Project
ESL, FSL Specialist
Toronto, Ontario

Dan Forbes
St. Anne Elementary School
St. Anne, Manitoba

Erminia Pedretti
Ontario Institute for Studies in Education of the University of Toronto
STSE Specialist
Toronto, Ontario

Pedagogical and Academic Reviewers

Dave Bekkers
Green Glade Senior Public School
Mississauga, Ontario

Pat Bright
University of Victoria
Victoria, British Columbia

Philip Capstick
Coldbrook & District School
Coldbrook, Nova Scotia

Audrey Cook
George Street Middle School
Fredericton, New Brunswick

Mike Elson
Dresden Area Central School
Dresden, Ontario

Professor Greg Finn
Brock University
St. Catharines, Ontario

Jenni Foss
Hilltop Middle School
Toronto, Ontario

Stephen Haberer
Trinity College School
Port Hope, Ontario

Bruce Hickey
Brother Rice High School
St. John's, Newfoundland

Roy Hughes
Breton Educational Centre
New Waterford, Nova Scotia

Marleen Kacevychius
Malcolm Munroe Memorial Junior High School
Sydney, Nova Scotia

David Knox
Canterbury High School
Ottawa, Ontario

Brenda Kusmenko
Humber Summit Middle School
Toronto, Ontario

Mia MacIntyre
Malcolm Munroe Memorial Junior High School
Sydney, Nova Scotia

Greg Mazanik
St. Cornelius School
Caledon, Ontario

Sean Marks
River Oaks Public School
Oakville, Ontario

Cedric McGrath
École Le Tremplin
Tracadie-Sheila, New Brunswick

Bob Mealey
Riverview Middle School
Riverview, New Brunswick

Bob Moulder
Montclair Public School
Oakville, Ontario

Terry Quinlan
Charles P. Allen High School
Bedford, Nova Scotia

Bill Reynolds
Morning Glory Public School
Pefferlaw, Ontario

Hazen Savoie
École Dr Marquerite-Michaud
Buctouche, New Brunswick

Taunya Sheffield
Central Kings Rural High School
Cambridge Station, Nova Scotia

Alison Smith
Bayside Middle School
Saint John, New Brunswick

John Smith
Green Glade Senior Public School
Mississauga, Ontario

Jay Sugunan
Beaumonde Heights Junior Middle School
Etobicoke, Ontario

Lindsay Thierry
St. Michael's University Middle School
Victoria, British Columbia

Gary Turner
Cunard Junior High School
Halifax, Nova Scotia

Elgin Wolfe
Ontario Institute for Studies in Education of the University of Toronto
Toronto, Ontario

Safety Reviewer

Margaret Redway
Fraser Scientific & Business Services
Delta, British Columbia

Field Testing Acknowledgements

Imagine asking teachers, just at the end of a busy teaching year, if they would like to be involved in field-testing a new science resource you are creating, not in beautifully polished final form, but as photocopied first draft, with stick figures sketched in by authors; now, more challengingly, try to imagine the generosity of the response we received. The following teachers were recommended to us by their boards. We asked them for their help and they provided it willingly and patiently, sharing with us their enthusiasm, their frustrations, and above all their invaluable recommendations for improvements prior to publication. We sincerely thank them and their students for helping us, through months of testing, to develop the most useful possible resource for you and your students in teaching and learning the grade 8 science curriculum.

We wish to extend special thanks to Dave Bekkers, who wrote the *Project* for Unit 1; to Sean Marks, who wrote the *Projects* for Units 2, 3, and 4; to Annelies Groen, for her excellent activity ideas and advice; and to Bob Moulder, for his invaluable assistance in providing props for set-up photographs.

Field Test Teachers

Mary Anderson
Our Lady of Peace School
Maple, Ontario

Vijaya Balchandani
Humberwood Downs Academy
Etobicoke, Ontario

Dave Bekkers
Green Glade Senior Public School
Mississauga, Ontario

Karen Bowers
Maplehurst School
Burlington, Ontario

Brian Buttery
Terry Fox Public School
Cobourg, Ontario

Nancy Di Nardo
Our Lady of Peace School
Maple, Ontario

Mike Elson
Dresden Area Central School
Dresden, Ontario

Bob Friesen
Maplehurst School
Burlington, Ontario

Annelies Groen
Deer Park Public School
Toronto, Ontario

Suzy Hall
Trinity College School
Port Hope, Ontario

Mike Hanson
St. Andrew's College
Aurora, Ontario

Joanne Harris
Centennial Middle School
Georgetown, Ontario

Adriana Hayden
Terry Fox Public School
Cobourg, Ontario

Renée Jensen
Westminster Public School
Thornhill, Ontario

Brenda Kusmenko
Humber Summit Middle School
Toronto, Ontario

Ed Malabre
Lawrence Heights Middle School
North York, Ontario

Sean Marks
River Oaks Public School
Oakville, Ontario

Bob Moulder
Montclair Public School
Oakville, Ontario

Jay Sugunan
Beaumonde Heights Junior Middle School
Etobicoke, Ontario

Contents

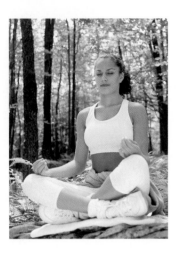

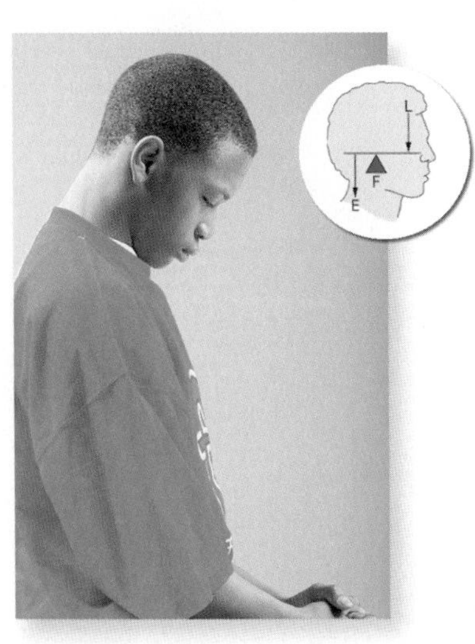

L
F
E

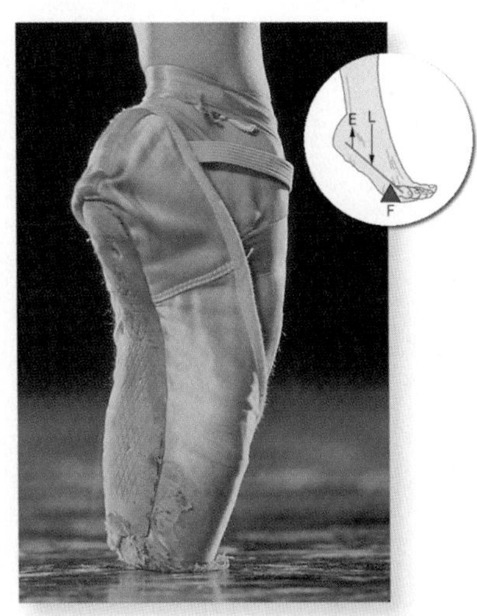

E
L
F

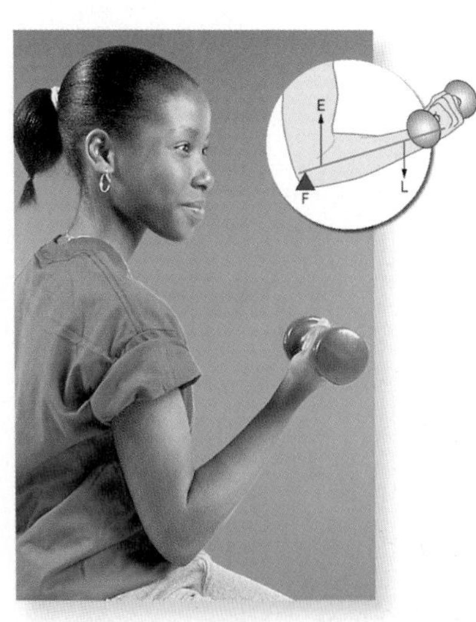

E
F
L

OF WINNIPEG
HYDRO
3

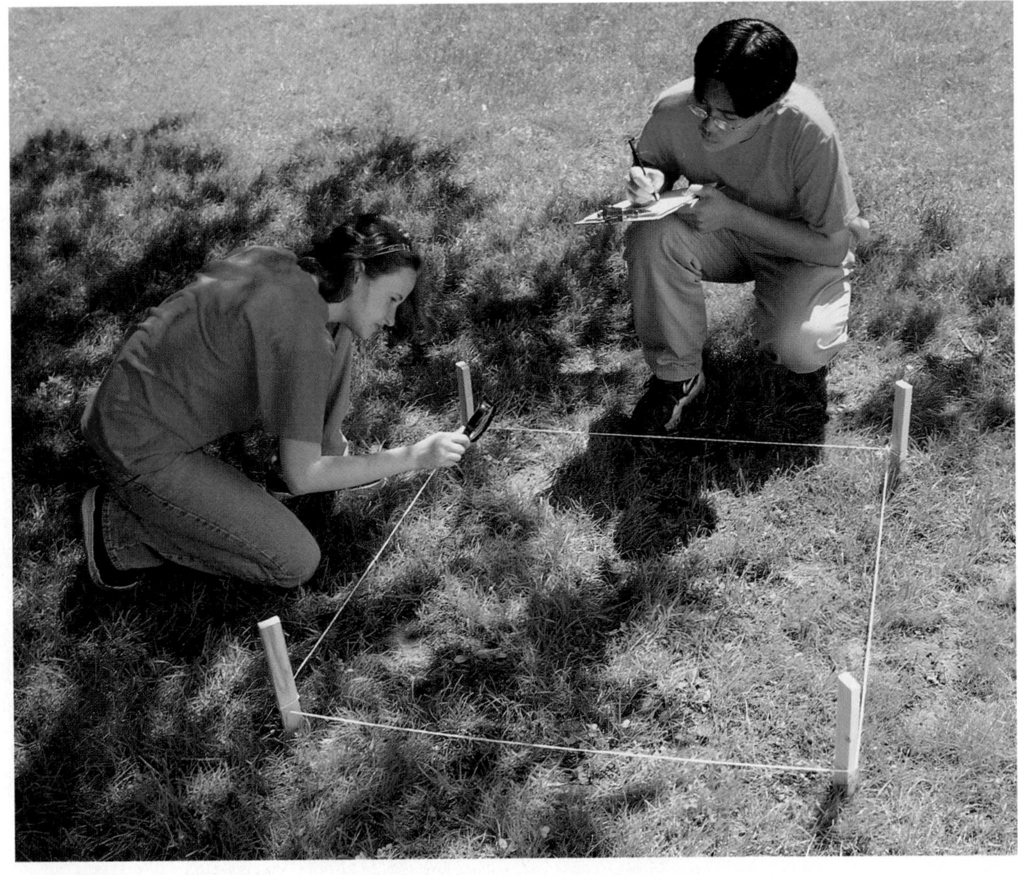

To the Teacher

We are very pleased to have been part of the team of experienced science educators and editors working together to bring you and your students this new program — the ***SCIENCEPOWER™ 7-10*** series of textbooks, and its French equivalent, ***OMNISCIENCES*** 7-10. The ***SCIENCEPOWER™*** and ***OMNISCIENCES*** student and teacher resources were specifically developed to provide 100 percent congruence with the new curriculum. As the titles ***SCIENCEPOWER™*** and ***OMNISCIENCES*** suggest, these resources are designed to foster an appreciation of the power of scientific explanation as a way of understanding our world and to empower students to critically examine issues and questions from a societal and environmental perspective.

SCIENCEPOWER™ 8/OMNISCIENCES 8 provide:

- A science inquiry emphasis, in which students address questions about the nature of science involving broad explorations as well as focussed investigations. Skill areas emphasized include: careful observing; questioning; proposing ideas; predicting; hypothesizing; making inferences; designing experiments; gathering, processing, and interpreting data; and explaining and communicating.
- A technological problem-solving emphasis, in which students seek answers to practical problems. Problem solving may either precede knowledge acquisition or provide students with opportunities to apply their newly acquired science knowledge in novel ways. Skill areas emphasized include: understanding the problem; setting and/or understanding criteria; developing a design plan, carrying out the plan; evaluating; and communicating.
- A societal decision-making emphasis, in which students draw upon those science and technology concepts and skills that will inform the question or issue under consideration. Students are encouraged to focus attention on sustainability and stewardship. Skill areas emphasized include: identifying the issue; identifying alternatives; researching, reflecting, and deciding; taking action; evaluating; and communicating.

The particular emphases within a unit are, in part, suggested by the topic itself. The primary and secondary emphases for ***SCIENCEPOWER™ 8*** and ***OMNISCIENCES 8*** are listed in the table opposite.

Scientific literacy has become the goal in science education throughout the world, and this goal has been given expression in Canada in the *Common Framework of Science Learning Outcomes, K-12: Pan-Canadian Protocol for Collaboration on School Curriculum* (Council of Ministers of Education, Canada, 1997).

> "Scientific literacy is an evolving combination of the science-related attitudes, skills, and knowledge students need to develop inquiry, problem-solving, and decision-making abilities, to become lifelong learners, and to maintain a sense of wonder about the world around them. To develop scientific literacy,

students require diverse learning experiences which provide opportunity to explore, analyze, evaluate, synthesize, appreciate, and understand the interrelationships among science, technology, society, and the environment that will affect their personal lives, their careers, and their future."

	***SCIENCEPOWER™ 8/ OMNISCIENCES* Unit**	**Primary Emphasis**	**Secondary Emphasis**
Life Systems Cells, Tissues, Organs, and Systems	Unit 1 Cells and Body Systems	Science and Science Inquiry	Societal Decision Making
Matter and Materials Fluids	Unit 2 Fantastic Fluids	Technology and Technological Problem Solving	Science and Science Inquiry
Energy and Control Optics	Unit 3 Light and Optical Instruments	Technology and Technological Problem Solving	Science and Science Inquiry
Earth and Space Systems Water Systems	Unit 4 Water Systems on Earth	Societal Decision Making	Science and Science Inquiry
Structures and Mechanisms Mechanical Efficiency	Unit 5 Mechanical Advantage and Efficiency	Technology and Technological Problem Solving	Science and Science Inquiry

Through varied text features, ***SCIENCEPOWER™ 8*** enables students to understand and develop skills in the processes of scientific inquiry, and in relating science to technology, society, and the environment.

Like the other textbooks in our series, ***SCIENCEPOWER™ 8*** builds on the three basic goals of the curriculum, and reflects the essential triad of knowledge, skills, and the ability to relate science to technology, society, and the environment (STSE). Science is approached both as an intellectual pursuit, and also as an activity-based enterprise operating within a social context.

Our extensive *Teacher's Resource Binder* provides essential planning and implementation strategies that you will find helpful and practical. Our *Blackline Masters* provide you with materials that you can use for vocabulary building, skill building, concept clarification, alternative activities for multiple learning styles, forms for performance task assessment of student achievement that are specific to the unit of study, and forms for assessment that focus on larger encompassing skills of science, technology, and societal decision making. Our *Computerized Assessment Bank* will assist you in your full implementation of the ***SCIENCEPOWER™ 8*** program.

We feel confident that we have provided you with the best possible program to help ensure your students achieve excellence and a high degree of scientific literacy through their course of study.

The Authors and Senior Program Consultants

A Tour of Your Textbook

Welcome to *SCIENCEPOWER™ 8*. This textbook introduces you to the wonders of cells, fluids, light, and Earth's water systems, as well as machines — some simple, some ingenious. To understand the book's structure, begin by taking the brief tour on the following pages. Then do the *Feature Hunt* on page xxiii to check your understanding of how to use this book.

Unit Opener

- *SCIENCEPOWER™ 8* has five major units.
- Each unit opener provides a clear overview of the unit's contents.
- The unit opener sparks interest in the topic by suggesting a problem to think about, presenting science ideas to consider, or highlighting a societal issue to explore.
- The unit opener identifies each of the three chapters in the unit.

Chapter Opener

- Each chapter opener gives you a clear idea of what the chapter is about.
- **Getting Ready** questions give you a chance to think about what you already know (or perhaps do not know) about the topics in the chapter.
- **Science Log** suggests various ways to answer the *Getting Ready* questions, and provides an opportunity to keep a record of what you learn, in the same way that scientists log their observations and the results of their findings. (Your teacher may call this record a *Science Journal* instead of a *Science Log*.)

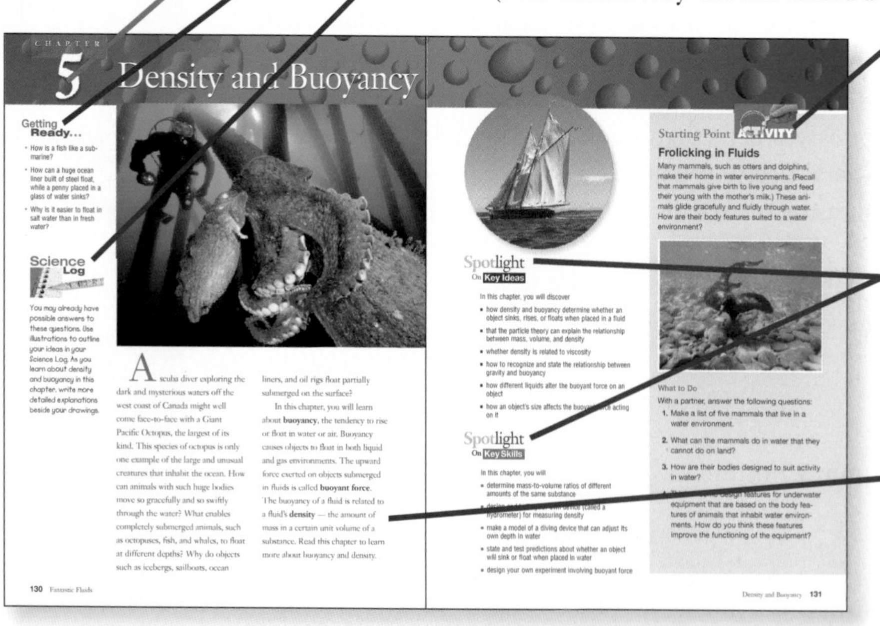

- **The Starting Point Activity** launches each chapter in a variety of ways. Like the *Getting Ready* questions, the *Starting Point Activity* helps you think about what you already know (or may not know) about the chapter's main topics.
- **Spotlight on Key Ideas** and **Spotlight on Key Skills** focus your attention on the major ideas and skills that you will be expected to know by the time you have completed the chapter.
- The **introductory paragraphs** of each chapter invite you to learn more about the topics and clearly tell you what you will be studying in the chapter.

Design & Do Investigation

- These hands-on activities set challenges to design and construct your own models, systems, or products. They teach design skills and blend science and technology in novel ways.
- The co-operative group work icon signals that you will be doing these investigations in a team.
- The Design Criteria provide a framework for evaluating your results.
- You and your team members are then on your own to design and construct!

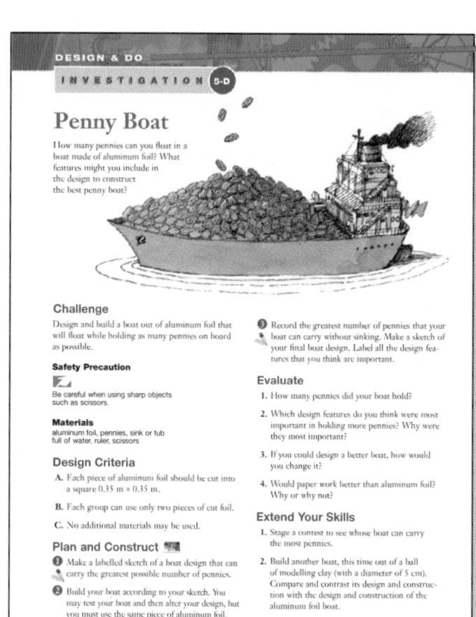

DESIGN & DO

INVESTIGATION 5-D

Penny Boat

How many pennies can you float in a boat made of aluminum foil? What features might you include in the design to construct the best penny boat?

Challenge

Design and build a boat out of aluminum foil that will float while holding as many pennies on board as possible.

Safety Precaution

Be careful when using sharp objects such as scissors.

Materials

aluminum foil, pennies, sink or tub full of water, ruler, scissors

Design Criteria

A. Each piece of aluminum foil should be cut into a square 0.35 m × 0.35 m.

B. Each group can use only two pieces of cut foil.

C. No additional materials may be used.

Plan and Construct

1. Make a labelled sketch of a boat design that can carry the greatest possible number of pennies.
2. Build your boat according to your sketch. You may test your boat and then alter your design, but you must use the same piece of aluminum foil.
3. Record the greatest number of pennies that your boat can carry without sinking. Make a sketch of your final boat design. Label all the design features that you think are important.

Evaluate

1. How many pennies did your boat hold?
2. Which design features do you think were most important in holding more pennies? Why were they most important?
3. If you could design a better boat, how would you change it?
4. Would paper work better than aluminum foil? Why or why not?

Extend Your Skills

1. Stage a contest to see whose boat can carry the most pennies.
2. Build another boat, this time out of a ball of modelling clay (with a diameter of 5 cm). Compare and contrast its design and construction with the design and construction of the aluminum foil boat.

Density and Buoyancy 147

Find Out Activity

- A short, informal inquiry activity that usually involves hands-on exploration.
- These activities call for simple, easy-to-obtain materials and equipment.
- In these activities, as well as in the investigations, you will use important science inquiry skills: predicting, estimating, hypothesizing, and so on.
- The pencil icon signals that you should make a written note of your predictions or observations.

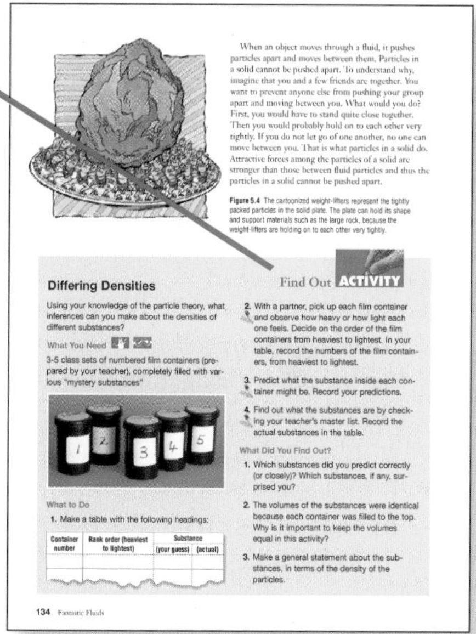

When an object moves through a fluid, it pushes particles apart and moves between them. Particles in a solid cannot be pushed apart. To understand why, imagine that you and a few friends are together. You want to prevent anyone else from pushing your group apart and moving between you. What would you do? First, you would have to stand quite close together. Then you would probably hold on to each other very tightly. If you do not let go of one another, no one can move between you. That is what particles in a solid do. Attractive forces among the particles of a solid are stronger than those between fluid particles and thus the particles in a solid cannot be pushed apart.

Figure 5.4 The cartoonized weight-lifters represent the tightly packed particles in the solid plate. The plate can hold its shape and support materials such as the large rock, because the weight-lifters are holding on to each other very tightly.

Find Out ACTIVITY

Differing Densities

Using your knowledge of the particle theory, what inferences can you make about the densities of different substances?

What You Need

3-5 class sets of numbered film containers (prepared by your teacher), completely filled with various "mystery substances"

What to Do

1. Make a table with the following headings:

Container number	Rank order (heaviest to lightest)	Substance (your guess)	Substance (actual)

2. With a partner, pick up each film container and observe how heavy or how light each one feels. Decide on the order of the film containers from heaviest to lightest. In your table, record the numbers of the film containers, from heaviest to lightest.
3. Predict what the substance inside each container might be. Record your predictions.
4. Find out what the substances are by checking your teacher's master list. Record the actual substances in the table.

What Did You Find Out?

1. Which substances did you predict correctly (or closely)? Which substances, if any, surprised you?
2. The volumes of the substances were identical because each container was filled to the top. Why is it important to keep the volumes equal in this activity?
3. Make a general statement about the substances, in terms of the density of the particles.

134 Fantastic Fluids

Conduct an Investigation

- One- to four-page "formal" labs provide an opportunity to develop science inquiry skills using various equipment and materials.
- These investigations provide a chance to ask questions about science, to make observations, and to obtain results.
- You then analyze your results to determine what they tell you about the topic you are investigating.
- Photographs showing each major step in the Procedure help you to carry out the investigation.
- Safety icons and Safety Precautions alert you to any special precautions you should take to help maintain a safe classroom environment.
- The pencil icon signals that you should make a written note of your predictions or observations.

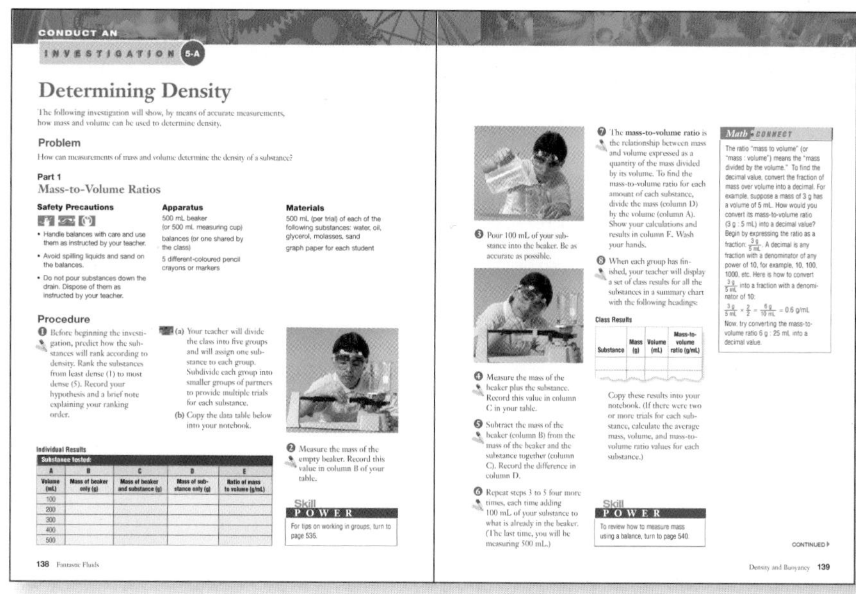

CONDUCT AN

INVESTIGATION 5-A

Determining Density

The following investigation will show, by means of accurate measurements, how mass and volume can be used to determine density.

Problem

How can measurements of mass and volume determine the density of a substance?

Part 1

Mass-to-Volume Ratios

Safety Precautions

- Handle balances with care and use them as instructed by your teacher.
- Avoid spilling liquids and sand on the balances.
- Do not pour substances down the drain. Dispose of them as instructed by your teacher.

Apparatus

500 mL beaker (or 500 mL measuring cup)

balances (or one shared by the class)

5 different-coloured pencil crayons or markers

Materials

500 mL (per trial) of each of the following substances: water, oil, glycerol, molasses, sand

graph paper for each student

Procedure

1. Before beginning the investigation, predict how the substances will rank according to density. Rank the substances from least dense (1) to most dense (5). Record your hypothesis and a brief note explaining your ranking order.
2. (a) Your teacher will divide the class into five groups and will assign one substance to each group. Subdivide each group into smaller groups of partners to provide multiple trials for each substance.
   (b) Copy the data table below into your notebook.

Individual Results

Substance tested:

A	B	C	D	E
Volume (mL)	Mass of beaker only (g)	Mass of beaker and substance (g)	Mass of substance only (g)	Ratio of mass to volume (g/mL)
100				
200				
300				
400				
500				

2. Measure the mass of the empty beaker. Record this value in column B of your table.

Skill POWER

For tips on working in groups, turn to page 536.

138 Fantastic Fluids

3. Pour 100 mL of your substance into the beaker. Be as accurate as possible.
4. Measure the mass of the beaker plus the substance. Record this value in column C in your table.
5. Subtract the mass of the beaker (column B) from the mass of the beaker and the substance together (column C). Record the difference in column D.
6. Repeat steps 3 to 5 four more times, each time adding 100 mL of your substance to what is already in the beaker. (The last time, you will be measuring 500 mL.)
7. The **mass-to-volume ratio** is the relationship between mass and volume expressed as a quantity of the mass divided by its volume. To find the mass-to-volume ratio for each amount of each substance, divide the mass (column D) by the volume (column A). Show your calculations and results in column E. Wash your hands.
8. When each group has finished, your teacher will display a set of class results for all the substances in a summary chart with the following headings:

Class Results

Substance	Mass (g)	Volume (mL)	Mass-to-volume ratio (g/mL)

Copy these results into your notebook. (If there were two or more trials for each substance, calculate the average mass, volume, and mass-to-volume ratio values for each substance.)

Skill POWER

To review how to measure mass using a balance, turn to page 540.

Math CONNECT

The ratio "mass to volume" (or "mass : volume") means the "mass divided by the volume." To find the decimal value, convert the fraction of mass over volume into a decimal. For example, suppose a mass of 3 g has a volume of 5 mL. How would you convert its mass-to-volume ratio (3 g : 5 mL) into a decimal value? Begin by expressing the ratio as a fraction: $\frac{3\ g}{5\ mL}$. A decimal is any fraction with a denominator of any power of 10, for example, 10, 100, 1000, etc. Here is how to convert $\frac{3\ g}{5\ mL}$ into a fraction with a denominator of 10:

$\frac{3\ g}{5\ mL} \times \frac{2}{2} = \frac{6\ g}{10\ mL} = 0.6\ g/mL$

Now, try converting the mass-to-volume ratio 6 g : 25 mL into a decimal value.

CONTINUED

Density and Buoyancy 139

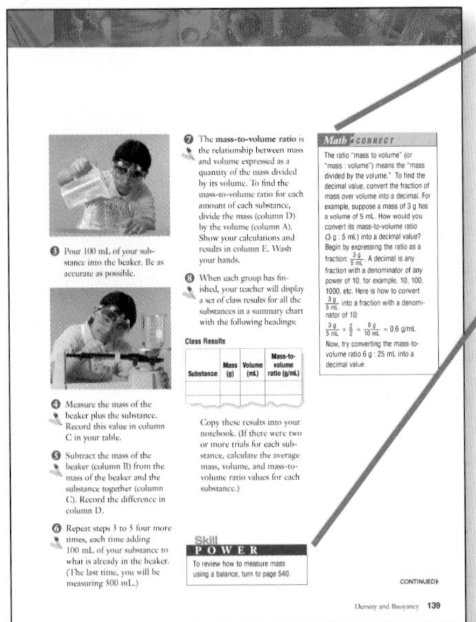

The mass-to-volume ratio is the relationship between mass and volume expressed as a quantity of the mass divided by its volume. To find the mass-to-volume ratio for each amount of each substance, divide the mass (column D) by the volume (column A). Show your calculations and results in column E. Wash your hands.

3. Pour 100 mL of your substance into the beaker. Be as accurate as possible.
4. Measure the mass of the beaker plus the substance. Record this value in column C in your table.
5. Subtract the mass of the beaker (column B) from the mass of the beaker and the substance together (column C). Record the difference in column D.
6. Repeat steps 3 to 5 four more times, each time adding 100 mL of your substance to what is already in the beaker. (The last time, you will be measuring 500 mL.)
7. The **mass-to-volume ratio** is the relationship between mass and volume expressed as a quantity of the mass divided by its volume. To find the mass-to-volume ratio for each amount of each substance, divide the mass (column D) by the volume (column A). Show your calculations and results in column E. Wash your hands.
8. When each group has finished, your teacher will display a set of class results for all the substances in a summary chart with the following headings:

Class Results

Substance	Mass (g)	Volume (mL)	Mass-to-volume ratio (g/mL)

Copy these results into your notebook. (If there were two or more trials for each substance, calculate the average mass, volume, and mass-to-volume ratio values for each substance.)

Skill POWER

To review how to measure mass using a balance, turn to page 540.

Math CONNECT

The ratio "mass to volume" (or "mass : volume") means the "mass divided by the volume." To find the decimal value, convert the fraction of mass over volume into a decimal. For example, suppose a mass of 3 g has a volume of 5 mL. How would you convert its mass-to-volume ratio (3 g : 5 mL) into a decimal value? Begin by expressing the ratio as a fraction: $\frac{3\text{ g}}{5\text{ mL}}$. A decimal is any fraction with a denominator of any power of 10, for example, 10, 100, 1000, etc. Here is how to convert $\frac{3\text{ g}}{5\text{ mL}}$ into a fraction with a denominator of 10:

$\frac{3\text{ g}}{5\text{ mL}} \times \frac{2}{2} = \frac{6\text{ g}}{10\text{ mL}} = 0.6\text{ g/mL}$

Now, try converting the mass-to-volume ratio 6 g : 25 mL into a decimal value.

CONTINUED

Density and Buoyancy 139

Mathconnect

- Reviews math skills as they are required to do activities.
- Makes connections between your science studies and your math studies.

SkillPower

- Skill development tips refer you to the *Science and Technology Skills Guide* at the back of the student text.
- These tips provide specific skill development strategies and activities as they are needed, for example, in the proper use of the microscope and in scientific drawing.

Think & Link Investigation

- These one- to two-page "thought" or "paper-based" investigations let you explore ideas or connections that might be impractical or dangerous in the science classroom.
- These investigations emphasize a variety of skills. These skills include analyzing data, interpreting diagrams or photographs, and forming ideas, opinions, or recommendations based on analysis of a societal issue.

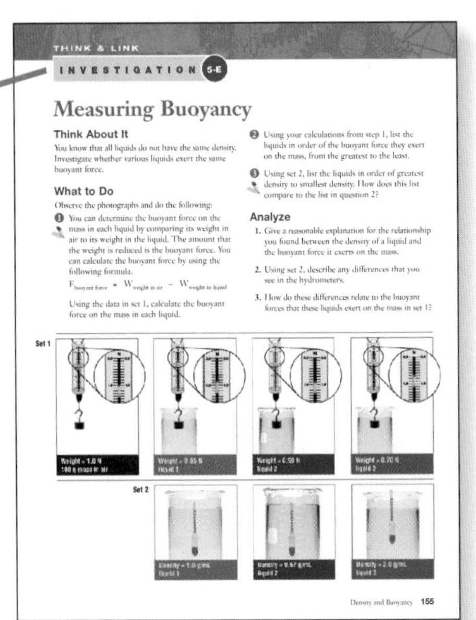

THINK & LINK

INVESTIGATION 5-E

Measuring Buoyancy

Think About It

You know that all liquids do not have the same density. Investigate whether various liquids exert the same buoyant force.

What to Do

Observe the photographs and do the following:

1. You can determine the buoyant force on the mass in each liquid by comparing its weight in air to its weight in the liquid. The amount that the weight is reduced is the buoyant force. You can calculate the buoyant force by using the following formula.

$F_{\text{buoyant force}} = W_{\text{weight in air}} - W_{\text{weight in liquid}}$

Using the data in set 1, calculate the buoyant force on the mass in each liquid.

2. Using your calculations from step 1, list the liquids in order of the buoyant force they exert on the mass, from the greatest to the least.
3. Using set 2, list the liquids in order of greatest density to smallest density. How does this list compare to the list in question 2?

Analyze

1. Give a reasonable explanation for the relationship you found between the density of a liquid and the buoyant force it exerts on the mass.
2. Using set 2, describe any differences that you see in the hydrometers.
3. How do these differences relate to the buoyant forces that these liquids exert on the mass in set 1?

Density and Buoyancy 155

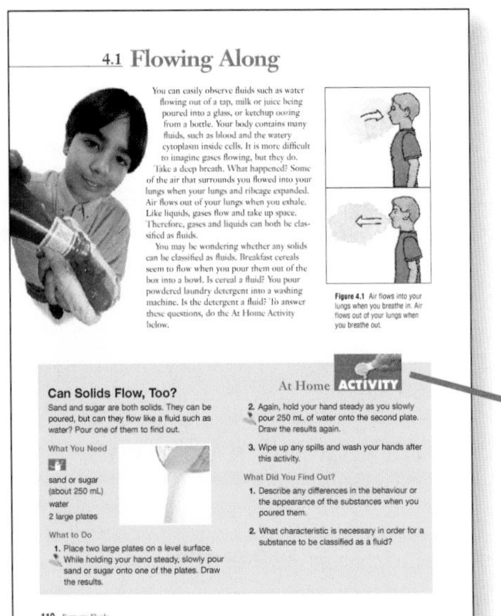

4.1 Flowing Along

You can easily observe fluids such as water flowing out of a tap, milk or juice being poured into a glass, or ketchup oozing from a bottle. Your body contains many fluids, such as blood and the watery cytoplasm inside cells. It is more difficult to imagine gases flowing, but they do. Take a deep breath. What happened? Some of the air that surrounds you flowed into your lungs when your lungs and ribcage expanded. Air flows out of your lungs when you exhale. Like liquids, gases flow and take up space. Therefore, gases and liquids can both be classified as fluids.

You may be wondering whether any solids can be classified as fluids. Breakfast cereals seem to flow when you pour them out of the box into a bowl. Is cereal a fluid? You pour powdered laundry detergent into a washing machine. Is the detergent a fluid? To answer these questions, do the At Home Activity below.

Figure 4.1 Air flows into your lungs when you breathe in. Air flows out of your lungs when you breathe out.

At Home ACTIVITY

Can Solids Flow, Too?

Sand and sugar are both solids. They can be poured, but can they flow like a fluid such as water? Pour one of them to find out.

What You Need

sand or sugar (about 250 mL)
water
2 large plates

What to Do

1. Place two large plates on a level surface. While holding your hand steady, slowly pour sand or sugar onto one of the plates. Draw the results.
2. Again, hold your hand steady as you slowly pour 250 mL of water onto the second plate. Draw the results again.
3. Wipe up any spills and wash your hands after this activity.

What Did You Find Out?

1. Describe any differences in the behaviour or the appearance of the substances when you poured them.
2. What characteristic is necessary in order for a substance to be classified as a fluid?

110 Fantastic Fluids

At Home Activity

- These short, informal inquiry activities can be done at home using simple, everyday materials.

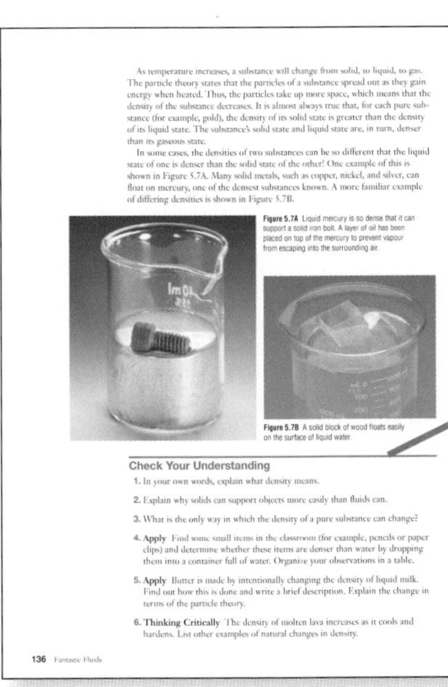

Check Your Understanding

- A set of review questions appears at the end of each numbered section in a chapter.
- These questions provide opportunities for ongoing self-assessment.
- "Apply," "Design Your Own," and "Thinking Critically" questions give you additional challenges.

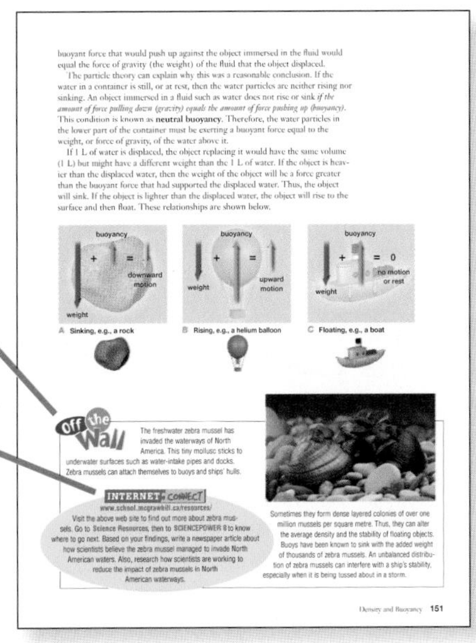

Off the Wall

- This item features intriguing situations, odd events, or weird facts.
- Ideas for connecting science with other curriculum areas are often included.

Internetconnect

- This feature encourages productive use of the Internet by offering content-appropriate sites.
- Web site suggestions will save you time as you do research.

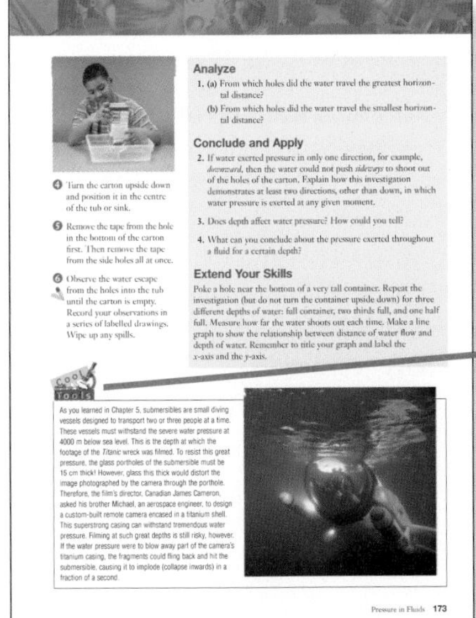

Cool Tools

- This feature provides information about some of the instruments and equipment invented to help humans explore the unknown.
- The information is often related to a variety of occupations and situations.

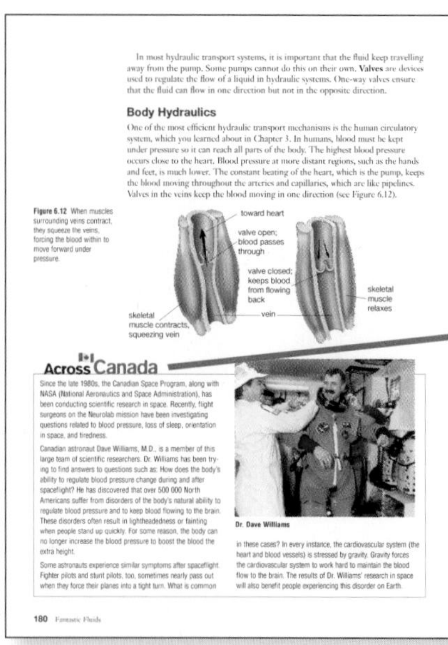

Across Canada

- These "mini-essays" feature information on Canadian scientists involved in important research and discoveries.
- *Across Canada* increases awareness and appreciation of the work of Canadian scientists. The essays also provide role models for those of you interested in careers or further study in science.

Chapter at a Glance

- Located at the end of each chapter, this page provides self-assessment opportunities as you look back at the chapter as a whole.
- It gives parents or guardians an overview of what you have accomplished.
- **Prepare Your Own Summary** encourages you to summarize your understanding in a variety of ways — using diagrams, flowcharts, concept maps, artwork, writing, or any approach you prefer.

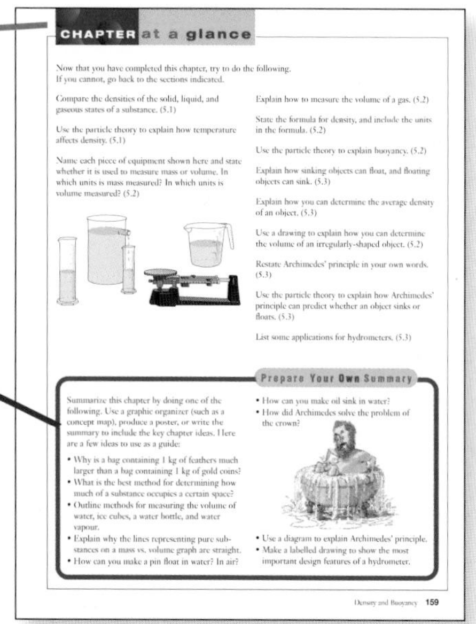

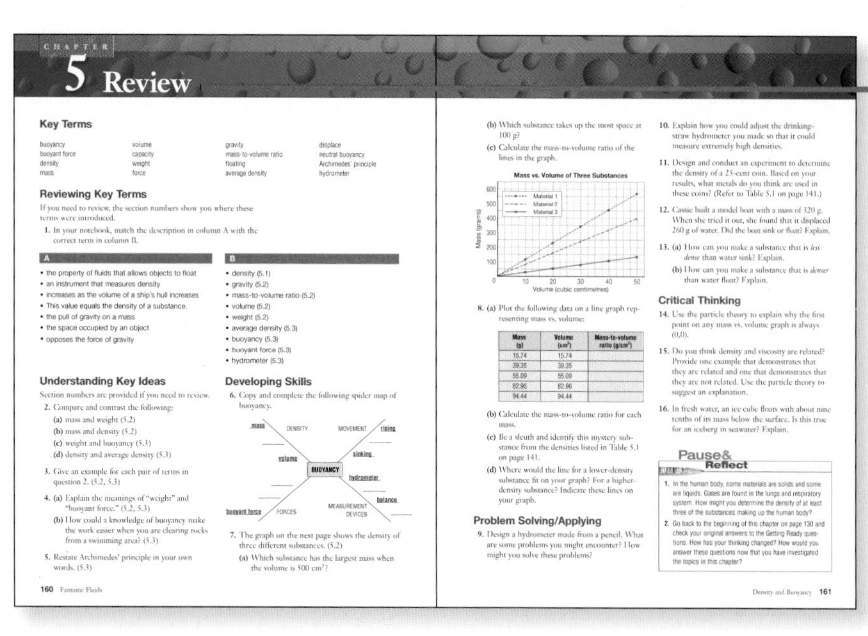

CHAPTER 5 Review
Key Terms
Reviewing Key Terms
Understanding Key Ideas
Developing Skills
Mass vs. Volume of Three Substances
Critical Thinking
Problem Solving/Applying
Pause& Reflect
160 Fantastic Fluids
Density and Buoyancy 161

Chapter Review

- This final wrap-up of each chapter reviews basic concepts, skills of inquiry and communication, and skills relating science to technology, society, and the environment.
- These questions help you recall, think about, and apply what you have learned.

End-of-Unit Features

Ask an Expert

- Experts in every area of science and technology are working to understand better how the world "works" and to try to find solutions to difficult problems. The *Ask an Expert* feature at the end of each unit is an interview with one of these many people.
- After the interviews, you will have a chance to do an activity that is related to the kind of work the expert does.

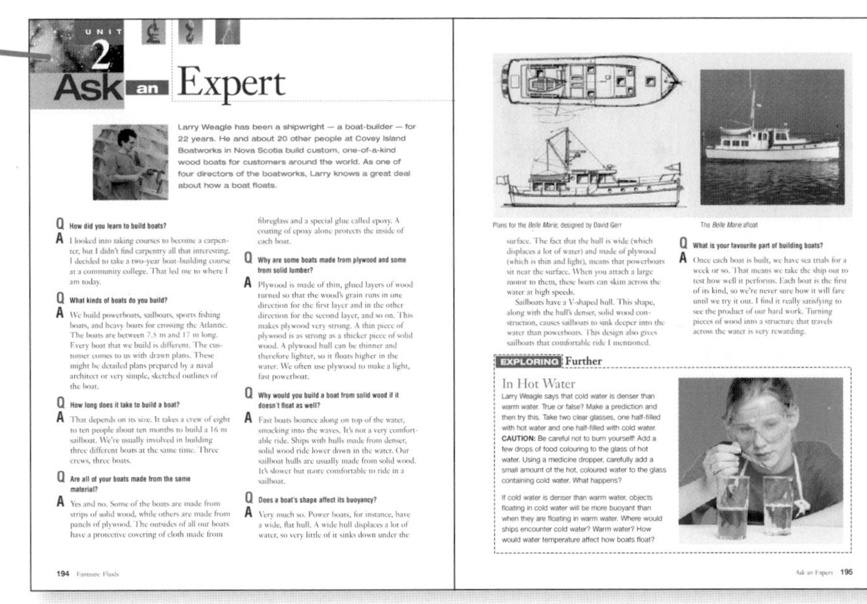

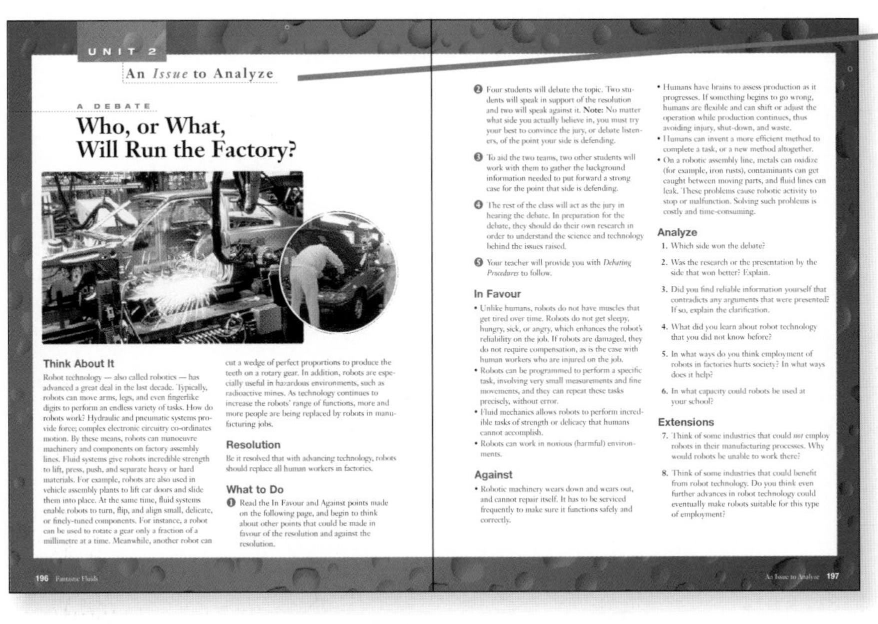

An Issue to Analyze

- You, your community, and society in general face complex issues in today's world. Understanding science and technology cannot provide a "correct" answer to the problems these issues present, but understanding will lead to more informed decisions. *An Issue to Analyze* gives you a chance to start thinking now about how you can help make the best decisions for yourself and your community, today and in the future.
- This feature takes the form of either a simulation or a debate.

Unit Project

- A *Unit Project* gives you a chance to use key ideas and skills from the unit to create your own device, system, model, or (in Unit 4) a research presentation.
- Your teacher might ask you to begin to consider early in the unit how you might design, plan, and complete your wrap-up project.
- You will complete the project as part of a team.
- *More Project Ideas* offers additional suggestions for enjoyable ways to demonstrate your learning.

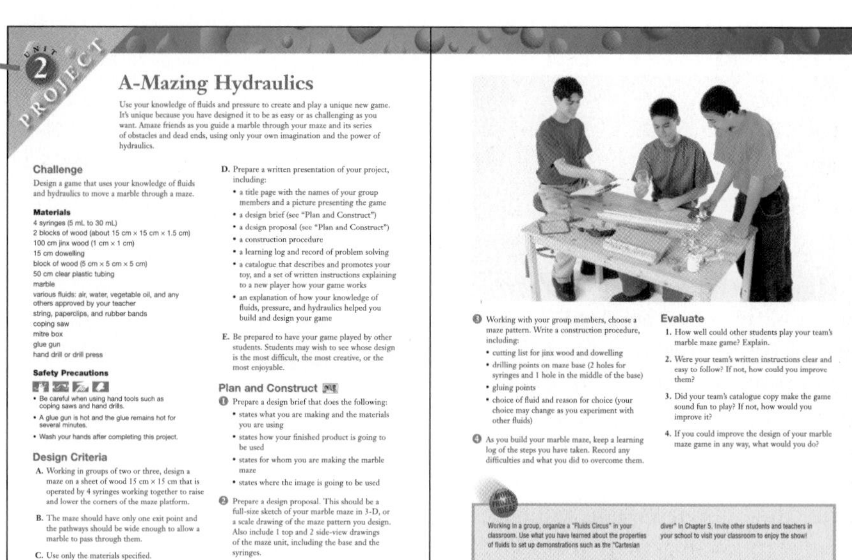
UNIT 2 PROJECT

A-Mazing Hydraulics

Use your knowledge of fluids and pressure to create and play a unique new game. It's unique because you have designed it to be as easy or as challenging as you want. Amaze friends as you guide a marble through your maze and its series of obstacles and dead ends, using only your own imagination and the power of hydraulics.

Challenge

Design a game that uses your knowledge of fluids and hydraulics to move a marble through a maze.

Materials

4 syringes (5 mL to 30 mL)
2 blocks of wood (about 15 cm × 15 cm × 1.5 cm)
100 cm jinx wood (1 cm × 1 cm)
15 cm dowelling
block of wood (5 cm × 5 cm × 5 cm)
50 cm clear plastic tubing
marble
various fluids: air, water, vegetable oil, and any others approved by your teacher
string, paperclips, and rubber bands
coping saw
mitre box
glue gun
hand drill or drill press

Safety Precautions

- Be careful when using hand tools such as coping saws and hand drills.
- A glue gun is hot and the glue remains hot for several minutes.
- Wash your hands after completing this project.

Design Criteria

A. Working in groups of two or three, design a maze on a sheet of wood 15 cm × 15 cm that is operated by 4 syringes working together to raise and lower the corners of the maze platform.

B. The maze should have only one exit point and the pathways should be wide enough to allow a marble to pass through them.

C. Use only the materials specified.

D. Prepare a written presentation of your project, including:
- a title page with the names of your group members and a picture presenting the game
- a design brief (see "Plan and Construct")
- a design proposal (see "Plan and Construct")
- a construction procedure
- a learning log and record of problem solving
- a catalogue that describes and promotes your toy, and a set of written instructions explaining to a new player how your game works
- an explanation of how your knowledge of fluids, pressure, and hydraulics helped you build and design your game

E. Be prepared to have your game played by other students. Students may wish to see whose design is the most difficult, the most creative, or the most enjoyable.

Plan and Construct

1. Prepare a design brief that does the following:
   - states what you are making and the materials you are using
   - states how your finished product is going to be used
   - states for whom you are making the marble maze
   - states where the image is going to be used
2. Prepare a design proposal. This should be a full-size sketch of your marble maze in 3-D, or a scale drawing of the maze pattern you design. Also include 1 top and 2 side-view drawings of the maze unit, including the base and the syringes.
3. Working with your group members, choose a maze pattern. Write a construction procedure, including:
   - cutting list for jinx wood and dowelling
   - drilling points on maze base (2 holes for syringes and 1 hole in the middle of the base)
   - gluing points
   - choice of fluid and reason for choice (your choice may change as you experiment with other fluids)
4. As you build your marble maze, keep a learning log of the steps you have taken. Record any difficulties and what you did to overcome them.

Evaluate

1. How well could other students play your team's marble maze game? Explain.
2. Were your team's written instructions clear and easy to follow? If not, how could you improve them?
3. Did your team's catalogue copy make the game sound fun to play? If not, how would you improve it?
4. If you could improve the design of your marble maze game in any way, what would you do?

Working in a group, organize a "Fluids Circus" in your classroom. Use what you have learned about the properties of fluids to set up demonstrations such as the "Cartesian diver" in Chapter 5. Invite other students and teachers in your school to visit your classroom to enjoy the show!

198 Fantastic Fluids

Unit 2 Project 199

Other Important Features

DidYouKnow?

- Presents interesting facts related to science, technology, nature, and the universe.

- These brain teasers are often mathematics-related.
- They draw upon your problem-solving skills and your imagination.

Pause& Reflect

- These items supply opportunities to reflect on what you know (or do not know) and to make connections among ideas throughout the text.
- This recurring feature encourages you to construct your own learning on an ongoing basis and to keep track of how your knowledge is building.

Word CONNECT

- Word origins and a variety of language activities provide links to language arts.

Career CONNECT

- Portrays people with various levels of education making practical use of science and technology in their jobs.

Computer CONNECT

- Highlights opportunities where using spreadsheets or data base applications would be helpful.

Design Your Own

- This signals opportunities in investigations, activities, Check Your Understanding sections, and Chapter Reviews for students to plan, design, and conduct their own experimental investigations.

Wrapping Up the Tour

At the back of ***SCIENCEPOWER™ 8***, you will find some additional features to help you review and develop skills and knowledge that you will need to be successful in this course. Are you having trouble with graphing? Would you like help setting up a data table? Have you forgotten how to make a concept map? Do you need a reminder about scientific notation or the metric system? The *Science and Technology Skills Guide* will help you review or improve your skills. A *Glossary* provides all the key vocabulary for the whole course, and an *Index* will help you find your way to a topic.

Science and Technology Skills Guide

530 Science and Technology Skills Guide

Special Icons

The co-operative group work icon alerts you to opportunities to work within a group, and the pencil icon reminds you to record your predictions or observations on paper. The safety icons are extremely important because they alert you to any safety precautions you must take, for example, the need for safety goggles or a lab apron. Other safety icons used in this book are shown on page 553. Make certain you become familiar with what they mean, and make sure that you follow their precautions.

Instant Practice — Feature Hunt

To acquaint yourself further with your textbook before you start using it, see if you can find some of the features it contains. Work with a partner.

1. Find the following features. Briefly tell what each feature is about and record the page number where you found it.
   - (a) a *Pause & Reflect* in Chapter 6
   - (b) an *Off the Wall* in Unit 2
   - (c) a *CareerConnect* in Chapter 5
   - (d) a *DidYouKnow?* in Unit 1
   - (e) a *Conduct an Investigation* entitled "Follow That Refracted Ray!" in Chapter 7
   - (f) a *Find Out Activity* entitled "How Far Up?" in Chapter 14
   - (g) an *Across Canada* in Chapter 3
2. Find these words and their meanings.
   - (a) buoyancy (Unit 2)
   - (b) spectrum (Unit 3)
   - (c) run-off (Unit 4)
   - (d) cytoplasm (Unit 1)
   - (e) turbine (Unit 5)
   - (f) Venn diagram (SkillPower 1)
   - (g) histogram (SkillPower 6)

Extensions

3. How did you find the words in question 2? Give one other way to find each word and its meaning.
4. Prepare your own feature hunt for a classmate to do.

Safety in Your Science Classroom

Become familiar with the following safety rules and procedures. It is up to you to use them and your teacher's instructions to make your activities and investigations in ***SCIENCEPOWER***™ ***8*** safe and enjoyable. Your teacher will give you specific information about any other special safety rules to be used in your school.

1. Working with your teacher . . .

- Listen carefully to any instructions your teacher gives you.
- Inform your teacher if you have any allergies, medical conditions, or other physical problems that could affect your work in the science classroom. Tell your teacher if you wear contact lenses or a hearing aid.
- Obtain your teacher's approval before beginning any activity you have designed yourself.
- Know the location and proper use of the nearest fire extinguisher, fire blanket, first-aid kit, and fire alarm.

2. Starting an activity or investigation . . .

- Before starting an activity or investigation, read all of it. If you do not understand how to do any step, ask your teacher for help.
- Be sure you have checked the safety icons and have read and understood the safety precautions.
- Begin an activity or investigation only after your teacher tells you to start.

3. Dressing for success in science . . .

- When you are directed to do so, wear protective clothing, such as a lab apron and safety goggles. Always wear protective clothing when you are using materials that could pose a safety problem, such as unidentified substances, or when you are heating anything.
- Tie back long hair, and avoid wearing scarves, ties, or long necklaces.

4. Acting responsibly . . .

- Work carefully with a partner and make sure your work area is clear.
- Handle equipment and materials carefully.
- Make sure stools and chairs are resting securely on the floor.
- If other students are doing something that you consider dangerous, report it to your teacher.

5. Handling edible substances . . .

- Do not chew gum, eat, or drink in your science classroom.
- Do not taste any substances or draw any material into a tube with your mouth.

6. **Working in a science classroom . . .**
   - Make sure you understand all safety labels on school materials or those you bring from home. Familiarize yourself, as well, with the WHMIS symbols and the special safety symbols used in this book, found on page 553.
   - When carrying equipment for an activity or investigation, carry the equipment carefully. Carry only one object or container at a time.
   - Be aware of others during activities and investigations. Make room for students who may be carrying equipment to their work station.
7. **Working with sharp objects . . .**
   - Always cut away from yourself and others when using a knife or razor blade.
   - Always keep the pointed end of scissors or any pointed object facing away from yourself and others if you have to walk with such objects.
   - If you notice sharp or jagged edges on any equipment, take special care with it and report it to your teacher.
   - Dispose of broken glass as your teacher directs.
8. **Working with electrical equipment . . .**
   - Make sure your hands are dry when touching electrical cords, plugs, or sockets.
   - Pull the plug, not the cord, when unplugging electrical equipment. Report damaged equipment or frayed cords to your teacher.
   - Place electrical cords in places where people will not trip over them.
9. **Working with heat . . .**
   - When heating an item, wear safety goggles and any other safety equipment that the text or your teacher advises.
   - Always use heatproof containers.
   - Do not use broken or cracked containers.
   - Point the open end of a container that is being heated away from yourself and others.
   - Do not allow a container to boil dry.
   - Handle hot objects carefully. Be especially careful with a hot plate that looks as though it has cooled down.
   - If you use a Bunsen burner, make sure you understand fully how to light and use it safely.
   - If you do receive a burn, inform your teacher, and apply cold water to the burned area immediately.
10. **Working with various chemicals . . .**
   - If any part of your body comes in contact with a substance, wash the area immediately and thoroughly with water. If you get anything in your eyes, do not touch them. Wash them immediately and continuously for 15 min, and inform your teacher.

- Always handle substances carefully. If you are asked to smell a substance, never smell it directly. Hold the container slightly in front of and beneath your nose, and waft the fumes toward your nostrils, as shown here.

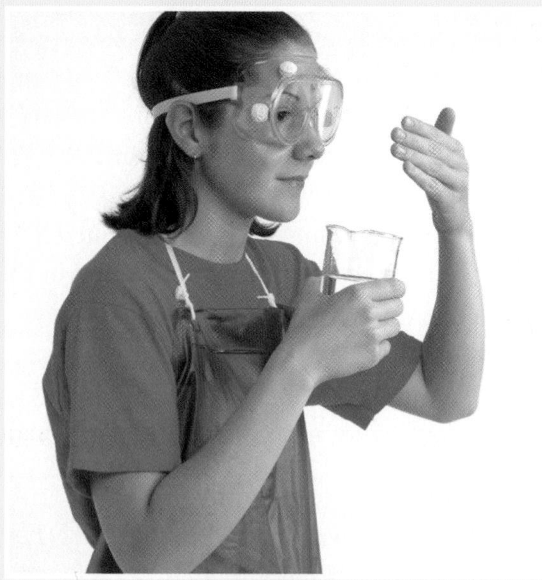

- Hold containers away from your face when pouring liquids, as shown here.

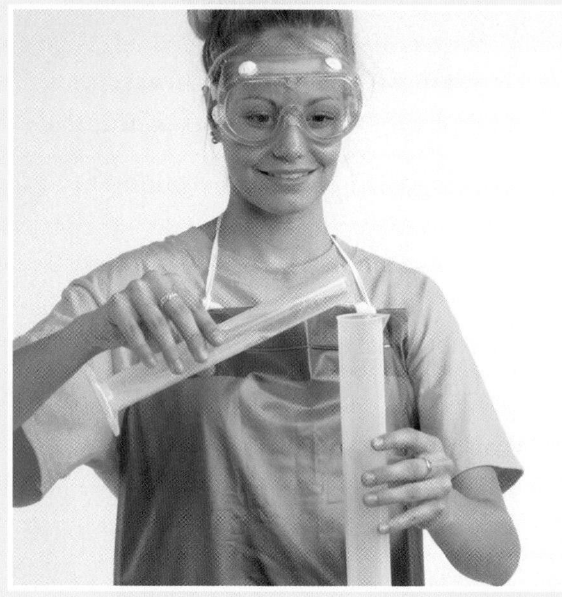

11. Working with living things . . .

On a field trip:

- Try not to disturb the area any more than is absolutely necessary.
- If you move something, do it carefully, and always replace it carefully.
- If you are asked to remove plant material, remove it gently, and take as little as possible.

In the classroom:

- Treat living creatures with respect.
- Make sure that living creatures receive humane treatment while they are in your care.
- If possible, return living creatures to their natural environment when your work is complete.

12. Cleaning up in the science classroom . . .

- Clean up any spills, according to your teacher's instructions.
- Clean equipment before you put it away.
- Wash your hands thoroughly after doing an activity or an investigation.
- Dispose of materials as directed by your teacher. Never discard materials in the sink unless your teacher requests it.

13. Designing, constructing, and experimenting with structures and mechanisms . . .

- Use tools safely to cut, join, and shape objects.
- Handle modelling clay correctly. Wash your hands after using modelling clay.
- Follow proper procedures when using mechanical systems and studying their operations.
- Use special care when observing and working with objects in motion (e.g., objects that spin, swing, bounce, or vibrate; gears and pulleys; elevated objects).
- Do not use power equipment such as drills, sanders, saws, and lathes unless you have specialized training in handling such tools.

Instant Practice

Select one of the numbered sections of the safety points listed. Use these safety points to create a poster that communicates the message of the importance of safety.

What You Need

poster paper or Bristol board
crayons/markers
glitter or other materials that will draw attention to your poster

What to Do

1. With your group, decide what safety point(s) you want to illustrate.
2. Decide how you are going to illustrate your safety point(s). Will you show a right way and a wrong way? Will you use humour? (If you do, you should make sure that your audience does not miss the point.)
3. Make a rough copy of your poster. Revise any part of it that you feel you can improve. Then make your final copy.
4. When you are satisfied with your poster, present it to your teacher for display.

Introducing the *SCIENCEPOWER*™ *8*

Almost everywhere you turn in today's world you see the effects of science and technology. Look up and see jets whisking people to places that once required long journeys by land or by sea. The photograph on the left below shows Toronto's busy Pearson International Airport. Air travellers can make the journey across the Atlantic to England, for example, in about six hours. In contrast, it would take early Canadian travellers, riding by horse and carriage, about six hours to cover the distance across a large modern city such as Toronto. Today, the Space Shuttle, a technological marvel, launches astronauts into orbit to conduct scientific research. In six hours the Shuttle can complete four orbits around Earth, travelling a total of nearly 160 000 km!

Do you know what science is? Do you know what technology is? Are they different? Many people think of them as one and the same, but they are not. How are science and technology related to each other and how do they affect us as individuals and as a society, especially in terms of the environment in which we live?

In your grade 8 course, you will participate in scientific and technological investigations. You will gain new skills and knowledge as you explore the relationships among science, technology, society, and the environment (abbreviated **STSE**).

Why is it important to do this? Science and technology affect us all, even though we are not always aware of it. For example, did you watch television or use a toaster this morning? Both involve science and technology. Go further and think about what makes these appliances function. Did you flip a light switch today? Electricity is so much a part of everyday life that we take it for granted, but science and technology are needed to produce and conduct electricity to power the many devices and appliances on which we depend.

Whether you decide to become a scientist or technologist or choose another area as a career, such as music, art, or teaching, you will need to know the basic ideas and skills that you will develop in studying science. Start with the *Instant Practice* below to find out what you already know about science and technology.

Instant Practice

1. As a class, make a list of the first things that come to mind when you think of
   (a) science **(b)** technology
2. Compare the definitions you produced. How are they different? How are they similar?
3. Based on the answers from questions 1. (a) and (b), decide together on the best definition for each of these terms.

Science, Technology, and Society

As a class, you probably came up with many good ideas about what "science" and "technology" mean. Everyone has a slightly different background, so there were probably various suggestions about what these words mean.

Some people think that science is a group of facts or knowledge. Science includes this, but, perhaps more importantly, science is a special way of thinking or asking questions. This way of thinking helps us find out about the natural world, from what the centre of Earth is made of to the size of the farthest stars. In a similar way, many people think that technology is using scientific knowledge to solve a practical problem. Technology can be this, but it can also be a process of discovery using your everyday experience. Give this some thought: People used technology long before they knew about the science involved in the processes they were using. For example, people learned how to make fire by rubbing pieces of wood together long before they understood the science of heat.

The key goal of science is to investigate and better understand the nature of the universe, which includes all the living and non-living things that we can know about. The goal of technology is to find solutions to practical problems by designing and developing devices, materials, systems, and processes. Science may investigate, for example, how heat and friction are related. On the other hand, humans made fire for tens of thousands of years without realizing that friction provides the initial energy to set off an exothermic, or energy-producing, reaction. The physical and chemical aspects of heat and friction were understood only about two centuries ago, after years of scientific investigation.

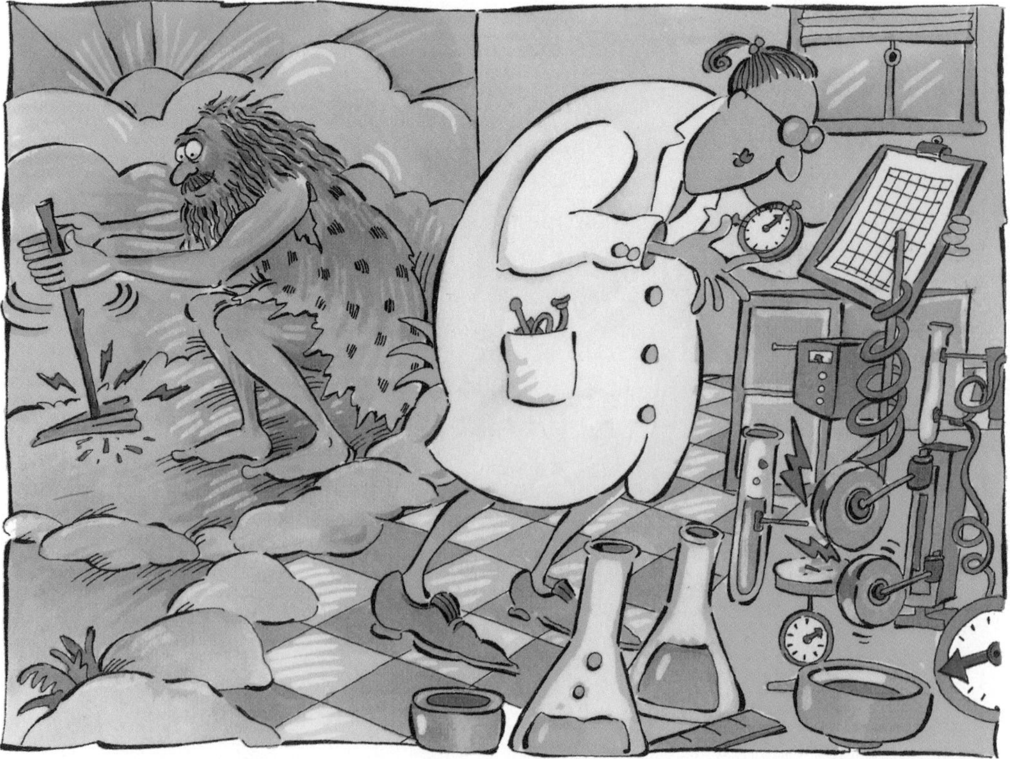

How can you easily grasp what science and technology are, especially in terms of how they relate to societal and environmental issues? The diagram below shows one way to think about what these words mean and how these ideas are related to each other.

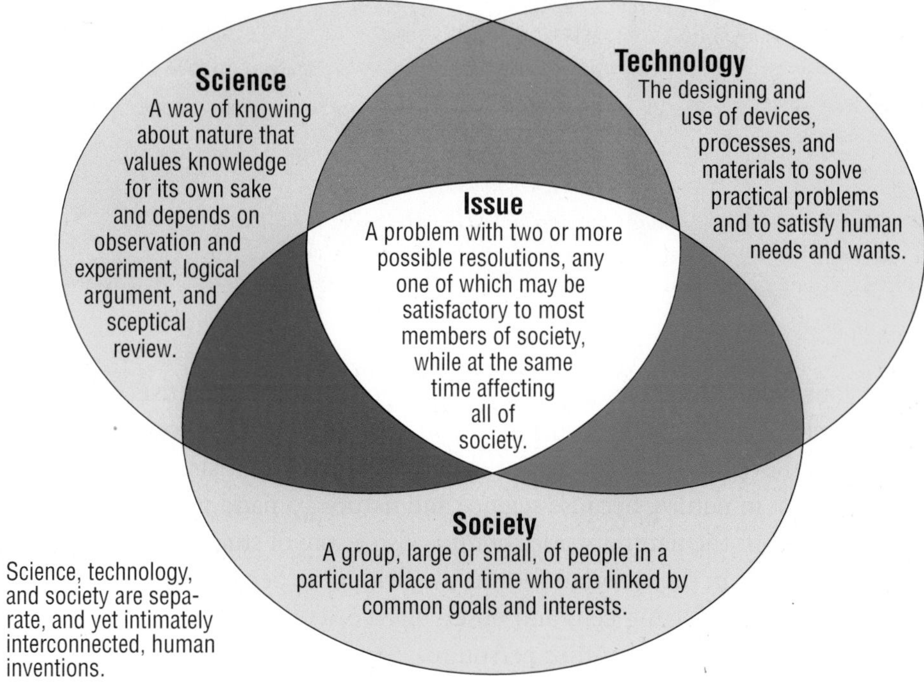

Science
A way of knowing about nature that values knowledge for its own sake and depends on observation and experiment, logical argument, and sceptical review.
Technology
The designing and use of devices, processes, and materials to solve practical problems and to satisfy human needs and wants.
Issue
A problem with two or more possible resolutions, any one of which may be satisfactory to most members of society, while at the same time affecting all of society.
Society
A group, large or small, of people in a particular place and time who are linked by common goals and interests.

Science, technology, and society are separate, and yet intimately interconnected, human inventions.

Instant Practice

1. Decide whether the following are questions of science, technology, or both.
   - **(a)** How does a bird fly?
   - **(b)** What is the most efficient shape for an airplane wing?
   - **(c)** What makes a sunset red?
   - **(d)** How does a car's engine turn the wheels of the car?
   - **(e)** How can you make an engine's use of gas more efficient?
2. How would you use technology to cross a shallow stream safely using only materials that you can find near the stream?
3. Does constructing a ladder involve technology, science, or both? Explain your answer.
4. Give two examples of modern technology that have never had an impact on you or someone you know.

Science and Science Inquiry

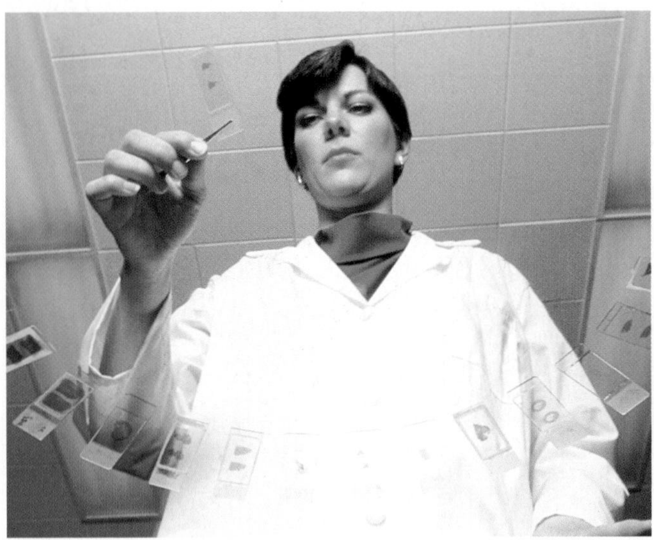

How would you interpret this photograph?

What does this photograph bring to mind?

What is your mental image of a scientist? How do you think scientists carry out their research? Would you be surprised if you were to learn that both people in the photographs above are scientists at work? Besides being scientists, they both have a genuine interest in nature, because science and nature go hand in hand. **Science** is a body of facts about the natural world, but it is also a way of studying and asking questions about nature, in the laboratory or in the field. One person may study nature by disrupting a sample of living cells and observing the action, in a test tube, of chemicals produced by the cells. Another person may study nature by probing distant galaxies by means of telescopes. Examine the following photographs to see the wide variety of places that scientists explore and the tools they use to learn about nature.

These astronomers may be looking at a star so far away that it took the light millions of years to reach Earth.

Sometimes scientists go into outer space to study nature. Canadian astronaut Julie Payette, shown here, flew on the Space Shuttle *Discovery* in May 1999 as part of a supply mission to the International Space Station. Payette was the first Canadian to visit and work aboard the Space Station. Her research activities include the application of interactive technology in space.

Dr. Biruté Galdikas, of Simon Fraser University in British Columbia, spent over 26 years studying orangutans in the jungles of Borneo. Her research has provided significant information not only about orangutans, but also about rain forests and biodiversity.

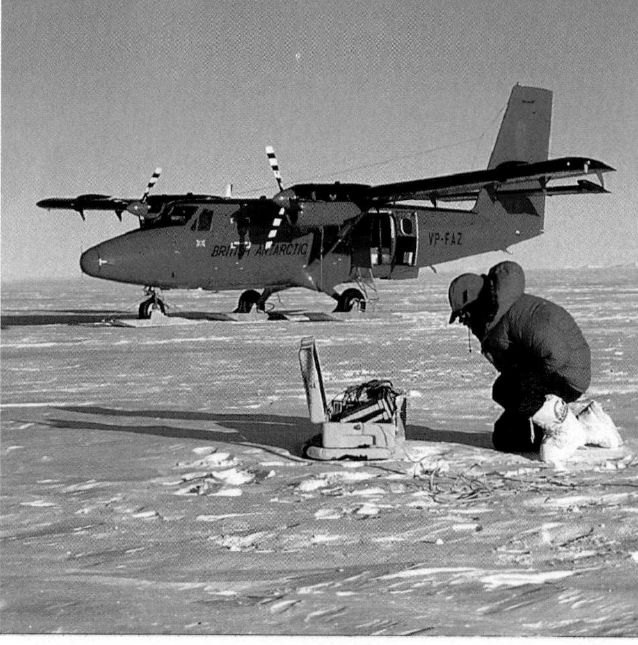

This scientist is on a glacier in Antarctica measuring the speed at which the glacier is moving. The Global Positioning System allows the scientist to obtain accurate position data by satellite, which can be used to calculate the glacier's speed. This information will help scientists make predictions about future climate changes.

As you can see from the photographs, scientists travel to every part of the world, as well as into outer space, to learn about the nature of the universe. With such diverse interests, what do these scientists have in common? No matter how their interests vary, scientists have an intense curiosity about some aspect of nature. Their curiosity compels them to seek answers wherever they can find them. However, along with nearly every answer comes a new question. Science is like a never-ending story. There is always more to discover.

When submersibles like "Alvin" were developed, marine scientists were able to descend farther below the surface of the ocean than ever before. They discovered forms of life, such as these tube worms, that are unlike anything that scientists had observed previously.

Instant Practice

What aspect of nature interests you? Would you travel into space or to the bottom of the ocean to look for answers to questions? Maybe you would rather read about things that other people have studied.

Take a few minutes to skim through the pages of this book. Think of questions about the natural world that you would like to investigate. Write down four or five questions that spark your curiosity. Exchange your questions with a partner and discuss how you might find answers to them.

Categories of Knowledge in Science

Suppose that your teacher gave you and your classmates an assignment to write a short essay that answers one of the questions you wrote for your *Instant Practice*. For example, assume that you saw a photograph of a microscope and wrote, "How does a microscope work? How can I see objects with a microscope that I cannot see with a magnifying glass?" Where would you look to find information that would answer this question? You probably know that there are thousands of books in libraries and bookstores on different topics in science. If you have searched the Internet, you have also discovered that there are a tremendous number of web sites on nearly every topic, including science. How could you narrow your search?

Topics in science are grouped into categories. Each category is then subdivided several times. The concept map below shows one way to start grouping science topics. Each topic in the last row can be subdivided many more times. For example, *animals* include many subdivisions, from insects such as ants, to mammals such as whales.

You can use the concept map to start searching for the answer to the question, "How do microscopes work?" The first three categories are *life topics*, *Earth topics*, and *physical topics*. Microscopes are not living things, nor are they part of Earth or space. Thus, you would look under *physical topics*. Microscopes use light, and light is a form of energy. This fact leads you to *physics*, the study of energy. If you looked for subdivisions of physics, you would find a topic called *optics*, the study of light as it passes through lenses and is reflected by mirrors. By exploring further in this way, you would eventually find information that explains how microscopes work.

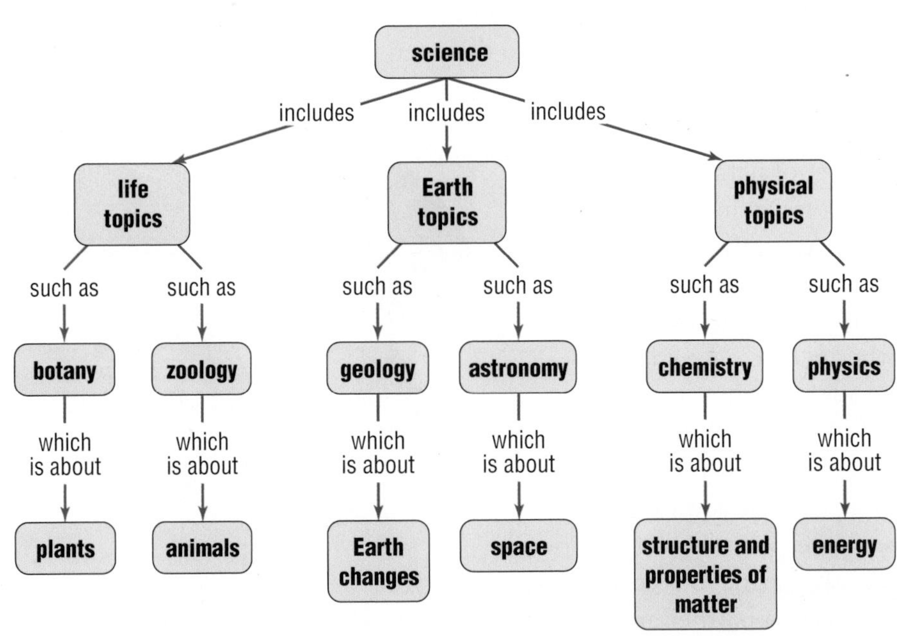

As you can imagine, no one person could possibly learn everything there is to know about science. However, everyone can learn enough about the major categories to gain a general understanding of the world around them. In this course, you will concentrate on the five different categories represented by these photographs.

Unit 1

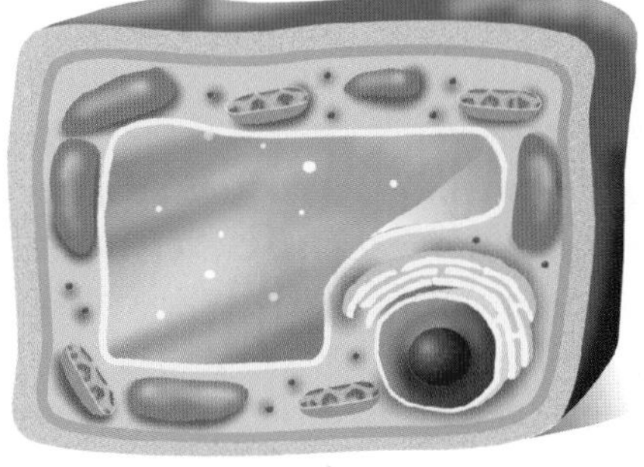

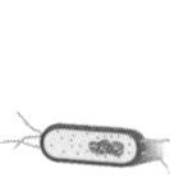

❶ First, you will learn about living cells, how cells function, and how they are organized into tissues, organs, and organ systems. Look again at the concept map opposite. To which category or categories would the topic of cells belong?

❷ Next, you will learn about fluids. Why do some fluids flow more rapidly than others? Why do some objects float in one fluid and sink in another? The study of fluids is a subdivision of which category?

Does one of these topics interest you more than the others? Your ideas may change as you learn more about each topic. You might find that a subject you knew nothing about is so interesting that you decide to pursue a career in that field. On the other hand, you may choose a career other than science. You may one day become a lawyer, a musician, a plumber, or an historian. In any case, the knowledge you gain in this course will help you to become scientifically literate. You will then be able to contribute to the decision-making processes that are so important in our highly technological society.

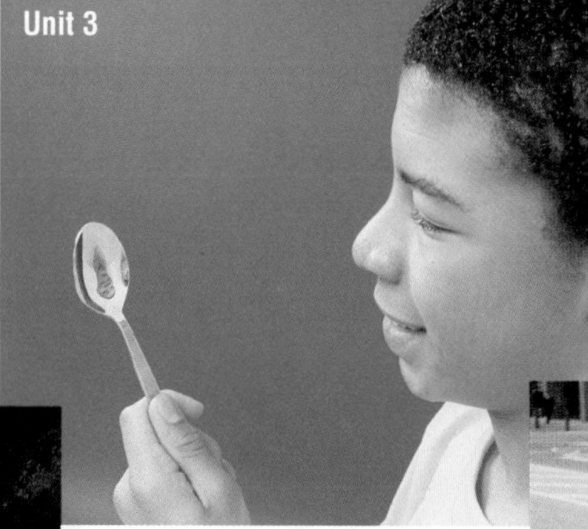

❸ Then you will study light and optical instruments. You will study reflection and other behaviours of light. You will also learn how a microscope, a telescope, and other optical devices work. You already know the category to which optics belongs.

❹ Following your study of optics, you will learn about water systems on Earth. Water is not a living organism, but without water there could be no life. How would you categorize Earth's water systems?

❺ You will conclude your studies with a unit on mechanical efficiency. What are the scientific principles that cause machines to work? How are the properties of fluids used in machines? Where do machines fit in the concept map on the previous page?

The Science Inquiry Process

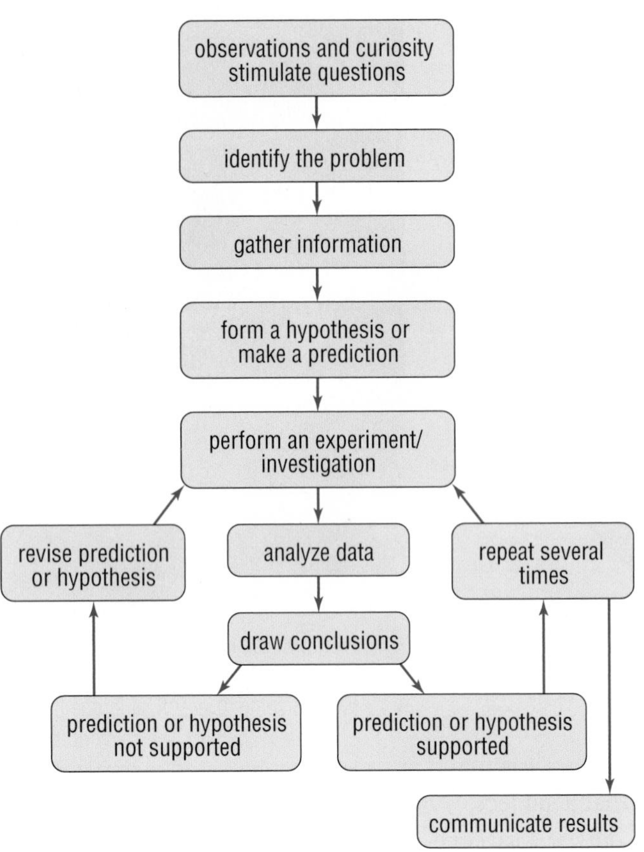

Science Inquiry

While looking at the photographs of scientists at work on pages IS-4 and -5, did you wonder how they carry out their experiments and observations? For example, if you were to ask the question "How do birds fly?", how would you find an answer? You have probably seen birds fly all your life, but you may not know exactly how they move through the air. To find out how birds fly or to answer any scientific question, you would have to design experiments carefully. Where would you begin?

Asking concise and focussed questions and designing experiments that will give clear answers to them is a very important part of science called *science inquiry*. While scientists over the centuries were accumulating the vast knowledge we now have about science, they were also developing an orderly process for carrying out science inquiry. One model of the science inquiry process is shown in the concept map here.

As you can see in the model, science inquiry begins with curiosity. Until someone asks a question, new knowledge cannot be gained. As well, someone must recognize the importance of asking the question in the first place. A basic **law** of science might have been overlooked if a curious person had not asked a specific question. For example, ever since humans inhabited Earth, people knew that all objects fell toward Earth — a fact that everyone took for granted.

law: an action or condition that has been observed so consistently that scientists are convinced it will always happen. A law has no theoretical basis. For example, scientists do not know exactly why gravity exists, but they understand how it works. No observations have ever contradicted the law of gravity.

A Question of Gravity

The behaviour of falling objects moved from common sense into science in the 1660s, thanks mostly to the famous British scientist Sir Isaac Newton. We can imagine Newton asking himself: "Why do objects fall to Earth at the same speed, no matter how massive they are? Could there be a universal way to describe how any object falls to Earth?" This led Newton to formulate his law of universal gravitation, also known as the law of gravity. Eventually, he showed that the motion of the planets in the solar system could be explained by the law of gravity. Today, scientists use Newton's concepts about gravity to plan the orbits of artificial satellites and to determine exactly how to guide them into the correct position.

This has never happened to you, has it? No observations have ever been made that contradict the law of gravity.

Long before humans were sending artificial satellites into orbit, they were using gravity. Clocks such as the "grandmother clock" in the photograph rely on gravity to keep time. When you pull a pendulum bob (the mass at the end of the string or rod) up to the side, gravity pulls it back down and causes it to swing with a rhythmic motion. The time it takes to make one complete swing back and forth is called the pendulum's period of motion.

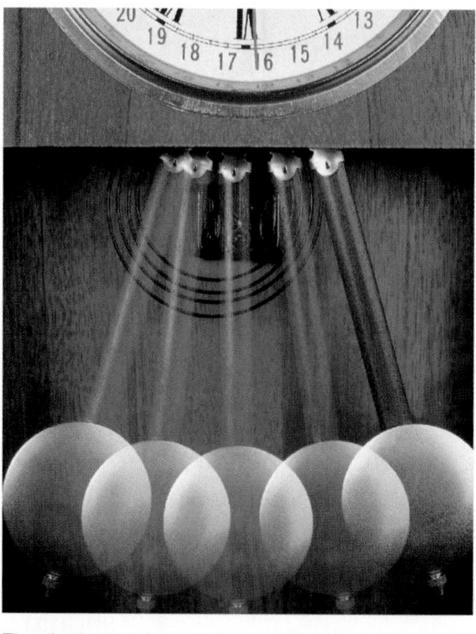
The rhythm of the pendulum determines the rate at which the hands of the clock move.

What Makes It Tick?

Since a pendulum is much less complicated than a satellite, it is a good example to use when learning about the science inquiry process. Consider a simple pendulum consisting of a mass tied to the end of a string. To begin the process of science inquiry, clearly state the problem in the form of a question. For example, "What property or properties of the pendulum determine the period of its motion?" Next, collect all the information you already have about the problem. This simple pendulum consists of only a length of string and a mass. In order to make the pendulum swing, you must pull the bob up to one side and release it. Now you are ready to state a **hypothesis**; for example, "The longer the string, the longer the period will be."

To perform a **fair test**, you need to identify what **variables** might affect the outcome, and then you have to control those variables. In the case of the pendulum, the possible variables are:

- the length of the string
- the angle at which you hold the string and the bob away from the vertical position
- the mass of the pendulum bob
- the period of the pendulum

Since you can choose the values of *length of string*, *mass of bob*, and *angle of swing*, these factors are called **independent** (or **manipulated**) **variables**. The period of the pendulum is the **dependent** (or **responding**) **variable**. The period will depend on (or respond to) your independent choice of length, mass, or angle.

To control the variables when performing an **experiment**, you must keep all the variables constant except the one you are testing. For example, if you are testing the effect of the length of the string on the period of a pendulum, you must keep the mass and the angle the same for every test.

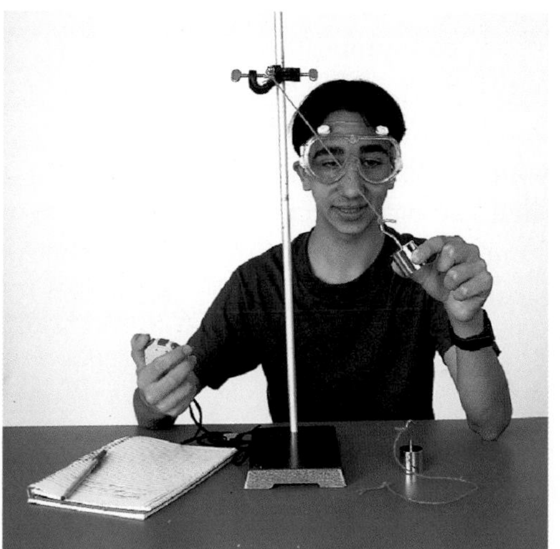

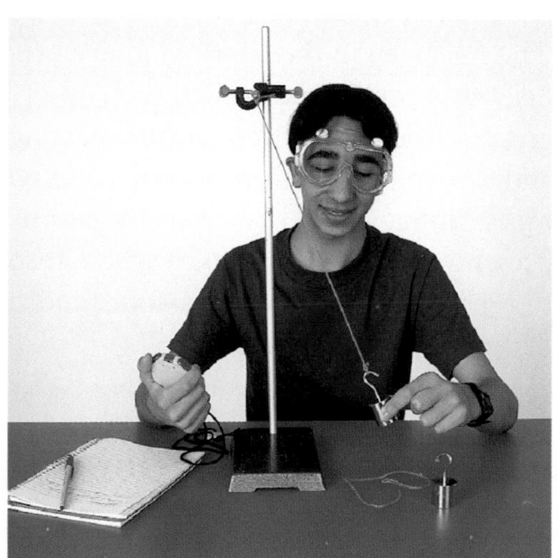
"I hypothesize that as the length of the string gets longer, the period will get longer."

hypothesis: a possible explanation for a question or observation, stated in a way that makes the explanation testable.

fair test: an investigation carried out under strictly controlled conditions to ensure accuracy.

variable: any factor that might influence the outcome of an experiment.

experiment: an activity or procedure designed to falsify a hypothesis. It may seem strange to attempt to prove something wrong. However, it is not possible to prove something to be absolutely true because there might always be one more experiment that will falsify the hypothesis. If you do not falsify a hypothesis, then your results support it.

observation: something seen and noted; the result of an experiment that you see and record

qualitative observation: an observation described using words only; for example, "When the length of the pendulum was increased, the period of the pendulum became longer."

quantitative observation: an observation that involves numbers; for example, "When the length of the pendulum was increased to 30 cm, the period increased to 1.1 s."

conclusion: an interpretation of the results of an experiment as it applies to the hypothesis being tested; for example, "Based on quantitative data for gasoline consumption, we found that regular gasoline is more efficient than premium gasoline."

When you carry out an experiment, you make **observations**, analyze them, and draw a **conclusion**. When investigating the pendulum, you would choose several different lengths of string. You would then attach the same mass to each length of string, hold each string at the same angle from the vertical, let the pendulum go, and measure the period of the pendulum with a stopwatch.

If your conclusion does not support your hypothesis, you have not failed. In fact, you may have made some observations that can lead to a new hypothesis. For example, you may have discovered that lengthening the string affected the period of the pendulum in a different way from what you had predicted. Thus, you can revise your hypothesis and try again. Even if your conclusion *does* support your hypothesis, you should not stop there. All experiments must be *reproducible*, that is, you should get the same results several times. Repeat the experiment, in exactly the same way in several *trials*, therefore, to confirm that your first results were not accidental. As well, if your experimental design has several possible variables, you should test them all one by one, keeping every variable — except the one you are testing — constant each time.

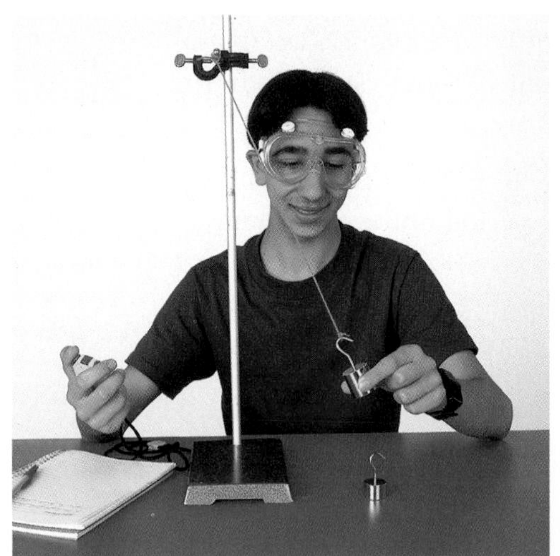

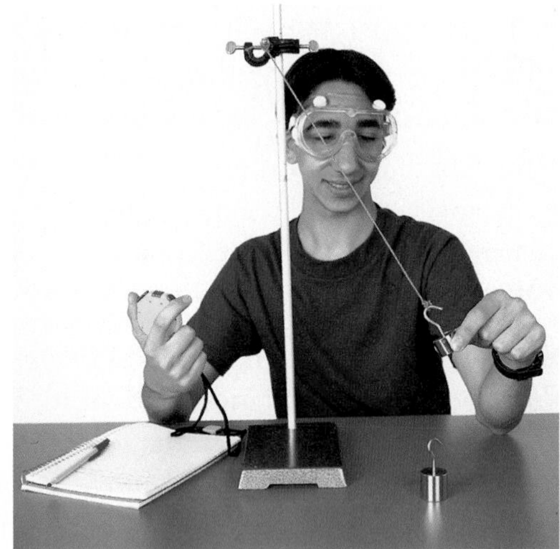

After you have made a conclusion about the effect of the length of the string on the period of the pendulum, you should test to see what effect, if any, differing masses might have on the period.

Communicating Results in Science

The final step in the science inquiry process is to communicate your results. Scientists communicate with each other by writing reports in scientific journals, by giving presentations at scientific meetings, and by discussing their research studies and experiments at these meetings. Often, your teacher will ask you to write a report about an activity or investigation. When you write a report, your goal is to present your procedure so clearly that someone reading the report would be able to repeat the experiment. In your report, you should also present your results and discuss your conclusion.

When many different scientists perform a variety of experiments to test the same hypothesis, and they all agree on the results, the hypothesis gains more and more support. Eventually, when a hypothesis has been thoroughly tested and all results support it, the hypothesis will gain the status of a **theory**.

The process that you have just been reading about is a **model** of science inquiry designed to show you how scientific knowledge is accumulated. (To learn more about models, turn to SkillPower 8 on page 554.) You will discover that scientists use models in many different ways. In this course, you will also be using models such as the one in the photograph.

theory: an explanation of an observation or event that has been supported by consistent, repeated experimental results, and has therefore been accepted by a majority of scientists. For example, you have probably heard of the theory of plate tectonics, also known as "continental drift."

model: a mental image, diagram, structure, or mathematical expression that attempts to explain a concept or hypothesis. A model of a volcano can be made from clay, a small amount of baking soda, vinegar, and a bottle cap. Other models can be devised on a computer.

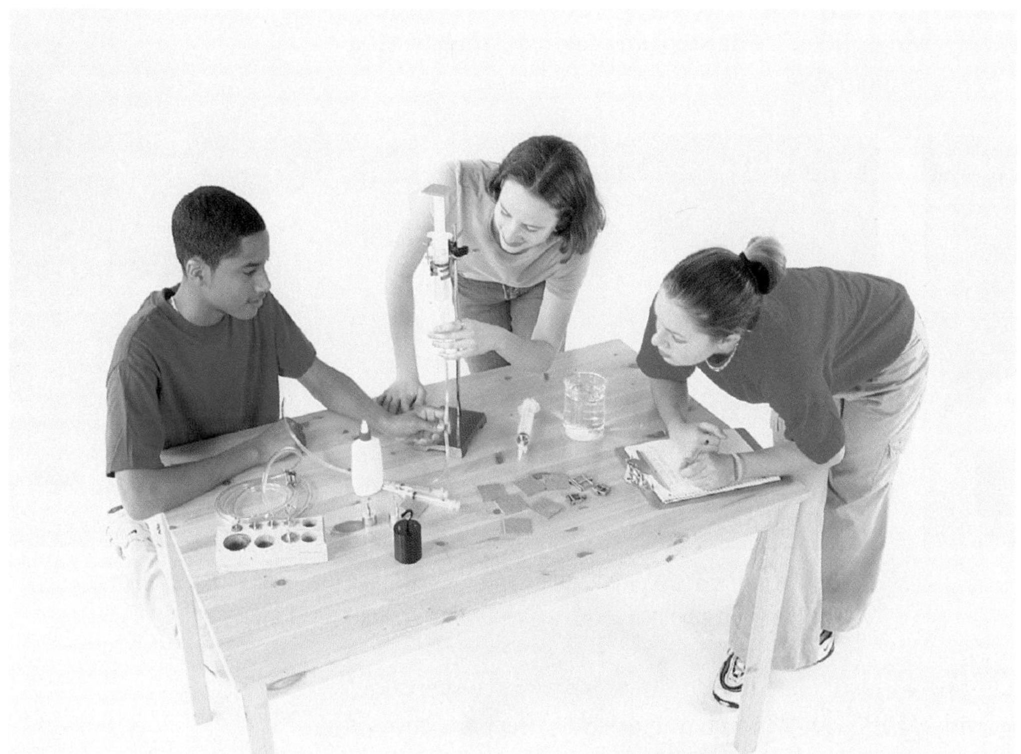

The students in the photograph are building a model of a hydraulic lift, similar to the lift that auto mechanics use to raise cars so they can work underneath them.

Now that you have learned about the process of science inquiry, it is time to try it out. By carrying out the investigation on the following pages, you will gain experience that will help you when you perform the activities and investigations in this book.

CONDUCT AN

INVESTIGATION IS-A

Observing Glider Characteristics

In this investigation, you will use the concepts you have just learned about the process of science inquiry. You will identify and test the variables that affect the motion of a device called a "ring-wing glider." The steps in making a ring-wing glider are shown in the photographs below.

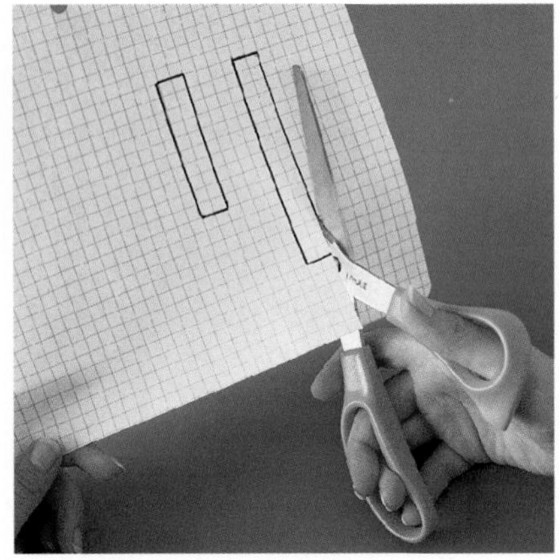

A Cut two strips of paper of different lengths from graph paper.

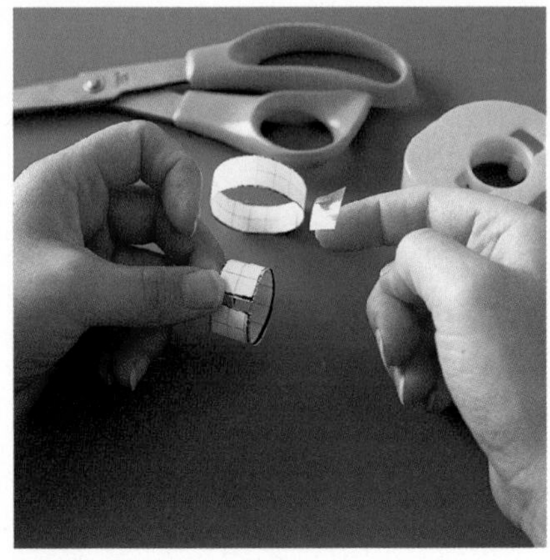

B Roll the strips of paper into loops and tape the ends together.

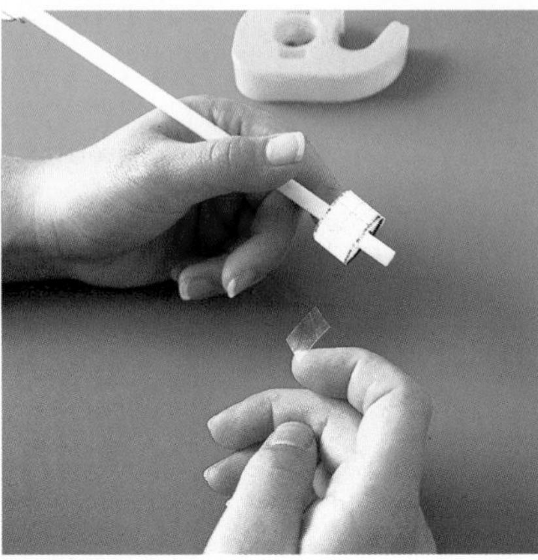

C Tape the loops to the ends of a drinking straw.

You have probably made many paper airplanes, but you may never have seen one that looked like a ring-wing glider. Find out what characteristics affect the motion of this type of glider. For example, how far apart should you position the small ring and the large ring? How does the width of the rings affect the flight? Does the glider fly better if the large ring or the small ring is in the front when you launch it? Should the straw be held at the top or the bottom when you fly the glider? Should you launch the glider rapidly or gently?

Problem

How do changes in two different characteristics (variables) of the ring-wing glider affect its flight patterns?

Safety Precautions

Be careful when using sharp objects such as scissors.

Apparatus

pencil
scissors
ruler

Materials

non-bendable drinking straws
graph paper
transparent tape

Procedure

1. To collect some general information about the ring-wing glider, each member of the group can make his or her own glider. Choose a size and shape for your glider and assemble it.
2. The whole group should observe the flight of each glider to gather information. Hold the gliders in different positions to find out if one way of launching them works better than another.

3. As a group, choose two different characteristics (independent variables) to test. For each variable, formulate and write down a hypothesis that predicts how the variable will affect the flight of the glider. For example, "We predict that placing the rings closer together will make the glider fly farther."
4. For each of the two variables you selected, decide on a basic design for all the glider characteristics except the one you are going to test. (Remember, to carry out a fair test, you must change only one variable at a time.) For example, if you decided to test the diameter of the rings, then the width of the rings and the distance between the rings must remain constant (the same). As well, you must hold the gliders in the same position and launch them with the same speed and amount of force when testing their flight patterns.
5. For each variable you have chosen, build at least four gliders that have differences in the variable you are testing.
6. Fly your gliders. To make good comparisons, test each glider several times. Have several different group members test each glider. Observe the flight characteristics carefully. Write down qualitative data (descriptions of flight patterns) for each of the two variables you have tested.

Analyze

1. Was there a relationship between either of the variables that you tested and the distance that the glider flew? If so, describe the relationship in writing.
2. Was there a relationship between either of the variables that you tested and the smoothness of the flight pattern of the glider? If so, describe the relationship in writing.

Conclude and Apply

3. For each of the two characteristics that you tested, did your results support or falsify your hypothesis? Explain.
4. Write a summary statement that describes any relationships that you observed between the characteristics you tested and the flight patterns of the ring-wing glider. Submit your summary to your teacher.

Technology and Technological Problem Solving

What would your life have been like if you had lived when farmers ploughed their fields with ploughs like this one?

How do modern tractors such as this one affect your everyday life?

The lives of ordinary people have changed a great deal over the last several hundred years. The two photographs hint at one of many reasons for these changes. Many years ago, one farmer could produce only enough food for his own family and a few other people. However, present-day farmers, with modern equipment, can provide food for thousands of people. Therefore, most people are free to pursue careers other than cultivating enough food to eat. Engineers and inventors have had time to develop electric lighting, telephones, televisions, trains, airplanes, automobiles, and computers, as well as many more modern conveniences. What is one word that can describe all these changes in our lives? Technology.

What is technology and how does it differ from science? **Technology** is the application of scientific knowledge and everyday experience to solve practical problems. Science and technology go hand in hand. When scientists discover new scientific principles, technologists can use the information to develop a new device or technique. For example, scientists studied the nature of sound and the behaviour of sound waves. On the basis of these findings, engineers and technologists developed ultrasound instruments. These instruments can diagnose conditions such as heart disease and can check the health of a fetus. Also, instruments using sound waves can measure the depth of the ocean.

Many forms of technology were developed before the scientific principles were understood. For example, humans made boats without knowing why they would float. They knew from experience that certain shapes would make boats more stable on water, although they did not know why. As well, new technologies allow scientists to carry out their research more efficiently. If you were to walk into any modern science research laboratory, you would see many advanced technological tools and instruments. Knowledge in science is growing more rapidly than ever, thanks in part to the technologically advanced instruments now available.

Technological Problem Solving

Just as scientists can reach their goals more efficiently by using an orderly science inquiry process, engineers and inventors can work more effectively when they follow an orderly problem-solving method. The concept map on the right shows a typical problem-solving procedure.

In this course, you will have many opportunities to apply the technological problem-solving method in *Design & Do Investigations* and in the end-of-unit *Projects*. In these projects and investigations, you will be given certain design criteria and then asked to develop and carry out a plan. You will test your product, system, or device, and evaluate how well it fulfilled the criteria. Finally, you will communicate your results and your evaluation of those results. Now, you will have a chance to practise the technological problem-solving process.

Solving a Technological Problem

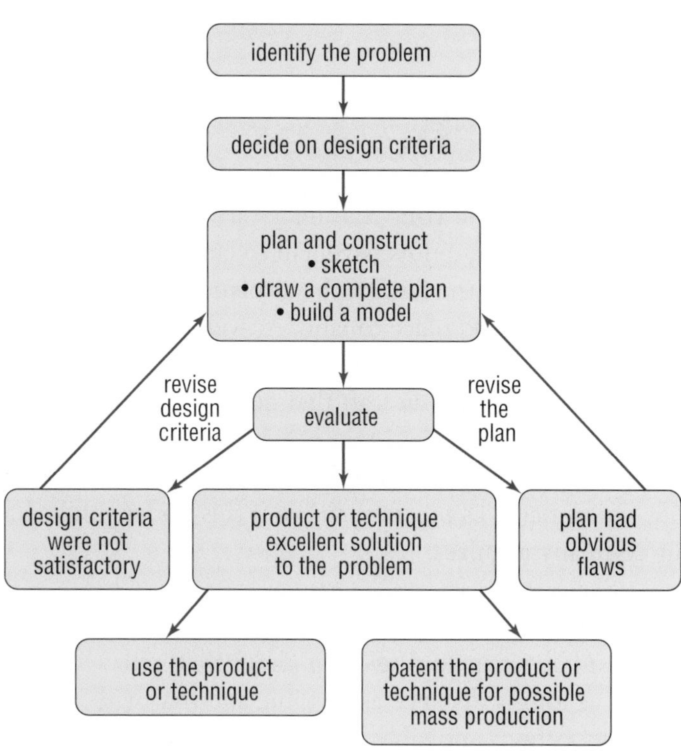

In your investigation of glider characteristics, you saw that strange-looking objects such as the ring-wing glider can actually fly. Your ring-wing glider may not be as strange as the airplane in the photograph, but it gives you an opportunity to use the information you discovered in your science inquiry to develop a new, and possibly improved, technology.

Who would ever have imagined that an airplane like this could fly?

Going the Distance

Imagine that you and your teammates are engineers for an airplane design company. You know that many pilots have had to land a plane that has lost all power. However, you also know that the farther the airplane can glide, the greater the chance that the pilot will find a good place to land. As well, the smoother the airplane glides, the easier it will be for the pilot to land it safely. Therefore, you and your teammates are working on designs for aircraft that glide as far as possible and that land as smoothly as possible. You have heard that some scientists are testing the gliding characteristics of a new ring-wing glider. You have decided to obtain as much information about those experiments as possible, to see if this glider might have the characteristics that your company requires.

Challenge

Design and construct a ring-wing glider that will glide for as long a distance as possible and that will land as smoothly as possible.

Apparatus

graph paper, pencil, scissors, ruler, long measuring tape

Materials

non-bendable drinking straws, transparent tape, masking tape

Safety Precautions

Be careful when using sharp objects such as scissors.

Design Criteria 

A. Your ring-wing glider must be designed and built by a team of students in no more than 20 min.

B. Your ring-wing glider must withstand five flights.

C. Your ring-wing glider must be no longer than two drinking straws.

D. You must draw and clearly label a sketch, with dimensions, before you begin constructing your ring-wing glider. Have your teacher approve the sketch.

Plan and Construct

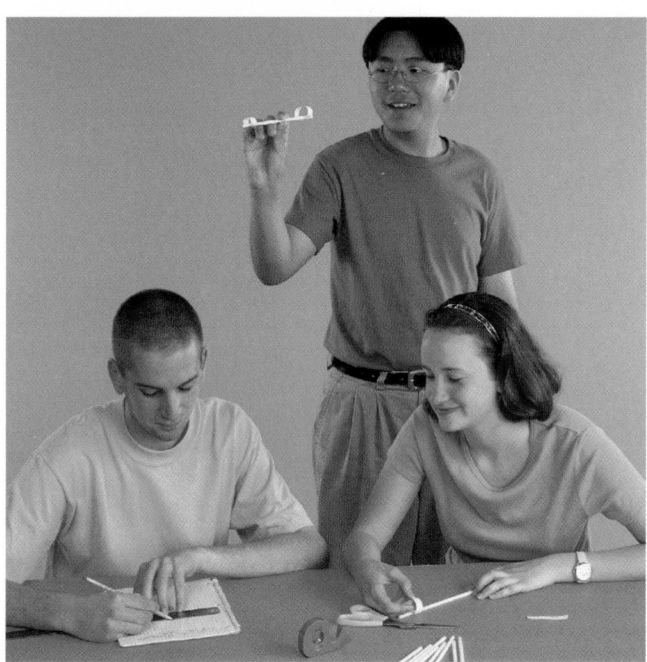

1. Your teacher will give you a summary of all the class data on the relationships between the variables of the ring-wing glider and the flight patterns. As a group, study the summary of the scientific data. Discuss the class data, your own experience with the ring-wing glider, and the design criteria.

2. Choose a specific design for your ring-wing glider. Draw the plan to scale. Study your design and discuss any possible changes you wish to make. Agree on a final design and show it to your teacher.

3. Assemble your ring-wing glider.

4. Test your ring-wing glider.
   - **(a)** Use masking tape to mark a spot for the "pilot" to stand when launching the glider.
   - **(b)** Launch the glider five times. Each team member may launch the glider or the team may elect a pilot.
   - **(c)** Using the measuring tape, measure and record the distance that the glider travelled during each trial.

5. Calculate the average distance that your glider travelled.

6. Post the class results and display the gliders.

Evaluate

1. Examine each of the gliders and their gliding distances. Evaluate your own design. If your glider did not travel the greatest distance, how did it differ from the glider that did? If your glider travelled the farthest, what characteristics do you think account for its success?

2. After considering all of the gliders, how would you revise yours to make it travel farther? If you have time, build and test your revised plan.

Societal Decision Making

For most people, this photograph of icebergs conjures up an image of a remote part of Earth, far removed from our everyday lives. Their potential importance may not seem immediately clear. Indeed, many people, if they think about icebergs at all, remember them in connection with the sinking of the *Titanic*. Glaciers and icebergs, however, contain most of the world's fresh-water supply, even though it is locked up in ice. This may seem surprising because of the amount of water present on Earth. As you will learn in more detail in Chapter 10, Earth is commonly referred to as the blue planet. It is easy to see why from the photograph of Earth as seen from space.

The blue comes from the water that covers more than 70 percent of Earth's surface. Water is vital to us. Without it, life could not exist. Yet most of it is unavailable because it is in the salt-water oceans. The water we drink comes mainly from fresh-water lakes and rivers, but these represent a tiny amount of the total water on Earth (see the circle graphs on the following page).

Think about what is happening to Earth's supply of drinking water. Look up statistics on the number of people born worldwide every minute. This tremendous increase in population has only been occurring over the last century. This, combined

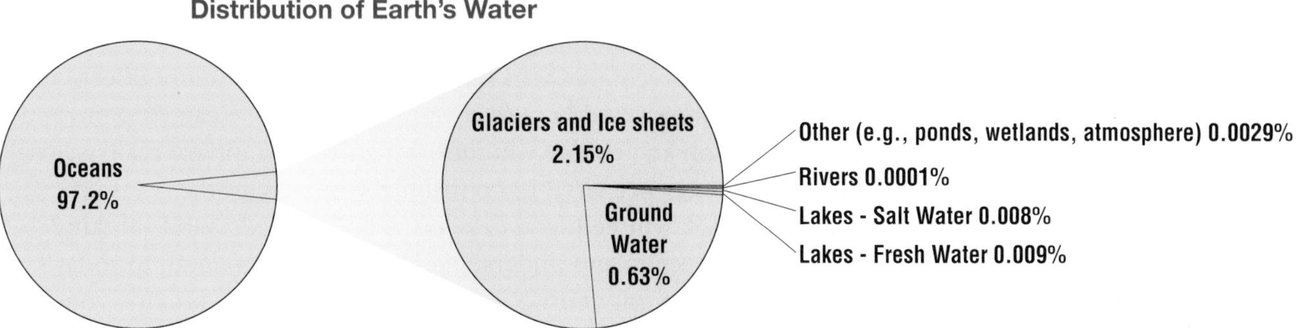

with pollution and other uses of water (in industry, for example), has placed a great demand on fresh-water supplies. Recent studies indicate that the supply of fresh water will soon be insufficient to meet the needs of Earth's growing population. This is already the case in some areas of the world, and some countries need to import fresh water. In Canada we are fortunate because of our very large supply of fresh water. Are we doing our best to use it wisely and to preserve it as a resource for the future? Think about this.

What options would we have if the water supply could not meet our needs? Should we consider "harvesting" the water locked up in ice? We could develop the technology to do this efficiently, but we would have many issues to consider.

- What effects would harvesting water from ice have on the environment?
- Would it affect the pattern of ocean currents, the weather, the balance of nature?
- If so, should we then abandon this idea?
- What alternatives should also be considered?
- What factors need to be taken into account before a decision is reached?

During your study of science this year, you will encounter various situations that involve science, technology, society, and the environment, sometimes abbreviated STSE. When these elements come together, it is important to consider how they interact and influence each other and the world we live in. It is important because no one lives alone. We all belong to a larger group, the society we live in, and what each of us does has an effect on others as well as on the environment.

The problem of fresh-water supply is only one example of how STSE interactions often raise a variety of issues. It is clear that there are many complex aspects to consider in deciding how best to resolve such issues. Technology can help meet societal needs, but its effects on the environment, which includes us as well, must be taken into account.

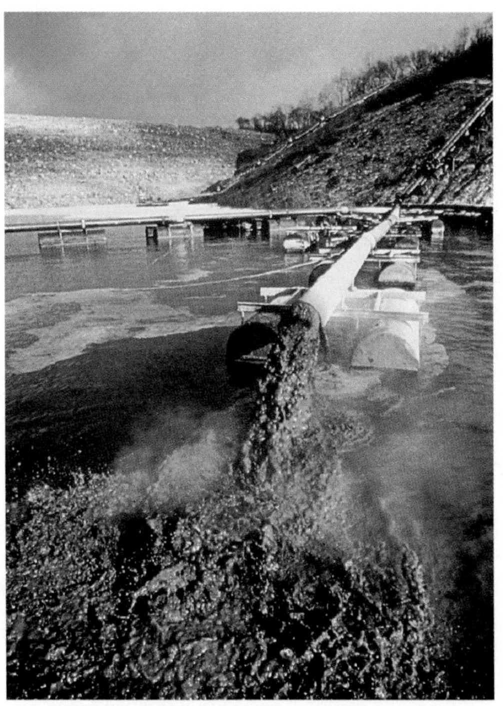
Human activity has an impact on our water supply.

The Role of Science

Science can guide us in reaching decisions because it provides us with basic information that helps us weigh the advantages and disadvantages of making any particular decision. However, it is important to realize that science alone cannot determine the decision, on either personal or public issues, for us. There are many other aspects that play a role in shaping how each person views the world, such as literature, history, philosophy, and cultural beliefs. All viewpoints should be considered with respect.

The roles of science and technology in resolving issues can be more effective if the citizens of a society understand how science works. If people can recognize valid scientific information and decide whether "scientific" claims are really scientific, then they will be better prepared to contribute positively to the decision-making process. This book will help you develop your knowledge and thinking skills so that you can appreciate and understand how science is related to today's societal issues and those that might arise in the future. It will also provide you with critical skills and concepts that will help you in future science courses. These same skills will be useful throughout your life in making important decisions about personal and community issues.

Use the "Developing Decision-Making Skills" flowchart shown here as a guide to analyzing issues.

Developing Decision-Making Skills

Identify the Issue

Gather Relevant Information

Identify All the Alternatives

Weigh Each Alternative by Clarifying its Consequences

Make a Decision

Evaluate the Decision

The decision is the best alternative based on risks/benefits and, thus, probable consequences.

One or more of the steps in the decision-making process were faulty. No action should be taken and the process should be repeated to ensure that the faulty steps are eliminated and replaced by improved thinking.

Errors of judgment may have been made at any of these steps in the decision-making process.

Take Action/Communicate the Decision

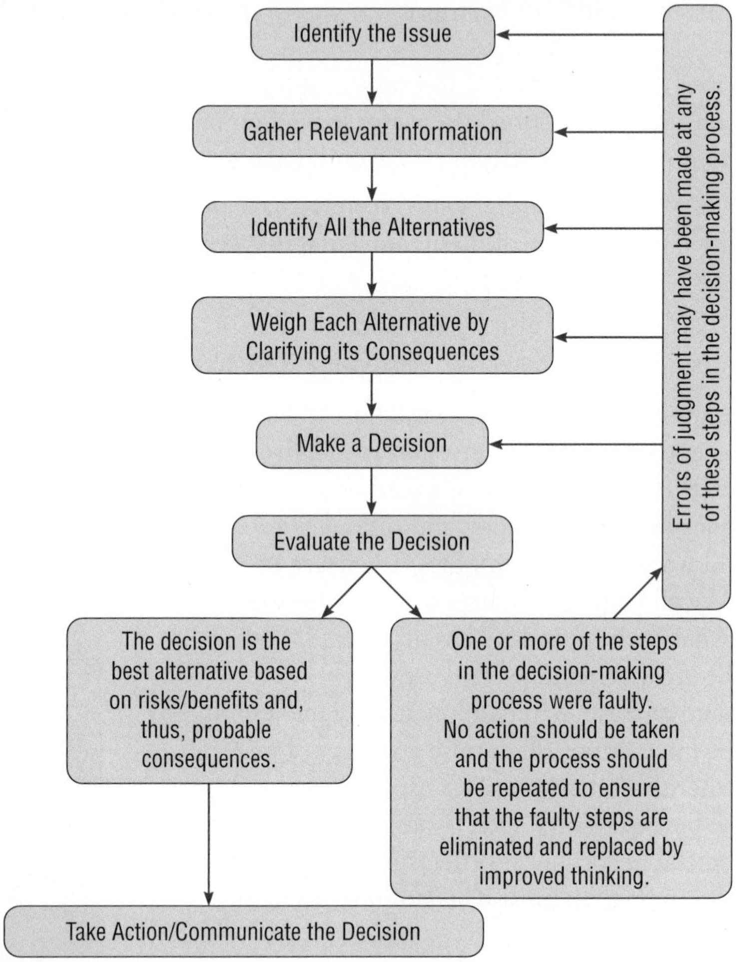

Your ***SCIENCEPOWER***™ ***8*** textbook promotes an understanding of issue analysis, but two special features are specifically designed to help you develop your skill at making decisions. Many of the *Think & Link Investigations* deal with issues. All of the activities in *An Issue to Analyze* in each unit provide an issue to debate or a simulation to role-play, providing you with the opportunity to participate in critical-thinking and decision-making situations.

Instant Practice

1. Describe two ways in which each of the following technological developments has affected society positively:
   - **(a)** dam building
   - **(b)** production of supertankers
   - **(c)** mass production of the automobile
   - **(d)** mass production of computers
   - **(e)** harnessing nuclear reactions

2. Now describe two ways in which each of the above technologies has affected or might affect society negatively.

3. How would you evaluate whether each of the technologies in question 1 is "good" or "bad" for society?

4. For each of the technologies in question 1, record ideas you have on the kinds of scientific knowledge that were necessary for its development.

5. An issue has two or more possible solutions, and a decision must be made in favour of one solution over the others. In weighing the advantages and disadvantages of the solutions, is it important to take into account:
   - **(a)** Who benefits from the solution?
   - **(b)** Who is affected by the costs or disadvantages?
   - **(c)** Should we be careful in selecting a solution based on who proposes it? Why or why not?

6. As a class, discuss the possible shortage of fresh water. Identify possible solutions and consider the benefits and risks of each solution. Do the solutions involve the use of technology? Do we have enough scientific knowledge to assess the risks attached to each solution? Is there a simple answer? Is there a solution that does not interfere with the environment at all? If so, is it a realistic solution?

UNIT 1

Cells and Body Systems

What does a whale have in common with a tiny pond organism too small to be seen with the unaided eye? To answer this question, you would have to study both the whale and some pond water with a microscope. It is amazing what the magnifying lenses of a microscope reveal. A spoonful of pond water is full of living creatures, many of them consisting of only single cells, like the amoeba shown here greatly magnified. The whale's enormous body is also made entirely of cells — trillions and trillions of them.

No one had observed cells before microscopes were invented in the seventeenth century. For the first time in history, scientists could study the detailed structures of living things. Scientists' observations led them to hypothesize that cells are the basic building blocks of all forms of life. From their observations came the theory that each cell is a miniature living system, able to carry out the same activities as a living organism.

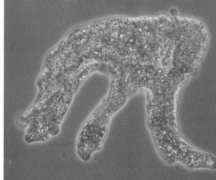

By learning what cells are, and how they work, you can understand better what makes all forms of life possible. You will also understand better the relationship between healthy cells and body systems, and healthy organisms as a whole — including yourself!

Unit Contents

CHAPTER

1 Observing Cells

Getting Ready...

- What do you think a cell looks like?
- A wooden baseball bat is made up of dead tree cells. What cell part gives the bat its hardness?
- Why is it not possible for a single-celled organism to grow to be as large as you?

Science Log

In your Science Log, sketch a diagram of your idea of what a cell looks like. Include as many of its parts as you know. Then answer the other two questions above as well as you can for now. Watch for answers to all these questions as you study this chapter.

Skill POWER

For tips on how to make and use a Science Log, turn to page 534.

This bird is alive! Movement is one of the signs of life. Is movement *always* a sign of life? An airplane moves, and it is certainly not alive. Describing the differences between living things and non-living things can be a little trickier than you might think at first. Many of the characteristics shown by living things, such as movement, growth, and reproduction, can also be seen in some non-living things. For example, crystals can grow, and blobs of oil floating on water can divide in two.

For biologists, however, one feature separates all forms of life from everything else. They have found that all living things are made of cells. The **cell** is the smallest unit scientists consider to be alive. Airplanes, crystals, and oil blobs are not made of cells. Thus they are not alive — even though they share some characteristics with living things. In this chapter, you will learn how scientists came to know about cells. You will use a microscope to observe cells for yourself and discover what is inside these remarkable structures.

Starting Point ACTIVITY

What Is Life?

One of the fascinating ideas you will explore in this chapter is that every living cell can carry out the same functions as every living organism. How would you decide if something is living or non-living?

What to Do 

1. With a partner or in a group, brainstorm all the characteristics you think are shared by living things.
2. Make a list of your ideas or draw a sketch or a cartoon to illustrate them. Later in this unit you will look back to your list or sketch.

Skill POWER

For tips on working in groups, turn to page 536.

Spotlight On Key Ideas

In this chapter, you will discover

- why the cell is considered to be the basic unit of life
- how a single cell in your body can be compared to your body as a whole
- how single-celled organisms meet their basic life needs
- how single-celled organisms differ from organisms made of many cells
- how plant cells differ in structure from animal cells
- what structures are inside cells and what they do
- why cells are small

Spotlight On Key Skills

In this chapter, you will

- learn how to use and care for a microscope
- prepare samples for viewing on microscope slides
- find, observe, and draw objects under a microscope
- estimate the size of microscopic objects
- calculate the relationship between volume and surface area
- design your own three-dimensional model of a cell using everyday materials

1.1 Microscopes and Cells

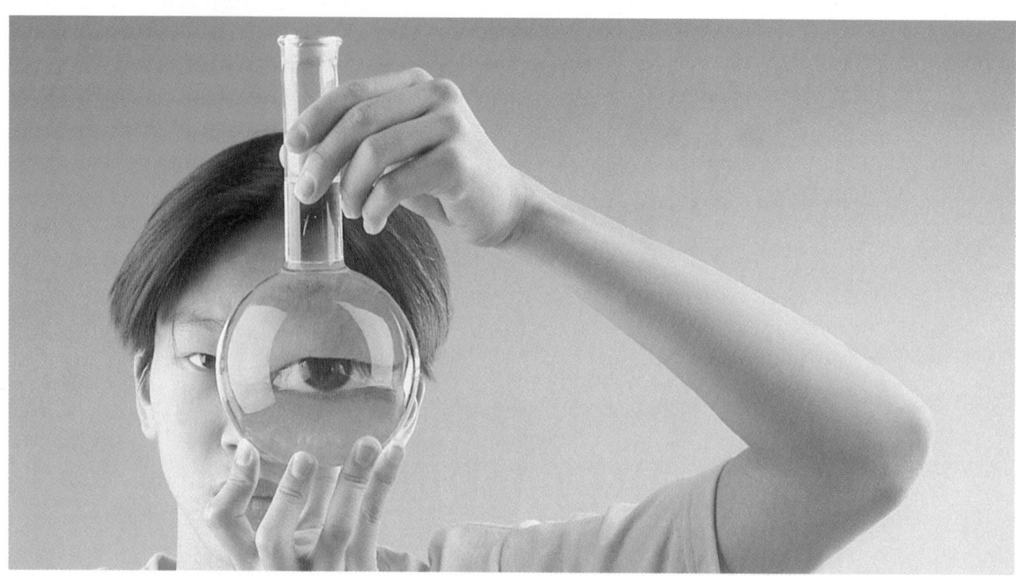

How do you enlarge something too small to be seen with the unaided eye? You do it in the same way that the student's face has been enlarged in the photograph shown here. **Magnifying** an object makes it appear larger.

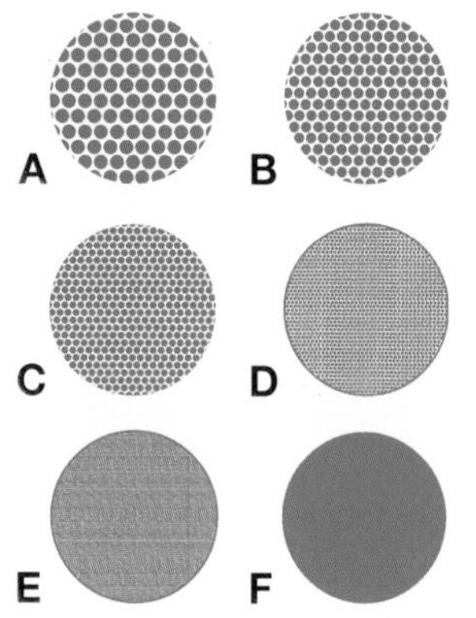

A World Too Small to See

It may be difficult to imagine, but there are living things all around you that you cannot see. The human eye can only see objects that are larger than 0.1 mm. Look at the circles of dots shown here. In the first circle, you can probably see individual dots. In which circle does the colour appear solid? Is the colour really "solid" or is it, too, made up of dots? For most of us to see separate dots, they must be more than 0.1 mm apart.

Find Out **ACTIVITY**

What Does It Take to Enlarge an Object?

For how long have people tried to magnify objects? Two thousand years ago, early Romans used waterfilled vessels to magnify objects that they were engraving. Test a similar magnifying technique in this activity.

What You Need

bottles, jars, and flasks of different shapes
water

What to Do

1. Predict which container shape will magnify the print in your textbook or in a newspaper the most.
2. Fill the containers with water and test your prediction.

What Did You Find Out?

If you were to design a lens to be placed in a magnifying glass for a detective to check out evidence of a crime, draw the shape of the lens you would use.

Early Microscopes

Early scientists found out exactly what you discovered in the Find Out Activity about the best shape for good magnification. One of the greatest explorers of all time discovered a whole new world without leaving his room. He was a Dutch linen merchant named Anton van Leeuwenhoek. His hobby was making magnifying lenses. With his great skill at grinding very small lenses, Leeuwenhoek made instruments called **microscopes**, which magnified objects up to 300 times (300×). Microscopes magnify objects by bending light through a lens. (You will learn more about lenses and magnification in Unit 3, Light and Optical Instruments.) Using his simple microscopes, Leeuwenhoek studied substances such as blood, pond water, and matter scraped from his teeth. He became the first person to observe organisms made of only one cell. He called these single-celled organisms "animalcules."

When Leeuwenhoek began writing about his discoveries in 1674, he created a sensation. Nobody before had suspected we are surrounded by a parallel world of living things so small they cannot be seen by the unaided eye. Leeuwenhoek's discoveries excited people's imaginations. Some people wondered if single-celled organisms might help answer the age-old question: What is life?

DidYouKnow?

Leeuwenhoek discovered micro-organisms by making lenses that could magnify objects as much as 300 times. By comparison, the palm of your hand magnified 300 times would be about the size of an average door.

Figure 1.1 Anton van Leeuwenhoek (1632–1723) used a simple homemade microscope to observe single-celled organisms and other objects.

Figure 1.2 Leeuwenhoek was the first to see red blood cells such as these (160×).

Pause&
Reflect

Imagine yourself in the place of Anton van Leeuwenhoek. What was life like at the time he lived? Use your library or a computer to research van Leeuwenhoek's life and times. Create a comic book, a play, a video, a talk, or some other means to present your ideas about this scientist and his experiences to your class.

Without microscopes, scientists could not have learned anything about the structure of cells. Our eyes cannot see such small objects unless the objects are magnified. Leeuwenhoek's microscopes had only one lens, similar to a magnifying glass. The sample (object) being studied with the microscope was held steady on a platform. A light placed behind the platform produced a bright image.

Microscopes Today

Improvements in technology and design gradually led to the development of modern compound light microscopes, such as the ones in your school. Compound light microscopes have two lenses, which give a greater power of magnification.

The best light microscopes can magnify objects as much as 2000×. This still is not enough, though, to see some of the smaller structures inside cells. For this, scientists use electron microscopes, which use beams of electrons instead of light. The electrons are bounced off the sample, then enlarged to form an image on a television screen or photographic plate. The first electron microscope was built in Germany in 1932. It could magnify up to 4000×. In 1938, the first practical electron microscope was developed by two Canadians at the University of Toronto: James Hillier of Brampton, Ontario, and Albert Prebus of Edmonton, Alberta. To test their valuable new laboratory tool, they first looked at the edge of a razor blade. Under a light microscope, the magnified blade edge appeared relatively smooth. Under their electron microscope, however, the same edge looked like a mountain range of rugged peaks and valleys! This electron microscope could magnify up to 7000×.

Both light and electron microscopes are used extensively today by scientists, engineers, and medical practitioners. Can you think of different microscopic objects each might want to observe?

Your investigations in this unit begin with an introduction to effective microscope use. With these skills, you, too, will be able to explore the microscopic world around you.

Figure 1.3A A compound light microscope

Figure 1.3B A scanning electron microscope (SEM)

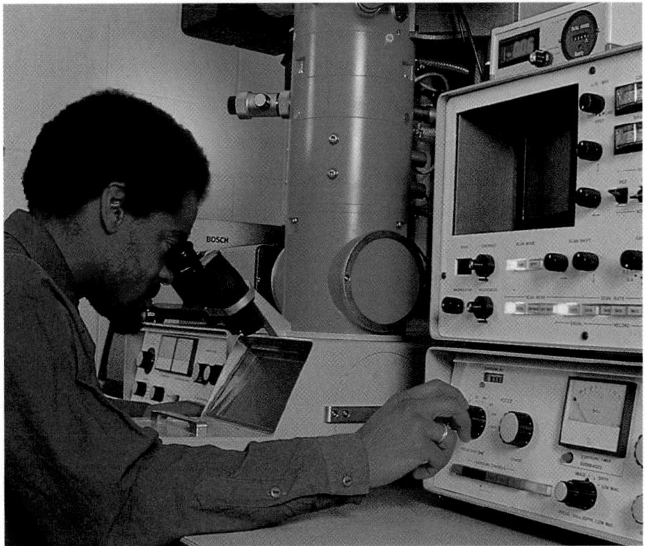

Figure 1.3C A transmission electron microscope (TEM)

Modern electron microscopes can magnify objects 2 000 000×. The image can be viewed on a television screen. However, it is usually photographed. The resulting image is known as an *electron micrograph*.

There are two main types of electron microscopes. In a transmission electron microscope (TEM), electrons are passed through very thin sections of a sample. Besides being sliced very thin, the object has to be placed in a vacuum. There is no air in a vacuum. Thus, only dead cells and tissues can be observed with a TEM. A scanning electron microscope (SEM) is used to observe the surfaces of whole objects. With an SEM, you can view and photograph living cells. In this type of electron microscope, electrons are reflected back from the surface of the sample, producing three-dimensional images.

Electron microscopes have helped scientists understand many microscopic structures such as the parts of a cell. The image on the left below was produced by a transmission electron microscope, while that on the right was produced by a scanning electron microscope.

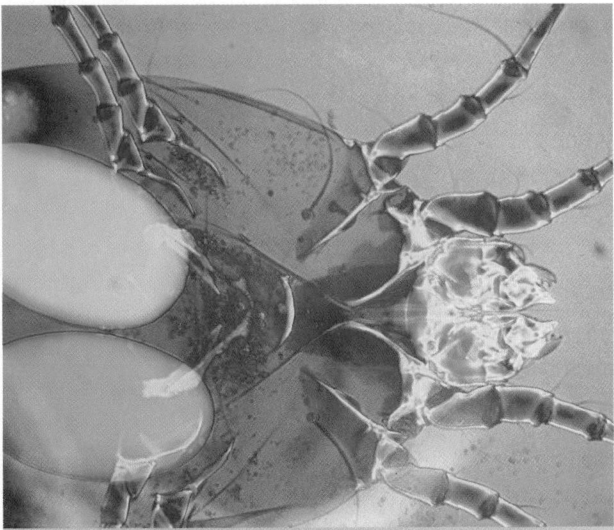

A This micrograph of a thin slice of a dust mite was taken by a transmission electron microscope.

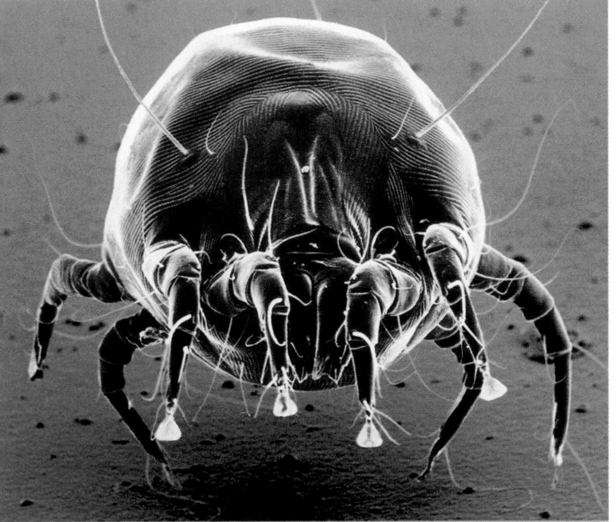

B Scanning electron microscopes show great detail on the surface of an organism. This is a tiny dust mite magnified 350×.

CONDUCT AN

INVESTIGATION 1-A

Using a Microscope

In this investigation, you will learn about the different parts of a compound microscope and how to use them. Then you will look at some prepared microscope slides provided by your teacher. Practise drawing what you observe through the microscope. In this investigation, you will also find a way to estimate the size of microscopic objects. With these skills, you will later be able to study cells from plants and animals, and observe live microscopic organisms such as those first seen by Leeuwenhoek.

Part 1

The Compound Light Microscope

Problem

What are the parts of a microscope?

Procedure

1. Study the photograph of the compound light microscope. Learn the names and functions of the different parts of the microscope.

2. Before going on to Part 2, close your book, and draw and label as many parts of the microscope as you can.

A Eyepiece (or ocular lens)
The part you look through. It has a lens that magnifies the object, usually by 10 times (10×). The magnifying power is engraved on the side of the eyepiece.

B Tube
Holds the eyepiece and the objective lenses at the proper working distance from each other.

C Coarse-adjustment knob
Moves the tube or stage up or down to bring the object into focus. Use it only with the low-power objective lens.

D Fine-adjustment knob
Use with medium- and high-power magnification to bring the object into sharper focus.

E Arm
Connects the base and tube. Use this for carrying the microscope.

F Revolving nosepiece
Rotating disk holds two or more objective lenses. Turn it to change lenses. Each lens clicks into place.

G Objective lenses
Magnify the object. Each lens has a different power of magnification, such as 10×, 40×, and 100×. The magnifying power is engraved on the side of each objective lens. Be sure you can identify each lens. For example, the low-power objective lens is usually 10×.

H Stage
Supports the microscope slide. Clips hold the slide in position. A hole in the centre of the stage allows the light from the light source to pass through the slide.

I Condenser lens
Directs light to the object being viewed.

J Diaphragm
Use this to control the amount of light reaching the object being viewed.

K Light source
Shining a light through the object being viewed makes it easier to see the details. (Your microscope might have a mirror instead of a light. If it does, you will adjust it to direct light through the lenses.)

Parts of a Compound Light Microscope

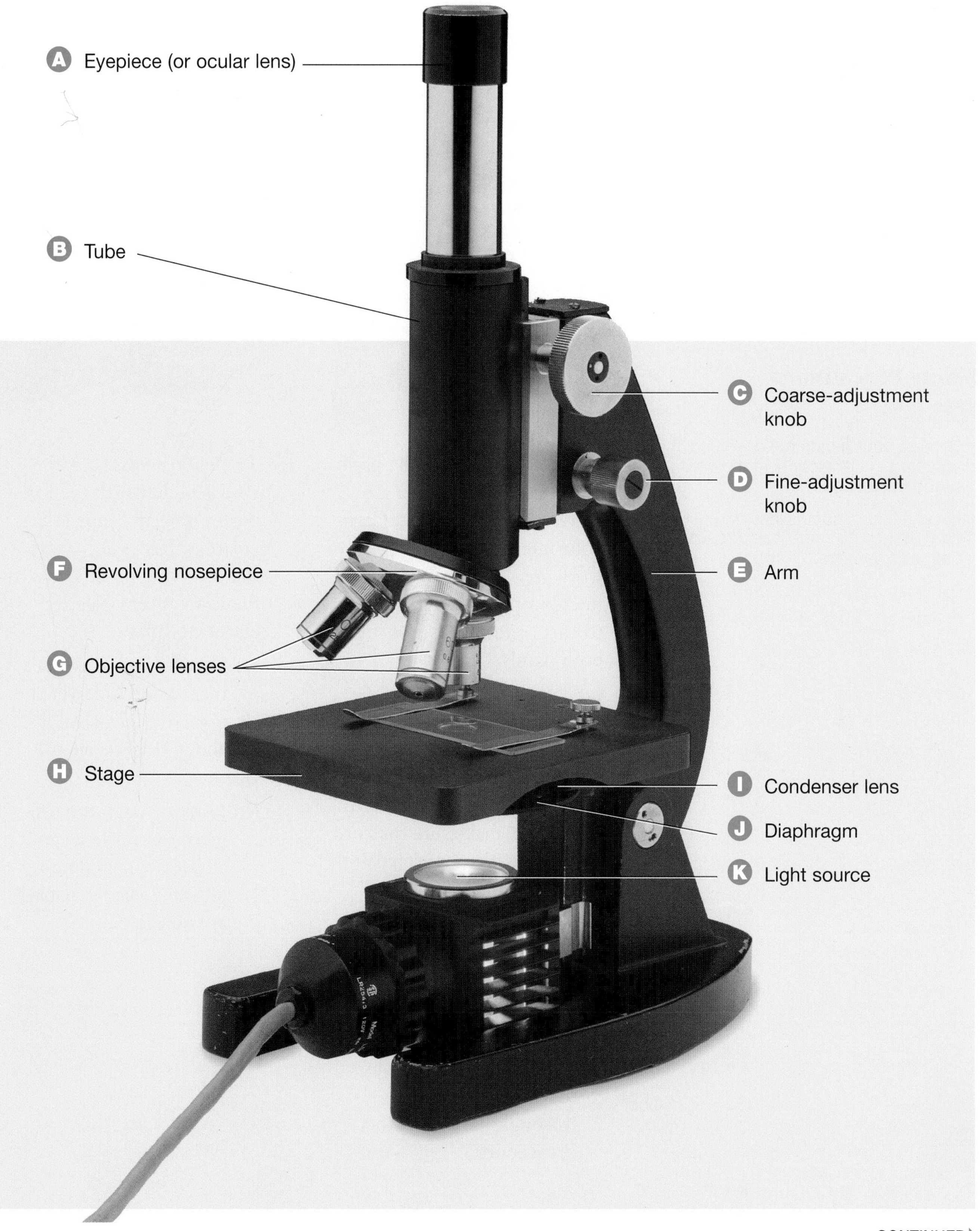

CONTINUED

Part 2
Using Your Microscope

Problem

What is the proper way to use and care for a microscope?

Apparatus

microscope

prepared microscope slides

Materials

lens paper

Safety Precautions

- Be sure your hands are dry when you plug in or disconnect the cord of the microscope.
- Handle microscope slides carefully so that they do not break or cause cuts or scratches.

Procedure

1. Now that you know the parts of the microscope, you are ready to begin using it. Carry your microscope to your work area. When carrying a microscope, hold it firmly by the arm and the base, using both hands.
   - **(a)** Position the microscope at your work area with the arm toward you. If the microscope has an electric cord for the light source, make sure the cord is properly connected and plugged in.
   - **(b)** Use lens paper to clean the lenses and the light source (or mirror). Do not touch the lenses with your fingers.
   - **(c)** Do not turn any knobs until you have read through the rest of the Procedure.

2. The microscope should always be left with the low-power objective lens in position. If it is not, rotate the revolving nosepiece until the low-power objective lens clicks into place.
   - **(a)** Use the coarse-adjustment knob to lower the objective lens until it is about 1 cm above the stage.
   - **(b)** Look through the eyepiece (ocular lens) and adjust the diaphragm until the view is as bright as you can get it.

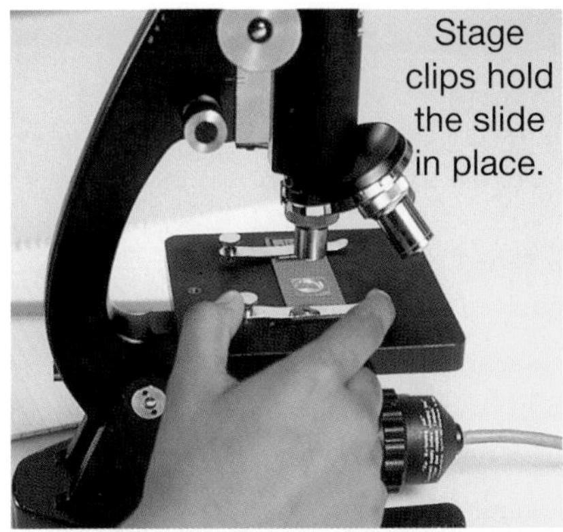
Stage
clips hold
the slide
in place.

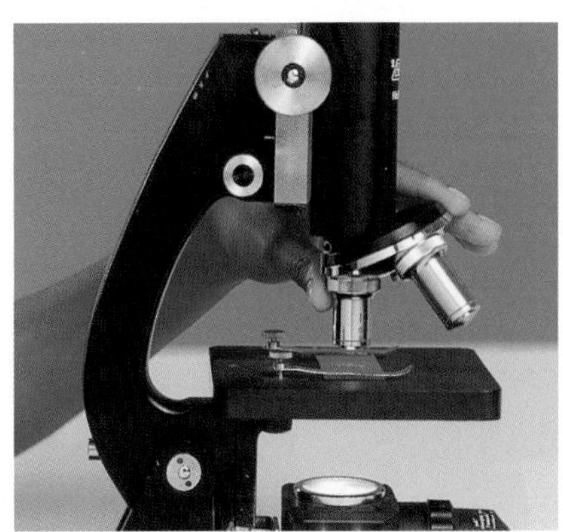

3. Place a prepared slide on the stage. Make sure the sample (object to be viewed) is centred over the opening.
   - **(a)** Look through the eyepiece and slowly turn the coarse-adjustment knob until the sample is in focus.
   - **(b)** Use the fine-adjustment knob to sharpen the focus.
   - **(c)** While looking through the eyepiece, move the slide a little to the left. In which direction does the image move? Move the slide a little away from you and then toward you. What happens to the image?

4. Find a part of the sample that interests you and, in your notebook, sketch what you see. Start by drawing a circle to represent the area you see through your eyepiece. This area is called the **field of view**. Make sure the details in your drawing fill the same space in the circle as they do when viewed through the microscope.
   - **(a)** Label your drawing to identify the sample.
   - **(b)** Calculate the magnification you are using. To do this, multiply together the magnifying power of the objective lens and the magnifying power of the eyepiece lens. Record this result on your drawing. **Example:** A 10× eyepiece and a 4× objective give a total magnification of 40×.

5. To see more details, rotate the revolving nosepiece to the next objective lens. **Do not change the focus first.** After the medium-power objective lens has clicked into place, adjust the focus using only the fine-adjustment knob.

   CAUTION: Do not use the coarse-adjustment knob with the medium- or high-power objective lens.
   - **(a)** When you have finished viewing and drawing the sample, remove the slide and return it to the proper container.
   - **(b)** If you do not continue to Part 3, carefully unplug the microscope and return it to the storage area.

Skill POWER

To learn how to make accurate scientific drawings, turn to page 557.

CONTINUED ▶

Part 3

Measuring the Field of View

Problem

How can the actual size of a microscopic object be determined?

Apparatus

microscope
prepared microscope slides
transparent plastic ruler

Materials

lens paper

Procedure

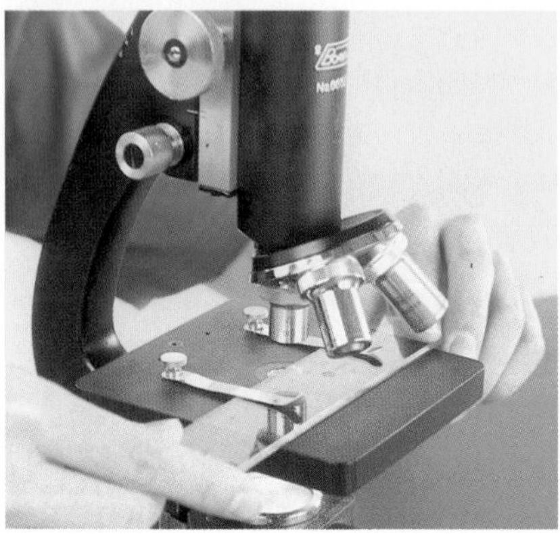

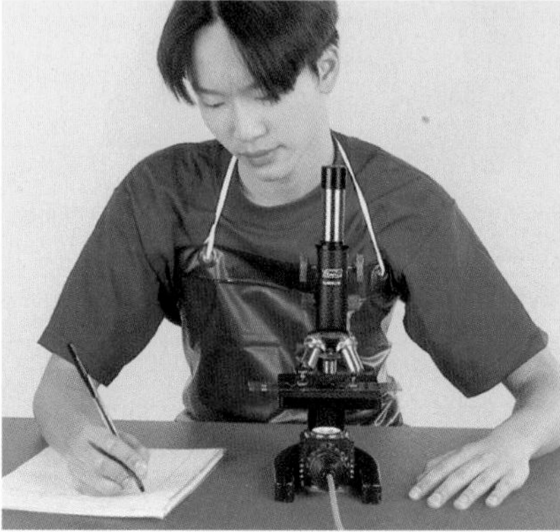

1. Set your microscope to the low-power objective and place a clear plastic ruler on the stage.

2. Focus on the ruler and move it so that one of the centimetre markings is at the left edge of the field of view.

3. Measure and record the diameter of the field of view in millimetres (mm). Millimetre markings on the ruler are too far apart to permit direct measurement of the field of view for lenses with magnification higher than 10×. You can, however, calculate the field of view for a higher magnification. To find out how to do this, go to "How to Calculate the Field of View" on the following page.
   (a) Unplug the microscope by pulling out the plug. Never tug on the electrical cord to unplug it.

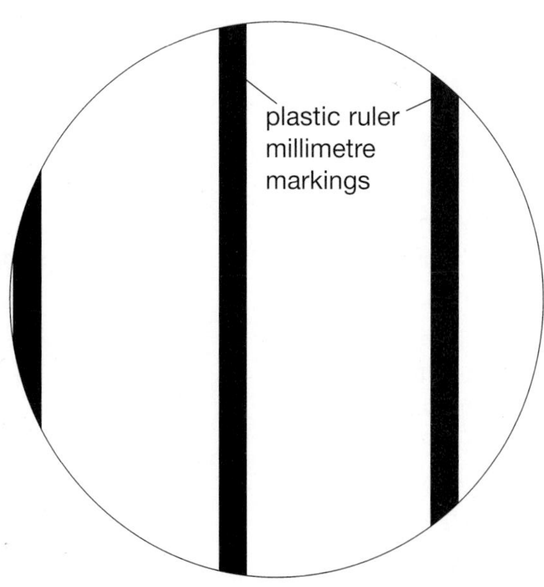

The diameter of the field of view illustrated here is 2.5 mm.

How to Calculate the Field of View

If you know the diameter of the field of view for the low-power lens, you can calculate the field of view for the other lenses. Use the following formula to do this.

$$\text{Medium-power field of view} = \text{Low-power field of view} \times \frac{\text{Magnification of low-power objective lens}}{\text{Magnification of medium-power objective lens}}$$

If, for example, your low-power objective lens is a 4× lens with a field of view of 4 mm, and your medium-power objective lens is a 10× lens, then your calculations would be:

$$\text{Medium-power field of view} = 4\text{ mm} \times \frac{4}{10}$$
$$= 4\text{ mm} \times 0.4$$
$$= 1.6\text{ mm}$$

Do a similar calculation to determine your high-power field of view. Record this value.

Analyze

1. How many lenses does the light pass through between the light source and your eye? Name them.
2. The coarse-adjustment knob is used only with which objective lens? Explain why.
3. How can you tell which objective lens is in the viewing position?
4. When you move a slide on the microscope stage away from you, in what direction does the object seen through the eyepiece move?
5. Make a table with two columns like the one shown here. Give your table a title. In the first column, list the parts of a microscope. In the second column, record the function of each part.

Microscope part	Function of microscope part

Conclude and Apply

6. Why should you never allow an objective lens to touch the slide?
7. Calculate the magnifying power of your school microscope when you use:
   (a) the medium-power objective lens;
   (b) the high-power objective lens.
8. Use your measurement in question 7, part (a) to calculate the diameter of the field of view under high power (in mm).
9. Why is the field of view under high-power magnification less than that under low-power magnification?

Skill POWER

If you need to review units of measurement in the metric system, turn to page 537.

Math CONNECT

Scientists measure the size of cells in units called *micrometres* (μm); 1000 μm = 1 mm. If you know that your field of view (its diameter) under an objective lens is 2.5 mm, how many micrometres is its diameter? If two cells of equal size occupy the entire field of view, what is the diameter of each cell in micrometres?

Preparing a Wet Mount

Now that you have learned how to use a microscope properly, you are ready to prepare some slides of your own, using a variety of materials.

Problem

How is a sample prepared on a microscope slide?

Safety Precautions

- Be careful when using sharp objects such as tweezers and scissors.
- Handle microscope slides and cover slips carefully so that they do not break and cause cuts or scratches.

Apparatus

microscope
microscope slides
cover slips
medicine dropper
tweezers
scissors

Materials

small piece of newspaper
tap water
other samples
lens paper

Procedure

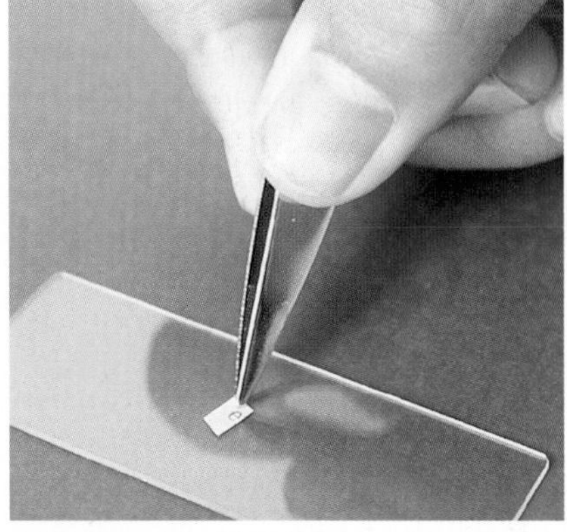

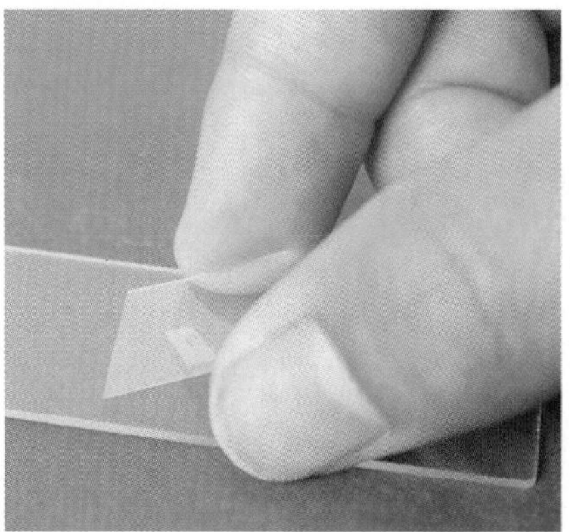

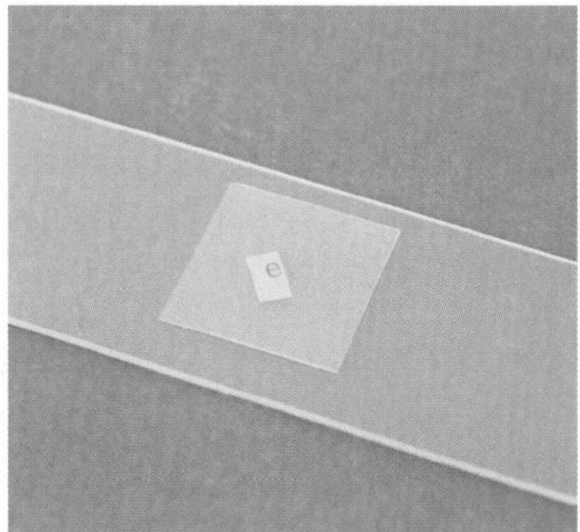

1. Cut out a small piece of newspaper containing a single letter. Use an *e*, *f*, *g*, *s*, or *h*. Pick up the letter with the tweezers and place it in the centre of a clean slide. **Note:** Always use clean slides and cover slips. Wash the slides with water and dry them carefully with lens paper when you have finished.
2. Use the dropper to place a very small drop of tap water on the newspaper sample. Then, hold a cover slip gently by its edges and place it at an angle of 45° on the surface of the slide. **Note:** Always hold microscope slides and cover slips by the edges to avoid fingerprints.
3. Slowly and carefully lower the cover slip over the sample. Make sure there are no air bubbles trapped underneath the cover slip. This type of sample preparation is called a **wet mount**.

Troubleshooting

- Do you see round or oval shapes on the slide? These are likely to be air bubbles. Move the cover slip gently with your finger to get rid of them, or study another area of the slide.
- Do moving lines and specks float across the slide? They are probably structures in the fluid of your eyeball that you see when you move your eyes. Don't worry; this is natural.
- Do you see a jagged line? This could be a broken cover slip.
- Do you close one eye while you look through the microscope with the other eye? You might try keeping both eyes open. This will help prevent eye fatigue. It also lets you sketch an object while you are looking at it.
- Always place the part of the slide you are interested in at the centre of the field of view before changing to a higher-power objective lens. The drawing on the right shows a view under low power. When you turn to medium and high power, you may not see A and parts of C. Why not?

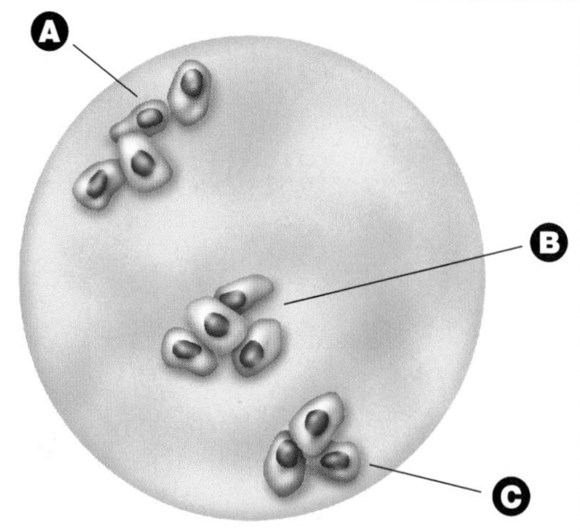

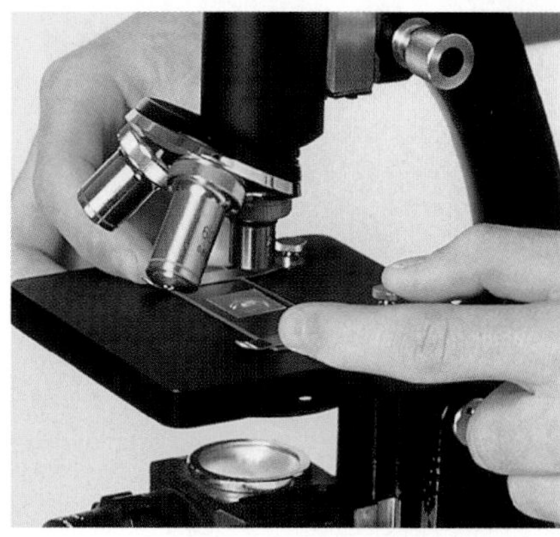

4. Set your microscope to the low-power objective lens. Place the slide with the letter on the microscope stage.
   (a) Look through the eye-piece and move the slide until you can see the letter. Adjust the coarse-adjustment knob until the letter is in focus.
   (b) Move the slide until you can see the torn edge of the piece of newspaper. Slowly turn the fine-adjustment knob about one-eighth turn either way. Do you see the whole view in sharp focus at one time?

Analyze

1. When scanning a microscope slide to find an object, would you use low, medium, or high power? Why?
2. Before rotating the nosepiece to a higher magnification, it is best to have the object you are examining at the centre of the field of view. Why?
3. To view the letter *e* through your microscope the right way up, how would you position the slide on the stage?
4. A student has made a wet mount of a piece of newspaper but sees several clear, round shapes in the field of view. What might these be? How could the student avoid having them appear on the slide?

Extend Your Skills

5. Prepare and examine microscope slides of different samples of materials, such as strands of hair, cotton, Velcro™, and grains of salt or sand. Obtain your teacher's approval of the material you select.

Road to Discovery

About the same time that Leeuwenhoek was making his observations in Holland, the English scientist Robert Hooke (1635–1703) was experimenting with microscopes he had built, like the one in Figure 1.4. Hooke looked through his microscope at a thin piece of cork that he had cut from the bark of an oak tree. Figure 1.5 shows what he saw — a network of tiny boxlike compartments that reminded him of a honeycomb. He described these little boxes as *cellulae*, meaning "little rooms" in Latin. Hooke's descriptions have given us our present-day word "cell."

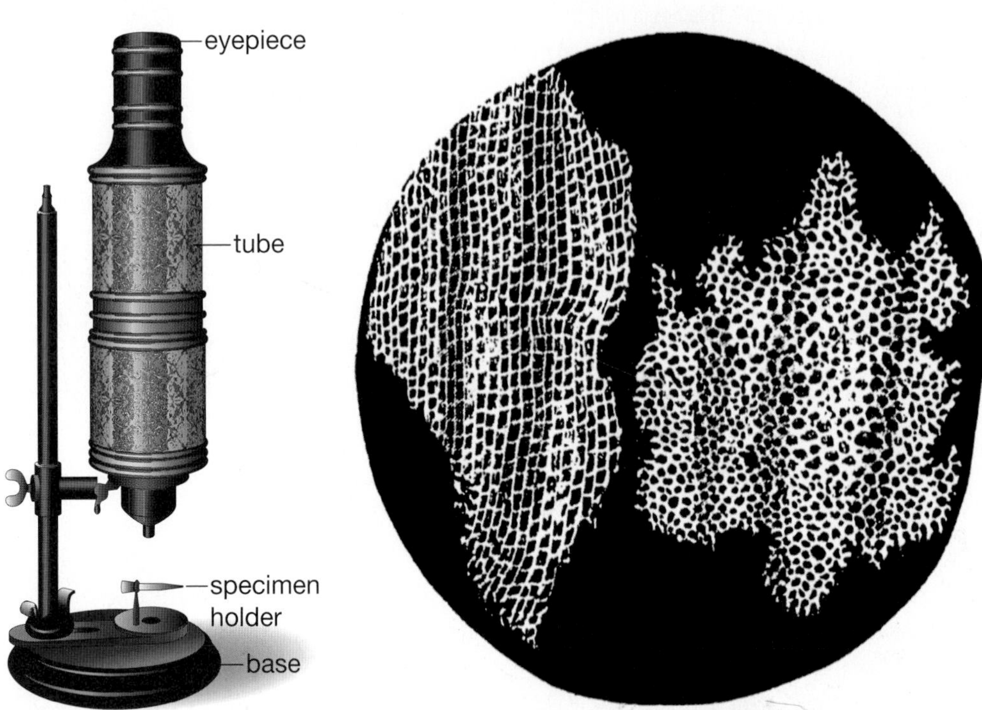

Figure 1.4 Hooke's microscope

Figure 1.5 The honeycomb appearance of cork cells seen by Robert Hooke. Hooke estimated that a cubic inch of cork (16.4 cm^3) would contain about 1259 million such cells.

DidYouKnow?

What are maggots? They are simply young flies, or the larvae of flies. Just as butterfly, mosquito, and dragonfly young look very different from the adults, so do the larvae (maggots) of flies such as deer flies and house flies.

Over the next century, many other scientists used microscopes to study microorganisms and to look at different parts of plants and animals. They saw cells in every living thing they examined. In 1839 German botanist Matthias Schleiden and zoologist Theodore Schwann combined their observations and made the hypothesis that all organisms are composed of cells. A cell is the basic unit of life, they suggested, because all the functions carried out by living things are carried out by their individual cells, as well.

If every organism is made of cells, what are cells made of? Where do they come from? In past centuries, it was commonly believed that some living organisms, such as maggots, flies, and even mice, could be produced from non-living matter such as air or water. This idea, called **spontaneous generation**, was used to explain such observations as maggots appearing suddenly in decaying carcasses. People believed that the maggots had appeared out of nowhere, as if by magic! In the late seventeenth century, an Italian physician, Francesco Redi, carried out the controlled experiment shown on page 19 to demonstrate that maggots come from tiny eggs laid on rotting meat by flies. (To learn more about controlled experiments in science, read pages IS-8–IS-13.)

Francesco Redi's Experiment

A Rotting meat is placed in two experimental jars and two control jars. Redi used two control jars to ensure accurate results.

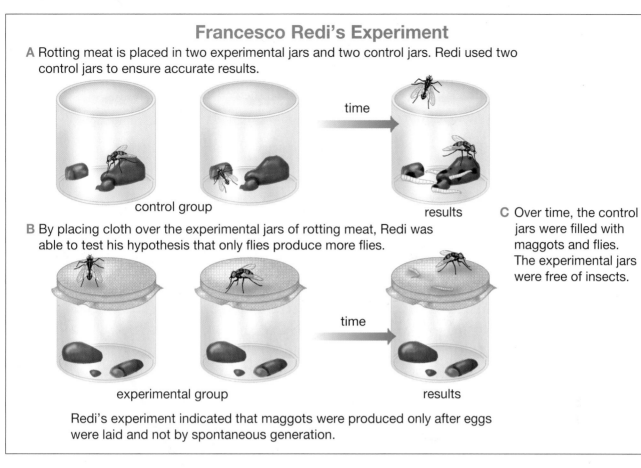

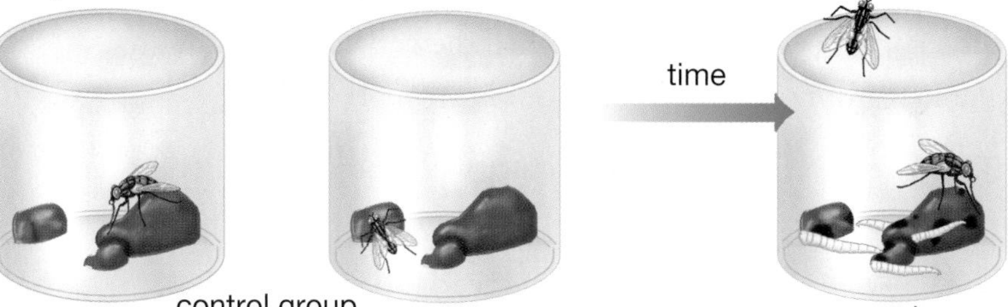
time
control group
results

B By placing cloth over the experimental jars of rotting meat, Redi was able to test his hypothesis that only flies produce more flies.

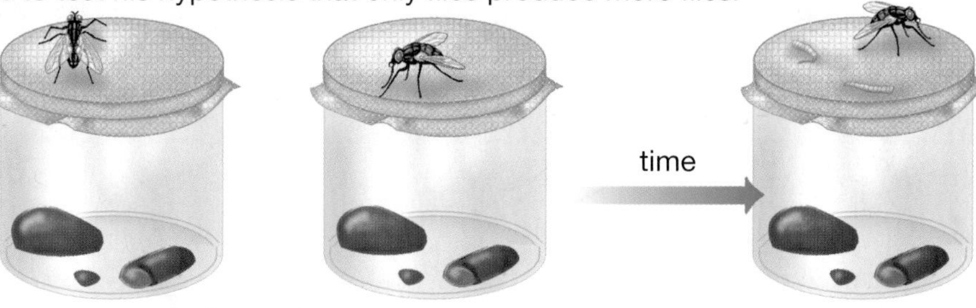
time
experimental group
results

C Over time, the control jars were filled with maggots and flies. The experimental jars were free of insects.

Redi's experiment indicated that maggots were produced only after eggs were laid and not by spontaneous generation.

Many people still believed in spontaneous generation until the mid-nineteenth century, however. They claimed that air itself contained an "active principle" that could give rise to micro-organisms. They thought that cells might form in liquids just as crystals form in solutions. Louis Pasteur, a French chemist, did controlled experiments in 1864 that showed that bacteria and other single-celled organisms are carried on particles of dust and water vapour in the air. Following experiments such as the one shown below, the old theories of spontaneous generation were finally abandoned.

Louis Pasteur's Experiment

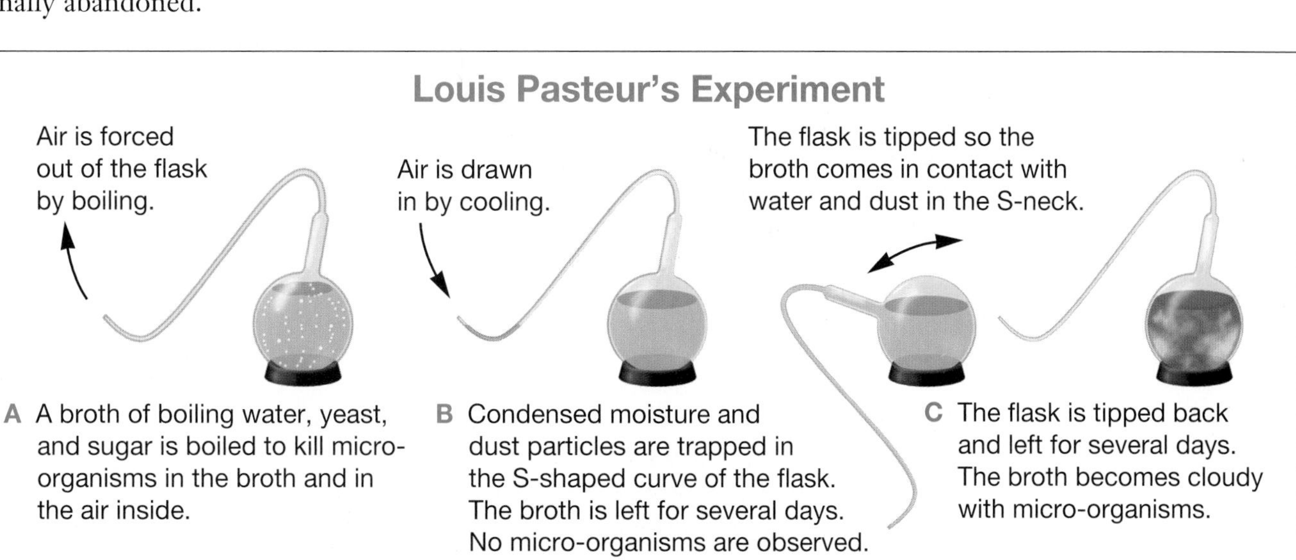
Air is forced out of the flask by boiling.
Air is drawn in by cooling.
The flask is tipped so the broth comes in contact with water and dust in the S-neck.

A A broth of boiling water, yeast, and sugar is boiled to kill micro-organisms in the broth and in the air inside.

B Condensed moisture and dust particles are trapped in the S-shaped curve of the flask. The broth is left for several days. No micro-organisms are observed.

C The flask is tipped back and left for several days. The broth becomes cloudy with micro-organisms.

Pasteur's experiment showed that micro-organisms come from other micro-organisms in the air and water. They do not appear by spontaneous generation.

The idea that life can be produced only by life had been suggested by a German scientist, Rudolf Virchow. He observed in 1858 that cells reproduce themselves and hypothesized that all cells can come only from already existing cells.

The general statements about living things made by Schleiden, Schwann, and Virchow form a set of hypotheses called the **cell theory**:

1. All living things are composed of one or more cells.
2. Cells are the basic units of structure and function in all organisms.
3. All cells come from previously existing cells.
4. The activity of an entire organism depends on the total activity of its independent cells.

You will explore more evidence supporting the cell theory in your grade 9 science course.

Check Your Understanding

1. In a compound light microscope, what is the function of
   (a) the eyepiece
   (b) the coarse-adjustment knob
   (c) the stage
   (d) the diaphragm?
2. (a) What is a wet mount?
   (b) How do you prepare a wet mount?
3. (a) When were micro-organisms first discovered?
   (b) Who discovered them?
4. Who was the first scientist to use the term "cell"?
5. Why was the development of the microscope important to the study of cells?
6. Why are cells considered to be the basic units of life?
7. **Thinking Critically** A biologist from another galaxy might think that automobiles are a dominant form of life on planet Earth. Automobiles move, consume gasoline and oil, and produce wastes. They are sheltered in garages and respond to stimuli. Automobiles age and break down, but new automobiles appear every year. They evolve, changing in appearance from year to year. What arguments would you use to persuade the alien visitor that cars are not alive?
8. **Thinking Critically** Over 2000 years ago, philosophers in ancient Greece tried to explain the nature of life based on their observations of the world around them. They proposed that living things combine the characteristics of water, heat, and air. Why might they have come to this conclusion? Do some research to learn more about the ideas of ancient thinkers such as Thales and Anaximander.

1.2 Unicellular and Multicellula

Strange as it might seem, a cell in your finger has characteristics in common with a microscopic organism and with the cells in an oak leaf. One way to understand the structure and function of cells in **multicellular** (many-celled) organisms, such as human beings, is to investigate the characteristics of **unicellular** (single-celled) organisms, like the ones shown in Figure 1.6.

The photographs in Figure 1.6 show a variety of microscopic pond organisms. Although unicellular, they are not simple. Each has a way of moving, obtaining food, and carrying out all other functions essential for life.

Figure 1.6

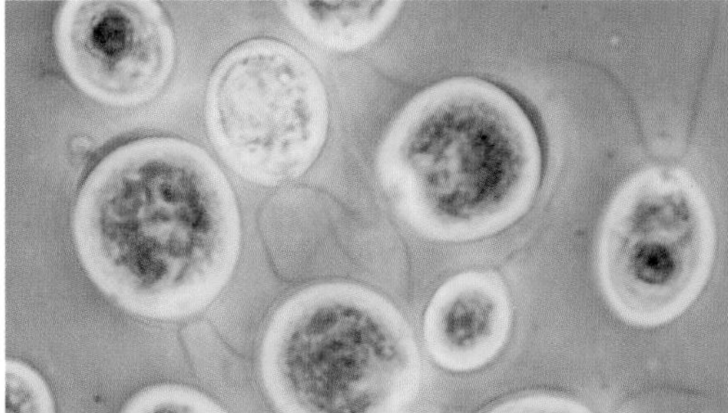

Chlamydomonas (180x) Makes its own food through photosynthesis, and moves by means of two long, whiplike structures called *flagella.*

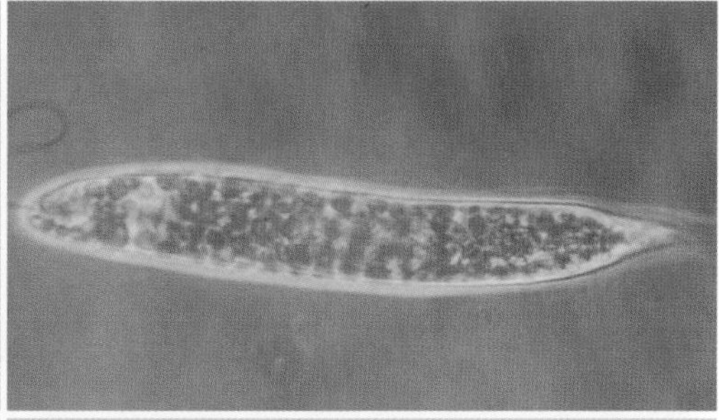

Euglena (100x) A common pond organism that also photosynthesizes, and moves by means of a single flagellum.

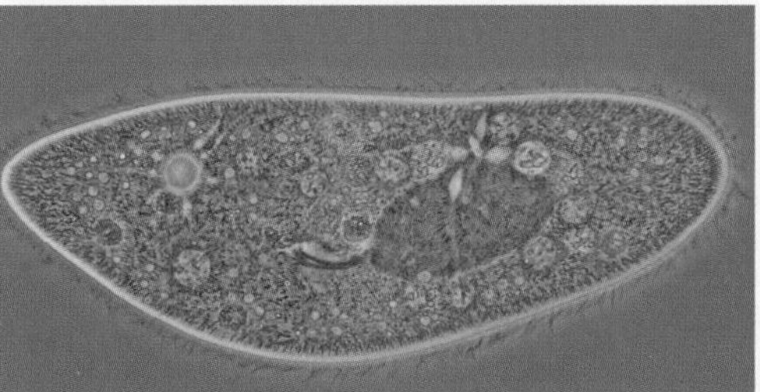

Paramecium (160x) Paramecia obtain their own food from the external environment. They are covered with short, hairlike structures called *cilia* that are used both for movement and to sweep food into a tiny groove that is similar to a mouth.

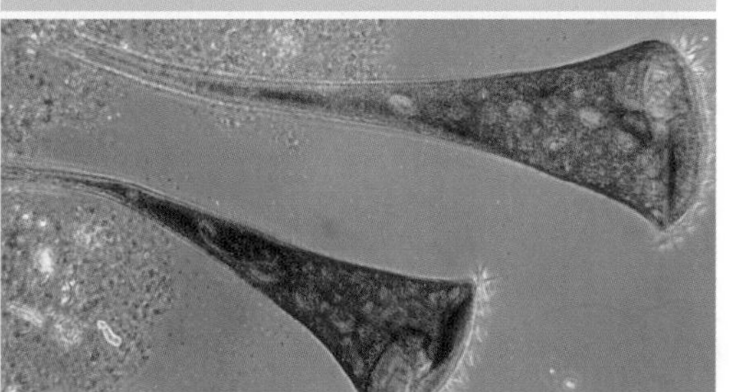

Stentor (125x) Stentor and some other unicellular organisms produce stalks to attach themselves to the bottom of ponds and streams. Stentor, like paramecium, has cilia, but these structures are used to bring in food rather than for movement.

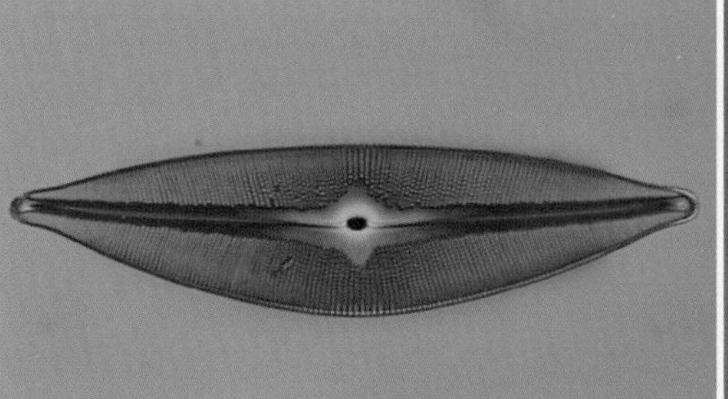

Diatoms (100x) Varied in shape and beautiful. Diatoms produce shells around themselves, make their own food through photosynthesis, and are free-floating.

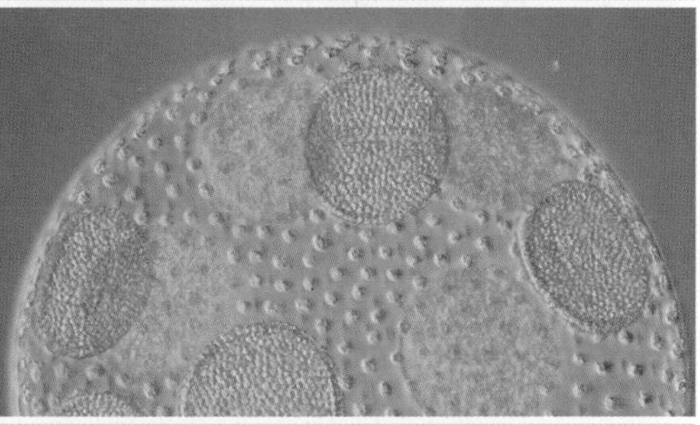

Volvox (30x) Living balls made of many volvox live together as a colony. Each has its own flagellum and makes its own food by photosynthesizing.

Pause& Reflect

In your Science Log, list or sketch some of the different types of homes that humans live in or have lived in. Include homes used in past periods of history, and homes used in different areas of the world. What parts do all these homes have in common? What are the functions of these parts? In what ways do you think a human home may be like a cell? Write your ideas in your Science Log.

Pond Water Safari

Here is what Leeuwenhoek wrote after looking through his microscope at a sample of lake water more than 300 years ago: "I found . . . many little animalcules. . . . And the motion of most of these animalcules in the water was so swift, and so various, upwards and downwards and round about, that it was wonderful to see."

In this investigation, you will observe and draw various micro-organisms found in pond water. Some of these tiny organisms are like animals, some are like plants. They move and feed in different ways. You will record what characteristics of living things you observe in unicellular organisms. You will probably also see small organisms made of more than one cell in the pond water you observe.

Problem

How do unicellular organisms meet their basic life needs?

Apparatus

microscope
microscope slides
cover slips
medicine dropper
tweezers

Materials

pond water
cotton wool

Procedure

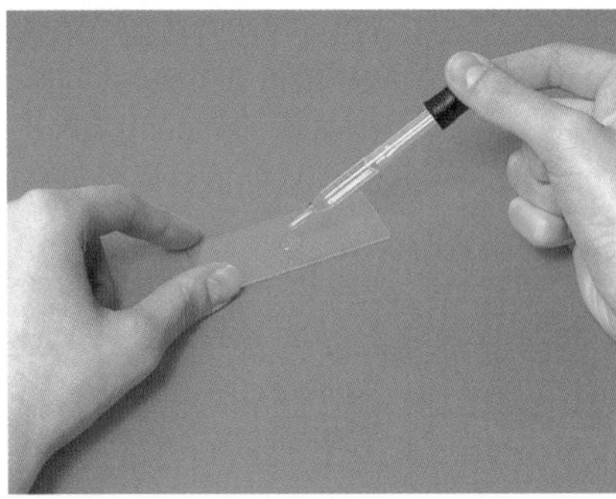

1. Obtain a sample of pond water from your teacher. Using a medicine dropper, place a drop of the pond water in the centre of a clean microscope slide.

Safety Precautions

- Be careful when using sharp objects such as tweezers
- Dispose of materials according to your teacher's instructions.

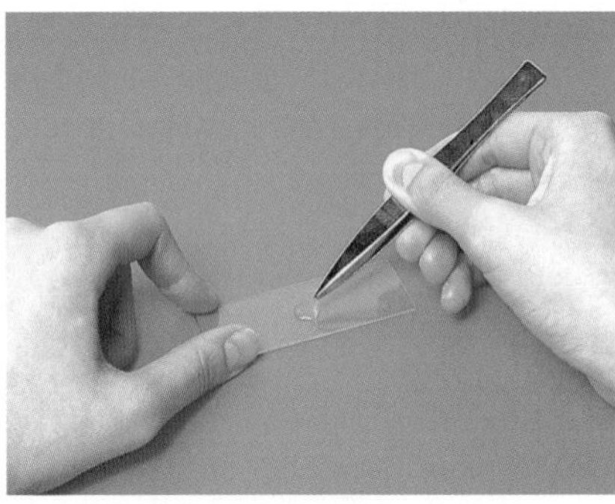

2. Pull two or three cotton fibres from the cotton wool and place them on the water drop.

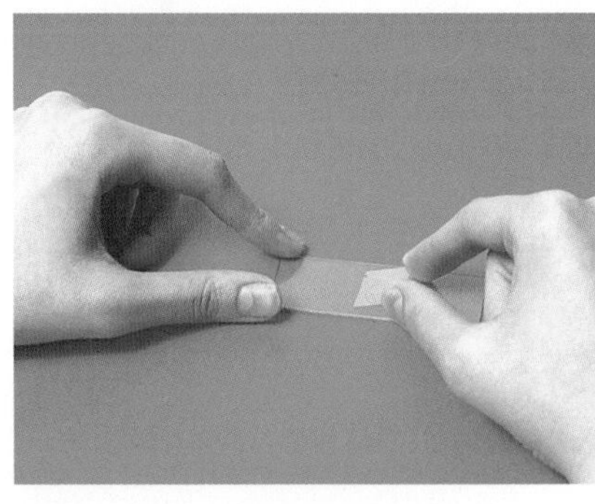

3. Place a cover slip on the sample.

4. Examine the slide under low power, looking for different unicellular organisms.

(a) Draw several different organisms, putting in as much detail as you can observe. Try to identify the organisms from the photographs in Figure 1.6 on page 21.

(b) Record which characteristics of living things you observe in unicellular pond organisms.

(c) Wash your hands after this investigation.

Amoeba Movement and Food-getting

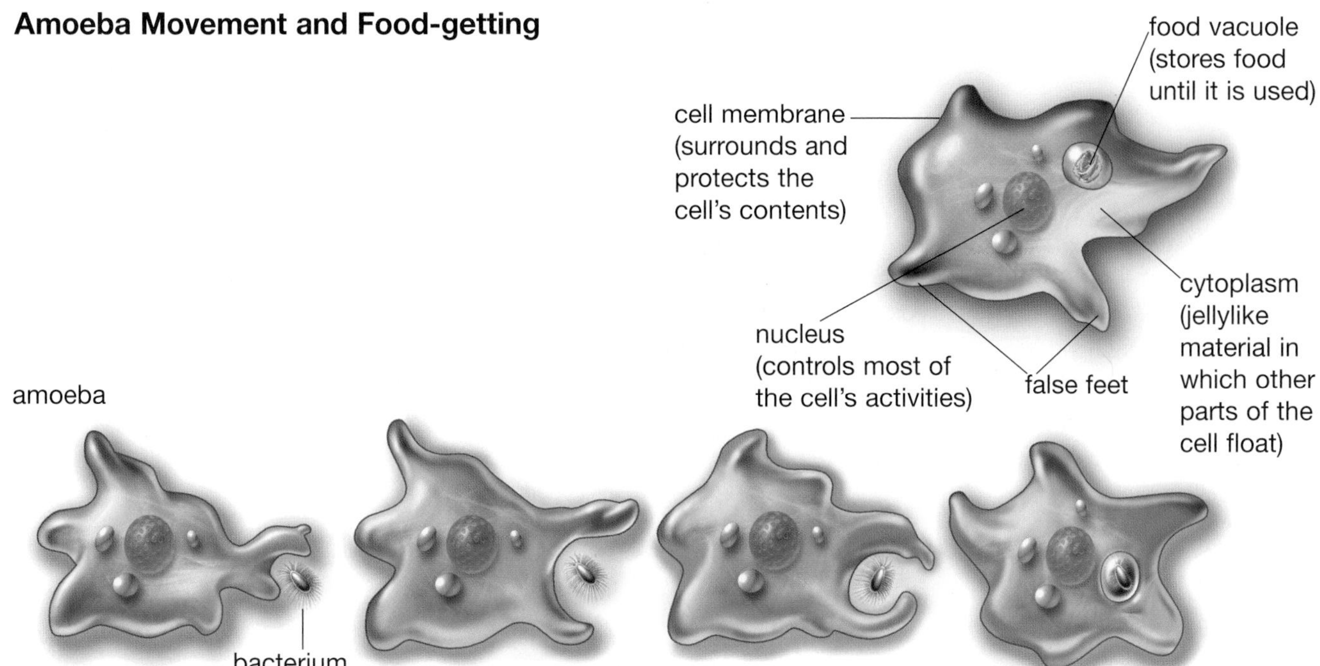

Look for some of the unicellular organisms shown on page 21. Look, too, for the amoeba, another common unicellular organism shown in the introduction to this unit. You might be fortunate enough to observe an amoeba in action. The amoeba moves by changing its shape. It pushes its cytoplasm against one part of its cell membrane, causing a bulge. This bulge is called a *false foot* (pseudopod). Then the amoeba shifts the rest of its cytoplasm in the same direction. The amoeba's ability to "flow" from place to place using its false feet also allows it to obtain food. The amoeba simply moves around a smaller organism such as a bacterium, trapping it inside a food vacuole.

Analyze

1. Suggest why you were asked to add cotton fibres to the water drop.
2. Describe what evidence you saw that unicellular organisms are able to feed. Recall that organisms may feed by ingestion — taking in substances, or by photosynthesis — producing food themselves (using energy from sunlight).

Conclude and Apply

3. What methods of movement did you observe?
4. Describe any evidence of growth or reproduction you saw.
5. Do unicellular organisms respond to stimuli (changes in their environment)? Explain.

Extend Your Knowledge

6. If you also observed multicellular organisms, describe how they differ in general from unicellular organisms. How are they similar to unicellular organisms?

Extend Your Skills

7. Keep a sample of pond water in a safe place exposed to sunlight for about a week before returning it to the pond. Use a microscope to observe the microscopic life in the water every day or two. Record any changes you see. Suggest an explanation for these changes.

CONDUCT AN

INVESTIGATION 1-D

Observing Plant and Animal Cells

You are now ready to begin observing the cells of multicellular organisms. When you first look at the cells of a plant or an animal through a microscope, they may look like rows of squashed boxes stacked together. There may appear to be little or nothing inside them. To observe the parts of a cell more clearly, scientists usually add coloured stains of various kinds to help.

In this investigation, you will continue to develop your skills using a microscope to investigate cells. You will prepare a wet mount of onion skin cells and look at a prepared slide of human skin cells.

Problem

What do plant and animal cells look like through a microscope?

Part 1

Observing Plant Cells

Safety Precautions

- Onion juice may sting your eyes. Wash your hands after handling the onion.
- Iodine solution may stain your hands or clothes. Avoid spilling it.
- Be careful when using sharp objects such as tweezers.

  CAUTION: If you get iodine on your skin or in your eye, inform your teacher and rinse the affected area with water. The eye should be rinsed for at least 15 min — iodine is an irritant and is toxic.

Apparatus

microscope
microscope slides
cover slips
medicine dropper
tweezers

Materials

small piece of onion
tap water
dilute iodine solution
lens paper

Procedure

1. Use tweezers to peel a single, thin layer from the inner side of a section of onion. If you cannot see light through your onion skin sample, try again.

2. Carefully place your onion skin sample in the centre of a clean slide. Make sure the onion skin does not fold over.

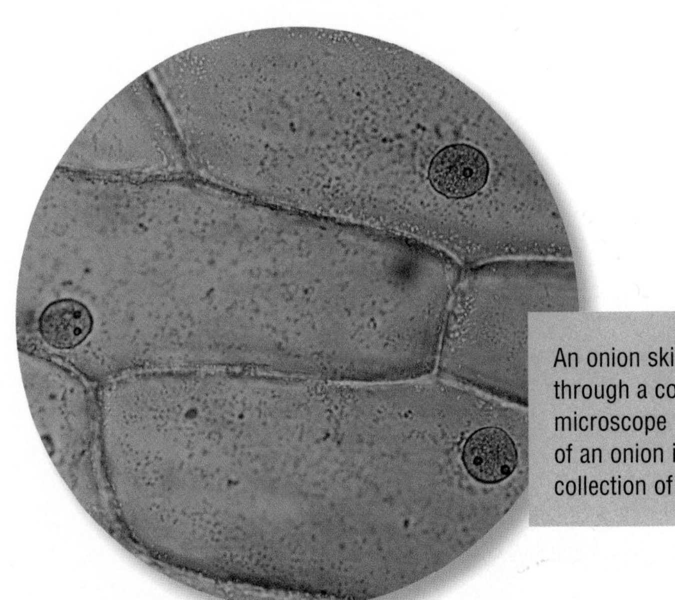

An onion skin sample viewed through a compound light microscope (100×). The skin of an onion is made of a collection of cells.

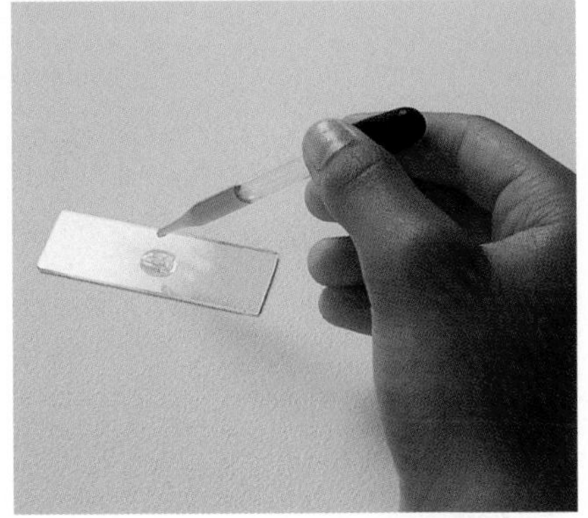

3 Add a small drop of water. Place a cover slip over the sample (see page 16).

(a) Examine the onion skin sample using the low-power objective lens on your microscope.

(b) Move the slide until you locate a group of cells that you wish to study. Centre these cells in your field of view and draw what you see.

4 Prepare another wet mount of the onion skin cells. This time, use a small drop of dilute iodine solution instead of water.

(a) Examine the stained cells, first using the low-power objective lens and then the medium-power objective lens.

(b) Carefully rotate the nose-piece to the high-power objective lens. Focus on the cells using the fine-adjustment knob. Draw what you see and label your diagram. (Hint: Do you see any parts that are similar to the parts of an amoeba shown on page 23?)

(c) Dispose of the onion material, clean your slides, and wash your hands. Set the microscope to the low-power objective lens.

CONTINUED

Part 2

Observing Animal Cells

Apparatus

microscope

prepared slide of human skin cells

Materials

lens paper

Procedure

1. Examine a prepared slide of human skin cells under different magnifications. Draw what you see and label your diagrams.

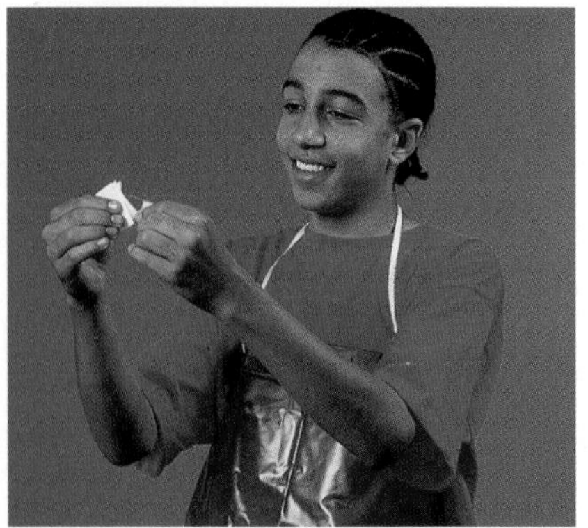

2. Clean your slide with lens paper, set the microscope to the low-power objective lens, and put the microscope away. Then wash your hands.

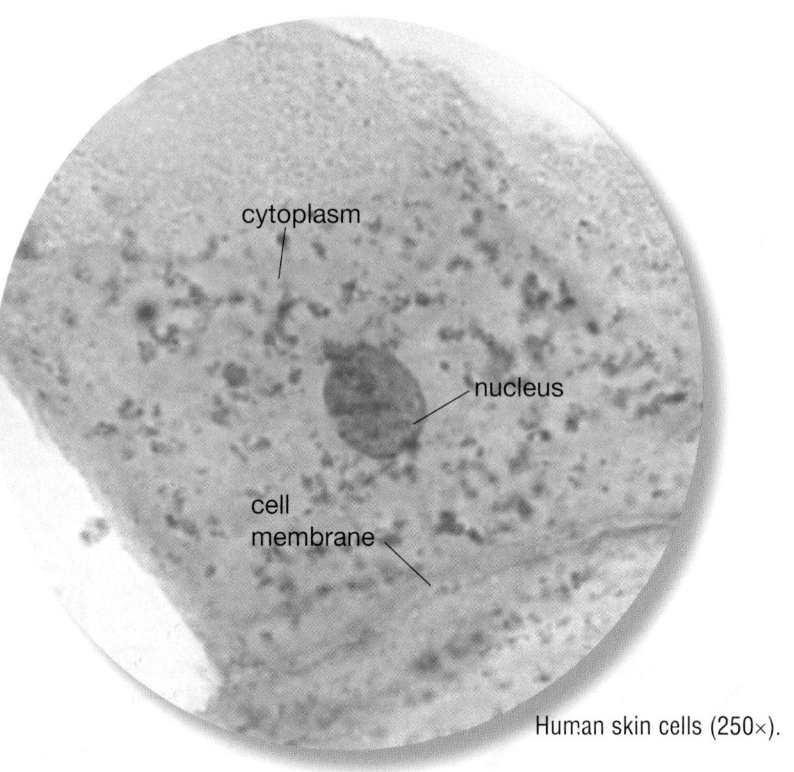

Human skin cells (250×).

Analyze

1. How was your study of cells affected by
   (a) using high-power magnification?
   (b) adjusting the diaphragm?
   (c) staining the cells?
2. List any differences you observed between onion skin cells and human skin cells. List any similarities.

Conclude and Apply

3. One function of skin is to protect and support the parts underneath it. How might the structure and arrangement of cells in the onion skin help do this?

Cell Parts Viewed With a Light Microscope

With your compound light microscope, you have been able to see the basic cell parts in unicellular organisms and in typical animal and plant cells. The cells of multicellular organisms fit together in much the same way as a building made of bricks (see Figure 1.7).

The cheek cells scraped from the inside of a person's mouth shown in Figure 1.8 again show you the major parts of an animal cell you should be able to see under a compound light microscope. Likewise, the cells scraped from the surface of a leaf (see Figure 1.9) show the major parts of a plant cell you can see with your own microscope. In the next section, you will learn more about these cell parts and what they do to keep each and every cell alive and functioning.

Figure 1.7 Cells fit together much like bricks in a wall.

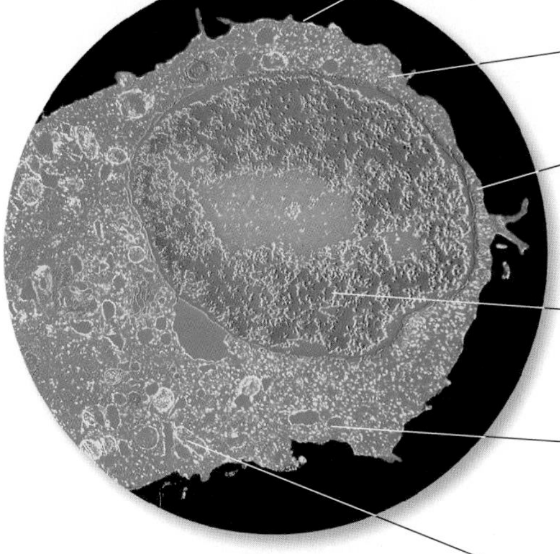

cell wall
(thick covering outside the cell membrane)

cell membrane
(surrounds the cell and protects the cell's contents)

cytoplasm
(jellylike material in which other parts of the cell float)

granules in the cytoplasm
(food storage; these are not a cell part)

nucleus
(controls most of the cell's activities)

vacuole
(liquid-filled part for storage; smaller and fewer in animal cells)

mitochondrion
(transforms energy for the cell)

chloroplast
(contains the green pigment, chlorophyll)

Figure 1.8 Colour-enhanced image of a typical animal cell magnified 250×.

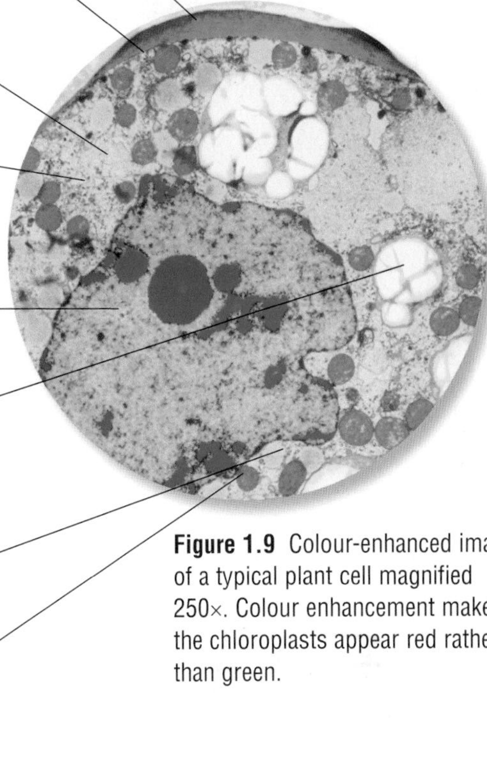

Figure 1.9 Colour-enhanced image of a typical plant cell magnified 250×. Colour enhancement makes the chloroplasts appear red rather than green.

Check Your Understanding

1. Give two examples each of (a) a unicellular organism, and (b) a multicellular organism.
2. List three key differences between a unicellular organism and a multicellular organism.
3. Describe two characteristics of life you have observed in a unicellular organism.
4. From your observations, list two structures that all cells seem to have in common.

1.3 Cell Parts and Cell Size

DidYouKnow?

The structures that Robert Hooke saw in a piece of bark were not living cells at all, but only the cell walls of dead, empty plant cells.

Cells are like factories in which the business of life is always going on. Every cell must carry out certain activities that keep it alive. These activities include obtaining materials and supplies of energy, making products, and getting rid of wastes. To carry out these functions, cells have some basic structures in common. Structures inside the cell are known as **organelles**. Each organelle has a role to play in the activities necessary for life. Many of the details of cell organelles have only been discovered since the invention of the electron microscope. Look closely at the diagrams on these two pages to see which organelles are found in both plant and animal cells. Which parts are found only in plant cells?

Animal Cel

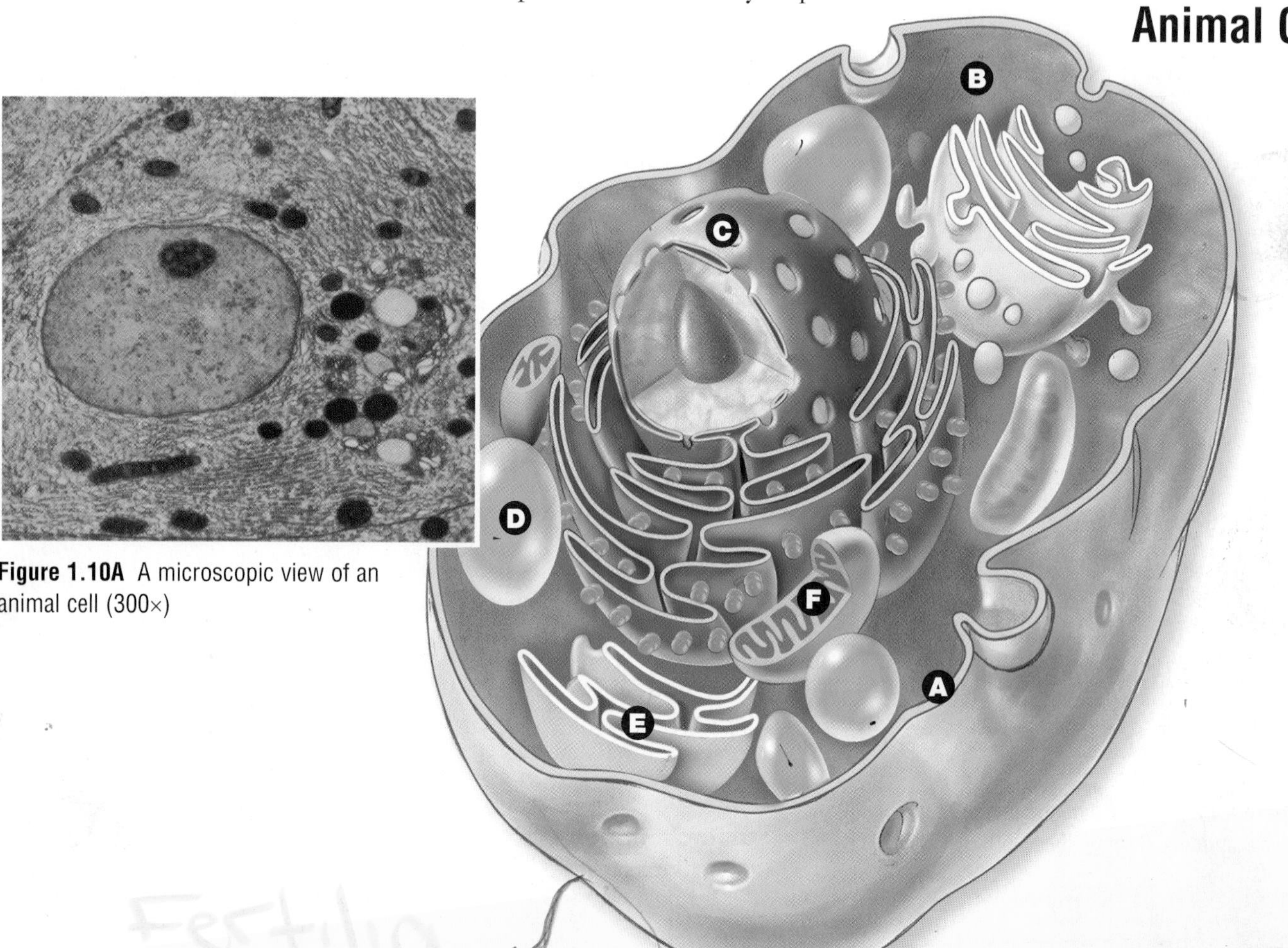

Figure 1.10A A microscopic view of an animal cell (300×)

A Cell membrane

Like the skin covering your body, the **cell membrane** surrounds and protects the contents of the cell. The cell membrane is not simply a container, however. Its structure helps control the movement of substances in and out of the cell.

B Cytoplasm

A large part of the inside of the cell is taken up by the jellylike **cytoplasm**. Like the blood flowing throughout your body, cytoplasm constantly moves inside the cell. The cytoplasm distributes materials such as oxygen and food to different parts of the cell. The cytoplasm also helps support all the other parts inside the cell.

C Nucleus

A large, dark, round **nucleus** is often the most easily seen structure in a cell. The nucleus controls the cell's activities. It contains the **chromosomes** — structures made of genetic material that direct a cell's growth and reproduction. The cell nucleus is enclosed by a **nuclear membrane**, which controls what enters and leaves the nucleus.

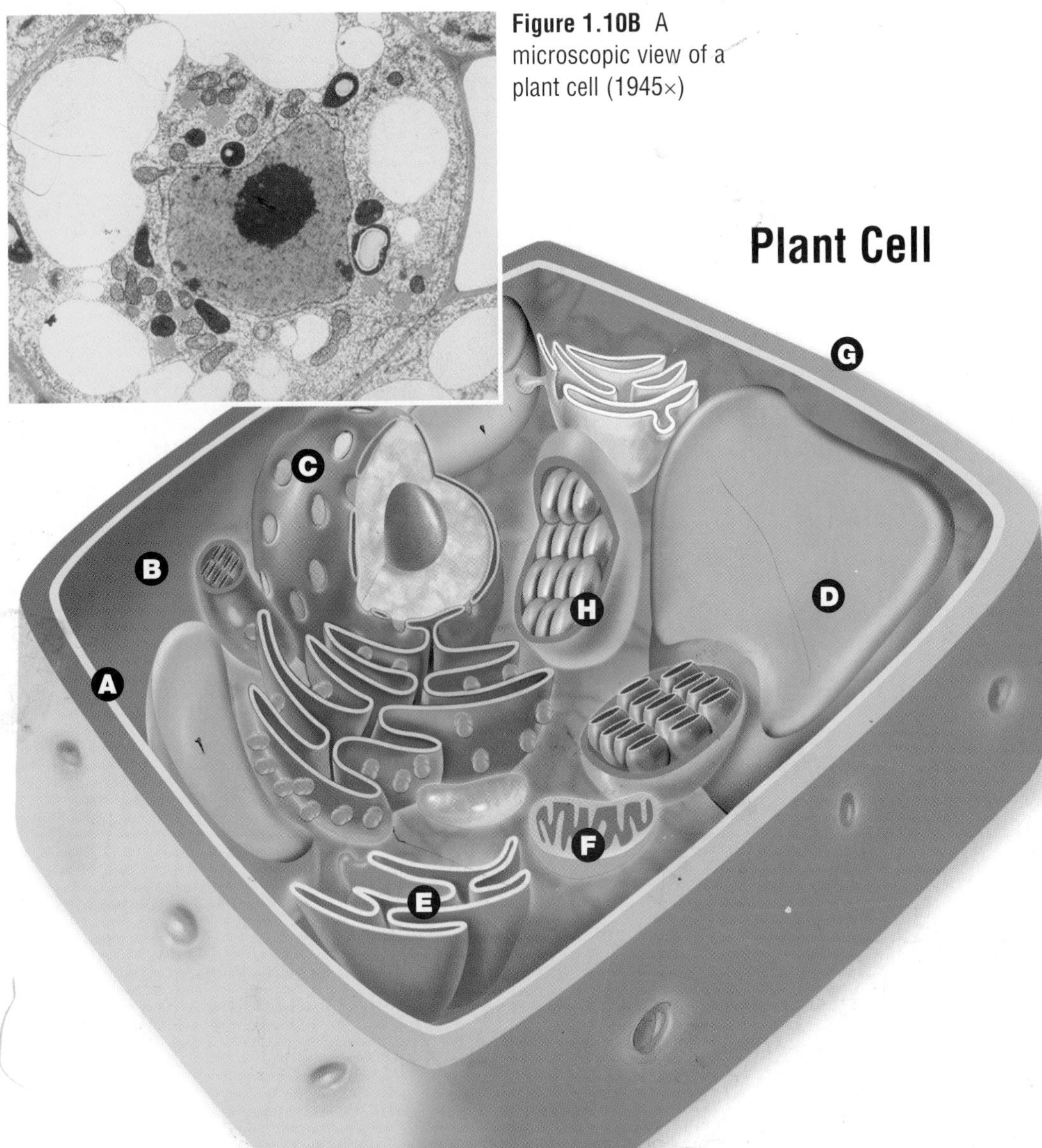

Figure 1.10B A microscopic view of a plant cell (1945×)

DidYouKnow?

A tree consists mainly of dead cells. The strength and rigidity of wood come from the cell walls. These cell walls remain stacked together solidly like bricks, long after the cells have ceased to carry out their living functions. The only living parts of a tree are the leaves, the growing tips of branches and roots, a thin layer of cells just under the bark, and the pith in the centre of the roots and the branches.

D Vacuoles
Balloonlike spaces within the cytoplasm are storage places for surplus food, wastes, and other substances that the cell cannot use right away. These structures, called **vacuoles**, are surrounded by a membrane.

E Endoplasmic reticulum
The **endoplasmic reticulum** is a folded membrane that forms a system of canals within the cytoplasm. Materials are transported through these canals to different parts of the cell, or to the outside of the cell.

F Mitochondria
Because cells do work, they need energy. Their energy is produced by oval-shaped organelles called **mitochondria** (singular: mitochondrion). Inside the mitochondria, tiny food particles are broken down to release their chemical energy for the cell's activities. Some cells, such as muscle cells, have more mitochondria than others because they need more energy to function.

G Cell wall
The **cell wall** occurs only in the cells of plants and fungi, and in some unicellular organisms. Cell walls are much thicker and more rigid than cell membranes, and are made mostly of a tough material called **cellulose**. They provide support for the cell.

H Chloroplasts
Chloroplasts are the structures in which the process of photosynthesis takes place. Photosynthesis uses energy from the Sun to make carbohydrates. Folded membranes inside each chloroplast contain the green pigment chlorophyll, which absorbs sunlight. Chloroplasts are found only inside cells in green plants and in some unicellular organisms. They are not found in animal cells.

INVESTIGATION 1-E

Build Your Own 3-D Cell

Making a three-dimensional model of a cell will help you remember all the different parts of a cell and how they fit together.

Challenge

Design and build a three-dimensional model of a cell that features all the organelles a cell needs in order to function.

Safety Precautions

- Never eat or drink anything in the science laboratory.
- Wash your hands after completing this activity.

Materials

Everyday items of your choice, for example, gelatin, modelling clay, shoe box, styrofoam, pipe-cleaners, plastic film, hard candies, dried pasta, craft items, etc.

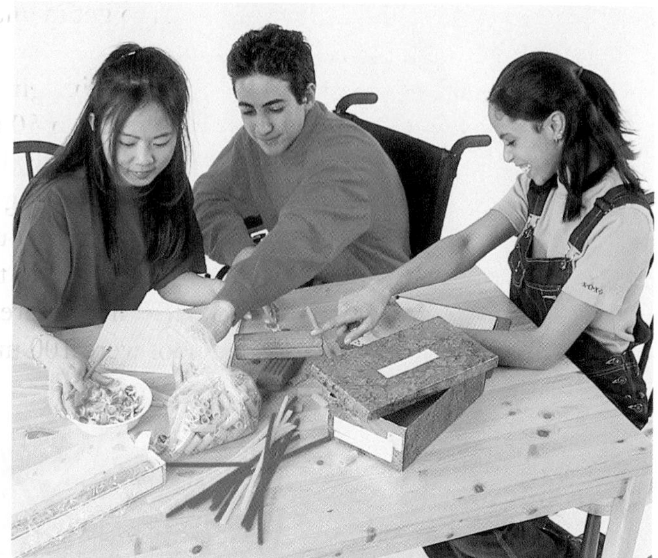

Design Criteria

A. Your model cell may be either a plant cell or an animal cell.

B. The organelles needed for the cell to function must be present.

C. Your model cell must contain all the right parts in the right proportions, and the parts must be clearly visible. It should be no larger than a shoe box or a basketball.

Plan and Construct

1. With your group, decide whether to build a plant cell or an animal cell.
2. List the organelles that your cell needs in order to function.
3. Decide which materials would best represent your cell and each organelle in your cell. Write each item beside the matching organelle in the list of organelles you made in step 2.
4. Make a neat labelled sketch of your design. Make sure you include and label all the organelles.
5. Start building your cell!

Evaluate

Work together to examine and compare the model cells constructed by the various groups. In what way or ways could you modify your design to improve it?

INTERNET CONNECT

www.school.mcgrawhill.ca/resources/

Do you want to take an imaginary journey through a cell? Go to the above web site, then to **Science Resources**, and on to **SCIENCEPOWER 8** to find out where to go next. You can zoom in, turn around, and check out different organelles inside a virtual cell.

Cell Size and Function

Why are cells so small? Why aren't larger organisms, such as the trees in Figure 1.11, made from one large cell instead of millions of microscopic cells? The explanation of why cells do not grow very large can be found in how cells function.

To carry out their work, cells need a constant supply of materials such as oxygen, water, and food particles. They also need to get rid of waste products. A larger cell would need more materials and would produce more waste products. However, the only way for materials to get in and out of the cell is through the cell membrane.

To have an idea of the problem this causes, imagine the cell as a round swimming pool with a diameter of 50 m. To keep this imaginary cell alive, you must swim to the centre of the pool carrying a beach ball (representing food particles), then swim back to the side carrying a lifebuoy (representing waste products). Suppose you must do this twelve times in a certain period of time. What differences would it make if the diameter of the pool were 100 m instead of 50 m?

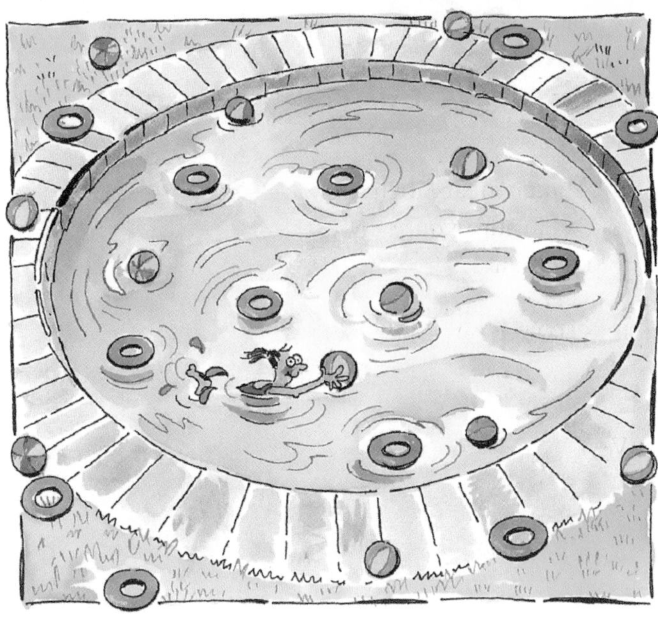

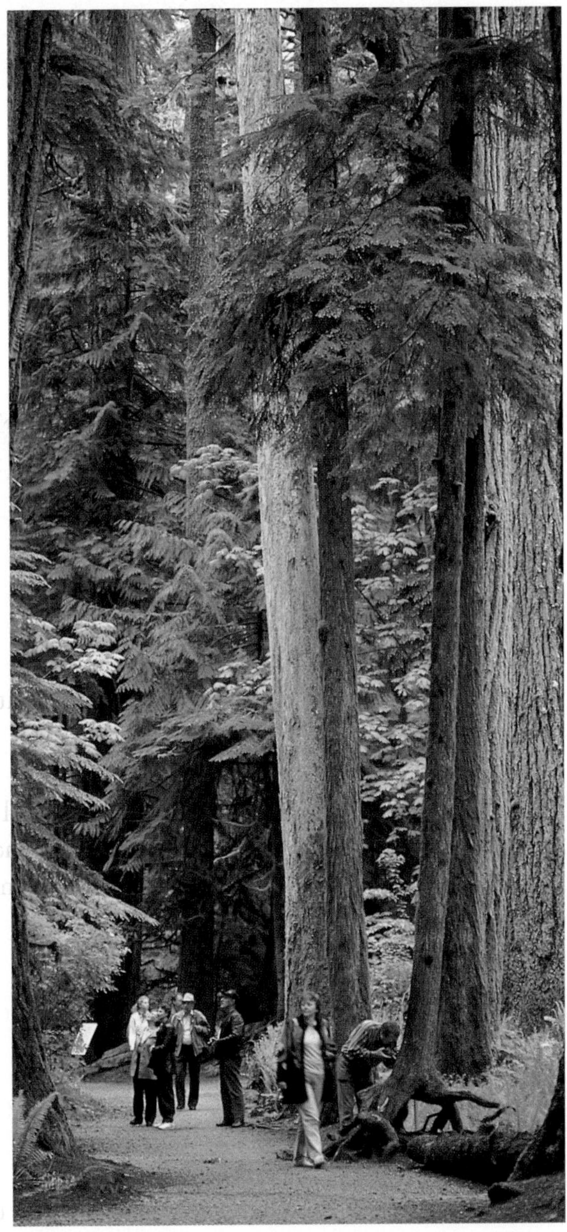

Figure 1.11 Why are all large organisms, including you, multicellular?

Pause& Reflect

Each organelle has a characteristic structure, and carries out the same function in every cell where it is found. How might the organelles of a cell be compared with the organs of your body, such as the stomach, lungs, and brain? In your Science Log, list some organelles. Beside each one, write what body organ they appear to be most like and why. Try this now, then after you have completed Chapter 3, check to see if you would like to change any of your comparisons.

DidYouKnow?

A single cell may contain many thousands of organelles. If the cell were the size of a large building, such as a school, the organelles would range in size from beachballs to classrooms. A mitochondrion would be about the size of a very large sofa.

INVESTIGATION 1-F

Is Bigger Better?

Think About It

The **volume** of any object, such as a cell, is the amount of space it takes up. The **surface area** of an object is the area of surface that encloses it. For a cell, the surface area is the amount of cell membrane. The relationship between volume and surface area changes as an object gets bigger or smaller. This relationship is called the **surface area-to-volume ratio**. In this investigation, you will calculate some of these relationships. From your results, you will make an inference about why small cells work more efficiently than large cells.

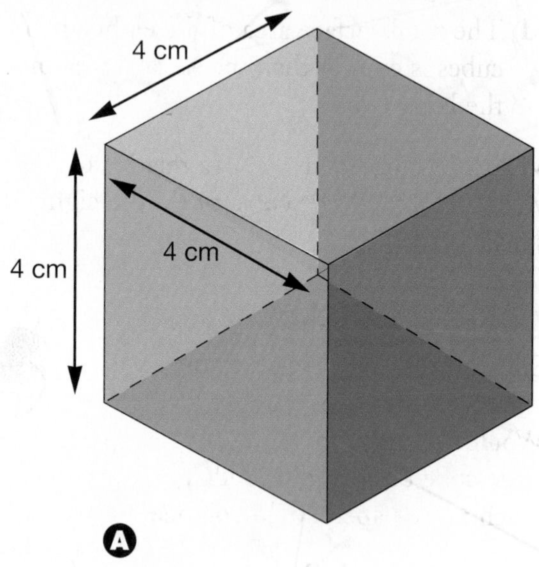

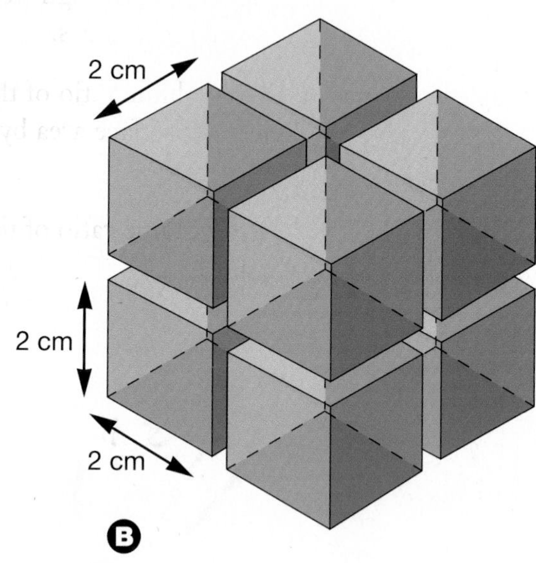

What to Do

1. In your notebook, draw a table with three column headings, such as the one shown below. Give your table a title.

Volume	Surface area	Surface area-to-volume ratio

2. Enter the results of the calculations on page 33 into this table as you complete them. You may use a calculator if you wish.

Math CONNECT

A ratio is a comparison of one number to another. For example, suppose a field contains 18 sheep and 3 cows. The ratio of cows to sheep is 3 to 18, or 1 : 6. If a banner is 2 m high and 6 m wide, what is the ratio of its height to its width?

Calculations

A. Calculate the volume of the cube in Diagram A. (Volume = height × width × depth.)

B. Calculate the volume of one of the small cubes in Diagram B. Multiply by eight to get the total volume of all eight cubes.

C. Calculate the surface area of the cube in Diagram A. To do this, calculate the area of one side, then multiply by six (the cube has six equal sides).

D. Calculate the total surface area of one of the small cubes in Diagram B. Multiply by eight to get the total surface area of all eight cubes.

E. Calculate the surface area-to-volume ratio of the cube in Diagram A. (Divide the surface area by the volume.)

F. Calculate the surface area-to-volume ratio of the eight small cubes in Diagram B.

Analyze

1. Answer each of the following questions by "True" or "False." If the statement is false, rewrite it to make it true.

- **(a)** The volume of the large cube is greater than the total volume of the eight small cubes.
- **(b)** The volume of the large cube is equal to the total volume of the eight small cubes.
- **(c)** The surface area of the large cube is greater than the total surface area of the eight small cubes.
- **(d)** The total surface area of the eight small cubes is greater than the surface area of the large cube.

2. Which has more surface area relative to volume: (a) the large cube, or (b) the eight small cubes?

Conclude and Apply

3. Suppose the large cube and the eight small cubes are gift boxes that you must wrap. Would you need more paper to wrap the large cube or all eight small cubes? How much more paper would you need?

4. You must store the boxes from question 3 in your locker. Which would take up more space, the large box or the eight small boxes? (Think carefully.)

5. Which size of cell would be more efficient at transporting substances in and waste products out — large or small? Explain why.

6. Why is it not possible for a single-celled amoeba to grow to be as large as you?

Skill POWER

To review how to measure volume, turn to page 540.

Small, Smaller, Smallest

Cells come in a variety of sizes and shapes. Most cells, however, fall into a narrow range of size — the size in which they function most efficiently. To grow bigger, organisms add more cells to their bodies rather than growing bigger cells. This occurs when cells divide, a process you will study in the next chapter.

Recall that cells are measured in micrometres (μm). Most cells in plants and animals have a diameter between 10 and 50 μm (see the examples in Figure 1.12). Bacterial cells are much smaller. They are only 1 μm to 5 μm across. That is about the same size as a mitochondrion.

Career CONNECT

The Art of Science

Many people with a passion for both art and science combine them to become scientific illustrators. Besides an eye for fine detail, most scientific illustrators have a degree or a diploma from an art college, as well as a university degree in a specific area of science.

Take a look at a few of the scientific illustrations in this unit. Which school subjects do you think would be important if you were planning to work in this field? What are some outside-school interests that might be useful? You can start thinking now about the types of drawings to include in your portfolio, if you think you might like to become a scientific illustrator.

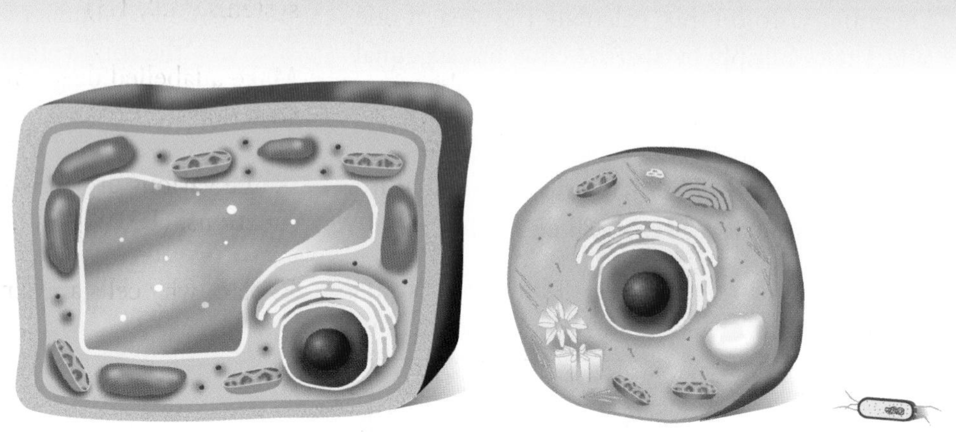

Figure 1.12 Relative sizes of a plant cell, animal cell, and bacterial cell

Check Your Understanding

1. Where would you find the genetic material in a cell?
2. What is the function of mitochondria?
3. Why does cytoplasm flow around inside a cell?
4. Which part of a cell stores food and waste materials?
5. Name two structures in a plant cell that are not found in an animal cell.
6. If you cut a cube in half, will the combined surface area of the two halves be more, less, or the same?
7. If you cut a cube in half, will the combined volume of the two halves be more, less, or the same?
8. Why is it an advantage for a cell to have a relatively large surface area-to-volume ratio?
9. Do organisms grow larger by (a) increasing the size of their cells, or (b) adding more cells?
10. **Thinking Critically** Why would you not expect to see chloroplasts in cells from an onion root?

CHAPTER at a glance

Now that you have completed this chapter, try to do the following. If you cannot, go back to the sections indicated.

Name the different parts of a microscope and describe the function of each part. (1.1)

List the characteristics of living things. (1.1)

Describe how to prepare a wet mount slide. You may show the procedure in a drawing. (1.1)

Calculate the size and magnification of an object seen through the microscope. (1.1)

Discuss how experiments were used to show that life comes from other life and not from spontaneous generation. (1.1)

Identify and draw various unicellular organisms. (1.2)

Distinguish between plant and animal cells. (1.3)

Explain why cells are considered to be living systems. (1.2, 1.3)

Make a labelled diagram showing basic cell structure. (1.3)

Name various organelles and describe their functions. (1.3)

Explain why cells are limited in size. (1.3)

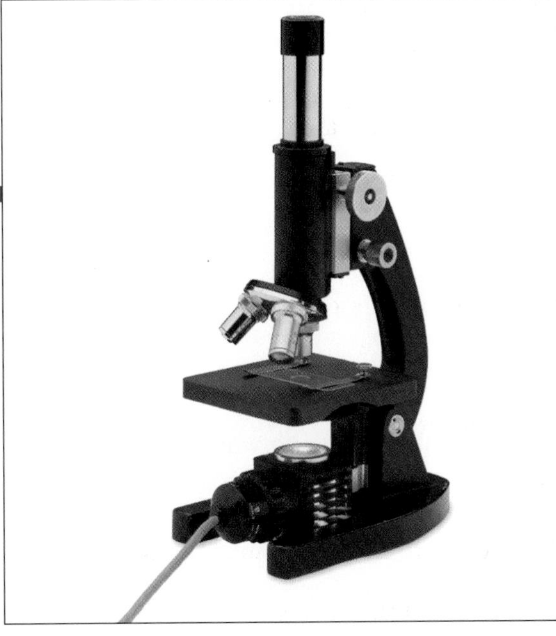

Prepare Your Own Summary

Summarize this chapter by doing one of the following. Use a graphic organizer (such as a concept map), produce a poster, or write the summary to include the key chapter ideas. Here are a few ideas to use as a guide:

- How can you distinguish between living and non-living things?
- How were cells first discovered?
- Make a drawing of the microscope shown above and label its parts.
- What is the cell theory?
- How do plant and animal cells differ?
- Copy the diagram of a cell shown below and label the cell parts.
- Describe the functions of some organelles.
- What is the difference between unicellular and multicellular organisms?
- Draw a diagram to illustrate the relationship between cell volume and surface area.

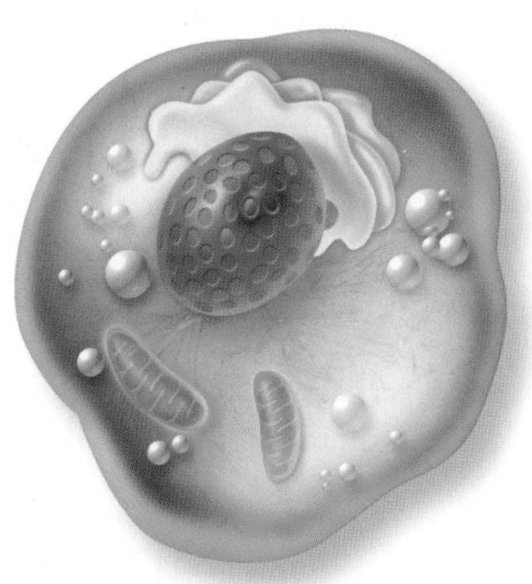

CHAPTER

1 Review

Key Terms

cell
magnifying
microscopes
field of view
wet mount
spontaneous generation
cell theory
multicellular
unicellular
organelles
cell membrane
cytoplasm
nucleus
chromosomes
nuclear membrane
vacuoles
endoplasmic reticulum
mitochondria
cell wall
cellulose
chloroplasts
volume
surface area
surface area-to-volume ratio

Reviewing Key Terms

If you need to review, the section numbers show you where these terms were introduced.

1. In your notebook, match each cell function in column A with a cell structure in column B.

A	B
• the "power-houses" of the cell	• nucleus (1.3)
• carries out photosynthesis in plant cells	• cell membrane (1.3)
• gives plant cells strength and support	• chloroplast (1.3)
• a moving fluid that distributes materials	• mitochondria (1.3)
• controls the cell's activities	• cytoplasm (1.3)
• a thin, protective "skin"	• vacuole (1.3)
• stores materials	• cell wall (1.3)
	• chromosome (1.3)

2. Is an earthworm unicellular or multicellular? Explain your answer. (1.2)

3. Describe two differences between the cell membrane and the cell wall. (1.3)

Understanding Key Ideas

Section numbers are provided if you need to review.

4. On a microscope, which adjustment knob must you use to focus the medium-power objective lens? Explain why. (1.1)

5. Which part of a cell allows it to exchange substances with its surroundings? (1.3)

6. Where would you find the substance chlorophyll in a cell? What is its function? (1.3)

7. Which of the following would you expect to find in an animal cell? Give a reason for each answer. (1.3)
   - **(a)** nucleus
   - **(b)** chloroplast
   - **(c)** vacuole

8. Is each of the following statements true or false? If a statement is false, write the correct statement. (1.3)
   - **(a)** Plant cells have a cell wall but no cell membrane.
   - **(b)** The cell nucleus contains chromosomes.
   - **(c)** Mitochondria and chloroplasts both absorb the energy of the Sun.

Developing Skills

9. Draw a Venn diagram like the one shown below. Where the circles overlap, list characteristics that living things and non-living things have in common. In the left circle, list characteristics shown by living things only. In the right circle, list characteristics shown by non-living things only.

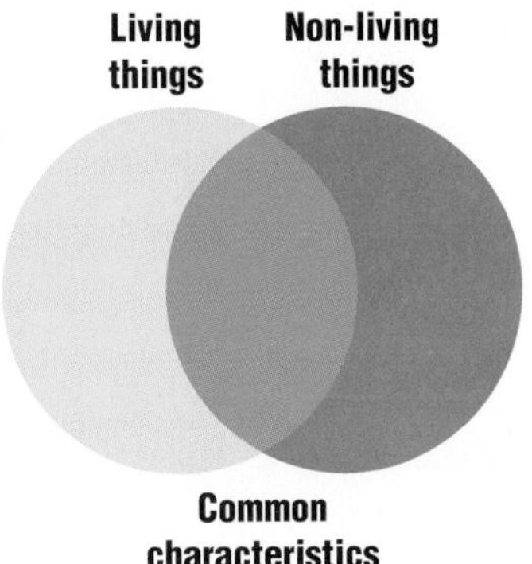

10. Draw another Venn diagram. Where the circles overlap, list parts that plant cells and animal cells have in common. In the left circle, list parts that only plant cells have. In the right circle, list parts that only animal cells have.

11. Copy and complete the following concept map of the basic units of life.

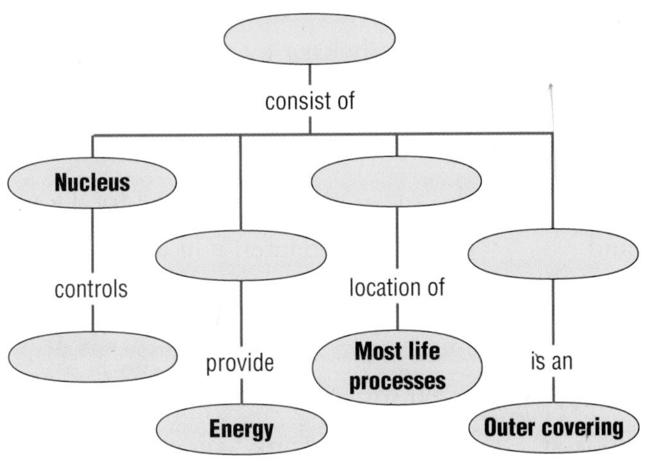

12. Suppose you are studying a slide of plant cells. You count 40 cells in a row across the diameter of the field of view. Make a flow chart describing or showing a technique you can use to estimate the average size of each cell.

Problem Solving/Applying

13. Imagine you are exploring another planet and you find a small, green, leaf-shaped object on the ground. How would you tell if it was a living, alien organism or a part of the non-living world?

14. Explain why animals and plants are made of billions or trillions of microscopic cells, rather than a few large cells. You may use a diagram in your answer.

Critical Thinking

15. A new science fiction movie shows giant amoebae from outer space invading Earth and ingesting people. Using your knowledge of cell structure and function, how would you explain to a younger student that this scene is physically impossible?

16. Chewing food into small bits helps digestion. Explain why, using the concept of surface area-to-volume ratio.

17. Suppose a new disease destroys the chloroplasts in plant cells. Explain what would happen to (a) the plant cells, (b) the plant, (c) other forms of life.

Pause& Reflect

1. A friend tells you that people and trees are two completely different forms of life. Explain why you would agree or disagree with your friend's comment.
2. In the eighteenth century, microscopes opened up a new world of learning for scientists. What instruments do you think are doing the same today?
3. Go back to the beginning of this chapter on page 4 and check your original answers to the Getting Ready questions. How has your thinking changed? How would you answer these questions now that you have investigated the topics in this chapter?

CHAPTER

2 The Functioning Cell

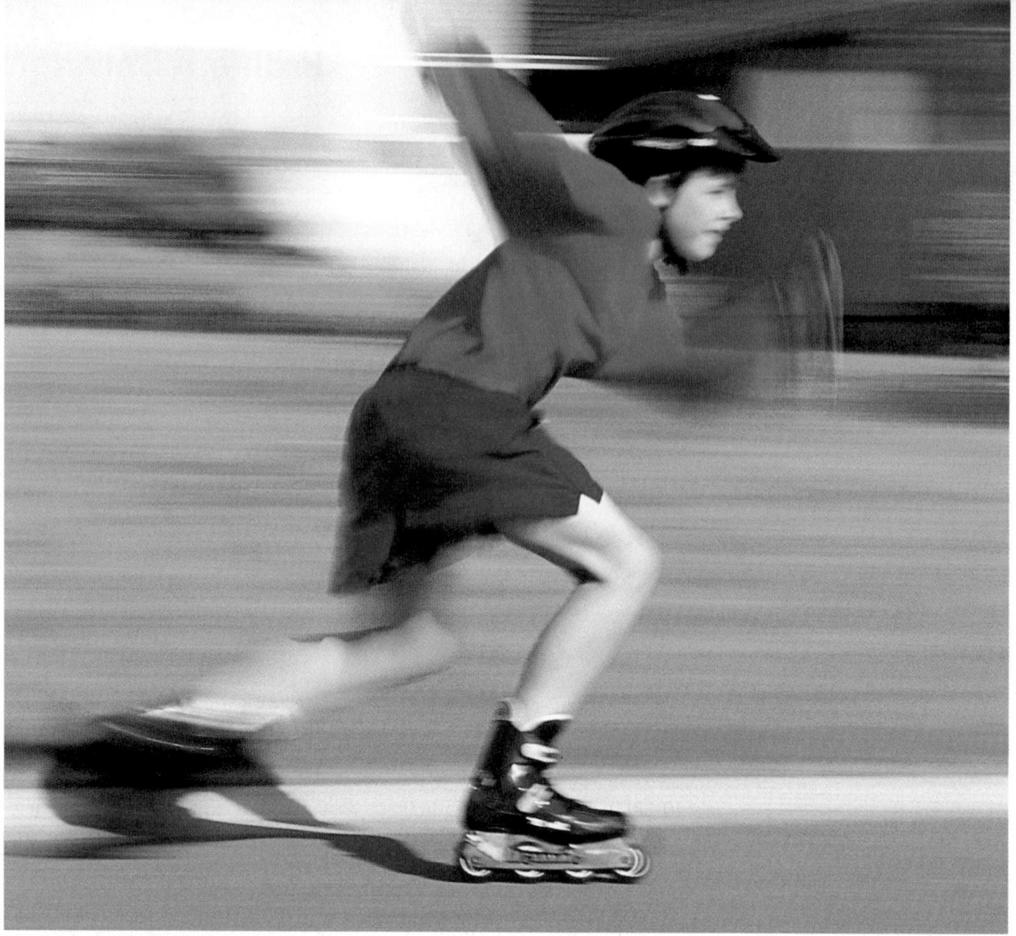

Getting Ready...

- How did you grow to your present size?
- How does your body heal a wound?
- Why can't you quench your thirst by drinking seawater?

Try to answer each question above using the word "cell." Record your answers in your Science Log. When you have finished studying this chapter, look back to see how your ideas have changed.

How long could you live without drinking? Without breathing? Without eating? These everyday activities are essential for life, but *why* do we need to drink, breathe, and eat? There are many ways to answer this question. Now that you have learned that you are made of cells, you can use this knowledge to look at your everyday activities in a new way. Individual cells carry out the same activities as whole organisms. When you drink, therefore, the water is eventually used by your cells, helping them to carry out their functions. Similarly, your cells also make use of the air you breathe and the food you eat. In this chapter, you will learn about some of the ways in which cells function.

If someone were to ask you why you need to drink, you could say it is because you are thirsty or hot. Now you could also say that you are drinking because your cells need the water.

Spotlight On Key Ideas

In this chapter, you will discover

- how substances move into and out of cells
- how cells obtain energy to carry out their functions
- how cells in multicellular organisms reproduce to make more cells for growth and repair
- how the structure of cells permits the movement of food, water, and gases needed by the cells
- the advantages of being multicellular over being unicellular

Spotlight On Key Skills

In this chapter, you will

- make predictions about the movement of particles into and out of cells
- make observations and measurements to test your predictions
- measure the amount of heat produced by active cells
- observe and measure the effect of cell division
- use a microscope to observe cells dividing
- design an experiment to show the effect of temperature on cell division
- design other experiments of your own

Starting Point ACTIVITY

Cells and Their Environment

Like the teenagers in the photographs, you constantly interact with your environment. Your body is active even when you are asleep. For example, you breathe in and out and process food from your last meal while you are sleeping. The individual cells in your body must carry out the same activities as your whole body. How do you think cells manage such functions as "breathing" and "eating"?

What to Do

1. List the key items that cells need from their environment in order to survive.
2. List the waste products that cells produce.
3.  You know that your body takes in materials for body growth and movement, and releases other materials as waste products. In a group, discuss how substances might enter and leave cells. Recall some of the structures and functions of cells that you learned about in Chapter 1. (You may wish to review the illustrations on pages 28-29.) You could illustrate some of your group's ideas in a sketch. Look at the sketch again when you have completed this chapter.

What Did You Find Out?

In what ways do you think individual cells function just like larger organisms such as you?

Skill POWER

For tips on working co-operatively in a group, turn to page 536.

2.1 The Cell Membrane: Gatekeeper of the Cell

At the border separating two nations, customs officers check the items that travellers are carrying. Some materials, such as firearms and plants, are not allowed to cross the border because of a country's laws. In a similar way, materials passing into and out of a cell are "checked" at the cell membrane. Like the customs checkpoint, a cell membrane allows some substances to enter or leave the cell, and it stops other substances. Because it allows only certain materials to cross it, the cell membrane is said to be **selectively permeable**. (A membrane that lets all materials cross it is **permeable**. A membrane that lets nothing cross it is **impermeable**.)

How does a cell membrane carry out this function? The answer is in the structure of the membrane. Imagine you have two small bags. One is made of plastic, the other of cheesecloth. Now imagine you pour water into both bags, as shown in Figures 2.1A and B. The plastic holds the water, but the cheesecloth lets the water run through. The plastic is impermeable to water, while the cheesecloth is permeable to water. This difference is due to differences in the structure of the materials from which the bags are made.

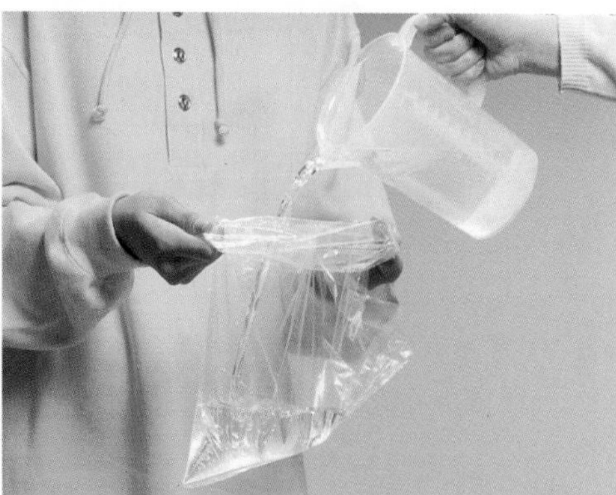

Figure 2.1A Plastic is impermeable to water.

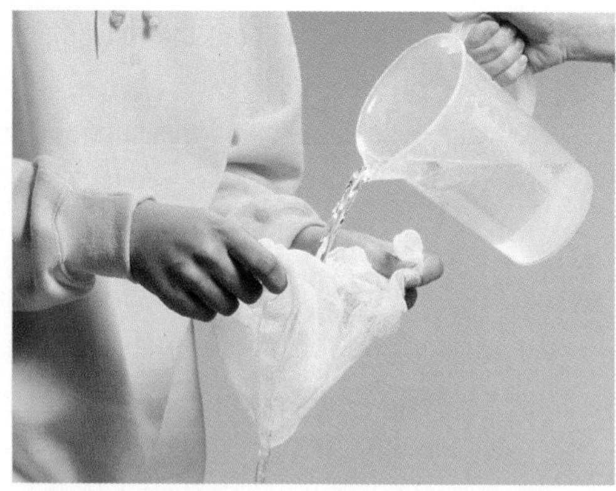

Figure 2.1B Cheesecloth is permeable to water.

Now imagine you are pouring a mixture of water and sand into both bags. Is each bag permeable, impermeable, or selectively permeable to the mixture? (If you are not sure, you could try carrying out a demonstration like the one in Figure 2.1 and observing what happens.)

Diffusion

The structure of the cell membrane controls what can move into and out of a cell. What causes substances to move in the first place? One clue is shown in Figure 2.2. What makes the blob of ink move outwards through the water in the container?

Figure 2.2 In time, the ink particles will become evenly dispersed with the water particles, and the whole solution will appear ink-coloured.

According to the particle theory, the particles in all liquids and gases are constantly moving in every direction and bumping into each other. (For more information on the particle theory, see page 111.) These collisions explain why particles that are concentrated in one area, such as the ink blob, spread apart into areas where there are fewer ink particles, and thus fewer collisions. This spreading-out process is called **diffusion**. Eventually, the ink particles will become evenly distributed throughout the container of water. At this time, individual ink particles continue to move, but there is no further change in the overall distribution of the ink in the water. Just like the ink, food colouring and the colour from certain crystals would also diffuse throughout the water, if they were left undisturbed for several minutes.

Pause& Reflect

Here are some situations in which diffusion occurs: A sugar cube is left in a beaker of water for a while. Fumes of perfume rise from the bottle when the top is removed. Give some other examples of diffusion. Can solids diffuse? Why or why not? Write your responses in your Science Log.

Recall that an average cell has a diameter of 20 to 30 μm. Suppose this cell were placed in a solution with a concentration of oxygen higher than in the cell's cytoplasm. The time required for diffusion to equalize the concentrations would be about 3 s at room temperature. If the cell were much larger — with a diameter of 20 cm, say — the same process would take about 11 years!

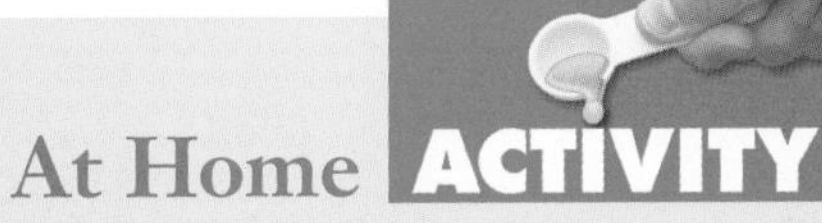

Observing Diffusion

You can observe diffusion by means of your sense of sight. Is there another way to observe this process? Try this. Have a friend or a family member stand at one end of a room with an orange while you stand at the other end facing the wall. Ask your friend to peel the orange. How do you know that particles have diffused from the orange throughout the air in the room?

Diffusion also plays a part in moving substances into and out of cells. For example, imagine an amoeba living in water. The concentration of dissolved carbon dioxide gas in the water is the same as the concentration of dissolved carbon dioxide gas in the amoeba's cytoplasm. Carbon dioxide particles therefore move into and out of the cell at the same rate, passing through small openings in the amoeba's selectively permeable membrane (see Figure 2.3A).

Now imagine the amoeba has been producing carbon dioxide as a waste product inside its single cell. The concentration of dissolved carbon dioxide particles in the amoeba's cytoplasm is now greater than the concentration of carbon dioxide in the surrounding water. As a result, more carbon dioxide particles move out of the cell by diffusion during a given time than move into the cell (see Figure 2.3B). The diffusion process continues until the concentration of the dissolved carbon dioxide gas on both sides of the cell membrane is once again equal.

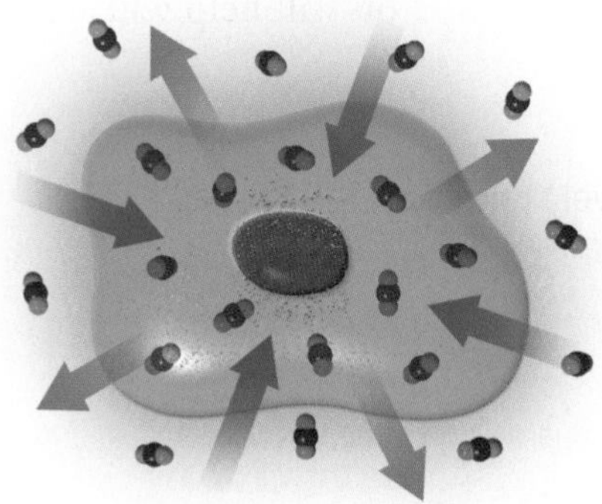

Figure 2.3A An equal concentration of carbon dioxide particles on both sides of the cell membrane. The particles move into and out of the cell at an equal rate.

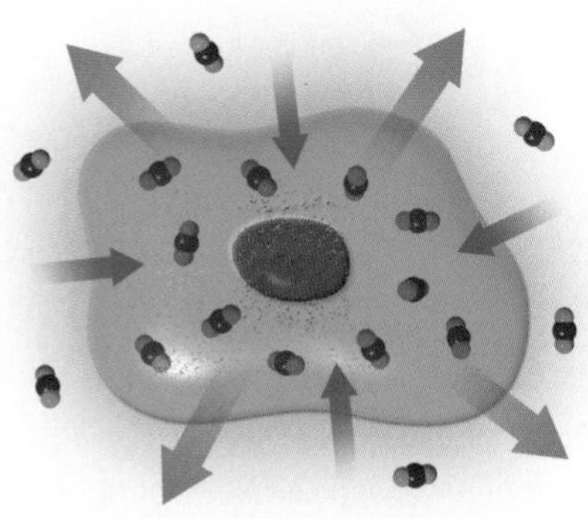

Figure 2.3B A greater concentration of carbon dioxide particles inside the cell. The particles move out of the cell at a greater rate than they move into the cell.

Figure 2.4A
Limp carrot sticks

Figure 2.4B
Carrot sticks 24 h later

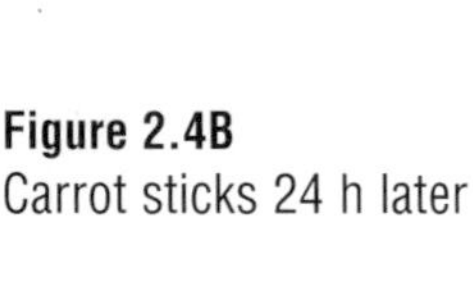

Osmosis

The most common substance found inside and around cells is water. About 70 percent of a cell's content is water, and most cells die quickly without a supply of this liquid. Water particles are small and can easily move into and out of cells by diffusion. The diffusion of water through a selectively permeable membrane is called **osmosis**.

You have probably already seen osmosis at work. Have you ever cut carrot sticks from a fresh carrot? You may have left some extra sticks in the refrigerator. By the next day, they have lost some of their moisture and they have gone limp. Suppose you place the sticks in a glass of water. Several hours later they are crisp again. What has happened? Water particles have moved from the water in the glass into the carrot cells by osmosis. (See Figures 2.4A and B.)

Now recall the idea introduced at the beginning of this chapter. It suggested that you drink water to help your cells carry out their functions. When you are very active, you lose moisture from your body in your breath and in sweat. Water is drawn out of your cells by osmosis. You need a new supply of water to restore the cell water content in your body to its normal level.

Water is important to living things because it dissolves many of the substances involved in cell processes. For example, glucose (which cells use for energy) dissolves in water to form a glucose solution. When water moves out of a cell, the dissolved substances inside the cell become more concentrated. When water moves into a cell, the dissolved substances inside the cell become more diluted.

Water tends to move by osmosis from a dilute solution to a more concentrated solution (see Figure 2.5). In other words, water moves from a region where it is in high concentration to one where it is in lower concentration. That is why water moves from tap water in a glass into dehydrated carrot cells. What do you think would happen if you put a fresh carrot stick into a glass containing a concentrated salt solution? Why might this happen? The next investigation will help you answer this question.

Pause& Reflect

What is the *solvent* in a salt solution? What is the *solute* in a salt solution? For that matter, what is a *solution*? If you remember these terms from your earlier studies, you are ready to do Conduct an Investigation 2-A. If you need review, look up the three terms in the Glossary at the back of this book. Write the definitions in your Science Log.

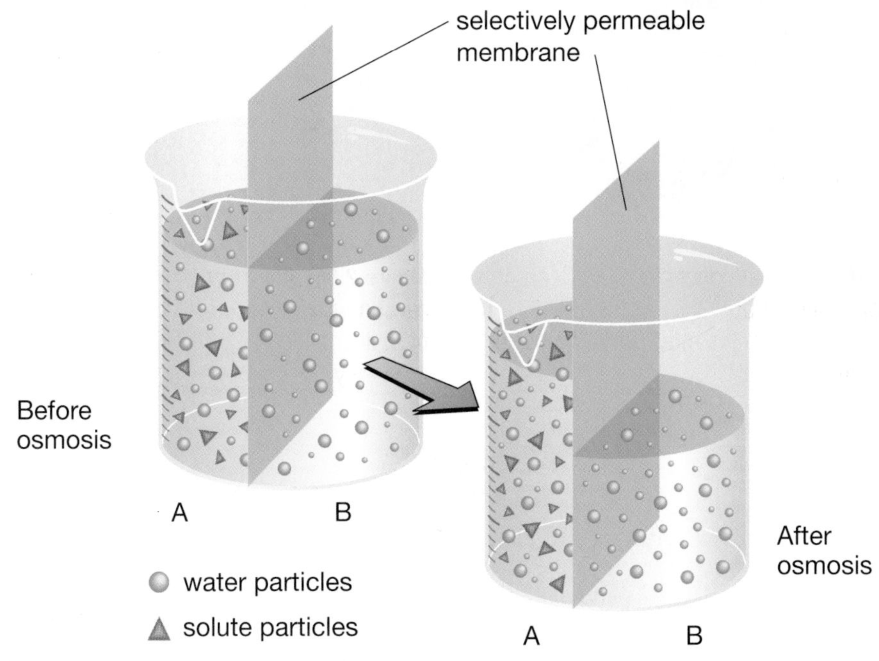

Figure 2.5 Water moves by osmosis from side B to side A inside the beaker. In this simplified diagram, which side represents a carrot stick and which side represents a glass of water?

DidYouKnow?

Can your doctor give you medicine without using pills, syrups, or needles? Yes, by using diffusion. Drugs can be put into a patch similar to a Band-Aid™ that is stuck onto the skin. There is a high concentration of drugs in the patch but a low concentration in the body. Therefore, the drug particles diffuse through the skin into the bloodstream.

Imagine you have a box of marbles. The box is divided in half by a strip of cardboard with an opening. All of the marbles are packed very tightly on one side of the cardboard strip. Some of the marbles roll to the other side of the box through the opening in the cardboard strip. The marbles keep rolling until each side of the box has about equal numbers of marbles. This is another example of osmosis.

CONDUCT AN INVESTIGATION 2-A

Measuring Osmosis

Just underneath the shell of an egg is a selectively permeable membrane. In this investigation, you will measure the movement of water by osmosis across this membrane. What conditions will cause water to move into the egg? What conditions will cause water to move out?

Note: You must start preparing your eggs for this investigation 24 h in advance. You will also need at least 24 h to observe the results of osmosis.

Problem

How can you measure the effects of osmosis?

Safety Precautions

- Handle all glassware carefully.
- Never eat or drink any substances in the science laboratory.

Apparatus

2 clean beakers (or glass jars) with lids
graduated cylinder
balance

Materials

2 uncooked eggs
white vinegar
pen or marker
labels
200 mL distilled water
200 mL salt solution
paper towel
water

Procedure

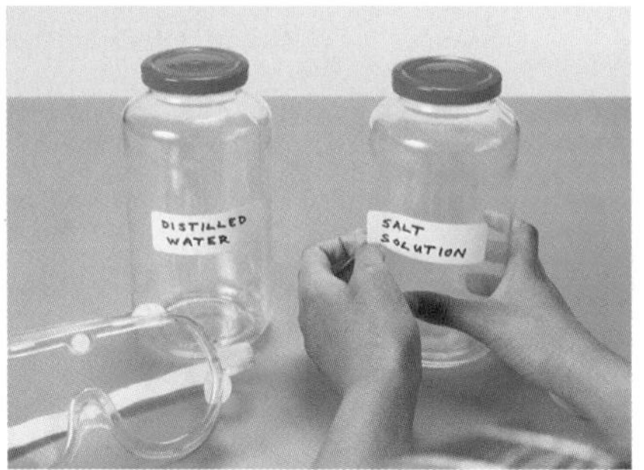

1. Prepare two raw eggs a day before your experiment by covering them with vinegar for 24 h.
   - **(a)** From your knowledge of osmosis, predict what will happen to the water content of an egg placed in distilled water, and one placed in salt solution. Record your predictions.
   - **(b)** Prepare a table like the one shown here. Give your table a title.

2. Label one jar "distilled water" and the other jar "salt solution."

3. Carefully remove the eggs from the vinegar, rinse them with water, and dry them with a paper towel. Record the appearance of the eggs.

Calculations	Egg in distilled water	Egg in salt solution
Original mass of egg		
Final mass of egg		
Change in mass (+ or –)		
Original volume of liquid		
Final volume of liquid		
Change in volume (+ or –)		

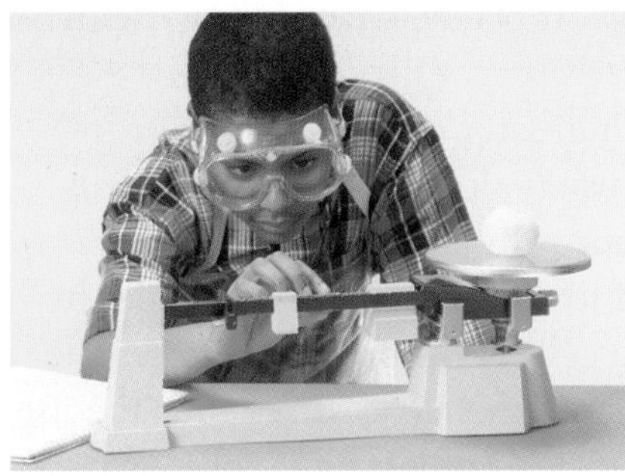

4 Measure and record the mass of each egg.

5 Place one egg in each jar. Pour 200 mL of distilled water into one jar and 200 mL of salt solution into the other. Cover the jars with lids and wait at least 24 h.

6 Carefully remove one egg and dry it. Measure and record its mass.

7 Using a graduated cylinder, measure and record the volume of liquid remaining in the jar.

(a) Repeat steps 6 and 7 for the other egg.

(b) Wash your hands after this investigation.

Skill POWER

To review how to measure the mass of an object and the volume of a liquid, turn to page 540.

Analyze

1. From your observations, make an inference about the effect of vinegar on egg shells. Why was this an important first step in the investigation?

2. What variables are held constant in this investigation? What variable is changed?

3. **(a)** What happened to the volume of liquid in the jar if the mass of the egg increased?
   (b) What happened to the volume of liquid in the jar if the mass of the egg decreased?
   (c) Explain these relationships.

Conclude and Apply

4. From your data, make an inference about the effect of osmosis on an egg placed in (a) distilled water, and (b) salt solution. In your answer, refer to the movement of water particles from a region where water is in high concentration to one where it is in lower concentration.

Extend Your Knowledge

5. Predict what results you might get if you repeated the experiment with a solution having double the concentration of salt. Explain your prediction.

6. Draw a flow chart to illustrate the sequence of events in this investigation, showing the movement of particles.

Active Transport

Small particles — such as water, carbon dioxide, and oxygen — diffuse freely into and out of cells through small openings in the cell membrane. This process depends only on the concentrations of the particles. It occurs without any use of energy by the cells.

However, cells also require certain substances in greater concentrations or in lower concentrations than can be obtained by diffusion alone. For example, cells need large amounts of glucose, which supplies them with energy. To meet this need, glucose particles must move from an area of low concentration (outside the cell) to an area of higher concentration (inside the cell). This process reverses the usual movement caused by diffusion. Unlike diffusion, this process requires the use of energy by the cells — like pushing a car uphill instead of letting it roll down to the bottom as it normally would.

The controlled movement of substances through the cell membrane is carried out by the membrane itself. To understand how it does this, scientists have studied the membrane structure in great detail (see Figures 2.6A and B). With the help of more powerful microscopes, they discovered large particles called **carrier proteins** embedded in the membrane. Like gates in a wall, carrier proteins control substances entering or leaving the cell. Each carrier protein attracts particles of a particular substance. The protein attaches to the substance, moves it through the membrane, and releases it on the opposite side, as shown in Figure 2.7. This energy-using process is called **active transport**.

Can cells break sidewalks? With the help of osmosis, they can! When cells take in water by osmosis, they tend to swell. The increasing pressure from the added volume of water may burst open animal cells. Plant cells, however, can withstand much greater pressure because they are surrounded by rigid cell walls. This pressure is called *osmotic pressure*. Have you ever seen weeds breaking through a paved sidewalk? They force their way through asphalt by osmotic pressure, generated by water in the cells of the shoot tip.

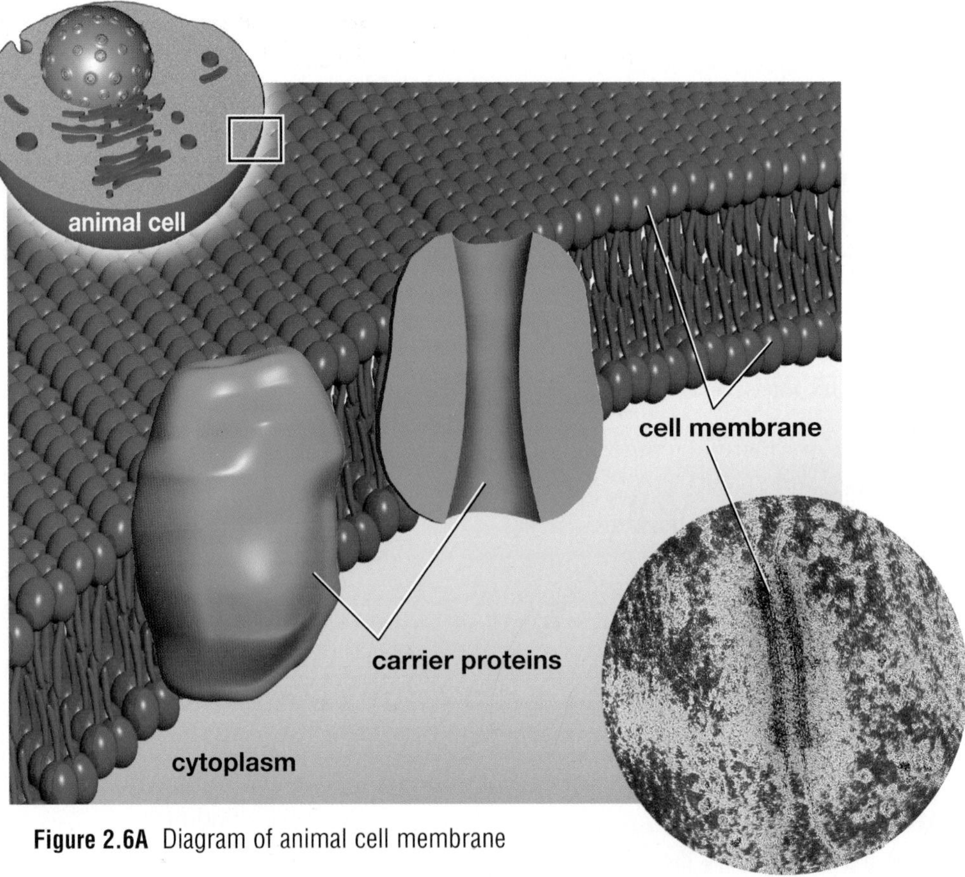

Figure 2.6A Diagram of animal cell membrane

Figure 2.6B Transmission electron microscope view of animal cell membrane (190 920×)

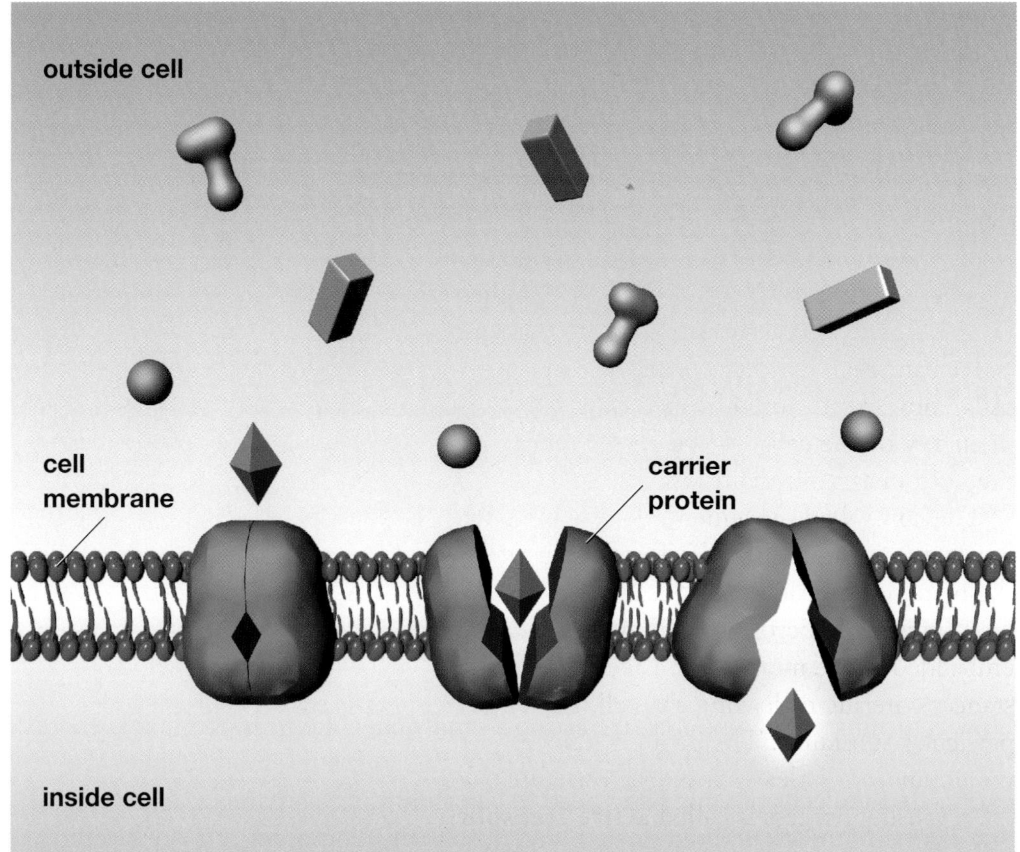

Figure 2.7 In active transport, carrier proteins in the cell membranes attract particles of particular substances, bind onto them, and release them on the opposite side of the membrane. These "gates" in the cell boundary can move particles either into or out of the cell.

In the Starting Point Activity on page 39, you were asked how cells might "breathe" and "eat." To carry out these activities, cells control the entrance and exit of key substances such as food particles. Active transport is one way this occurs. In the next section of this chapter, you will see how the food that you eat eventually gets turned into energy by your cells.

Check Your Understanding

1. What process causes water to enter or leave a cell?
2. How are osmosis and diffusion alike? How are they different?
3. How does active transport differ from diffusion? Give an example of each.
4. If your teacher opens a bottle of ammonia at the front of the classroom, you will smell ammonia at the back of the room a short time later. Explain what has occurred.
5. **Apply** Why do grocery stores spray their fresh vegetables with water?
6. **Thinking Critically** Why will a goldfish die if it is placed in salt water?

Word CONNECT

In the poem "The Rime of the Ancient Mariner" by Samuel Taylor Coleridge, a sailor stranded on a calm ocean says: "Water, water, everywhere, nor any drop to drink." Suggest what would happen to someone who drinks seawater. Explain why.

DidYouKnow?

Some diseases are caused by a defect in the transport of chemicals across the cell membrane. For example, kidney stones are produced by the buildup of crystals in kidney cells. Normally, the chemicals that form these crystals are transported across the cell membranes from kidney cells into the blood vessels and carried to other parts of the body.

2.2 How Cells Get Energy

How is turning on a computer like eating a sandwich? The common link is energy. Just as electricity supplies the energy that runs the computer, food supplies the energy that the cells of your body need to carry out their activities.

Cells cannot work without energy any more than a computer can work without being plugged in and turned on. The energy that cells use comes from food. What is food? For animals, food may be a sandwich, a mouse, or a blade of grass. For plants, food is carbohydrates made in their leaves by the process of photosynthesis. What all foods have in common, however, is particles that contain chemical energy.

The energy in food can only be released after food particles have entered the cells and have been broken down by a chemical reaction. The process that releases food energy is called **cellular respiration**.

You probably think of respiration as breathing in and out. That is what you and all other air-breathing animals do to obtain oxygen from the air and to get rid of carbon dioxide. Recall that cells carry out all the functions of living things. Your cells use the oxygen that you breathe in for cellular respiration, and they produce the carbon dioxide that you breathe out. Cellular respiration occurs in nearly all living cells of every organism — in plants and micro-organisms, as well as in animals.

Inside cells, oxygen combines with food particles (such as carbohydrates) in cellular respiration. The word equation for this chemical reaction is:

carbohydrates + oxygen → carbon dioxide + water + energy

DidYouKnow?

The word *photosynthesis* comes from two Greek words. *Photo* means "light." *Synthesis* means "putting together." Recall that in the process of photosynthesis, carbon dioxide and water are put together to make carbohydrates. The energy for this chemical reaction comes from sunlight. The word equation for photosynthesis is:

carbon dioxide + water + light energy → carbohydrates + oxygen

This chemical change can be compared to the burning of fuel. Like burning, much of the energy from the reaction ends up as heat. Think about what happens when your body demands more energy, such as when you run a race (see Figure 2.8). First, you need a good meal of energy-rich carbohydrates. As you run, you breathe more quickly, pumping in more oxygen for your cells to use. The oxygen and the food particles react inside your cells, producing energy for your muscles. At the end of the race, you feel warm.

No wonder! Every one of the millions of cells in your muscles has been burning food particles at a higher rate, not to mention the trillions of other cells in your body all working to carry out the respiration reaction.

Figure 2.8 Cellular respiration provides the energy to run a race.

Powerhouses of the Cell

Cellular respiration does not take place everywhere inside the cell. It occurs mainly inside the mitochondria (see page 29). Because energy is produced within the mitochondria, these organelles are often called the "powerhouses" of the cell. Different cells use different amounts of energy and have different numbers of mitochondria. Active cells, such as those in muscles, may contain several hundred mitochondria. The energy produced inside the mitochondria can be used by other parts of the cell (see Figure 2.9).

Why do cells need energy? Cell membranes need energy to move materials into and out of cells by active transport. Muscle cells need energy to contract. Sperm cells use energy to swim. Nerve cells use energy to send signals. Most cells also use energy to grow and reproduce. You will study cell reproduction in the next section.

DidYouKnow?

Have you ever felt stiff and sore after working your muscles harder than usual? This happens because your muscle cells cannot get oxygen fast enough to meet their demand for energy. So, instead of using cellular respiration to release energy from glucose, the muscle cells use a chemical reaction called fermentation. A product of this reaction is a chemical called lactic acid. It is this chemical that makes your muscles feel stiff. However, regular exercise strengthens your muscles and improves the efficiency of your heart and lungs. This helps get more oxygen to your cells faster and reduces the risk of soreness after exercise.

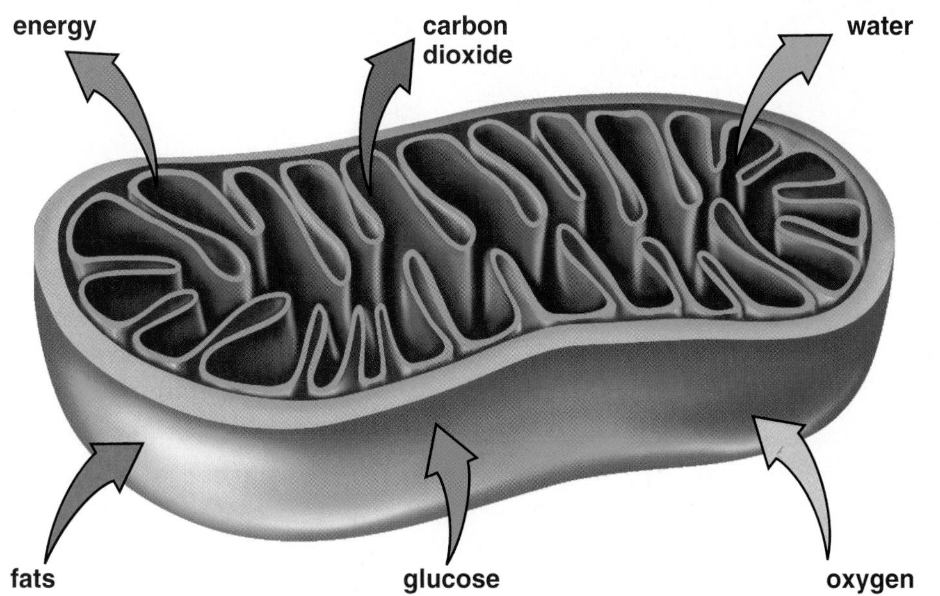

Figure 2.9 Cutaway diagram of a mitochondrion. Fats and glucose are broken down into smaller particles before entering the mitochondria, where cellular respiration occurs.

Measuring Cellular Respiration

You know that your cells are carrying out respiration because your body produces carbon dioxide and water vapour. These substances are two products of the respiration equation. As well, respiration produces heat. In this investigation, you will find out whether active plant cells also produce heat. If so, this will provide evidence of respiration in plant cells.

Problem

Do active plant cells produce heat?

Safety Precautions

Handle the thermometers carefully to avoid breaking them.

Apparatus

2 clean beakers (or glass jars)
2 alcohol thermometers
watch or clock

Materials

2 labels
2 large wads of cotton
50 dry kidney beans
50 kidney beans soaked in water overnight

Procedure

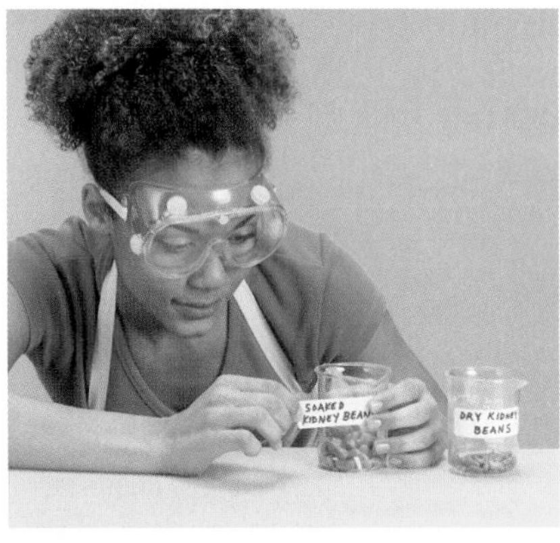

1. Place the dry kidney beans in a beaker and label it. Place the soaked beans in the other beaker and label it.
2. Place a thermometer in each beaker.
3. Cover the beans with a large wad of cotton, and place both beakers in a sheltered spot at room temperature.
   (a) Predict what difference you might expect between the beans in the two beakers.
    (b) Record your prediction, with an explanation.

Skill POWER

To review how to make a line graph, turn to page 546.

	Temperature (°C)						
Beaker contents	**At start**	**After 30 min**	**After 1 h**	**After 1.5 h**	**After 2 h**	**After 2.5 h**	**After 3 h**
Dry kidney beans							
Soaked kidney beans							

4 Read the temperature on each thermometer every half hour for 3 h. Record your results in a data table such as the one above. Record any other changes you observe in either beaker of kidney beans. Wash your hands after this investigation.

Analyze

1. Plot your temperature data for both beakers on a line graph. Put the time scale on the horizontal axis (x-axis) and the temperature scale on the vertical axis (y-axis).
2. Why do you think both beakers were covered with a large wad of cotton?
3. How did soaking one set of beans affect the temperature?

Conclude and Apply

4. From your observations, make an inference about what was occurring in each beaker. Give reasons for your inferences.

Extend Your Skills

5. **Design Your Own** How would you modify this investigation to show that growing kidney beans produce carbon dioxide gas as a product of cellular respiration? How would you test for this gas? What variable would be held constant in your investigation? What variable(s) would change? Make a sketch to show your experimental set-up.

At Home **ACTIVITY**

The Cell as a Factory

Now that you have investigated the cell, its organelles, and their functions, think about this. It has been said that a cell is much like a factory. See how well you can develop a storyboard or a flow chart using this comparison. How can you match the parts of a cell and their functions with what happens in a factory? The illustration shown here may give you some ideas for your storyboard or flow chart.

What to Do

1. Select a cell in your body, for example, a muscle cell in your big toe or a nerve cell in your brain. List the cell's organelles and their functions.
2. Select a type of factory, for example, a company that manufactures orange juice. List what comes into the factory, what goes out of the factory, and what happens inside the factory during the manufacturing process.
3. Create a labelled storyboard or a flow chart in which the functions of a cell are compared to the processes involved in manufacturing a finished product.
4. Bring your storyboard or flow chart to class for presentation or display.

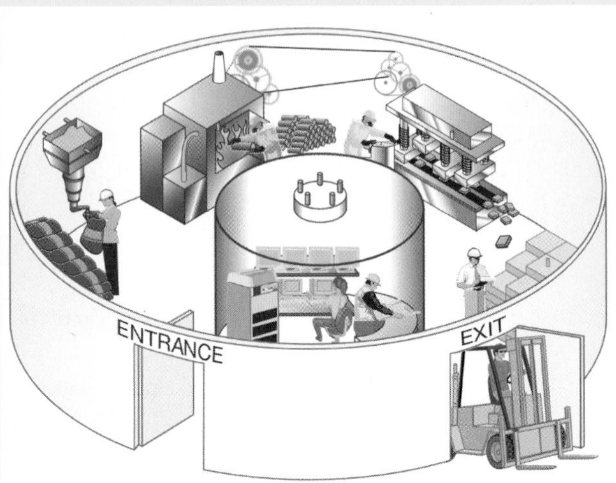

Check Your Understanding

1. What is cellular respiration?
2. In what part of a cell does cellular respiration occur?
3. Why do cells need energy?
4. **Apply** Create a diagram to show the steps by which energy from the Sun is turned into energy that your cells can use.
5. **Thinking Critically** How are the equations for cellular respiration and photosynthesis related?
6. **Design Your Own** Design and conduct one of the following experiments:
   (a) make and test a hypothesis about the effect of chemicals on a unicellular organism
   (b) test the effectiveness of different substances in preventing flowers from wilting
7. **Design Your Own** Formulate your own question about some aspect of cell functioning, and design your own experiment to explore possible answers.

2.3 How Cells Reproduce

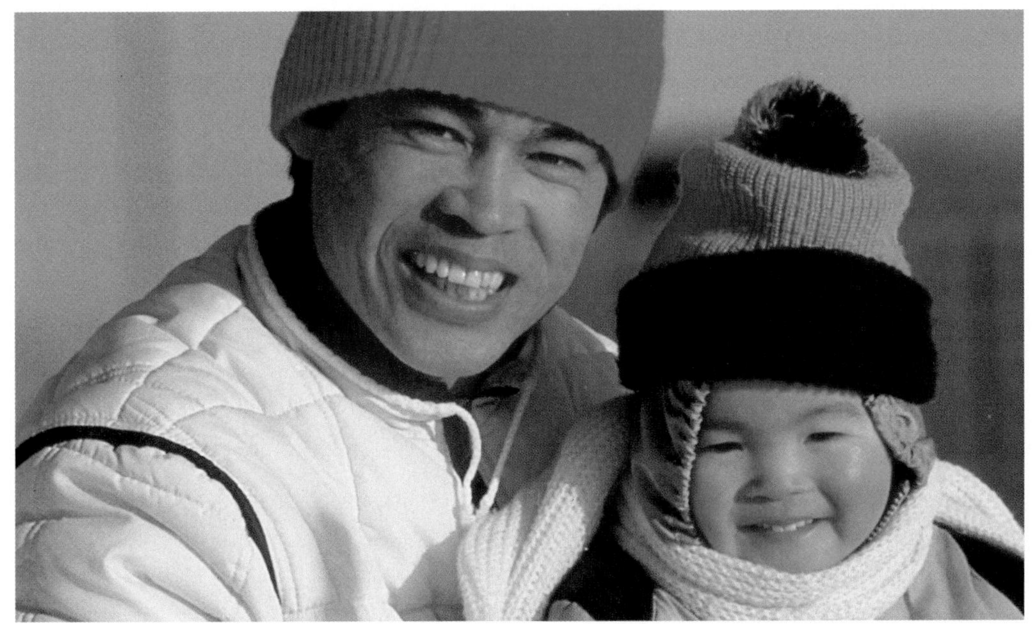

All organisms, including you, begin life as a single cell. How does one cell produce a multicellular organism? A cell divides itself over and over again to make new cells. Remember that cells can only come from previously existing cells. During cell division, one cell divides to become two cells. Then each of these two cells divides into two more cells, and so on. In this way, with amazing speed, the single fertilized egg cell from which a human develops eventually produces a baby consisting of trillions of cells.

The process does not stop when a baby is born. How do you think you grew to your present size? More cells have been added to your body continually since you were born, making you taller and heavier. Even when you are a fully grown adult, many of your cells continue dividing.

Pause& Reflect

Think of reasons why your cells do not stop dividing when you are fully grown. Write your ideas in your Science Log.

Regeneration

There is a story told in Greek mythology about Hydra, a many-headed monster. If one head was cut off, two heads grew in its place. There are no real monsters like Hydra, but many living things are able to grow new body parts to replace damaged or missing parts (see Figure 2.10). This ability to replace body parts is called **regeneration**.

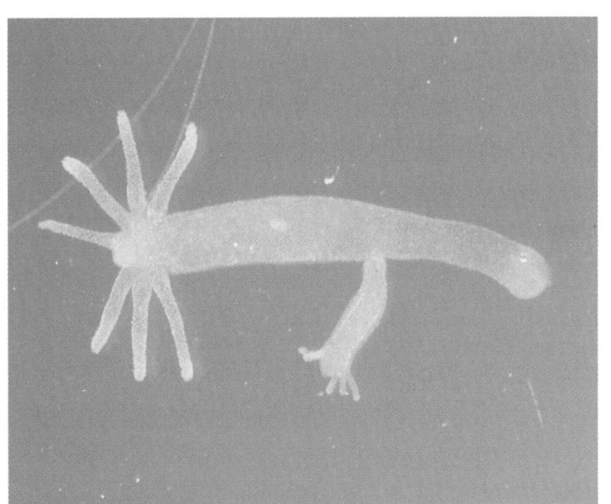

Figure 2.10 The tiny pond organism called hydra was named after the mythical monster. This organism is able to regenerate missing parts.

A fertilized egg divides in two. The two cells both divide to make four. The four cells divide to make eight, the eight cells become sixteen, and so on. If a cell divides once every hour, how many hours will it take to make one million cells? Estimate, then calculate!

DidYouKnow?

Your body produces about one million new red blood cells every second. They are made by dividing cells in the bone marrow (the inner core of the bone). In a few weeks' time, all the red blood cells now in your body will have died and been completely replaced by new cells.

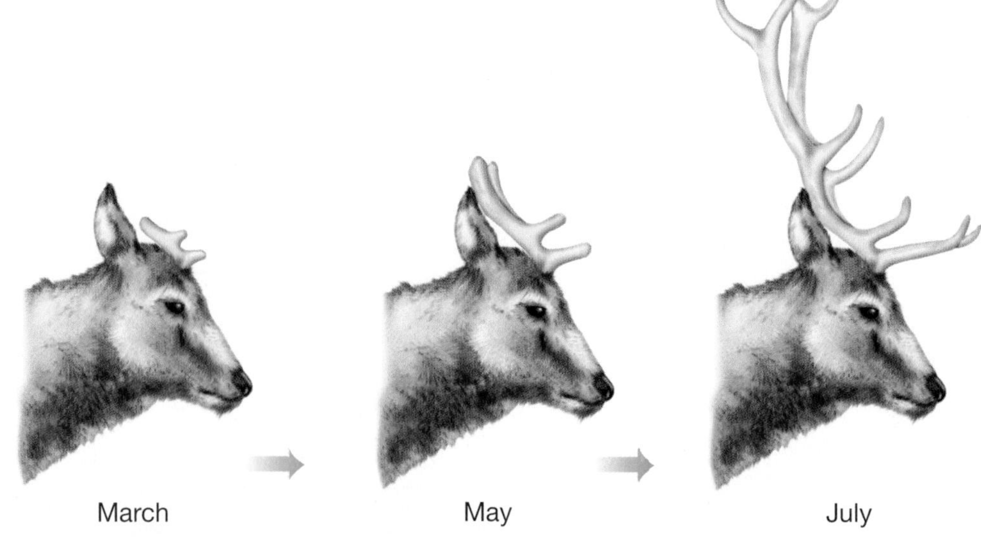

Figure 2.11 Some kinds of deer, such as elk, lose their antlers each year and grow new ones. The new antlers are built up by dividing cells. This is an example of regeneration.

People cannot grow new arms or legs, but we do show a kind of regeneration. Have you ever scraped off some skin, or broken a bone? The wound is able to heal because dead and damaged cells are replaced by new ones. The new cells, produced by cell division, grow over the injury and repair the wound.

Like organisms, individual cells have a life cycle — they grow, divide, and eventually die. Different cells grow and divide at different rates, depending on their structure and function. For example, the cells in your skin wear out quickly and last only days or weeks. The longest-lived cells in your body are nerve cells, which can last a lifetime.

Regeneration is evidence that organisms can produce new cells throughout their lives. For example, some lizards can grow a new tail after a predator snatches away their old one. Lobsters, newts, tadpoles, and insects can grow new legs. Deer shed their antlers each year, then grow new antlers to replace them (see Figure 2.11). If you cut off a plant stem and place it in water, it will grow new roots, while the part of the plant left behind will grow a new stem. These are all examples of regeneration. Can you think of any other examples?

Mitosis

How does one cell become two? If you look at cells under a microscope, you may be able to observe them in the process of dividing. The best places to look for dividing cells are in parts of an organism that are growing rapidly, such as the root tips of young seedlings (see Figure 2.12).

Notice that some cells in Figure 2.12 do not have a solid, round nucleus. Instead, they have clusters of dark, threadlike objects. These objects are the **chromosomes**, which become visible only when a cell is about to divide. Chromosomes contain the cell's genetic material, or DNA (**d**eoxyribo**n**ucleic **a**cid) — instructions for producing new cells with the same characteristics as the parent cell.

During cell division, the genetic material duplicates and then divides into two identical sets of chromosomes. This process is called **mitosis**. It is very similar in all forms of life, whether the organism is a unicellular micro-organism or a multicellular plant or animal. The two new cells formed by this division are called **daughter cells**. Each daughter cell gets one set of chromosomes. You will learn more about mitosis in later studies.

How Rapidly Do Cells Divide?

All healthy cells have regular rates of dividing. For example, certain bacterial cells divide once every 20 min. Frog embryo cells divide in about an hour, cells lining your intestine take about 48 h to divide, and liver cells divide only once every 200 days.

How can you determine how rapidly different kinds of cells divide? In the next investigation, you will be able to estimate the average rate of cell division in a young plant.

Figure 2.12 Cells in a growing root tip. Roots grow longer as more cells are added at the tip. Root cap cells protect the growing root. They are easily rubbed off by soil particles, but they are constantly replaced by new cells.

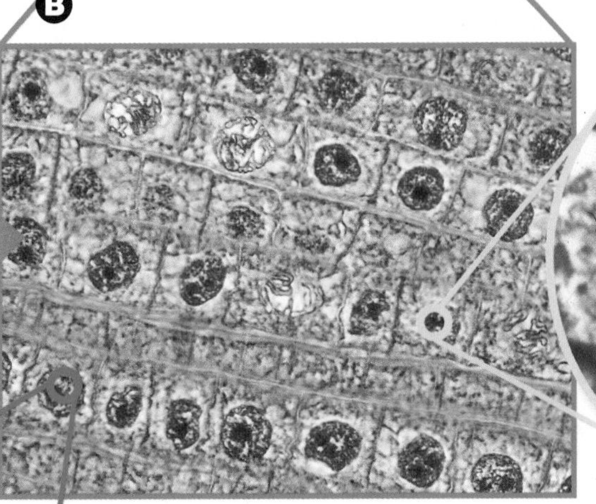

Many cells are dividing by mitosis in this enlarged area of the root tip. Note how the appearance of the nucleus changes in cells that are dividing.

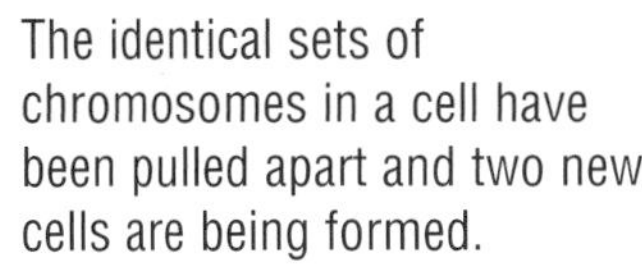

The identical sets of chromosomes in a cell have been pulled apart and two new cells are being formed.

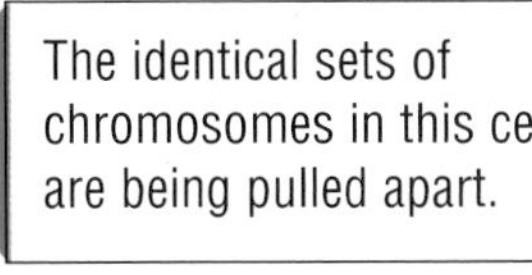

The identical sets of chromosomes in this cell are being pulled apart.

CONDUCT AN

INVESTIGATION 2-C

Observing Cell Division

You have learned that growth in organisms occurs as a result of cell division. In this investigation, you will observe and measure the growth of root seedlings. You can use your results to estimate the average rate of cell division in a growing root tip.

Problem

What is the average hourly growth rate of a young root tip?

Safety Precautions

Apparatus

5 small plastic bags
waterproof marking pen
ruler

Materials

5 young corn seedlings
5 labels
paper towels
water

Procedure

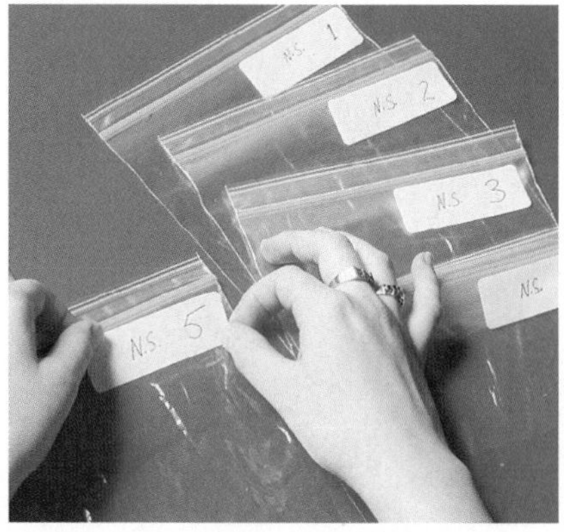

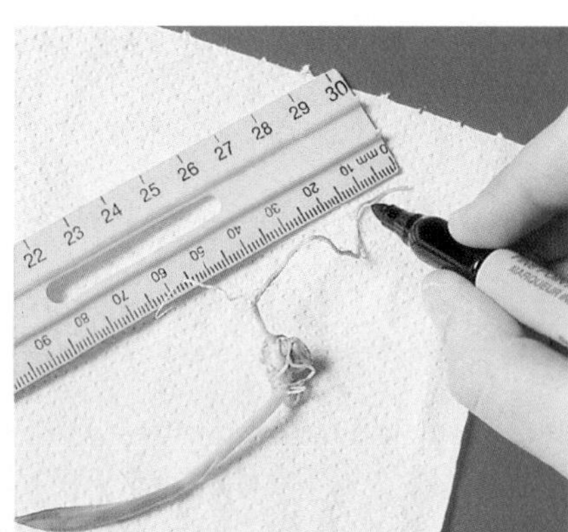

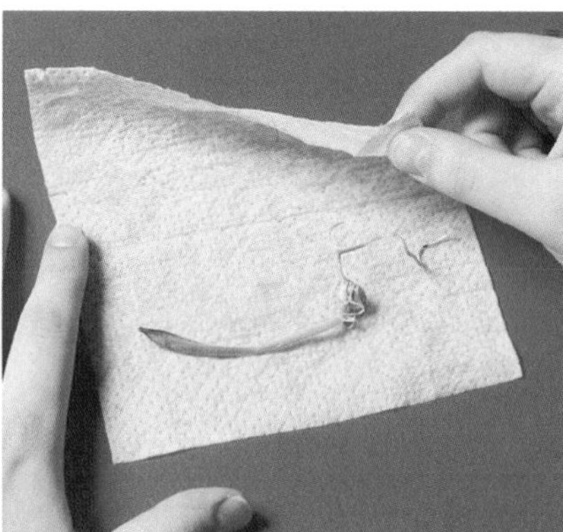

1. Number the five labels from 1 to 5. Add your initials to each label (for identification). Then place one label on each plastic bag.
   - (a) Prepare a data table like the one shown on the next page.
   - (b) Give your table a title.

2. Examine one of your seedlings and identify the growing root. With the marking pen, carefully place a dot on the root 10 mm from the root tip. Record this distance in your data table under "Initial length (mm)."

3. Fold the seedling in a square of damp paper towel.

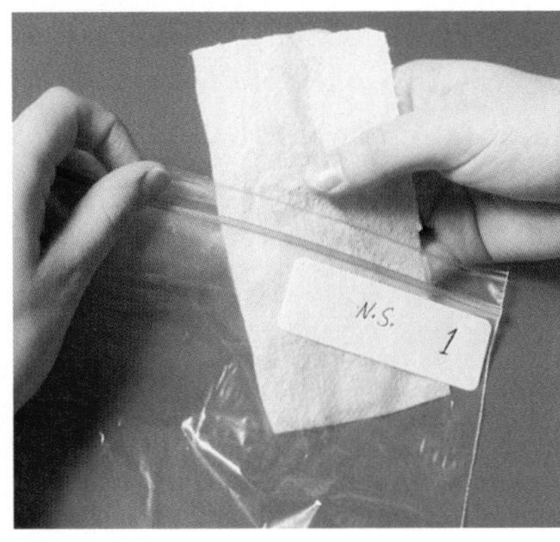

4 Place the wrapped-up seedling carefully into one of the labelled plastic bags.

(a) Repeat steps 2 to 4 for the other four seedlings.

(b) Leave all five bags of seedlings in a warm, dark place where they will not be disturbed for two days.

5 After 48 h, measure the length of each root from its tip to the mark you made. Record this data in your data table under "Final length (mm)."

(a) Calculate the amount by which each root has grown. Record these figures under "Total growth (mm)."

(b) Calculate the average total growth in your sample of five roots. (Add all the numbers from the last column and divide by 5.) Record the result in your data table.

(c) Wash your hands after this investigation.

6 Compile the growth data for the entire class in a stem-and-leaf plot and then find the class average.

Seedling number	Initial length (mm)	Final length (mm)	Total growth (mm)
1			
2			
3			
4			
5			
Average total growth			

Analyze

1. What evidence do you have that mitosis occurs in root tip cells?
2. What was the average increase in root length in 48 h?
3. What is the average growth rate of a young root tip per hour?
4. Make a flow chart showing the steps you would use to estimate the number of new cells added to the corn root each hour.
5. Predict the average amount of growth that would occur in a corn root in one week.

Conclude and Apply

6. Why was each seedling wrapped in a damp paper towel rather than a dry one?
7. Suppose you found that some seedlings in your investigation showed no root growth. Hypothesize some possible explanations for this result.
8. **Design Your Own** Design an experiment, similar to this one, to illustrate the effect of temperature on the rate of mitosis in root tips. Write out the steps of your experiment and the data you would record. Remember to include a control.

Extend Your Skills

9. Prepare and stain a wet mount of cross-sections of a root tip, or obtain a prepared slide of a root tip. Examine the slide under a microscope and identify cells undergoing mitosis.

Cells and Cancer

Cell division is essential for body growth and repair. What happens if cells begin to multiply and spread in an uncontrolled way? That is what happens in the bodies of people with cancer. Cancer cells divide at a far greater rate than normal cells and they spread to other parts of the body. Cancer has been described as "mitosis gone wild." As the abnormal cancer cells continue to multiply, they spread to other parts of the body and damage them. Cancer is one of the leading causes of death in Canada today.

Cancer can affect many parts of the body and may be caused by many different factors. Factors that produce cancer are called *carcinogens*. They include some types of chemicals, radiation, inherited (genetic) factors, certain viruses, and repeated damage to the body.

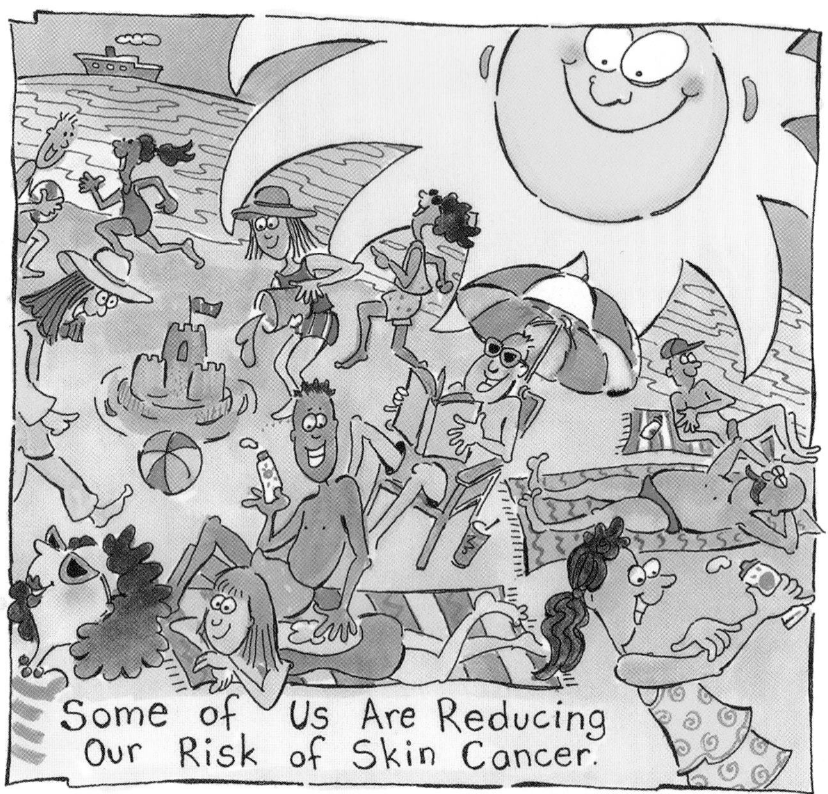

Some of Us Are Reducing Our Risk of Skin Cancer.

People working in some jobs may be exposed to particular types of carcinogens. For example, people working in industries using asbestos have had high rates of lung cancer because they inhaled fine particles of this substance over long periods of time. Some farmworkers have had high rates of cancer after improperly using certain pesticides. The rates of industry-related cancers have been reduced by use of protective clothing, air filters, and banning some harmful chemicals.

Some cancers can be prevented by changing lifestyle habits to reduce exposure to carcinogens. One example is the link between smoking and lung cancer. Smokers are far more likely to die of lung cancer than non-smokers and they can reduce this risk by not smoking. Another example is the connection between Sun exposure and skin cancer. People who spend many hours in the Sun without protective clothing or sunblock have a high risk of developing skin cancer, caused by ultraviolet radiation from the Sun. The incidence of skin cancer in Canada is on the rise. About 8000 cases of skin cancer are diagnosed each year in Canada.

Although prevention is better than cure, there are some treatments that can slow or stop the spread of cancer in patients who already have the disease. The techniques consist of destroying the cancerous cells while leaving normal cells intact. This can be done by chemicals (chemotherapy) or by radiation treatment — using high-energy particles to kill cells. These treatments are most successful if the cancer is diagnosed in an early stage, before the abnormal cells have spread widely through the body.

New techniques may give better methods of curing cancer in the future. One method is gene therapy, the altering of genes that cause cells to divide and produce cancer. Alternative therapies focus on ways to boost the body's own natural immune system. For example, people may be able to use vaccines or drugs that stimulate their bodies to destroy cancer cells, making them immune to cancer.

Career CONNECT

Just Right for the Job

Read this list of ten personal traits carefully. Then read the career advertisement for a medical technologist. Which traits do you think would be most important if you were interested in this type of career? Why? Rank the personal traits in order from 1 to 10, from most important (1) to least important (10).

Personal Traits

strong artistic abilities
good memory
steady hands
good athletic ability
dramatic talent
concern for accuracy
sense of humour
excellent math skills
good communication skills
love of the outdoors

Add any other personal traits to the list that you think would suit the job of medical technologist.

Mount Kinmore Hospital

has an immediate opening for a medical technologist. The successful applicant must have completed a three-year program in medical laboratory technology and be familiar with advanced laboratory equipment. He or she must be skilled in performing the following tests to assist doctors in diagnosing a patient's condition:

- CBC – a complete blood count of white blood cells, red blood cells, hemaglobin, and platelets
- urinalysis – testing urine samples for blood, protein, white blood cells, sugar, etc.
- microbiological culturing – growing bacteria from throat swabs, wound swabs, urine samples, and stool samples to identify the cause of infections or food poisoning
- tissue typing – matching of donor and recipient tissue types for organ transplants
- blood banking – matching of donor and patient blood types for blood transfusions

Check Your Understanding

1. What are chromosomes?
2. What is mitosis?
3. Give an example of regeneration.
4. Why do the cells of a multicellular organism continue to reproduce even after the organism is fully grown?
5. How is a daughter cell like the original cell from which it formed?
6. How can you tell from a cell's appearance alone whether the cell is undergoing mitosis?
7. What were some of your qualitative observations in Conduct an Investigation 2-C? What were some quantitative observations?
8. **Apply** What evidence do you have that some cells in your body are dividing?

2.4 Specialized Cells

Imagine an orchestra made up of only a hundred trumpet players or a hundred violins. Such an orchestra would be very limited! To play every kind of music, an orchestra needs a variety of musical instruments — some flutes, some oboes, a piano, drums, and so on. In the same way, a multicellular organism cannot be made up only of identical cells. As Figure 2.13 shows, although multicellular organisms grow from single cells that repeatedly divide, their cells are not all the same.

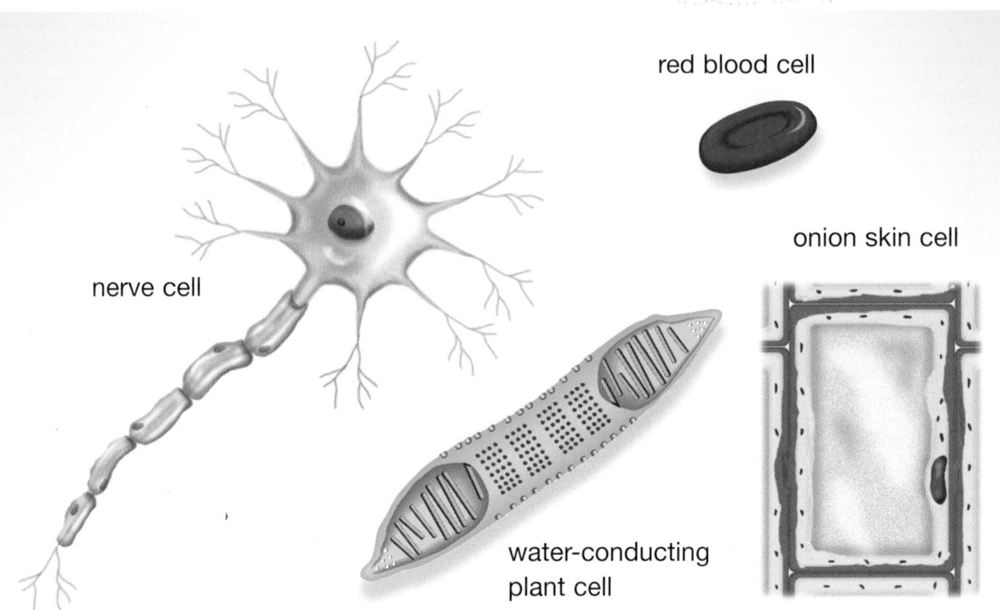

Figure 2.13 Different cells have different shapes and functions.

Like the instruments in an orchestra, different cells have different appearances and perform different jobs. They are said to be **specialized** for particular tasks. For example, your muscle cells are shaped to move parts of your body, and your skin cells are built to protect your body from the drying rays of the Sun. Humans have about a hundred different types of cells, each with their own particular structure and functions.

Look at the examples of plant and animal cells in Figure 2.13. How do their shapes relate to their functions? Nerve cells have long, branched fibres running from the main part of the cell, shaped to carry nerve signals from one part of the body to another. Red blood cells, which carry oxygen in the bloodstream, have a thin, disk-like shape. This gives them a large surface area to pick up large amounts of oxygen. The water-conducting cells of a plant are tubelike, with thick walls and a network of holes that lets water pass easily through them. Onion skin cells are flat and brick-shaped, so they can fit closely together to form a continuous protective layer.

A nerve fibre in the neck of a giraffe can be up to 1 m in length. However, the main part of the cell from which it comes is about the same size as a human nerve cell.

INVESTIGATION 2-D

Different Cells, Different Jobs

If you were to design a human body, what different cells would you need? What jobs must these cells do in your body? What characteristics would be needed for each job? In this investigation, you will work with groups in your class to discover the variety of different cells and to think about their functions and importance to the body as a whole.

Think About It

What is the structure and function of various cells in the human body?

What to Do

1. The class will divide into five groups. Each group will represent a different part of the body from the following list:

blood in heart	bone in leg
nerve in toe	skin on head
muscle in arm	

2. Imagine your classroom is a giant body. Go with your group to the approximate location of your part of the body.

3. Prepare a brief presentation about the specialized cells you might expect to find there. Your report should answer the following questions:
   (a) What is the main type of cell found in this part of the body?
   (b) What is its main function?
   (c) What type of structure would help it carry out this function? (For example, consider the cell shape, size, use of energy, and other characteristics.)
   (d) Could the body as a whole survive if it had none of these cells?

Analyze

The photographs below show several types of human cells with descriptions of their characteristics. Match each photograph to one of the cell types described by your class. (Answers are found on page 560.)

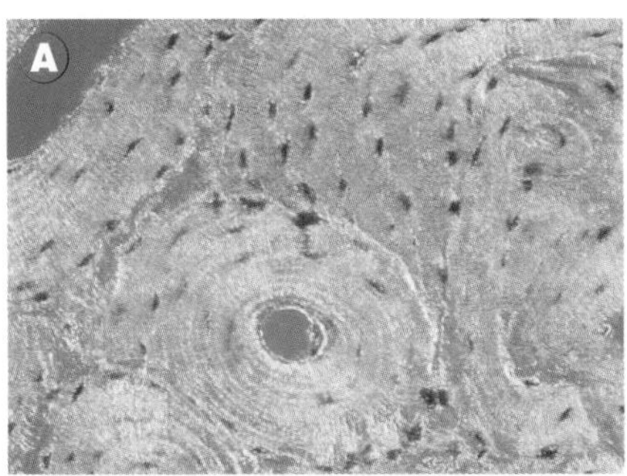

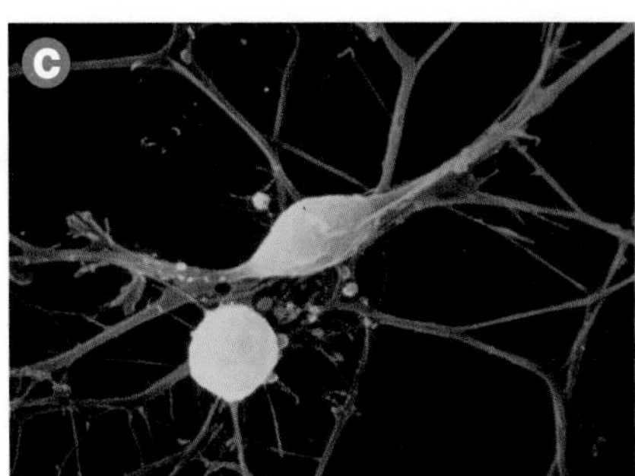

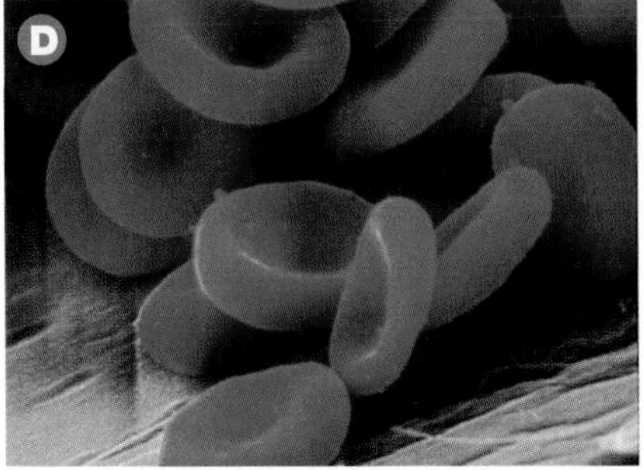

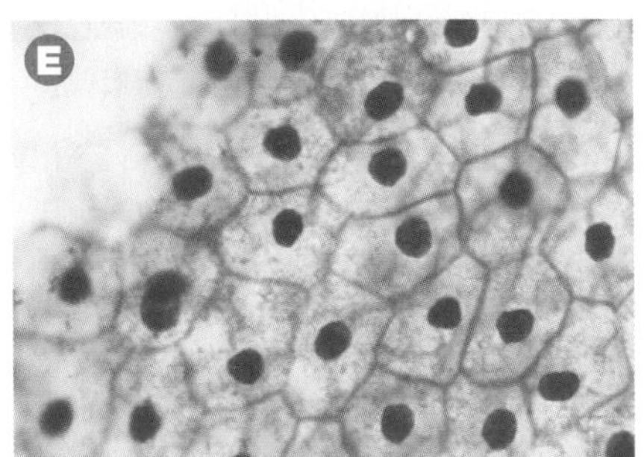

DidYouKnow?

Most household dust is made up of dead human skin cells. You and everyone around you are continually shedding parts of the thin outer layer of skin. Your entire outer layer of skin is completely replaced by the growth of new cells approximately every 28 days.

The Advantages of Being Multicellular

Imagine you are a microscopic, unicellular organism. Your whole body is one cell. This one cell must carry out all the functions needed to keep you alive. It must be able to move, obtain food, reproduce, and respond to the environment. There are many living organisms that consist of only one cell. What disadvantages do you think they have, compared with multicellular organisms?

In the last chapter, you learned of one disadvantage. Unicellular organisms cannot grow very large. Also, because they must take in all the materials they need through their cell membranes, most unicellular organisms can only live in watery, food-rich surroundings.

Multicellular organisms have several advantages compared to unicellular living things. They can live in a wide variety of environments. They are able to grow very large — as large as a whale or a Douglas fir. Multicellular animals can obtain their energy from a wide variety of foods. Their bodies are more complex. By specializing in particular functions, each cell in a multicellular organism can work much more efficiently than the cell of a unicellular organism, which must do every job itself.

In multicellular organisms, specialized cells of a similar kind work closely together, and are usually found grouped closely together in the body. Groups of specialized cells, in turn, work in harmony with other groups. You will investigate this organization of cells in the next chapter.

Check Your Understanding

1. Why do cells in your body need to be specialized?
2. Why do nerve cells have long fibres, whereas red blood cells are thin and disklike?
3. Why do unicellular organisms live mainly in a watery environment?
4. **Apply** Most people think of the skin as just a body covering. How do you think skin cells are important to other body cells?

CHAPTER at a glance

Now that you have completed this chapter, try to do the following. If you cannot, go back to the sections indicated.

Identify which part of the cell controls the movement of substances into and out of cells. (2.1)

Write a definition of (a) diffusion, and (b) osmosis. (2.1)

Draw a diagram to illustrate the process of active transport. (2.1)

Write a word equation for the process of cellular respiration. (2.2)

Describe how your body obtains each of the two substances that react together in the process of cellular respiration. (2.2)

Name the organelle in which cellular respiration takes place. (2.2)

Give examples of processes that depend on cell division. (2.3)

Explain where you would look for actively dividing cells in a plant. (2.3)

Describe an investigation you could carry out to show that plant cells divide. (2.3)

Name some different types of specialized cells in plants and animals. (2.4)

Explain how the structure of a specialized cell is related to its function in the body of a multicellular organism. (2.4)

List some advantages that multicellular organisms have over unicellular organisms. (2.4)

Prepare Your Own Summary

Summarize this chapter by doing one of the following. Use a graphic organizer (such as a concept map), produce a poster, or write the summary to include the key chapter ideas. Here are a few ideas to use as a guide:

- How do a permeable membrane and a selectively permeable membrane differ?
- How does the particle theory help explain how substances move into and out of cells?
- Copy and complete the diagram of osmosis on the top right.
- Describe a simple method of illustrating osmosis in a plant.
- Why do cells need energy?
- How do cells get energy?
- Why do cells need to divide?
- Sketch the appearance of a cell about to divide.
- Identify the specialized cells shown on the right. Describe how each cell is suited to its role.

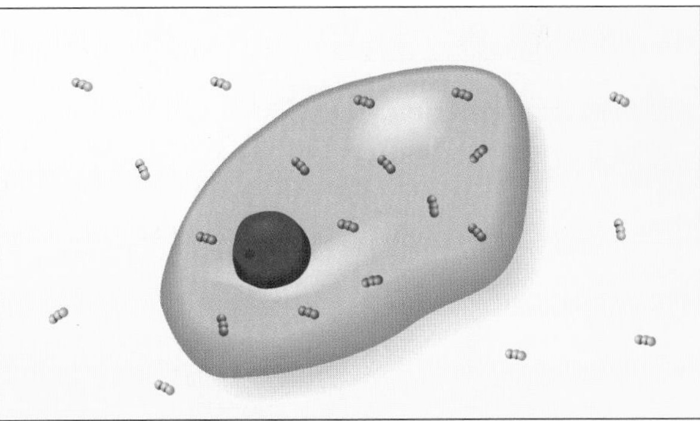

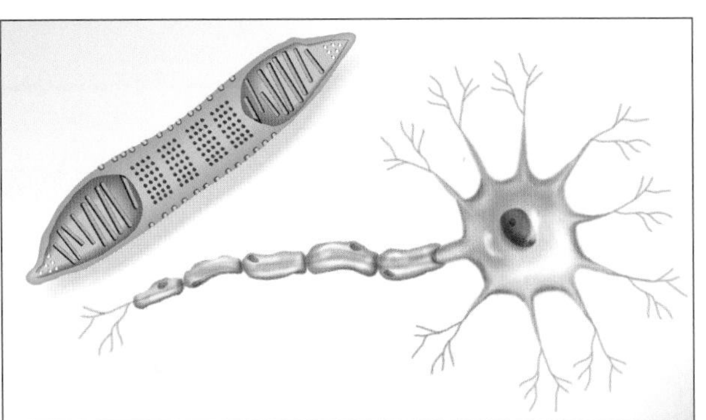

CHAPTER

2 Review

Key Terms

selectively permeable
permeable
impermeable
diffusion
osmosis
carrier proteins
active transport
cellular respiration
regeneration
chromosomes
mitosis
daughter cells
specialized

Reviewing Key Terms

If you need to review, the section numbers show you where these terms were introduced.

1. In your notebook, complete each sentence from column A with the correct ending from column B.

A
• A cell membrane is
• Cell nuclei divide by
• Cell nuclei contain
• Water enters or leaves cells by
• Oxygen enters or leaves cells by
• A dividing cell produces two

B
• osmosis (2.1)
• diffusion (2.1)
• daughter cells (2.3)
• mitosis (2.3)
• selectively permeable (2.1)
• chromosomes (2.3)
• carrier proteins (2.1)

2. Describe what is meant by regeneration. Give some examples. (2.3)

Understanding Key Ideas

Section numbers are provided if you need to review.

3. Compare cell membranes with the screen doors used on houses in summer. Explain why neither can be completely impermeable or completely permeable. (2.1)
4. How are osmosis and diffusion different? (2.1)
5. If a cell is placed in a concentrated solution of glucose, would you expect water to move into or out of the cell? Explain. (2.1)
6. Describe the role of carrier proteins in transporting particles into and out of a cell. Does this process use energy? Why or why not? (2.1)
7. Explain how (a) animals, and (b) plants obtain carbohydrates. (2.2)
8. Explain why cells need (a) energy, (b) oxygen, and (c) food. (2.2)
9. Give two reasons why cell division is necessary for the development and survival of multicellular organisms. (2.3)
10. Why are cells specialized in multicellular organisms? Give two examples of specialized cells. (2.4)

Developing Skills

11. Copy and complete the spider map on the work of cells. Use the following words and phrases: movement through a cell membrane; mitosis; selectively permeable; growth; active transport; carbon dioxide; energy; osmosis; water; impermeable; types of membranes; regeneration.

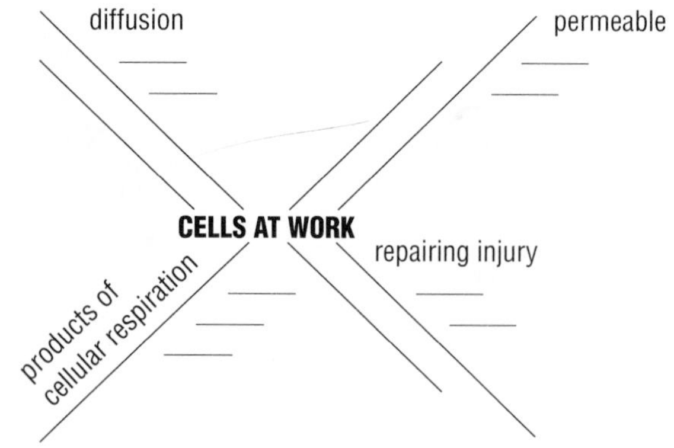

12. The table on the right shows the results of an experiment to find the effect of osmosis on potato cells. Two cubes of potato were weighed and placed in distilled water and two cubes were weighed and placed in salt solution. The mass of each potato cube was determined at 15 min intervals.

	Salt water			Distilled water		
Time (min)	**Cube 1 mass (g)**	**Cube 2 mass (g)**	**Average (g)**	**Cube 1 mass (g)**	**Cube 2 mass (g)**	**Average (g)**
0	59	60		51	52	
15	58	58		51	52	
30	50	55		52	53	
45	50	54		53	54	
60	50	53		55	53	

(a) Calculate the average mass of the cubes in water and the average mass of the cubes in salt solution at each time interval.

(b) Plot the data on a graph, showing average mass along the *y*-axis (the vertical axis) and time along the *x*-axis (the horizontal axis).

(c) Briefly interpret the results. What has happened to the mass of potato in water? What has happened to the mass of potato in salt solution? Why?

Problem Solving/Applying

13. Study the two illustrations of red blood cells. One set of cells has been placed in distilled water, the other set in strong salt solution. Make an inference about which set was in which solution, giving your reasons.

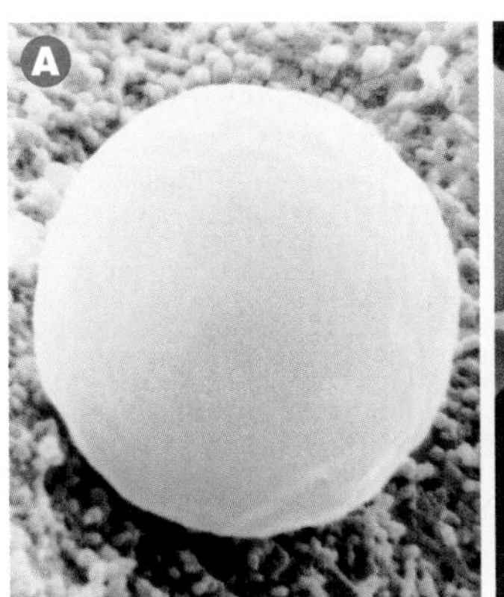

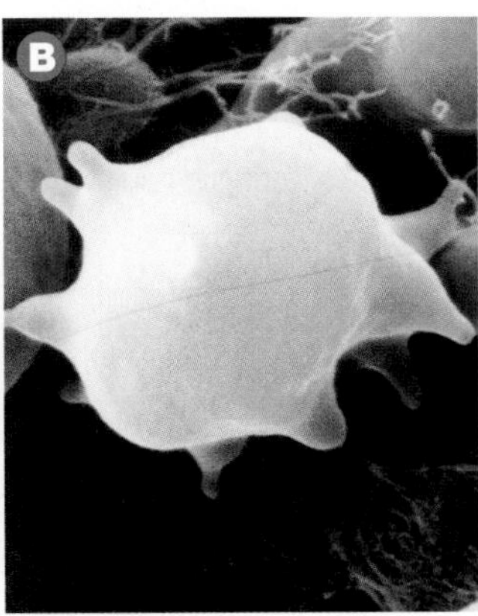

14. What cell process causes a suntan on your skin to fade?

15. Some of your body cells were dividing rapidly when you were 5 years old and will still be dividing when you are 25 years old. However, some of the cells that were dividing when you were a small child will divide slowly or stop dividing when you are an adult. Suggest an example of each type of cell and give a reason for this difference.

Critical Thinking

16. In what ways is the process of cellular respiration similar to the burning of a piece of paper?

17. What problem might athletes face if they drink only distilled water after a race? Explain your reasoning and suggest how the problem can be avoided.

Pause& Reflect

1. Three items you cannot live without are water, food, and air. Why does a lack of air lead to death much faster than a lack of the other two items?
2. Go back to the beginning of this chapter on page 38 and check your original answers to the Getting Ready questions. How has your thinking changed? How would you answer these questions now that you have investigated the topics in this chapter?

CHAPTER

3 Cells Working

Getting Ready...

- Why do plants lack muscles?
- How does the food you eat provide energy to cells in your toes?
- Why do doctors measure their patients' blood pressure?

Science Log

In your Science Log, describe how a horse obtains water and how a tree obtains water. Then describe the system you think each organism has inside its body to deliver water to all its cells. At the end of this chapter, look back to see how your ideas have changed.

Most cells of multicellular organisms such as a horse or a tree or yourself are not in direct contact with the outside environment. So how do the cells in the hind legs of a horse get oxygen? How do cells in the roots of a tree get food (sugars)?

To think about this problem, imagine that all the cells of an animal or plant are organized into different systems. Each system has a particular function to perform. For example, one system carries oxygen throughout the body to every cell. Other systems make sure that each cell receives food, and so on. Together, the systems keep all the cells of the body functioning properly.

In this chapter, you will learn how biologists think of cells as being organized into systems. The interactions of systems of cells maintain a stable environment inside multicellular organisms, even though conditions may change outside their bodies. What happens when a system does not work properly? Problems in the organization and function of cells and systems can cause illness. This chapter also explores how healthy lifestyle choices can help prevent the onset of illness and disease. Start this chapter by taking your pulse under different conditions. The pulse is one guide to how well your circulatory system is working.

Together

Starting Point ACTIVITY

Changing Your Pulse Rate

Your pulse is produced by blood surging through your arteries each time your heart "beats." What factors affect the rate at which your heart pumps blood? Measure your pulse rate to find out.

What to Do

1. Locate one of your radial arteries, on the inside of your wrist in line with your thumb (see the photograph).

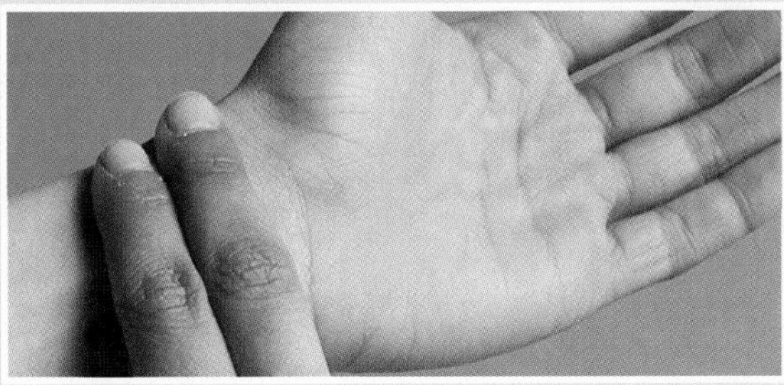

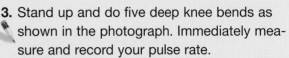

2. Using a watch or timer, count the number of pulses you feel in 15 s while you are sitting comfortably at rest. Multiply the number by 4 to obtain your heart rate per minute. Record your results.
3. Stand up and do five deep knee bends as shown in the photograph. Immediately measure and record your pulse rate.
CAUTION: Do not do this activity if you have any health problems that may put you at risk.

4. Rest for 1 min and again measure and record your pulse rate.

What Did You Find Out?

From your results, what is the relationship between exercise and pulse rate? Suggest an explanation for this relationship.

Spotlight On Key Ideas

In this chapter, you will discover

- how cells in multicellular organisms are organized into tissues, organs, and systems
- how systems of cells interact to maintain cell functions throughout the body
- how organ systems in animals and plants compare with each other
- how the structures of plants allow them to live in specific environments
- how good health depends on the proper functioning of cells, tissues, organs, and systems
- how research into cells has brought about improvements in human health and nutrition

Spotlight On Key Skills

In this chapter, you will

- identify main types of animal tissue from photographs
- research and describe factors that contribute to the efficient functioning of human body systems
- make a model to simulate how lungs function
- illustrate how blood transports substances to and from cells
- design your own experiments to investigate cell processes

3.1 How Cells Are Organized

Pause& Reflect

Select an animal such as a bird, insect, fish, or mammal. In your Science Log, make a labelled drawing of the levels of organization within that animal's body.

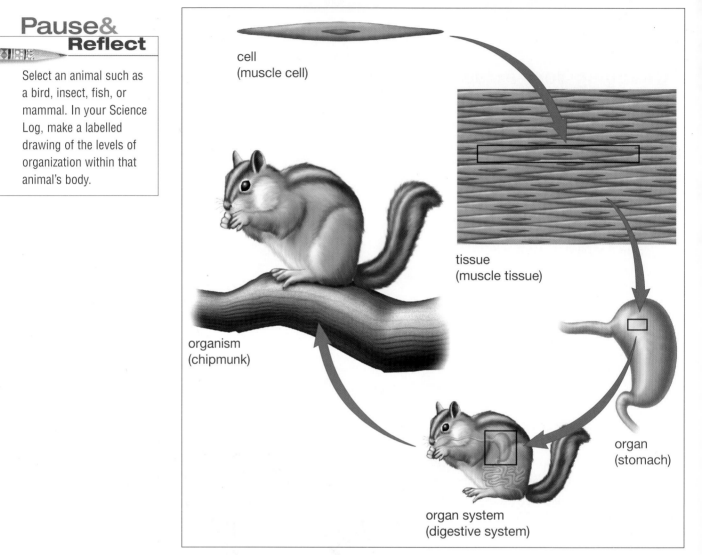

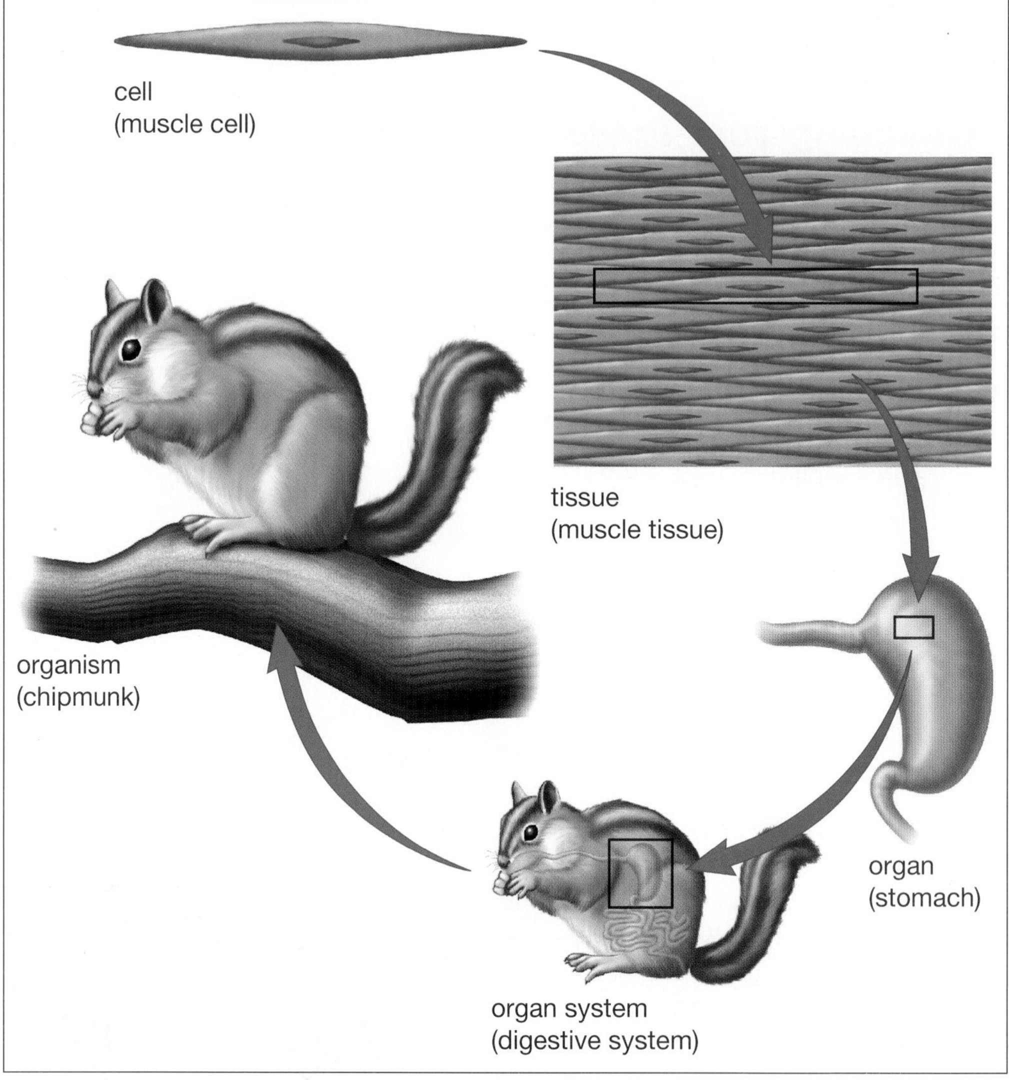

Figure 3.1 How cells are organized in living things

Many animals and plants are made of trillions of cells. To learn how these cells are organized in the body, compare the organization of cells with the way students are organized in a school district. First, students in the same grade are grouped together in classes. Then, different classes of students together make up a school. Finally, a number of schools are organized into a single school district.

Similarly, cells with the same structure and function are grouped into **tissues** (such as muscle, nerves, and skin). Groups of different tissues form **organs** (such as the heart, stomach, and liver). Organs work together in **organ systems** (such as the respiratory system and the digestive system). This arrangement of cells, tissues, organs, and systems forms several different **levels of organization** in living things, as shown in Figure 3.1. Each level can be studied on its own, or in relation to the levels above or below it. You have studied the cell level in Chapters 1 and 2. Now you are ready to examine each of the other levels.

Tissues

Tissues are groups of similar cells. Onion skin is a tissue made of sheets of similar, thin, tightly-packed cells. You have already observed the structure of onion skin under a microscope. These specialized skin cells form a layer that covers and protects the onion. Figures 3.2 and 3.3 show the main types of tissues found in animals and plants. These tissues are classified according to the functions they perform. You will examine these tissues in greater detail later in the chapter.

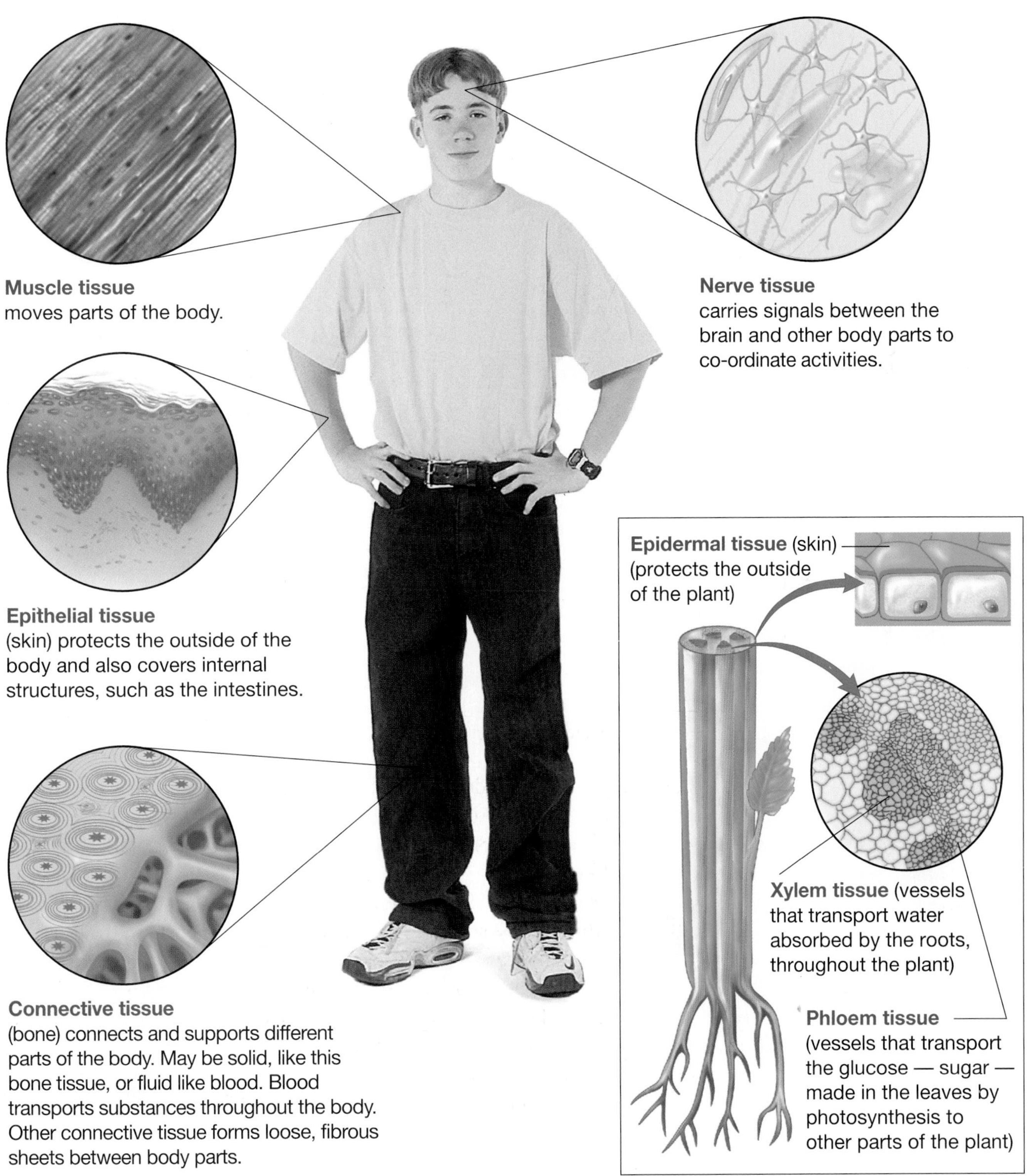

Muscle tissue
moves parts of the body.
Nerve tissue
carries signals between the brain and other body parts to co-ordinate activities.
Epithelial tissue
(skin) protects the outside of the body and also covers internal structures, such as the intestines.
Connective tissue
(bone) connects and supports different parts of the body. May be solid, like this bone tissue, or fluid like blood. Blood transports substances throughout the body. Other connective tissue forms loose, fibrous sheets between body parts.
Epidermal tissue (skin) (protects the outside of the plant)
Xylem tissue (vessels that transport water absorbed by the roots, throughout the plant)
Phloem tissue (vessels that transport the glucose — sugar — made in the leaves by photosynthesis to other parts of the plant)

Figure 3.2 Main types of tissues found in animals

Figure 3.3 Main types of tissues found in plants

Organs

Suppose you feel hungry, see a juicy apple, and eat it. This simple action would not be possible without the next level of organization in the body — the organs. Organs are distinct structures in the body that perform particular functions. You used your eyes (to sense the apple), your brain (to plan and co-ordinate your actions), and your mouth and stomach (to start digesting the apple). Each organ is made of several tissues working together. For example, your stomach is made of four main types of tissues, as shown in Figure 3.4. Other examples of organs in your body are the lungs, the heart, and the kidneys. Plants have organs, as well. Plant organs include roots, stems, leaves, and flowers.

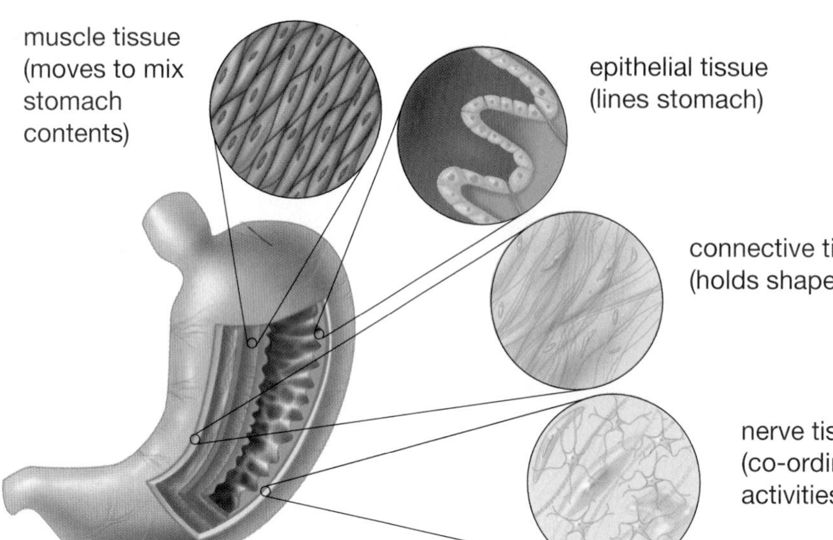

Figure 3.4 The stomach is an organ made of different tissues.

Find Out ACTIVITY

Looking at Animal Tissues

The photographs below show tissues observed under a compound light microscope similar to the one you have used.

What to Do

1. Look closely at the tissues shown in A, B, and C. The tissues are bone tissue, nerve tissue, and skeletal muscle (the kind of muscle you use, for example, to bend your arm).
2. Based on what you have read about these tissues, try to identify each type. Record your decision for each tissue.

What Did You Find Out?

1. Which tissue did you find easiest to identify and why? (Answers are found on page 560.)
2. List three different careers in which a person might need to examine tissues. Suggest at least one way a person could use information obtained by observing tissues.

A

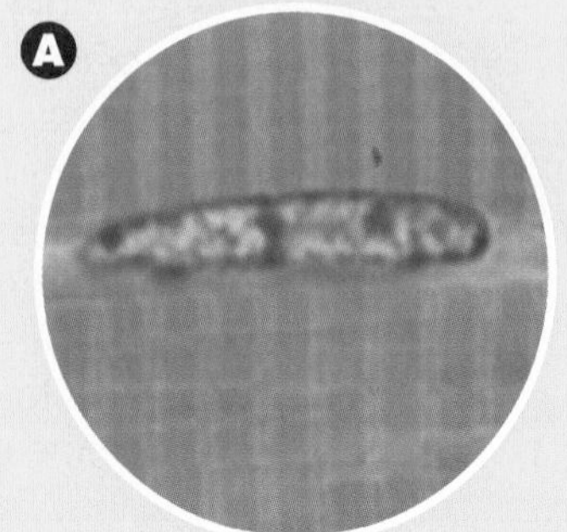

B

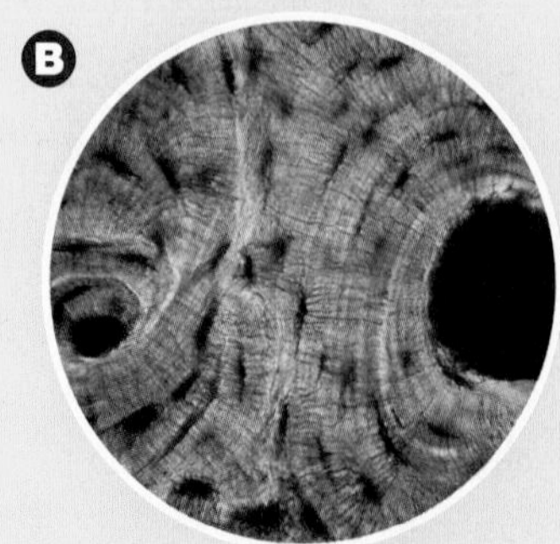

C

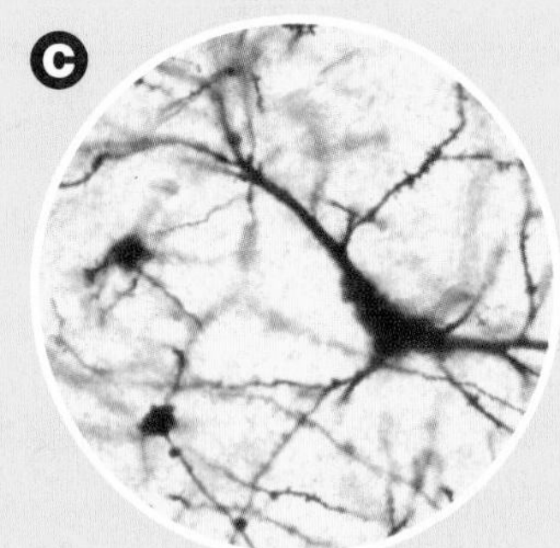

INVESTIGATION 3-A

Teamwork!

Think About It

Why do you need a liver? A heart? A pair of lungs? What do these and other organs do? What are their main parts? Are they part of an organ system? Your class will divide into small groups to investigate these questions and each group will present its findings to the class. How you complete this task is up to your imagination and your research skills!

What to Do

1. As a class, brainstorm a list of organs found in the human body. Divide into groups and assign one organ to each group.

2. Decide how the class will evaluate each group's presentations. As a minimum, each group's presentation must answer these three questions:
   - What is the organ's function?
   - What is the organ's structure?
   - To which organ system does the organ belong?

3. Use your library, the Internet, and any other resources to research the structure and function of your group's organ. Here are some more ideas to start you thinking:
   - What happens if the organ does not work properly?
   - Can the organ be transplanted? Can it be replaced by an artificial organ?
   - Which other animals have this organ? Are there some interesting differences or similarities compared with the human organ?

4. Decide how your group will co-ordinate and present the results of its research to the class. Group members may choose to present a scientific lecture with charts and graphs. They may write a play and act the roles of tissues, cells, or organs. Group members might devise a quiz modelled on games such as "20 Questions" or "Jeopardy," they might invent a board game, or they might construct a three-dimensional model.

Analyze

Use the criteria decided in advance (see step 2) to evaluate each group's presentation.

Skill POWER

For tips on doing research in science, turn to page 559.

INTERNET CONNECT

www.school.mcgrawhill.ca/resources/

For information about organs in the human body and diseases affecting them, visit the above web site. Go to **Science Resources**, then to **SCIENCEPOWER 8** to know where to go next. Search for facts or diagrams that you might use in this investigation.

Organ Systems

As you have seen, organs work together just as cells and tissues do. Organs form organ systems to perform activities that help the body function as a whole. For example, your stomach is part of a group of organs that form your **digestive system**. Other organs in this system include your tongue, pancreas, and small intestine. The role of the digestive system is to digest food and eliminate waste. No organ alone can perform all these functions. However, all the organs of the digestive system can do the job together.

Check Your Understanding

Choose the correct answer and write each complete sentence in your notebook:

1. A tissue is made from groups of (a) organs, (b) cells, (c) organelles.
2. Muscle is an example of (a) a system, (b) an organ, (c) a tissue.
3. The heart is an example of (a) an organ, (b) a system, (c) epithelial tissue.
4. One example of connective tissue is (a) nerve tissue, (b) bone tissue, (c) epithelial tissue.
5. An example of an organ system in animals is (a) the brain, (b) the respiratory system, (c) skin.
6. The type of tissue that protects the outside of a plant is called (a) epithelial tissue, (b) epidermal tissue, (c) connective tissue.
7. **Thinking Critically** The two sets of data on the right represent the heart rates of 30 individuals — 15 athletes and 15 non-athletes — after ten minutes of intense physical activity. The sample frequency table shows the number of athletes in five different heart rate ranges. Follow this model to make a frequency table for the non-athletes. Compare the data in the two tables. What can you conclude about the effect of physical training on heart rate? How could you display the data to make it easier to analyze?

Heart rates of 15 trained athletes after physical activity: 128, 131, 120, 127, 132, 125, 129, 122, 127, 133, 135, 130, 123, 128, 124

Heart rates of 15 non-athletes after physical activity: 143, 139, 144, 132, 138, 135, 141, 137, 128, 139, 140, 136, 133, 143, 135

Sample Frequency Table Heart Rates of Athletes After 10 min of Activity

Range of heart rates	Tabulate data	Total number of athletes in range (frequency)
119-122	//	2
123-126	///	3
127-130	//////	6
131-134	///	3
135-138	/	1

Many animals have specialized organs for certain functions. One example is the electric eel. This South American fish has electric organs in its tail region. The organs are made from muscle tissue connected to spinal nerves. The fish uses these organs to produce an electric shock to stun its prey. The shock from a large electric eel, say 2.75 m long, can measure over 600 volts — enough to stun a human.

3.2 Organ Systems in Plants

Why don't plants have muscles? The function of muscles is movement. One reason animals need to move is to find food. Plants do not need to move from place to place, however. They obtain their food by photosynthesis, using substances in the air (carbon dioxide) and soil (water). Plants can get these materials while remaining in one place, with their leaves in the air and their roots in the soil.

Because of differences in how plants and animals survive, plants have fewer types of tissues and organ systems than animals have. Movement is not the only difference between plants and animals. Unlike animals, plants do not need sense organs (such as eyes and ears) to locate their food. They do not need a digestive system to break down large pieces of food into small particles for their cells to use. Also, plants do not need a nervous system to send rapid signals throughout the body and to co-ordinate movement.

What do plants need in order to survive, and how do plant structures ensure the plant's survival? Plants have only two main organ systems: a root system below ground and a shoot system (the stems and leaves) above ground, as shown in Figure 3.5. The functions of the root system are to obtain water and minerals from the soil and to anchor the plant in the ground. The function of the shoot system is to make food for the plant. At certain times, flowering plants produce a third system, for reproduction. The main organs of the reproductive system are the flowers.

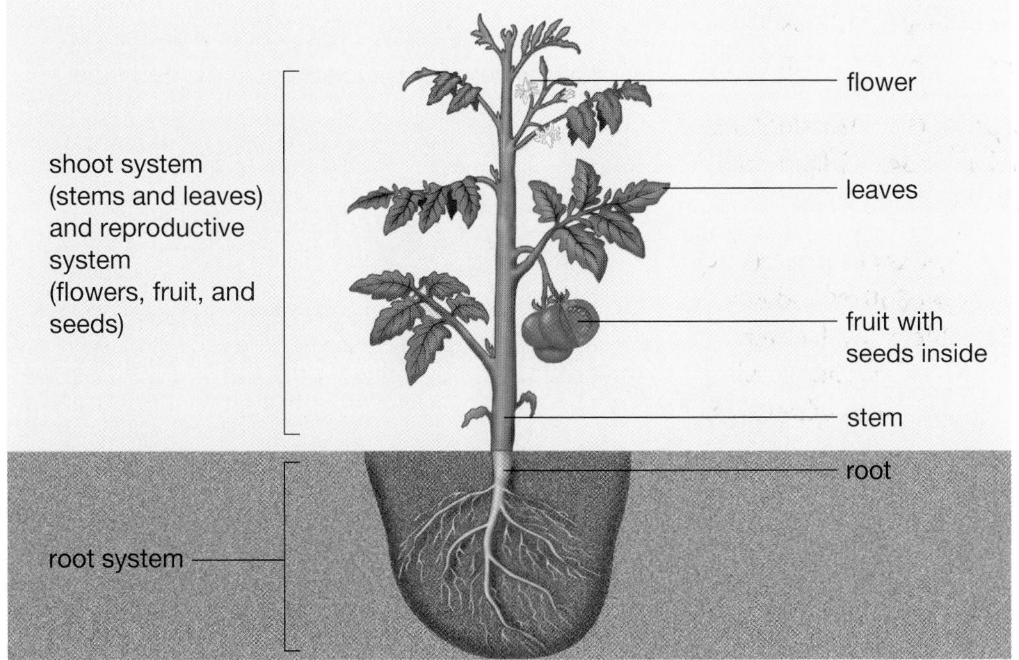

Figure 3.5 Organ systems in a plant

DidYouKnow?

Plants never stop growing in the way that many animals do. Instead, plants continue to add new cells, tissues, and organs at the ends of their shoots and roots. They do this by means of a permanent area of unspecialized cells that can develop into new plant structures and replace any parts that are damaged. Thus, some trees are estimated to be thousands of years old. Some patches of prairie grass are thought to be single individuals that have been growing in one place since the last ice age ended — 10 000 years ago!

Connecting the Systems

At the beginning of this chapter, you were asked how cells in the roots of a tree obtain sugars made in the leaves. Transport of nutrients is the role of a plant's tissues. Inside the plant, two types of tissues, called **vascular tissues**, connect the root system and the shoot system. **Phloem tissue** transports sugars manufactured in the leaves to the rest of the plant. **Xylem tissue** conducts water and minerals absorbed by the root cells to every cell in the plant (see Figure 3.6). Recall from Chapter 2 that all cells need food and water, together with oxygen, to carry out their functions. Xylem and phloem tissue usually occur together, along the length of the plant stems and roots. Both types of tissue are surrounded and supported by other tissue that gives the plant strength and has large vacuoles for storing food and water.

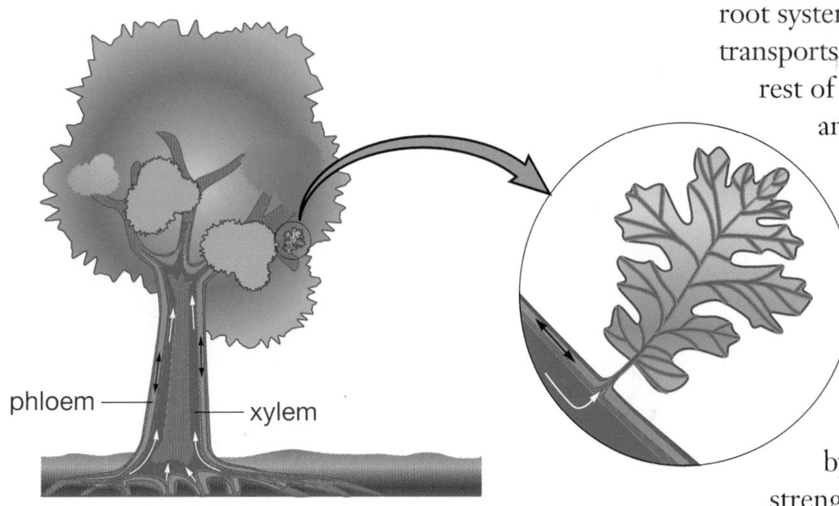

Figure 3.6 Xylem tissue conducts water from the root system to the shoot system in a plant. Phloem tissue carries sugars from the leaves to the rest of the plant.

Find Out **ACTIVITY**

What Tissues Are in Celery Stalks?

When you take a bite of celery, your teeth are crunching through the tissues in the celery stalk. The stalk is a plant stem, with vascular tissues for transporting water and sugars. To identify the xylem tissue, place the bottom end of a celery stalk in a jar of water to which red food colouring has been added. The next day, follow the steps below.

What You Need

celery stalk placed in red food colouring
single-edged razor blade or sharp knife
microscope, microscope slide, and cover slip
medicine dropper
water
cutting board or other cutting surface

What to Do

1. Hold the celery stalk firmly on a suitable cutting surface and use a sharp blade to cut off a very thin slice across the stalk, as shown. Try several times, making your slices as thin as possible. **CAUTION:** Never cut an object held in your hand. Cut with the blade moving away from you. Cut the celery stalk with your teacher present.
2. Take your thinnest celery slice and prepare a wet mount of it (see page 16).
3. Observe the celery cross-section under the low power of a microscope. Draw and label what you see.

What Did You Find Out?

1. What do the red "dots" in the celery section indicate?
2. Where are the xylem tissues in the stem? What other tissues did you observe?

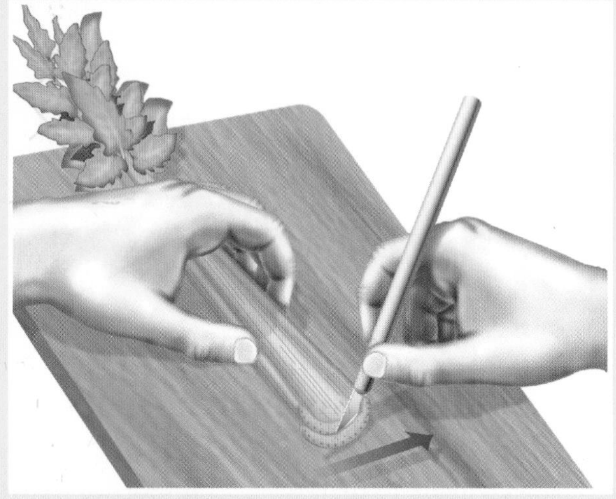

From Root to Leaf

One way to understand how the different levels of organization (cells, tissues, organs, and systems) work together is to follow the path of water through a plant. Most plants need a large supply of water. Plants require water to make sugars in the process of photosynthesis (see page 48). Plants obtain water from the soil. How does water get from the soil into the plant?

If you examine the structure of a root system, you will see that its growing tips are covered with fine **root hairs**. These "hairs" are, in fact, extensions of single epidermal cells (see Figure 3.7). When the concentration of water in the soil is greater than the concentration of water in the root cells, water enters these root hairs by osmosis (see page 42). The process of water uptake is therefore a function of the first level of organization — the cells.

From the root hairs, water passes from cell to cell by osmosis until it reaches the xylem tissue. The tube-shaped cells making up xylem tissue have thick walls with holes in their ends (see Figure 3.8). Stacked end to end, they form bundles of hollow vessels similar to drinking straws. Water can flow easily through these vessels. As more water enters the root hairs, it creates pressure that pushes water up the plant through the xylem tissue — the second level of organization.

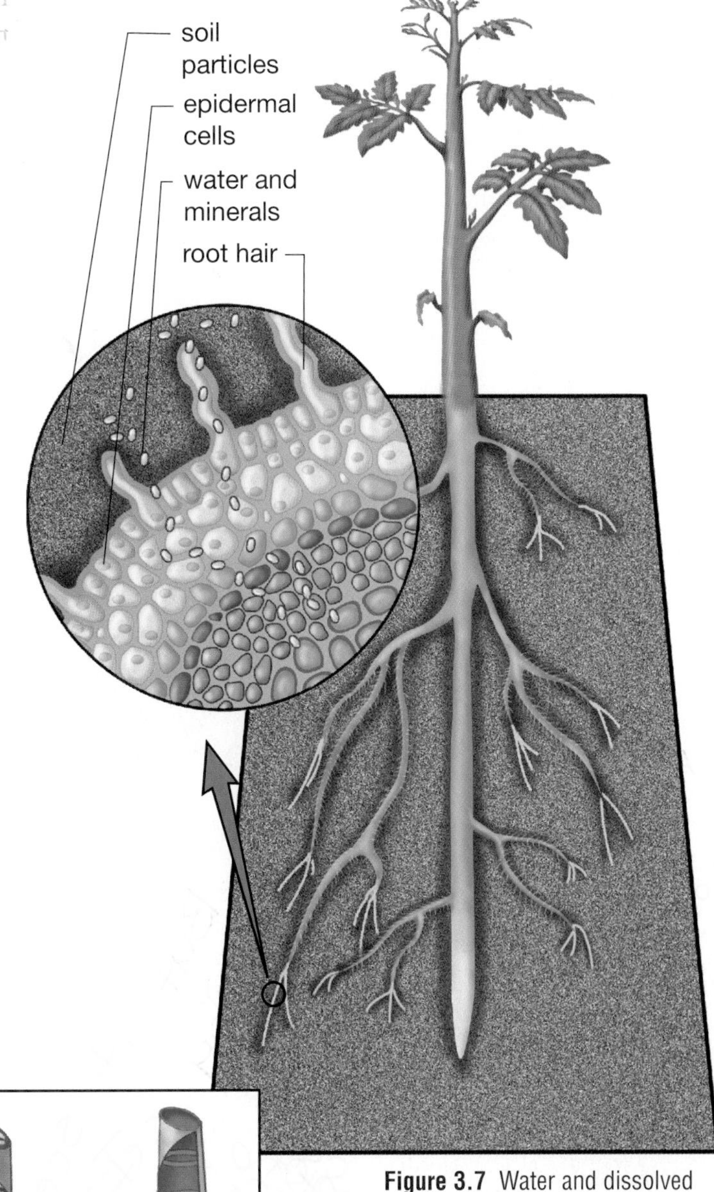

Figure 3.7 Water and dissolved minerals enter the plant by osmosis through the root hairs.

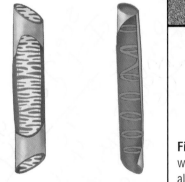

Figure 3.8 Xylem cells have thick walls for strength. Their open ends allow water to pass through freely.

DidYouKnow?

The phloem tissue of a tree lies close to the outer surface of the trunk, just below the bark. Because of this, sugar tappers can easily draw sugar solution from the trunks of maple trees. This is done by boring a small hole through the bark and pushing small tubes into the phloem tissue. The best time to tap maple trees is early in the spring, when large amounts of sap are flowing to provide energy for new growth.

Pause& Reflect

In your Science Log, make a full-page labelled drawing that shows all the levels of organization in a plant: cells, tissues, organs, and systems.

Water is transported by xylem tissue into the stems and the leaves. Leaves are the plant's food-producing organs — the third level of organization. Recall that photosynthesis manufactures sugars from water, carbon dioxide, and sunlight. Most photosynthesis takes place in a layer of cells in the leaf that are filled with chloroplasts. These cells are called **palisade cells**. Why are many leaves typically flat and thin? This shape provides a large surface area to absorb sunlight, and it makes it easy for gases to diffuse into the leaf cells (see Figure 3.9).

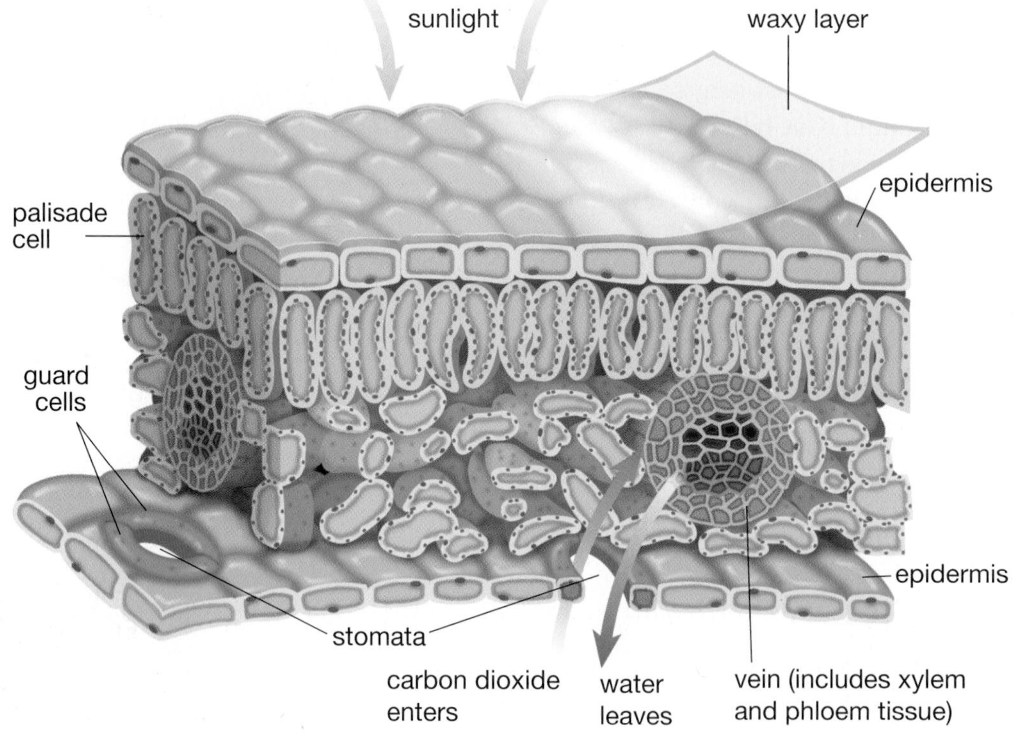

Figure 3.9 Tissues in a leaf represent the second level of organization.

Notice the tiny openings on the underside of the leaf. These openings are called **stomata** (singular **stoma**). They allow air to enter the leaf, supplying the oxygen the cells need for respiration and the carbon dioxide they need for photosynthesis. Spaces between leaf cells allow the air to flow around each cell. Surrounding each stoma are **guard cells**, which can expand to close off the stoma. Why do the stomata in a leaf open and close? To answer this question, recall that water first enters a plant through its root system. Then it moves into its shoot system. These are two of the plant's organ systems — the fourth level of organization. What happens next? The water does not continually circulate like the blood in our bodies. It does not go back into the root system. Instead, it exits the plant — through the open stomata in the leaves.

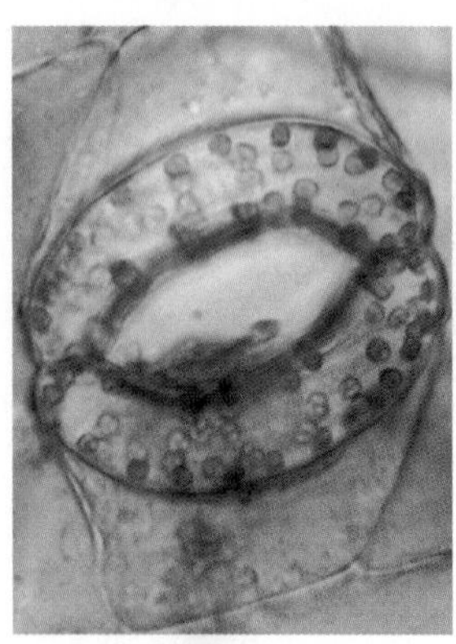

Figure 3.10A Guard cell with open stoma

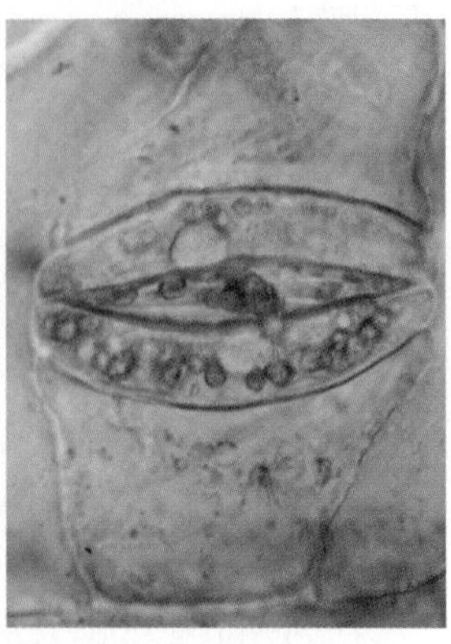

Figure 3.10B Guard cell with closed stoma

This loss of water from a plant through evaporation is called **transpiration**. The loss of water is not a problem as long as it is replaced by more water that enters the plant through the roots. In periods of drought and in deserts, however, water loss from a plant can be a serious problem. Figures 3.10A and B show open and closed stomata on the underside of a leaf.

Pulling and Pushing

If all the tissues of a plant were to magically disappear, leaving only the water in them behind, you would see a ghostly outline of the plant in a weblike network of water. There is no break in this water system. Fine columns of water connect every cell, from the leaves to the roots. The network extends even beyond the root hairs — it connects root hairs to channels of water in the soil.

According to the particle theory, individual water particles are held together by bonds of attraction, which make the plant's water network behave as a single unit. Water drawn into the root hairs by osmosis *pushes* slender water columns up the plant. At the same time, water lost from the leaves by transpiration *pulls* water up the xylem tissues all the way from the roots. Both these actions — pushing and pulling — are necessary to raise the water up to the top of very tall trees. In this way, trees can transport water without having a pumping organ similar to the human heart.

It is estimated that over 3.5 million litres of water are lost by transpiration from a single hectare of corn during one growing season. What would be the depth in centimetres of this volume of water if it were to flood the hectare of corn?

Transpiration and Leaves

What is the relationship between leaves, transpiration, and the movement of water through a plant? You can find out by doing this activity.

What You Need

2 leafy stalks of celery
beaker (or jar)
red food colouring
single-edged razor blade or sharp knife
cutting board or other cutting surface

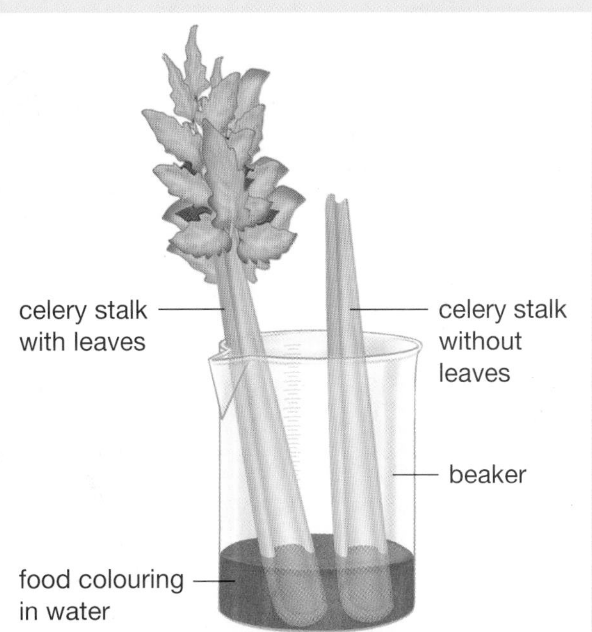

What to Do

1. Take two leafy stalks of celery of about the same length.
2. Remove all the leaves from one stalk.
3. Place both stalks in a beaker or jar of water to which red food colouring has been added, as shown in the diagram.
4. Leave the jar in a sunny spot or under a bright light for at least 3 h.
5. Place the stalks on a cutting board. Beginning near the bottom end of the stalks, cut across the stems at short intervals to determine how far the coloured water has risen up each stalk. **CAUTION:** Never cut an object held in your hand. Place the object on a cutting surface and cut with the blade moving away from you. Cut the celery stalk with your teacher present.
6. Based on your observations, what inference can you make about the effect of leaves on the rate of transpiration?

Organ Adaptations in Plants

Figure 3.11A Cactus spines and stems are adaptations to dry conditions.

Figure 3.11B These pine needles are an adaptation to dry conditions, as well.

Like animals, plants have adaptations that help them grow and survive in different environments. You can see some of these adaptations in the structure of roots, stems, and leaves. For example, many plants growing in deserts have small, fleshy leaves with a heavy wax coating that helps reduce water loss. Cactus spines are, in fact, narrow, waxy leaves. To compensate for their reduced leaf area, cacti such as the one shown in Figure 3.11A carry out photosynthesis in their stems. The leaves of coniferous (cone-bearing) trees, such as pines and other evergreens, are also adapted to dry conditions. The needlelike shape of the leaf reduces evaporation from the surface of the leaf (see Figure 3.11B).

Plants that grow in water, such as water lilies, could have a problem obtaining the air they need to survive (see Figure 3.12). To ensure their underwater roots obtain the oxygen they need for cell respiration, the root tissues of these plants have large air spaces in them. Still other plants have roots in the air! Orchids such as the ones shown in Figure 3.13 grow high above the ground on the branches of trees in tropical forests. Their root tissues are specially adapted to absorb moisture from the warm, humid air.

Figure 3.12 The roots of water lilies have adapted to a water environment. What other organ adaptations can you identify?

Figure 3.13 The roots of these tropical orchids are exposed to the air. What is one advantage of this adaptation?

Check Your Understanding

1. Which tissues conduct water in plants?
2. Which tissues conduct sugars in plants?
3. What is the function of guard cells?
4. Give two examples of organ systems in plants and explain their functions.
5. Describe some organ adaptations in plants.
6. **Design Your Own** Is it better to water plants in the evening or during the day? Make a hypothesis that answers this question, and then design an experiment to test it. Remember to include a control in your experiment.

3.3 Organ Systems in Humans

The woman shown in the photograph is practising the Indian philosophy of yoga. Although she appears very relaxed, there is a lot going on inside her body. As you learned in Chapter 2, every cell in the body needs a steady supply of food and oxygen to give it energy. Three different organ systems must work together to make this possible. Do you know what they are?

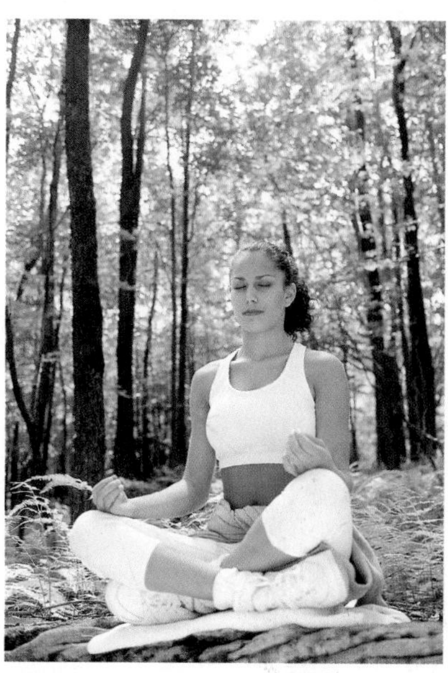

Food first enters the body through the mouth, then passes to the stomach and the intestine. It is broken down along the way into small, soluble particles that can be used by cells. Unused food is expelled from the body as waste. As you read on page 72, the organs involved in these processes form the digestive system, shown in Figure 3.14.

The woman practising yoga is taking slow, deep breaths. Breathing in (inhalation) fills her lungs with oxygen-containing air. Breathing out (exhalation) rids her body of waste carbon dioxide. The organs involved in this gas exchange form the **respiratory system**, as shown in Figure 3.15.

The Digestive System

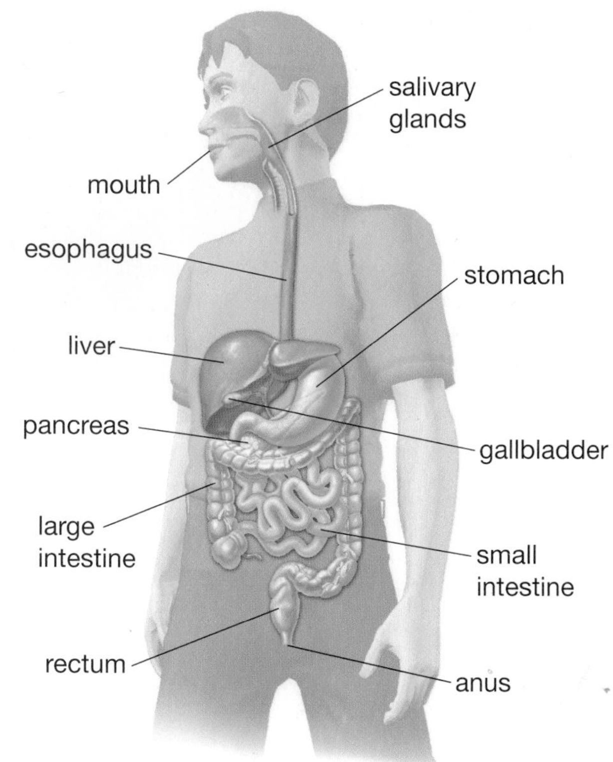

Figure 3.14 This organ system breaks down food by digestion.

The Respiratory System

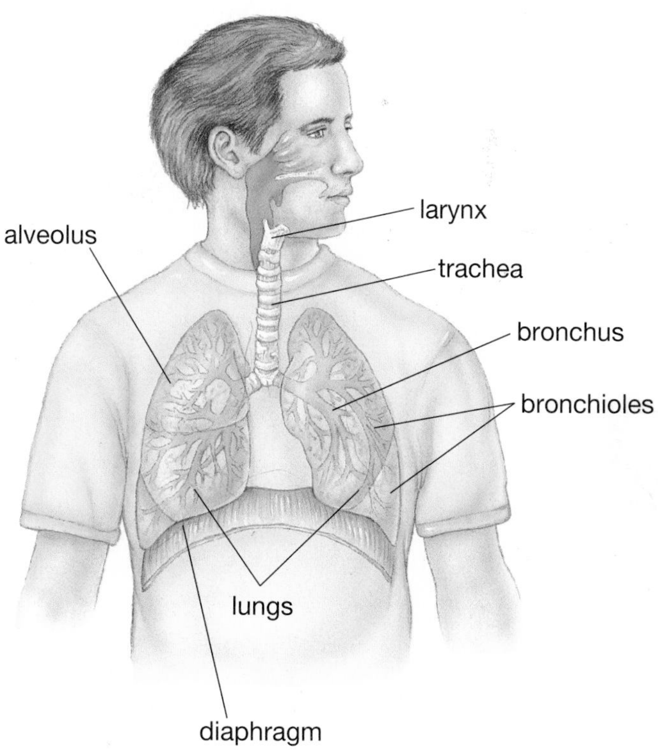

Figure 3.15 This organ system moves air in and out of the body. This in-and-out movement of air supplies oxygen for cells and removes waste carbon dioxide.

The digestive system puts food into the intestine and the respiratory system puts oxygen into the lungs. How do particles of food and oxygen eventually get from these systems to cells in the toes, the brain, and other parts of the body? A third system transports particles of food and oxygen. The **circulatory system** consists of the heart, blood, and blood vessels (see Figure 3.16). This system circulates blood around the body, delivering food particles, dissolved gases, and other materials to every cell and carrying away cell wastes.

There are about 11 different systems in the human body. Each system has a major function. The systems are co-ordinated into the total living organism, and all the systems depend on one another.

The Circulatory System

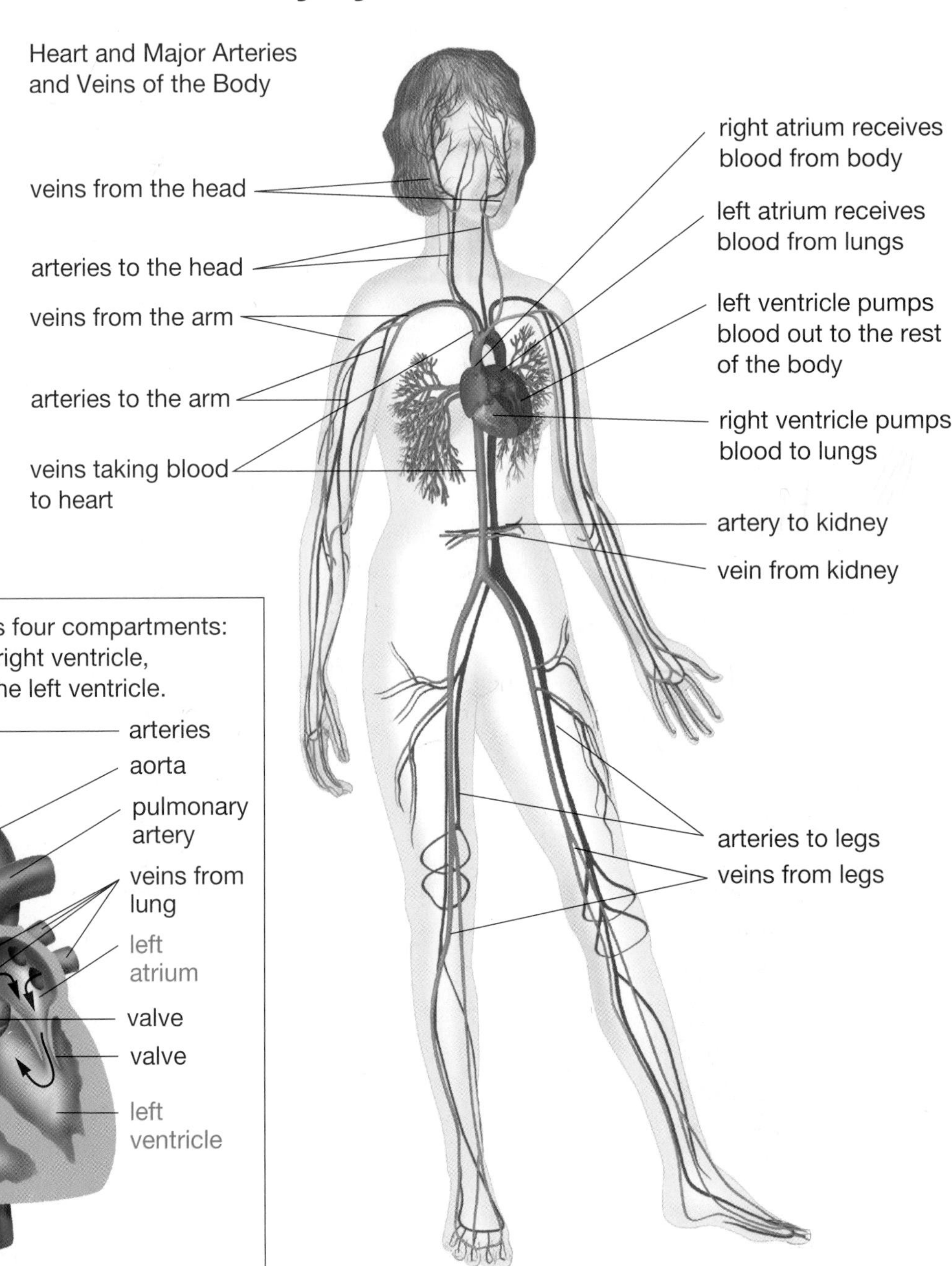

Figure 3.16 The circulatory system's function is to carry materials to and from all the cells in the body.

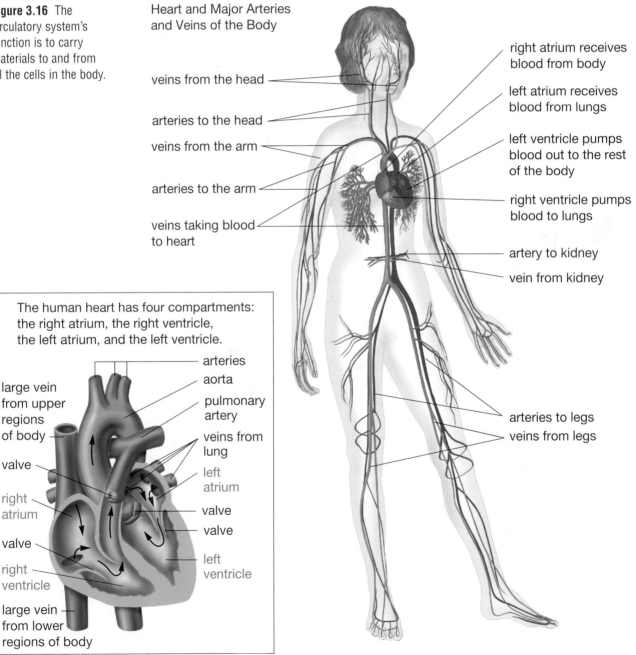

A Tale of Two Systems

To connect all the individual cells throughout your body with the air around you, the respiratory system and the circulatory system work together.

The function of the respiratory system is to exchange oxygen and carbon dioxide, while the circulatory system transports those gases throughout the body. How do the gases pass from one system to the other? Look for an answer where the two systems come into closest contact — among the tissues of the lungs.

Examine the structure of the respiratory system in Figure 3.15 on page 79. After air enters the nose, it passes to the lungs through a series of smaller and smaller tubes. The **trachea** (windpipe) is about 20 mm in diameter. It divides into a right and a left **bronchus**, each about 12 mm across. Each bronchus tube branches into thousands of small, narrow **bronchioles**, with diameters of 0.5 mm. Finally, the bronchioles divide and end in millions of tiny air sacs called **alveoli** (singular **alveolus**), only 0.2 mm in diameter.

The circulatory system also involves a series of tubes — the blood vessels. Like the air tubes of the respiratory system, blood vessels branch and divide into smaller and smaller channels. The three main types of blood vessels are shown in Figure 3.17. The smallest blood vessels are the **capillaries**.

DidYouKnow?

A capillary may be so narrow that only one red blood cell can pass through it at a time.

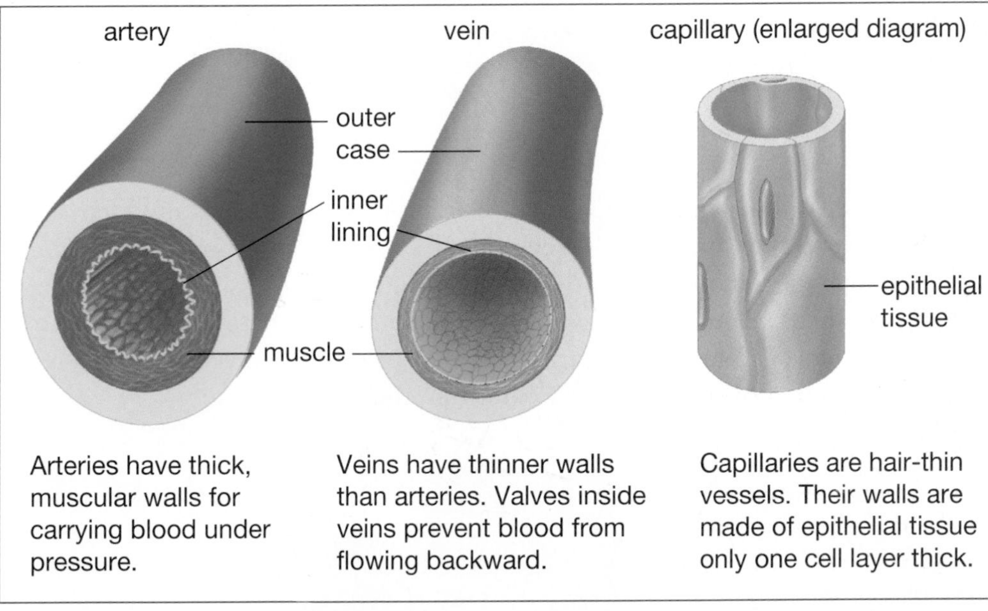
artery
vein
capillary (enlarged diagram)
outer case
inner lining
muscle
epithelial tissue
Arteries have thick, muscular walls for carrying blood under pressure.
Veins have thinner walls than arteries. Valves inside veins prevent blood from flowing backward.
Capillaries are hair-thin vessels. Their walls are made of epithelial tissue only one cell layer thick.

Figure 3.17 Cross-sections of human blood vessels

Now look closely at the detailed structure of the lungs (see Figures 3.18A and B on page 82). Each alveolus is surrounded by a web of capillaries. It is here, between these delicate tubes, that gases are exchanged. Oxygen and carbon dioxide pass back and forth between the air in the alveoli (part of the respiratory system) and the blood in the capillaries (part of the circulatory system).

In Chapter 2, you learned how substances move into and out of cells by diffusion. Diffusion also causes oxygen to pass from the alveoli into the capillaries. Why? The air in your alveoli has the same composition as the air in the atmosphere — about 20 percent oxygen. This is a much higher concentration of oxygen than the concentration found in the blood in your capillaries. The oxygen first dissolves in a thin film of moisture covering the walls of the alveoli. Then it diffuses from the alveoli through the thin capillary walls into the bloodstream.

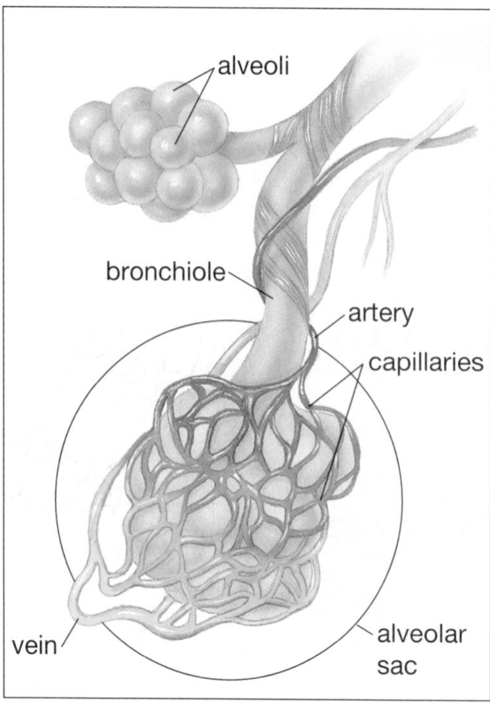

Figure 3.18A An enlarged alveolus. There are about 300 million alveoli in the human lungs.

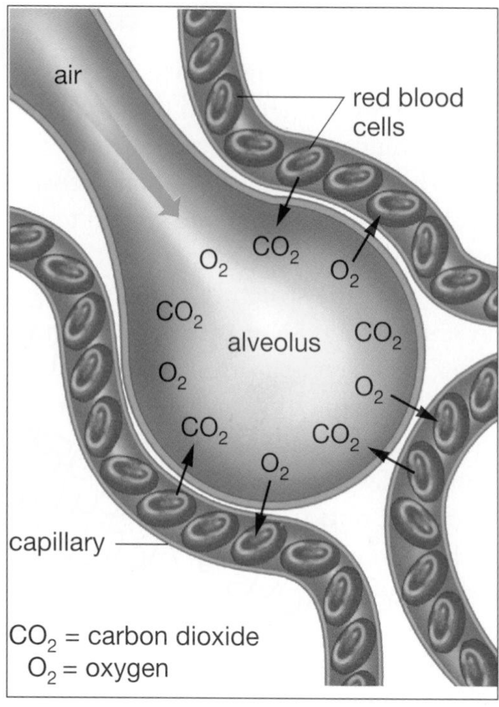

Figure 3.18B Gases move back and forth between the alveoli and the surrounding blood vessels.

Carbon dioxide diffuses in the opposite direction. Air normally contains only a very low concentration of carbon dioxide (about 0.03 percent). Blood in the capillaries carries all the dissolved carbon dioxide collected from cells throughout the body. (Recall that carbon dioxide is a waste product of cellular respiration.) This gas therefore moves from the capillaries into the alveoli. When you exhale, you release the carbon dioxide and water vapour into the air.

Off the Wall

Many seals, with lungs no bigger than a human adult's, can easily stay underwater without breathing for 20 min or more. Even more curious, they breathe out before they dive. The explanation for this puzzle lies in how the seal's blood and circulatory system function. All the oxygen a seal needs while underwater is stored in its blood and muscle tissue, rather than in its lungs. To be able to store this large amount of oxygen, a seal has about one-and-a-half times to twice as much blood in its body as other mammals of similar size.

As soon as a seal dives underwater, a series of changes takes place in its body. Its heartbeat slows at once, from about 100 beats per minute to about 10 beats per minute. Blood flow to some parts of its body, such as the kidneys and the muscles, slows or stops. The seal is also able to tolerate a high level of carbon dioxide in its blood, as this gas builds up during the dive.

When the seal returns to the surface, it can breathe in and out very rapidly, almost completely emptying its lungs of waste gas. With each breath, a seal can exchange 90 percent of the air in its lungs. By comparison, with each breath, humans exchange only about 20 percent of the air in their lungs.

Find Out ACTIVITY

Make a Model of the Lungs

If you look at lung tissue under a microscope, you will see that it has no muscle cells. So what makes your lungs expand and contract? You can make a model of your lungs and chest cavity to find out.

What You Need

large plastic pop bottle
2 plastic straws
2 small balloons
2 elastic bands
modelling clay

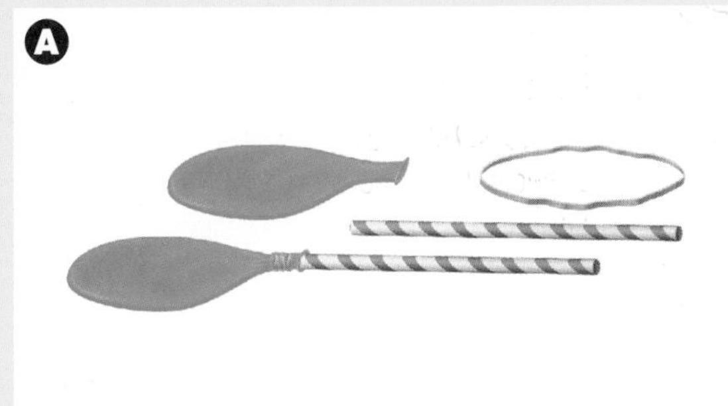

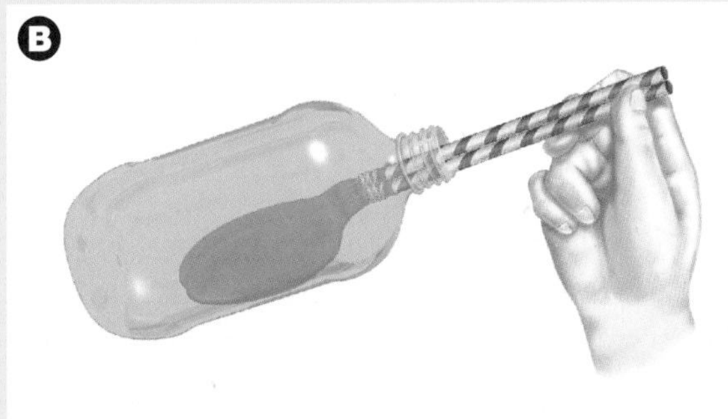

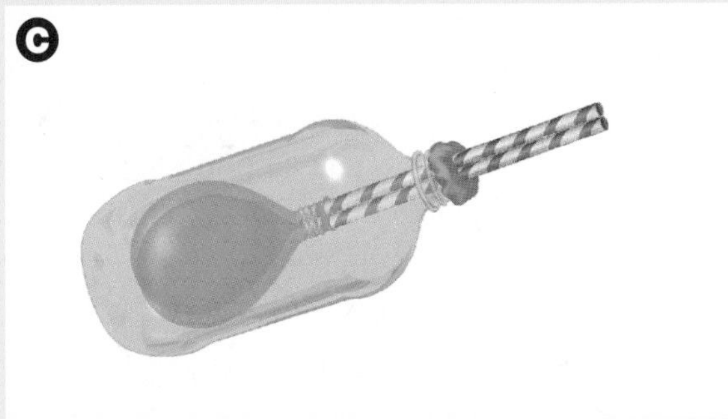

What to Do

1. Inflate the balloons to stretch them, then let the air out. Insert a straw into the neck of each balloon and fasten the balloon and straw tightly together, using elastic bands (see Diagram A).
2. Insert the straws into the bottle so the balloons are hanging inside the bottle, and the other ends of the straws are sticking above the neck, as in Diagram B.
3. Completely seal the neck of the bottle around the straws with modelling clay, as in Diagram C.
4. Squeeze and release the sides of the bottle and watch what happens to the balloons.

What Did You Find Out?

1. What happened to the balloons when the bottle was squeezed? What happened when the bottle was released?
2. Write an explanation for your observations. Try to use the term "air pressure."
3. In what way is the squeezing and releasing of the bottle similar to inhaling and exhaling?
4. Sketch your model and indicate with labels which parts represent the following structures: trachea, lungs, chest cavity, diaphragm (a band of muscle just under your breastbone that separates your chest cavity from your abdomen).
5. Put your hands on your ribs while taking deep breaths. Then feel the area near your diaphragm. Make an inference about which muscles are used to help you breathe. Describe what happens when the muscles contract and what happens when they relax.
6. In this model, squeezing and releasing the bottle is compared to normal breathing. Suppose you inflate the balloons by blowing into the straws. With what could you compare this?

Getting Food to Body Cells

You have discovered how your bloodstream obtains oxygen from your lungs. Your bloodstream also carries food particles, which it gets from your digestive system. The transfer of food from the digestive system to the circulatory system takes place at the inner lining of the small intestine, as shown in Figure 3.19. Covering the surface of this lining are millions of tiny, fingerlike projections called **villi** (singular **villus**). Each villus contains a network of capillaries. Dissolved food particles pass from the intestine into the capillaries by a process called **absorption**. The food particles are now small enough to enter your body's cells to supply them with the food they need. The arteries of your circulatory system provide the transportation network.

Do you see any similarities between the villi in the intestine and the alveoli in the lungs? Like alveoli, villi have thin walls through which particles can pass into the circulatory system. Both alveoli and villi consist of tiny projections, and both occur in huge numbers. This arrangement greatly increases the surface area that is in contact with capillaries, without taking up a large amount of space in the body.

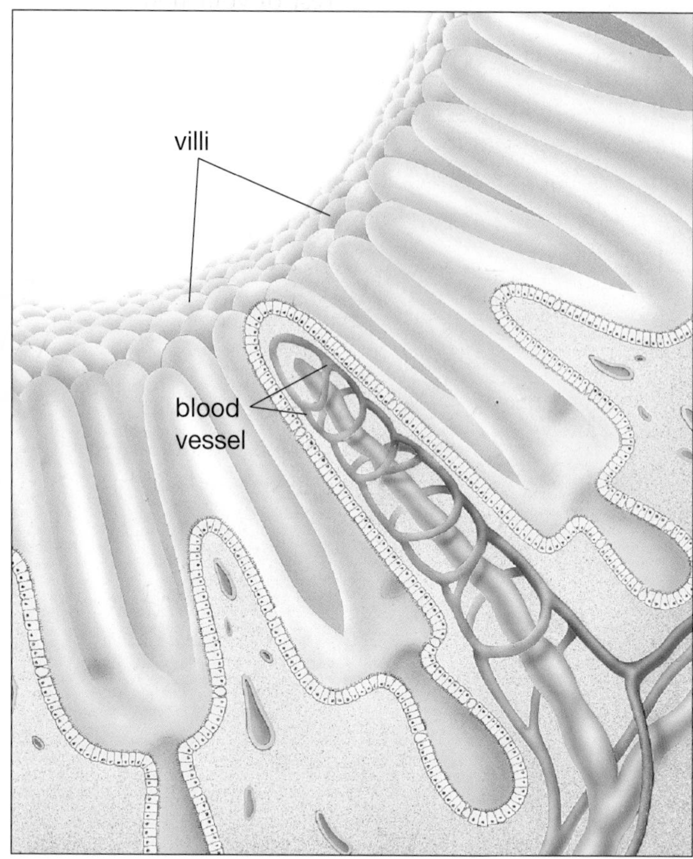

Figure 3.19 The villi in the small intestine. These structures increase the surface area of the small intestine for more efficient absorption of nutrients.

Getting Rid of Waste from the Blood

Getting food and oxygen to your cells is only one half the equation for good health. Your body must also get rid of wastes. Filtering waste materials from the blood is the main function of another system — the **excretory system**. The key organs in this system are your two kidneys. You can research this system for yourself in the Find Out Activity on the next page.

Word CONNECT

Ludwig von Bertalanffy, a Canadian biologist, developed the idea of homeostasis in living things. He suggested that organisms are like waterfalls: they appear to be constant while the individual parts that make them up are always changing and being replaced. Write a haiku or other poem to express in your own words this idea of constant change combined with stability in living organisms.

It's All Under Control

Feeling tired? Maybe you need a nap. Feeling cold? Why not put on a sweater? Feeling hungry? It's time to eat. In these and other ways, you respond to changing external conditions and make appropriate adjustments. Your body systems also make constant adjustments to maintain a stable internal environment for your cells. This process is known as **homeostasis**. It occurs automatically, usually without your even being aware of it. For example, no matter whether it is hot or cold outside, the inside of our bodies remains at an amazingly constant temperature of about 37°C all year round. A change in body temperature of as little as 0.5°C can make us feel either feverish or chilled. How is this steady body temperature maintained?

Nearly 90 percent of your body heat is lost through the skin. Most of the rest of your body heat is lost through your lungs. When you get cold, you may shiver. Your quivering muscles generate heat. You may also get "gooseflesh" — small bumps on your skin. The bumps are produced by the contraction of small muscles in the skin that make your hairs stand on end. In animals with a thick coat of hair, and in our hairier prehistoric ancestors, fluffing up the body hair helps reduce heat loss by improving insulation.

When you are hot, your body tries to cool you down. Do you get flushed and red after hard exercise? This happens because tiny blood vessels in your skin expand. This increases blood flow near the body surface where heat can be lost to the outside. Sweating helps cool your body as the moisture evaporates from your skin surface.

Employment — Excretion!

Imagine you are a body in search of kidneys. You must write a job description to get the right organs for the job. What exactly must kidneys do? What qualifications (structure) do they need? Will they be expected to work in co-operation with other organs? Present your job description in class.

What to Do

1. Do some research on kidneys and the excretory system to answer the questions above.
2. Present your findings in the form of a job advertisement with the heading: "Wanted: Highly Qualified Kidneys." The ad should give details of kidney structure and functions, and it should include a labelled sketch.

What Did You Find Out?

1. What waste materials do kidneys remove from the body, where do these wastes come from, and why must they be removed?
2. Which other system is most closely connected with the excretory system?

Extensions

3. In what ways can the millions of tubules (called nephrons) in the kidneys be compared with the alveoli in the lungs and the villi in the small intestine?
4. In the past, kidney failure often led to death because toxic wastes built up in the bloodstream. Today, dialysis machines may be used to filter a patient's blood. These machines use an artificial membrane. Do you think the membrane is permeable or selectively permeable? Explain your answer.

Your body's responses to stimuli are co-ordinated by the **nervous system** (the brain, spinal cord, and nerves) and the **endocrine system** (a set of glands that produce chemical messengers called **hormones**). These systems can be compared to the traffic lights and speed limits that keep city traffic moving smoothly. Suppose drivers could go in any direction they liked at any time, or drive at any speed. There would soon be chaos. In the body, too, a number of factors can affect the smooth working of the organ systems (listed in Table 3.1). Diet, exercise, drugs, injury, and disease can affect body systems and disrupt homeostasis. In some people, an inherited disorder leaves the body unable to control particular functions.

Table 3.1 Major Body Systems

System	Functions
Digestive	Breaks down food, absorbs food particles, and eliminates wastes.
Respiratory	Exchanges oxygen and carbon dioxide.
Circulatory	Circulates blood. Transports food particles, dissolved gases, and other materials.
Nervous	Controls and co-ordinates body activities. Senses internal and external changes.
Excretory	Regulates blood composition and excretes waste fluids.

Find Out ACTIVITY

A Simple Reflex

Homeostasis depends on a feedback system. The knee-jerk reaction is a simple example of a feedback system in the body. A sharp tap (stimulus) at the knee causes a signal to be sent to the spinal cord. A return signal (the feedback) causes the leg to react to the stimulus (see the diagram below). You can observe this feedback system for yourself.

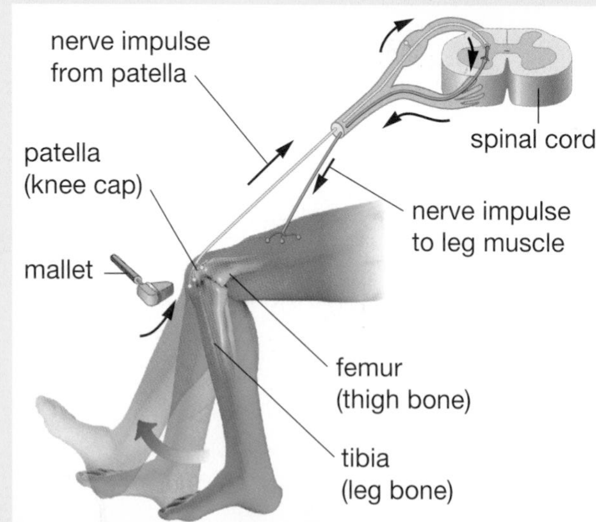

Note: This activity will be done as a controlled volunteer demonstration at the front of the classroom. Two students will be guided by the teacher.

What You Need

rubber reflex hammer

What to Do

1. Your teacher will ask for two volunteers to come to the front of the classroom. One student sits on the edge of a bench or table so his or her legs hang freely and do not touch the floor.
2. The second student locates the first student's kneecap and feels the position of the tendon at its lower edge. With the rubber hammer, or the edge of a hand, the second student (guided by the teacher) quickly and firmly taps the tendon. Note the automatic response of the lower leg.
3. The students will change roles and repeat steps 1 and 2.

What Did You Find Out?

1. Which muscles contract to lift the leg upwards?
2. Draw a flow chart showing the sequence of events from tapping the tendon to the response of the leg.

To keep your body temperature stable, your nerves, muscles, and blood all function together. Your nervous system monitors conditions outside the body through temperature receptors in your skin. (If you stand in a shower and quickly change the temperature of the water flow, you will know how sensitive your skin is to hot and cold.) Information from temperature receptors in your skin goes to the heat-regulating centre of your brain, called the **hypothalamus**. Responding to this information, the brain sends nerve signals to your muscles, skin, and blood vessels. Working together, your muscles, skin, and blood vessels adjust your blood flow and muscle activity, causing your body to increase its heat production or reduce its heat loss.

DidYouKnow?

Drug addiction or dependence is like an adjustment of homeostasis. When the body is first exposed to a drug (whether caffeine, nicotine, opium, or another chemical), the cells respond so as to maintain their normal function in the presence of the new chemical. After a time, the cells become so tolerant that the drug loses its effect. The person may then need to take larger quantities of the drug to produce the desired effect. If the substance is withdrawn, the cell response is disturbed for a time — sometimes severely — until a new re-adjustment is made.

Across Canada

In 1921, Canadian researchers Frederick Banting and Charles Best, working at the University of Toronto, discovered the hormone insulin. This hormone, produced by cells in the pancreas, sends a message to other cells in the body when there is a lot of glucose in the blood. The cells respond by processing the glucose, which lowers the blood's glucose level. Hormone production is constantly adjusted by the pancreas as glucose levels rise and fall. (The amount of glucose in a person's blood depends on their eating habits and amount of physical activity.)

Some people's bodies are unable to control the glucose levels in their blood. This leads to the disorder called diabetes. Before the discovery of insulin, diabetes was fatal. Today, diabetics are able to live full lives by controlling their diets and by injecting insulin, if their bodies cannot make it for themselves.

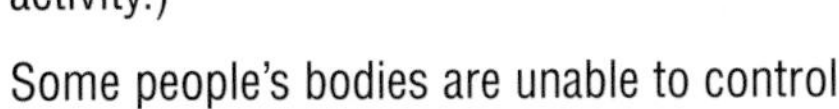

Check Your Understanding

1. Why does the body need oxygen?
2. Describe how carbon dioxide from a cell in your hand leaves your body.
3. The lungs can expand and contract because they have muscular walls. Is this statement true or false? Explain.
4. What structures help increase the absorption of food in the small intestine? How do they do this?
5. What is meant by homeostasis? Give an example.
6. How might your muscular system help you stay warm?
7. How might your circulatory system help you stay cool?
8. **Thinking Critically** Homeostasis keeps your home from getting too hot or too cold. What device maintains home temperature?

3.4 Body Systems and Your Health

DidYouKnow?

Iron (used in red blood cells) is found in certain foods, such as liver, egg yolks, beans, nuts, dried fruits, and leafy green vegetables. A shortage of iron in the body may result in anemia. This is a condition in which the ability of the blood to carry oxygen is greatly reduced. Symptoms of anemia include dizziness, fainting, and shortness of breath. The problem can be remedied by an improved diet that is rich in iron.

Put your hand over your heart and feel the beat. That rhythmic pulse is evidence of a pump at work, pushing your blood in a continuous flow through all the vessels of your body. A healthy heart and blood vessels are essential to overall good health. Without this pumping action, the blood will not circulate, and your cells will not get the supplies of oxygen and food they need for their activities. The circulating blood also carries wastes produced by the cells, taking them to other organ systems that break them down or excrete them from the body.

Moving substances around the body is a necessary job in all multicellular organisms. In a unicellular organism, materials are exchanged directly between the cell and its external environment, as shown in Figure 3.20. In a larger organism, however, most cells are not in direct contact with the external environment. Substances must be brought to cells and taken away from them by an ever-moving system of transport — the circulatory system. The blood vessels of this system run throughout the human body like a complex network of highways, roads, and footpaths.

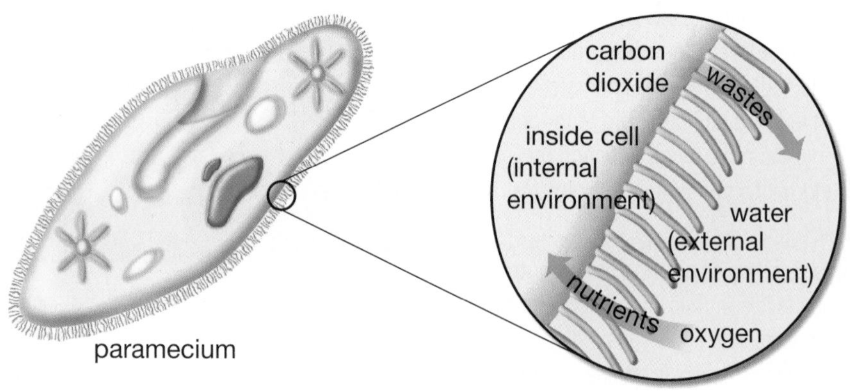

Figure 3.20 Paramecium exchanging materials with its external environment

In humans, the vehicle that transports substances is **blood**. About 8 percent of an adult's body weight is blood. What exactly is blood made of? The main components of this fluid and their functions are listed in Table 3.2. Plasma and red blood cells make up 99 percent of the volume of blood. **Plasma** is the liquid portion of the blood. It transports most of the carbon dioxide produced during cellular respiration. The red blood cells are specialized to carry oxygen. They contain an iron-rich chemical called **hemoglobin**, which attracts oxygen. This allows the blood to carry much more oxygen than it could otherwise.

Table 3.2 Blood Components and Their Functions

Component	Percentage of blood (by volume)	Main function
plasma	55%	carries nutrients, waste products, hormones, and blood cells
red blood cells	44%	carry oxygen
white blood cells	less than 1%	defend body against infection and disease
platelets	less than 1%	cause blood to clot (thicken) at site of wounds to prevent blood loss

INVESTIGATION 3-B

Have a Heart!

Think About It

The circulatory system seems complicated at first. Think of it as a number 8, with the heart at the centre where the lines of the 8 cross. This simplified diagram and activity will help you learn how blood flows through this system.

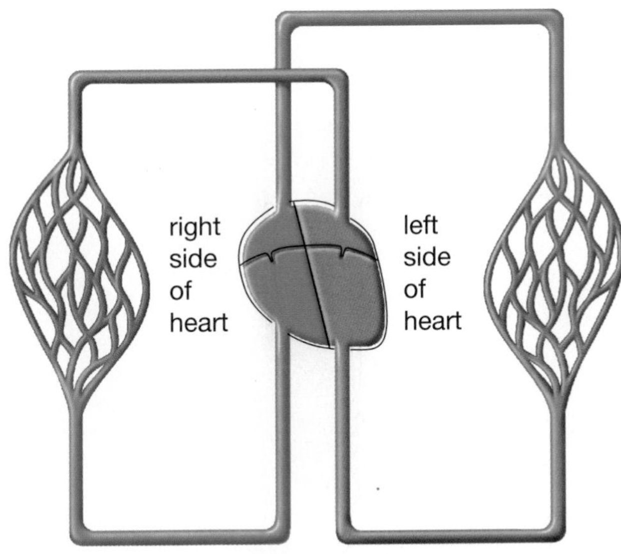

less oxygenated blood

oxygenated blood

What to Do

1. Read "Journey of a Red Blood Cell" on pages 90–91 before you go on to step 2.

2. Copy the above diagram. It represents the human circulatory system with some details and labels missing.

3. Using the diagram, and the illustration of the circulatory system in Figure 3.16, add the following labels to your drawing:

left atrium	capillaries
left ventricle	right atrium
arteries	right ventricle
arteries	heart
veins	body
veins	lungs
capillaries	

4. Draw arrows to show the direction of blood flow in the different blood vessels and through the four chambers of the heart.

Analyze

1. Which side of the heart collects blood from the body and pumps it to the lungs?

2. Which chamber receives blood from the lungs?

Extend Your Knowledge

3. Valves between the heart chambers prevent blood flowing from the ventricle to the atrium. Suggest what might happen to the pressure of blood leaving the ventricles if these valves do not function properly.

Pause& Reflect

When a person suffers a heart attack, severe chest pain is felt and the heart may stop working. In your Science Log, explain what you think could cause a heart attack. Compare your responses with those of your classmates. If possible, show your ideas to your family doctor or to your school nurse and ask them to make comments on your explanation.

Journey of a Red Blood Cell

The blood in your body travels in a double circuit, going through your heart twice before it completes one full circulation. The first circuit is from the heart to all tissues and organs except the lungs. The second circuit is from the heart to the lungs, where carbon dioxide diffuses out of the capillaries and oxygen diffuses into the capillaries. Follow a group of red blood cells on their journey around the body, then complete the investigation on page 89.

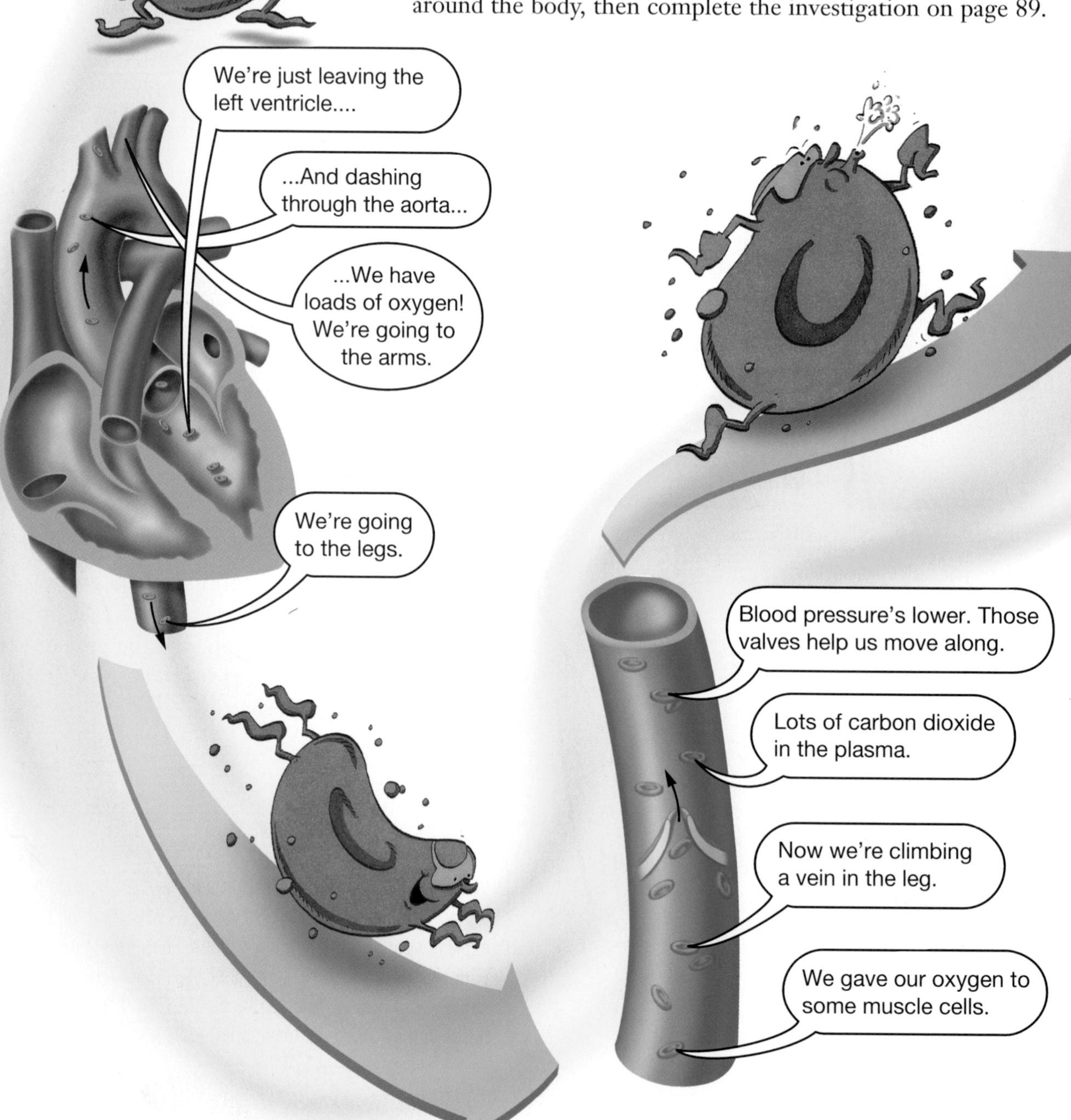

We're just leaving the left ventricle....
...And dashing through the aorta...
...We have loads of oxygen! We're going to the arms.
We're going to the legs.
Blood pressure's lower. Those valves help us move along.
Lots of carbon dioxide in the plasma.
Now we're climbing a vein in the leg.
We gave our oxygen to some muscle cells.

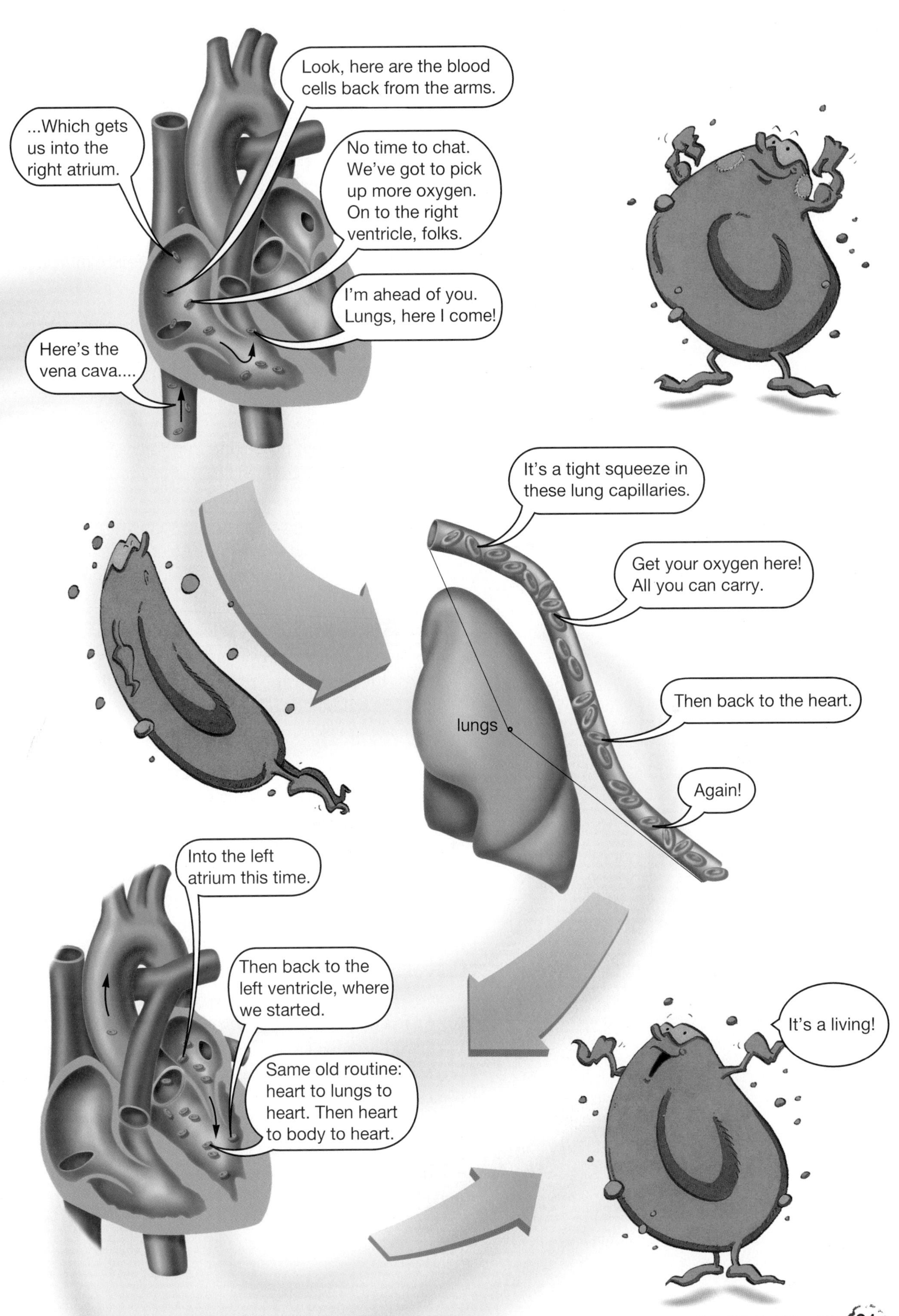
Here's the vena cava....
...Which gets us into the right atrium.
Look, here are the blood cells back from the arms.
No time to chat. We've got to pick up more oxygen. On to the right ventricle, folks.
I'm ahead of you. Lungs, here I come!
It's a tight squeeze in these lung capillaries.
Get your oxygen here! All you can carry.
lungs
Then back to the heart.
Again!
Into the left atrium this time.
Then back to the left ventricle, where we started.
Same old routine: heart to lungs to heart. Then heart to body to heart.
It's a living!

Blood Pressure

The device used to measure blood pressure is called a **sphygmomanometer** (see Figure 3.21A). It consists of an inflatable cuff that is wrapped around the arm. Air is pumped into the cuff, squeezing it against the artery in the arm and restricting the blood flow. Air is then slowly let out of the cuff to the point where the blood pressure matches the cuff pressure, letting blood force its way back through the artery. A doctor can listen for the sound of the blood using a **stethoscope** (see Figure 3.21B).

Why do doctors measure a patient's blood pressure? Blood pressure indicates several things about the health of the circulatory system:

- The volume of blood. If a person has lost a lot of blood through injury, their blood pressure will be low.
- Heart rate. A fast-beating heart pushes blood rapidly through the arteries, building up blood pressure.
- Artery size. Large, open arteries conduct larger volumes of blood, producing low blood pressure. Small, narrow, or partly clogged arteries produce high blood pressure.
- Artery elasticity. Flexible arteries can easily expand, letting more blood through. Loss of elasticity results in "hardening" of the arteries, producing higher blood pressure.
- Blood viscosity. Viscosity refers to the thickness of the blood. Thick fluids flow less easily than thin, watery fluids. Blood viscosity is a measure of the balance between red blood cells and plasma. You will learn more about viscosity in Chapter 4.

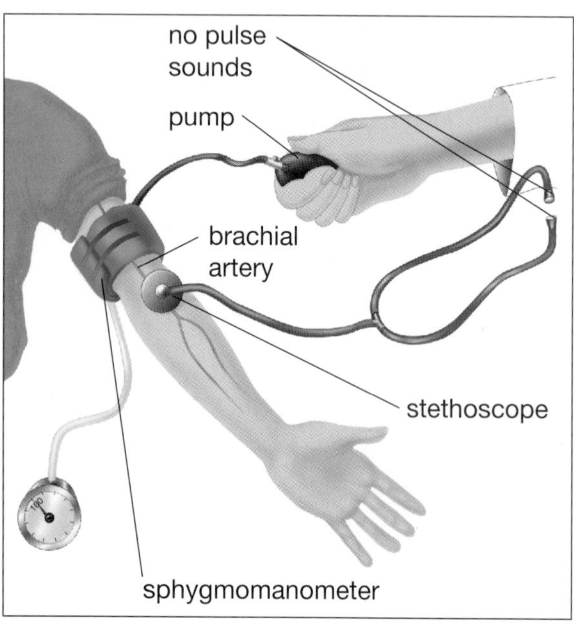

Figure 3.21A Measuring blood pressure

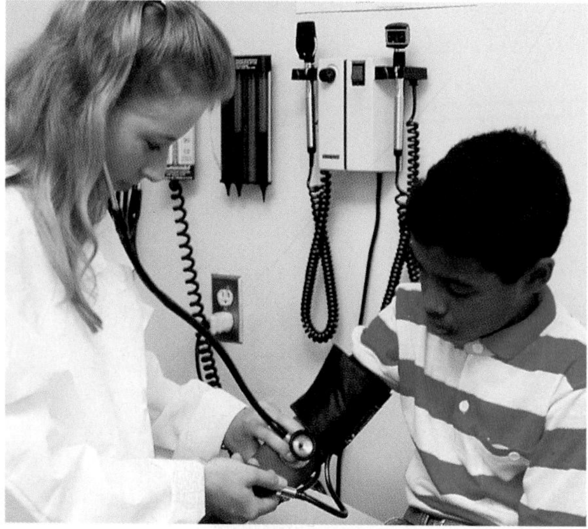

Figure 3.21B A doctor uses both a sphygmomanometer and a stethoscope to measure blood pressure.

Doctors measure blood pressure as a simple first step to assess the health of the circulatory system. High blood pressure can become a deadly disease, sometimes known as the "silent killer" because people with high blood pressure may not feel ill.

Disorders of the heart and circulatory system are a leading cause of death in North America. Many of these disorders are related to lifestyle or pollution. Thus, the risks of these disorders can be reduced or avoided by choosing healthy lifestyle habits. For example, cigarette smoke is a double threat to the circulatory system. Nicotine in cigarette smoke causes blood vessels to constrict, which increases blood pressure. Also, smokers may experience difficulty breathing. Why? Carbon monoxide in the smoke competes with oxygen in the lungs. This reduces the blood's ability to carry oxygen to the cells. The circulatory system is also harmed by a poor diet, especially one that includes too much fat or salt.

You Are What You Eat

Do you eat foods that help your body or harm it? Too much or too little of particular foods over long periods of time can eventually have a negative impact on one or more of your body systems. Different foods contain different combinations of substances that your body needs. Substances in foods that provide energy and materials for cell development, growth, and repair are often referred to as **nutrients**. **Carbohydrates** are the cells' main source of quick energy. (You may have heard athletes talk about "carbohydrate loading" — eating plenty of carbohydrates before they compete.) **Fats** are another source of energy. Unlike carbohydrates, however, fats can be stored by the body. When you eat more food than the body needs for its activities, the surplus is stored in your tissues as fats. **Proteins,** found in foods such as meat, fish, and eggs, are essential for growth and repair of body tissues. In addition to carbohydrates, fats, and proteins, a complete, healthy diet of nutrients includes minerals, vitamins, and water.

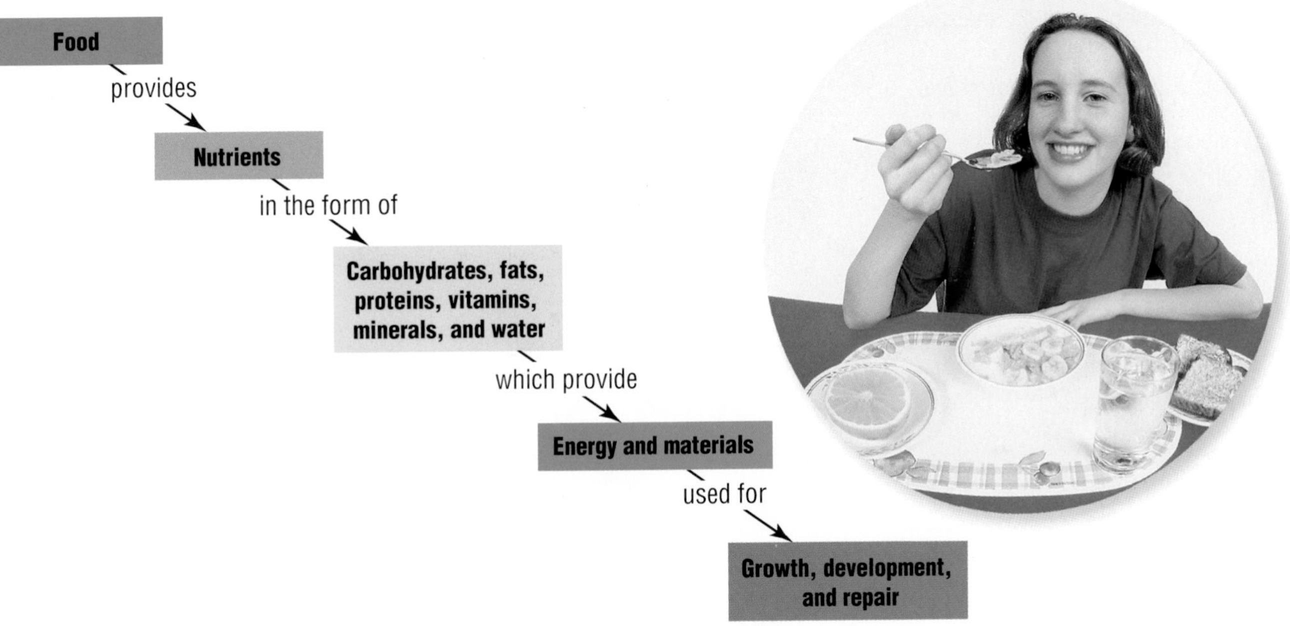

Food
provides
Nutrients
in the form of
Carbohydrates, fats, proteins, vitamins, minerals, and water
which provide
Energy and materials
used for
Growth, development, and repair

Career CONNECT

Different Foods for Different Sports

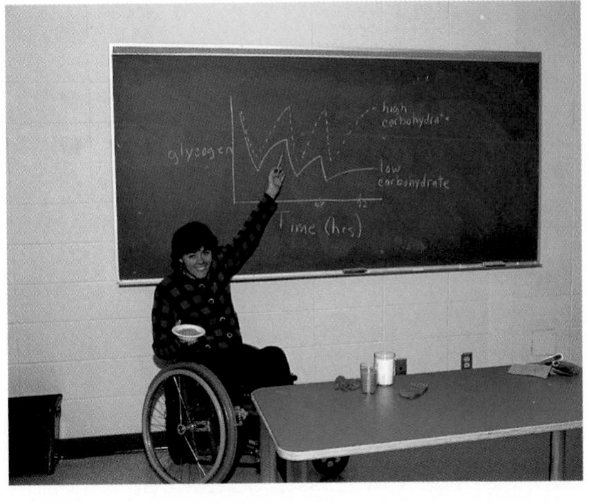

As a sports nutritionist, Helga Rempel advises athletes about the effect of the foods they eat on their athletic performance. She knows how the different parts of the body respond to nutrients such as fats, vitamins, or sugars. She also knows that different sports require different diets. For example, because of how our muscle cells use energy, a marathon runner needs to eat more proteins than a weight lifter.

A dietitian is another professional who knows how food affects the human body. Talk to a teacher or a guidance counsellor about the difference between a dietitian and a nutritionist. Check your phone book, local medical centre, or hospital to find a nutritionist or a dietitian in your area. Perhaps you could arrange to speak to them about the work that they do. Take notes on your conversation and present your findings in class.

Testing for Carbohydrates

Carbohydrates include starches and sugars. In this investigation, you will test a variety of food items to determine which of them contain starch.

Problem

Which foods contain starch?

Safety Precautions

- Be careful when handling sharp objects such as knives.
- Iodine is a toxic irritant that stains. Do not drip any of this liquid on your skin or clothing.

Apparatus

5 petri dishes
2 medicine droppers
knife

Materials

iodine solution
white paper
potato
apple
carrot
additional food samples from your lunch

Procedure

1. Prepare a data table in your notebook like the one below.

Testing Foods for Starch		
Food substance	Colour change	Is starch present?

2. Cut a small slice of each solid food sample. Predict which of the food samples will contain starch. Record your prediction.

3. Place a food slice, or a few drops of a liquid sample, in each dish. Set the dishes on a sheet of white paper to observe the results more clearly.

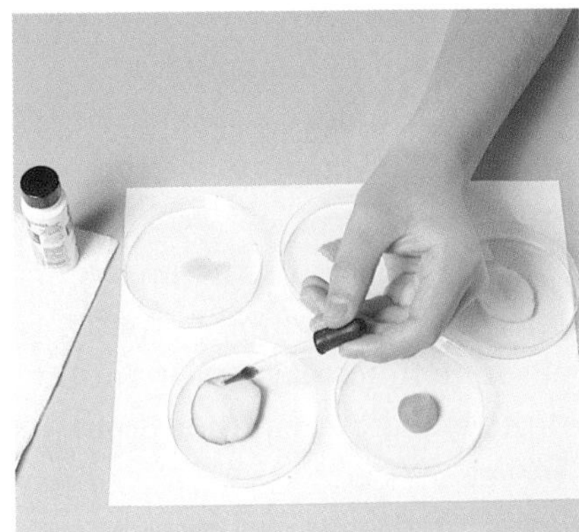

4 Add a few drops of iodine solution to each sample. Wait a few minutes and examine the samples for any colour changes. A colour change indicates the presence of starch.

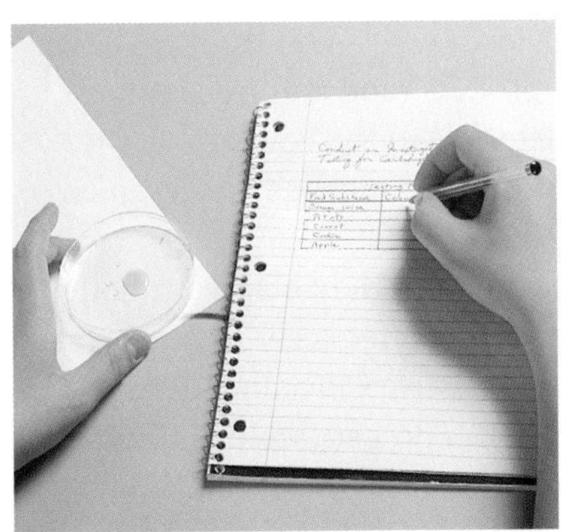

5 Record the results in your table. Wash your hands after this investigation.

CAUTION: If you get any iodine in your eye, rinse your eye for at least 15 min and inform your teacher.

Analyze

1. Which foods contain starch?
2. Are all starch-containing foods solid?
3. Based on results from tests of lunch items, what are the most common sources of carbohydrates eaten by you and your classmates?

Conclude and Apply

4. What is a balanced diet? Write out a menu for a meal that includes all the substances (nutrients) your body needs to function properly. Then write another menu for a meal that might damage your body if it were eaten regularly over a long period. Explain what harm this diet might do and why.

Extend Your Knowledge

5. After you eat, the amount of sugar in your blood increases. Find out how the body controls blood sugar levels. What common disorder is caused by a failure of this control, resulting in large amounts of glucose in the blood and urine? How is this disorder treated?

You and Your Body

What would you think if someone advised you to sit on a couch for at least eight hours a day in a smoky room, eat plenty of candy bars and potato chips, drink lots of pop, and get no more than three or four hours of sleep each night? You would probably assume that you would not feel too well after a few weeks of this lifestyle. Like any complex structure, your body needs proper care to function properly. However, people sometimes pay less attention to the health of their bodies than they do to maintaining a bicycle or car.

To maintain healthy organs and systems, everyone has the same essential needs: clean air and water, nutritious foods, exercise, and sleep.

Clean air means oxygen for your cells, which use the oxygen to produce energy. Pollution from smoke and traffic fumes decreases the ability of oxygen to get into your body. A balanced diet provides your cells with the food materials they need for growth and activities. Lack of essential materials makes the body grow weaker, while too much of some substances such as fats, sugar, and salt can place a strain on certain organs and systems.

Exercise is important because it helps the body process food and oxygen more efficiently. A healthy heart and lungs help carry materials to the cells and get rid of wastes. Strong muscles help protect the body from injury. Your body is designed to work, and you feel better when you are active.

Not only do healthy lifestyle habits make you feel better, they help your body resist diseases. Your immune system is able to work more effectively when you are well fed and rested. Your immune system attacks and destroys invading germs and helps break down harmful materials in your body. If you do get a cold, a disease, or an injury, you are likely to recover faster if you are basically healthy.

If you are always feeling tired or lacking zest or catching colds, think about what you can do to improve your overall health and energy level. This unit has given you some ideas of the variables to consider in making choices that will benefit your cells, tissues, organs, organ systems — and you!

Check Your Understanding

1. Why do humans need a complex circulatory system while an amoeba does not?
2. What are the main components of blood?
3. Why do arteries have thick walls?
4. What are the three main types of food?
5. **Apply** Your doctor finds that you have high blood pressure. What two features of your circulatory system might produce this result?

CHAPTER at a glance

Now that you have completed this chapter, try to do the following. If you cannot, go back to the sections indicated.

Name the five levels of organization in multicellular organisms. Give an example of each level of organization. (3.1)

Copy the sketch of the plant below and label the three organ systems shown. (3.2)

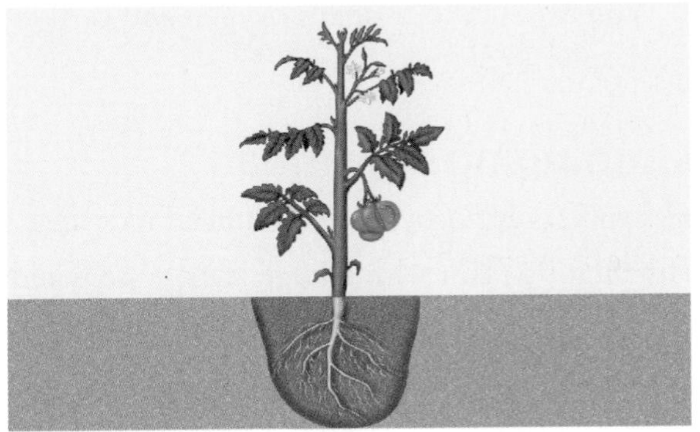

Explain the functions of each of the three organ systems of a plant. (3.2)

Give reasons why plants have fewer types of tissues and organ systems than animals have. (3.2)

Name two tissues that connect roots with shoots. (3.2)

Describe the process by which water (a) enters a plant, and (b) leaves a plant. (3.2)

Give an example of how plant tissues are adapted to either a dry or a wet environment. (3.2)

List five organ systems found in humans. (3.3)

Name the organs in which substances pass between the circulatory system and (a) the respiratory system, (b) the digestive system, (c) the excretory system. (3.3)

Name the structures in each of the above organs through which substances pass into or out of capillaries. (3.3)

Define homeostasis and give an example. (3.4)

Explain how a fatty diet could affect your circulatory system. (3.4)

Prepare Your Own Summary

Summarize this chapter by doing one of the following. Use a graphic organizer (such as a concept map), produce a poster, or write the summary to include the key chapter ideas. Here are a few ideas to use as a guide:

- How are cells organized in multicellular organisms? (3.1)
- What are the main tissues found in plants? In animals? (3.1, 3.2)
- What organ system supplies food in plants? In animals? (3.2, 3.3)
- Describe a simple experiment you could do to show how water moves through stem tissues in a plant. (3.2)
- Copy the diagram on the right showing lung structure. Add labels to identify the main parts. Briefly explain how exchange of gases takes place in these structures. (3.3)

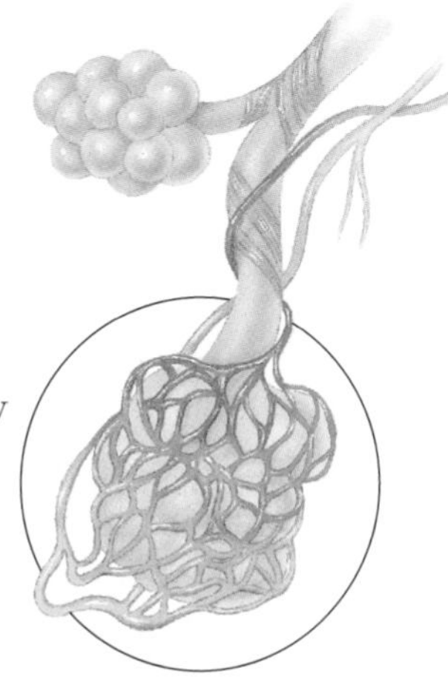

- What is the function of the villi lining the small intestine? (3.3)
- Draw a sketch to show the pattern of blood circulation between the heart, the lungs, and the rest of the body. (3.3, 3.4)
- Write a paragraph that includes the words "doughnuts," "exercise," "carbohydrates" and "homeostasis." (3.4)

CHAPTER 3 Review

Key Terms

tissues
organs
organ systems
levels of organization
epithelial tissue
epidermal tissue
digestive system
vascular tissues
phloem tissues
xylem tissue
root hairs
palisade cells
stomata (singular stoma)
guard cells
transpiration
respiratory system
circulatory system
trachea
bronchus
bronchioles
alveoli (singular alveolus)
capillaries
villi (singular villus)
absorption
excretory system
homeostasis
nervous system
endocrine system
hormones
hypothalamus
blood
plasma
hemoglobin
sphygmomanometer
stethoscope
nutrients
carbohydrates
fats
proteins

Reviewing Key Terms

If you need to review, the section numbers show you where these terms were introduced.

1. Draw a flow chart illustrating the following terms in the correct order:

   organs, cells, tissues, organism, organ systems (3.1)

2. Describe one difference between
   (a) xylem and phloem (3.2)
   (b) an artery and a vein (3.3)
   (c) a vein and a capillary (3.3)

3. In your notebook, write the correct term to complete the following sentences:
   (a) Water loss through a leaf is ________. (3.2)
   (b) ________ in cells causes stems and leaves to stay firm. (3.2)
   (c) Blood pressure is measured with a ________. (3.4)

4. What do the digestive system and the respiratory system have in common? (3.3)

5. **(a)** Which system in the human body regulates blood composition and gets rid of waste fluids? (3.3)
   (b) Which system exchanges oxygen and carbon dioxide? (3.3)

Understanding Key Ideas

Section numbers are provided if you need to review.

6. Why might a plant with a huge stem system and a tiny root system have difficulty surviving? (3.2)

7. Why might a plant with a huge root system and a tiny stem system have difficulty surviving? (3.2)

8. Why is a plant growing in the shade less likely to wilt than one growing in full sunlight? (3.2)

9. What effect might pruning (trimming) a tree's branches have on the transport of water up the tree trunk? (3.2)

10. Why is such a large percentage of your blood volume made up of red blood cells? (3.3)

11. Carbon dioxide is transported through the body in solution in plasma, but oxygen is not. Explain why there is a difference in the way these two gases are transported by the circulatory system. (3.3)

12. Suppose you receive a sudden surprise, such as your teacher announcing a surprise test. Your heart may beat faster and your breathing may become irregular. After a short time, your breathing and heart rate return to normal.
    (a) Which two systems are interacting in this initial reaction? (3.3)
    (b) Which system controls and co-ordinates their interaction? (3.4)
    (c) What process returns the systems to normal functioning? (3.4)

Developing Skills

13. Make a concept map to show how the following are related in living organisms. Link the concepts together using a few key words or descriptions. Add any other terms you need.

cellular respiration	small intestine
food	cells
external environment	circulatory system
oxygen	carbon dioxide
lungs	

14. Draw a large square on a sheet of paper. Divide the square into four equal squares. This diagram represents the four-chambered heart. Label each chamber. Add arrows in red or blue showing the direction of blood flow to or from each chamber and between chambers. Use red arrows for blood that is rich in oxygen and blue arrows for blood that is low in oxygen. Finally, write on your diagram where blood entering each chamber is coming from and where blood leaving each chamber is going.

Problem Solving/Applying

15. Suggest what adaptations might be found in (a) the leaves, and (b) the roots of a plant living on the tundra, where conditions are cold and dry. Give reasons for your ideas.

16. Design Your Own Why might a plant wilt in hot weather? Why might a different plant not wilt in hot weather? Design your own experiment to compare the responses of two different kinds of plants to an increase in temperature. Sketch the steps of your procedure.

17. Some babies are born with a hole in the heart between the left ventricle and the right ventricle. Explain what problems this would cause for the baby.

18. The normal breathing rate of an infant is faster than that of a teenager. Why do you think this is so?

19. How can diet affect your circulatory system?

Critical Thinking

20. When you exercise on a hot day, you sweat and become thirsty. Explain how sweating and thirst are examples of homeostasis.

21. Imagine you are a science fiction writer describing a planet where plants can move from place to place. Why might plants do this? What tissues would they need apart from muscle tissues? Why don't plants move on Earth in the same way that animals do? Write a short science fiction story that takes place on this imaginary planet.

Pause& Reflect

1. The ageing and death of an organism can be described as a failure of homeostasis. In your Science Log, write why you agree or disagree with this description.
2. Go back to the beginning of this chapter on page 66 and check your original answers to the Getting Ready questions. How has your thinking changed? How would you answer these questions now that you have investigated the topics in this chapter?

UNIT 1 Ask an Expert

Diabetes is a disease that affects more than 1.5 million Canadians. That number is expected to double by the year 2010. Research into how our cells function is being done in hopes of finding a cure — or at least better treatment — for this disease. That is where people like Dr. Amira Klip come in. She is a biochemist and research scientist at the Hospital for Sick Children in Toronto, Ontario. Dr. Klip's work is known around the world.

Q How did you become interested in science?

A When I was in high school, I had a terrific teacher of organic chemistry, the chemistry of living things. I learned from this teacher that although we can't always see what is happening inside our cells, we can still understand what is happening because of the reactions that we *can* see. It's a bit like looking at the wind — you can't see it but you know it's blowing because you can see the trees moving.

Q When did you begin doing scientific research for a living?

A I completed my bachelor's and master's degrees in biochemistry at the Centre for Advanced Studies in Mexico City, where I lived. Then I worked in laboratories running experiments for other scientists. Most of the research I did involved studying the membranes of human cells. I was trying to understand how materials pass in and out through cell membranes.

Q Does your early research have much to do with the research you do now?

A Yes. I study one particular kind of protein in the cell. This is the protein that takes glucose from the bloodstream and brings it into the cells of our bodies, especially our fat and muscle cells. This protein is called a "glucose transporter" because it transports, or carries, the glucose from outside the cell membrane to the inside of the cell.

outside cell
glucose in blood
1.
2.
3.
4.
5.
6.
cell membrane
inside cell
extra glucose transporters

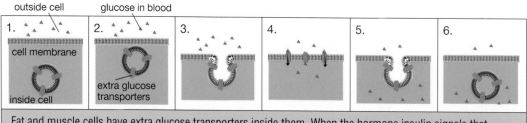

Fat and muscle cells have extra glucose transporters inside them. When the hormone insulin signals that there is a lot of glucose in the blood, these extra transporters come to the membrane and let glucose pass through the membrane and into the cell. These diagrams show what scientists think happens as those extra transporters come to the membrane and then return inside the cell once their job is done.

Q How does your research relate to diabetes?

A When a person has diabetes, the glucose in their blood does not get processed by the cells. In our lab, we study how the glucose transporter does its job and we try to figure out what goes wrong in the bodies of some diabetic people. We know that people with type 1 diabetes can't produce their own insulin. They need daily insulin injections just to stay alive. Many more people have type 2 diabetes. For some reason, insulin doesn't activate their cells to process the glucose. If we can find out more about how the insulin and the parts of the cells work together, maybe we can find better medical treatment for all diabetics.

Q How do you study these cells? Can you observe them with a microscope?

A It's not quite that simple. The cell parts we study are too small to be visible, even with a microscope. One way researchers study what's happening is by tagging the glucose transporters with something that we are able to see. That means we attach something to them, such as a tiny bit of chemical that gives off fluorescent light. Then we can examine samples of cells at different points in the experiment. The location of the fluorescence tells us how the glucose transporter has responded to various stimuli. If the bit of fluorescent light shows up in the centre of the cell, then we know that that's where the glucose transporter is. If the fluorescent tag is at the cell membrane, we know the transporter is at the membrane to do its job.

Q Do you spend most of your day at the microscope?

A Not really. I hire research technicians and train graduate students to run many of the experiments. I look at the experimental results and try to understand what they mean.

The rest of my time is spent teaching university classes and lecturing at conferences all over the world. Conferences keep me in touch with my fellow researchers so we can share what we have learned, either new data concerning glucose transporters or new techniques for studying them.

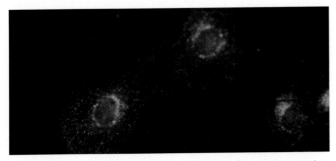

In this micrograph showing muscle cells, the glucose transporter proteins have been tagged with a fluorescent chemical so that they appear as bright green spots clustered around the nucleus of the cell.

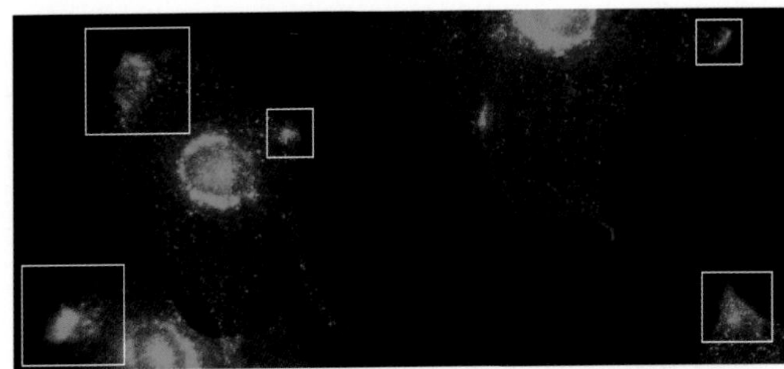

These same muscle cells have been stimulated with insulin for 30 min. The highlighted green spots are glucose transporters that have responded to the insulin and have come to the cell membrane.

EXPLORING Further

Watch Some Cells at Work

Researchers have developed a technique for videotaping special fat cells cultured (grown) in the laboratory, as they take glucose from the bloodstream. Visit the McGraw-Hill Ryerson web site at www.school.mcgrawhill.ca/resources/ to see some green fluorescent-tagged glucose transporters moving about in a cell as they are stimulated.

Search for other Internet sites related to diabetes to find information on the disease itself and to learn what it is like to live with diabetes. Write a brief summary of the information you find at each of the sites.

UNIT 1

An *Issue* to Analyze

A DEBATE

Science Versus Alternative Medicine

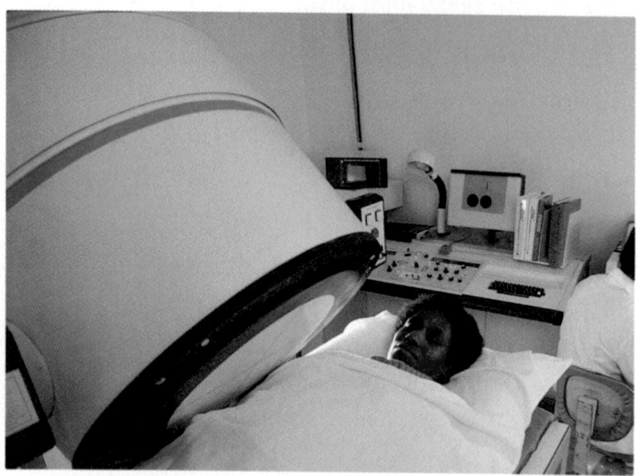

The large device on the left, called a gamma camera, is used in nuclear medicine to treat cancer.

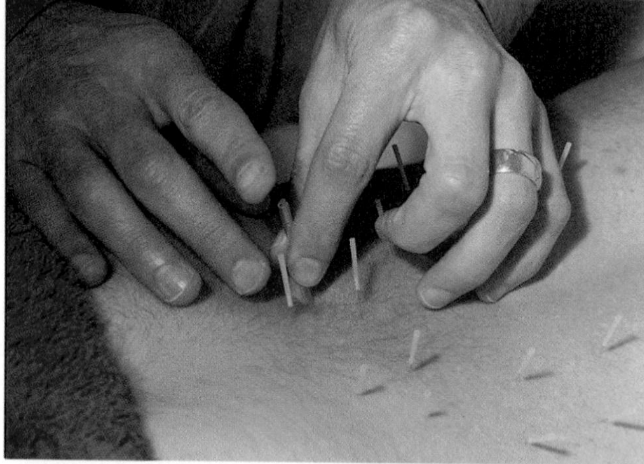

An acupuncturist at work. Acupuncture is a method of pain relief developed in China many centuries ago.

Think About It

During the past century, scientific medicine has developed vaccines, antibiotics, and surgical techniques that have almost eliminated some diseases, for example, smallpox and polio. Scientific medicine has also reduced the risk of death from other disorders, such as heart attacks and cancer. As a result of medical advances, together with improved nutrition and sanitation, Canadians born today have a much higher life expectancy than did people born a hundred years ago.

Many medical advances are based on better understanding of how cells and organs function in the body to help maintain normal health. For example, analysis of cells in the immune system helped researchers develop methods of controlling the once deadly disease, AIDS. The value of new techniques in scientific medicine is measured by quantitative results. Quantitative results include improved rates of recovery and increased survival rates, for example.

Despite the great success of medical treatments based on scientific understanding, many types of alternative medicine have become popular. Some of these alternative therapies are centuries old, and come from non-Western cultures. For example, acupuncture, a method of relieving pain by inserting special needles into the skin, was developed by the ancient Chinese. Herbal remedies, based on the use of various plants, have been used by most cultures around the world since prehistoric times.

Other alternative practices, such as homeopathy, chiropractic, faith healing, and naturopathy, were developed in the nineteenth century to compete with scientific medicine. More recent alternative practices include aromatherapy, therapeutic touch, magnet therapy, reflexology, and various dietary supplements.

Alternative treatments are not taught in most medical schools in Canada. People using these techniques vary in their training and knowledge.

Some practitioners of alternative medicine are licensed and regulated, for example, chiropractors, but others are not. In general, alternative techniques have not been tested according to the same standards used to establish the value of scientific therapies.

Resolution

Be it resolved that methods used in alternative medicine are unproven as effective ways of treating illness.

What to Do

1. Read the In Favour and Against points listed on this page, and begin to think about other points that could be made in favour of the resolution and against the resolution.

2. Four students will debate the topic. Two students will speak in support of the resolution and two will speak against it. **Note**: No matter which side of the debate you actually believe in, you must try your best to convince the jury, or debate listeners, of the point your side is defending.

3. To aid the two teams, two other students will work with them to gather the background information needed to put forward a strong case for the point that side is defending. (To make the research more manageable, each team may choose to find detailed information about one of the alternative therapies mentioned on page 102.)

4. The rest of the class will act as the jury in hearing the debate. They, too, should do their own research in order to understand the science and technology behind the issues raised. Possible areas to research include:
   - the placebo effect
   - the safety and effectiveness of various treatments
   - the difference between statistical evidence and anecdotal evidence
   - the difference between correlation and causation
   - the ethics of charging fees for unproven therapies

5. Your teacher will provide you with *Debating Procedures* to follow.

In Favour

- There is little or no statistical evidence based on controlled clinical trials that demonstrate that alternative therapies produce cures.
- Many alternative therapies contradict or ignore what is known about how cells and organs in the body carry out their functions.
- The apparent success of some therapies may be a result of faulty diagnosis. In other words, a patient appears to recover because he or she did not really have the illness that was diagnosed in the first place.

Against

- Good health is a result of many factors affecting organ systems as a whole, and alternative therapies may affect this complex balance rather than one factor alone.
- Some alternative therapies are based on practices that have been used to treat illnesses successfully for hundreds or thousands of years.
- Alternative therapies may use some principles not yet discovered by scientific research.

Analyze

1. Which side won the debate, based on a class vote?

2. Did the side that won produce better research or make a better presentation? Explain.

3. Did your initial viewpoint change as a result of this issue analysis? Explain why or why not.

UNIT 1 PROJECT

An ExCELLent Pop-Bottle Environment

Here is an opportunity to see biology in action and to design your own investigation. Use your knowledge of science inquiry processes, the microscope, cells, and micro-organisms to investigate "life in a bottle." In Part A, you will construct an apparatus using pop bottles, as shown in the diagrams. Your apparatus represents a model of an ecosystem. Then, in Part B, you will use this apparatus to test a hypothesis in an experiment that you design yourself.

Safety Precautions

- Be careful when using sharp objects such as scissors and hand tools such as a back saw.
- Use caution when handling plastic bottles that have been cut; the edges may be sharp.
- A glue gun is hot and the glue remains hot for several minutes.
- Wear safety goggles and take proper safety precautions when using the drill press.
- Wash your hands after completing this project.

Part A

Construct Your Pop-Bottle Environment 

Materials

2 L plastic pop bottles (2)
sharp scissors
100 cm jinx wood (1 cm × 1 cm)
coping saw or back saw
mitre box
glue gun
hand drill or drill press
drill press vice or hand screw
clear packing tape
ruler or metal scale
garden soil

Skill POWER

For information on the safety symbols used throughout this book, turn to page 553.

What to Do

1. With a team member's help, carefully cut bottle 1 just below the shoulder, as shown. Cut off the bottom of the bottle as well, leaving about 2 cm of the hip on the cylinder. Set aside the newly cut top and base (sections A and B).

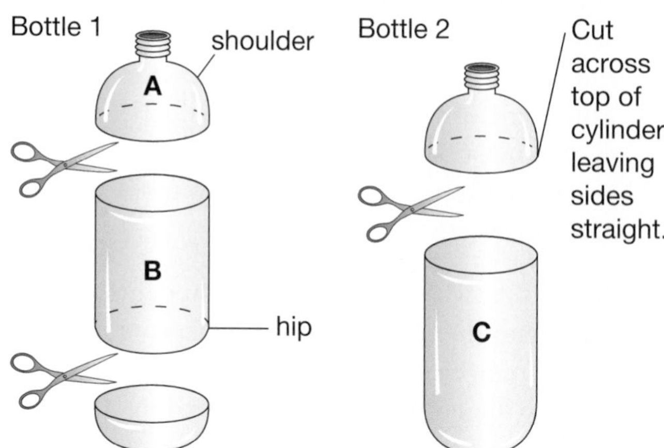

2. Repeat step 1 using bottle 2, but this time do not cut off the bottom. Retain only the newly created cylinder, section C.
3. Remove the twist-off cap from section A. Place it in a drill press vice or a hand screw, and drill 5 or 6 small holes. Replace the cap on the threaded bottle portion.
4. Measure the diameter of the base of bottle 1, and construct a simple wooden frame to stabilize the pop-bottle environment.
5. Pour about 1 L of pond water, aquarium water, or water from a stream into section C. You can also add some aquatic plants, but they will need a small quantity of soil for anchoring.

6 Invert section A into section B. Slide this A/B unit into section C, as shown.

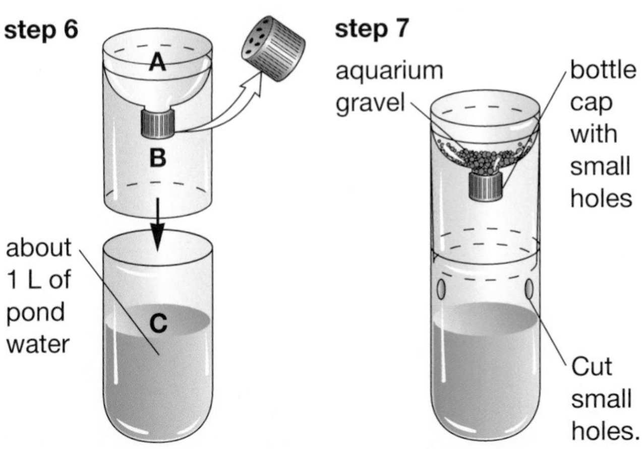

7 Add a small amount of aquarium gravel to the neck of the bottle. Gently tape the bottle sections together. Cut small holes in section C, as shown.

8 Add garden soil to section A, as shown below.

9 Plant some seeds in the soil and water it with aquarium or pond water until the soil is damp.

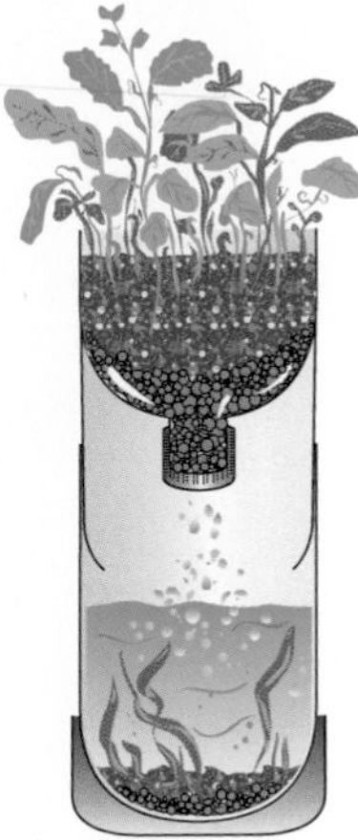

Your completed pop-bottle environment is a model of an ecosystem.

In a group, brainstorm the materials you would need to build a model of a neuron (nerve cell). Then design and construct a model that is no larger than a bread box and no smaller than a shoe box.

Part B

Design Your Own Investigation

Materials

microscope
microscope slides
cover slips
medicine dropper
tweezers
test tubes or collection bottles

What to Do

1 List the various factors (independent variables) that would affect the life forms inhabiting your model ecosystem. Select one dependent variable to investigate. (For example, you might wish to investigate how a certain type of micro-organism living in the pop-bottle environment is affected by varying temperatures.) While investigating your dependent variable, you must keep all the other independent variables other than temperature constant (the same). This will ensure that your experiment is a "fair test."

2 After you have selected your variable, make and record a hypothesis related to that variable.

3 Use your apparatus to design and conduct an investigation to test your hypothesis:

(a) Plan and write a step-by-step procedure. Show your procedure to your teacher.

(b) Determine and set up a control for your experiment.

(c) Decide what you will measure in your experiment, and how you will record your measurements.

(d) Make a conclusion, based on analyzing the results you obtain when you conduct your experiment.

(e) Write a detailed report of your findings.

Evaluate

1 Did the results you obtained provide an answer to the problem or question you were investigating? If not, what factors or steps might you change or improve in order to obtain an answer?

2 Did your results support your hypothesis? Explain.

UNIT 2

Fantastic Fluids

Imagine travelling in a miniature submarine through a blood vessel in a human body, and facing an attack by gigantic white blood cells! This is one of many exciting episodes in scientist Isaac Asimov's science fiction novel, *Fantastic Voyage*. These two pages show a scene from the Hollywood film on which Asimov based his book.

Inside your body, fluids help carry gases, food materials, wastes, and chemical messages back and forth between cells. Blood and the air you breathe act as transportation fluids. The fluid in the lining surrounding your lungs protects the delicate air sacs from rubbing against your ribs. Tears and mucus lubricate sensitive body tissues.

The properties of fluids also affect your external environment. Many organisms inhabit the surface of Earth and move through air, while thousands of species live in water environments. Humans have designed many different crafts and vessels to travel through these two fluid environments, from kayaks to speedboats, from hot-air balloons to supersonic jets. Explore this unit to find out more about fluids — their viscosity, density, buoyancy, and pressure — and how we make use of these properties.

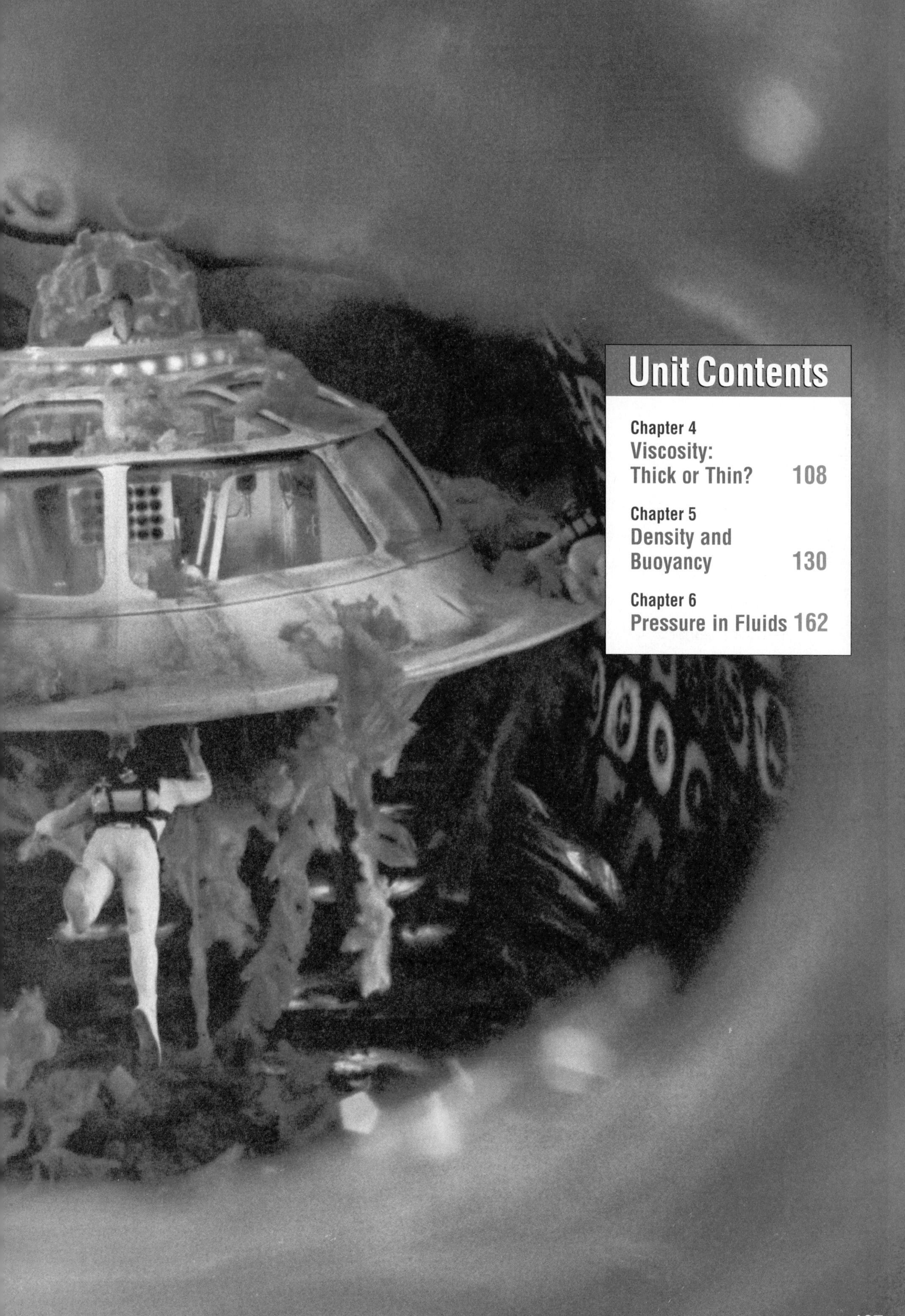

Unit Contents

CHAPTER

4 Viscosity: Thic

Getting Ready...

- Why does maple syrup flow more slowly than water?
- What causes ice cream to dribble down a cone on a hot summer day?
- Why are different types of motor oil used in a car, depending on the season?

Science Log

Try to answer the questions above in your Science Log. Instead of describing the liquids as thick or thin, use the word "viscosity" in your explanations. Then read this chapter to find the answers.

Skill POWER

For tips on how to make and use a Science Log, turn to page 534.

Smooth milk chocolate... creamy toffee...thick, rich fudge. Do these sound tempting? Does your mouth water at the thought of gooey caramel topping, pancakes drizzled with syrup, or melted chocolate chips in cookies hot from the oven?

Some of the most delicious treats are liquids that flow thickly and smoothly. The property that describes a liquid's thickness or thinness is called **viscosity**. In this chapter, you will discover how high-viscosity (thick) liquids differ from low-viscosity (thin) ones. You will also use the particle theory to explain some behaviours and properties of fluids — liquids and gases. A **fluid** is any substance that flows. The particle theory can explain why some substances flow while others do not. Investigate the viscosity of fluids in this chapter.

r Thin?

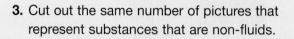

Fluid or Non-fluid?

How does a fluid differ from a non-fluid?

What You Need

large sheet of paper for class chart
newspapers and magazines
scissors
masking tape

What to Do 

1. Your teacher will make and display a large class chart with the headings: "Fluids" and "Non-fluids."
2. In your group, select and cut out three to five newspaper or magazine pictures that represent substances that are fluids.

   CAUTION: Be careful when using sharp objects such as scissors.
3. Cut out the same number of pictures that represent substances that are non-fluids.
4. Each group will take turns taping their pictures under the appropriate heading on the wall chart.

What Did You Find Out?

1. Look at the pictures that students have classified as fluids and as non-fluids. Do you agree with the classifications? Why or why not?
2. Compare and contrast a fluid and a non-fluid.

Extension

3. What is a simple test you could do to determine whether a substance is a fluid or a non-fluid?

Spotlight On Key Ideas

In this chapter, you will discover

- why liquids and gases are classified as fluids but solids are not
- what viscosity is and how to measure it
- how and why the viscosity of a fluid can vary
- how to compare the viscosity of various liquids
- what factors affect a fluid's flow rate
- how viscosity can make substances useful

Spotlight On Key Skills

In this chapter, you will

- classify, compare, and contrast fluids and non-fluids
- measure the flow rate of various liquids
- assemble apparatus according to a diagram
- make a bar graph based on given data
- use the particle theory to explain differences in the viscosity of fluids
- design your own experiments

4.1 Flowing Along

You can easily observe fluids such as water flowing out of a tap, milk or juice being poured into a glass, or ketchup oozing from a bottle. Your body contains many fluids, such as blood and the watery cytoplasm inside cells. It is more difficult to imagine gases flowing, but they do.

Take a deep breath. What happened? Some of the air that surrounds you flowed into your lungs when your lungs and ribcage expanded. Air flows out of your lungs when you exhale. Like liquids, gases flow and take up space. Therefore, gases and liquids can both be classified as fluids.

You may be wondering whether any solids can be classified as fluids. Breakfast cereals seem to flow when you pour them out of the box into a bowl. Is cereal a fluid? You pour powdered laundry detergent into a washing machine. Is the detergent a fluid? To answer these questions, do the At Home Activity below.

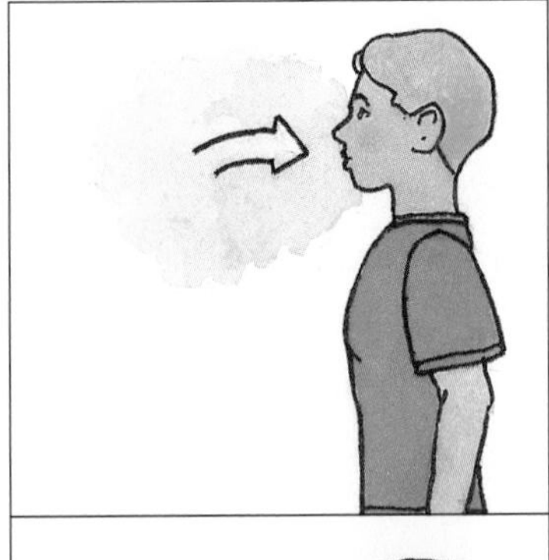

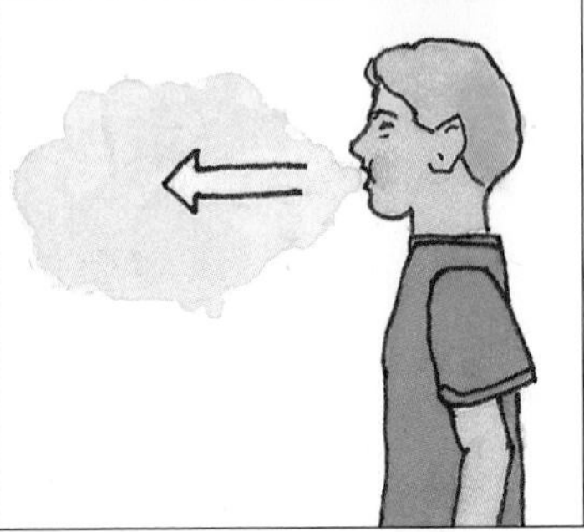

Figure 4.1 Air flows into your lungs when you breathe in. Air flows out of your lungs when you breathe out.

At Home ACTIVITY

Can Solids Flow, Too?

Sand and sugar are both solids. They can be poured, but can they flow like a fluid such as water? Pour one of them to find out.

What You Need

sand or sugar (about 250 mL)
water
2 large plates

What to Do

1. Place two large plates on a level surface. While holding your hand steady, slowly pour sand or sugar onto one of the plates. Draw the results.
2. Again, hold your hand steady as you slowly pour 250 mL of water onto the second plate. Draw the results again.
3. Wipe up any spills and wash your hands after this activity.

What Did You Find Out?

1. Describe any differences in the behaviour or the appearance of the substances when you poured them.
2. What characteristic is necessary in order for a substance to be classified as a fluid?

The States of Matter and the Particle Theory

The essentials of life — food, water, and air — are examples of substances that occur in the three different states of matter: solid, liquid, and gas. As you learned in earlier science studies:

- **Solid** is the state of matter of a substance that has a definite shape and volume (for example, a sugar cube).
- **Liquid** is the state of matter of a substance that has a definite volume, but no definite shape (for example, water).
- **Gas** is the state of matter of a substance that has neither a definite shape nor a definite volume (for example, oxygen).

By using the particle theory, you can explain why liquids and gases flow but solids do not.

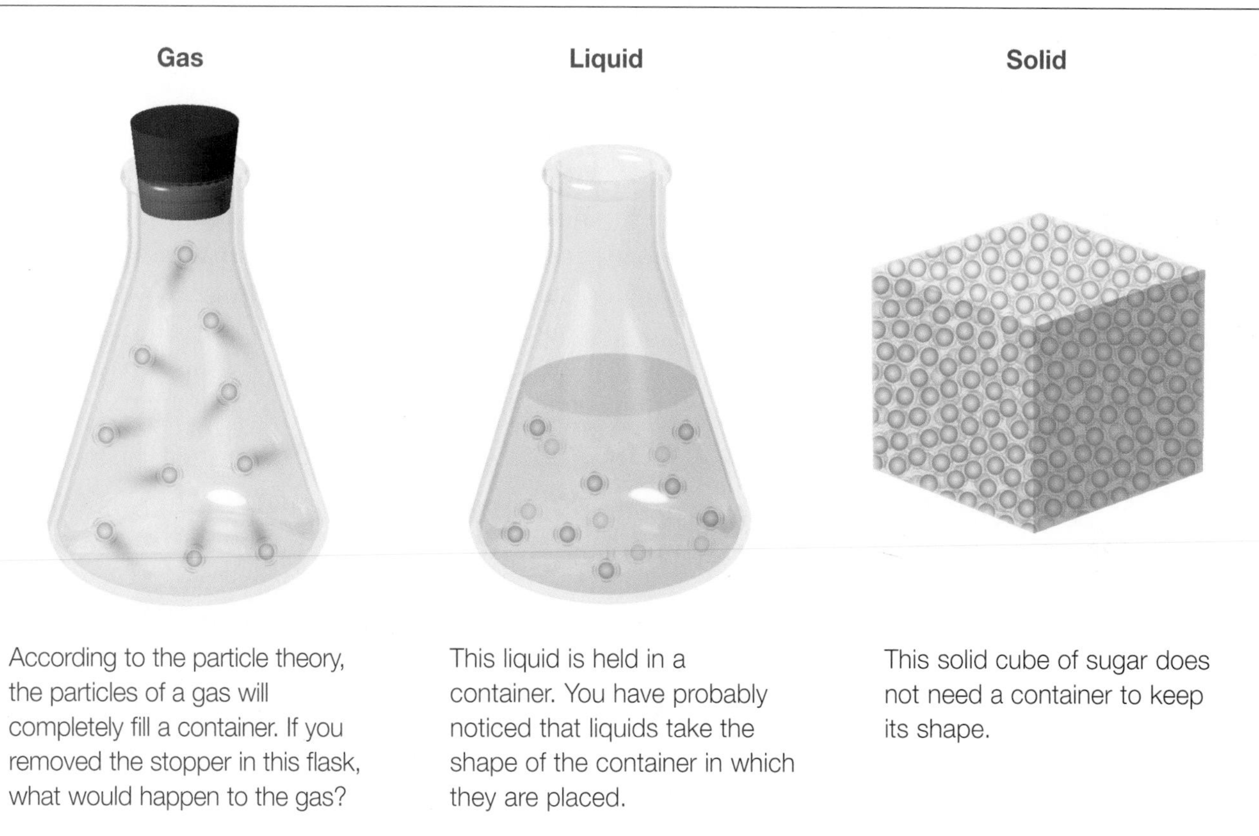

According to the particle theory, the particles of a gas will completely fill a container. If you removed the stopper in this flask, what would happen to the gas?

This liquid is held in a container. You have probably noticed that liquids take the shape of the container in which they are placed.

This solid cube of sugar does not need a container to keep its shape.

The five major points of the particle theory of matter are:

1. All matter is made up of very small particles.
2. All particles in a pure substance are the same. Different substances are made of different particles.
3. There is space between the particles.
4. The particles are always moving. As the particles gain energy, they move faster.
5. The particles in a substance are attracted to one another. The strength of the attractive force depends on the type of particle.

Particles in Solids

According to the particle theory, we can think of solids as being made up of particles that are tightly packed together, like bees in a hive. This way of thinking about the particles of a solid can explain why solids are greatly affected by gravity. It explains why a solid will tumble toward the lowest surface when suspended in the air and then dropped. The particle theory suggests that the particles of a solid are so close together that they cannot move around freely — they can only vibrate.

Many solids can be ground into such small pieces that they can slip past each other when they are poured out of their containers. Sugar, salt, flour, powdered cleansers and detergents, and many other crystals and powders that we use every day are examples of solids that can be poured. However, according to the particle theory, each tiny fragment of these solids contains billions of even smaller particles that are tightly packed together. Thus, each tiny fragment is also like a miniature solid in itself. This explains why solids form a pile when they are poured and why they do not keep flowing apart from each other. You observed this behaviour of solids in the previous At Home Activity.

Particles in Liquids

Again, according to the particle theory, the particles that make up liquids have enough energy to pull away from each other and slide around each other, while at the same time vibrating close together in small clusters. Another way to think about what is happening on the level of the particles is to imagine groups of guests talking and dancing at a party. The party guests can move around by shifting as a group, or by flowing in between the other groups of partygoers. Similarly, liquid particles can slip past each other. Unlike the particles in solids, they do not form rigid clumps. As a result, the particles of a liquid cannot hold their shape; instead, they fill a container and take the shape of that container.

As in solids, liquid particles are so tightly packed together that they are easily affected by the downward pull of gravity. Therefore, liquids always flow to the lowest possible level, like the water flowing over a waterfall (see Figure 4.2). As well, liquids form a level surface when they are at rest. Some foods, like chocolate and ice cream, can be melted to form a gooey liquid. Many other substances, such as water, oil, syrup, and perfume, are liquids at room temperature.

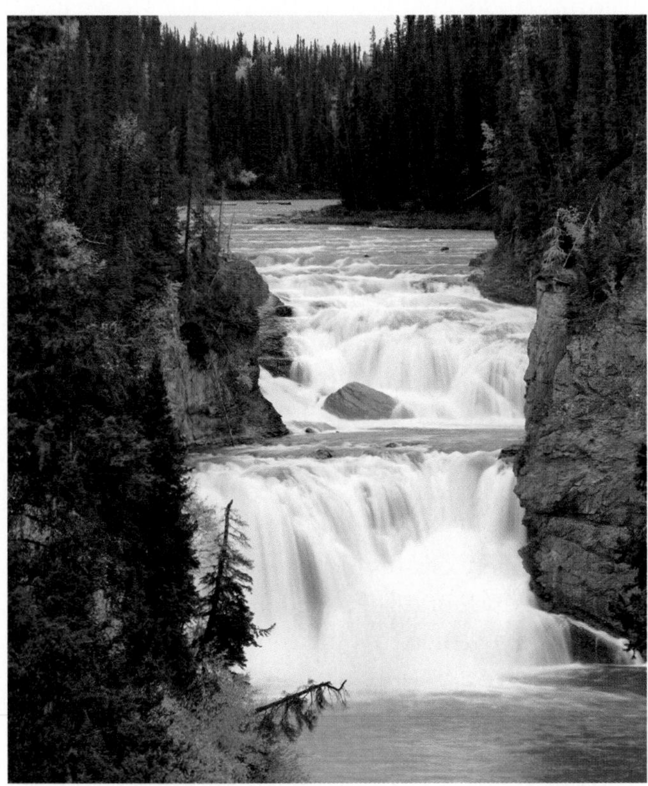

Figure 4.2 Liquids always flow to the lowest point possible.

Particles in Gases

All liquids can be transformed into their gaseous state when the liquids are heated. Many substances are gases at room temperature; for example, the air around you is a gas. According to the particle theory, gas particles are so far apart from each other that there is an enormous amount of empty space between them. Imagine that you and a friend are as far apart from each other as possible in a baseball stadium, and no one else is there. This is similar to what it would be like to be a gas particle. In fact, the particle theory explains that most gases seem invisible to you because you are observing mostly empty space. The particle theory also explains why gas particles have no difficulty moving past each other, and why they flow very easily.

Find Out **ACTIVITY**

Magic Mud

How thick is too thick? Can a liquid become so viscous that it can no longer flow?

What You Need

75 mL cornstarch
45 mL water
food colouring
mug
sturdy spoon

What to Do

1. Spoon the cornstarch into the mug.
2. Add the water and stir. When the water and the cornstarch are completely mixed, the Magic Mud mixture should be difficult to stir.
3. Mix in a few drops of food colouring and stir. (If the mixture becomes runny, add a little more cornstarch.)
4. Squeeze some of the Magic Mud between your fingers. Describe what happens when you stop squeezing. Try rolling the Magic Mud into a ball, a sausage, or any other shape you choose. What happens to the shape?

What Did You Find Out?

1. Is Magic Mud a liquid or a solid? Decide how to classify it, and record your answer. What makes it more like one than the other?
2. Wash your hands after this activity.

The particle theory suggests that gas particles are so free to move that they do so in every direction, and they have a great deal of energy to move extremely far apart. Therefore, gas particles spread out so much that in a brief time, they fill up the space of an entire container or room. For this reason, gases, like liquids, take on the shape of the container in which they are sealed. However, gases do not flow to the lowest possible level as do liquids. Because gas particles are not clustered or packed tightly together, the energy of gas particles allows them to move in all directions, sometimes against gravity, and to remain suspended. For example, water vapour forms clouds that float in the sky. Unlike what happens to liquids, when the lid is taken off a container of gas, the gas particles will start to spread apart again, until they have filled the entire room or building. The particle theory can be used to explain how gases always occupy all the space that they can fill — up, down, or sideways.

The distance between yourself and a friend at opposite sides of a stadium can be likened to the distance between gas particles.

You have learned that the behaviour of gas particles as they spread out in all directions is known as diffusion. Diffusion is the movement of particles from an area of high concentration to an area of low concentration of the same substance. For example, the particles in perfume vapour begin to move away from the perfume bottle (where there is a high concentration of perfume particles) when it is opened, until the perfume can be smelled across the room (where there is a low concentration of perfume). As you know, diffusion can occur in liquids as well. You have probably seen cream added to coffee. In seconds, the coffee and cream are mixed together — even without stirring. Stirring with a spoon simply allows the cream to diffuse more quickly.

Solids, liquids, and gases are not the most common forms of matter in the universe. The most common form of matter exists in a fluid state called *plasma*. Plasma is a gaslike mixture of positively and negatively charged particles. When matter is heated to extremely high temperatures, the particles begin to collide violently and to break apart into smaller particles that can conduct electricity. Glowing plasmas occur naturally, for example, in stars such as the Sun, and in lightning.

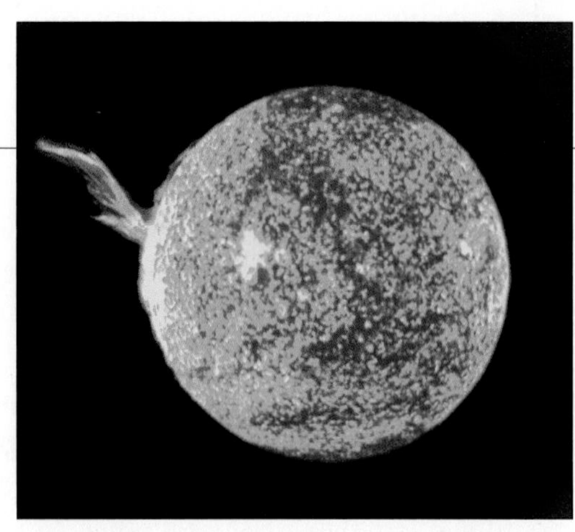

Changes of State

As you may recall from earlier studies, the process of melting is an example of a **change of state**, which occurs when the physical state of a substance is transformed into another state. Figure 4.3 shows changes of state. The change from solid to liquid is called **melting**, and the change from liquid to gas is called **vaporization**. These changes of state occur when the substance is heated and the particles of the substance gain energy. If you were to cool the substance, the reverse changes of state would occur because the particles lose energy. The change from gas to liquid is called **condensation**, and the change from liquid to solid is called **freezing**.

An unusual change of state occurs when a solid turns into its gaseous state without going through the liquid state. This change of state is known as **sublimation**. An example of sublimation occurs when dry ice is used at a rock concert — a chunk of frozen carbon dioxide (a solid) gains energy and gives off a thick cloud of fog (carbon dioxide gas). Figure 4.4 shows this change of state. The change from a gas directly to a solid is called sublimation as well. An example of this occurs when frost forms on windows on bitterly cold days (water vapour in the air loses energy rapidly and forms snowy ice).

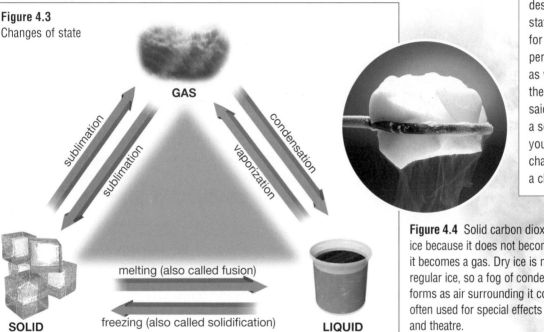

Figure 4.3
Changes of state

GAS
sublimation
sublimation
condensation
vaporization
melting (also called fusion)
freezing (also called solidification)
SOLID
LIQUID

Figure 4.4 Solid carbon dioxide is called dry ice because it does not become a liquid before it becomes a gas. Dry ice is much colder than regular ice, so a fog of condensed water vapour forms as air surrounding it cools. This result is often used for special effects in television, film, and theatre.

Pause& Reflect

How do raindrops form? If clouds contain water vapour and remain suspended in the sky, try to explain what you think happens inside the clouds to form rain. Using the word "condensation," write an explanation in your Science Log. You may draw a sketch to illustrate your answer.

Word CONNECT

What if you wanted to describe the change of state from liquid to solid for a metal, such as copper, or a non-metal, such as wax? Neither one of these substances can be said to "freeze" back into a solid. What verb would you use to describe this change of state? Look for a clue in Figure 4.3.

DidYouKnow?

What is the difference between a gas and a vapour? A substance is called a gas if it exists as a gas at room temperature (for example, oxygen gas or carbon dioxide gas). The same substance is called a vapour if it normally exists as a solid or a liquid at room temperature (for example, water vapour or perfume vapour).

Off the Wall

Did you know that science fiction script writers often feature changes of state in imaginary ways? "Morphing" has become one of the most popular special effects developed for science fiction movies. By means of computer-generated graphics, characters on screen appear to metamorphose, or morph, into someone or something else. In some movies, for example, solid beings appear to morph into a liquid that can creep under doors, or slip through cracks, and then quickly resolidify. Morphing has also become a popular feature of music videos, television programs, and commercials.

Evaporation is slow vaporization. It occurs over a wide range of temperatures. A wet towel will dry even if the air temperature is not high. On a cool day it will simply take longer for the water to evaporate from the towel.

Boiling is rapid vaporization. It occurs at a specific temperature, called the **boiling point**. The boiling point of water is 100°C (at sea level). Similarly, every substance has its own **freezing point** and **melting point**. The freezing point of water, for example, is 0°C (at sea level). This is the temperature at which liquid water freezes. It is also the temperature at which ice melts — its melting point. Figure 4.5 illustrates the melting point of two other substances: paraffin (wax) and silver. The pictures show how the properties of a substance, such as the melting point, can have valuable uses. When normally solid substances are melted, the liquid can be poured into moulds of various shapes. When the substance is cooled, it solidifies and takes the shape of the mould. The result can be a wax candle, a silver teapot, and many other products. The processes pictured here make good use of flowing liquids and viscosity.

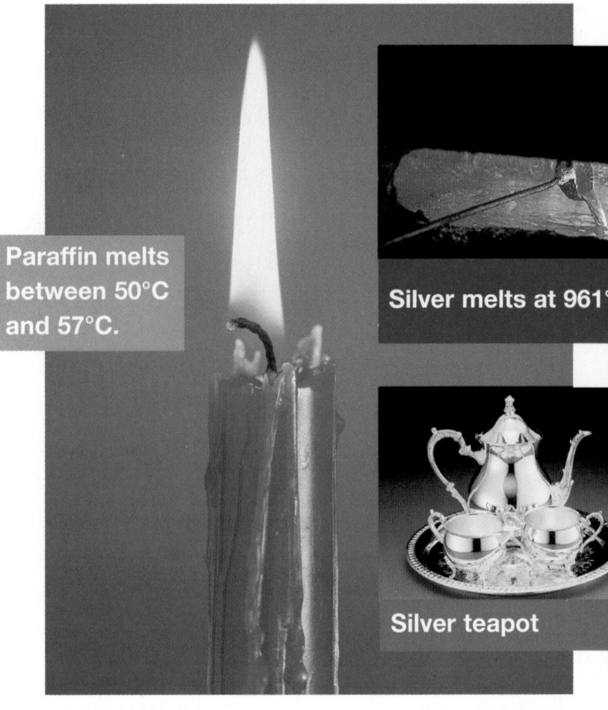
Paraffin melts between 50°C and 57°C.
Silver melts at 961°
Silver teapot

Figure 4.5 Every substance has its own melting point.

Check Your Understanding

1. In your own words, what does "to flow" mean?

2. Classify the following items as fluids or as non-fluids.

shampoo	pencil	honey	thumbtacks	hairspray
thread	sap	paper	nail polish	balloon
blood	air	molten lava	smoke	perfume
sugar	natural gas	ash	gravel	snow

3. How could you test whether or not a substance is a fluid?

4. A substance has a definite volume but an indefinite shape. Is the substance a solid, a liquid, or a gas?

5. **Apply** Use the particle theory to explain why ice cubes form in your freezer.

6. **Thinking Critically** Describe a place or a situation in which you could find water as a solid, a liquid, and a gas all at the same time.

7. **Thinking Critically** Which substances could you use to demonstrate "morphing"? Are these substances fluids? For each substance, explain your answer.

Word CONNECT

In your Science Log, write a short story that involves "morphing." Include an illustration of the morphing process that occurs in your story.

4.2 How Fast Do Fluids Flow?

Pause& Reflect

In your Science Log, describe what it is like to blow bubbles in different kinds of drinks. Use the word "viscous" to explain why some drinks bubble more easily than others.

Some liquids can flow faster than others. Orange juice flows freely from a jug or a carton, but how would you describe the flow of chocolate syrup from a bottle or the flow of honey from a jar? Look at the fluids shown above. Predict which fluid will flow to the end of the pan first. Then predict in which order the other fluids will flow to the end of the pan.

If you wanted to know how fast you could run, you might ask a friend to time how long it would take you to run, say, 100 m. This measurement would be your speed. In a similar way, you could measure how fast a fluid "runs." You would measure the time it takes for the fluid to flow from one point to another (its distance). This characteristic is called the fluid's **flow rate**.

Just as certain variables can slow you down in a race, certain variables can also slow down the flow of fluids. What property of fluids might cause a liquid to slow down? Explore this question further in the next investigation.

Figure 4.6 Donovan Bailey of Canada won the gold medal for the 100 m dash in the 1996 Summer Olympic Games. What factors, or variables, might reduce a runner's speed?

INTERNET CONNECT

www.school.mcgrawhill.ca/resources/

Go to the above web site to find out Donovan Bailey's time when he won the gold medal for the 100 m dash at the 1996 Summer Olympics in Atlanta, Georgia. Go to **Science Resources**, then to **SCIENCEPOWER 8** to know where to go next. What was Bailey's speed? Has anyone beat his speed? If so, by how much has his speed been improved? Who beat his speed? When and where did this occur?

CONDUCT AN

INVESTIGATION 4-A

The Flow Rate of Liquids

Viscosity is a difficult property to measure directly. However, the flow rate of a liquid is a good indicator of viscosity. You can determine how fast a fluid flows by measuring the amount of time it takes for a certain amount of the fluid to flow past a specific point.

Problem

How can you compare the flow rates of various liquids?

Safety Precautions

- Keep your hands away from your face and mouth. Do not eat or drink any substances in the science laboratory.
- Wipe any spills immediately. Do not leave floors wet.
- Dispose of materials properly, according to your teacher's instructions.

Apparatus

ramp made of smooth plastic or glass (minimum 0.5 m × 0.3 m)
stack of books (0.25 m – 0.3 m high)
thermometer
measuring spoon (15 mL) with rounded bottom
watch with second hand or stopwatch
rubber gloves

Materials

waterproof marker or wax pencil
tape
paper towels
soap for cleaning ramp surface
water
15 mL of any two of the following liquids (at room temperature):

cooking oil, honey
motor oil, liquid detergent
molasses, ethyl alcohol
corn syrup

Procedure

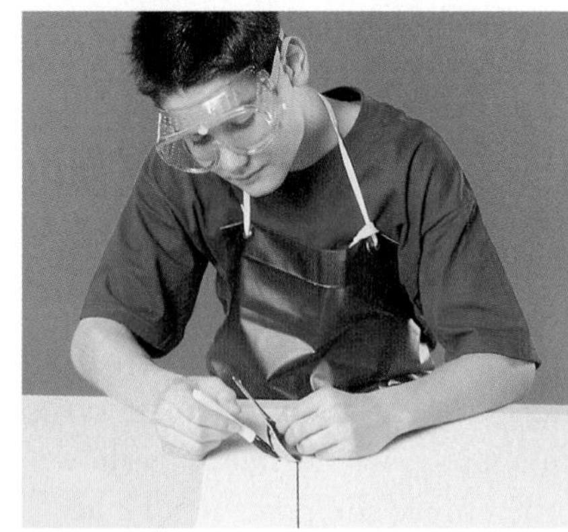

1. Draw a line across the width of the ramp, approximately 10 cm from the top. Draw a dot in the centre of the line and just above it. When you place the measuring spoon on the dot and rock the spoon forward, the lip of the spoon will touch the top line. This is the start mark.

2. Draw a finish line 10 cm below the top line. Then assemble the apparatus as shown here.
   (a) Measure the temperature of the room and record it in your notebook.
   (b) Make a data table with the following column headings:
   Liquid [water plus 2 others]
   Time (s)
   Flow rate (cm/s)
   Ranked flow rate
   Ranked viscosity
   (c) Hypothesize which of the three liquids will flow fastest.

Skill POWER

To review how to make a data table, turn to page 546.

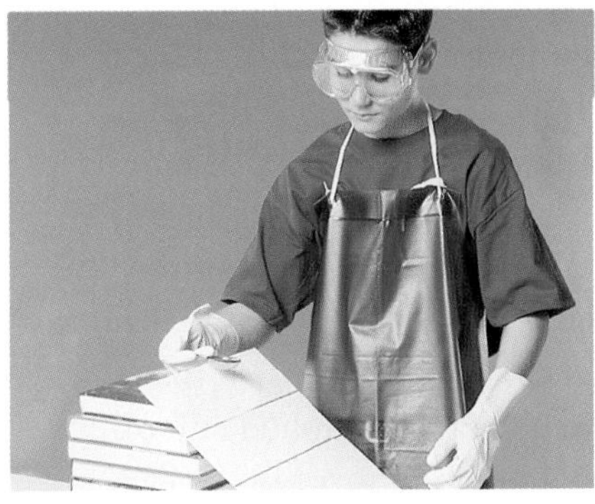

3 Test one liquid at a time, starting with water. The "spoon student" will pour enough water into the clean, dry measuring spoon so that it is level, then place the spoon at the start mark on the ramp.

(a) Another group member, the timer, will work the stopwatch.

(b) A third group member, the marshall, will say "go" and will call "stop" when the liquid reaches the finish line.

4 When the marshall says "go," the spoon student rocks the spoon quickly but carefully to pour its contents down the ramp. At the same time, the timer starts timing.

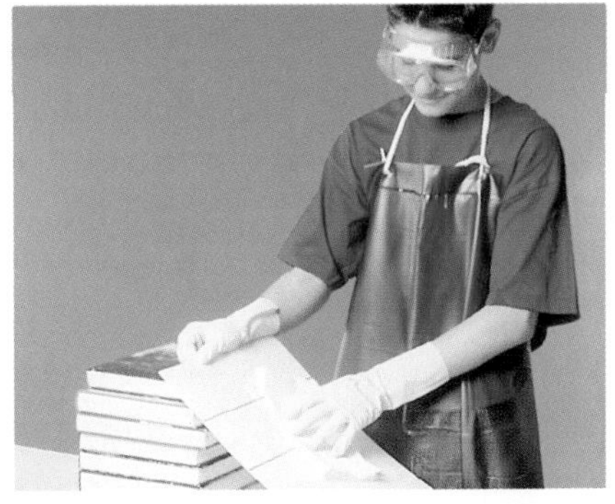

5 When the marshall says "stop," the timer stops timing and records the time in the data table. Students clean and dry the ramp and the measuring spoon to get ready for the next liquid.

6 Repeat steps 3 to 5 for the other two liquids. (Do this at least twice for each liquid; note that alcohol should be tested last because it might erase the start and finish lines.)

(a) Wash your hands after this investigation.

(b) Keep the flow rate ramp assembled so you can do the Find Out Activity on page 121.

Analyze

1. Determine the flow rate (in cm/s) for each substance. Do this by dividing the distance travelled (10 cm) by the time recorded for each substance (in seconds). Record each result in your data table.
2. Rank the liquids from fastest flow rate (1) to slowest flow rate (3). Record these rankings in the fourth column of your data table ("Ranked flow rate"). Was your hypothesis correct?
3. Rank the viscosities in the table from lowest (1) to highest (3). Record these values in your data table under "Ranked viscosity."
4. Describe two sources of error that might affect your results. Are these errors due to the equipment or to human factors? How could you reduce or eliminate these errors?

Conclude and Apply

5. Compile the class data in a stem-and-leaf plot. From your results, what is the median flow rate of the liquids you tested? How does it compare with the mean (average) value?
6. How is the flow rate of a liquid related to its viscosity?
7. Which liquids were more difficult than others to measure with the viscosity ramp? What could you have done to the ramp to make it easier to measure these liquids?
8. Make a bar graph showing flow rate (in cm/s) along the vertical axis (y-axis), and the various liquids along the horizontal axis (x-axis). Plot the data for each liquid on this graph, using a different colour for each liquid. Include a legend on your graph.

Skill POWER

To review how to make a stem-and-leaf plot, turn to page 546.

Figure 4.7 Paint, varnishes, and other such liquids must have just the right viscosity.

Applications of Viscosity in Liquids

Why might it be important to know how to determine the flow rate and, therefore, the viscosity of a liquid? The viscosity of liquids is an important property that must be measured precisely in some industries. For example, the viscosity of paints, varnishes, and similar household products is closely regulated so that the paints and varnishes can be applied smoothly and evenly with a brush or a roller. In fact, antique dealers and many householders are glad that furniture stripping liquid has finally been thickened. Previously, this thin, smelly liquid was difficult to use because it tended to run down and off the furniture before it had a chance to remove old paint and finishes. Now, however, the viscosity has been increased to produce an almost gel-like texture, so that the product is easier to apply and sticks well to the surface of the furniture.

The viscosity of some medications, such as the various liquids used to remove warts, has also been modified for easier application. Drug companies manufacture medicines, such as cough syrup, that have a high viscosity yet are still drinkable, in order to coat and soothe the throat.

People in many occupations need to know how to adjust the viscosity of a substance to suit specific applications. For example, chefs need to know how to make gravies thinner than sauces and frostings thicker than icings. Mechanics must choose an engine oil that is the right viscosity for the season. Artists need to know how to thin or thicken oil paints or acrylics. Technicians must control the viscosity of various chemicals in chemical processing plants.

Pause& Reflect

Motor oils are manufactured in a variety of viscosities to suit weather conditions. The viscosity of a motor oil decreases as it is heated and increases as it cools. Predict when a mechanic would choose to put a higher-viscosity oil in a vehicle, and write your prediction in your Science Log. Then visit a service garage or a hardware store and find out under what conditions specific viscosities of motor oil are used.

Product Appeal and Viscosity

Your mouth is highly sensitive to viscosity, so food manufacturers ensure that ice-cream toppings, pasta sauces, soups, gravies, salad dressings, and other products are just the right consistency (thickness) to suit consumers' tastes. Food manufacturers must also know how to regulate the effect of heat on the viscosity of a substance.

Figure 4.8 Viscosity affects how foods taste.

For example, chocolate coating for candy bars must be at precisely the right consistency and temperature in order to cover the bar completely with the same amount of chocolate each time. Some candy coatings are especially sensitive to temperature. If the candy were to stay too hot for too long, it might become so hard that it would be unpleasant to the bite. The candy would then have to be remade.

Figure 4.9 Heating the sap from sugar maple trees changes its viscosity.

Canada's maple syrup industry depends on controlling the viscosity of a liquid. Maple syrup comes from the sap of sugar maple trees. The sap itself is thin, runny, and not sweet to the taste. However, when the sap is heated and most of the water has evaporated, the result is a thick, sweet, and flavourful syrup, irresistible on pancakes and waffles!

Career CONNECT

Viscosity at Work

As a class, brainstorm how fluid viscosity might be used in each of these occupations: candy maker; glass blower; beekeeper; baker; motor mechanic; maple syrup manufacturer. Contact someone in your community who works in one of these occupations, and ask if you might job shadow him or her for half a day. (When you job shadow someone, you observe and assist the person at work.) Take notes on what you learn about the role of viscosity in the job. Present an in-class report describing your experience, and write a letter to thank the person you visited.

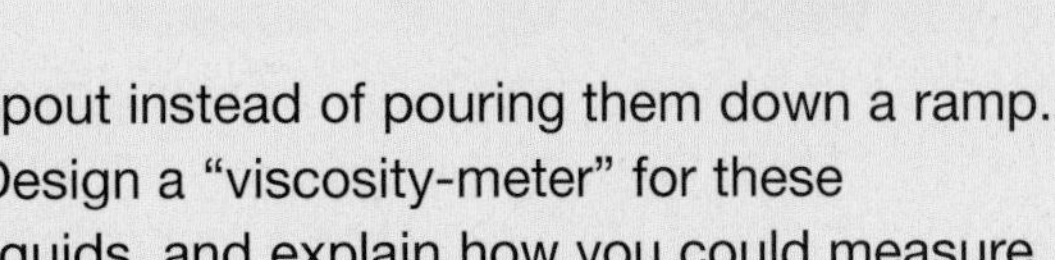

Cool It!

Go back to Conduct an Investigation 4-A on page 118. Make a hypothesis stating the effect that temperature has on the flow rate of a liquid. Cool the liquids to exactly the same low temperature and repeat the flow rate procedure.

What Did You Find Out?

1. What differences in flow rate did you observe?
2. How can the particle theory be used to predict the effect of temperature on a liquid's flow rate?

Extensions

3. It might be easier to measure the flow rate of low-viscosity liquids by dripping them from a spout instead of pouring them down a ramp. Design a "viscosity-meter" for these liquids, and explain how you could measure flow rate using your new apparatus.
4. It is common practice for scientists to repeat an experiment two or three times to reduce the effect of any errors during the procedure. Each repetition is called a *trial*. The average of three or four trials is reported as the final result. Repeat this activity twice so that you have three trials for each liquid. Report the flow rate as the average flow rate for each liquid.

Find Out **ACTIVITY**

The Value of Viscosity

Have you ever bought a product such as a milk-shake and wished it were thicker? Did you ever buy a product such as nail polish that, over time, became too thick to be useful? Can the viscosity of a product make it more valuable to you? Would you pay more for just the right viscosity in a product?

What to Do

1. With a group of classmates, choose a product from the following list:

hair conditioner	ice cream topping
shampoo	hand lotion
water-based paint	liquid detergent
yogurt	face lotion
salad dressing	ketchup
nail polish	liquid bubble bath

   Then obtain three or four different brands of this product, making sure that the viscosities are obviously different (for example, runny salad dressings versus thick salad dressings).

2. Make a table for the different brands of your product. Write the following headings:

Name brand of product	Cost ($)	Volume (mL)	Cost per volume (¢/mL)	Relative viscosity (high, medium, low)

3. Give your table a suitable title. Complete the table, filling in the information for each product that you examine. (If a difference in viscosities is not immediately obvious, do a flow rate test to determine relative, or comparative, viscosities. Wear an apron and safety goggles if you do this test.) Wash your hands after this activity.

What Did You Find Out?

1. Is there a relationship between cost per volume and viscosity? If so, describe it. How is viscosity related to the product's usefulness?

2. As a class, list three products that are more useful because their viscosity is high and three products that are more useful because their viscosity is low.

Check Your Understanding

1. List two substances that have a low viscosity and two substances that have a high viscosity.
2. What is the relationship between the viscosity of a liquid and its flow rate?
3. How can you test the viscosity of a liquid?
4. (a) What is the effect of temperature on the flow rate — and thus the viscosity — of a liquid?
   (b) Explain this effect using the particle theory.
5. **Apply** Asphalt is the black, sticky material that binds gravel in the pavement that covers streets and highways. Explain why paving is almost always done during the summer months.

Math CONNECT

Go to the mathematics books you are using to find out how to calculate cost per volume.

4.3 Why Viscosity Varies

Imagine that you and a group of friends are moving through a crowded shopping mall. Could you make your way through the crowd more quickly if your group were small rather than large? Small groups can move through a crowd more quickly than large groups because they can fit into the empty spaces between other groups more easily. This scenario is shown in Figure 4.10A. In a similar way, the particle theory suggests that small particles can move past each other more easily than large particles can. Large particles take up more space.

Figure 4.10A Small groups of people can move through crowds more easily than large groups. Similarly, small particles can flow more easily than large particles.

Also imagine that you, your friends, and everyone in the shopping mall are carrying huge parcels and knapsacks, as shown in Figure 4.10B. It would be even more difficult to squeeze through the crowd because everyone's bulkiness would be taking up even more space. Similarly, some particles are bulkier than others because of their shape. For example, oil particles are bulkier than water particles. Thus, oil is more viscous than water.

Figure 4.10B Bulkiness can slow down the flow of pedestrian traffic in a shopping mall. Similarly, bulky particles flow past each other more slowly than less bulky ones.

Pause& Reflect

Each one of the three cartoons represents a fluid having difficulty moving because of the nature of its particles. As a class, act out each of these scenarios. In your Science Log, describe what happens.

Next, imagine that everyone in the mall is wearing shoes made entirely of Velcro™. Every time you would try to walk by someone, your shoes would stick to theirs. With every step, you would have to stop and pull your foot away from another person's foot. No matter how large or small the people in the crowd, it would take a very long time to walk through the mall. In a manner similar to people wearing Velcro™ shoes, all particles attract each other. However, certain types of particles attract and hold on to each other more tightly than other types. It is very hard for these particles to flow past each other, so they do so very slowly. The strength of the attraction that particles have for each other is the most important factor in determining a fluid's viscosity.

Finally, imagine that everyone in the shopping mall is moving very, very slowly and will not move out of the way as you and your friends try to pass. In this situation, shown in Figure 4.10C, it would take an even longer time to move through the crowd. This is similar to what happens to particles when they are cooled — they slow down.

Even though all fluids flow smoothly, they flow at different rates because they have different viscosities. Another way to define viscosity is **resistance to flow**. Resistance to flow means that the particles can move around, but it may be difficult for them to pass by each other; this resistance generates **internal friction**. Another word for friction is "rubbing." It is easier to skate on ice than on pavement because the friction between the skates and the ice is less than the friction between skates and pavement. Similarly, it is easier for some fluid particles to move past each other, compared to other fluid particles.

A liquid's ability to flow also depends on the energy that the particles have to move around. The warmer the liquid becomes, the more energy the particles have to move out of the way and make room for other particles to pass. However, as the temperature drops, the particles have less energy to move around, so the empty spaces between them get smaller and smaller. In general, a fluid's viscosity *decreases as the fluid is heated* and *increases as the fluid is cooled*.

Figure 4.10C It is hardest to move through a crowd of people who are moving very, very slowly. When particles are cooled, they take much longer to flow around each other because all the particles have slowed down.

INVESTIGATION 4-B

Flowing Fluid Floods City

Think About It

You may have heard the expression "as slow as molasses in January." Read the following true story and then answer the questions.

January 15, 1919, was an unusually warm day. The fine weather lured the citizens of Boston, Massachusetts, outside to enjoy the springlike temperatures. It hardly seemed like the setting for a disaster.

The workers in Boston's industrial North End were enjoying lunch and the pleasant weather. Suddenly, they heard a low rumbling and then an explosive crack. A 30-m wide cast-iron tank, standing 15 m above street level on the property of the United States Industrial Alcohol Company, burst apart! Like lava spilling from a volcano, crude molasses flowed into the street. The result was a "flash flood" consisting of 10 million litres of sweet, sticky, deadly goo.

The "wall of molasses" — some witnesses say as high as 5 m — poured through the streets at a speed of almost 60 km/h. It demolished buildings, ripping them off their foundations. It flipped vehicles over and buried horses. People tried to outrun the gooey tidal wave, but they were overtaken and either hurled against solid objects, or drowned where they fell. Within minutes, 21 people were killed and more than 150 injured.

The clean-up took weeks. Lawsuits were filed against the United States Industrial Alcohol Company, charging it with negligence. After six years, the court made a final ruling against the Company. The court's findings showed that the tank had been overfilled and that it was not properly reinforced. The United States Industrial Alcohol Company had to pay more than one million dollars in damages.

Analyze

1. Gather some clues from the story:
   (a) What was the date?
   (b) What was unusual about the weather?
   (c) What was the first clue that something disastrous was about to happen?
   (d) How fast did the molasses pour out of the tank?
   (e) Who was accused of being responsible for the accident?
2. Use the particle theory to explain why the tank burst.
3. Energy is reponsible for making things move. Use the particle theory to try to account for why something as viscous as molasses could move as quickly as it did on that particular day, at that particular moment.

Critical Thinking

4. For what purpose do you think the company used the molasses?

Helium is one of the strangest substances in the universe. Although helium is commonly used in its gaseous state, it has unusual properties as a liquid. At a few degrees above absolute zero (about -270°C), helium changes from a "normal" fluid to what is known as a superfluid, a substance that can flow without friction. Superfluids have zero viscosity and no internal resistance to flow, which allows them to pass through a hole of any size and to flow up and over the sides of containers. Scientists are researching future uses of superfluids.

What Factors Affect the Viscosity of Gases?

Although gases, in general, flow much more easily than liquids, the viscosity of gases can vary, too. The size and shape (or bulkiness) of gas particles and temperature are factors that affect the viscosity of gases as well as the viscosity of liquids. Just like liquid particles, as gas particles get larger and bulkier, their viscosity increases.

The effect of temperature on gas particles is different, however, from the effect on liquid particles. While an increase in temperature reduces viscosity in liquids, the opposite is true for gases. Why? Gas particles do not depend on an increase in energy (a rise in temperature) to move farther apart, as is the case for liquids. The particle theory suggests that gas particles are already very far apart. Extra energy increases the internal friction of gas particles because the particles speed up and collide with each other more frequently. Cooler temperatures in gases keep internal friction (and therefore viscosity) low.

Summary

Now that you have investigated the property of viscosity, you can distinguish clearly between a fluid and a non-fluid. You can also explain, using the particle theory, why liquids and gases flow, but solids do not. Finally, you have learned that adjusting factors such as temperature can change the viscosity of a fluid. The ability to change the viscosity of fluids has many applications in manufacturing, health care, food preparation, and other areas.

Check Your Understanding

1. Define internal friction.
2. Name three factors that affect the internal friction of a fluid.
3. List two ways to (a) increase, and (b) decrease the viscosity of a liquid.
4. **Apply** Keeping in mind the factors that can affect viscosity, design a container for an imaginary liquid that is half as viscous as water and that must be used in small quantities during emergency surgery.
5. **Design Your Own** Bubble gum can be thought of as a liquid with a high viscosity. Design an experiment to determine the best way to remove a wad of gum from the sole of a shoe.
6. **Thinking Critically** Examine the ingredients of two different brands of the same product that have different viscosities. Research which ingredients are responsible for the product's viscosity.

CHAPTER at a glance

Now that you have completed this chapter, try to do the following. If you cannot, go back to the sections indicated.

Define the term "fluid." (4.1)

Use the particle theory to explain why liquids and gases are fluids, but solids are not. (4.1)

Give some characteristics of liquids and solids that are the same. What are some characteristics that are different? (4.1)

Explain how liquids and gases are similar. How are they different? (4.1)

Gas

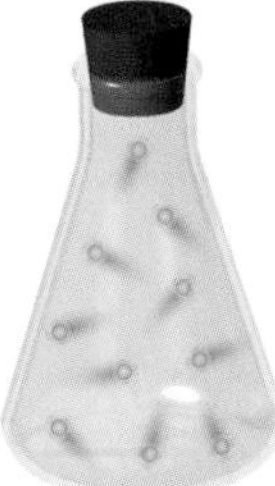

Liquid

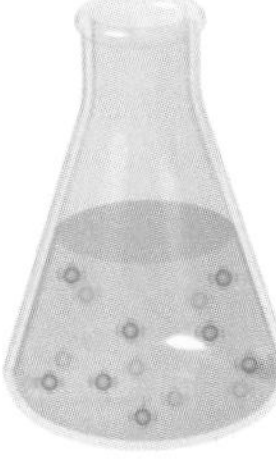

Solid

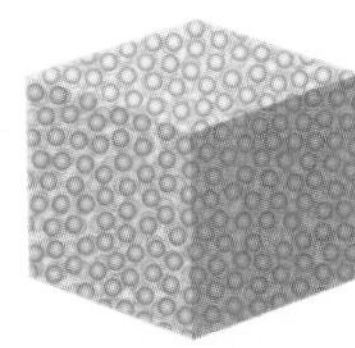

Describe how to measure viscosity. (4.2)

Explain how to determine the flow rate of a liquid. (4.2)

Name some industries in which measuring and controlling viscosity are important. (4.2)

Name some jobs in which workers regulate the viscosity of substances. (4.2)

Summarize the main factors that affect the viscosity of liquids and gases. (4.3)

Use the particle theory to explain how each factor affects viscosity. (4.3)

Explain why the effect of temperature on the viscosity of gases differs from the effect of temperature on the viscosity of liquids. (4.3)

Prepare Your Own Summary

Summarize this chapter by doing one of the following. Use a graphic organizer (such as a concept map), produce a poster, or write the summary to include the key chapter ideas. Here are a few ideas to use as a guide:

- Why do some liquids kept in the refrigerator flow more slowly than liquids kept in a cupboard?
- What are some ways to change the viscosity of a liquid?
- How can the particle theory explain how fluids behave?
- How are a liquid's flow rate and its viscosity related?
- How can viscosity be used to improve the performance of products such as motor oil and medications?
- Why is oil more viscous than water?

- Why are warm gases more viscous than cool ones?
- Which everyday household products do you prefer to have a high viscosity? A low viscosity? Why?

CHAPTER 4 Review

Key Terms

viscosity
fluid
solid
liquid
gas
change of state
melting
vaporization
condensation
freezing
sublimation
evaporation
boiling
boiling point
freezing point
melting point
flow rate
resistance to flow
internal friction

Reviewing Key Terms

If you need to review, the section numbers show you where these terms were introduced.

1. Use the key terms to complete the following statements. Do not write in the textbook.
   - **(a)** ________ chocolate causes it to flow. (4.1)
   - **(b)** A ________ always has an even surface. (4.1)
   - **(c)** A ________ can be poured only if it is ground into a fine powder. (4.1)
   - **(d)** ________ occurs when dishes "air dry" on a rack. This change of state is called ________. (4.1)
   - **(e)** The particles of a ________ are spread extremely far apart. (4.1)
   - **(f)** Liquid water can change into a gas at its ________ (100°C). (4.1)
   - **(g)** A fluid's resistance to flow is known as its ________; a good way to measure this property is by determining the ________ of a liquid. (4.2, 4.3)
   - **(h)** There is ________ between the particles of a fluid that causes viscosity. (4.3)

Understanding Key Ideas

Section numbers are provided if you need to review.

2. How do you distinguish between a fluid and a non-fluid? (4.1)
3. Compare and contrast liquids and gases. (4.1)
4. How can you demonstrate that finely ground solids are not fluids? (4.1)
5. What is the relationship between a liquid's viscosity and its flow rate? (4.2)
6. Some foods taste better if their viscosity is high; others taste better if their viscosity is relatively low. Name two foods that you would place in each category. (4.2)
7. What is the relationship between viscosity and internal friction? (4.3)
8. How do the size and the shape of the particles of a fluid affect internal friction? (4.3)
9. Use the particle theory to explain why and how temperature affects the viscosity of a liquid. You may use a sketch in your answer. (4.3)

Developing Skills

10. Copy and complete the following concept map using key ideas you have learned in this chapter.

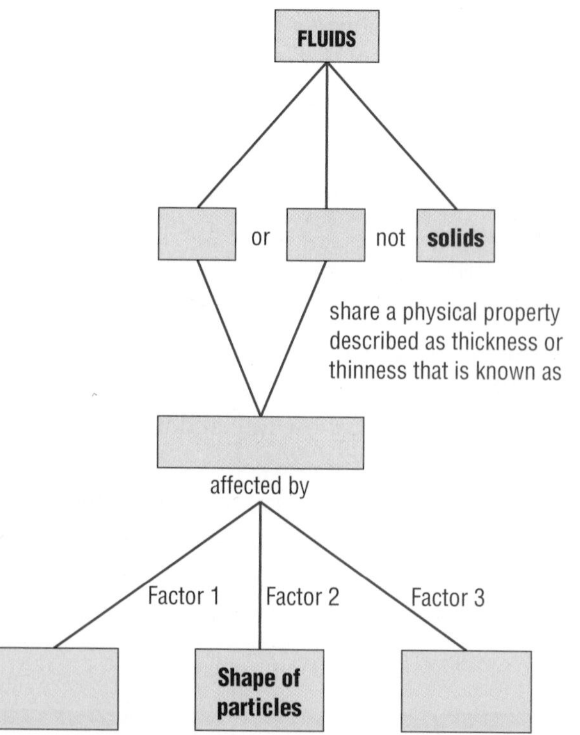

11. Make a bar graph for the following data, with "Flow rate" along the vertical axis (y-axis) and "Temperature" along the horizontal axis (x-axis).

(a) Which substance is the most viscous? Which one is the least viscous?

Substance	Flow rate at 10°C (cm/s)	Flow rate at 25°C (cm/s)	Flow rate at 50°C (cm/s)
1	2.0	4.0	9.0
2	13.0	13.0	14.0
3	0.0	0.0	2.0
4	5.0	8.0	13.0
5	0.0	1.0	4.0

(b) Which substance is a solid at room temperature (about 20°C)?

(c) Infer why the values for substance 2 are so similar.

12. Suppose you were sent into a department store and told to group all the liquids according to viscosity. Explain how you would classify the liquids. List three items that you would place in each category.

13. Make a collage of the fluids that are used in one of the following industries:

cosmetics
cooking
construction
petroleum refining
graphic design

14. Formulate your own question related to viscosity, and design a fair test to explore possible answers.

Problem Solving/Applying

15. Design a simple test that a chef or a medical lab technician might use that could provide an estimate of the viscosity of a substance. Prepare a step-by-step procedure to explain how this test might be done and how the viscosity of a substance can be estimated as high, medium, or low. Which variable(s) would you keep constant in your test? Which variables would you change?

16. Design a time-keeping device based on the flow rate of a liquid.

Critical Thinking

17. Many products, such as paint, nail polish, and glue, can work effectively only when they have a certain viscosity. What do manufacturers do to the products' containers to help prevent any changes in viscosity? What are three things you can do once you have opened the product to keep its viscosity intact over time?

18. Sometimes substances are sensitive to heat treatment. For example, if they are heated for too long or to very high temperatures, they will become much thinner than they were originally. Keeping internal friction in mind, can you suggest how the substance might have changed at the particle level?

19. Various kinds of oil are transported to places all over the world. Name as many forms of transportation that you can think of that are used to move oil from one location to another. Make a list of any special considerations involved in the design of each method of transportation. Make another list of the environmental damage that might result in each case if an accident or a leak were to occur.

Pause& Reflect

1. Besides temperature and the shape and size of particles, think of another factor or factors that might affect the viscosity of a fluid.
2. Imagine you have been asked to create a milkshake for a well-known fast-food restaurant. How would you design a straw to match the viscosity of your new milkshake? What factors (variables) would you take into account?
3. Go back to the beginning of this chapter on page 108 and check your original answers to the Getting Ready questions. How has your thinking changed? How would you answer these questions now that you have investigated the topics in this chapter?

CHAPTER

5 Density and Buoyanc

Getting Ready...

- How is a fish like a submarine?
- How can a huge ocean liner built of steel float, while a penny placed in a glass of water sinks?
- Why is it easier to float in salt water than in fresh water?

Science Log

You may already have possible answers to these questions. Use illustrations to outline your ideas in your Science Log. As you learn about density and buoyancy in this chapter, write more detailed explanations beside your drawings.

A scuba diver exploring the dark and mysterious waters off the west coast of Canada might well come face-to-face with a Giant Pacific Octopus, the largest of its kind. This species of octopus is only one example of the large and unusual creatures that inhabit the ocean. How can animals with such huge bodies move so gracefully and so swiftly through the water? What enables completely submerged animals, such as octopuses, fish, and whales, to float at different depths? Why do objects such as icebergs, sailboats, ocean liners, and oil rigs float partially submerged on the surface?

In this chapter, you will learn about **buoyancy**, the tendency to rise or float in water or air. Buoyancy causes objects to float in both liquid and gas environments. The upward force exerted on objects submerged in fluids is called **buoyant force**. The buoyancy of a fluid is related to a fluid's **density** — the amount of mass in a certain unit volume of a substance. Read this chapter to learn more about buoyancy and density.

Spotlight On Key Ideas

In this chapter, you will discover

- how density and buoyancy determine whether an object sinks, rises, or floats when placed in a fluid
- that the particle theory can explain the relationship between mass, volume, and density
- whether density is related to viscosity
- how to recognize and state the relationship between gravity and buoyancy
- how different liquids alter the buoyant force on an object
- how an object's size affects the buoyant force acting on it

Spotlight On Key Skills

In this chapter, you will

- determine mass-to-volume ratios of different amounts of the same substance
- design and build your own device (called a hydrometer) for measuring density
- make a model of a diving device that can adjust its own depth in water
- state and test predictions about whether an object will sink or float when placed in water
- design your own experiment involving buoyant force

Starting Point 

Frolicking in Fluids

Many mammals, such as otters and dolphins, make their home in water environments. (Recall that mammals give birth to live young and feed their young with the mother's milk.) These animals glide gracefully and fluidly through water. How are their body features suited to a water environment?

What to Do

With a partner, answer the following questions:

1. Make a list of five mammals that live in a water environment.
2. What can the mammals do in water that they cannot do on land?
3. How are their bodies designed to suit activity in water?
4. Think of some design features for underwater equipment that are based on the body features of animals that inhabit water environments. How do you think these features improve the functioning of the equipment?

5.1 Density and the Particle Theory

Figure 5.1 There is very little space between the people on a crowded elevator — they are densely packed together.

Have you ever been on a crowded elevator? As Figure 5.1 shows, it is definitely uncomfortable when too many people are jammed together tightly, or densely, on an elevator.

Using everyday words, density can be described as the "crowdedness" of the particles that make up matter. In scientific terms, density is the amount of a substance that occupies a particular space. When you describe a substance as being "heavy" or "light," you are referring to the property of density.

According to the particle theory, different substances have different-sized particles. The size of the particles determines how many particles can "fit into" a given space. Therefore, each substance has its own unique density, based on particle size. Furthermore, the particle theory suggests that there is empty space between the particles of matter. Could as many people fit onto an elevator if each person were surrounded by a large "spacing box"? Would larger spaces among the people increase or reduce the density (crowdedness) of the people travelling on the elevator? The answer is shown in Figure 5.2.

Density of Solids, Liquids, and Gases

How is the density of a substance related to the substance's physical state? Imagine filling a film container with liquid water and another film container with water vapour. Both liquid water and water vapour are the same substance and therefore have particles of the same size. According to the particle theory, gas particles have more space between them than do liquid particles. Therefore, the water vapour in the container would have fewer particles than the liquid water. It would be reasonable to conclude that the density of the water vapour is less than the density of liquid water.

Skill POWER

For tips on how to make and use a Science Log, turn to page 534.

Pause& Reflect

Imagine you were to fill three balloons to the same size: one with plaster, which would harden solid; the second with water, a liquid; and the third with air, a mixture of gases. In your Science Log, draw these three balloons. Draw O's or X's to represent the particles and to show how the particles are spread out inside each balloon. Use the particle theory to try to explain how and why the density of the substances inside each balloon varies.

Figure 5.2 Increasing the spaces between people on an elevator will reduce the density.

DidYouKnow?

Could the solid state be less dense than the liquid state of the same substance? The answer is yes. You have probably seen ice cubes floating in a glass of water. As solid ice forms from liquid water, the ice *expands*. This means that there is more empty space trapped in a chunk of ice than in the same amount of water. Thus, solid ice is less dense than liquid water.

How are density and state of matter related to the physical properties of a substance? Solid objects can move easily through liquids and gases. For example, dolphins can leap through the air and then dive back underwater so smoothly that the activity appears almost effortless, as shown in Figure 5.3. According to the particle theory, the fluid properties of water and of air allow water particles and air particles to move out of the way of the firmer, non-fluid bodies of marine animals. Why do solid particles tend to hold together while fluid particles tend to move apart?

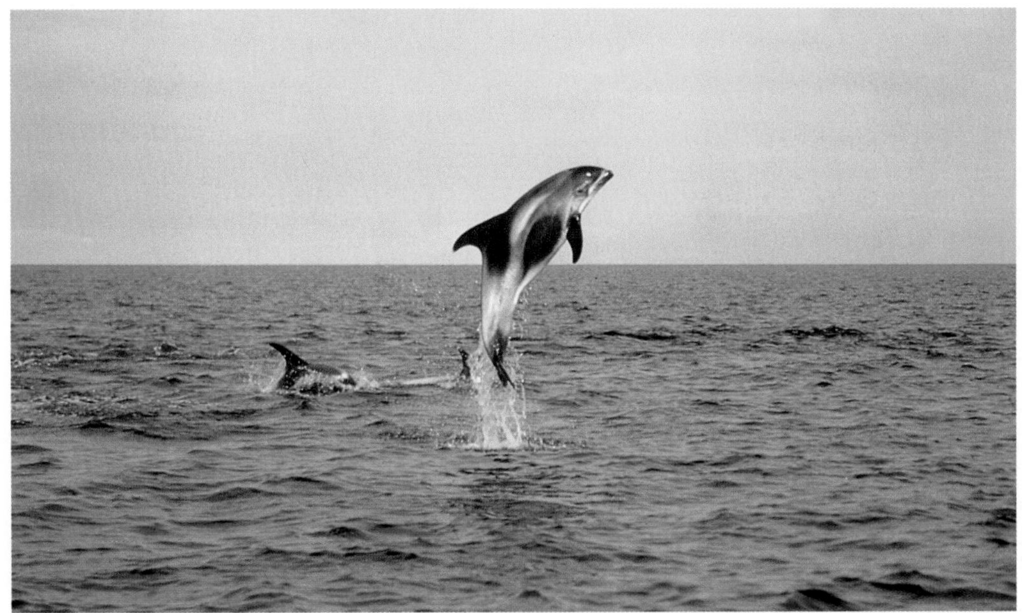

Figure 5.3 The particles of a fluid move apart easily when a solid object, such as a dolphin, travels through the fluid.

When an object moves through a fluid, it pushes particles apart and moves between them. Particles in a solid cannot be pushed apart. To understand why, imagine that you and a few friends are together. You want to prevent anyone else from pushing your group apart and moving between you. What would you do? First, you would have to stand quite close together. Then you would probably hold on to each other very tightly. If you do not let go of one another, no one can move between you. That is what particles in a solid do. Attractive forces among the particles of a solid are stronger than those between fluid particles and thus the particles in a solid cannot be pushed apart.

Figure 5.4 The cartoonized weight-lifters represent the tightly packed particles in the solid plate. The plate can hold its shape and support materials such as the large rock, because the weight-lifters are holding on to each other very tightly.

Find Out **ACTIVITY**

Differing Densities

Using your knowledge of the particle theory, what inferences can you make about the densities of different substances?

What You Need  

3-5 class sets of numbered film containers (prepared by your teacher), completely filled with various "mystery substances"

What to Do

1. Make a table with the following headings:

Container number	Rank order (heaviest to lightest)	Substance	
		(your guess)	(actual)

2. With a partner, pick up each film container and observe how heavy or how light each one feels. Decide on the order of the film containers from heaviest to lightest. In your table, record the numbers of the film containers, from heaviest to lightest.

3. Predict what the substance inside each container might be. Record your predictions.

4. Find out what the substances are by checking your teacher's master list. Record the actual substances in the table.

What Did You Find Out?

1. Which substances did you predict correctly (or closely)? Which substances, if any, surprised you?

2. The volumes of the substances were identical because each container was filled to the top. Why is it important to keep the volumes equal in this activity?

3. Make a general statement about the substances, in terms of the density of the particles.

Figure 5.5 Although liquid particles are sometimes quite closely packed together, they cannot support objects in the same way that solids can, because the particles do not have a strong enough attraction for each other. Thus, liquid particles move apart easily.

DidYouKnow?

"Empty space" does not mean "air." Empty space means a separation between two objects with nothing between those objects, not even air particles. Simply because empty space and air *look* the same does not mean they *are* the same! For example, outer space is mostly empty space, but it has no air. Astronauts would suffocate if they were to venture outside their spacecraft without masks and air tanks, because there is no oxygen in outer space.

If you were to step onto the surface of a lake, the water would not support your foot. Instead, your foot would go right through the water. In fact, you would continue to fall through the water, pushing the water particles out of the way, as shown in Figure 5.5. Liquids cannot support objects in the same way that solids can, because the particles of a liquid move apart easily, allowing a dense, solid object, such as your foot, to pass through the liquid. The attractive forces between liquid particles are not strong enough to prevent your foot from pushing them apart.

Similarly, you cannot walk on air, because gases are even less dense than solids or liquids. When you move through air, you are moving through mostly empty space. You do not have to move as many particles of air out of the way as you do in water (see Figure 5.6). This explains why running through air is much easier and faster than running through water. In general, *gases are less dense than liquids.*

Figure 5.6 When you move through air, you do not have to move as many particles of air out of the way as you do water particles in water.

As temperature increases, a substance will change from solid, to liquid, to gas. The particle theory states that the particles of a substance spread out as they gain energy when heated. Thus, the particles take up more space, which means that the density of the substance decreases. It is almost always true that, for each pure substance (for example, gold), the density of its solid state is greater than the density of its liquid state. The substance's solid state and liquid state are, in turn, denser than its gaseous state.

In some cases, the densities of two substances can be so different that the liquid state of one is denser than the solid state of the other! One example of this is shown in Figure 5.7A. Many solid metals, such as copper, nickel, and silver, can float on mercury, one of the densest substances known. A more familiar example of differing densities is shown in Figure 5.7B.

Figure 5.7A Liquid mercury is so dense that it can support a solid iron bolt. A layer of oil has been placed on top of the mercury to prevent vapour from escaping into the surrounding air.

Figure 5.7B A solid block of wood floats easily on the surface of liquid water.

Check Your Understanding

1. In your own words, explain what density means.
2. Explain why solids can support objects more easily than fluids can.
3. What is the only way in which the density of a pure substance can change?
4. **Apply** Find some small items in the classroom (for example, pencils or paper clips) and determine whether these items are denser than water by dropping them into a container full of water. Organize your observations in a table.
5. **Apply** Butter is made by intentionally changing the density of liquid milk. Find out how this is done and write a brief description. Explain the change in terms of the particle theory.
6. **Thinking Critically** The density of molten lava increases as it cools and hardens. List other examples of natural changes in density.

5.2 Density: How Are Mass and Volume Related?

How can you measure the density of a substance? You could determine a substance's density if you knew how much of the substance occupies a certain space. To find out how much of a substance occupies a space, you can first of all measure the mass of the substance. **Mass** is the amount of matter in a substance (see Figure 5.8). **Volume** is a measurement of the amount of space occupied by the substance. Figures 5.9A and B show how the volume of a solid can be measured either directly or indirectly, depending on the shape of the solid. As you know, the volume of a liquid can be measured using a measuring cup or a graduated cylinder, for example. The volume of a gas can be determined by measuring the volume of the container that holds it. The greatest amount of fluid that a container can hold is called its **capacity**. Capacity is usually measured in litres or millilitres.

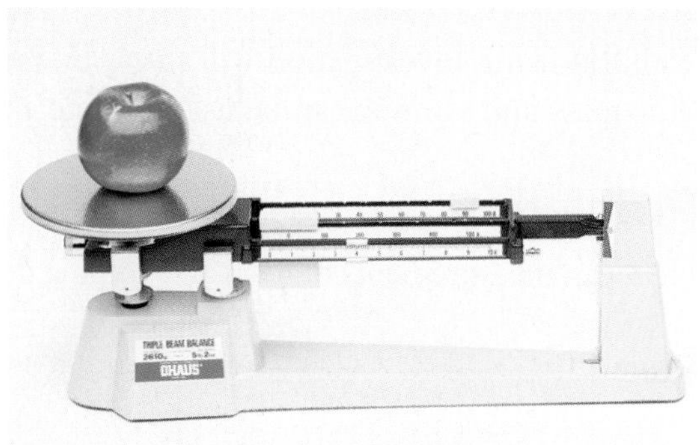

Figure 5.8 A balance is used to measure mass in grams (g) or in kilograms (kg). This apple has a mass of 102 g. It weighs about 1 N.

Keep in mind that mass and weight are not the same. **Weight** is the force of gravity exerted on an object. As you may recall from earlier studies, a **force** is a push or a pull, or anything that causes a change in the motion of an object. **Gravity** is the natural force that causes an object to move toward the centre of Earth. All forces, including weight, are measured in newtons (N). The pull of gravity everywhere on Earth's surface is essentially the same. On Earth, gravity pulls on an object with a downward force of 9.8 N for every kilogram of its mass. Thus, a bag of sugar with a mass of 2.26 kg weighs 22.1 N on Earth.

In the next investigation, find out how measuring mass and volume can determine the density of a substance.

The mass of an object depends on the amount of matter that makes up the object. The weight of an object changes as gravitational forces change. The pull of gravity on the Moon is about one sixth of Earth's gravity, or approximately 1.6 N/kg. On the Moon, what would your mass be? What would your weight be?

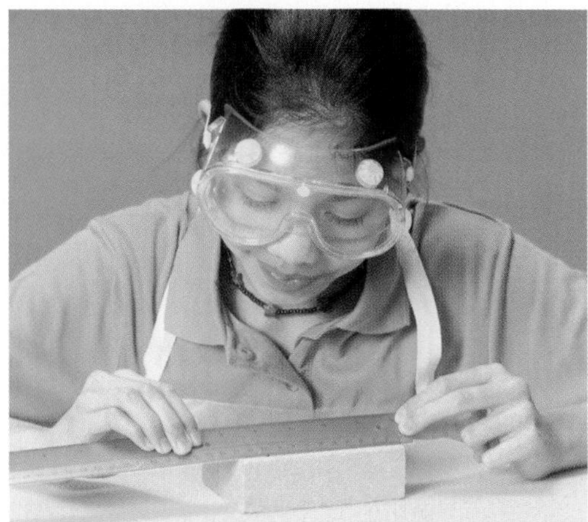

Figure 5.9A If an object has a regular shape — for example, if it is a block of wood — you can use a mathematical formula to calculate the object's volume: $V = l \times w \times h$.

Figure 5.9B The volume of an irregularly shaped object can be found by measuring the volume of the water that spills out of an overflow can.

Skill POWER

For tips on how to measure volume, turn to page 540.

CONDUCT AN

INVESTIGATION 5-A

Determining Density

The following investigation will show, by means of accurate measurements, how mass and volume can be used to determine density.

Problem

How can measurements of mass and volume determine the density of a substance?

Part 1

Mass-to-Volume Ratios

Safety Precautions

- Handle balances with care and use them as instructed by your teacher.
- Avoid spilling liquids and sand on the balances.
- Do not pour substances down the drain. Dispose of them as instructed by your teacher.

Apparatus

500 mL beaker (or 500 mL measuring cup)

balances (or one shared by the class)

5 different-coloured pencil crayons or markers

Materials

500 mL (per trial) of each of the following substances: water, oil, glycerol, molasses, sand

graph paper for each student

Procedure

1. Before beginning the investigation, predict how the substances will rank according to density. Rank the substances from least dense (1) to most dense (5). Record your hypothesis and a brief note explaining your ranking order.

(a) Your teacher will divide the class into five groups and will assign one substance to each group. Subdivide each group into smaller groups of partners to provide multiple trials for each substance.

(b) Copy the data table below into your notebook.

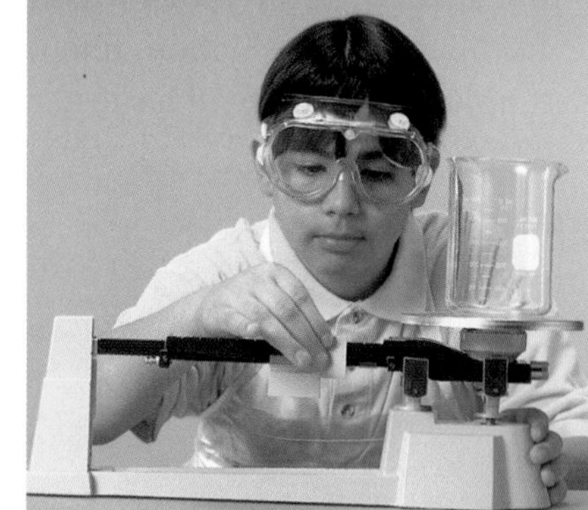

2. Measure the mass of the empty beaker. Record this value in column B of your table.

Individual Results

Substance tested:				
A	B	C	D	E
Volume (mL)	Mass of beaker only (g)	Mass of beaker and substance (g)	Mass of substance only (g)	Ratio of mass to volume (g/mL)
100				
200				
300				
400				
500				

Skill POWER

For tips on working in groups, turn to page 536.

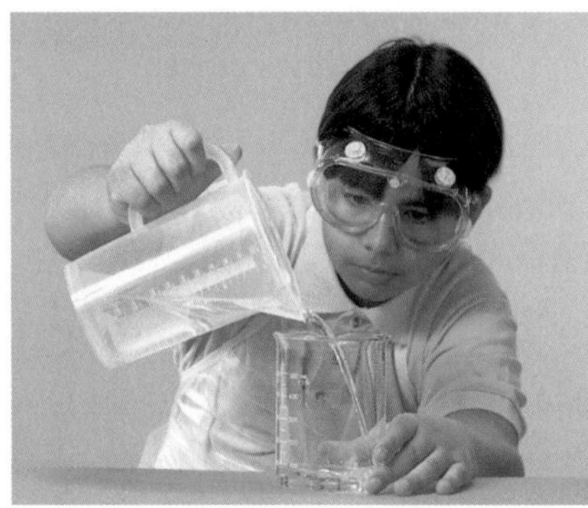

3 Pour 100 mL of your substance into the beaker. Be as accurate as possible.

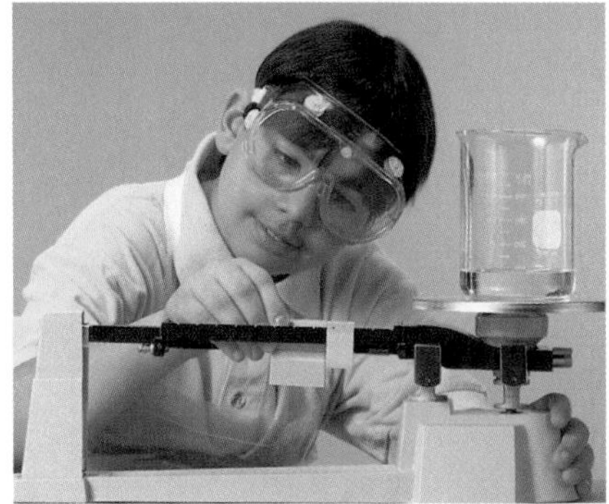

4 Measure the mass of the beaker plus the substance. Record this value in column C in your table.

5 Subtract the mass of the beaker (column B) from the mass of the beaker and the substance together (column C). Record the difference in column D.

6 Repeat steps 3 to 5 four more times, each time adding 100 mL of your substance to what is already in the beaker. (The last time, you will be measuring 500 mL.)

7 The **mass-to-volume ratio** is the relationship between mass and volume expressed as a quantity of the mass divided by its volume. To find the mass-to-volume ratio for each amount of each substance, divide the mass (column D) by the volume (column A). Show your calculations and results in column E. Wash your hands.

8 When each group has finished, your teacher will display a set of class results for all the substances in a summary chart with the following headings:

Class Results

Substance	Mass (g)	Volume (mL)	Mass-to-volume ratio (g/mL)

Copy these results into your notebook. (If there were two or more trials for each substance, calculate the average mass, volume, and mass-to-volume ratio values for each substance.)

Skill POWER

To review how to measure mass using a balance, turn to page 540.

Math CONNECT

The ratio "mass to volume" (or "mass : volume") means the "mass divided by the volume." To find the decimal value, convert the fraction of mass over volume into a decimal. For example, suppose a mass of 3 g has a volume of 5 mL. How would you convert its mass-to-volume ratio (3 g : 5 mL) into a decimal value? Begin by expressing the ratio as a fraction: $\frac{3\text{ g}}{5\text{ mL}}$. A decimal is any fraction with a denominator of any power of 10, for example, 10, 100, 1000, etc. Here is how to convert $\frac{3\text{ g}}{5\text{ mL}}$ into a fraction with a denominator of 10:

$$\frac{3\text{ g}}{5\text{ mL}} \times \frac{2}{2} = \frac{6\text{ g}}{10\text{ mL}} = 0.6\text{ g/mL}$$

Now, try converting the mass-to-volume ratio 6 g : 25 mL into a decimal value.

CONTINUED

Part 2

Graphing

Procedure

1. Make a line graph of the class results recorded in Part 1. Place the volume scale along the horizontal axis (x-axis), and the mass scale along the vertical axis (y-axis).
2. Plot the (average) results for the first substance on the graph. Draw a line through these points in one colour. Record this colour in a legend on the graph, and write the name of the substance beside it.
3. *On the same graph*, plot the results for the next substance. Draw a line through these points using another colour. Record this colour in the legend, and write the name of the substance beside it.
4. Repeat step 3 for the three remaining substances.
5. Give your graph a title.

Skill POWER

To review how to make a line graph, turn to page 546.

Computer CONNECT

Create a computer file for a data base of all the "raw data" in the class. Use the program features to sort the data, find averages, and generate computer graphs. Make a poster to display your results.

Analyze

1. Describe the lines on your graph. Are they straight or curved? Do they have the same slope? If not, are some lines closer together than others?
2. Look back to the data table you made for your substance. What happens to the mass-to-volume ratio for each volume measurement of your substance? Why do you think this happens?
3. Compare your hypothesis to the final results.
4. There is a chance of error in every experiment. Suggest ways to improve (a) how you performed the investigation (more accurate measurement, avoidance of spilling, etc.), (b) how you calculated results (possible math errors), and (c) how you graphed your results.

Conclude and Apply

5. Why are some lines in the graph similar to each other while some are different?
6. How can you tell from your mass-to-volume ratios and your graph which substance is the least dense? Which substance is the most dense?
7. Look at the mass-to-volume ratios in the Class Results table in Part 1. Compare these values with the slope of the lines in the graph that correspond to them. How does the slope of a line change as the mass-to-volume ratio changes?
8. Add a sixth line to your graph for a substance that is denser than water but less dense than sand. Between which values would its mass-to-volume ratio be?
9. Use the particle theory to explain the relationship between the mass, volume, and density of the substances you examined in this investigation.
10. From your observations, do you think that density and viscosity are related? Explain your answer.

Extend Your Skills

11. Use the particle theory to predict the effect of temperature on mass-to-volume ratios.

A Formula for Density

The density of a substance can be determined by calculating its mass-to-volume ratio. You can do this by dividing the object's mass by its volume. Therefore, the formula for density is:

$$\text{Density } (D) = \frac{\text{Mass } (m)}{\text{Volume } (V)} \text{ or simply, } D = \frac{m}{V}$$

For example, the density of an object having a mass of 10 g and a volume of 2 cm^3 is 5 g/cm^3. The density of solids is usually given in g/cm^3 (grams per cubic centimetre). The density of liquids and gases is often given in g/L (grams per litre) or g/mL (grams per millilitre). Using pure water as an example, you could express its density as either 1 g/cm^3 or as 1 g/mL (1 cm^3 = 1 mL).

As long as the temperature and pressure stay the same, the mass-to-volume ratio, or density, of any pure substance is a *constant*, which means it does not change.

***Word* CONNECT**

Density is an example of an *intrinsic property* of a pure substance, because density depends only on the particles that make up the substance. Intrinsic properties can be used to identify pure substances, because each pure substance has its own specific set of intrinsic properties. Therefore, you could measure the density of a pure substance to help determine its identity. Name two other intrinsic properties of a pure substance.

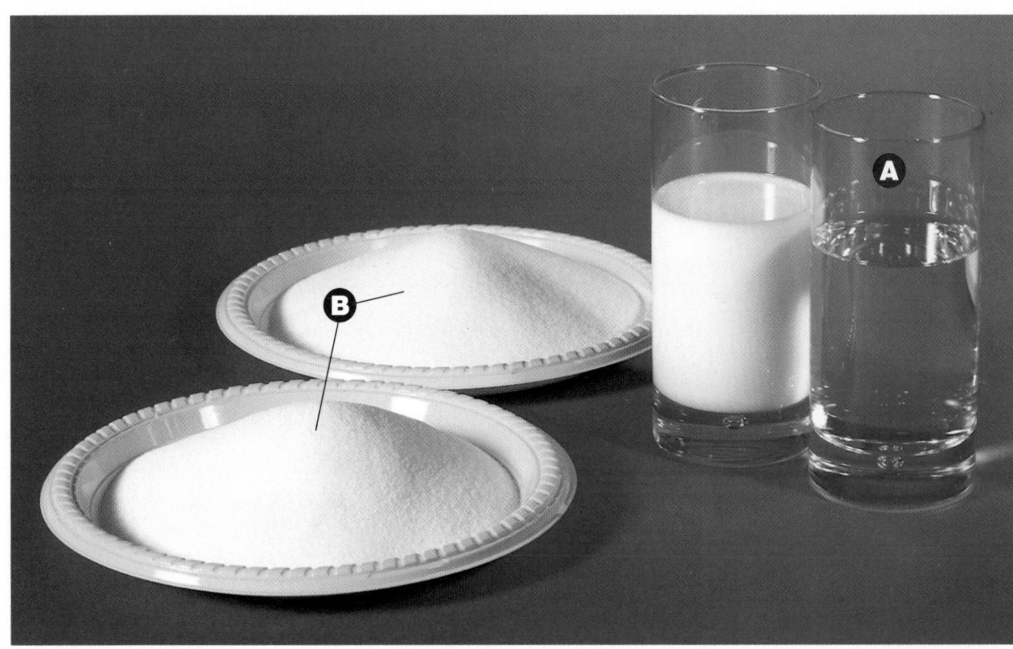

A Seawater may look like regular water, but its density is closer to that of milk — 1.03 g/mL.

B The salt and sugar shown here both have a mass of 0.5 kg and are the same colour. However, their densities differ.

You observed the property of density in Conduct an Investigation 5-A. When you increased the volume of each substance, the mass increased. Therefore, the mass-to-volume ratio, represented as a decimal value, remained constant. If the mass of a pure substance increases, the volume will increase. Similarly, if the volume of a pure substance increases, the mass will also increase. According to the particle theory, the size of the particles of a substance does not change when the mass or volume of the substance changes. A certain number of particles of a particular size will always occupy a certain space. As the number of particles increases from substance to substance, the space required for those particles must also increase. Therefore, density is a property of matter that is unique to a specific pure substance. Table 5.1 lists the approximate densities of some common substances at 0°C and 101.3 kPa of pressure.

Table 5.1 Approximate Densities of Common Substances

Fluid	Density (g/mL)	Solid	Density (g/cm^3)
hydrogen	0.00009	styrofoam	0.005
helium	0.0002	cork	0.24
air	0.0013	oak	0.70
oxygen	0.0014	sugar	1.59
carbon dioxide	0.002	salt	2.16
ethyl alcohol	0.79	aluminum	2.70
machine oil	0.90	iron	7.87
water	1.00	nickel	8.90
seawater	1.03	copper	8.92
glycerol	1.26	lead	11.34
mercury	13.55	gold	19.32

INVESTIGATION 5-B

Comparing Densities

Think About It

When you compare the masses of equal volumes of different kinds of matter, you are comparing their densities. Scientists have recorded the densities of many substances. Does knowing how the density of a substance compares with the densities of other substances tell you something about the characteristics or behaviour of the substance? Find out in this investigation.

What to Do

Use the information in Table 5.1 on page 141 to answer the following questions:

1. Which substance in the table is the most dense? Is it a solid, a liquid, or a gas at room temperature?
2. Which substance is the least dense? What is its physical state at room temperature?
3. Write a conclusion about which physical state tends to be the most dense.
4. Identify one combination of two substances in which a liquid is more dense than a solid.
5. Name the substance that is denser than mercury.

Analyze

1. Which substances would float in water?
2. Which substances would sink in water?

Extension

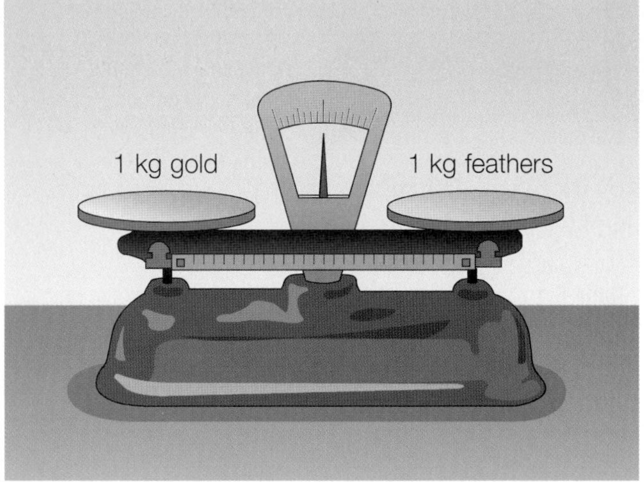

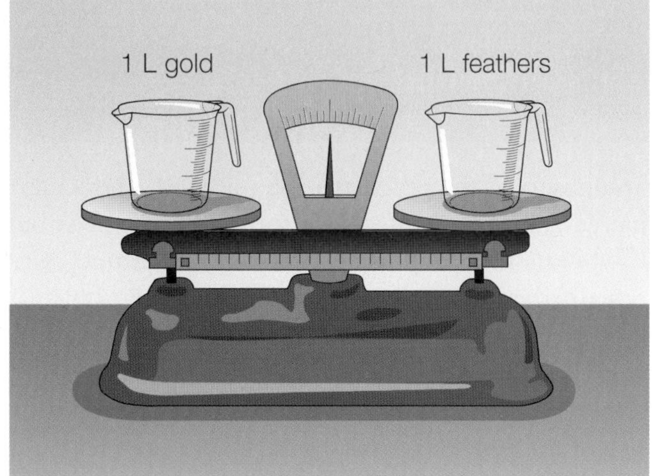

1. Copy and complete the diagram above by adding a sack of gold and a sack of feathers. If both sacks have the same mass, which one will have the larger volume?

2. Copy and complete the diagram above. Add the gold and the feathers to the measuring cups, and show how the balance would tip as a result. If both substances have the same volume, which one will have the larger mass? Record your answer. Comment on how a knowledge of density helped you with your answers.

What Is the Density of a Pencil?

You need two measurements to calculate the density of a pencil. What are they?

What You Need

100 mL graduated cylinder
balance
pencil
water

What to Do

1. Use a balance to measure the mass of a pencil in grams.
2. Pour 90 mL of water into a 100 mL graduated cylinder.
3. Lower the pencil, eraser end down, into the cylinder. Continue to push the pencil down until it is completely underwater, but be sure your finger is not also submerged.
4. Read and record the new volume of water.
5. Calculate the pencil's density by dividing its mass by the change in volume of the water level when the pencil was completely underwater.

What Did You Find Out?

Is the density of the pencil greater or less than the density of water? How do you know?

Extension

Use the same method to find out the density of another object, such as a rubber stopper or a cork. Make a prediction, then carry out the activity to see if you were correct.

Check Your Understanding

1. (a) Define mass.
   (b) How do you measure mass?
   (c) In which units is mass measured?
2. (a) Define volume.
   (b) How do you measure the volume of a liquid? Of a solid? Of a gas?
   (c) In which units is volume measured?
3. What does the mass-to-volume ratio tell you about a substance?
4. **Apply** Using information from Table 5.1 on page 141, copy the table below and fill in the missing information.

Substance	Mass (g)	Volume (cm^3)	Density (g/cm^3) (Mass-to-volume ratio)
aluminum	5.40		
	6.48	3.0	
		5.0	8.92
oak	0.33		
salt		4.0	

5. **Thinking Critically** Make a paper airplane. Then make an identical airplane out of aluminum foil and observe which one flies better. Explain your observations.

What shiny solid has a mass of 356 g and a volume of 40 cm^3? (Hint: The substance is listed in Table 5.1.)

5.3 Buoyancy: The "Anti-Gravity" Force

Playing in water is fun because there are many things you can do in water that you cannot do on land. For example, as Figure 5.10 shows, doing handstands in water is easy. The water exerts an upward force that helps support you when you do a handstand. Buoyancy refers to the ability of a fluid to support an object floating in or on the fluid. The particles of a fluid exert a force in a direction opposite to the force of gravity. The force of gravity *pulls down*, toward the centre of Earth. Buoyant force — the upward force on objects submerged in or floating on fluids — *pushes up*, away from Earth. Like all other forces, buoyant force is measured in newtons (N).

Floating occurs when an object does not fall in air or sink in water, but remains suspended in the fluid. People cannot walk on water, but many people can swim, and they can also float in boats. Hot-air balloons, like the one shown in Figure 5.11, can float at one altitude for a long time. Why do you think fluids can support certain objects? Explore this question in the next investigation.

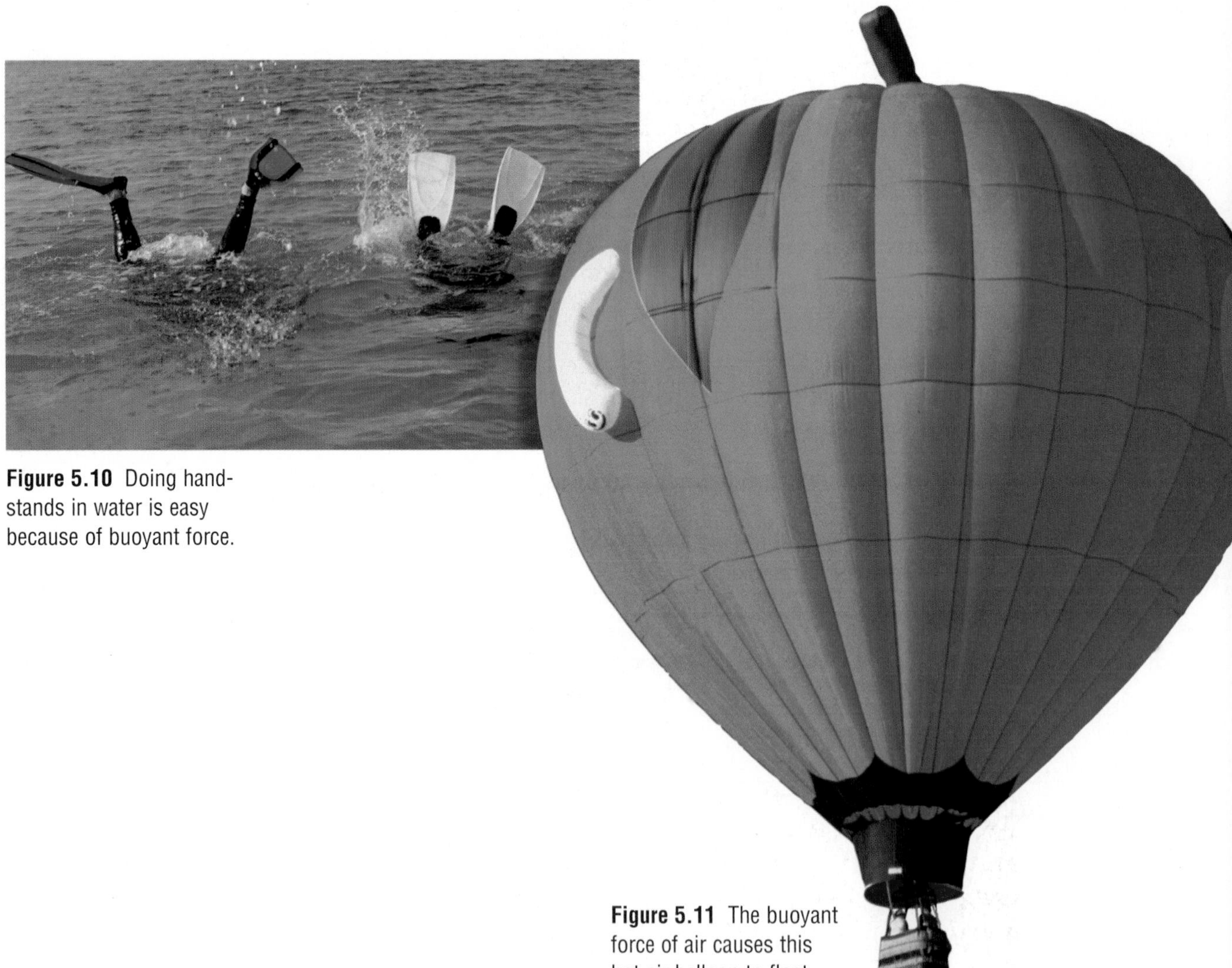

Figure 5.10 Doing handstands in water is easy because of buoyant force.

Figure 5.11 The buoyant force of air causes this hot-air balloon to float.

CONDUCT AN

INVESTIGATION 5-C

Build a Density Tower

Do you think density plays a role when a fluid supports an object? Find out in this investigation.

Problem

How can you build a tower out of liquids that support each other as well as solids?

Safety Precautions

Do not pour substances down the drain. Dispose of them as instructed by your teacher.

Apparatus

tall plastic jar or cup (or transparent container) with lid
cork
toothpick or wood chip
paper clips
rubber gloves

Materials

water, with food colouring added
vegetable oil

Procedure

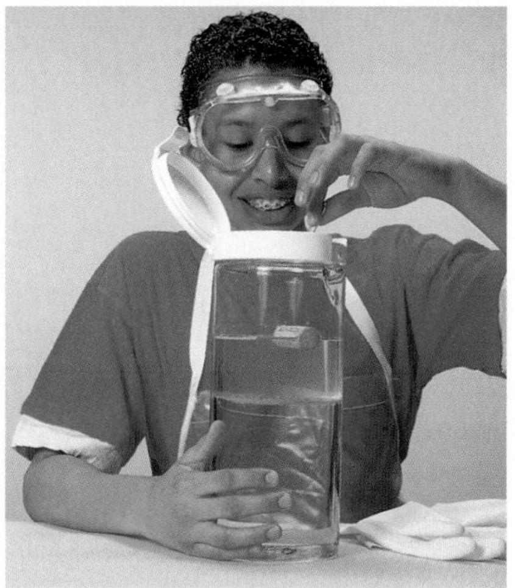

1. Combine the water, oil, cork, wood chip, and paper clips in the container. Allow the substances to settle (stop moving). Sketch and label the tower and its contents.

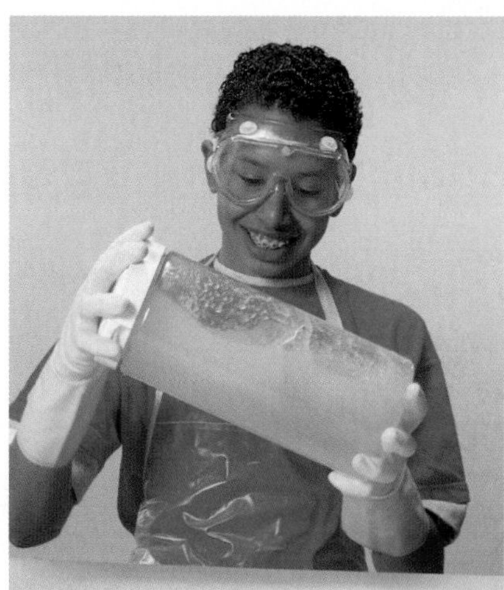

2. Shake the tower and allow the substances to settle again. If the shaken tower appears different, draw a new labelled sketch. Wash your hands after this investigation.

Analyze

1. Make a data table and rank the substances in the density tower in order from least dense (1) to most dense (5).
2. Which substances are denser than water? Which substances are less dense than water?

Conclude and Apply

3. Can a solid be less dense than a liquid? Use the particle theory to explain your answer.
4. Does the volume of an object determine its density?

Extend Your Skills

5. Add more items of your choice to the density tower, for example, a rubber stopper, a small rubber duck, a candle stub, a small plastic toy, and a safety pin. Predict where you think these objects will settle in the tower. Then test your prediction.

Sinkers and Floaters

How can people travel in the air and on water if the density of their bodies is greater than the density of both these fluids? Why don't they sink? Is density the only factor that explains why fluids can support certain substances? Look at Figures 5.12A and B. It seems that water can support objects that have densities greater than water — as long as the weight of the object is spread over a large enough area.

Figure 5.12B The mass of this straight pin is slightly less than 1 g. However, the pin sinks when placed in water.

Figure 5.12A The *Hibernia* oil rig in Atlantic Canada has a mass of more than 14 000 t, yet it floats on water.

Average Density

Ships can be built of steel (density = 9.0 g/cm^3) as long as they have large, hollow hulls. A large, hollow hull ensures that the **average density** of the ship (that is, the total mass of all substances on board divided by the total volume) is less than that of water. Similarly, life jackets are filled with a substance of very low density. Thus, life jackets lower a person's average density, allowing the person to float (see Figure 5.13). Experiment with average density in the next investigation.

Figure 5.13 A life jacket lowers the average density of the person wearing it.

DESIGN & DO

INVESTIGATION  5-D

Penny Boat

How many pennies can you float in a boat made of aluminum foil? What features might you include in the design to construct the best penny boat?

Challenge

Design and build a boat out of aluminum foil that will float while holding as many pennies on board as possible.

Safety Precaution

Be careful when using sharp objects such as scissors.

Materials

aluminum foil, pennies, sink or tub full of water, ruler, scissors

Design Criteria

A. Each piece of aluminum foil should be cut into a square 0.35 m × 0.35 m.

B. Each group can use only two pieces of cut foil.

C. No additional materials may be used.

Plan and Construct 

1. Make a labelled sketch of a boat design that can carry the greatest possible number of pennies.
2. Build your boat according to your sketch. You may test your boat and then alter your design, but you must use the same piece of aluminum foil.
3. Record the greatest number of pennies that your boat can carry without sinking. Make a sketch of your final boat design. Label all the design features that you think are important.

Evaluate

1. How many pennies did your boat hold?
2. Which design features do you think were most important in holding more pennies? Why were they most important?
3. If you could design a better boat, how would you change it?
4. Would paper work better than aluminum foil? Why or why not?

Extend Your Skills

1. Stage a contest to see whose boat can carry the most pennies.
2. Build another boat, this time out of a ball of modelling clay (with a diameter of 5 cm). Compare and contrast its design and construction with the design and construction of the aluminum foil boat.

It took 3 h for submersibles (underwater vessels) to free-fall 4 km through the icy waters 650 km off the coast of Newfoundland to reach the wreck of the *Titanic*. Would it take as long to free-fall the same distance from an airplane? Why or why not? What accounts for the difference?

Pause& Reflect

Imagine a contest in which you must drop a sheet of paper from a tower five storeys high, into a small box below. This contest would take place in an auditorium, so there would be no wind to affect the results. In your Science Log, explain why you would probably not succeed in winning this contest. Why can you not rely on an object such as a sheet of paper to land on a small target directly below you? Paper is made from wood. Do you think a wooden block would fall straight down? In your Science Log, comment on the difference between dropping a sheet of paper and dropping a wooden block.

Benefits of Average Density

Average density is useful because it enables objects that would otherwise sink — such as large ships and oil rigs — to float. Average density also helps floating objects to sink. For example, most fish have an organ called a swim bladder (also called an air bladder). This organ, a large sac near the spine of the fish, contains a mixture of air and water (see Figure 5.14A). The fish's depth in the water depends on how much air is inside the sac. As the amount of air decreases, the fish sinks lower. As the amount of air increases, the fish rises closer to the surface. This depth-control structure has been adapted in the submarine, allowing the submarine's crew to adjust its depth underwater (see Figure 5.14B).

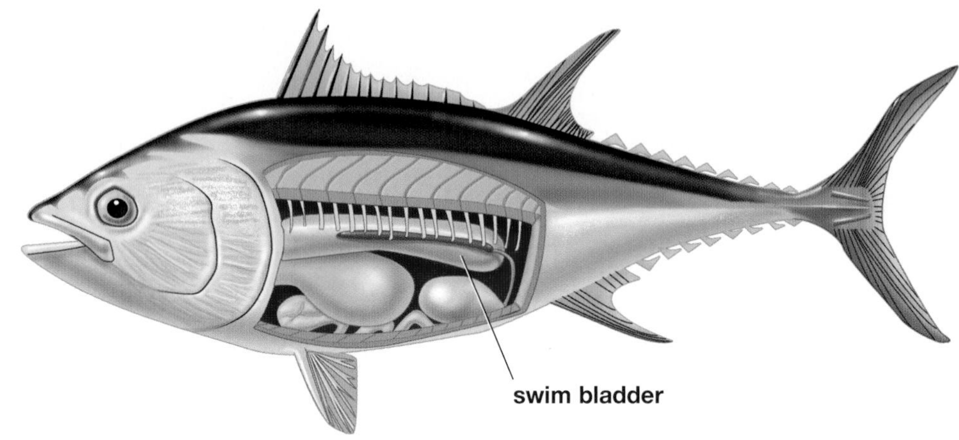

Figure 5.14A Cutaway drawing showing the swim bladder inside a fish

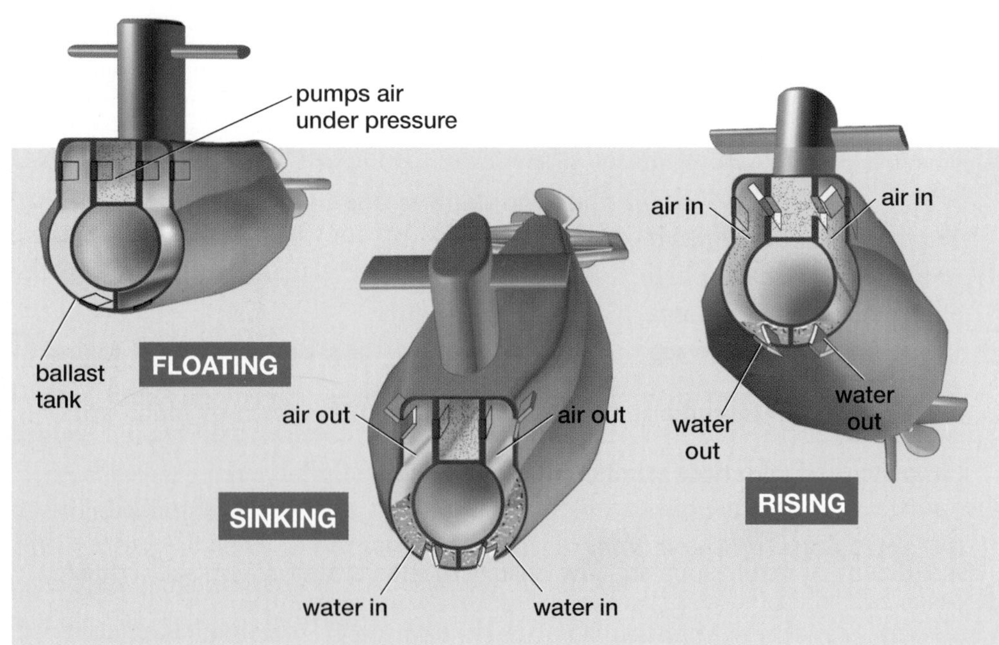

Figure 5.14B Cross-section of a submarine

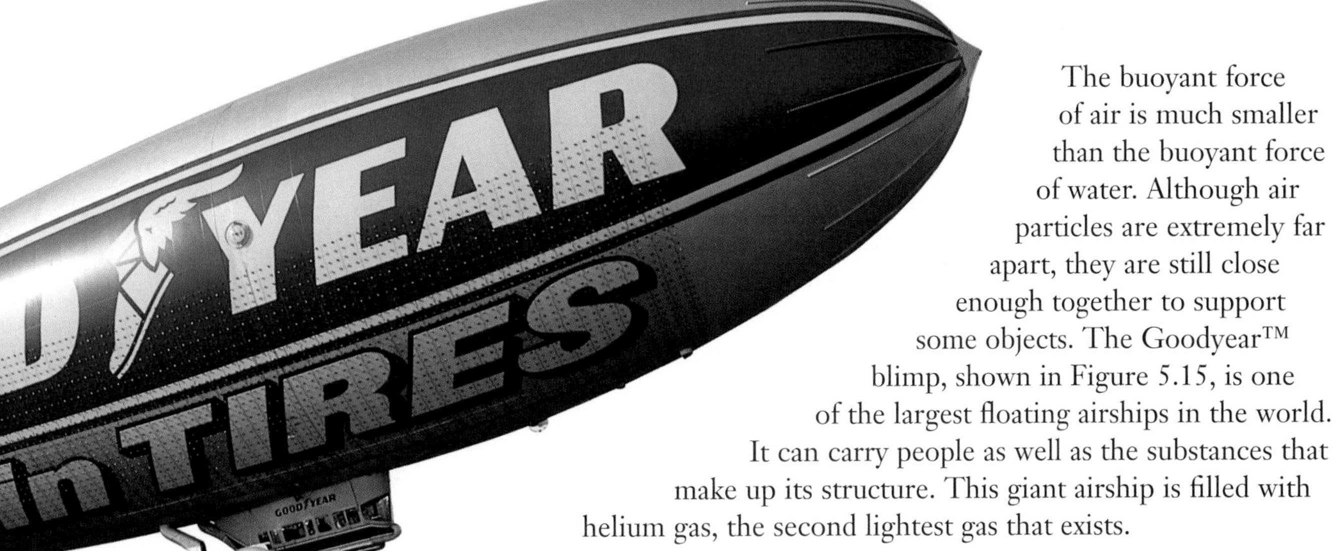

The buoyant force of air is much smaller than the buoyant force of water. Although air particles are extremely far apart, they are still close enough together to support some objects. The Goodyear™ blimp, shown in Figure 5.15, is one of the largest floating airships in the world. It can carry people as well as the substances that make up its structure. This giant airship is filled with helium gas, the second lightest gas that exists.

An airship such as a blimp can float because its mass is relatively small compared to its enormous volume. Thus, its average density is slightly less than the density of the air surrounding it. Ocean-going ships, hot-air balloons, and blimps all have huge volumes. The relationship between the size of an object and the buoyant force exerted on it was established long ago by a scholar named Archimedes.

Figure 5.15 The Goodyear™ blimp is often seen hovering over open stadiums during sports events. It captures a unique "bird's-eye view" of the action below.

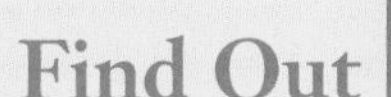

Cartesian Diver

Make your own model of a diving device that can adjust its own depth.

What You Need

1 L plastic pop bottle and cap
water
medicine dropper

What to Do

1. Fill the pop bottle three quarters full with water.
2. Fill the medicine dropper half full with water.
3. Drop the medicine dropper, or "diver," into the pop bottle. Put the cap on the bottle.
4. Squeeze the bottle hard, then release it. Record your observations and include a sketch.

What Did You Find Out?

What happens to the amount of water in the "diver" as you squeeze the bottle? What happens to the water level in the "diver" when you release the bottle?

Extension

Describe the links between cause and effect that you observed.

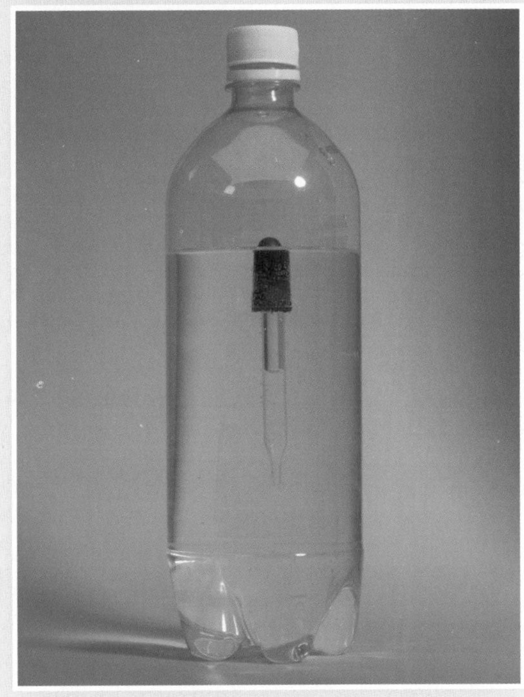

Archimedes' Principle

The Greek scientist Archimedes made a brilliant discovery around 212 B.C.E. Hiero II, ruler of Syracuse, suspected that the royal goldsmith had not used pure gold to make his crown. The king asked Archimedes to determine whether the crown was made entirely of gold.

Archimedes knew that all he had to do was determine whether the density of the crown matched the density of gold. Recall that the formula for density requires only two values: mass and volume. Archimedes could measure the mass of the crown easily with a balance. How could he measure the volume of an object as irregularly shaped as a crown?

Word CONNECT

Apparently, when Archimedes suddenly realized how to find the volume of the crown, he jumped out of the bath and ran through the streets shouting "Eureka!" What does this Greek word mean in English?

Archimedes solved the problem while at the public baths. He stepped into the almost-full bath, and water gushed all over the floor. The solution to the problem came to Archimedes in a flash — a solid object can **displace** water out of a container. That is, a solid object can move water out of the way, as shown below.

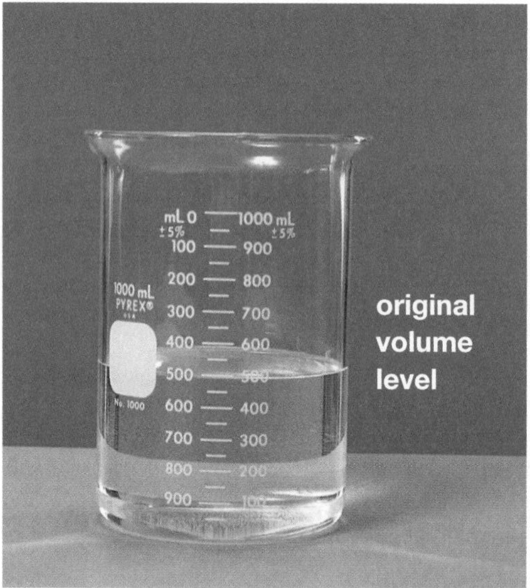

Archimedes reasoned that the water that was displaced in the tub must have exactly the same volume as the volume of his body. Therefore, to find the volume of the crown, Archimedes would simply submerge the crown in a container full of water. He would then collect and measure the volume of the water that spilled out. When Archimedes carried out this test, he showed that the crown was made of a mixture of gold and silver. He concluded that the goldsmith who had made the crown had tried to cheat the king.

Archimedes applied his new ideas to another property of fluids. He believed that the displaced fluid held the key to whether the object placed in the fluid would sink or float. He wondered why he would sink if he stepped into a bathtub, but he would float if he stood in a boat on the water. He concluded that the amount of buoyant force that would push up against the object immersed in the fluid would equal the force of gravity (the weight) of the fluid that the object displaced.

The particle theory can explain why this was a reasonable conclusion. If the water in a container is still, or at rest, then the water particles are neither rising nor sinking. An object immersed in a fluid such as water does not rise or sink *if the amount of force pulling down (gravity) equals the amount of force pushing up (buoyancy).* This condition is known as **neutral buoyancy**. Therefore, the water particles in the lower part of the container must be exerting a buoyant force equal to the weight, or force of gravity, of the water above it.

If 1 L of water is displaced, the object replacing it would have the same volume (1 L) but might have a different weight than the 1 L of water. If the object is heavier than the displaced water, then the weight of the object will be a force greater than the buoyant force that had supported the displaced water. Thus, the object will sink. If the object is lighter than the displaced water, the object will rise to the surface and then float. These relationships are shown below.

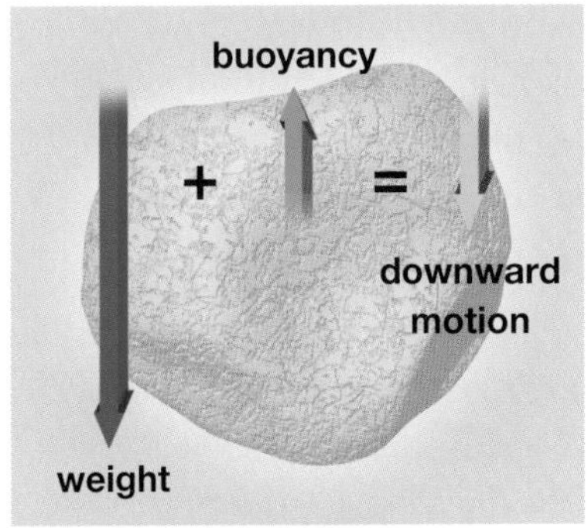

A Sinking, e.g., a rock

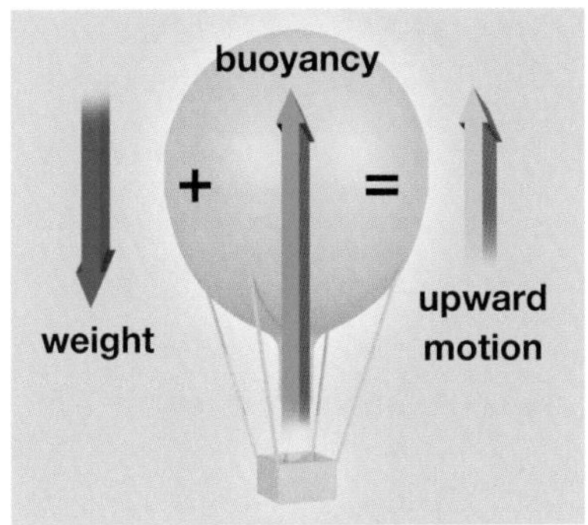

B Rising, e.g., a helium balloon

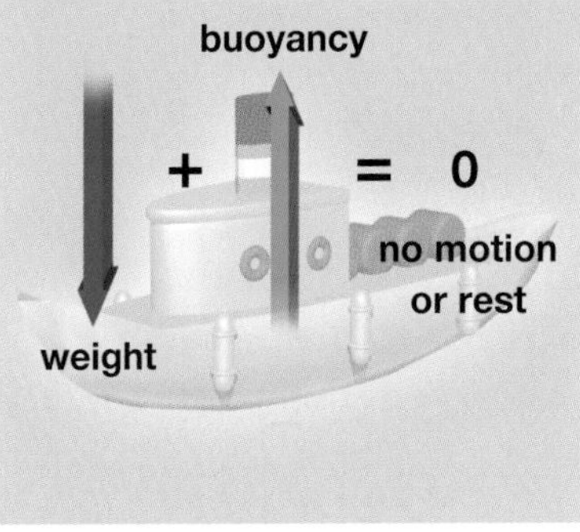

C Floating, e.g., a boat

The freshwater zebra mussel has invaded the waterways of North America. This tiny mollusc sticks to underwater surfaces such as water-intake pipes and docks. Zebra mussels can attach themselves to buoys and ships' hulls.

Sometimes they form dense layered colonies of over one million mussels per square metre. Thus, they can alter the average density and the stability of floating objects. Buoys have been known to sink with the added weight of thousands of zebra mussels. An unbalanced distribution of zebra mussels can interfere with a ship's stability, especially when it is being tossed about in a storm.

INTERNET CONNECT

www.school.mcgrawhill.ca/resources/

Visit the above web site to find out more about zebra mussels. Go to **Science Resources**, then to **SCIENCEPOWER 8** to know where to go next. Based on your findings, write a newspaper article about how scientists believe the zebra mussel managed to invade North American waters. Also, research how scientists are working to reduce the impact of zebra mussels in North American waterways.

When Archimedes stepped into the bath, he sank because the amount of water that he displaced weighed less than he did. When he stepped into a boat, however, a larger volume of water was displaced. The weight of this water far exceeded the weight of the boat and Archimedes combined. Therefore, the buoyant force was greater and the boat, with Archimedes in it, floated on the surface.

Why would Archimedes and his boat not continue to rise, with such a large buoyant force pushing it upward? At the surface of the water, the fluid supporting the object is air. As mentioned earlier, the buoyancy of air is much less than that of water. Therefore, the upward motion stops at the water's surface.

Archimedes made the following conclusion, now known as **Archimedes' principle:** *The buoyant force acting on an object equals the weight (force of gravity) of the fluid displaced by the object.* Archimedes' principle is useful in

Find Out **ACTIVITY**

The Amazing Floating Egg

Do you think that different liquids exert a similar buoyant force? Find out for yourself in this activity.

What You Need

glass
fresh egg
teaspoon
water
salt

What to Do

1. Place an egg in a glass half-full of water and observe what happens. Record your observations.
2. Stir salt into the water one teaspoonful at a time; stop when the egg floats. Try to explain why the egg floats.
3. When the egg is floating, carefully pour more tap water into the glass until it is almost full. Add the water *slowly* and near the side of the glass so the fresh water and the salt water do not mix. Where does the egg float now? Sketch a labelled diagram of your floating egg. Suggest an explanation for your observation.

predicting whether objects will sink or float. Note that the buoyant force does not depend on the weight of the submerged object, but rather on the weight of the displaced fluid. A solid cube of aluminum, a solid cube of iron, and a hollow cube of iron, all of the same volume, would experience the same buoyant force!

How Buoyancy and Density Are Related

Think back to your density tower (page 145). Both water and oil are liquids, but they did not support the same objects. Therefore, the buoyant force of a liquid does not depend on *physical state*, but rather on *density*. (This is also true for buoyancy in gases.)

As you observed in the previous Find Out Activity, objects float more easily in salt water than in fresh water. Seawater (salt water) has a density of 1.03 g/mL and fresh water has a density of 1.00 g/mL. The density of salt water is greater than that of fresh water, which means that the particles of salt water are packed together more tightly. Therefore, salt water can support more weight per volume than fresh water. A buoy floating in salt water extends out of the water more than a buoy in fresh water. The next time you have a chance to swim in the ocean, observe how much more easily you can float on your back!

ocean buoy **fresh-water buoy**

The relationship between buoyancy and density is the basis for the **hydrometer**, an instrument designed to measure liquid density. Figure 5.16 on the next page shows two kinds of commercial hydrometers. A hydrometer will extend farther out of a liquid if the liquid has a higher density, for example, water (1 g/mL). A hydrometer will sink lower if the liquid has a lower density, such as vegetable oil (0.9 g/mL).

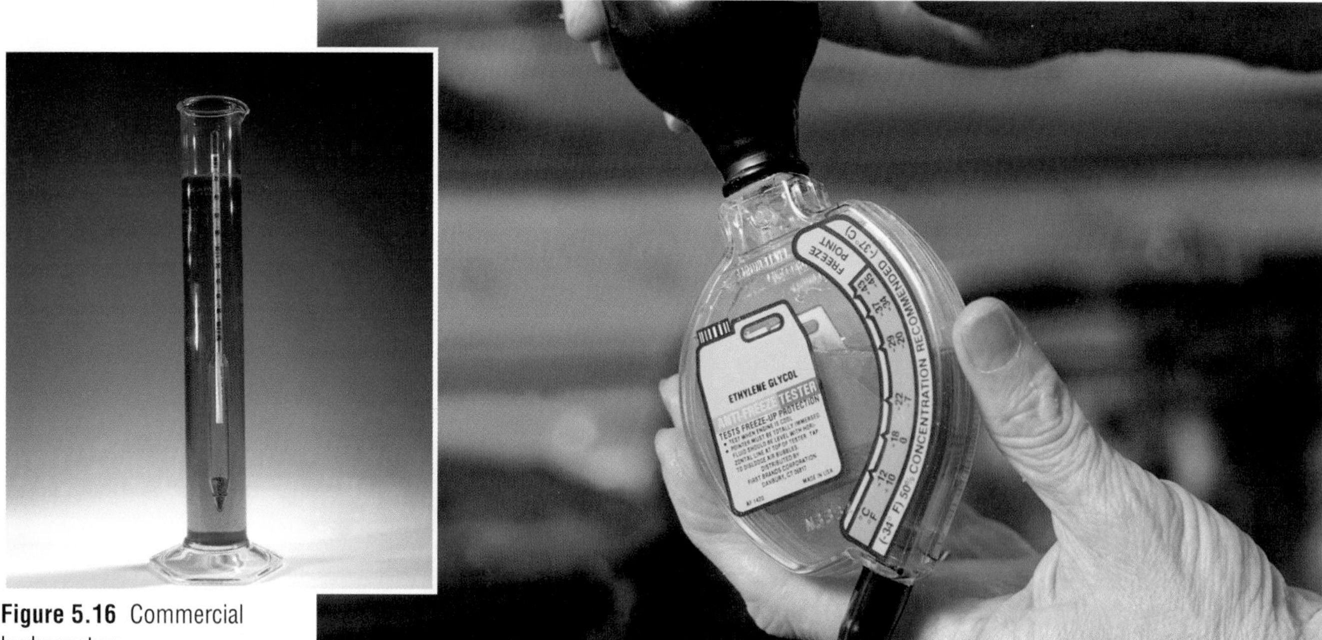

Figure 5.16 Commercial hydrometers

Many different hydrometers are available commercially, all designed for specific uses. Hydrometers are widely used in the food and beverage industries. Although they measure density, these instruments can be used to determine other values indirectly. For example, they can be used to determine the sugar content of canned fruit syrup, or the alcohol content of wine. Analyze how a hydrometer works in the next investigation. Then design and build your own hydrometers using various materials, in Conduct an Investigation 5-F on page 156.

Career CONNECT

Working Underwater

A number of careers involve scuba-diving in lakes, oceans, or other bodies of water. One example is underwater welding. As another example, the divers shown here are conducting research on marine life. How many other underwater careers can you think of? Brainstorm in a group to see how many diving-related careers you can list. The scuba gear that divers use includes masks, fins, air tanks, and weight belts. The word "scuba" comes from "**s**elf-**c**ontained **u**nderwater **b**reathing **a**pparatus." Search the Internet for information on scuba-diving or look in the Yellow Pages of your telephone book for companies that offer scuba training. Try to find out:

- how scuba gear helps you sink instead of float on top of the water
- how you are able to rise to the surface after your dive
- what determines how long a diver can stay underwater
- how much training is necessary before a beginning diver can dive without an instructor
- what careers involve scuba diving

INVESTIGATION 5-E

Measuring Buoyancy

Think About It

You know that all liquids do not have the same density. Investigate whether various liquids exert the same buoyant force.

What to Do

Observe the photographs and do the following:

1. You can determine the buoyant force on the mass in each liquid by comparing its weight in air to its weight in the liquid. The amount that the weight is reduced is the buoyant force. You can calculate the buoyant force by using the following formula.

   $$F_{\text{buoyant force}} = W_{\text{weight in air}} - W_{\text{weight in liquid}}$$

   Using the data in set 1, calculate the buoyant force on the mass in each liquid.

2. Using your calculations from step 1, list the liquids in order of the buoyant force they exert on the mass, from the greatest to the least.

3. Using set 2, list the liquids in order of greatest density to smallest density. How does this list compare to the list in question 2?

Analyze

1. Give a reasonable explanation for the relationship you found between the density of a liquid and the buoyant force it exerts on the mass.
2. Using set 2, describe any differences that you see in the hydrometers.
3. How do these differences relate to the buoyant forces that these liquids exert on the mass in set 1?

1

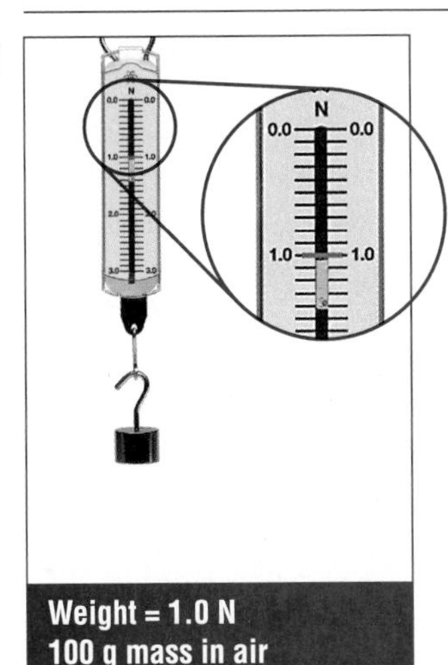

Weight = 1.0 N
100 g mass in air

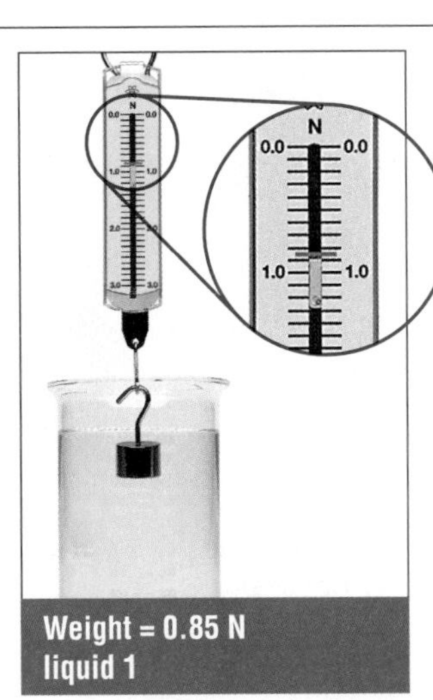

Weight = 0.85 N
liquid 1

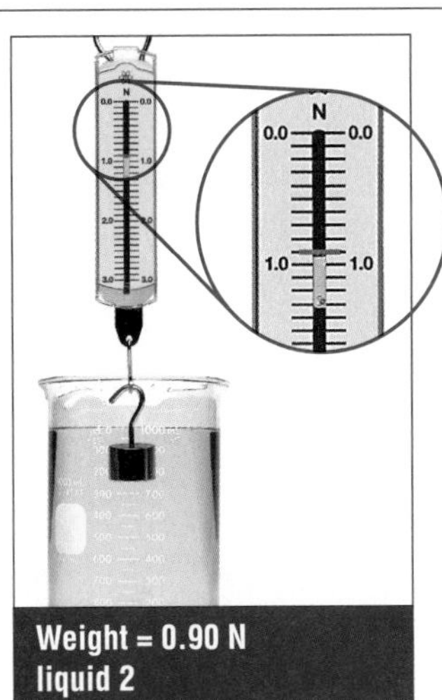

Weight = 0.90 N
liquid 2

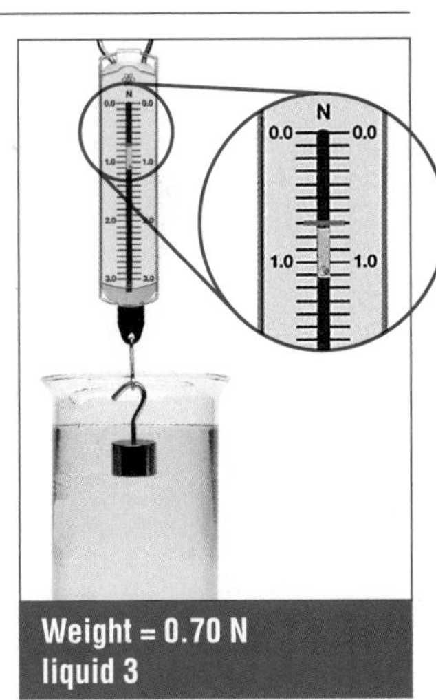

Weight = 0.70 N
liquid 3

Set 2

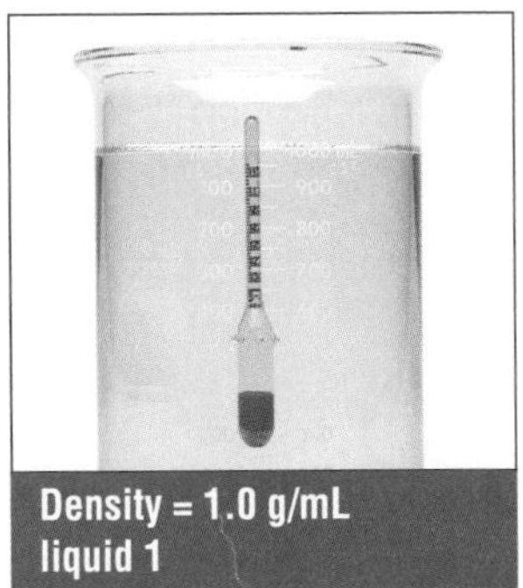
Density = 1.0 g/mL
liquid 1

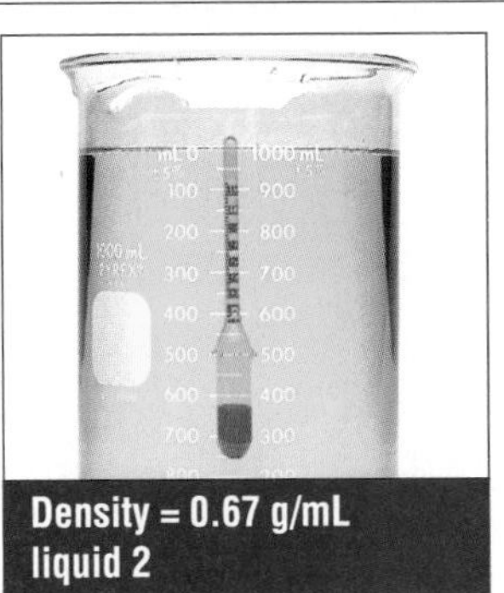
Density = 0.67 g/mL
liquid 2

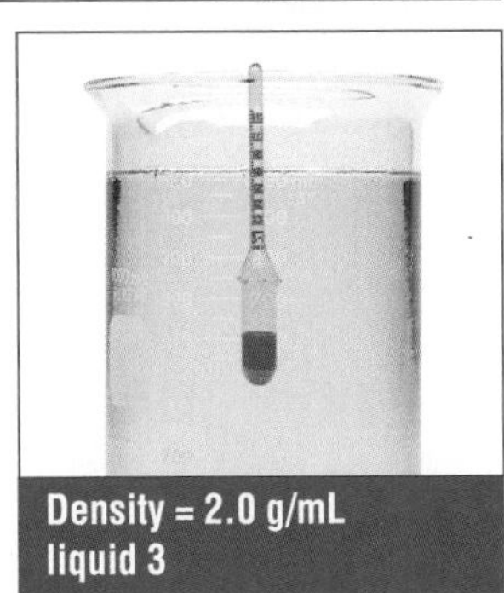
Density = 2.0 g/mL
liquid 3

Build Your Own Hydrometer

To make simple hydrometers, you can use a variety of objects, for example, pencils, rulers, and even candles. Anything that floats can function as a hydrometer. In this investigation, you will use a drinking straw to make your own hydrometer.

Problem

How can you build a homemade hydrometer?

Safety Precautions

Do not pour substances down the drain. Dispose of them as instructed by your teacher.

Apparatus

4 test tubes (or tall glasses)
test-tube rack
pencils or markers in different colours
plastic drinking straw
modelling clay
scissors

Materials

sand
paper funnel
water
salt
vegetable oil
ethyl alcohol
cloth or tissue

Procedure

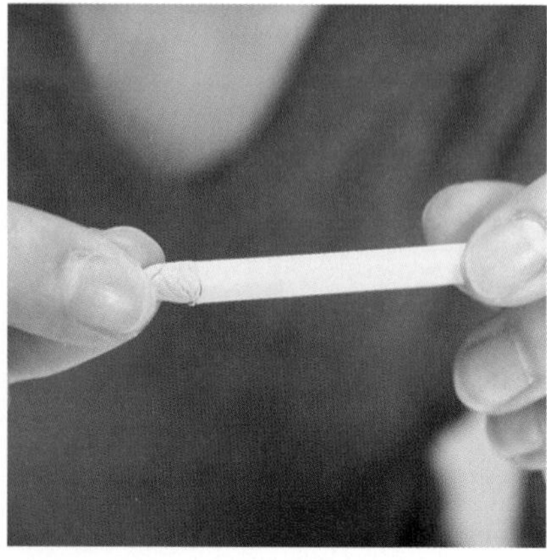

1. Cut the straw to a length of 10 cm. Plug one end of the straw with a piece of modelling clay.

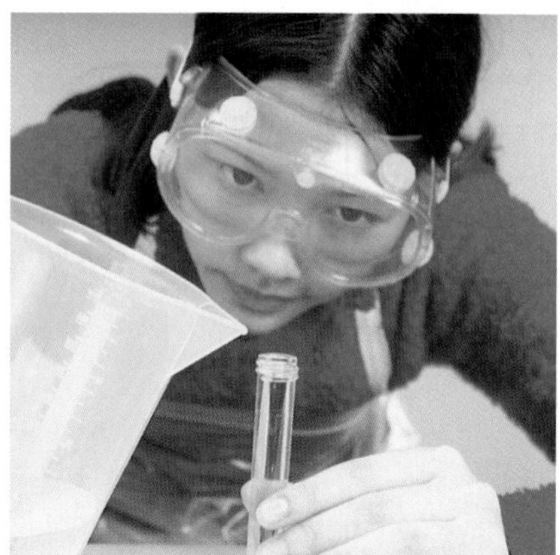

2. Slowly fill a test tube with water until the water bulges slightly above the rim.

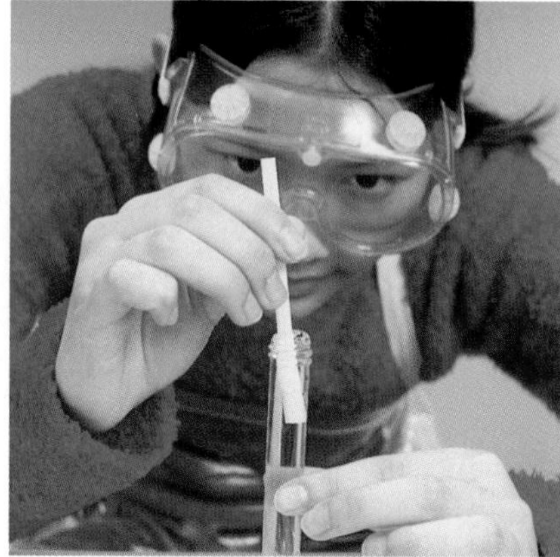

3. Place the full test tube in the rack. Put the straw, plugged end down, into the test tube filled with water (some water will spill out).

❹ Using a paper funnel, add sand to the straw until the straw floats upright in the test tube, with approximately 4 cm of the straw above the water. Notice that the bulge causes the straw to float in the centre of the test tube.

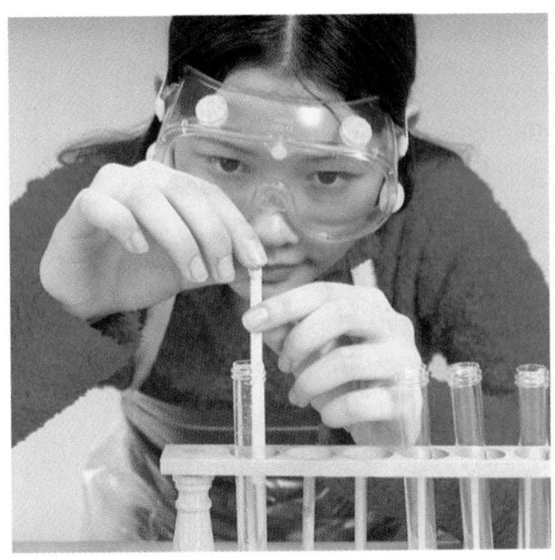

❺ Plug the open end of the straw with modelling clay.

❻ Mark a line for "water level" on the straw as it floats in the water. Your hydrometer is now ready to use.

❼ Place the hydrometer into liquids that have different densities: salt water, vegetable oil, and ethyl alcohol. Mark its floating level in each liquid, using a different colour for each. Wipe the hydrometer dry between uses. Wash your hands after this investigation.

Conclude and Apply

1. How does the hydrometer behave when it is placed in a liquid with a density that is less than the density of water? How does it behave when placed in a liquid whose density is greater than the density of water?
2. Can you use your hydrometer to rank liquids in order according to their densities? Explain your answer.
3. How might you calibrate or set your hydrometer to give a numerical value for the densities of these liquids? You might need to draw a sketch to explain your answer.
4. Use the word "density" to explain how a hydrometer works.

Extend Your Knowledge

5. Dissolve 100 mL of sugar in a glass of water. Place your hydrometer in the solution. Mark the surface line on the hydrometer. Repeat this step, each time adding and dissolving 100 mL of sugar. How does the density of the sugar solution change? Comment on how a hydrometer might be useful in determining the sugar content of a liquid in the food industry. (Remember to wash your hands after completing this extension activity.)

Across Canada

Dr. Verena Tunnicliffe

Did you know that ocean water covers about 70 percent of Earth's surface? As a high school student, Verena Tunnicliffe was torn between studying this fascinating fluid — or Russian history. The ocean won out. Today, Dr. Tunnicliffe is a professor at the School of Earth and Ocean Sciences, located at the University of Victoria in British Columbia.

Currently, Dr. Tunnicliffe is studying fish communities off the coast of British Columbia. She is especially interested in how global warming and other climate changes affect these communities. She is also investigating ecosystems surrounding deep-sea thermal vents. Thermal vents are ocean-floor openings that release hot water and gases from deep inside Earth. Certain species of bacteria thrive around deep-sea thermal vents. The bacteria serve as food for other species of clams, crabs, and sea worms, which are among the creatures Dr. Tunnicliffe studies.

Dr. Tunnicliffe conducts much of her research 2 or 3 km below the surface of the ocean. To descend to the bottom of the sea, she and her colleagues climb into a research submersible, which holds three people in a steel ball about 2 m across. Conditions in the submersible are cramped and cold, but most of the time Dr. Tunnicliffe doesn't mind. She is too interested in her research and what it may reveal. "In a rapidly changing world of great environmental challenges," she says, "even the distant ocean depths may hold important lessons for humanity."

Check Your Understanding

1. Copy and complete the table shown here.

Property	Measuring instrument	Unit
		g
density		
	graduated cylinder	
		N

2. A block of an unknown metal measures . The block has a mass of 235 g. Of what metal do you think the block is made? Would this metal float or sink in mercury?

3. (a) What does buoyant force mean?
   (b) What does displace mean?

4. Explain how you can make plastic sink and steel float.

5. What is meant by average density?

6. State Archimedes' principle.

7. (a) If the buoyant force is less than the weight of an object immersed in a fluid, what will happen to the object?
   (b) If the buoyant force equals the object's weight, what will happen to the object?
   (c) Give an example of what can happen when the buoyant force on an object is greater than the weight of the object.

8. **Design Your Own** Design an experiment to determine whether liquids can exert a buoyant force on other liquids. Make sure your procedure is approved by your teacher before you carry out your experiment.

CHAPTER at a glance

Now that you have completed this chapter, try to do the following. If you cannot, go back to the sections indicated.

Compare the densities of the solid, liquid, and gaseous states of a substance. (5.1)

Use the particle theory to explain how temperature affects density. (5.1)

Name each piece of equipment shown here and state whether it is used to measure mass or volume. In which units is mass measured? In which units is volume measured? (5.2)

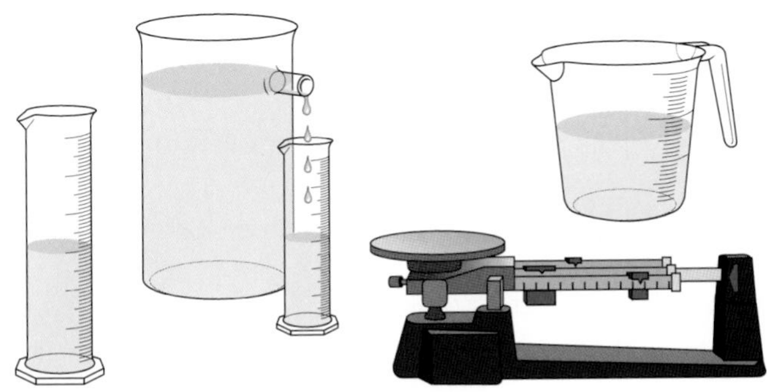

Explain how to measure the volume of a gas. (5.2)

State the formula for density, and include the units in the formula. (5.2)

Use the particle theory to explain buoyancy. (5.2)

Explain how sinking objects can float, and floating objects can sink. (5.3)

Explain how you can determine the average density of an object. (5.3)

Use a drawing to explain how you can determine the volume of an irregularly-shaped object. (5.2)

Restate Archimedes' principle in your own words. (5.3)

Use the particle theory to explain how Archimedes' principle can predict whether an object sinks or floats. (5.3)

List some applications for hydrometers. (5.3)

Prepare Your Own Summary

Summarize this chapter by doing one of the following. Use a graphic organizer (such as a concept map), produce a poster, or write the summary to include the key chapter ideas. Here are a few ideas to use as a guide:

- Why is a bag containing 1 kg of feathers much larger than a bag containing 1 kg of gold coins?
- What is the best method for determining how much of a substance occupies a certain space?
- Outline methods for measuring the volume of water, ice cubes, a water bottle, and water vapour.
- Explain why the lines representing pure substances on a mass vs. volume graph are straight.
- How can you make a pin float in water? In air?
- How can you make oil sink in water?
- How did Archimedes solve the problem of the crown?

- Use a diagram to explain Archimedes' principle.
- Make a labelled drawing to show the most important design features of a hydrometer.

CHAPTER

5 Review

Key Terms

buoyancy
buoyant force
density
mass
volume
capacity
weight
force
gravity
mass-to-volume ratio
floating
average density
displace
neutral buoyancy
Archimedes' principle
hydrometer

Reviewing Key Terms

If you need to review, the section numbers show you where these terms were introduced.

1. In your notebook, match the description in column A with the correct term in column B.

A

- the property of fluids that allows objects to float
- an instrument that measures density
- increases as the volume of a ship's hull increases
- This value equals the density of a substance.
- the pull of gravity on a mass
- the space occupied by an object
- opposes the force of gravity

B

- density (5.1)
- gravity (5.2)
- mass-to-volume ratio (5.2)
- volume (5.2)
- weight (5.2)
- average density (5.3)
- buoyancy (5.3)
- buoyant force (5.3)
- hydrometer (5.3)

Understanding Key Ideas

Section numbers are provided if you need to review.

2. Compare and contrast the following:
   (a) mass and weight (5.2)
   (b) mass and density (5.2)
   (c) weight and buoyancy (5.3)
   (d) density and average density (5.3)

3. Give an example for each pair of terms in question 2. (5.2, 5.3)

4. **(a)** Explain the meanings of "weight" and "buoyant force." (5.2, 5.3)
   (b) How could a knowledge of buoyancy make the work easier when you are clearing rocks from a swimming area? (5.3)

5. Restate Archimedes' principle in your own words. (5.3)

Developing Skills

6. Copy and complete the following spider map of buoyancy.

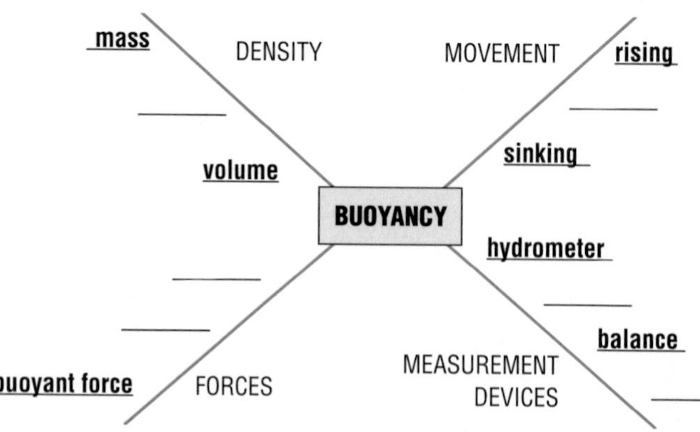

7. The graph on the next page shows the density of three different substances. (5.2)
   (a) Which substance has the largest mass when the volume is 50 ?

(b) Which substance takes up the most space at 100 g?

(c) Calculate the mass-to-volume ratio of the lines in the graph.

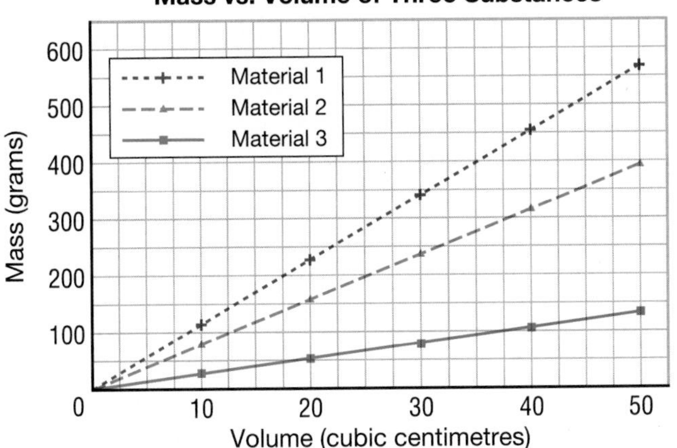

8. (a) Plot the following data on a line graph representing mass vs. volume:

Mass (g)	Volume (cm^3)	Mass-to-volume ratio (g/cm^3)
15.74	15.74	
39.35	39.35	
55.09	55.09	
82.96	82.96	
94.44	94.44	

(b) Calculate the mass-to-volume ratio for each mass.

(c) Be a sleuth and identify this mystery substance from the densities listed in Table 5.1 on page 141.

(d) Where would the line for a lower-density substance fit on your graph? For a higher-density substance? Indicate these lines on your graph.

Problem Solving/Applying

9. Design a hydrometer made from a pencil. What are some problems you might encounter? How might you solve these problems?

10. Explain how you could adjust the drinking-straw hydrometer you made so that it could measure extremely high densities.

11. Design and conduct an experiment to determine the density of a 25-cent coin. Based on your results, what metals do you think are used in these coins? (Refer to Table 5.1 on page 141.)

12. Cassie built a model boat with a mass of 320 g. When she tried it out, she found that it displaced 260 g of water. Did the boat sink or float? Explain.

13. (a) How can you make a substance that is *less dense* than water sink? Explain.

(b) How can you make a substance that is *denser* than water float? Explain.

Critical Thinking

14. Use the particle theory to explain why the first point on any mass vs. volume graph is always (0,0).

15. Do you think density and viscosity are related? Provide one example that demonstrates that they are related and one that demonstrates that they are not related. Use the particle theory to suggest an explanation.

16. In fresh water, an ice cube floats with about nine tenths of its mass below the surface. Is this true for an iceberg in seawater? Explain.

Pause& Reflect

1. In the human body, some materials are solids and some are liquids. Gases are found in the lungs and respiratory system. How might you determine the density of at least three of the substances making up the human body?
2. Go back to the beginning of this chapter on page 130 and check your original answers to the Getting Ready questions. How has your thinking changed? How would you answer these questions now that you have investigated the topics in this chapter?

CHAPTER 6

Pressure in Fluids

Getting Ready...

- Why does pop become more fizzy when you shake it?
- How does water get to the top floor of an apartment building?
- How do the Jaws of Life work?

Science Log

Ask friends or family members to tell you what they know about water pressure and air pressure. Together, try to answer the questions above in your Science Log.

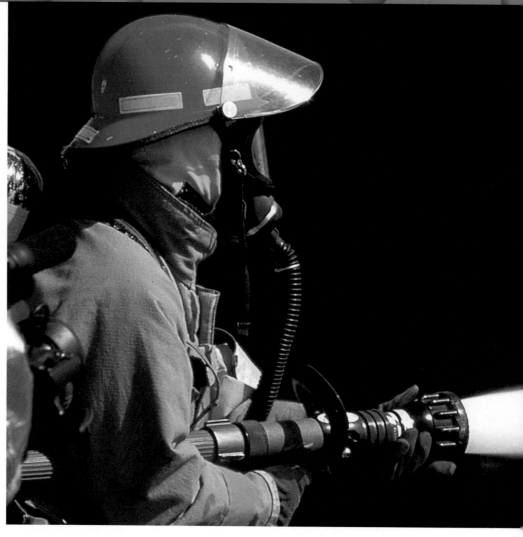

Fighting fires is a high-pressure job, in more ways than one. **Pressure** is the force acting on a certain area of surface. Without fluid pressure, the water from the hoses could not reach the fire, nor could the air in the firefighters' tanks reach their lungs to protect them from the choking smoke. The pressure gauges on the fire trucks are carefully checked every day so that firefighters can do their jobs properly.

Everyone, not just firefighters, relies on fluid pressure in many different ways. For example, many foods and beverages are canned and bottled under pressure to preserve their quality and to keep them safe to eat. Vacuum cleaners, automobile brakes, toilets, dentists' chairs, inflatable mattresses and toys — even our bodies — all depend on the pressure of fluids in order to function.

Spotlight On Key Ideas

In this chapter, you will discover

- how external pressure affects the behaviour of fluids
- why pressure in fluids varies with depth and with altitude
- how fluid pressure is affected by temperature
- why gases are compressible but liquids are incompressible
- how liquids compare to air as transmitters of force
- how pneumatic and hydraulic systems depend on fluid pressure

Starting Point

Balloon Lift

See if you can use a balloon to lift some books.

What You Need

straw
balloon
strong tape
textbooks
strong elastic band

What to Do

1. Insert a straw into the mouth of a balloon. Seal the straw to the balloon with tape. Reinforce the seal by winding the elastic band around the balloon seal.

2. Place a book on the balloon. Blow steadily into the straw and observe what happens.
3. Place more books on top of the first one, one book at a time. Record how many books you can lift by blowing into the straw.

What Did You Find Out?

1. What happened to the balloon as you blew into it? Which part of the balloon inflated first?
2. Was it harder to blow up the balloon with two books on top of it rather than one?
3. Which part of the air-filled balloon is doing the work to lift the books?

Spotlight On Key Skills

In this chapter, you will

- use a formula to calculate pressure
- compare the compressibility of solids, liquids, and gases
- predict and test the effect of depth on water pressure
- construct your own barometer to measure atmospheric pressure
- observe how pressure is transmitted in a simple pneumatic system
- design and build your own simple pneumatic and hydraulic systems

6.1 All About Pressure in Fluids

Figure 6.1 Blaise Pascal (1623–1662)

Every time you lean against a wall, you are exerting pressure on the wall. Pressure is a measure of the force acting perpendicular to a unit area. When you press your hand against a wall, you are *applying pressure* on that particular *area* of the wall. If the wall were made of whipped cream, your hand would push right through the whipped cream, leaving a hand shape that is the outline of the area over which the force was applied. If the force is increased, the pressure will increase. What happens if the area is increased? Do the Find Out Activity on page 165 to see for yourself.

A Formula for Pressure

Pressure can be calculated by using the following formula:

$$\text{Pressure } (P) = \frac{\text{Force } (F)}{\text{Area } (A)} \text{ or } P = \frac{F}{A}$$

Force is measured in newtons (N) and area is often measured in square metres (m^2). The unit for pressure, therefore, is newtons per square metre (N/m^2). This unit is also called a **pascal** (Pa), named after the French scientist Blaise Pascal (1623–1662) in honour of his pioneering work with pressure. A **kilopascal** (kPa) is equal to 1000 Pa.

Imagine a cubic aquarium that is 1 m × 1 m × 1 m. Suppose it is full of water, and a pressure gauge (an instrument for measuring pressure) is attached to the bottom of the aquarium (see Figure 6.2). What pressure reading would the gauge show? The answer to this question would depend on two factors: the weight of the water, and the area of the base of the aquarium.

What is the weight, or force of gravity, of this amount of water? First, find the volume of the cubic aquarium:

$$\text{Volume} = \text{length} \times \text{width} \times \text{height}$$
$$V = l \times w \times h$$
$$= 1 \text{ m} \times 1 \text{ m} \times 1 \text{ m}$$
$$= 1 \text{ m}^3$$

Math CONNECT

Imagine a cube of water that is 2 m × 2 m × 2 m. What pressure would this imaginary cube exert on a table? Use the formula for pressure in your calculations.

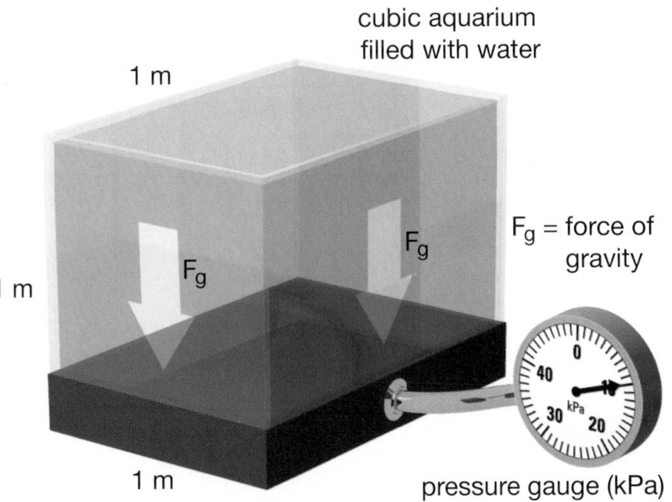

Figure 6.2 The water in the aquarium weighs 10 000 N. In other words, the water is exerting a force of 10 000 N on the area of the base of the cubic aquarium.

Recall from Chapter 5 that the density of water is 1 g/cm³, or 1000 kg/m³ (that is, a 1000 kg mass of water occupies a 1 m³ volume of water). If 1 m³ of water has a mass of 1000 kg, and if the force of gravity is approximately 10 N for every kilogram (9.8 N/kg), then one cubic metre of water would exert a force of approximately 1000 kg × 10 N/kg. This equals a weight of 10 000 N.

When the 10 000 N "cube" of water in the aquarium rests on a surface, its weight pushes down over a certain area. Determine the area of the base of the cubic aquarium:

$$\begin{aligned} \text{Area} &= \text{length} \times \text{width} \\ A &= l \times w \\ &= 1\ \text{m} \times 1\ \text{m} \\ &= 1\ \text{m}^2 \end{aligned}$$

The area of the base of the cube is 1 m². You can now calculate the pressure exerted by the cube of water:

$$P = \frac{F}{A}$$

If $F = 10\ 000$ N and $A = 1\ \text{m}^2$, then

$$\begin{aligned} P &= \frac{10\ 000\ \text{N}}{1\ \text{m}^2} \\ &= 10\ 000\ \text{N/m}^2 \\ &= 10\ 000\ \text{Pa} \\ &= 10\ \text{kPa (kilopascals)} \end{aligned}$$

Find Out **ACTIVITY**

Pop 'em Quick!

Suppose you were in a contest to see who could pop the greatest number of balloons in 1 min. What could you do to pop the balloons as quickly as possible?

What You Need

2 balloons   straight pin

Safety Precautions

Be careful when using sharp objects such as straight pins.

What to Do

1. Blow up both balloons to approximately the same size. Knot the end of each one.
2. Set one balloon on a table. Push your index finger into the balloon until it pops. (You may need to steady the balloon with your other hand.)
3. Repeat step 2 using the straight pin instead of your finger.

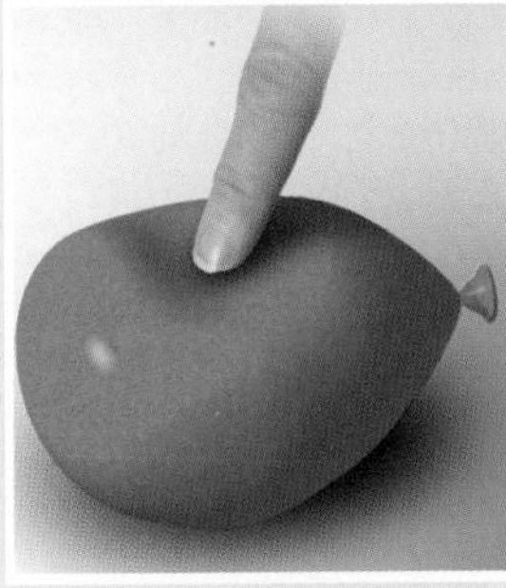

What Did You Find Out?

1. Which method required less force to pop the balloon? Which method was faster?
2. Which "popping tool" had the smaller surface area: your finger or the straight pin?
3. Which popping method required more pressure?

Pause& Reflect

How does water behave when you fill a bathtub or a swimming pool? Does the end of the tub near the faucet or the part of the pool nearest the hose fill up first? What about an inner tube for a tire? Does the part nearest the spout fill up before the rest of the tire? Think about these questions and write your answers in your Science Log.

Pressure and the Particle Theory

The particle theory states that particles in solids, liquids, and gases are constantly moving. Particles move quickly if they have a great deal of energy. They move more slowly if they have less energy. When particles move, there is always a chance that they will bump into each other, like bumper cars. When a collision occurs, the particles move apart, leaving only empty space between them.

Why do fluids, such as juice in a cup or air in a tire, appear to be at rest if all their particles are moving? Moving particles exert a force in the direction of their motion. Fluid particles are moving in all directions at all times. Thus, most of the forces cancel each other out, but some are not cancelled. These forces are exerted against the walls of the container, causing pressure. What happens if there is a crack in a cup or a hole in a tire? In which direction does the fluid flow? Regardless of where the crack or hole is, the result is always the same: the fluid *flows out*. This indicates that the *pressure of a fluid is exerted equally in all directions*, as shown in Figure 6.3. As the particles in the inner walls of the container apply pressure on the fluid to stay inside the container, the particles of the fluid press against the container with an equal force.

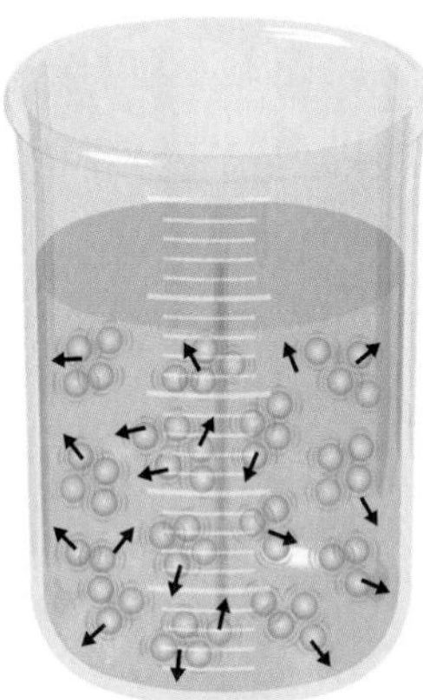

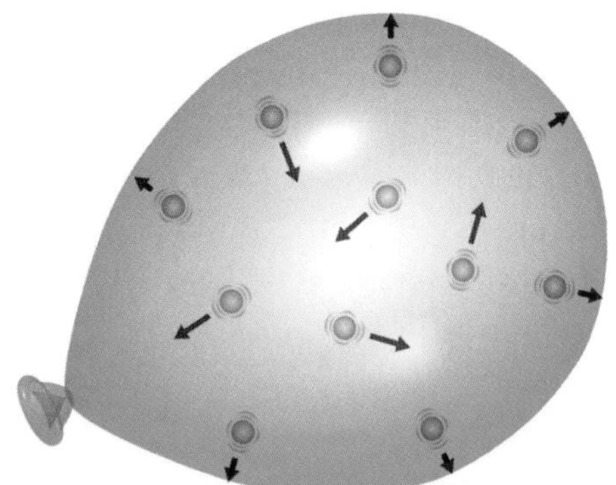

Figure 6.3 The pressure of a fluid is exerted equally in all directions.

DidYouKnow?

Squeeze a plastic ketchup or mustard container, and your hamburger bun is coated. Why? Because of a useful property of fluids discovered by Blaise Pascal and now known as Pascal's law. You will learn more about how we make use of this property in Unit 5.

Compressibility

The particles in a fluid push against each other until something more rigid, for example, the walls of their container, exerts a force in the opposite direction. What would happen if a fluid were completely enclosed and pressure were exerted on the walls of a flexible container? For example, what happens when you squeeze a bottle filled with water? Is it possible for the particles of the fluid to move closer together when this pressure is applied?

According to the particle theory, the amount of empty space between particles depends on two factors:

1. the physical state of the substance (whether the substance is a solid, liquid, or gas)
2. the amount of energy that the particles have

In general, the amount of empty space between solid particles and between liquid particles is small, but the amount of empty space between gas particles is huge.

An interesting property of gases, then, is **compressibility**, the ability to be squeezed into a smaller volume. Gases are **compressible** because gas particles are extremely far apart (see Figure 6.4A). However, the particles remain far enough apart to behave like a gas, even if the gas is compressed (see Figure 6.4B).

Although there is empty space between the particles of solids and liquids, the spaces are already almost as small as possible. Thus, when a force is applied to a solid or a liquid, the particles cannot move much closer together. Because solids and liquids cannot be squeezed into a smaller volume, they are said to be almost **incompressible**. Instead of changing the volume of either the solid or the liquid, the applied force is **transmitted** (passed along), from one particle to the next, throughout the substance, somewhat like falling dominoes (see Figure 6.5).

In the investigation on the next page, you will examine compressibility in greater detail.

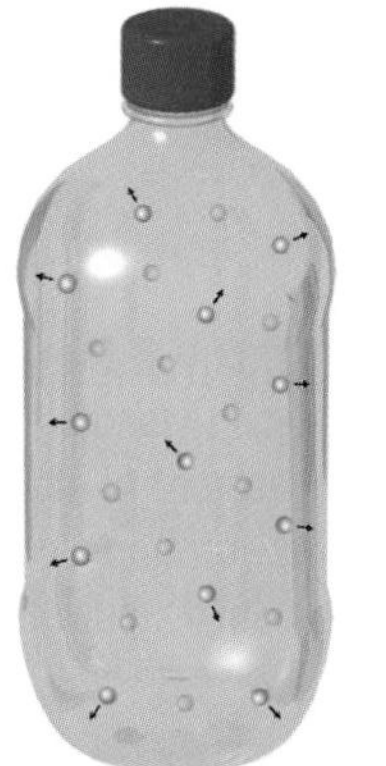

Figure 6.4A The particles of gas in this closed, flexible bottle are extremely far apart. Thus, the gas can be compressed (squeezed) into a smaller volume.

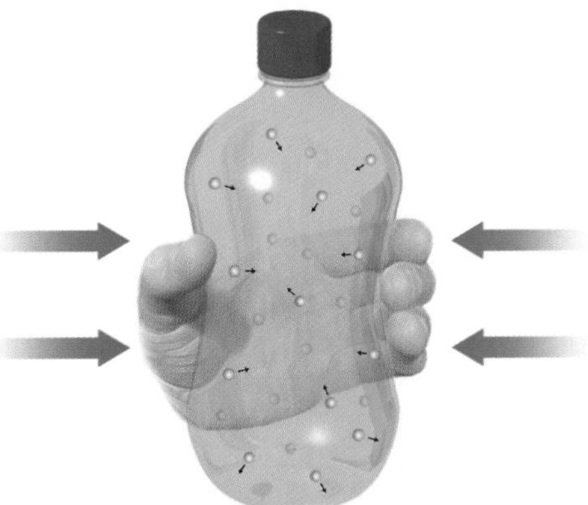

Figure 6.4B When external pressure is applied to the bottle, the gas particles are compressed, or forced closer together.

Figure 6.5 An applied force is transmitted from one particle to the next, in both solids and liquids. This is similar to dominoes falling in a row.

INVESTIGATION 6-A

Bottle Squeeze

In this investigation, you will test the compressibility of solids, liquids, and gases.

Problem

How do the compressibilities of solids, liquids, and gases compare?

Safety Precaution

Dispose of materials as instructed by your teacher.

Apparatus

500 mL beaker (or measuring cup)

rubber gloves

3 empty plastic pop bottles with twist-off caps (500 mL each)

Materials

water

sand

Procedure

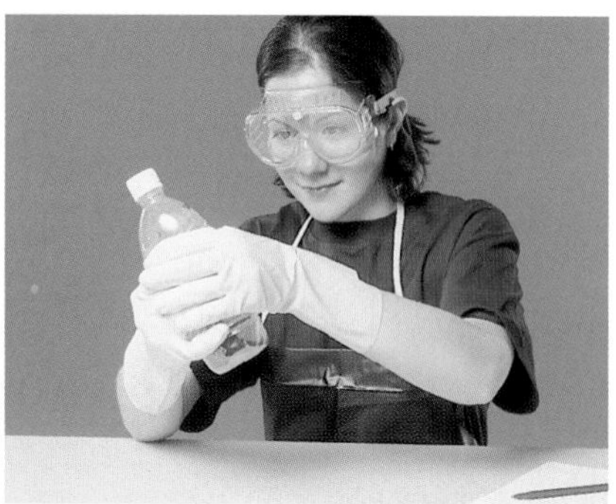

1. Twist the cap *tightly* on an empty pop bottle.

2. Squeeze the bottle as hard as you can. Estimate how much of the original volume you could compress — one quarter, one third, one half, more? Record your estimate.

3. Find the precise volume of the bottles. Although these bottles are meant to hold 500 mL of pop, the bottle might actually have a larger capacity. Fill the bottle right to the top with water. Then use the beaker or the measuring cup to measure the total actual volume of the bottle. Record the total capacity of the bottle.

4. Half-fill the other two bottles, one with water and one with sand. Repeat step 2 for each of these bottles.

Skill POWER

To review how to use a graduated cylinder, turn to page 540.

Skill POWER

To review how to organize data in a table, turn to page 546.

5. Completely fill each bottle: the water bottle with more water, the sand bottle with more sand. *Make sure no air is left inside the bottles.*
   (a) Fill the water bottle until the water rises to mid-neck. Wait about 5 min.
   (b) Then, very carefully, pour more water into the bottle until the water bulges at the top.

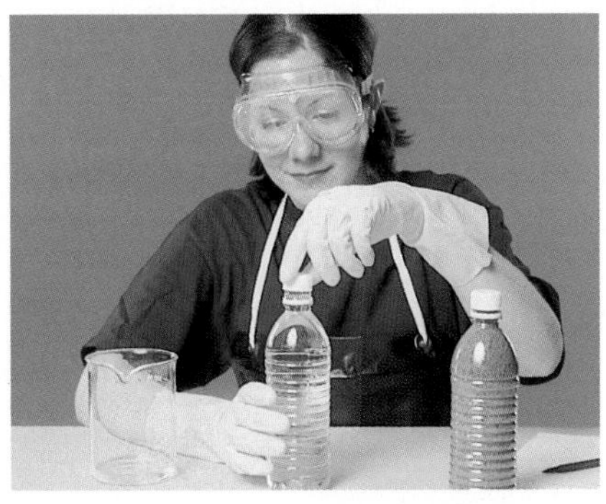

6. Twist the cap tightly on the bottle of sand, then on the water bottle. You should see some water leaking out as you do this. *Be careful not to squeeze the bottle while you are twisting the cap shut.*
   (a) Repeat step 2 for each bottle.
   (b) Organize all the data you have collected in a table and give your table a title.

Analyze

1. (a) How did your ability to compress the bottle containing water change as the amount of water increased?
(b) How did your ability to compress the bottle containing sand change as the amount of sand increased?

Conclude and Apply

2. (a) How does the compressibility of a gas compare to the compressibility of a solid?
(b) How does the compressibility of a gas compare to the compressibility of a liquid?

3. How does the compressibility of a liquid compare to the compressibility of a solid?

4. If a car ran over the water-filled bottle, what do you think would happen? Would the water inside the bottle compress as the bottle flattened, or would the bottle burst as the plastic gave way? Explain your answer.

Extend Your Knowledge

5. Would a force greater than the force exerted by your two hands be able to compress the water-filled bottle and the sand-filled bottle? Find or design a device that could exert a greater, controlled force on your sample bottles.

Pause& Reflect

Think about equipment used in your home, at school, in grocery stores, and in hospitals, and make a list of devices, mechanisms, or situations in which compression can occur. In each case, decide whether the compression is desirable or undesirable. How might compression be prevented when it is not needed? Record your answers in your Science Log.

Remember the Cartesian diver from Chapter 5? When you squeezed the bottle, the diver sank to the bottom. Explain how the Cartesian diver works in terms of air compression.

Under Pressure

Imagine that you and several of your friends are imitating the behaviour of gas particles under pressure, and you are "forced" to squeeze into a small space, such as a telephone booth. What would your main goal be soon after the door of the telephone booth closed? You would probably want to get out! Similarly, gases under pressure are ready to expand again, because the particles have so much energy. If gases under pressure find a way to escape from a container — for example, through a nozzle or a hole — they exit the container with a great deal of force. That force can be used in many applications to push or move objects, as you saw in the Starting Point Activity on page 163. The compressibility of gases is useful for storing gases in a small volume (for example, in oxygen tanks). Compressibility is also useful in helping people do work. Later in this chapter, you will see how certain tools depend on the compressibility of gases, as well as the incompressibility of liquids.

Pressure Changes with Depth

You experience pressure exerted by fluids every day. Wind is moving air that can lift your hair or push against you. Water running out of a faucet can rinse away bits of food from a plate. This type of pressure is called **flow pressure**, or pressure that causes motion because the fluid is moving. Fluid pressure can also exert a force on an object even if the fluid is not moving. This is known as **static pressure**. The plumbing system in your home uses both types of pressure, as shown in Figures 6.6A and B. When a water faucet is turned off, the water under pressure in the pipes is an example of static pressure. As soon as you turn the faucet on, water flows out of the tap. This is an example of flow pressure.

You can feel the effects of static pressure in a swimming pool. You may have noticed your ears begin to hurt if you swim deep underwater. This discomfort is caused by water pressure on your eardrums. Why do you feel more pressure the deeper you go underwater? The weight of all the water — and the air — above you pushes down on you and the water below. Since gravity pulls everything toward the

Did You Know?

Air pressure is even greater at locations on Earth's surface that are below sea level, for example, in Death Valley in California.

Figure 6.6A Static pressure of water sitting in pipes

Figure 6.6B Flow pressure of water running out of a faucet

centre of Earth, all particles have weight. As suggested by the particle theory, fluid particles exert their weight on the particles beneath them. Therefore, water pressure is greater the deeper you go underwater. The same is true for gases, such as air. Air pressure is greater at sea level than at higher altitudes, and it decreases as you rise in the atmosphere. You will learn more about air pressure in the next section. Examine the relationship between fluid pressure and depth in the next investigation.

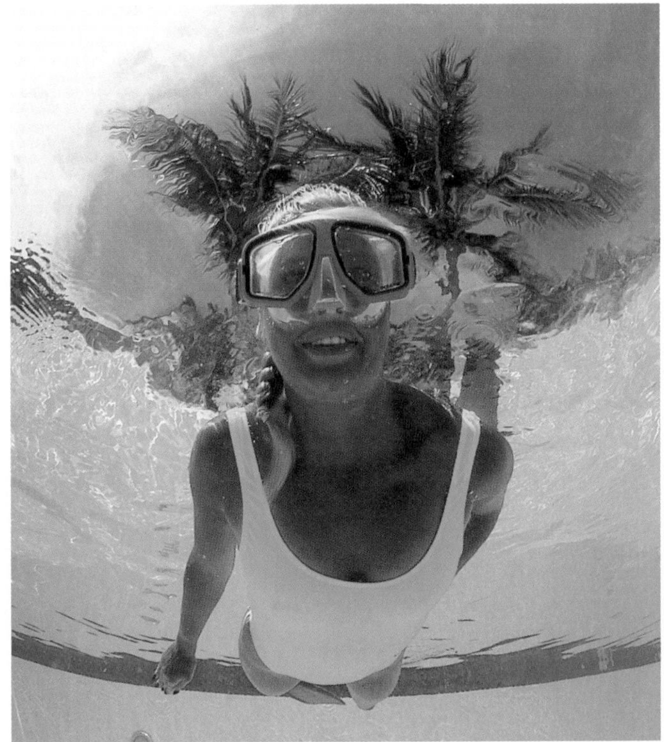

Figure 6.7A Have you felt a pain in your ears when you swam underwater?

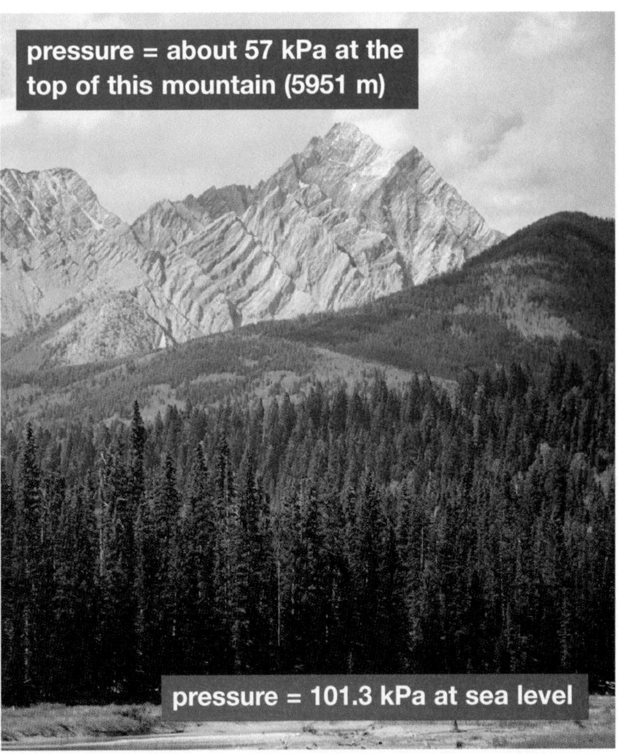

Figure 6.7B Air pressure decreases with increasing altitude.

DidYouKnow?

Bottled pop on store shelves does not look carbonated because it has been bottled under pressure. To manufacture pop, carbon dioxide gas is forced to dissolve into the flavoured and sweetened water solution called the "syrup." The bottle cap is secured before the pressure is released, and the strength of both bottle and cap prevents a loss of pressure. If you open a bottle of pop quickly, however, the pressure inside the bottle is released suddenly and the carbon dioxide escapes rapidly from the solution, forming bubbles, or fizz. The more quickly you open the bottle, the more fizz is produced. Shaking the pop ahead of time, or warming the bottle in the Sun, gives the carbon dioxide particles more energy. Thus, more carbon dioxide escapes when the pressure is released than if the bottle were not shaken, or if the bottle were taken out of the refrigerator.

CONDUCT AN

INVESTIGATION 6-B

Pressure Puzzle

In this investigation, water will be allowed to escape out of a milk carton through holes made at different heights on one side of the carton. Water pressure will cause the water to shoot out of the holes a certain distance horizontally. Predict whether all the distances the water travels will be the same or different. Record your prediction and then test it.

Problem

How does depth affect water pressure?

Safety Precautions

- Be careful when you poke holes in the milk carton, and when using sharp objects such as scissors.
- Do this investigation outdoors or in a sink or plastic tub to prevent a watery mess.

Apparatus

large beaker or pitcher

pen or other object for poking holes

wide tape (duct tape or packing tape)

plastic tub or sink

scissors for cutting tape

Materials

water

empty milk carton with a twist-off cap or juice carton with a twist-off cap

Procedure

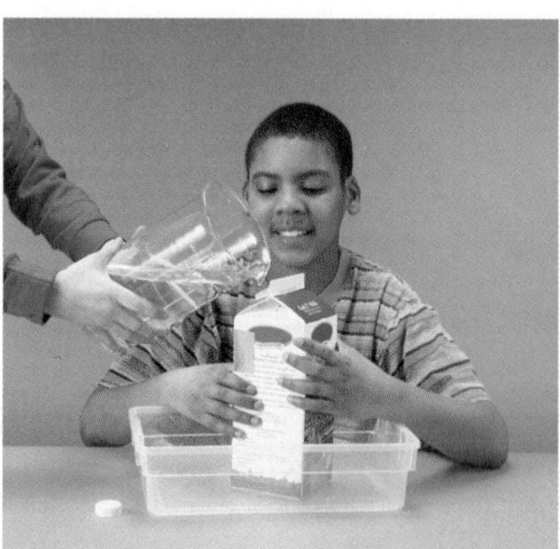

1. Using the tip of a pen or some other tool, poke several holes (about 2 mm in diameter) into one side of the empty carton near the top. Poke one more hole in the bottom of the carton.
2. Cover the holes in the carton with tape.
3. With the help of a partner, fill the carton to the top with water. Then replace the twist-off cap.

Skill POWER

For tips on scientific drawing, turn to page 557.

4 Turn the carton upside down and position it in the centre of the tub or sink.

5 Remove the tape from the hole in the bottom of the carton first. Then remove the tape from the side holes all at once.

6 Observe the water escape from the holes into the tub until the carton is empty. Record your observations in a series of labelled drawings. Wipe up any spills.

Analyze

1. (a) From which holes did the water travel the greatest horizontal distance?

   (b) From which holes did the water travel the smallest horizontal distance?

Conclude and Apply

2. If water exerted pressure in only one direction, for example, *downward*, then the water could not push *sideways* to shoot out of the holes of the carton. Explain how this investigation demonstrates at least two directions, other than down, in which water pressure is exerted at any given moment.

3. Does depth affect water pressure? How could you tell?

4. What can you conclude about the pressure exerted throughout a fluid for a certain depth?

Extend Your Skills

Poke a hole near the bottom of a very tall container. Repeat the investigation (but do not turn the container upside down) for three different depths of water: full container, two thirds full, and one half full. Measure how far the water shoots out each time. Make a line graph to show the relationship between distance of water flow and depth of water. Remember to title your graph and label the *x*-axis and the *y*-axis.

Cool Tools

As you learned in Chapter 5, submersibles are small diving vessels designed to transport two or three people at a time. These vessels must withstand the severe water pressure at 4000 m below sea level. This is the depth at which the footage of the *Titanic* wreck was filmed. To resist this great pressure, the glass portholes of the submersible must be 15 cm thick! However, glass this thick would distort the image photographed by the camera through the porthole. Therefore, the film's director, Canadian James Cameron, asked his brother Michael, an aerospace engineer, to design a custom-built remote camera encased in a titanium shell. This superstrong casing can withstand tremendous water pressure. Filming at such great depths is still risky, however. If the water pressure were to blow away part of the camera's titanium casing, the fragments could fling back and hit the submersible, causing it to implode (collapse inwards) in a fraction of a second.

Pressure and Temperature

The particle theory suggests that particles move faster when they are heated because they gain more energy. Keeping the particle theory in mind, compare Figures 6.8A and B, representing the behaviour of air particles inside two balloons.

- In which balloons are the particles of air colliding with each other and with the inner walls of the balloon with greater force?
- Pressure is force measured over a certain area. In which balloon are the particles of air exerting greater pressure against the inner walls of the balloon?
- Does an increase in temperature cause an increase or a decrease in pressure?

Figure 6.8A The particles of air inside this chilled balloon are moving slowly and are colliding infrequently with the particles that make up the inner walls of the balloon.

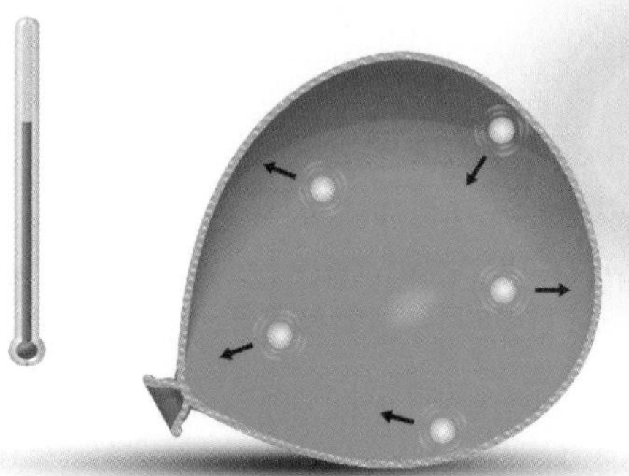

Figure 6.8B The particles of air inside this balloon are moving faster and are colliding frequently with the particles that make up the inner walls of the balloon.

Check Your Understanding

1. Define pressure.
2. (a) What unit is used to measure pressure?
   (b) What is another way to express this unit?
3. Explain why fluid pressure varies with depth in (a) liquids, and (b) gases.
4. **Apply** Why do you think airplane cabins are "pressurized"? If an airplane door were to open at a high altitude, in which direction would the air move — into or out of the airplane? Why?
5. **Thinking Critically** The atmosphere is a layer of gases that surrounds Earth and is approximately 1000 km high. Normal atmospheric pressure at sea level, at 25°C, is about 101 300 Pa. As you rise through the air, atmospheric pressure decreases steadily. Imagine that you are in a hot-air balloon, and you need to measure the air pressure regularly, from the time the balloon starts to rise off the ground. Without doing any calculations, what would you predict the reading to be when you reach the outermost limit of Earth's atmosphere? Explain your reasoning.

6.2 Water Pressure and Hydraulic Systems

When water and other liquids are at rest, the particles push at 90° against, or *perpendicular* to, the walls of the container, as shown in Figure 6.9. Due to static pressure, water would gush out of a crack that occurred anywhere along the pipe.

In Conduct an Investigation 6-A, you saw that as much as you tried to compress a sealed liquid by squeezing the walls of its container, you could not do it. Liquids are incompressible. What happens to the force that you exert? As long as the liquid is *continuous* and *enclosed* in a tube or a pipe, the force will be transmitted along the liquid until something moves or bulges. The pressure produced in this way is the same everywhere in the liquid and is exerted in all directions equally. Observe this for yourself in the following Find Out Activity.

Figure 6.9 Static pressure of water in pipes causes the water particles to push at 90° against the inside walls of the pipes.

Find Out **ACTIVITY**

Balloon Bulger

Find out how squeezing a plastic bottle in one place can cause a reaction somewhere else.

What You Need

rubber gloves
strong elastic bands
small, flexible plastic bottle with a narrow spout
water
balloon
deep pail

What to Do

1. Fill the pail with water. Submerge the bottle and the balloon. Make sure all the air has escaped from both objects and has been replaced with water.
2. While holding both objects underwater, stretch the opening of the balloon over the opening of the bottle. You should have a completely enclosed and continuous liquid inside (no air). Use one or two strong elastic bands to make sure the balloon is securely fastened to the bottle.
3. Press the bottle and observe what happens. Record your observations.
4. Now press the balloon and observe what happens to the bottle. Record your observations.

What Did You Find Out?

1. What makes the balloon bulge?
2. The amount of pressure you put on the bottle in step 3 may affect your results in step 4. In step 4, under what conditions does the bottle bulge? When does it not bulge?

Hydraulic Systems

The Balloon Bulger uses the incompressibility of water to cause motion. The force that you apply on the bottle is transmitted through the liquid to make the water particles move. When the water particles move into the balloon, they make the flexible walls of the balloon move.

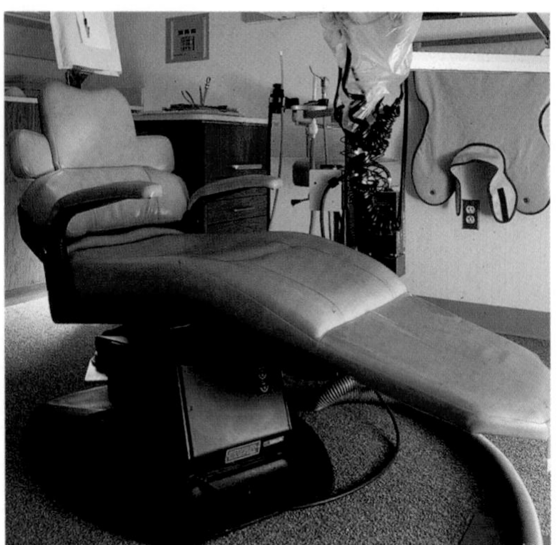

Hydraulics is the study of pressure in liquids. Devices that transmit applied forces through a liquid to move something else, because of pressure, are called **hydraulic systems.** In most hydraulic systems, a force is exerted on a continuous, enclosed liquid. This applied force creates pressure that moves the liquid through a series of tubes, pipes, or hoses, which causes motion at the other end of the system. The Balloon Bulger is a simple model of a hydraulic system. It behaves like the lifting mechanism that moves a barber's or dentist's chair, for example (see the photograph on the left). Try making a simple model of a hydraulic system in the following Find Out Activity.

Find Out **ACTIVITY**

Simple Hydraulics

Modified syringes filled with water and joined with plastic tubing provide a simple model of a hydraulic system.

What You Need

2 modified syringes
short piece of plastic tubing
glass of water

What to Do

1. Fill the cylinder of one syringe (the "main cylinder") with water by inserting the cylinder tip into a glass filled with water and pulling back the plunger (see Diagram A).
2. Attach a piece of tubing to this syringe. Push the plunger until the tubing is filled with water (see Diagram B).
3. Attach the cylinder of the other syringe (the "reacting cylinder") to the other end of the plastic tubing (see Diagram C). Make sure the plunger of the reacting cylinder is completely pushed in before connecting the tubing!
4. Push the plunger of the main cylinder in all the way. **CAUTION:** Never point the tubing or syringe toward anyone when expelling excess fluid.

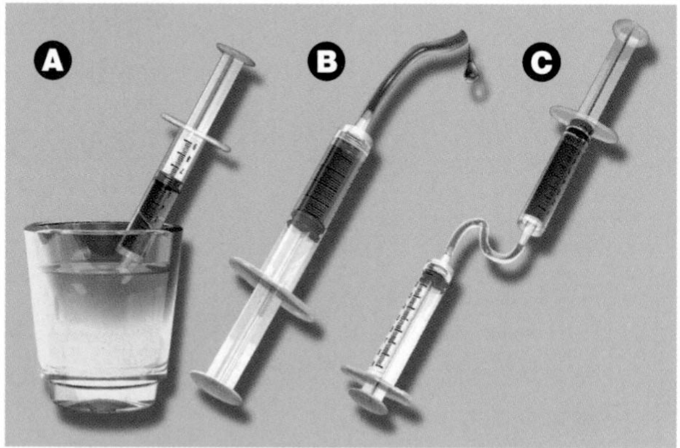

What Did You Find Out?

What happens when you apply a force on the plunger of the main cylinder? Explain your observation using the term "hydraulic system."

Hydraulics in Action

Rescue workers use hydraulic-powered devices to cut away or move heavy metal at the scene of automobile accidents. One such device is called the Jaws of Life. A powerful pump is used to move the hydraulic fluid continuously through the hoses to produce a strong force in each attached rescue tool. The three types of tools that can be attached to the Jaws of Life are shown directly below. The spreaders can be used to pry open a door, for example. However, their spread is limited. When an opening is made large enough, a ram can be fitted into the space to spread the opening even farther apart. Rams come in different sizes, depending on how large an opening is needed. The cutters act like huge pruning shears and can cut through thick metal and other materials.

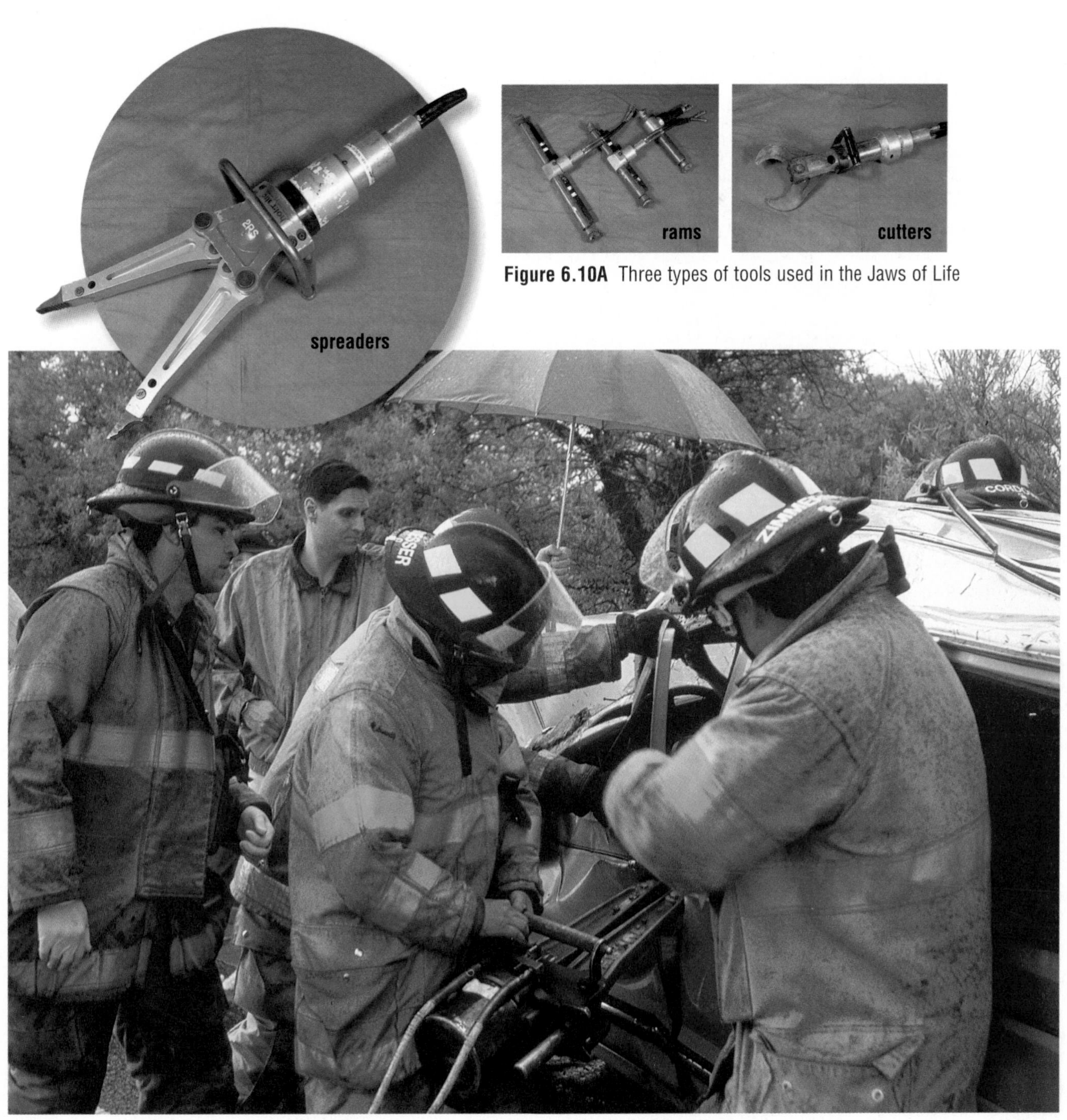

Figure 6.10A Three types of tools used in the Jaws of Life

Figure 6.10B Firefighters are using the Jaws of Life to rescue an accident victim from a crushed car.

INVESTIGATION 6-C

Make a Model of a Dentist's Chair

A dentist's chair can be lowered, raised, and tilted by means of a hydraulic system. Use your knowledge of hydraulics and some everyday materials to design and build your own working model of a dentist's chair.

Challenge

Create a working model that simulates the movement of a dentist's chair.

Materials

modified syringes; plastic tubing; water; jinx wood, plywood, stiff cardboard or corrugated plastic; masking tape; elastic band; glue gun; butterfly fasteners or paper fasteners; dowelling; thread spool; scissors or cutting knife; pencil or skewer for punching holes

Safety Precautions 

- Be careful when using sharp tools such as scissors, knives, or skewers.
- A glue gun is hot and the glue remains hot for several minutes.

Design Criteria 

A. Your team's model must use hydraulics to exert a large force, using minimal space.

B. Your model hydraulic system must transmit motion smoothly.

C. You must be able to operate your model with no breakage of parts.

Plan and Construct

1. Work with your team members to make initial concept sketches showing how your model will work. Draw a side view and a top view. Include all measurements and dimensions. Label the parts of your model.
2. Decide which materials to use as components. Select the materials to use as connecting parts or hinges in your model.
3. Assemble your model and test it to see if it works.

Evaluate

1. Does your team's model meet all the specified design criteria?
2. Did the members of your group work successfully as a team in designing and building the prototype? If not, how might problems or conflict be avoided next time?
3. Compare your team's model with those constructed by other groups. How would you rate your model compared to the others? Not as effective? As effective? Superior? Explain your rating.
4. In what ways might you improve your design?

Hydraulics to Transport Fluids

Just as water gushes out of an open faucet, liquids under pressure flow away from the applied force in all directions. Hydraulic systems can be used to transport fluids over large distances. The ancient Romans constructed huge aqueducts to transport water from lakes to distant cities. Today, water, natural gas, and oil are typical examples of fluids transported in extensive pipelines (see Figure 6.11A). Pumps provide the force that pushes the fluid through the pipes. Why do the travelling fluids need to be placed under pressure?

Think about the water that comes out of your faucet. Where did this water come from? Drinking water can come from lakes, rivers, and underground wells deep in Earth's crust. Unless you live underground, water must flow up in order to get from its source to most people's homes. How does water travel up to reach homes in highrise apartment buildings? To travel so high, water must be placed under pressure in order to give the water particles the energy to move against gravity. Otherwise, the water would simply rest in puddles in the lowest parts of the pipe! The amount of pressure transmitted in the pipes must be enough to transport the liquid over a large distance, but not too much to make the pipes burst.

Friction in the pipes — caused by rough surfaces or numerous bends in the pipeline — can affect fluid pressure. The particle theory suggests that particles lose energy as they brush past each other in confined spaces and as they bump into the walls of the pipelines. Therefore, pumping stations are frequently needed to restore the pressure lost on long routes (see Figure 6.11B). The water reaching your home has probably been to a nearby pumping station to maintain a constant water supply to your home. Imagine how much time you would spend hauling water to your home from a lake, well, or river if water pumps and faucets were not available to us!

DidYouKnow?

The type of firetruck known as the "pumper" acts just like a pumping station. When the hose is attached to a fire hydrant, it directs the water to the pumper truck first, so that the water pressure can be increased as much as possible. High pressure is needed so that a powerful jet of water can shoot out to extinguish the blaze.

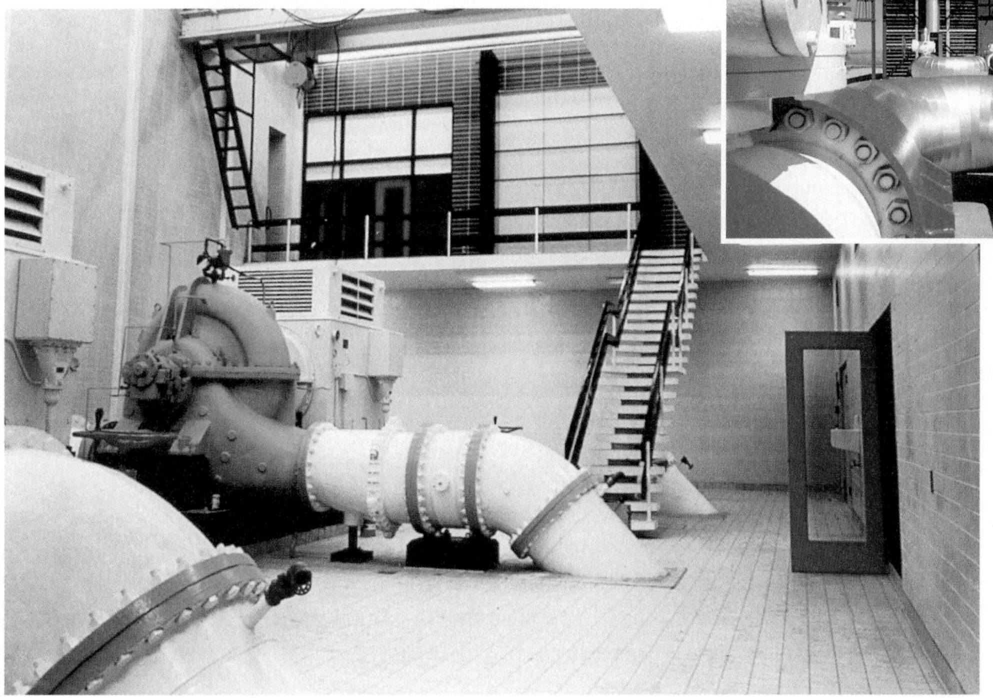

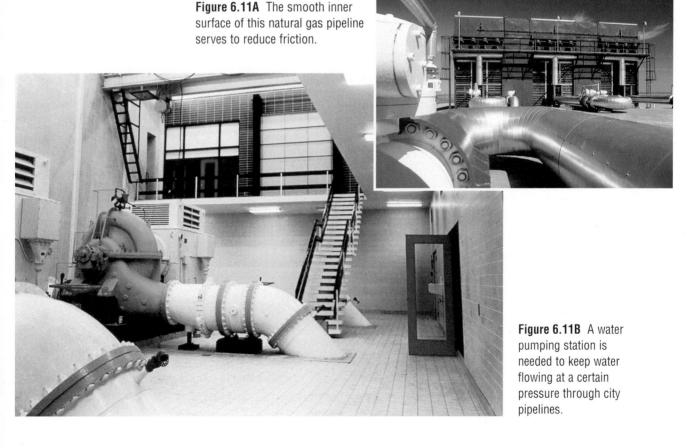

Figure 6.11A The smooth inner surface of this natural gas pipeline serves to reduce friction.

Figure 6.11B A water pumping station is needed to keep water flowing at a certain pressure through city pipelines.

In most hydraulic transport systems, it is important that the fluid keep travelling away from the pump. Some pumps cannot do this on their own. **Valves** are devices used to regulate the flow of a liquid in hydraulic systems. One-way valves ensure that the fluid can flow in one direction but not in the opposite direction.

Body Hydraulics

One of the most efficient hydraulic transport mechanisms is the human circulatory system, which you learned about in Chapter 3. In humans, blood must be kept under pressure so it can reach all parts of the body. The highest blood pressure occurs close to the heart. Blood pressure at more distant regions, such as the hands and feet, is much lower. The constant beating of the heart, which is the pump, keeps the blood moving throughout the arteries and capillaries, which are like pipelines. Valves in the veins keep the blood moving in one direction (see Figure 6.12).

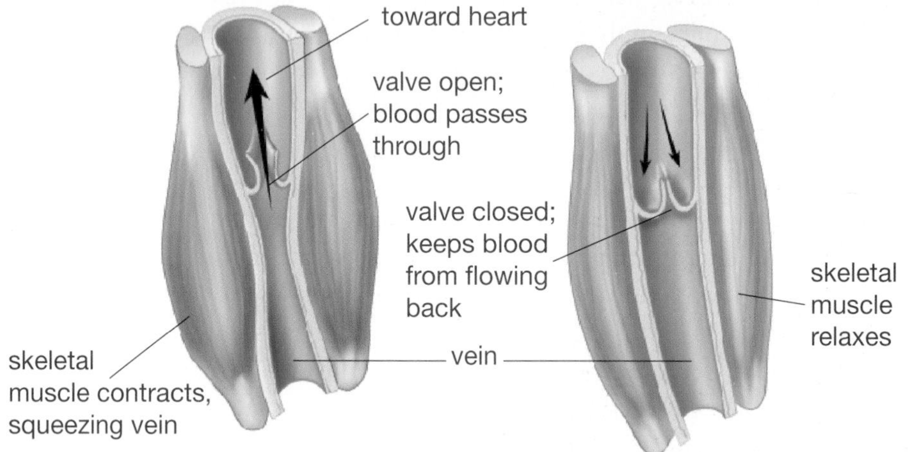

Figure 6.12 When muscles surrounding veins contract, they squeeze the veins, forcing the blood within to move forward under pressure.

Across Canada

Since the late 1980s, the Canadian Space Program, along with NASA (National Aeronautics and Space Administration), has been conducting scientific research in space. Recently, flight surgeons on the Neurolab mission have been investigating questions related to blood pressure, loss of sleep, orientation in space, and tiredness.

Canadian astronaut Dave Williams, M.D., is a member of this large team of scientific researchers. Dr. Williams has been trying to find answers to questions such as: How does the body's ability to regulate blood pressure change during and after spaceflight? He has discovered that over 500 000 North Americans suffer from disorders of the body's natural ability to regulate blood pressure and to keep blood flowing to the brain. These disorders often result in lightheadedness or fainting when people stand up quickly. For some reason, the body can no longer increase the blood pressure to boost the blood the extra height.

Some astronauts experience similar symptoms after spaceflight. Fighter pilots and stunt pilots, too, sometimes nearly pass out when they force their planes into a tight turn. What is common in these cases? In every instance, the cardiovascular system (the heart and blood vessels) is stressed by gravity. Gravity forces the cardiovascular system to work hard to maintain the blood flow to the brain. The results of Dr. Williams' research in space will also benefit people experiencing this disorder on Earth.

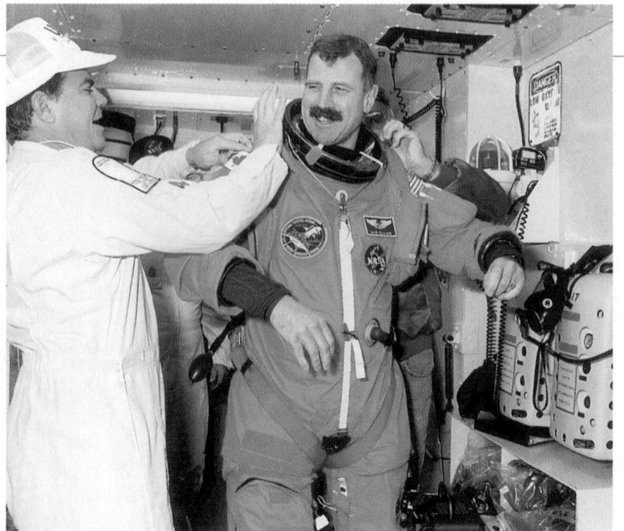

Dr. Dave Williams

Blood pressure increases and decreases between heartbeats. Immediately after the heart contracts, a surge of blood causes high pressure in the arteries. Then, before the next heartbeat, the pressure falls, only to increase again at the next contraction. You learned in Chapter 3 that blood pressure is measured with a sphygmomanometer, more commonly referred to as a blood pressure cuff. Both the maximum pressure, called *systole*, and the minimum pressure, called *diastole*, are measured using this device (see Figures 6.13A and B). A normal blood pressure reading is 120/70. This means that the systolic pressure is 120 mm Hg (millimetres of mercury), and the diastolic pressure is 70 mm Hg (millimetres of mercury).

Did You Know?

It is important for all mammals to keep blood flowing to the brain to remain conscious. This is probably most difficult for an extremely tall animal, such as a giraffe. In fact, a giraffe's body has to maintain an enormous systolic blood pressure to stay conscious as it raises its head 4 m after having a drink of water.

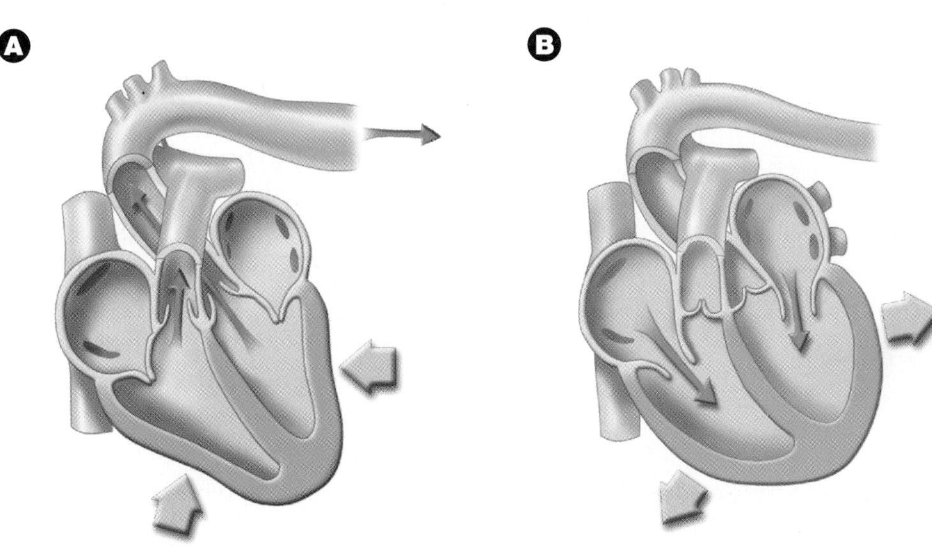

Figure 6.13A Heart during contraction (systole)
Which valves are open and which valves are closed?

Figure 6.13B Relaxed heart, between contractions (diastole)
Which valves are open and which valves are closed?

Check Your Understanding

1. In which direction is pressure exerted in a fluid at rest?
2. Pipelines are used to transport liquids such as natural gas. Why are these pipelines made with few bends and kept free of dirt and rust?
3. What are the names of the two blood pressure readings? What does each reading indicate?
4. **Apply** The fuel line in a car is an example of a hydraulic transport system. Look up *Chilton's Auto Repair Manual* in your library to find a diagram of such a system. Draw a simplified version of it. Label the pump, valves, pipeline, source, and destination of the fuel.
5. **Thinking Critically** Doctors always remove all the air from a needle before inserting the tip of the needle into a patient's blood vessel. Why do you think an air bubble in the bloodstream would be dangerous?

Pause & Reflect

Can you guess why the sphygmomanometer is placed on the upper arm? Discuss this with your classmates and write the most likely explanation in your Science Log.

6.3 Air Pressure and Pneumatic Systems

Figure 6.14 Air pressure exists everywhere and is exerted in all directions.

Just like other fluids, air exerts pressure on everything that it surrounds, as shown in Figure 6.14. The study of pressure in gases is called **pneumatics** (pronounced new-mat-ics). Why don't you feel weighed down by air? The pressure of Earth's atmosphere is so well balanced by your body, both inside and out, that you hardly ever notice air pressure.

Just as water pressure changes with depth, air pressure changes with altitude. If you were driving up a mountainside, your ears would "pop" as your eardrums adjusted to the change in air pressure. According to the particle theory, as you climb higher in the atmosphere, fewer air particles press against you on the outside of your body. The air pressure inside your body does not change as quickly, however. Thus, the number of particles pressing from the inside out is still the same as it was when you were at the base of the mountain. Your eardrum is a very thin membrane that can move in response to a difference in air pressure (see Figure 6.15). If the difference in pressure on either side of the eardrum becomes great, you experience a "pop" inside your ear as the pressure equalizes.

On average, the atmosphere exerts a force of 10 N (about the weight of a 1 kg object) on every square centimetre of Earth's surface at sea level. This is equal to about 101 300 Pa, or 101.3 kPa. See for yourself in the next investigation just how strong this pressure is.

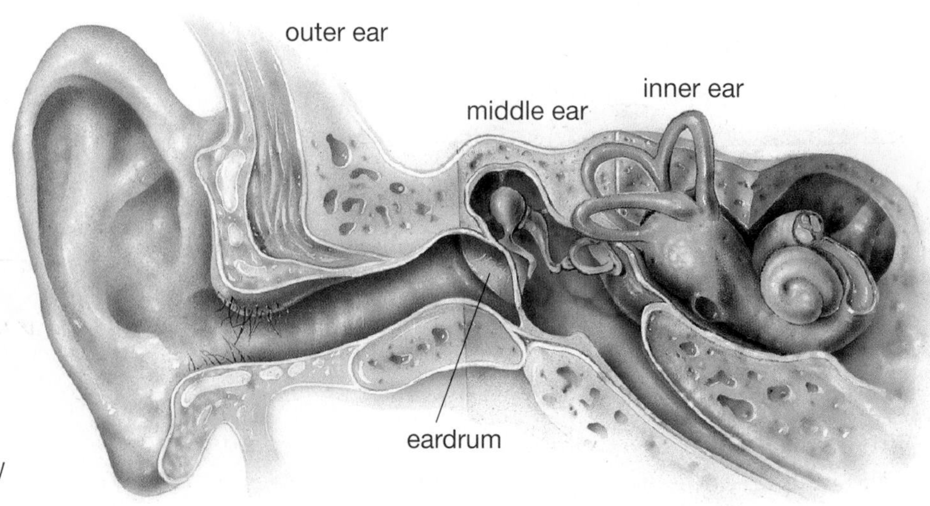

Figure 6.15 The eardrum is a very thin membrane that is sensitive to changes in air pressure.

Imploding Pop Can: A Teacher Demonstration

Pop cans are made of metals strong enough to withstand the high pressure of a carbonated soft drink. Even when a pop can is empty, a large force is needed to crush the can. Find out if a pop can is able to withstand the forces of normal atmospheric pressure.

Safety Precautions

Care must be taken when a heat source is used during this investigation.

Apparatus

empty pop can
barbecue tongs (or other metal tongs)
oven mitts
medium-sized bowl
hot plate
stopwatch or watch with a second hand

Materials

ice water

Procedure

Your teacher will do the following:

1. Fill the bowl with ice water. Set the bowl beside the hot plate.
2. Rinse the pop can with water. Leave a small amount of water inside (about 5–10 mL).
3. Set the hot plate to high, and place the pop can on it.
4. Watch for steam to rise out of the can. Observe the steam escaping for 15 s.
5. Wearing oven mitts, place the tongs around the steaming can and quickly turn the can upside down in the bowl of ice water.

Analyze

1. What was inside the can just before it was turned upside down?
2. Describe what happened to the pop can after it contacted the cold water.
3. What did you observe when the can was lifted out of the water?
4. Infer what happened to the contents of the can when it was cooled by the water.

Conclude and Apply

5. What caused the pop can to implode?
6. Why would this experiment not work as well with a dry can?

Word CONNECT

Look up the meaning of "implosion" in a dictionary. What is the opposite of an implosion?

Balanced and Unbalanced Forces

If the inside of a closed container experiences a lower air pressure than the air pressure pushing on the outside, the walls of the container will buckle and cave in. In other words, the lower air pressure inside the container does not balance the higher air pressure outside the container. This results in an unbalanced force, the force of atmospheric pressure, which pushes on the walls *toward the inside* of the container. You may have noticed this imbalance when drinking juice from a juice box. The straw makes such a tight seal that as you draw the juice up the straw and reduce the air pressure inside the juice box, the box buckles inward. The air pressure outside the juice box pushes the walls of the box together.

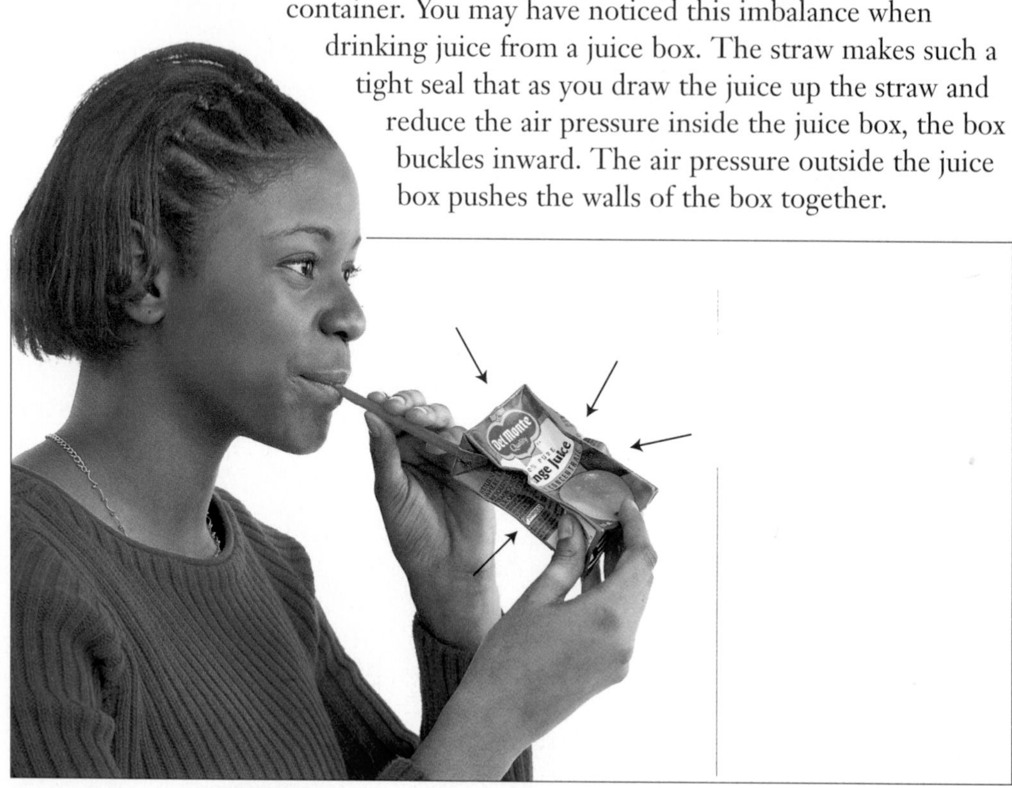

Measuring Air Pressure

The most common device for measuring air pressure is a **barometer**. The earliest barometers were made with mercury and are still used by many weather stations. Figure 6.16 shows how a mercury barometer works. First, a thin, strong-walled glass tube, sealed at one end and open at the other, is completely filled with mercury. It is then inverted (turned upside down) in a pool of mercury, allowing no air to enter. The mercury falls to a lower level in the tube, due to gravity, leaving a vacuum (no air, only empty space) at the top of the tube. Why does some mercury stay in and some come out?

Pause& Reflect

Drink bottles are often made of flexible plastic. Have you ever taken a big sip from a plastic water bottle and seen the sides buckle? What about glass bottles? Do the sides of a glass bottle buckle when you take a big sip? At home, take sips from various types of bottles in which the openings are sealed around a straw, and record your findings in your Science Log. Try to explain your observations in terms of "low pressure," "high pressure," and "unbalanced force."

As the air pushes down on the mercury in the pool, forcing it up the column, the mercury pushes down through the column and into the pool because of its weight. The mercury will stop moving when the force of the air pressure pushing it up equals the force of gravity pulling it down. Air pressure can support 760 mm of mercury (Hg) at sea level.

A more convenient barometer for use at home is the **aneroid barometer** (shown in Figure 6.17), which contains no liquid at all. Make your own aneroid barometer in the following investigation.

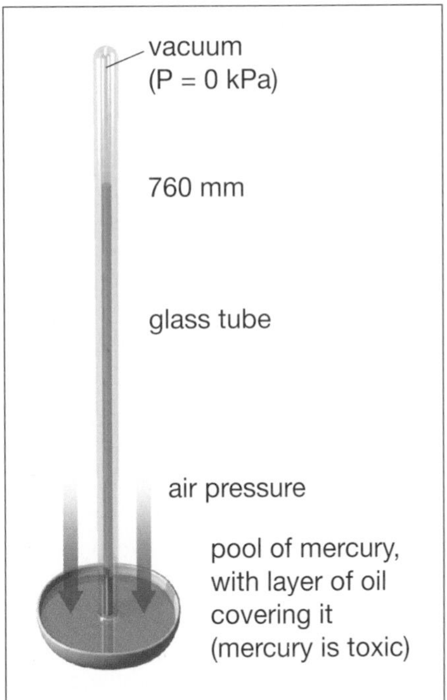

Figure 6.16 A mercury barometer

Pause& Reflect

Water is much less dense than mercury. How high do you think a column of water might rise in a barometer? Discuss this question with your classmates, and record your answer in your Science Log.

Figure 6.17 Aneroid barometers contain a sealed drum under a partial vacuum (some of the air has been removed). As outside air pressure changes, the lid moves in and out. Levers magnify the movement and transfer it to the pointer, which moves along a scale. This scale has been calibrated (set) to read pressure values in kilopascals.

Measuring Air Pressure

Why do weather forecasters report high-pressure systems and low-pressure systems on the weather maps? High air pressure generally means fair weather, and low air pressure often means clouds and rain. Although changes in atmospheric pressure can be very small, they can signal the coming of a storm. How can you make a device to detect small changes in atmospheric pressure?

Problem

How can a homemade barometer detect changes in atmospheric pressure?

Safety Precautions

- Be careful of sharp edges as you construct your barometer.
- Be careful when using the scissors.

Apparatus

small coffee can with plastic lid (or any wide-mouthed container)

strong elastic band

scissors

Materials

piece of rubber from a balloon

piece of lined paper

tape and straw

Procedure

1. Construct your barometer as shown in the diagram, or make any adjustments that you think might improve the design. Remove the creases in the piece of rubber by pulling it gently through the elastic band seal. This will also help to tighten the rubber.

2. Once you have built your barometer, place it inside the classroom, in a location where the temperature will remain constant. Mark an initial reading on the lined paper, and record it in a data table. The table should contain columns for date, time, air pressure reading, and weather conditions at the time of each reading. Give your table a title.

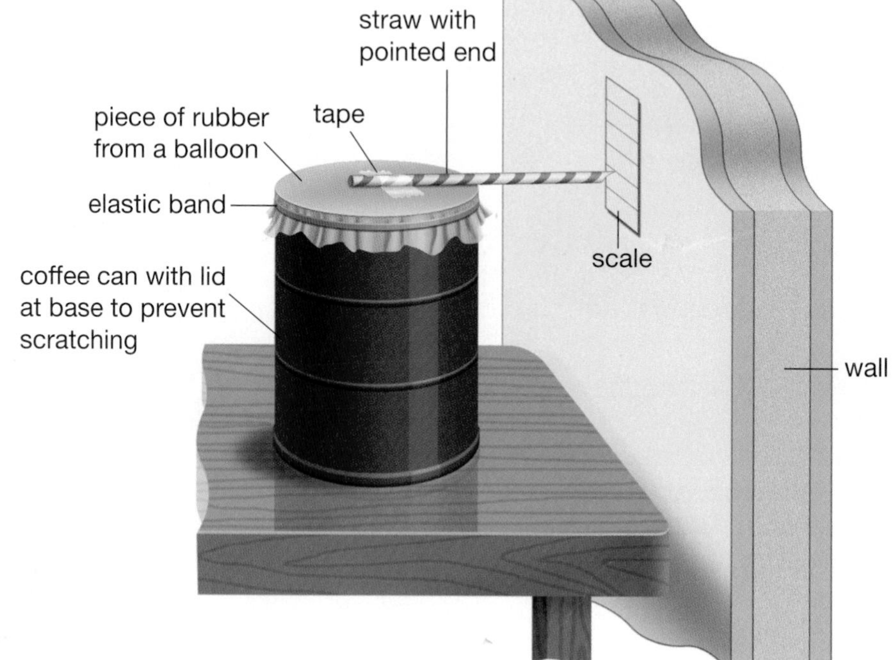

3. You will need to take readings each day for at least a week before you can construct a scale for your barometer. (If you do not have a commercial barometer in your classroom, consult weather reports for air pressure at the time that you check your barometer.) Mark each reading on the lined paper with a dash, and label it with the official air pressure value. Record all information in your table.

4. At the end of a week, answer the questions on the right.

Skill POWER

To review how to make a data table, turn to page 546.

Analyze

1. What were the highest and lowest atmospheric pressures that you recorded during your test period? Did the air pressure change significantly during your test period?
2. Was your barometer sensitive enough to show the air pressure differences clearly? Compare your scale readings with those on your classmates' barometers. Were any of their readings more sensitive than yours? If so, discuss the possible reasons for this difference, and try to summarize the possible variables involved.

Conclude and Apply

3. Compare your readings with the weather conditions at the time of each reading. Is there any relationship between atmospheric pressure and daily weather? Describe any evidence in your results that supports your answer.
4. Propose changes to your design. What could you do to (a) increase your barometer's sensitivity, (b) increase the distance between readings, and (c) attach the scale to the barometer?

Career CONNECT

Up, Up, and Away

How would you like to blow up balloons every day as part of your job? Tom Kudloo is an aerologist. He works at a weather station in the Arctic, taking air pressure readings that help forecasters predict the next day's weather. Every 12 h, he pumps hydrogen into a thick weather balloon until it is about 2 m in diameter. Then he checks to see if the balloon is inflated enough to lift the 2 kg box of instruments it must carry up into the atmosphere.

The box contains a radio transmitter and a sensor that measures air pressure, temperature, and humidity. Tom attaches the box of instruments to the balloon, then takes it outside to launch it. As the balloon rises, the radio transmitter relays data about the air through which it passes to a computer at the weather station.

Part of Tom's job is to interpret the data transmitted from the balloon and translate it into weather information. Weather forecasters across North America depend on information from Tom and from many aerologists in other areas to predict what the next few days' weather will probably be like.

Listen to a weather report on your local television station. What can you learn from it about weather and air pressure? Arrange to talk to a weather reporter at your local television station to find out how air pressure is involved in weather forecasting. Ask about other careers that involve the weather. You can also search for "weather careers" on the Internet.

Using a straw to drink will not work if the straw is too long, for example, 12 m long. Refer to what you learned about the barometer to help you explain why this statement is true.

Pressure and Pneumatic Systems

Pneumatic systems are similar to hydraulic systems, except gases are used instead of liquids. In **pneumatic systems**, a gas transmits a force exerted on the gas in an enclosed space.

The operation of most pneumatic systems is based on the fact that gases can be compressed. Therefore, **compressors** — devices that compress air — are needed for pneumatic devices to operate. Air pressure builds up in these devices. As the pressure in the device is released, the compressed air **decompresses**. In other words, the particles start to move apart suddenly, creating a strong, steady force that can perform powerful tasks. Many tools use pneumatics, from large tampers used to pack down dirt and gravel when building a road, to tiny precision drills used by dentists. As well, heavy trucks and buses rely on pneumatic brakes (also called air brakes) to stop quickly and smoothly. The pneumatic brakes in the logging truck shown in Figure 6.18 are powerful enough to bring this huge vehicle safely to a halt.

Figure 6.18 This logging truck is equipped with pneumatic brakes.

Find Out **ACTIVITY**

Exploring Pneumatics

Observe how pressure is transmitted in a simple pneumatic system.

What You Need

modified syringes (various sizes)
short piece of plastic tubing

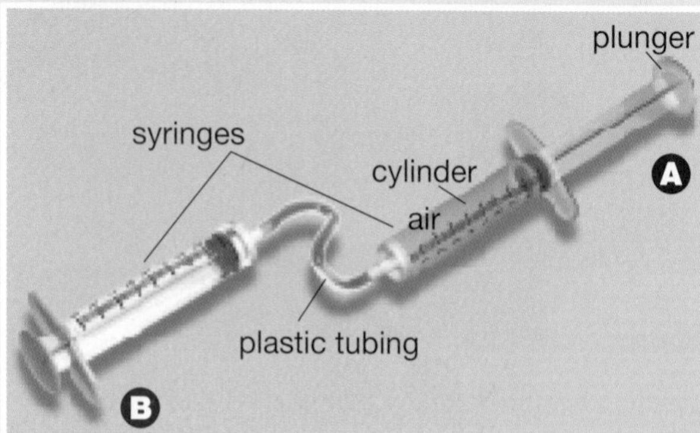

What to Do

1. Connect two identical syringes together with a piece of plastic tubing.
2. Before you make the final connection, make sure that the syringe you will push on has its plunger pulled completely out. Make sure that the plunger that will react is pushed completely in.
3. Press on A. What happens at B?
4. Press on A while your other thumb is on B. What can you feel? Why don't you feel the pressure immediately?
5. Try different-sized syringes at each end. How far does the plunger at B move compared to the distance A moves?

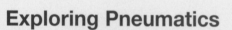

What Did You Find Out?

1. Is the force exerted at B always the same?
2. Do you think this simple pneumatic system could be used to make work easier? How?

INVESTIGATION 6-F

Giant Jaws

Apply what you have learned about pneumatic systems to design and build a set of "Giant Jaws" that can perform a simple task.

Challenge

Using the materials listed, create a device that can grasp an object with a mass of approximately 250 g.

Materials

2 modified syringes (any size)
plastic tubing
jinx wood, stiff cardboard, or corrugated plastic
split pins or butterfly fasteners
double-sided foam tape or masking tape
glue gun
clothes hanger wire
scissors or cutting knife
pencil or skewer for punching holes

Safety Precautions

- Be careful when using sharp tools such as scissors, knives, or skewers.
- A glue gun is hot and the glue remains hot for several minutes.

Design Criteria

A. The reach of your "Giant Jaws" must be no less than 30 cm and no more than 50 cm.

B. Your device must be able to withstand its own weight (it should not bend or break during performance).

C. Your "Giant Jaws" must be powered by pneumatic pressure.

D. Your device must work smoothly and efficiently to grasp an object having a mass of approximately 250 g.

Plan and Construct

1. Make an initial concept sketch or sketches showing how your "Giant Jaws" will function.

2. How will you make sure that the "Giant Jaws" will be strong enough to grasp an object having a mass of 250 g?

3. Decide how to prepare and assemble the pieces that will make up your "Giant Jaws." How will the components be held together?

4. When your "Giant Jaws" is assembled, test it out!

Evaluate

1. Were your team's drawings helpful in designing an effective device? Were they sufficiently clear and detailed?

2. Did your team's "Giant Jaws" meet all the specified design criteria? Could the device grasp the object without bending or breaking?

3. How could you improve your team's design or procedure?

More Uses for Pneumatics

Another useful property of gases is their ability to exert a force *back* (a counterforce) when they are compressed. This property can be used to cushion shocks. For example, the air in a car tire pushes back against the force exerted by the weight of the car. Otherwise, the car would simply sink to the ground. If the car hits a bump, the extra force compresses the air in the tires even farther. This allows the effect of the force to be spread out over the entire tire, rather than being transmitted directly to the body of the car and its passengers. When the extra force is removed, the air returns to its original volume, and the tire resumes its original shape.

An air bag in an automobile is another device used to cushion shocks. Air bags are designed to be used in addition to seat belts. An air-bag system consists of one or more crash sensors, an ignitor and gas generator, and an inflatable nylon bag. The nylon bag for the driver is stored in the steering wheel, and the bag for the front-seat passenger is usually stored inside the dashboard. If a car hits something with sufficient force (speeds in excess of 15 – 20 km/h), impact sensors trigger the flow of electric current to an ignitor. The ignitor causes an explosive chemical reaction to occur, producing harmless nitrogen gas. The nitrogen gas propels the air bag from its storage compartment just in time to apply a restraining force that protects the driver from serious injury. The bag then immediately deflates. The entire process takes only 0.04 s!

Check Your Understanding

1. Why do the sides of a juice box buckle when you suck the juice out through a straw?
2. Name two devices used to measure the pressure of a gas.
3. (a) On what property of gases is a pneumatic system based?
   (b) How does this property help pneumatic systems do work?
4. **Apply** Which do you think would produce more fizz when opened: a bottle with a twist-off cap or one that has a bottle cap? Why?
5. **Thinking Critically** Would you be able to drink through a straw in outer space? Could you drink through a straw on the Moon? Explain your answers.

CHAPTER at a glance

Now that you have completed this chapter, try to do the following.
If you cannot, go back to the sections indicated.

Define pressure, both in words and as a formula. (6.1)

State how pressure is exerted in all fluids. (6.1)

Compare the compressibility of solids, liquids, and gases. (6.1)

Describe how pressure changes with depth or with altitude. (6.1)

Use the particle theory to explain how temperature affects pressure. (6.1)

Define hydraulic system and give some common examples of hydraulics. (6.2)

Explain why pumps and valves are necessary in hydraulic systems. (6.2)

Describe how the body's circulatory system is similar to a hydraulic system. (6.2)

Use the particle theory to explain how air exerts pressure. (6.3)

Explain how a barometer measures atmospheric pressure. (6.3)

Define pneumatic system and give some examples of pneumatic devices. (6.3)

Examine the simple hydraulic device shown below and explain how it works. Use the words "force," "cylinder," "pressure," "transmitted," and "hydraulics" in your answer.

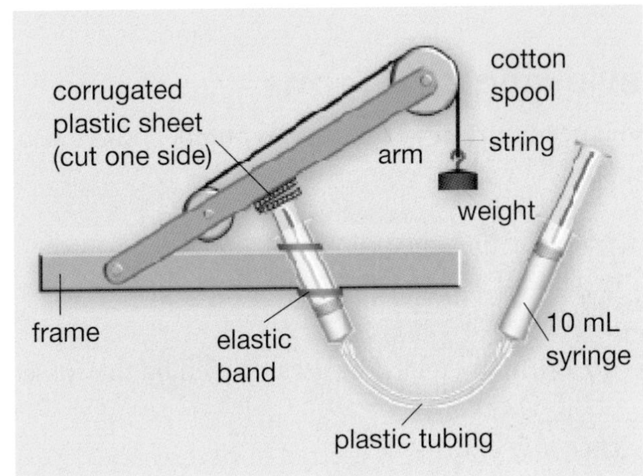

Prepare Your Own Summary

Summarize this chapter by doing one of the following. Use a graphic organizer (such as a concept map), produce a poster, or write the summary to include the key chapter ideas. Here are a few ideas to use as a guide:

- Use the concept of pressure to explain why popping a balloon with a pin is easier than stepping on it.
- How does the particle theory explain compressibility of gases and incompressibility of liquids?
- Use the particle theory to explain why your ears hurt when you swim deep underwater and "pop" when you travel up in a rapidly moving elevator.
- What are the basic requirements for a hydraulic system, such as the Jaws of Life?
- What is the main purpose of a hydraulic transport system?
- Identify the main components of the human circulatory system as a hydraulic system. State the purpose of each component.
- What does a sphygmomanometer tell you?
- How do mercury barometers and aneroid barometers measure atmospheric pressure?
- Why is a compressor necessary in pneumatic devices?
- What are the basic design requirements for a simple hydraulic or pneumatic device?

CHAPTER

6 Review

Key Terms

pressure
pascal
kilopascal
compressibility
compressible
incompressible
transmitted
flow pressure
static pressure
hydraulics
hydraulic systems
valves
pneumatics
barometer
aneroid barometer
pneumatic systems
compressors
decompresses

Reviewing Key Terms

If you need to review, the section numbers show you where these terms were introduced.

1. State whether the following statements are true or false. If a statement is false, rewrite it to make it true.
   - (a) Pressure equals the area divided by the force. (6.1)
   - (b) Fluid pressure is exerted in all directions. (6.1)
   - (c) Temperature and depth can affect fluid pressure. (6.1)
   - (d) Hydraulic systems depend on air pressure. (6.2)
   - (e) Pipelines with many bends lose pressure more quickly than straighter pipelines. (6.2)
   - (f) Liquids must be continuous and in an open system to transmit forces. (6.2)
   - (g) Hydraulics is the study of pressure in liquids. (6.2)
   - (h) A sphygmomanometer measures body temperature. (6.2)
   - (i) A dentist's chair is an example of a pneumatic system. (6.3)
   - (j) A pop can will explode if you remove all the air from it and then cool it suddenly. (6.3)
   - (k) A barometer measures air pressure. (6.3)
   - (l) An air bag in an automobile uses hydraulics. (6.3)

Understanding Key Ideas

Section numbers are provided if you need to review.

2. Explain, using the particle theory, why gases can be compressed but liquids and solids cannot. (6.1)
3. How do (a) depth, and (b) temperature affect pressure? (6.1)
4. What are the requirements for a hydraulic system to work? (6.2)
5. Explain why the blood pressure is higher in your arteries than in your veins. (6.2)
6. Explain how a mercury barometer works. (6.3)

Developing Skills

7. Copy and complete the following concept map of the basic principles of fluid pressure.

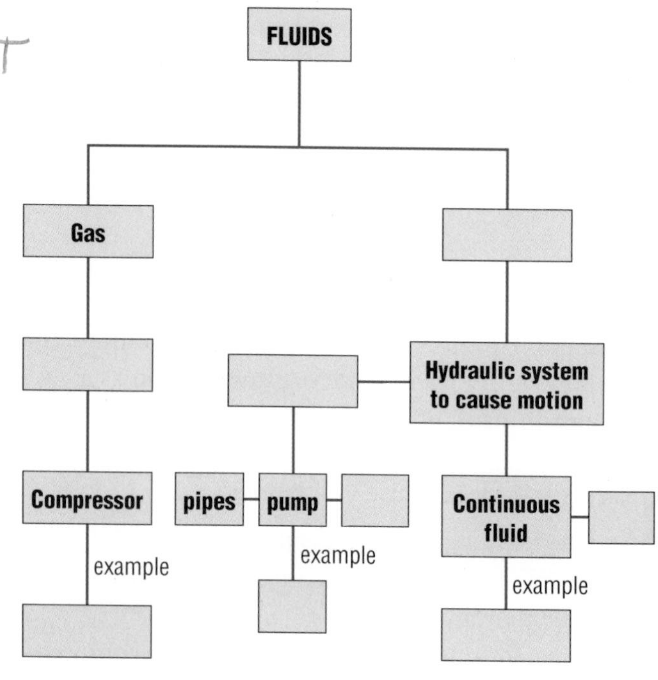

8. When you press the button on a drinking fountain, a jet of water shoots upward into the air. Formulate your own question about this observation, based on what you know about the properties of fluids. Design an experiment that could provide an answer to your question.

9. Arrange to visit an automotive service garage. Make note of all the hydraulic and pneumatic mechanisms a mechanic uses, and which ones are needed and must be checked regularly in a car. What do these mechanisms look like? How do they work?

10. (a) Observe a toy water pistol. Make a hypothesis about how its mechanism works. Take it apart and see if your hypothesis was correct.

    (b) Make a hypothesis about how a large, "pumper" water gun works. Why does it not spray water while you are pumping? Why can it shoot water extremely far?

Problem Solving/Applying

11. Michelle weighs 500 N. How much pressure does she apply when she stands on her head, if the part of her head that is touching the floor has a radius of 2 cm?

12. Why did you need the hole in the bottom of the milk carton in the Pressure Puzzle investigation on page 172?

13. Breanna's friends got her a helium balloon for her birthday with "Happy Birthday, Breanna!" printed on it. By accident, it slipped out of her grasp and rose higher and higher into the air until she could not see it anymore. A few days later, she and her friends found it in a park. It had burst. Why do helium balloons eventually burst, or explode, as they rise higher and higher into the atmosphere?

14. Design and make one or more of the following using simple hydraulic systems:
    - a jack-in-the box
    - a door closer
    - a toy with blinking eyes
    - a moving miniature billboard
    - a simple puppet with moving parts
    - a robot arm that moves up and down

Critical Thinking

15. Explain why your body is sensitive to 30 kPa of water pressure at the bottom of a swimming pool, but not sensitive to 101.3 kPa of normal atmospheric pressure.

16. Marco always struggles to open a pickle jar, or any food jar, for the first time. He has tried many methods, including hitting the lid with a knife (which always damages the lid and the knife), and using a rubber pad specially designed to give a good grip on the lid. In all cases, however, he still had to wrench the lid open with as much strength as possible. One day his brother told him to use a small spoon to simply pry the lid slightly apart from the glass lip. Explain why this simple method works.

17. What do you think would happen if a container such as an aerosol spray can that has been sealed under pressure were exposed to extremely high temperatures? (This is similar to what happened during the Great Molasses Flood described in Chapter 4.) Explain your answer.

Pause& Reflect

1. Go back to the Starting Point Activity on page 163 and use the particle theory to explain how the air inside the balloon "pushes" the books upward.
2. Go back to the beginning of this chapter on page 162 and check your original answers to the Getting Ready questions. How has your thinking changed? How would you answer these questions now that you have investigated the topics in this chapter?

UNIT 2 Ask an Expert

Larry Weagle has been a shipwright — a boat-builder — for 22 years. He and about 20 other people at Covey Island Boatworks in Nova Scotia build custom, one-of-a-kind wood boats for customers around the world. As one of four directors of the boatworks, Larry knows a great deal about how a boat floats.

Q How did you learn to build boats?

A I looked into taking courses to become a carpenter, but I didn't find carpentry all that interesting. I decided to take a two-year boat-building course at a community college. That led me to where I am today.

Q What kinds of boats do you build?

A We build powerboats, sailboats, sports fishing boats, and heavy boats for crossing the Atlantic. The boats are between 7.5 m and 17 m long. Every boat that we build is different. The customer comes to us with drawn plans. These might be detailed plans prepared by a naval architect or very simple, sketched outlines of the boat.

Q How long does it take to build a boat?

A That depends on its size. It takes a crew of eight to ten people about ten months to build a 16 m sailboat. We're usually involved in building three different boats at the same time. Three crews, three boats.

Q Are all of your boats made from the same material?

A Yes and no. Some of the boats are made from strips of solid wood, while others are made from panels of plywood. The outsides of all our boats have a protective covering of cloth made from fibreglass and a special glue called epoxy. A coating of epoxy alone protects the inside of each boat.

Q Why are some boats made from plywood and some from solid lumber?

A Plywood is made of thin, glued layers of wood turned so that the wood's grain runs in one direction for the first layer and in the other direction for the second layer, and so on. This makes plywood very strong. A thin piece of plywood is as strong as a thicker piece of solid wood. A plywood hull can be thinner and therefore lighter, so it floats higher in the water. We often use plywood to make a light, fast powerboat.

Q Why would you build a boat from solid wood if it doesn't float as well?

A Fast boats bounce along on top of the water, smacking into the waves. It's not a very comfortable ride. Ships with hulls made from denser, solid wood ride lower down in the water. Our sailboat hulls are usually made from solid wood. It's slower but more comfortable to ride in a sailboat.

Q Does a boat's shape affect its buoyancy?

A Very much so. Power boats, for instance, have a wide, flat hull. A wide hull displaces a lot of water, so very little of it sinks down under the

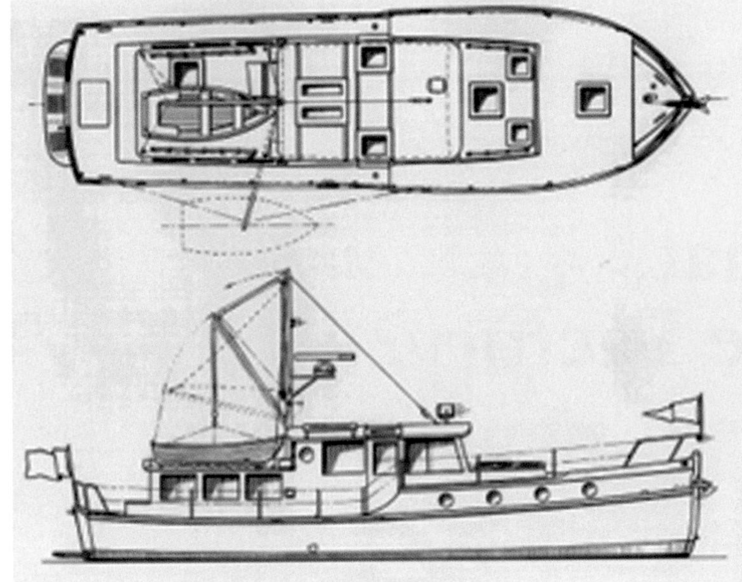

Plans for the *Belle Marie*, designed by David Gerr

The *Belle Marie* afloat

surface. The fact that the hull is wide (which displaces a lot of water) and made of plywood (which is thin and light), means that powerboats sit near the surface. When you attach a large motor to them, these boats can skim across the water at high speeds.

Sailboats have a V-shaped hull. This shape, along with the hull's denser, solid wood construction, causes sailboats to sink deeper into the water than powerboats. This design also gives sailboats that comfortable ride I mentioned.

Q **What is your favourite part of building boats?**

A Once each boat is built, we have sea trials for a week or so. That means we take the ship out to test how well it performs. Each boat is the first of its kind, so we're never sure how it will fare until we try it out. I find it really satisfying to see the product of our hard work. Turning pieces of wood into a structure that travels across the water is very rewarding.

EXPLORING Further

In Hot Water

Larry Weagle says that cold water is denser than warm water. True or false? Make a prediction and then try this. Take two clear glasses, one half-filled with hot water and one half-filled with cold water. **CAUTION:** Be careful not to burn yourself! Add a few drops of food colouring to the glass of hot water. Using a medicine dropper, carefully add a small amount of the hot, coloured water to the glass containing cold water. What happens?

If cold water is denser than warm water, objects floating in cold water will be more buoyant than when they are floating in warm water. Where would ships encounter cold water? Warm water? How would water temperature affect how boats float?

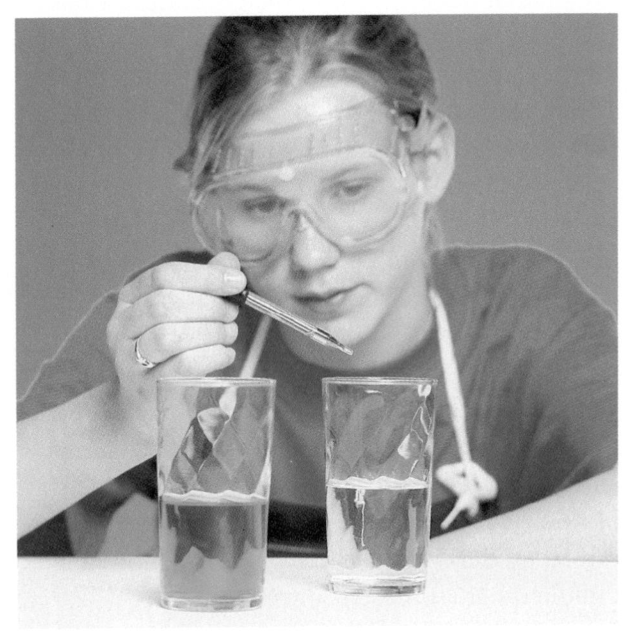

An *Issue* to Analyze

A DEBATE

Who, or What, Will Run the Factory?

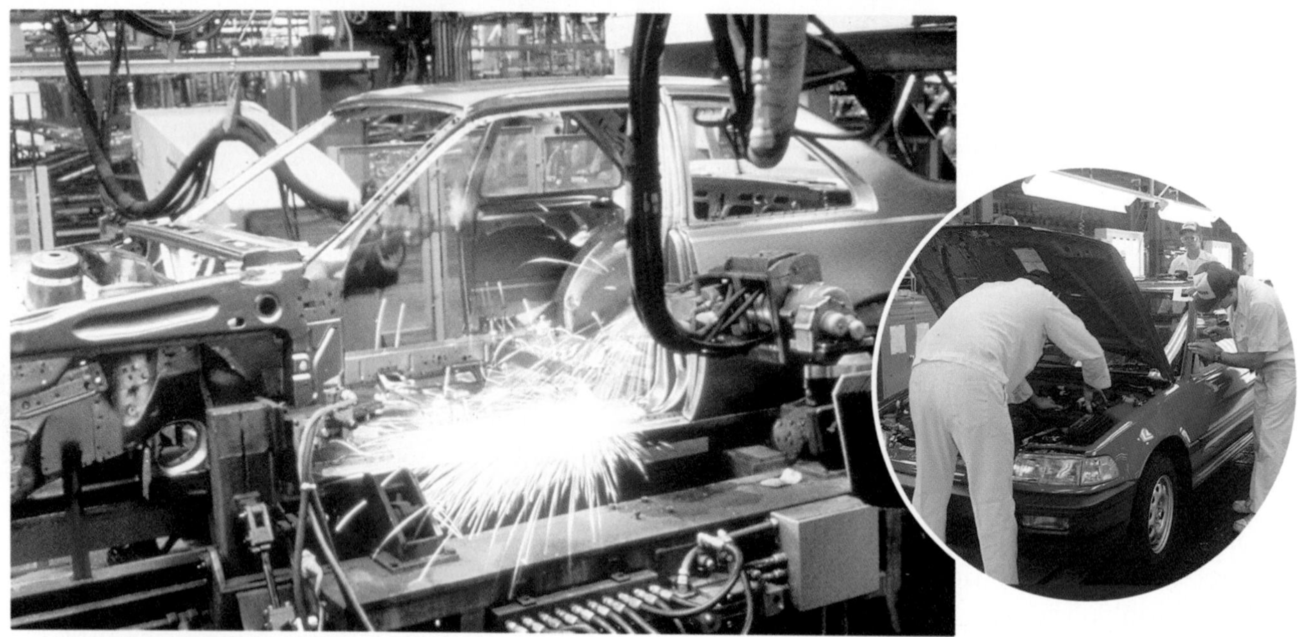

Think About It

Robot technology — also called robotics — has advanced a great deal in the last decade. Typically, robots can move arms, legs, and even fingerlike digits to perform an endless variety of tasks. How do robots work? Hydraulic and pneumatic systems provide force; complex electronic circuitry co-ordinates motion. By these means, robots can manoeuvre machinery and components on factory assembly lines. Fluid systems give robots incredible strength to lift, press, push, and separate heavy or hard materials. For example, robots are also used in vehicle assembly plants to lift car doors and slide them into place. At the same time, fluid systems enable robots to turn, flip, and align small, delicate, or finely-tuned components. For instance, a robot can be used to rotate a gear only a fraction of a millimetre at a time. Meanwhile, another robot can cut a wedge of perfect proportions to produce the teeth on a rotary gear. In addition, robots are especially useful in hazardous environments, such as radioactive mines. As technology continues to increase the robots' range of functions, more and more people are being replaced by robots in manufacturing jobs.

Resolution

Be it resolved that with advancing technology, robots should replace all human workers in factories.

What to Do

1. Read the In Favour and Against points made on the following page, and begin to think about other points that could be made in favour of the resolution and against the resolution.

2. Four students will debate the topic. Two students will speak in support of the resolution and two will speak against it. **Note:** No matter what side you actually believe in, you must try your best to convince the jury, or debate listeners, of the point your side is defending.

3. To aid the two teams, two other students will work with them to gather the background information needed to put forward a strong case for the point that side is defending.

4. The rest of the class will act as the jury in hearing the debate. In preparation for the debate, they should do their own research in order to understand the science and technology behind the issues raised.

5. Your teacher will provide you with *Debating Procedures* to follow.

In Favour

- Unlike humans, robots do not have muscles that get tired over time. Robots do not get sleepy, hungry, sick, or angry, which enhances the robot's reliability on the job. If robots are damaged, they do not require compensation, as is the case with human workers who are injured on the job.
- Robots can be programmed to perform a specific task, involving very small measurements and fine movements, and they can repeat these tasks precisely, without error.
- Fluid mechanics allows robots to perform incredible tasks of strength or delicacy that humans cannot accomplish.
- Robots can work in noxious (harmful) environments.

Against

- Robotic machinery wears down and wears out, and cannot repair itself. It has to be serviced frequently to make sure it functions safely and correctly.
- Humans have brains to assess production as it progresses. If something begins to go wrong, humans are flexible and can shift or adjust the operation while production continues, thus avoiding injury, shut-down, and waste.
- Humans can invent a more efficient method to complete a task, or a new method altogether.
- On a robotic assembly line, metals can oxidize (for example, iron rusts), contaminants can get caught between moving parts, and fluid lines can leak. These problems cause robotic activity to stop or malfunction. Solving such problems is costly and time-consuming.

Analyze

1. Which side won the debate?
2. Was the research or the presentation by the side that won better? Explain.
3. Did you find reliable information yourself that contradicts any arguments that were presented? If so, explain the clarification.
4. What did you learn about robot technology that you did not know before?
5. In what ways do you think employment of robots in factories hurts society? In what ways does it help?
6. In what capacity could robots be used at your school?

Extensions

7. Think of some industries that could *not* employ robots in their manufacturing processes. Why would robots be unable to work there?
8. Think of some industries that could benefit from robot technology. Do you think even further advances in robot technology could eventually make robots suitable for this type of employment?

UNIT 2 PROJECT

A-Mazing Hydraulics

Use your knowledge of fluids and pressure to create and play a unique new game. It's unique because you have designed it to be as easy or as challenging as you want. Amaze friends as you guide a marble through your maze and its series of obstacles and dead ends, using only your own imagination and the power of hydraulics.

Challenge

Design a game that uses your knowledge of fluids and hydraulics to move a marble through a maze.

Materials

4 syringes (5 mL to 30 mL)
2 blocks of wood (about 15 cm × 15 cm × 1.5 cm)
100 cm jinx wood (1 cm × 1 cm)
15 cm dowelling
block of wood (5 cm × 5 cm × 5 cm)
50 cm clear plastic tubing
marble
various fluids: air, water, vegetable oil, and any others approved by your teacher
string, paperclips, and rubber bands
coping saw
mitre box
glue gun
hand drill or drill press

Safety Precautions

- Be careful when using hand tools such as coping saws and hand drills.
- A glue gun is hot and the glue remains hot for several minutes.
- Wash your hands after completing this project.

Design Criteria

A. Working in groups of two or three, design a maze on a sheet of wood 15 cm × 15 cm that is operated by 4 syringes working together to raise and lower the corners of the maze platform.

B. The maze should have only one exit point and the pathways should be wide enough to allow a marble to pass through them.

C. Use only the materials specified.

D. Prepare a written presentation of your project, including:

- a title page with the names of your group members and a picture presenting the game
- a design brief (see "Plan and Construct")
- a design proposal (see "Plan and Construct")
- a construction procedure
- a learning log and record of problem solving
- a catalogue that describes and promotes your toy, and a set of written instructions explaining to a new player how your game works
- an explanation of how your knowledge of fluids, pressure, and hydraulics helped you build and design your game

E. Be prepared to have your game played by other students. Students may wish to see whose design is the most difficult, the most creative, or the most enjoyable.

Plan and Construct

1. Prepare a design brief that does the following:
   - states what you are making and the materials you are using
   - states how your finished product is going to be used
   - states for whom you are making the marble maze
   - states where the image is going to be used

2. Prepare a design proposal. This should be a full-size sketch of your marble maze in 3-D, or a scale drawing of the maze pattern you design. Also include 1 top and 2 side-view drawings of the maze unit, including the base and the syringes.

3. Working with your group members, choose a maze pattern. Write a construction procedure, including:
   - cutting list for jinx wood and dowelling
   - drilling points on maze base (2 holes for syringes and 1 hole in the middle of the base)
   - gluing points
   - choice of fluid and reason for choice (your choice may change as you experiment with other fluids)

4. As you build your marble maze, keep a learning log of the steps you have taken. Record any difficulties and what you did to overcome them.

Evaluate

1. How well could other students play your team's marble maze game? Explain.
2. Were your team's written instructions clear and easy to follow? If not, how could you improve them?
3. Did your team's catalogue copy make the game sound fun to play? If not, how would you improve it?
4. If you could improve the design of your marble maze game in any way, what would you do?

Working in a group, organize a "Fluids Circus" in your classroom. Use what you have learned about the properties of fluids to set up demonstrations such as the "Cartesian diver" in Chapter 5. Invite other students and teachers in your school to visit your classroom to enjoy the show!

UNIT 3

Light and Optical Instruments

What is light? You know it helps you to see, but do you know how? Do you know what causes the shimmering colours in the northern lights? Why do we see things in colour? For centuries, scientists around the world have tried to find answers to questions such as these. Gradually, they have developed explanations for light and colour, and with their growing understanding has come an ability to use light for many technological purposes.

In the past, people could obtain light energy only from fire and the Sun. Today, we use laser lights, radio waves, infrared light, and other forms of light energy in many applications. During laser eye surgery, for example, powerful laser light can vaporize tissue from part of the eye to correct vision.

Our ability to produce and control visible light and radiation benefits our daily lives in countless ways. In this unit, you will learn about visible light and other kinds of radiation and the many ways in which we apply them.

Unit Contents

CHAPTER

7 Understanding Light

Getting Ready...

- How is the light from a light bulb similar to the light from the Sun?
- Why do you have to stand directly in front of a bathroom mirror to see your face?
- Why do objects look different when they are immersed in water?

Science Log

In your Science Log, list some of the ways you used light today. Make sure you include situations involving reflections. Then try to answer the questions above as well as you can. You will look back to your answers when you have completed this chapter.

Think about all the ways in which you have used light during the past few days. How many times did you control the brightness of light or make a shadow? How often did you use a mirror or see reflections off a shiny surface? Do you or anyone in your family wear glasses or contact lenses? Perhaps you used a camera, a pair of binoculars, or another optical device. This chapter will discuss some common sources of light, both natural and artificial. It will also introduce some interesting properties of light to help you understand how light travels and how it behaves in predictable ways.

Skill POWER

For tips on how to make and use a Science Log, turn to page 534.

Starting Point ACTIVITY

Switched On

You probably turn on a light bulb several times every day, but have you ever looked at one closely? How does the bulb produce light?

What You Need

convex lens or magnifying glass
lamp with a clear showcase bulb
light-coloured wall or screen

What to Do

1. Using the lens or magnifying glass, look closely at the clear light bulb. Draw a diagram showing the wires inside the bulb and how they are supported.
2. Turn on the light bulb. Hold the lens between the bulb and a light-coloured wall or screen. Move the lens back and forth until you see a sharp image of the glowing part of the bulb on the wall or screen. An **image** is the likeness of an object. Make a drawing of what you see, including the bulb, lens or magnifying glass, and wall.

What Did You Find Out?

1. The part of the bulb that lights up is called the **filament**. What shape does the filament have? Why do you think it is shaped like this?
2. The light bulb is giving off more than light. What other kind of energy does the bulb release?
3. Why does a glowing spot on the wall appear when you hold the lens up to the bulb?

Skill POWER

For information on the safety symbols used throughout this book, turn to page 553.

Spotlight On Key Ideas

In this chapter, you will discover

- how various sources produce light
- two basic properties of visible light
- how to use the laws of reflection to describe the behaviour of light
- what happens to light when it strikes different types of surfaces
- how transparent objects can change the path of light

Spotlight On Key Skills

In this chapter, you will

- use a protractor to measure angles of reflection and refraction
- calculate the cost efficiency of lighting
- analyze experimental results using diagrams
- design your own investigation of refraction

7.1 What Is Light?

Light from the Sun is produced by nuclear fusion of hydrogen particles. The Sun is composed of about 70 percent hydrogen. When hydrogen particles fuse (combine), they form another particle called helium. During this process, an enormous amount of energy is released as heat and light. The temperature inside the Sun is about 16 000 000°C!

In the simplest terms, **light** is the form of energy that you can see. The Sun is a **natural light source**, the most abundant and least expensive light source in the world. Fire is another natural source of light (see Figure 7.1).

The Sun is a star; all stars in the universe are sources of light. Light spreads out, or **radiates**, from the Sun and other stars, in all directions, like the spokes of a bicycle wheel. This type of energy transfer does not require matter; it is known as **radiation**. Energy such as light that travels by radiation is often called **radiant energy**.

Less than one tenth of one millionth of a percent of the Sun's energy actually reaches Earth, but our lives are totally dependent upon this energy. Plants, people, and other animals could not live without light from the Sun. Because sunlight is not always available, people have developed light-producing technologies, or artificial lights. A light bulb is an example of an **artificial light source**. Like the Sun, light from a bulb radiates in all directions.

What else can produce light? Think about what happens when you strike a match. Chemicals on the tip of the match react to produce heat and light. Once the chemical energy is used up, the match is no longer useful. Like the match, all other sources of light require energy. Flashlights use electrical energy from batteries. Light bulbs glow when you switch on electricity. The light that leaves the Sun is formed through a process called *nuclear fusion*.

Figure 7.1 Besides the Sun and the stars, flames and sparks are natural sources of light.

Find Out **ACTIVITY**

Is Light Energetic?

It takes energy to do work. Just getting out of bed some mornings takes what seems like a lot of energy! It takes energy to produce light, as well. Do you think light, like other forms of energy, can cause a change in an object?

What You Need

solar-powered calculator
2 identical black film canisters
aluminum foil
bright light source, such as a 100 W bulb

What to Do

1. Find the solar cells on a calculator. Enter some digits, then completely cover the solar cells with your finger to block the light. What happens to the digits? If nothing happens, the calculator has "dual power." What do you think this means?
2. Wrap one of the film canisters with a single layer of aluminum foil, shinier side out. Place both canisters in a bright light, such as sunlight or light from a 100 W bulb. Wait a few minutes, then remove the top of each canister and feel the inside surface of the containers.

What Did You Find Out?

1. What happened when you prevented the light from reaching the solar cells on the calculator?
2. What difference did you observe between the two film canisters?
3. What evidence do you have that light caused a change in both steps 1 and 2 of "What to Do"?

Extensions

4. When light energy is absorbed by solar cells, into what form of energy does it change so that the calculator can use the energy?
5. In your notebook, complete the following sentence: "Light can be changed into energy forms such as...."
6. In step 2 of "What to Do," what is the independent variable? What is the dependent variable? What variables should be controlled to get meaningful results?

The First Basic Property of Light

You have seen that light is a form of energy. This is the first basic property of light. When light is absorbed by a surface, it can be transformed into several different forms of energy. Light can be transformed into thermal energy, electrical energy, or chemical energy. For example, the absorption of sunlight by a black sweater causes the garment to gain thermal energy. Solar cells change light into electricity. Trees in your neighbourhood absorb sunlight to make chemical energy (sugars).

Did You Know?

Satellites use solar cells to power their electronic equipment. Someday, we might all use sunlight to produce the electrical energy we need. In 1987, the Sunraycer, a test car covered with solar cell panels, drove across Australia powered only by energy from the Sun.

Figure 7.2 Sunlight is absorbed by the pavement on this runway and transformed into thermal energy. You can see heated air rising from the pavement on a road or a runway on a hot, sunny summer day.

The brightness, or **intensity**, of light indicates how much energy a surface will receive. A surface can absorb more energy if the brightness of the light intensifies. For instance, pavement may feel hot to the touch on a sunny summer day (see Figure 7.2). However, the pavement will feel only warm if the clouds block out the sunlight. In the activity below, explore further the concepts of light, intensity, and radiant energy.

At Home

Reading with Intensity

Light intensity is determined by how much energy is received on a unit of area. In this activity, you will observe how distance affects the intensity (brightness) of light striking an object.

What You Need

book
lamp with the shade removed
60 W bulb
100 W bulb
measuring tape

What to Do

1. Ask an adult to place a 60 W bulb in the lamp. **CAUTION:** Turn off the electricity before the bulb is changed.
2. Darken the room. Turn on the lamp and stand about 60 cm away from it while holding this book. Read a sentence from the book at this distance.
3. Move about 3 m from the lamp. Read a sentence from the book at this distance.
4. Repeat steps 1 to 3 using a 100 W bulb.

What Did You Find Out?

1. How does increasing the distance from the bulb affect the intensity of the light striking the book's pages?
2. Describe the difference between reading the book using the 60 W bulb and reading the book using the 100 W bulb.
3. Draw two diagrams, one showing light leaving the 60 W bulb and one showing light leaving the 100 W bulb. Think of a way to represent the amount of energy striking the book at each distance you measured. You might try drawing different numbers of lines to represent different intensities of light. Remember that light radiates in all directions from the bulb, just as it radiates from the candle flame shown in the diagram below.

Sources of Light

How would your life be different if the Sun and stars were the only sources of light available to you? You would probably go to bed very early, especially in the winter, because there would not be much that you could do after dark. Without artificial sources of light, there would be no television, no lamps for reading, no computers. All the rooms in buildings would probably have windows or skylights.

We are lucky to have many sources of light available to us. In earlier times, once the Sun had set, people found their way around outside with the aid of torches and lanterns. Candles and oil lamps were commonly used indoors. Imagine trying to study by the light of a candle!

Today, we have so much light in our cities that light pollution can wash out our view of the skies at night. That is why many observatories, such as the one shown in Figure 7.3, are located far from urban areas. However, some communities are taking steps to conserve light energy. For example, new types of streetlights are designed to direct their light downward, so that they illuminate the ground or the street and not the sky. In addition, these lights are comparatively energy-efficient. For example, the yellow sodium vapour lights shown in Figure 7.4 on the next page are much more efficient than white lights. The following sections will examine and compare different types of light sources, both natural and artificial.

Pause& Reflect

In your Science Log, describe several ways in which plants and animals respond to changes in the intensity of light. For example, how do your eyes react to a bright light? How do roosters behave when the Sun rises? What do you think birds do during an eclipse of the Sun?

Figure 7.3 This photograph is a time exposure image of star trails over the dome of the Mayall telescope at the Kitt Peak National Observatory in Arizona, USA. The telescope's high-altitude location (over 2000 m) and the clear desert skies reduce atmospheric interference to incoming light.

Figure 7.4 These bright yellow lights contain sodium vapour. Electricity makes the gas glow, producing a very intense yellow light.

Incandescent Sources

An object can be heated to such a high temperature that it emits visible light. Such an object is called an **incandescent source** of light. The emission of visible light by a hot object is called **incandescence**. Both candle flames and light bulbs are examples of incandescent sources. As you saw in the Starting Point Activity, in the light bulbs used most commonly in our homes, electricity heats a metal wire filament in the bulb (see Figure 7.5). This filament becomes so hot that it glows white. The change from electrical energy to visible light energy involves the following energy transformation:

Electrical energy → Thermal energy → Visible light energy

Figure 7.5 An incandescent light bulb

Have you ever touched an incandescent bulb right after you turned off the light? If so, you probably burned your fingers! About 95 percent of the energy given off by incandescent light bulbs is released as heat. In a way, an incandescent source of light is like having a small electric heater in the room.

DidYouKnow?

The filament in an incandescent light bulb is usually made of the element tungsten.

Fluorescent Sources

You may have noticed that when you stand under a so-called "black light," some of your clothing glows, especially white socks! In this process, high-energy, invisible ultraviolet light is absorbed by the particles in the fabric. (You will learn more about ultraviolet light in Chapter 9.) These particles then emit some of this energy as light that you can see, making the clothing glow. This glow is called **fluorescence**. You can summarize this energy transformation as follows:

Ultraviolet light energy → Energy absorbed by particles → Visible light energy

A **fluorescent source** of light makes use of this energy transformation. Figure 7.6 shows the typical parts of a fluorescent tube. An electric current from the lead-in wires and electrodes cause the mercury vapour inside the tube to give off ultraviolet radiation. A phosphor coating on the inside of the tube absorbs the ultraviolet energy. This causes the coating to glow, thus producing light that you can see. The energy pathway for a fluorescent source is summarized as follows:

Electrical energy → Energy absorbed by mercury particles →

Ultraviolet light energy → Energy absorbed by phosphor particles → Visible light energy

Did You Know?

Ultraviolet light can cause eye damage. Never stare directly at an ultraviolet light source.

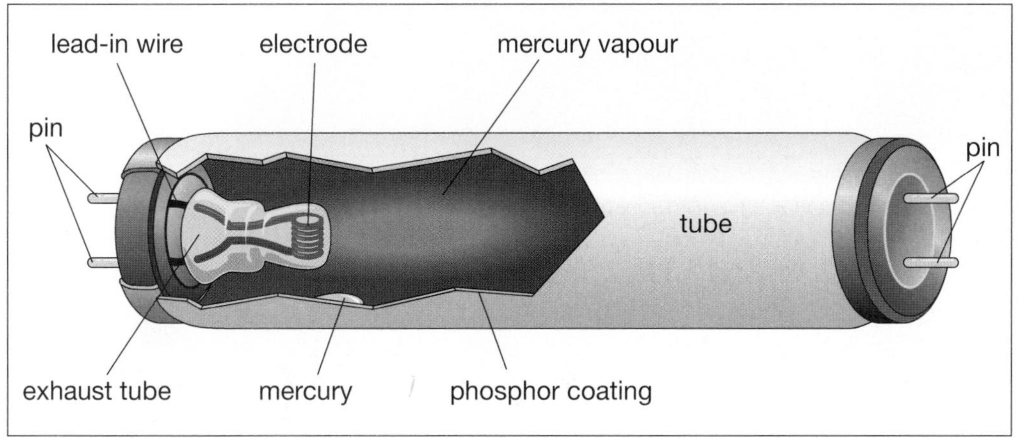

Figure 7.6 A fluorescent tube

Fluorescent tubes have a few disadvantages compared to incandescent light bulbs. They are much more expensive to manufacture and more difficult to dispose of than incandescent bulbs. Also, both the phosphor coating and the mercury vapour of fluorescent tubes are toxic.

However, if you compare the energy pathways for the fluorescent tube and the incandescent light bulb, you will notice a definite advantage for fluorescent sources. Thermal energy is not involved in the operation of a fluorescent light source. You can even touch the tubes when they are lit. As a result, fluorescent lighting wastes much less energy as heat than incandescent lighting. In other words, fluorescent lighting is more energy-efficient. Most schools and businesses use fluorescent tubes rather than incandescent bulbs to conserve energy and thus save money.

Did You Know?

Fluorescent tubes use a device called a ballast resistor. One function of this device is to limit the amount of electricity flowing to the tube. After a few years of use, parts of this device can become loose and begin to vibrate. These vibrating parts cause the annoying hum that you sometimes hear in your classroom.

Phosphorescent Sources

A **phosphorescent source** of light is similar to a fluorescent source. Light energy is absorbed by certain particles that can store this energy for a while. The stored energy is later released as visible light. The original light energy may be either in the form of high-energy ultraviolet light (as in fluorescent tubes) or in the form of visible light. The persistent emission of light following exposure to and removal of a source of radiation is known as **phosphorescence**.

The main difference between a fluorescent source and a phosphorescent source is that particles in the fluorescent source release their light energy immediately. Phosphorescent particles take longer to emit light, and they continue to glow for a while after the energy source has been removed (see Figure 7.7).

Figure 7.7 Phosphorescent materials are often used in photographic darkrooms. This phosphorescent dial on a darkroom timer glows to indicate how long a photograph should remain in various solutions.

Find Out **ACTIVITY**

Recycling Fluorescent Tubes

Disposal of fluorescent tubes poses a challenge. The mercury vapour and the phosphor coating in these tubes are toxic, so we cannot simply throw the tubes away. Is there a way to recycle fluorescent tubes?

What to Do

Conduct research on how fluorescent tubes should be recycled. You might contact a lighting store and arrange to speak to a salesperson or a manager. You could also speak to your school board's health and safety officer. Make sure you write down your questions ahead of time. Another research strategy would be to search the Internet using key terms such as "fluorescent lighting" and "disposal."

What Did You Find Out?

Present your findings to your class in a question-and-answer format. (Use visual aids if possible.) Include your recommendations for the safe disposal of fluorescent tubes.

Chemiluminescent Sources

Light can also result from the energy released in chemical reactions. The chemical reaction produces particles that give off visible light energy. This process is called **chemiluminescence**. The energy pathway for a **chemiluminescent source** can be represented as follows:

Chemical energy → Visible light energy

Glow sticks, often used as emergency signal lights, produce light by chemiluminescence. In a glow stick, a breakable barrier separates two liquids. Bending the stick causes the barrier to break. The liquids mix and cause a chemical reaction that releases light, as shown in Figure 7.8.

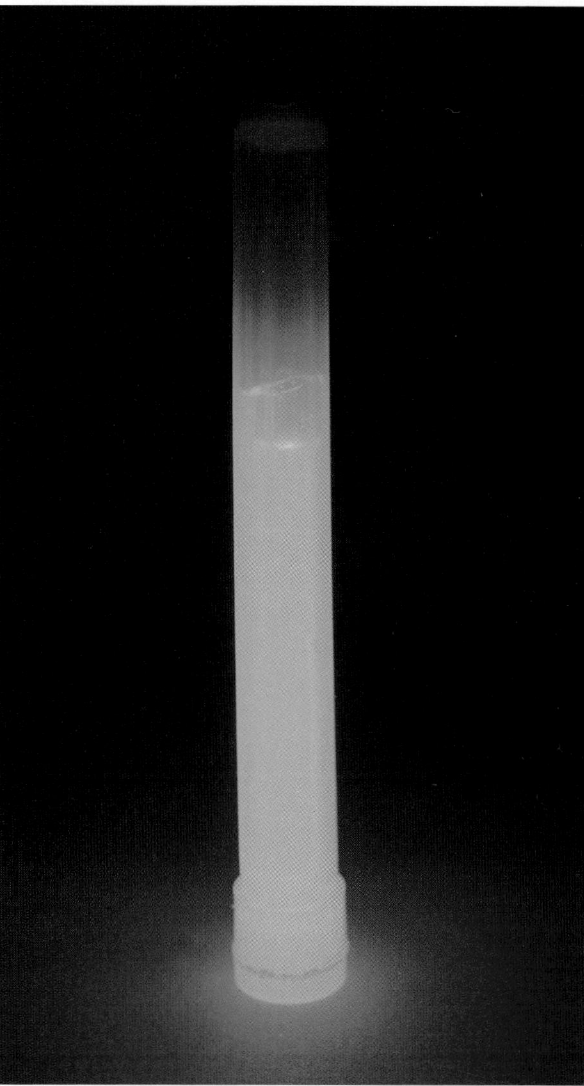

Figure 7.8 A mixture of chemicals releases light when a glow stick is bent. The stick will glow for several hours until the chemical energy is used up.

Bioluminescent Sources

If you were moving through the darkest depths of the ocean in a research submarine, you might be surprised to see glowing creatures swimming past your porthole (see Figure 7.9). These animals cannot be incandescent or fluorescent sources. Instead, they rely on chemical reactions inside their bodies to provide the energy for light. This special type of light produced in living creatures is called **bioluminescence**, and the result is known as a **bioluminescent source** of light. Many organisms that live deep in the ocean use bioluminescence because so little sunlight reaches far below the surface. Some fish produce bioluminescence to attract prey. Certain fungi in caves also produce bioluminescence, as do fireflies. Fireflies glow to attract mates.

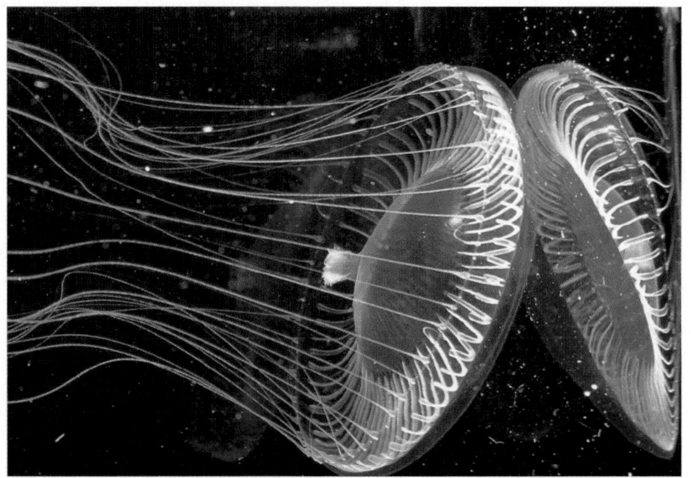

Figure 7.9 How might bioluminescence be helpful to these jellyfish?

Pause& Reflect

In your Science Log, record the various sources of light you encounter in the next few days. Make a table and classify the light sources as incandescent, fluorescent, phosphorescent, chemiluminescent, or bioluminescent. If possible, and if it is safe to do so, bring an unusual light source to class and explain how it works.

The Cost of Lighting

So far, you have looked at how various light sources produce light. Now consider the cost of using different sources of light. Since incandescent and fluorescent sources are the most common around the home, you can compare their costs.

Electrical energy costs about eight cents per kilowatt hour. A **watt** is a unit of electrical power. A **kilowatt hour** is one thousand watts of electrical power operating for one hour. The symbol for watt is W and the symbol for kilowatt hour is kW•h. To understand how to calculate the cost of lighting, look at the following example.

Example: How much will it cost to leave a 60 W bulb on for 10 h if electrical energy costs 8¢/kW•h?

Solution:

1. Convert 60 W to kilowatts by dividing by 1000.
   $60\ \text{W} \div 1000 = 0.06\ \text{kW}$

2. Calculate the number of kilowatt hours by multiplying the power (in kW) by the number of hours.
   $\text{Number of kW•h} = 0.06\ \text{kW} \times 10\ \text{h} = 0.6\ \text{kW•h}$

3. Calculate the cost of leaving the light on for 10 h by multiplying the number of kilowatt hours by the cost per kilowatt hour.
   $\text{Cost (in cents)} = \text{Amount of energy (in kW•h)} \times \text{Unit price (in ¢/kW•h)}$

   $\text{Cost} = 0.6\ \text{kW•h} \times 8\text{¢/kW•h} = 4.8\text{¢}$

   Therefore, the cost of leaving the light on for 10 h is 4.8¢.

Fluorescent tubes are more energy-efficient than incandescent bulbs. A fluorescent tube with a power of 12 W can produce the same amount of light as a 60 W incandescent bulb. The cost of operating the 12 W fluorescent tube for 10 h would be 0.96¢. This is only one fifth the cost of operating the 60 W incandescent bulb for the same amount of time!

The Second Basic Property of Light

Understanding the first basic property of light — that light is energy — helps explain some aspects of light. However, it does not explain everything. When someone stands in front of you during a movie, part of your view is blocked, as shown in Figure 7.10. This hints at another basic property of light. The light from the screen will not bend around the person to reach your eyes because light travels in straight lines from its source. This is the second basic property of light. Knowledge of how light travels allows us to predict how light will behave. To show the path of light, you can draw a straight line with an arrowhead to show the direction in which light is travelling. This type of drawing is called a **ray diagram** (see Figure 7.11).

Figure 7.10 A shadow results because light travels in straight lines from its source and does not bend around objects.

Until light strikes something, it will continue to travel in straight lines away from the source, as shown in Figure 7.11. When light strikes clear substances such as air and water, it passes through them. These media are **transparent**. Window glass is transparent, and so are the lenses in your eyes. You can look through a transparent surface to see the light source clearly on the other side. Other substances let some light pass through them, but the light is scattered from its straight path. These substances are **translucent**. Wax paper is translucent. You can sometimes see the source of light on the other side of a translucent surface, but the image is not clear. Many materials will not allow any light to pass through them. These materials are **opaque** — they block the path of light. A book is opaque, for example. As you learned in previous science studies, opaque objects produce shadows when light strikes them.

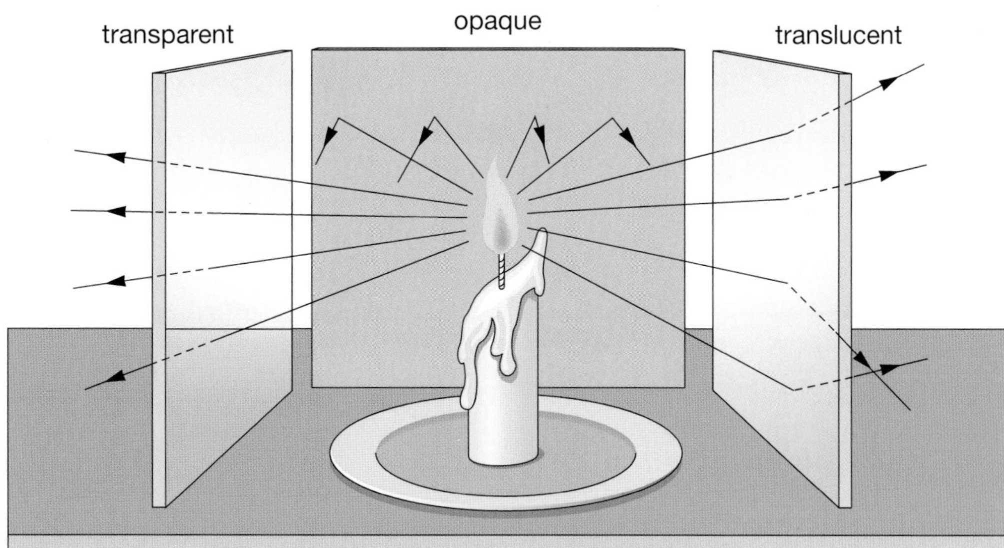

Figure 7.11 Light will travel in straight lines until it strikes something.

Find Out **ACTIVITY**

Make a Large Pinhole Camera

To see for yourself that light travels in straight lines, try making a large pinhole camera.

What You Need

sharp knife

large cardboard box, about twice as large as your head

aluminum foil

bright object, such as a light bulb

masking tape

What to Do

1. The box will serve as a camera body. Cut a 3 cm square hole in one side of the box, 2–3 cm from the closed end of the box. **CAUTION:** Be careful when using sharp objects such as scissors.
2. Fold a 5 cm square piece of aluminum foil in half. Cut a semicircle 1 cm wide into the fold.
3. Unfold the foil and tape it over the square opening in the box. The hole will let light into your camera.
4. Turn your back on the bright object (light bulb) and lower the box over your head with the hole behind you.
5. Fold the flaps of the box against your neck and head to make the inside of the box as dark as possible. Slowly move the box around until you can see a bright patch of light on the inside of the box.
6. Observe the patch of light at several distances from the bright object. **CAUTION:** Have a partner move the object for you. Do not walk while wearing the box on your head.

What Did You Find Out?

1. What happens to the brightness of the image when an object is closer to the camera?
2. What happens to the size of the image when an object is closer to the camera?
3. What happens to the sharpness of the image when an object is closer to the camera?
4. What evidence have you observed that light travels in straight lines?

Extension

Try a similar activity using a tissue box with a small hole at one end of the box and wax paper (as the screen) taped to the other end.

Career CONNECT

Stage Lighting

Have you ever watched a play in a theatre? Did you notice how lighting can change the mood of a scene on stage? Bright lights shining from all sides create a warm, relaxed atmosphere. Dim lights and shadows create an atmosphere of danger and suspense.

The lighting technician in the theatre must know which of the many lights needs to be lit for each scene and just how quickly or slowly the lights should fade out or come on. In the past, the job often required two people: one to make each lighting change on cue and another to set the switches in preparation for the change. Now, most large theatres have a computerized lighting board. The technician can program the lighting changes for the show ahead of time. During the show, only one person is needed to shift from one lighting cue to the next. The equipment is more complicated, but it allows the show to run more smoothly, with fewer errors.

Ask your teacher's permission to interview a person in a light-related career, such as a photographer, home-security installer, photo-lab manager, or videographer. Ask about the technological changes that have taken place in the industry and how these changes affect the work the person does. What changes are likely to take place in the future? Write a report comparing and contrasting past methods of doing the job with today's methods.

Check Your Understanding

1. What is light?
2. Write the energy pathway for
   (a) an incandescent source
   (b) a fluorescent source
   (c) a chemiluminescent source
3. State one advantage that incandescent bulbs have over fluorescent tubes.
4. State one advantage that fluorescent tubes have over incandescent bulbs.
5. If electrical energy costs 7¢/kW•h, calculate the cost of running a 15 W scanner for 10 min. (You will need to convert 10 min into hours.)
6. Describe what happens when light strikes a translucent surface, a transparent surface, and an opaque surface. Give one specific example of an object that has each type of surface.
7. What would happen to the intensity of sunlight if Earth were twice as far from the Sun?
8. **Apply** The diagram on the right shows the relative positions of Earth, Moon, and Sun during a solar eclipse, as well as the path of the light during an eclipse. Is much of Earth's surface in complete shadow? Use the idea that light travels in straight lines to explain how a solar eclipse occurs. (The motions of Earth and Moon are factors, as well.)

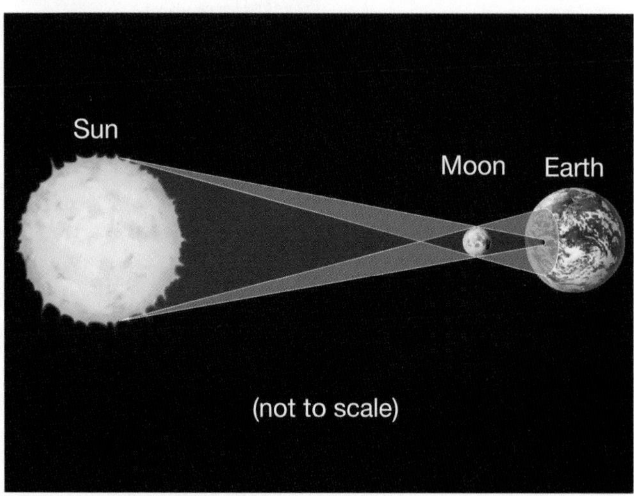

Pause& Reflect

Start a list of careers that require knowledge of the behaviour of light. Share what you have written with at least two class members. Add to your list of careers as you learn more about light in this chapter.

7.2 Reflecting on Reflections

You see some objects because they emit their own light. The Sun, a glowing light bulb, and a burning candle all produce their own light energy. These objects are **luminous**. Most objects, however, do not produce their own light energy — they are **non-luminous**. They can be seen only when light from a luminous source strikes the object and then reflects off the object into your eyes. **Reflection** occurs when light bounces off a surface. When a room is poorly lit, you see less because less light is reflecting off the objects around you. If you shine a flashlight on an object, you can see the object clearly again because you have increased the amount of light reflecting off its surface.

All the light striking an object is not reflected — some light energy is absorbed. Dark surfaces tend to absorb most of the light that strikes them. That is why wearing dark clothes in the summer Sun can make you feel hot. That is also why dark objects are more difficult to see at night. Very little light is reflected from dark surfaces, even in full daylight. On the other hand, light-coloured objects reflect most of the light that strikes them.

What you see depends on the images your eyes form using the light reflected off an object, and also on your brain's interpretation of these images. Sometimes the brain misinterprets an image. Optical illusions fool our brains into making false conclusions (see Figures 7.12 and 7.13). Try to create your own optical illusion.

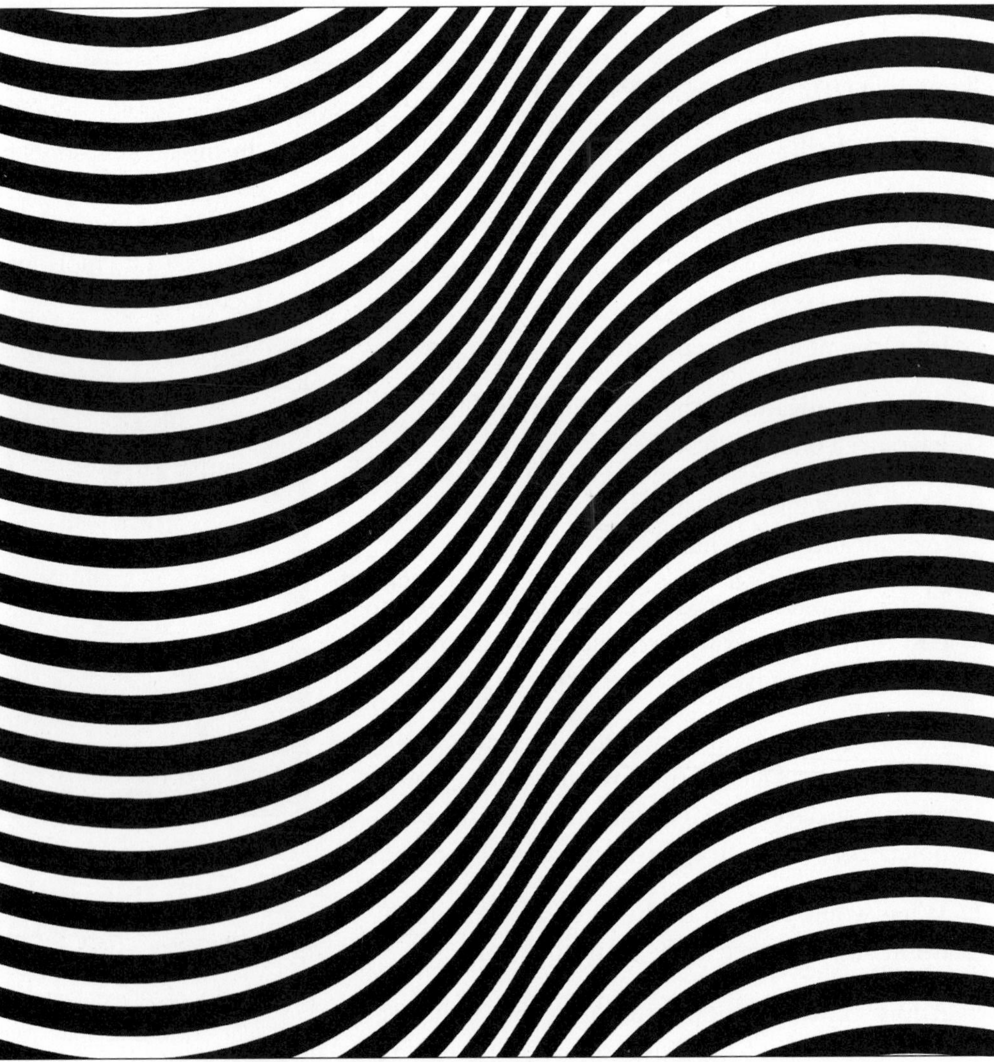

Figure 7.12 Observe this image through one eye while you slowly lift the page up to the level of your chin. Did you see the lines on the flat page appear to rise into a gently sloping hill on the right?

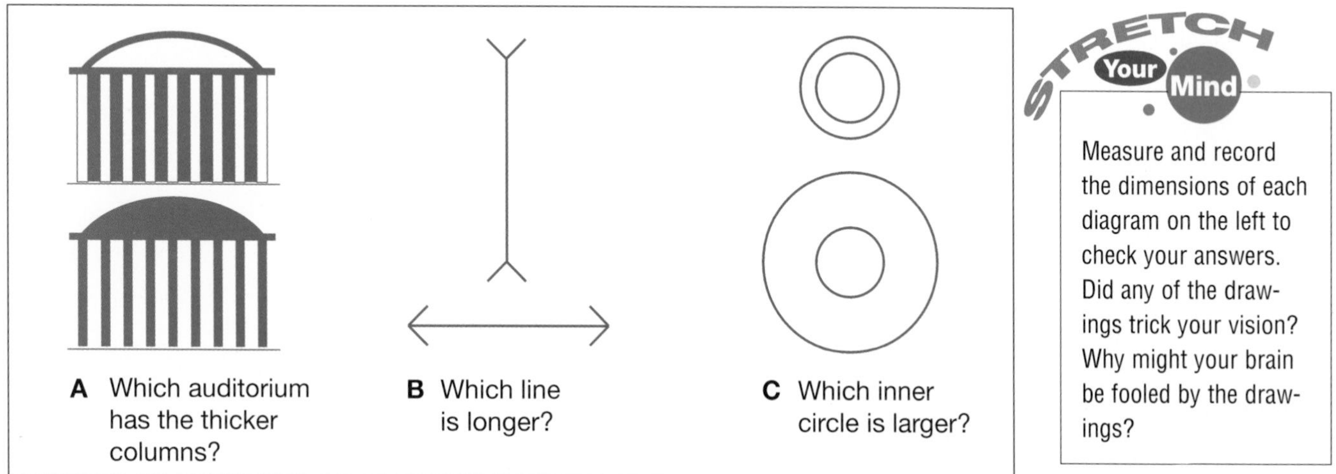

A Which auditorium has the thicker columns?
B Which line is longer?
C Which inner circle is larger?

Figure 7.13 Study these optical illusions and answer the questions.

STRETCH Your Mind

Measure and record the dimensions of each diagram on the left to check your answers. Did any of the drawings trick your vision? Why might your brain be fooled by the drawings?

When light strikes a reflecting surface, it bounces off the surface. A light ray that strikes a surface is called an **incident ray**. The light that is reflected from a surface is called a **reflected ray** (see Figure 7.14). To find out in which direction each ray of light is travelling, you can measure an angle. To do this, draw a reference line that is perpendicular (90°) to the reflecting surface at the point where the incident ray strikes the surface. This line is called the **normal**. The angle between the incident ray and the normal is called the **angle of incidence**, *i*. The angle between the normal and the reflected ray is called the **angle of reflection**, *r*. This approach provides a standard way to describe the direction in which light travels during reflection. You will investigate reflection in the next two investigations.

INTERNET CONNECT

www.school.mcgrawhill.ca/resources/

To see more optical illusions, and to learn about sensory systems and perception, visit the above web site. Go to **Science Resources**. Then go to **SCIENCEPOWER 8** to find out where to go next.

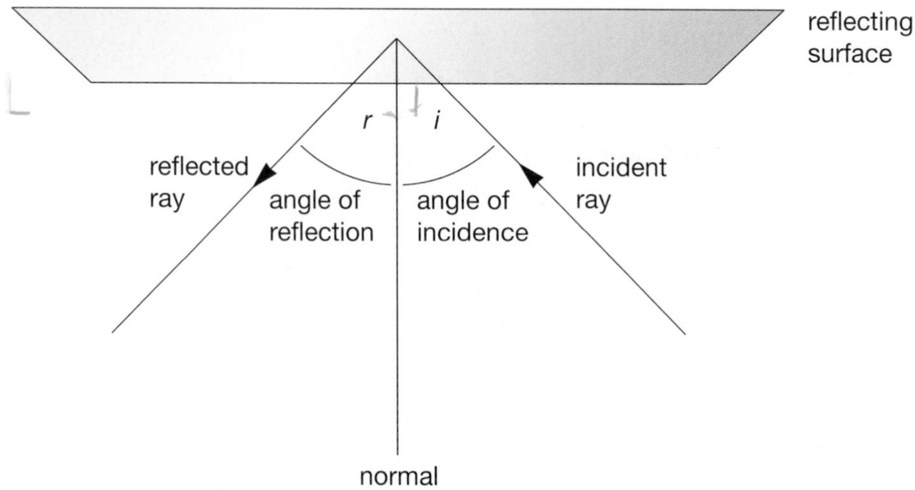

Figure 7.14 The normal is a reference line that is drawn perpendicular to the reflecting surface at the point where an incident ray strikes a reflecting surface. The angle of incidence, *i*, is the angle between the normal and the incident ray. The angle of reflection, *r*, is the angle between the normal and the reflected ray.

When Light Reflects

Light travels in straight lines until it reaches a surface. What happens when light is reflected by a surface? Where does the light energy go? In this investigation, you will observe the behaviour of reflected light.

Problem

What happens to light when it reflects off a surface?

Apparatus
clear plastic cup
wooden pencil
ruler

Materials
water
paper

Procedure

1. Fill the cup about three quarters full of water. Place the cup of water on a level surface.

2. Observe the surface of the water. Move your head around until you can see a reflection of the lights overhead, or a reflection of a window.

3. Make a simple ray diagram to show the direction in which light travels before it reaches your eye. Show and label the positions of the light source, the surface of the water, and your eye. This drawing should show the situation as someone would observe it from the side.

4. Move the cup of water to the edge of the desk or table. Wait until the water stops jiggling. Crouch down so that you can look up at the bottom of the water's surface.

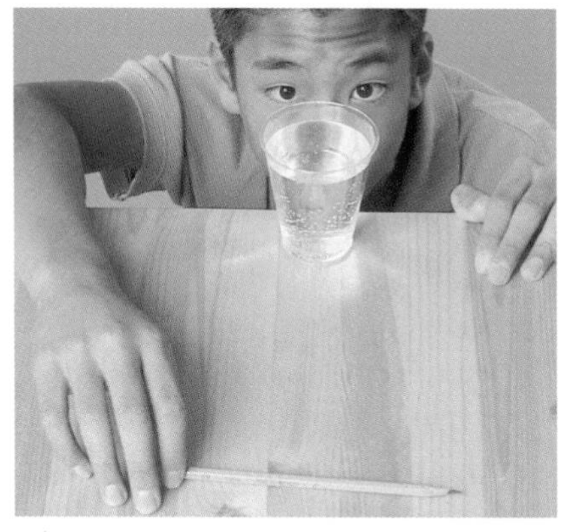

5. Slide a pencil across the desk, toward the cup and your eye.

6. Move the pencil along the desk surface until you can see a reflection of the pencil in the lower surface of the water. Make a simple ray diagram showing the path of the light from the pencil to your eye.

7. Look at the reflection of the pencil as you did in step 6, but now gently tap on the rim of the glass. What do you see now?

Wipe up any spills as wet floors are slippery.

Analyze

1. In steps 4 and 6, what happened to some of the light that struck the lower flat surface between the air and the water? What common device depends on this behaviour of light?
2. In step 7, what change occurred in the surface of the water when you tapped on the glass? Could you still see a reflection of the pencil?

Conclude and Apply

3. During reflection, what happens to the direction in which light travels?

CONDUCT AN

INVESTIGATION 7-B

Laws of Reflection

When you look in your bathroom mirror, light from a window or a bulb reflects off your face in all directions. Some of the light from your face reflects off the mirror into your eyes. This reflected light must follow a consistent pattern because you always see the same image of your face when you look in a mirror.

In this investigation, you will demonstrate the two laws that describe the path of light when it bounces off a surface (in this case, a **plane mirror** — one that has a flat surface). In science, a **law** is a statement of a pattern that has been observed again and again, with no exceptions. A law describes an action or condition that has been observed so consistently that scientists are convinced it will always happen.

Problem

How does light behave when it reflects off a flat surface?

Safety Precaution

The edges of the mirror may be sharp. Be careful not to cut yourself.

Apparatus

ray box

plane mirror (about 5 cm × 15 cm) with support stand

small object such as a short pencil, a common nail (about 5 cm), or a toothpick that is thicker at one end than the other (the object should not be longer than the mirror)

protractor

ruler

pencil (for drawing)

Materials

sheet of blank paper (letter size)

Skill
POWER

For tips on making tables, turn to page 546. For tips on measuring angles with a protractor, turn to page 540.

Procedure

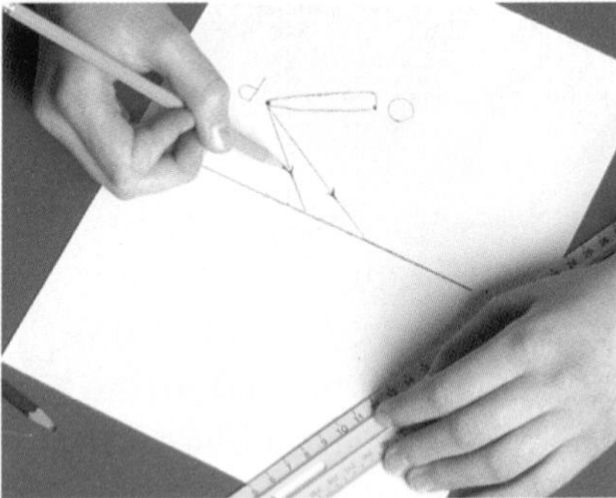

1. Near the middle of the blank sheet of paper, draw a straight line to represent the reflecting surface of the plane mirror. (This is usually the back surface of the mirror because the front surface of the mirror has a protective glass pane.)

2. Lay the small object on its side on the paper about 5–10 cm in front of the line you just drew, and at an angle to the line. Trace the shape of the object. Mark one end of the object **P** and the other end **O**.

3. Remove the object. Draw two straight lines that start from **P** and that end on the straight line that represents the plane mirror. On each line, neatly draw an arrowhead pointing toward the reflecting surface. These lines represent the paths of two rays of light that come from the object and reflect off the mirror. How many of these incident rays could you draw?

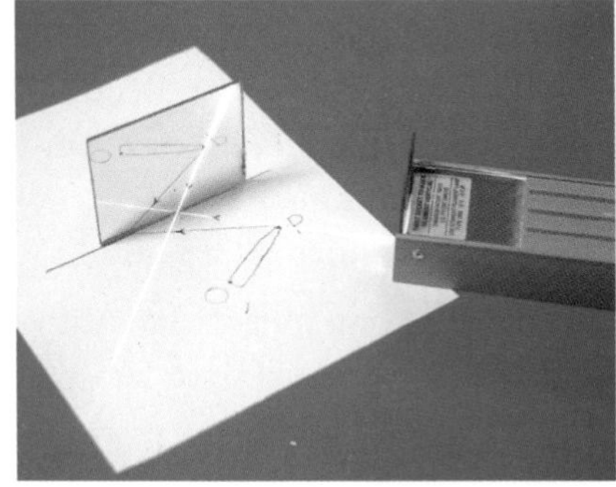

4 Carefully place the mirror in its stand on the sheet of paper so that the reflecting surface of the mirror is exactly along the line you drew in step 1.

5 Use the ray box to shine a thin beam of light along each of the incident rays you drew. Where does the reflected light go?

6 Accurately draw a straight line to show the path of each ray after the light strikes the mirror. On each line, draw an arrowhead pointing away from the mirror. Each line shows a reflected ray.

7 Place the mirror and the object back on the sheet of paper. Observe the reflection of the object and the reflected rays that you drew. What do you notice about the reflected rays? Indicate on your drawing the point from which the two reflected rays seem to come.

Analyze

1. On your sheet of paper, use a protractor to draw the normal at the point where each incident ray strikes the mirror. (You may refer to Figure 7.14 on page 217 as a guide.) Use the normal as a reference line to measure angles.

2. Make a table with three columns that have these headings: "Incident ray," "Angle of incidence," and "Angle of reflection." Give your table a suitable title. Accurately measure the angle of incidence, *i*, and the angle of reflection, *r*, for each of the two light rays. Record the data in your table.

3. Extend each reflected ray behind the mirror, using a dotted line. Measure the perpendicular distance from the mirror to the point where the reflected rays meet (see the diagram below). Compare this distance to where the incident rays started at point **P**.

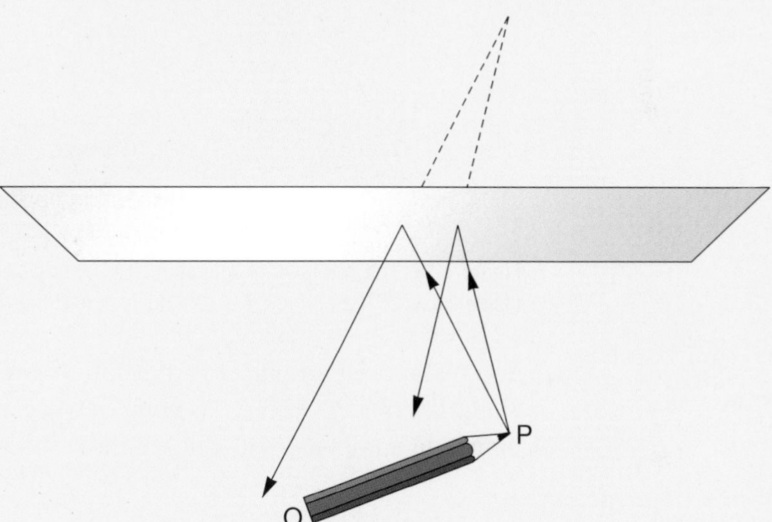

Conclude and Apply

4. From your data table, describe the pattern relating the angle of incidence and the angle of reflection. This pattern is the first law of reflection. Find two other groups in your class that found a similar pattern. Did any group discover a different pattern?

5. Hold a pencil in front of you. It can be moved in three dimensions — toward or away from you, to your left or right, or up and down. How many dimensions are necessary to contain the incident ray, the normal, and the reflected ray? What is another name for this type of space? (Hint: Think about the name given to the mirror in this investigation.) The second law of reflection describes this arrangement of the three lines — the incident ray, the normal, and the reflected ray. State the second law of reflection.

Predictable Behaviour

Light that is reflected from a surface behaves in two predictable ways. You demonstrated this in Conduct an Investigation 7-B. The two predictable behaviours of light are called the **laws of reflection** (see Figure 7.15). The first law of reflection states that the angle of reflection, *r*, is equal to the angle of incidence, *i*. For example, if the angle of incidence is 30°, then the angle of reflection is also 30°.

The second law of reflection states that the incident ray, the normal, and the reflected ray are all in the same **plane** (an imaginary flat surface). This is why you can draw all three lines on a flat sheet of paper.

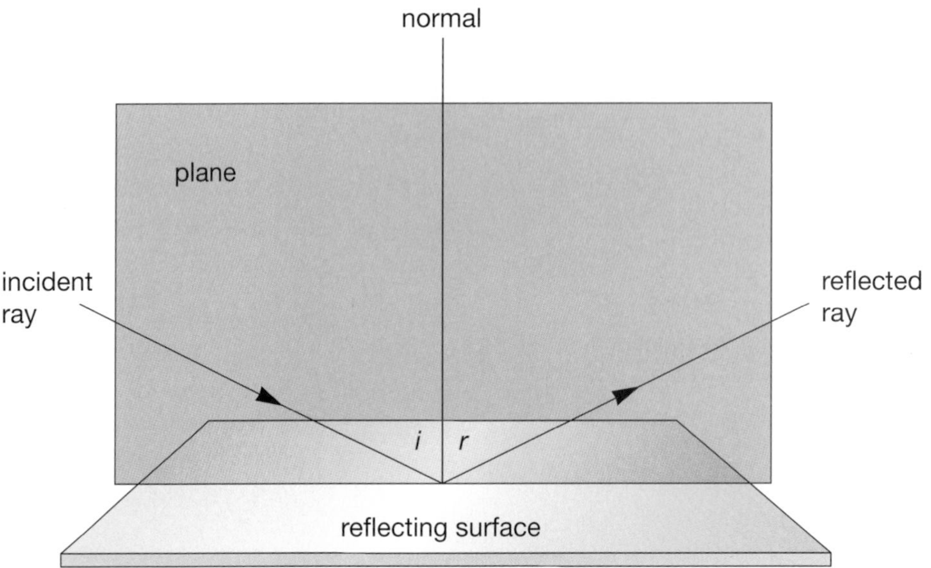

Figure 7.15 The two laws of reflection: 1. The angle of reflection, *r*, is always equal to the angle of incidence, *i*. 2. The incident ray, the normal, and the reflected ray are always in the same plane.

You realize that an object is in front of you only because light is spreading out from that object. As long as your eyes receive light as if it were spreading out from points on an object, you will see that object.

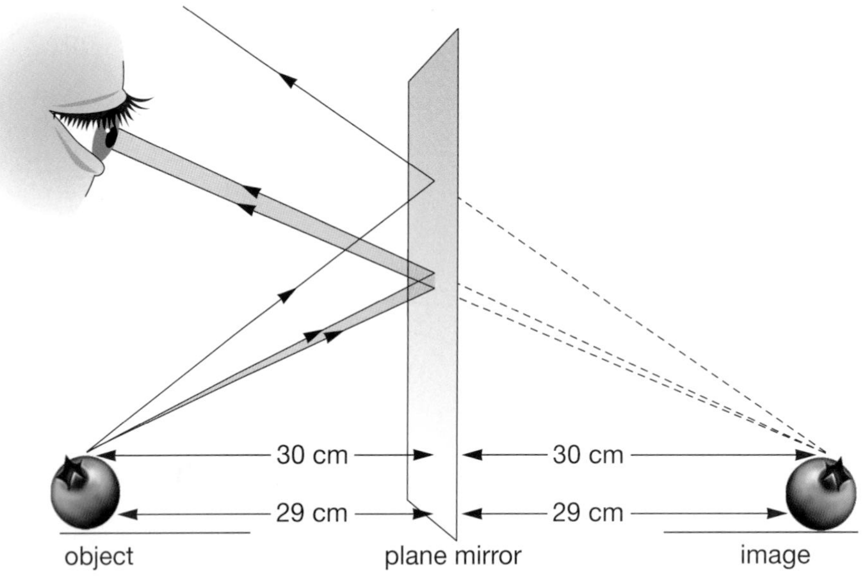

Figure 7.16 Only a small fraction of the light reflecting from an object enters your eyes.

Mirrors can trick your brain. Light is reflected off an object toward the mirror, which reflects the light back to you. The object appears to be located at the point from which the light seems to come — a position behind the mirror's surface. The image appears to be the same size and shape as the real object, and it seems to be the same distance behind the mirror's surface as the real object is in front (see Figures 7.16 and 7.17).

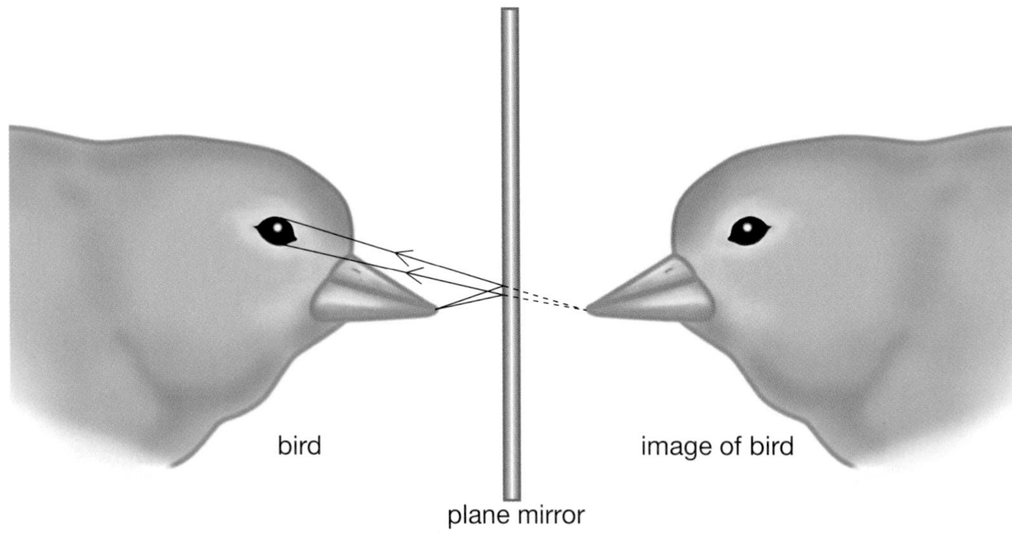

Figure 7.17 We know that what we see in a mirror is just an image. However, a pet bird will chatter for hours to a "friend" in the mirror.

To confirm this, choose a point on a diagram of a reflected object and measure the perpendicular distance between that point and the reflecting surface, as shown in Figure 7.18. Mark a point at the same distance behind the reflecting surface. Do this for several other points on the object until you can see the object's shape and size emerge from the points. If you do this for enough points, you could connect the dots to draw the location and shape of the image formed by the reflection.

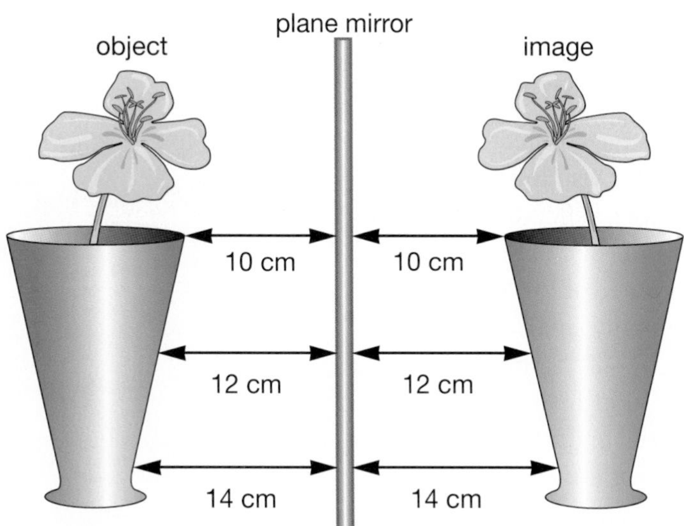

Figure 7.18 The image of an object reflected by a plane mirror appears to be at the same distance behind the mirror's surface as the object itself is in front.

A smooth, flat reflecting surface always produces an image that has the same size and shape as the object. All the normals related to light reflected from a bathroom mirror point in the same direction (see Figure 7.19A). Clear images are created from light reflected by a plane mirror.

Compare the normals related to light reflected from a mirror to the normals related to light reflected from a wall or a piece of paper (see Figure 7.19B). Although the wall and the paper surfaces might appear to be smooth, both are much rougher than the mirror's surface. The normals related to light reflected from rough surfaces will point in random directions, depending on exactly where the incident rays strike the surface. When light reflects off a rough surface, **diffuse reflection** occurs and no image results, as shown in Figure 7.19B.

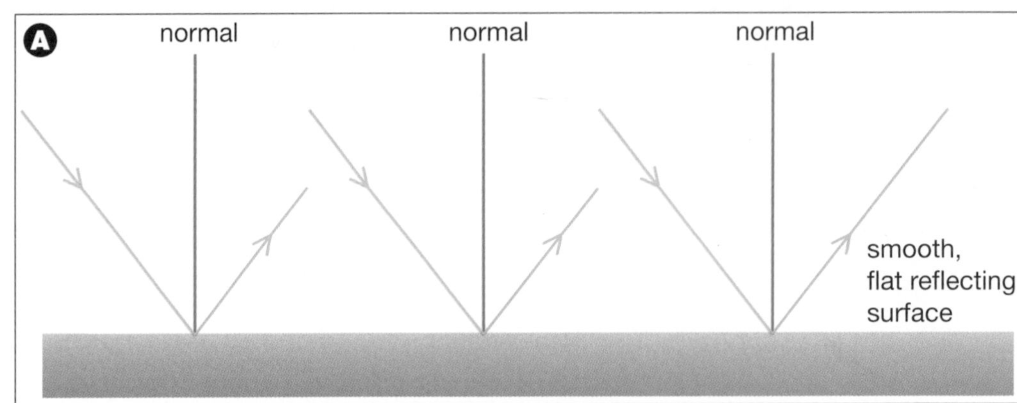

Figure 7.19A Reflected light from a smooth surface

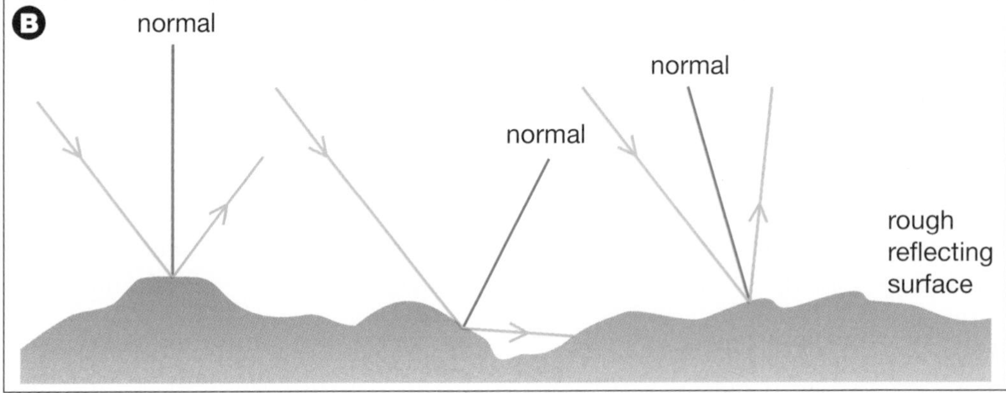

Figure 7.19B Reflected light from a rough surface

Astronauts have placed corner reflectors on the Moon. Scientists on Earth then aimed pulses of laser light at these reflectors. By measuring the time it took for the light to return to Earth, the scientists determined the distance to the Moon's surface to within a few centimetres!

Using Reflections

Cars and bicycles have reflectors to make these vehicles visible at night. Figure 7.20 shows a reflector in which hundreds of tiny, flat reflecting surfaces are arranged at 90° to one another. These many small surfaces are packed side by side to make the reflector. When light from another vehicle hits the reflector, the light bounces off the many tiny surfaces back toward the source of the light. The driver in the other vehicle sees the reflection and realizes that something is ahead.

Figure 7.20 A bicycle reflector

Pool players can use the laws of reflection to improve their game. Like a light ray, a pool ball travels in a straight line until it strikes something. In a "bank shot," the white cue ball bounces off a cushion before it strikes the target ball. To decide where to aim the cue ball against the cushion, the player chooses a spot that is the same distance behind the cushion as the target ball is in front (see Figure 7.21). This spot is the "image" of the target ball. The player now shoots the cue ball toward the image. Because the ball bounces off the cushion at the same angle at which it strikes the cushion, the cue ball bounces off the cushion and strikes the target ball.

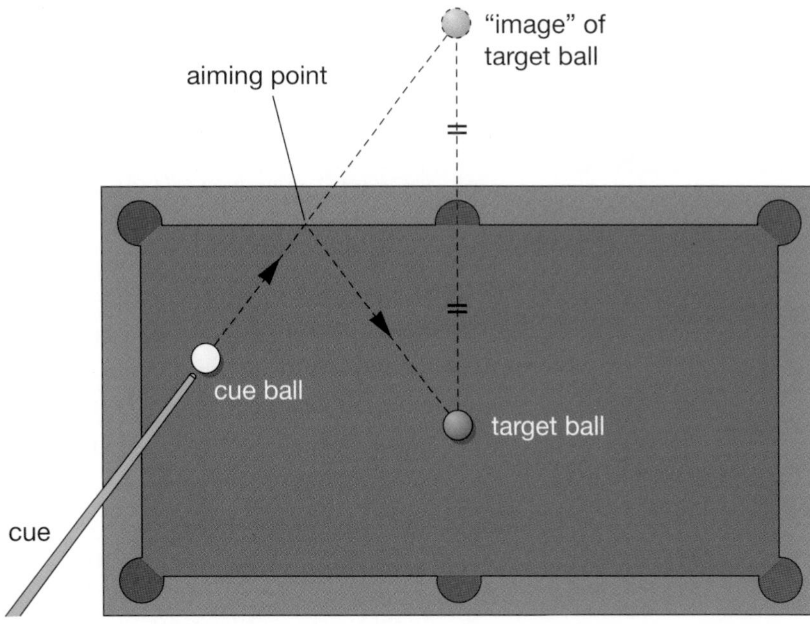

Figure 7.21 You can improve your pool game by applying the laws of reflection.

Check Your Understanding

1. Make a simple, accurate drawing in which you show and label: an incident ray, a reflected ray, the normal, the angle of incidence, and the angle of reflection. Write a definition for each term.
2. State the two laws of reflection.
3. When you see the reflection of the tip of your nose in a plane mirror, from where do the reflected rays of light appear to be coming? If you move twice as far away from the mirror, what happens to the position of the image of your nose?
4. In your notebook, trace each diagram below. Make the measurements and draw the missing parts.

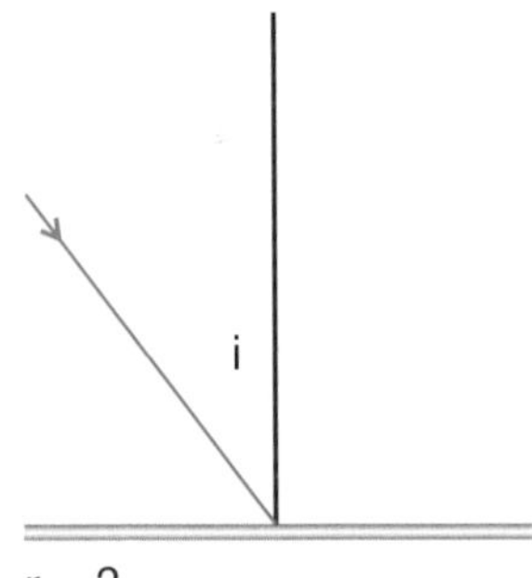

r = ?
Draw the reflected ray.

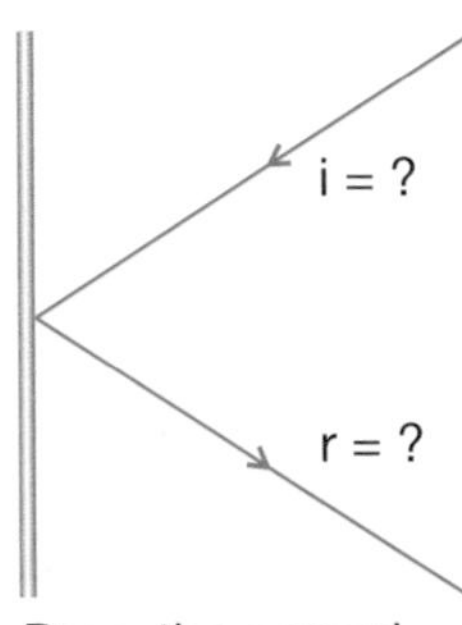

Draw the normal.

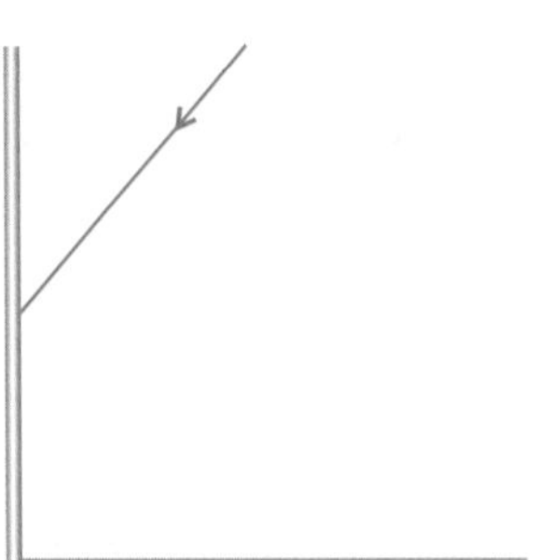

Draw the two reflected rays. Compare the directions of the light striking and bouncing off the mirror.

7.3 Refraction

DidYouKnow?

In air, light travels at 300 000 km/s. It slows down to 200 000 km/s in glass and 165 000 km/s in diamond.

You know that reflection occurs when light rays bounce off objects, and you can now accurately predict the direction in which reflected light rays travel by using the laws of reflection. You can also predict where an image will be located in a plane mirror.

What happens when light moves from air into a medium such as water? If you have ever stood on the side of a pool and tried to dive for an object on the bottom, you may have been surprised that the object was not where you expected it to be. **Refraction** is the bending of light when it travels from one medium to another. Light bends because it changes speed when it moves between materials that have different densities. Light usually travels more slowly in comparatively dense material. The bending of light makes the object's image appear to be in a different position from where the object really is (see Figure 7.22). Explore refraction in the next two investigations.

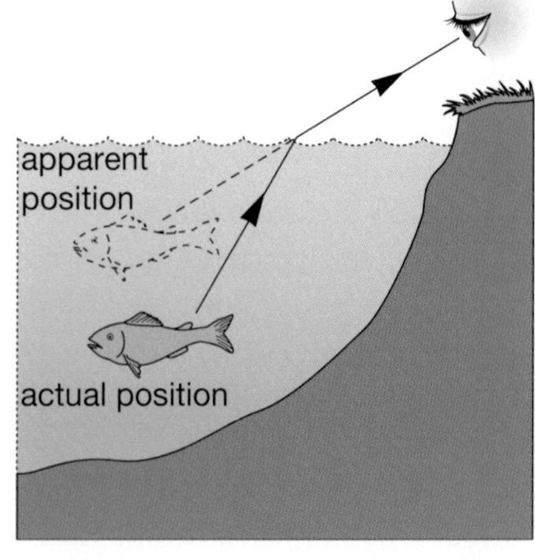

Figure 7.22 The bending of light can make it difficult to see where an object is located in the water.

Find Out ACTIVITY

The Re-appearing Coin

You can observe refraction in the following activity.

What You Need

cup or bowl with opaque sides
water
coin

What to Do

1. Work with a partner. Place the coin in the middle of the empty cup or bowl. Look down on the coin with one eye, then lower your head until the edge of the cup blocks your view of the coin. Do not move your head.
2. Your partner now slowly pours water into the cup until you can see the coin again. If the coin moves because of the flow of water, start again. Use a pencil to temporarily hold the coin in place.
3. Wipe up any spills and wash your hands after this activity.

What Did You Find Out?

When water was poured into the cup, you could see the coin, even though the straight-line path of the light was blocked by the cup. Copy the diagram below, indicate the water line, and draw rays to show the light's path. What happened to the rays of light when light passed from air to water?

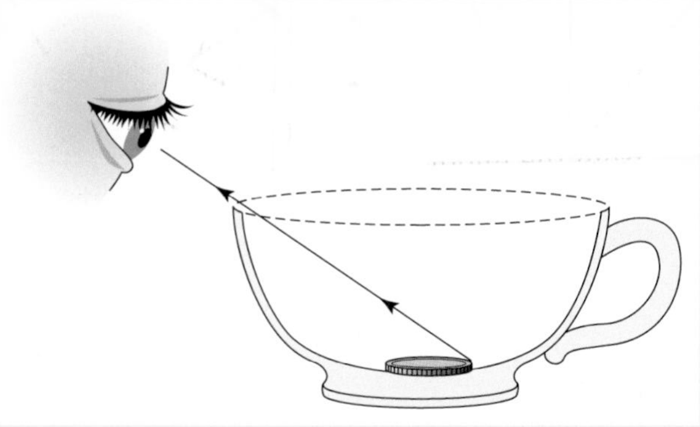

CONDUCT AN

INVESTIGATION 7-C

When Light Refracts

Observe how refraction affects light travelling through different materials.

Problem

What happens to light when it travels from one medium into another?

Apparatus

clear plastic cup, three quarters full of water

wooden pencil

ruler

pencil (for drawing)

Materials

water

paper

Procedure

1. Lay the pencil on the desk so that the middle of the pencil touches the back edge of the cup. Look at the pencil through the front of the cup. Draw what you see.

2. Lean the pencil, pointed end down, in the cup of water. Lower your eyes so that they are level with the surface of the water. When you look at the pencil from the side, what seems to happen to it at the surface of the water? Draw what you see.

3. Look straight down along the pencil that is standing in the water. What seems to happen to the pencil at the surface between the water and the air? Move your head slightly to the side. Draw the pencil as it appears to you. Wipe up any spills after this investigation.

Analyze

1. In step 1, through what medium did light from the ends of the pencil travel before reaching your eyes? Through what medium did light from the middle of the pencil have to travel? Did this light travel in a straight line all the way? How do you know?

2. In steps 2 and 3, through what medium does the light from the bottom part of the pencil travel before reaching your eyes? Did this light travel in a straight line all the way? What happens to the path of the light when it moves from the water to the air?

Conclude and Apply

3. Through how many different media does light travel during refraction in this activity?

4. What can happen to the path of light during refraction? What can happen to the image you see, compared with the real object?

CONDUCT AN

INVESTIGATION 7-D

Follow That Refracted Ray!

Your class will work in groups to design an investigation to study the behaviour of light as it passes through different substances. When you examined the path of light rays that reflected off a plane mirror, you discovered a pattern. The angle of reflection always equals the angle of incidence. Understanding this pattern helps you predict how light reflects off a flat surface. Does refracted light behave according to a consistent pattern as well? Make a hypothesis and then test it.

Problem

Is there a pattern that describes the path of light during refraction?

Apparatus

ray box (placed on the edge of a sheet of white paper)

transparent plastic, watertight tray (box top from greeting cards, candies, etc.)

ruler

protractor

Materials

sheet of blank white paper (letter size)

water

liquid other than water (vegetable oil, liquid soap, etc.)

Procedure

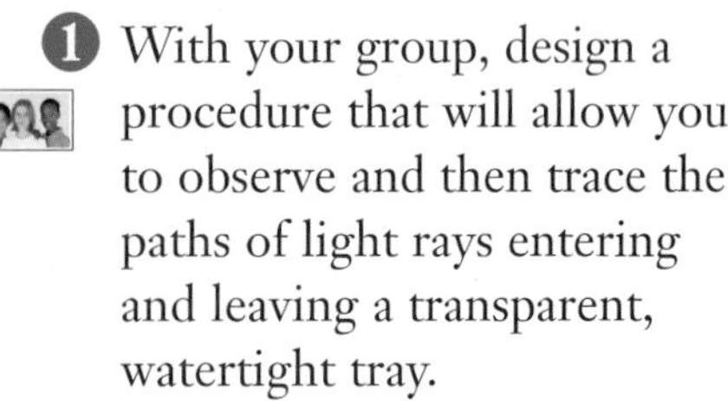

1. With your group, design a procedure that will allow you to observe and then trace the paths of light rays entering and leaving a transparent, watertight tray.
2. Write a brief description of the procedure you plan to follow. Have your teacher approve your procedure.
3. After conducting your investigation, prepare actual-size diagrams on which you can record your observations.
4. Each of your diagrams should show the following:
   - **(a)** the path of a light ray going from the air, through the empty tray, and back into the air through the opposite side of the tray, for at least two different angles of incidence
   - **(b)** the path of the same incident rays when there is water in the tray
   - **(c)** the path of light that travels through one corner of the tray, through the water, and out the adjacent side of the tray back into the air
   - **(d)** the paths of light travelling through some liquid other than water for the same incident rays you used in (a) and (b)

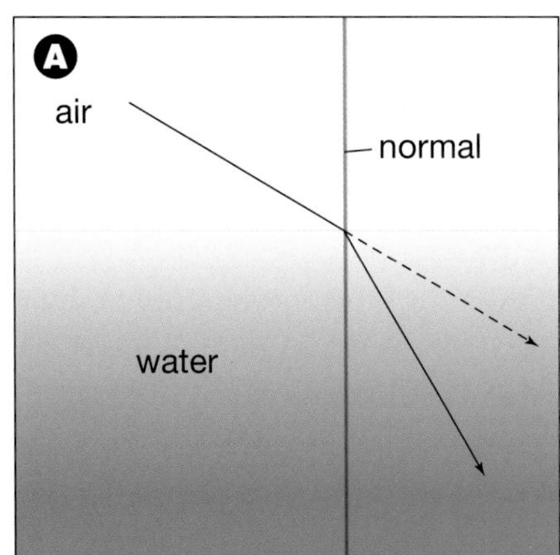

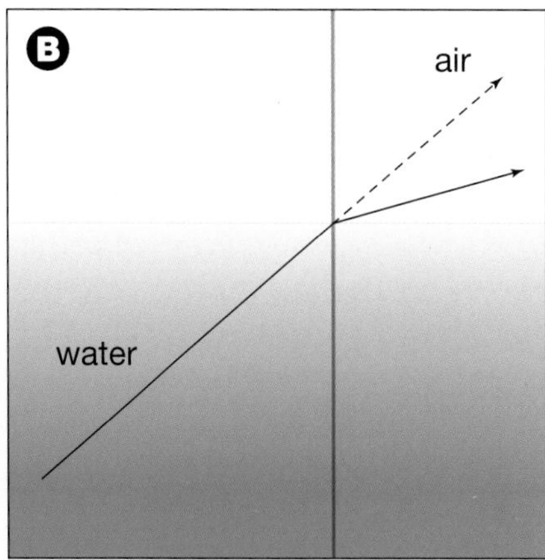

Analyze

1. On your diagrams, draw the normal at each point where a light ray travels from one medium into another.

2. Measure and record the size of each angle on your diagrams. Record the angles using the symbols *i* and *R*.

Conclude and Apply

3. Does light bend toward or away from the normal when it travels from air into another medium such as water? Did your results support your hypothesis? Did other groups get similar results using their procedures?

4. What happens to the size of the angle at which the light bends when the angle of incidence increases?

5. What happens to the size of the angle at which the light bends when a liquid other than water is used and the angle of incidence is the same?

6. Does light move toward or away from the normal when it travels from a medium such as water into air?

7. Is there an angle of incidence for which there is no change in the direction of the light? Draw a diagram for this situation for light travelling through a rectangular shape.

8. Write a statement that answers the problem on page 228.

5 If necessary, you can refer to the diagrams on this page as a guide.

(a) Wipe up any spills as wet floors are slippery.

(b) Wash your hands after this investigation.

Around a Bend with Light

When light travels from one medium into a denser one — for example, when it moves from air into water — it will bend *toward* the normal. When light exits a denser substance, its direction of travel bends *away from* the normal. How much bending occurs depends on the type of substance through which the light travels. The new direction of the light is called the **angle of refraction,** *R* (see Figure 7.23). When the angle of incidence, *i*, increases, the angle of refraction, *R*, also increases. However, doubling the angle of incidence does not mean that the angle of refraction also doubles.

Refraction can also occur when light travels through air at different temperatures. Warm air is less dense than cold air. Light bends as it travels through different densities of air. The refraction of light through air can result in a mirage. Have you ever been driving along a highway on a hot summer day and noticed what looked like pools of water lying ahead? However, when you got close to the pools, they mysteriously disappeared. You were seeing a mirage. The air close to the ground is hotter and less dense than air higher up. As a result, light from the sky directed at the ground is bent upward as it enters the less dense air. The "pools of water" were actually images of the sky refracted by warm air near the ground (see Figure 7.24).

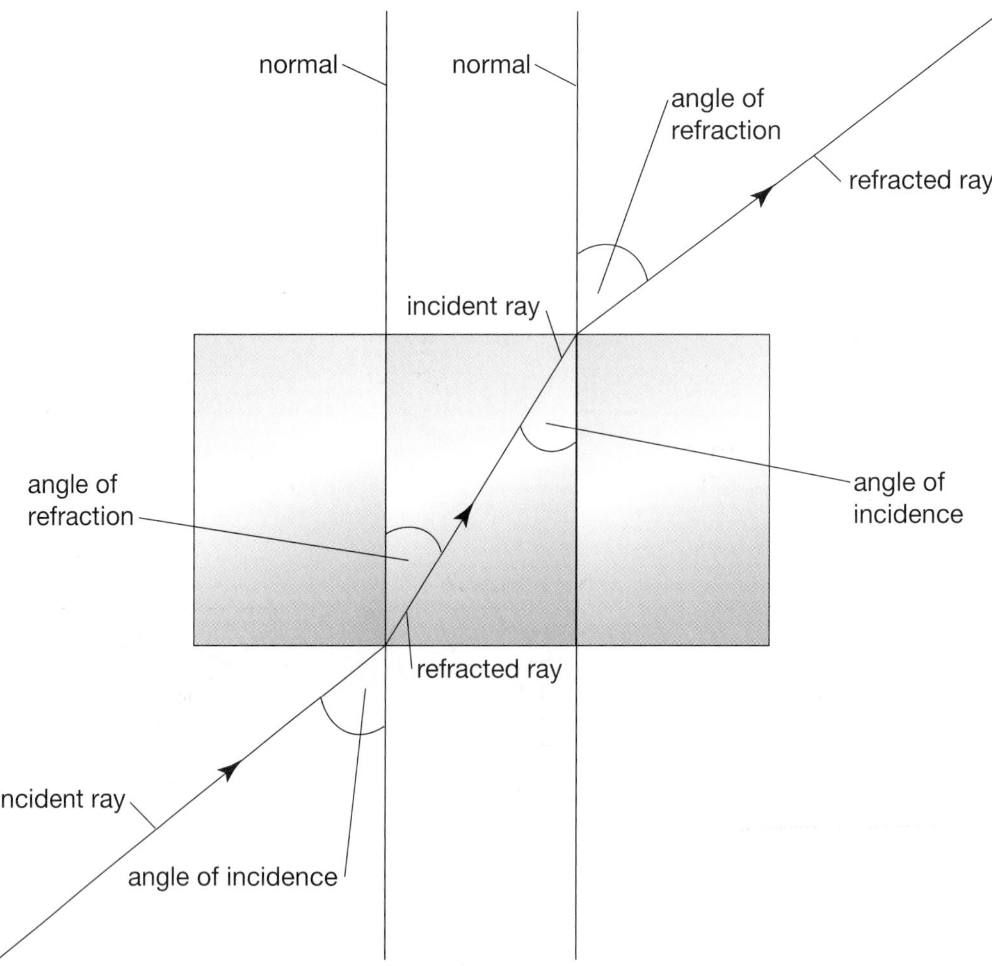

Figure 7.23 Light is refracted as it passes through one medium into a denser medium.

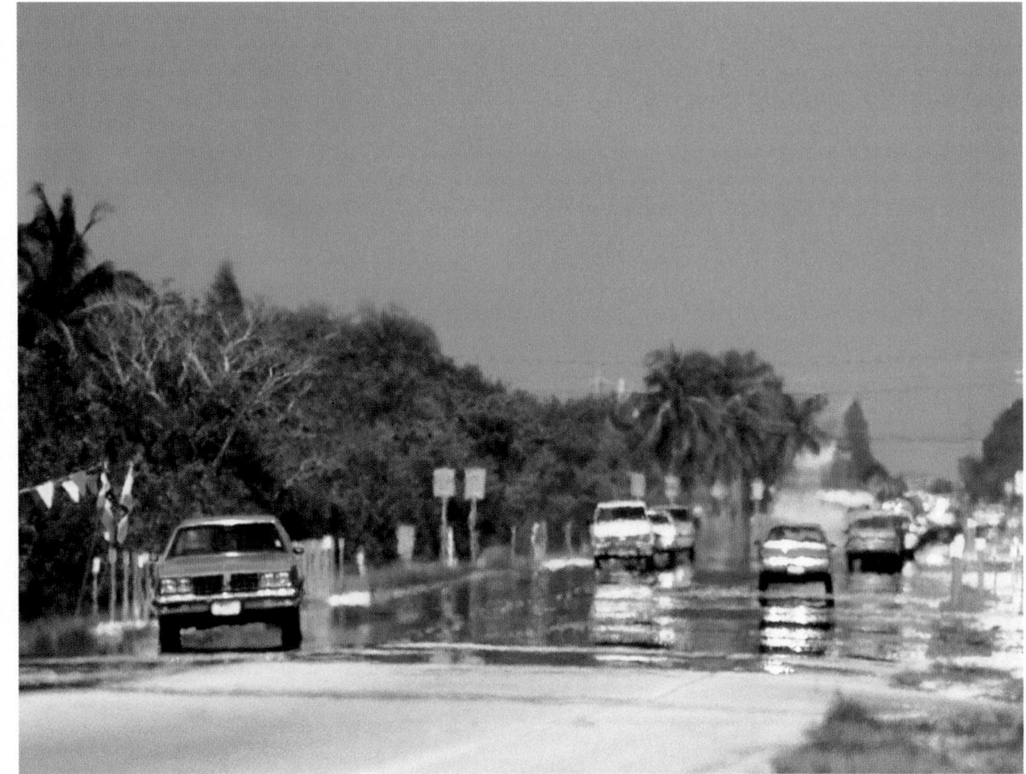

Figure 7.24 Refracted light is responsible for creating mirages. When air near the ground warms up, the light from objects at a distance is refracted into a curved path. This causes the illusion of a water surface, which is really an image of the sky refracted by warm air near the ground.

Is That All There Is to Light?

Table 7.1 summarizes what happens to light when it strikes different surfaces. Sometimes all three light behaviours happen at once. Light from the Sun can reflect off the still surface of a lake to produce a mirrorlike reflection. At the same time, the water can absorb light, transforming the light energy into thermal energy. The water warms during the day and cools off at night. If you are looking down into the water from shore, you might be uncertain about the location of objects on the bottom. This happens because light is refracted as it travels from water into the air.

Table 7.1 What Happens When Light Strikes a Surface?

Type of behaviour	What happens to light striking a surface?	Nature of surface	What else happens?
Absorption	Changes into some other kind of energy.	Occurs mostly on rough, dark, opaque surfaces.	Some light is usually reflected off the surface.
Reflection	Bounces off the surface and travels in a new direction.	Occurs best when light hits a smooth, shiny surface.	Some light is usually absorbed.
Refraction	Travels through the surface, often in a new direction.	Occurs when light strikes a different, transparent medium.	Some light is usually reflected off the surface.

Check Your Understanding

1. Distinguish between reflection and refraction.

2. Give two specific examples of materials that mostly refract light. Is light also reflected or absorbed by these materials?

3. What happens to light when it is refracted?

4. Using the normal as a reference line, describe the change in direction of a light ray that travels from
   (a) air into glass
   (b) water into air

5. **Apply** A student chops a piece of ice out of a frozen lake and holds its smooth, parallel sides tilted toward the Sun. Show the path of a ray of sunlight through the ice. (Hint: Light travels more slowly in ice than in air.)

6. Use a ruler to trace the following diagrams. Use the laws of reflection to locate and draw the image. Measure accurately.

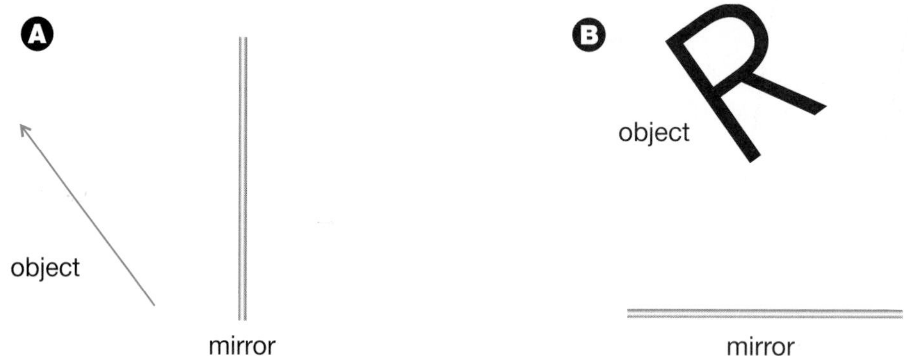

7. **Thinking Critically** What if light behaved differently from what you have learned? Describe the changes you would notice if
   (a) black surfaces and white surfaces reflected the same amount of light (that is, the incident light falling on both surfaces is the same)
   (b) human skin absorbed all the visible light that struck it
   (c) instead of travelling at 300 000 km/s in air, light travelled only at the speed of a car on a highway, about 0.03 km/s

8. **Design Your Own** Make a hypothesis about how various liquids will refract light, and design an investigation to test your hypothesis. Carry out your investigation (with your teacher's approval). Describe the angle of refraction for each liquid.

9. **Thinking Critically** Formulate your own question about the behaviour of light and explore possible answers and solutions.

CHAPTER at a glance

Now that you have completed this chapter, try to do the following. If you cannot, go back to the sections indicated.

Give a simple explanation of what light is. (7.1)

Discuss how light is produced and how light can be transformed into other forms of energy. (7.1)

Name some natural and some artificial sources of light. (7.1)

Explain why fluorescent tubes are more energy-efficient than incandescent bulbs. (7.1)

Describe how light travels. (7.1)

Describe what happens to light when it strikes different surfaces. (7.1)

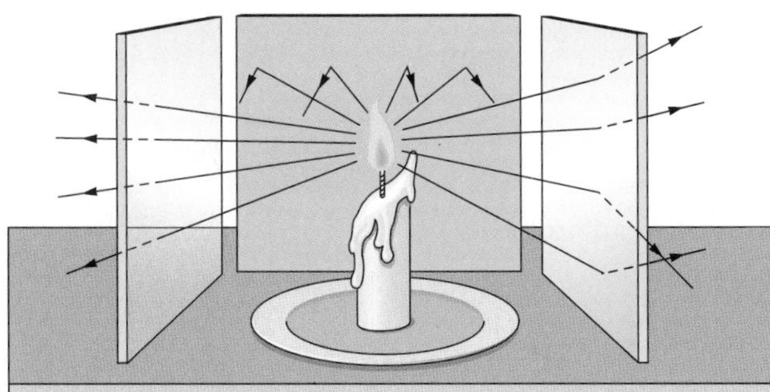

Give some examples of luminous and non-luminous objects. (7.1)

State the laws of reflection. (7.2)

Make a drawing to show how a plane mirror produces an image that is the same size as the reflected object. (7.2)

In your notebook, copy the diagram below and add the correct labels. (7.2)

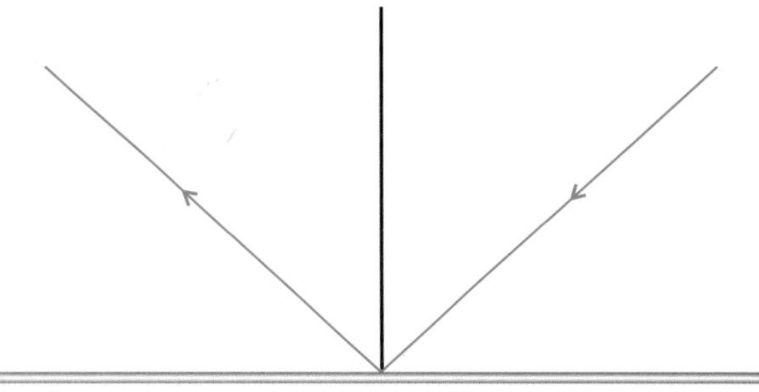

Explain what causes refraction of light rays. (7.3)

Give some factors that can affect the angle of refraction of a light ray. (7.3)

Prepare Your Own Summary

Summarize this chapter by doing one of the following. Use a graphic organizer (such as a concept map), produce a poster, or write the summary to include the key chapter ideas. Here are a few ideas to use as a guide:

- Define radiant energy.
- List some sources of light and the type of energy used by each one.
- Explain why disposal of fluorescent tubes poses an environmental problem.
- State the two basic properties of light.
- Give some examples of differing intensities of light.
- Contrast the energy efficiency of a fluorescent tube and an incandescent bulb.
- Distinguish between the terms “transparent,” “translucent,” and “opaque.”
- Define reflection and refraction.
- Explain how you would draw a simple ray diagram.
- Give some examples of applications of the laws of reflection.
- Describe how refraction affects light travelling through different materials.
- Name some careers requiring a knowledge of the behaviour of light.

CHAPTER

7 Review

Key Terms

image
filament
light
natural light source
radiates
radiation
radiant energy
artificial light source
intensity
incandescent source
incandescence
fluoresence
fluorescent source
phosphorescent source
phosphorescence
chemiluminescence
chemiluminescent source
bioluminescence
bioluminescent source
watt
kilowatt hour
ray diagram
transparent
translucent
opaque
luminous
non-luminous
reflection
incident ray
reflected ray
normal
angle of incidence
angle of reflection
plane mirror
law
laws of reflection
plane
diffuse reflection
refraction
angle of refraction

Reviewing Key Terms

If you need to review, the section numbers show you where these terms were introduced.

1. In your notebook, match the description in column A with the correct term in column B.

A

- the bending of light as it passes from one medium to another
- can be transformed into chemical energy, electrical energy, or thermal energy
- allowing no light to pass through
- an artificial light source
- a type of diagram explaining how light travels
- occurs when light bounces off a surface
- describes objects that produce their own light

B

- incandescent bulb (7.1)
- light (7.1)
- image (7.1)
- refraction (7.3)
- plane (7.2)
- ray diagram (7.1)
- opaque (7.1)
- luminous (7.2)
- reflection (7.2)

Understanding Key Ideas

Section numbers are provided if you need to review.

2. Why do many people prefer to wear light-coloured clothing outdoors in the summer? (7.1)
3. What are some things you could say or do to convince a younger sibling that light is a form of energy that travels in straight lines? (7.1)
4. Name the sources of light that you have learned about in this chapter, and give an example of each. (7.1)
5. What properties must a surface have if the light that reflects from it forms a clear image of an object? (7.2)
6. Your vision involves more than just your sense of sight. Explain. (7.2)
7. State the two laws of reflection. Draw a simple diagram that shows these laws. (7.2)
8. Describe a situation in which each of the following could occur:
   (a) Almost all of the light energy that strikes a surface is absorbed. (7.1)
   (b) Very little light is absorbed by the surface. Most of the light energy is reflected to produce images. (7.2)

Developing Skills

9. Copy and complete the following concept map.

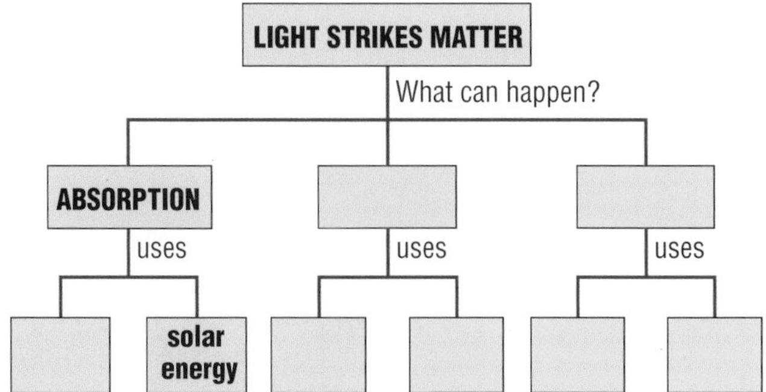

10. A student who is 140 cm tall stands in front of a vertical plane mirror that is 200 cm away. The mirror is 180 cm high and stands on the floor.
    (a) Make a scale drawing of this situation, reduced to one twentieth the actual size.
    (b) Use the laws of reflection to locate the image of the head and feet of the student. Draw the image of the student in faint lines.
    (c) How short could the mirror be and still allow the student to see her/his full length?

Problem Solving/Applying

11. The diagram below shows the top view of a fish tank full of water. In your notebook, draw a larger diagram similar to this one. A student looks through the tank at a small light bulb on the other side of the tank. Show the refraction of the two incident rays as they travel into the water, then back into the air. Will the student see the bulb to the left or the right of its actual position? Will the bulb seem closer or farther away?

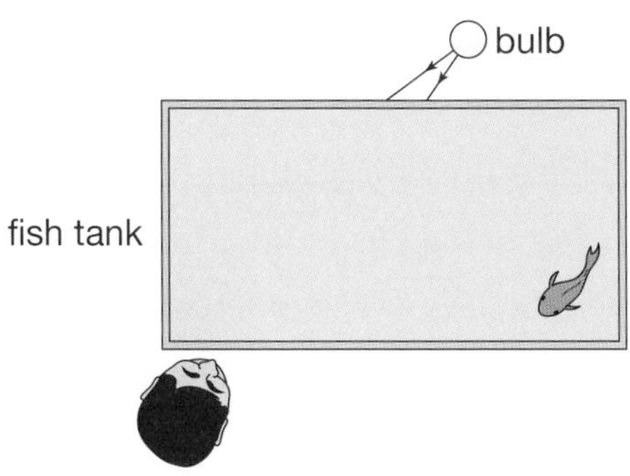

12. Many optical devices are designed to create an image. To create a sharp, clear image, every ray of light that leaves a point on an object should arrive at one point on the image. Draw a simple pinhole camera with a small source of light in front of it. Draw all the rays that can leave the object and strike the back of the camera. Now draw the same pinhole camera but give it a much larger hole, like one punched by a pencil. Again, draw all the rays that can leave the object and strike the back of the camera. What is one critical requirement for a pinhole camera that makes a sharp image?

13. What is the cost of burning a 100 W light bulb for a day if electrical energy costs 8¢/kW•h?

Critical Thinking

14. When sunlight travels from the vacuum of space into Earth's atmosphere, it slows down slightly. Why? What behaviour of light occurs?

15. Copy the following diagram and show the bending that occurs when light enters our atmosphere. Because of this, does the Sun appear higher or lower in the sky at sunset and sunrise?

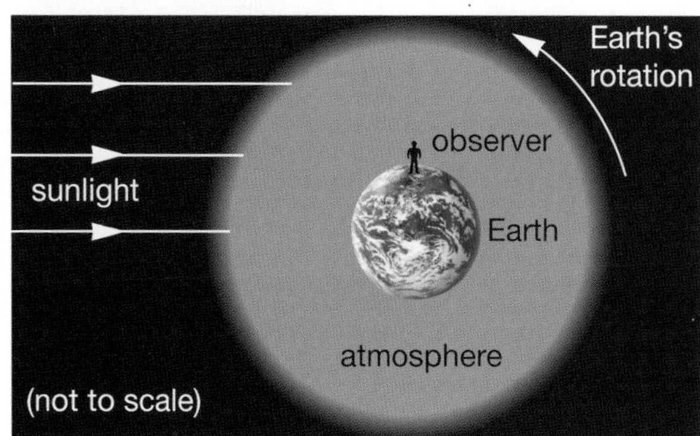

Pause& Reflect

Go back to the beginning of this chapter on page 202 and check your original answers to the Getting Ready questions. How has your thinking changed? How would you answer those questions now that you have investigated the topics in this chapter?

CHAPTER

8 Light Technologies

Getting Ready...

- Why does your face look distorted when it is reflected in a shiny silver spoon?
- How do we see things?
- How does a telescope work?

Look around the room. In your Science Log, make a list of as many different items as you can that change the path of light. Then answer the questions above in your Science Log as best you can for now. You will discover detailed answers to these questions as you read this chapter.

In the distant reaches of space, stars are being born, slowly forming out of clouds of gas and dust. The Orion Nebula, shown here, is one of the stellar birthplaces. Young, hot stars release huge amounts of energy into the cloud in which they have formed, causing the cloud to glow. With the help of powerful telescopes, astronomers can see these clouds of gases as a glowing mist or nebula spreading out through space.

Other scientists keep their focus at very close range. Compound light microscopes allow biologists to study tiny structures such as the bacteria shown in the micrograph on the next page. In this chapter, you will find out how we see things and how our understanding of light and its behaviour has helped us build optical devices that extend human vision.

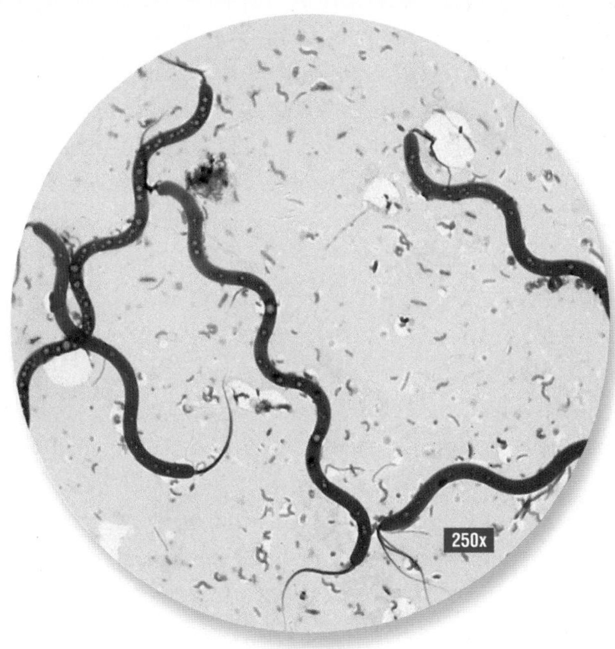

Spotlight On Key Ideas

In this chapter, you will discover

- how light behaves when it strikes curved, reflecting surfaces such as concave and convex mirrors
- what a lens is and how it works
- the difference between a real and a virtual image
- how an optics system works
- how human eyes, cameras, telescopes, binoculars, and microscopes form images

Spotlight On Key Skills

In this chapter, you will

- create different kinds of images using lenses and mirrors
- determine how lenses are arranged in telescopes, microscopes, and other optical devices
- role-play the parts of a system that involves input, output, and feedback
- design and build simple optical devices
- design an experiment to determine whether an image is real or virtual

Starting Point ACTIVITY

Out of the Looking Glass

The telescope that is used to view the Orion Nebula uses a large mirror that curves inward. Such a shape is **concave**. Another type of mirror bulges outward. This shape is **convex.** Compare the images formed by curved mirrors with the images you see in plane (flat) mirrors.

What You Need

concave mirror
convex mirror
white cardboard or Bristol board (about 20 cm × 30 cm)

What to Do

1. Stand near a window and aim the concave mirror toward the outside.
2. Hold the cardboard screen in front of the mirror but a bit off to the side.
3. Angle the mirror so that the reflected light strikes the screen. Move the screen back and forth until you see a sharp image. Describe the image.
4. Hold the mirror close to your face and look at yourself. Describe the image.
5. Repeat steps 1 to 4 with the convex mirror.

What Did You Find Out?

1. How are the images formed by a concave mirror different from the ones you see in a plane mirror? Describe any similarities you observe between concave mirror images and plane mirror images.
2. How are the images formed by a convex mirror different from the ones you see in a plane mirror? Describe any similarities you observe between convex mirror images and plane mirror images.

8.1 Concave and Convex Mirrors

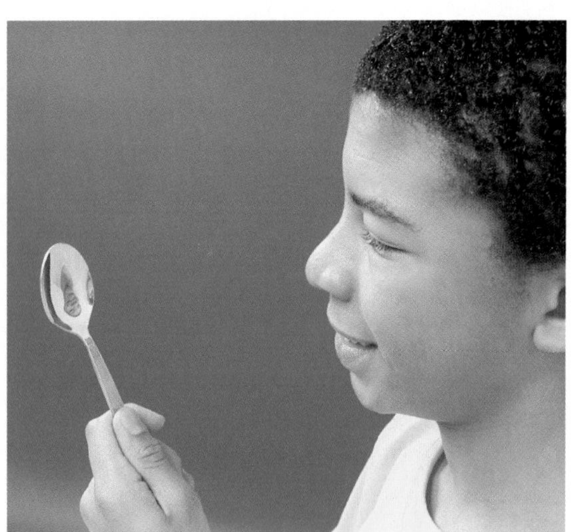

Figure 8.1 What kind of image is reflected by the bowl of a spoon?

Have you ever looked at yourself in the bowl of a spoon (see Figure 8.1)? How did the image look? Was it upright or upside down (inverted)? The **attitude** of an image describes whether the image is upright or upside down in relation to the object. Was the image large or small? Did the image change when you looked at the underside of the spoon?

Figures 8.2A and B show the two main types of curved mirrors. Each type can form images with different attitudes and sizes. You used both types of mirror in the Starting Point Activity. The concave mirror curves in like a cave entrance. The bowl of a spoon is concave. So is the palm of your hand. Astronomers' telescopes often use concave mirrors. A shiny sphere and the underside of a spoon, on the other hand, are examples of convex mirrors.

When you did the Starting Point Activity, you saw that a curved mirror could create images in different places. Sometimes the mirror would project an image in front of it. At other times, it appeared to form images of various sizes behind the mirror. Why does a curved shape form an image so different from the ones that you see in a plane mirror? Find out in the next investigation.

Figure 8.2A A concave mirror is curved like the inside of a shiny bowl.

Figure 8.2B A convex mirror bulges outward like the surface of a shiny, helium-filled party balloon.

CONDUCT AN

INVESTIGATION 8-A

Concave Mirror Images

Just as for plane surfaces, the laws of reflection can be used to determine what happens to the light that reflects off curved surfaces.

Problem

How can you determine how a concave mirror forms images?

Apparatus

metric ruler
mathematical compass
protractor
pencil

Materials

2 sheets of paper

Procedure

Part 1

Objects Far from the Mirror

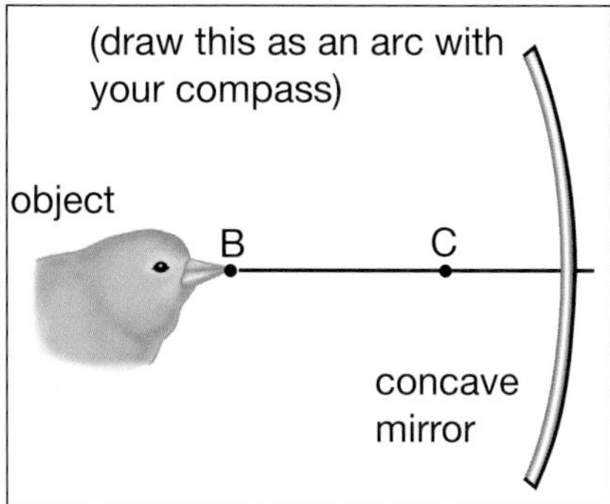

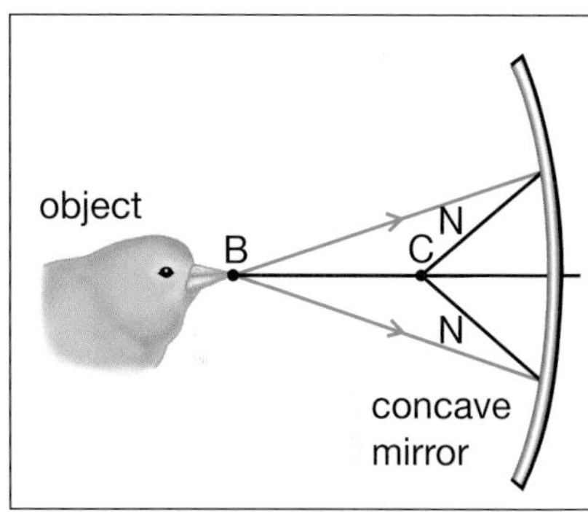

1 Draw a line across the width of one sheet of paper, as shown.

(a) Mark a point on the line about halfway across the paper. Label this point **C**.

(b) Using this point as the centre, draw an arc with a radius of 5 cm to the right of **C**.

(c) Mark a point on the line 7 cm to the left of **C**. This point will represent the beak of the bird that is representing your object. Do not bother sketching the bird unless you really enjoy drawing. Label the point **B** for beak.

2 Draw two incident rays from the tip of the beak, **B**, to the mirror. Use arrows to show the direction in which the rays are travelling.

(a) Now draw radii from the centre, **C**, to the points where the incident rays meet the mirror. Because a radius meets the circumference of the circle at right angles, any radius is also a normal.

(b) Label these radii with an **N** for normal.

CONTINUED

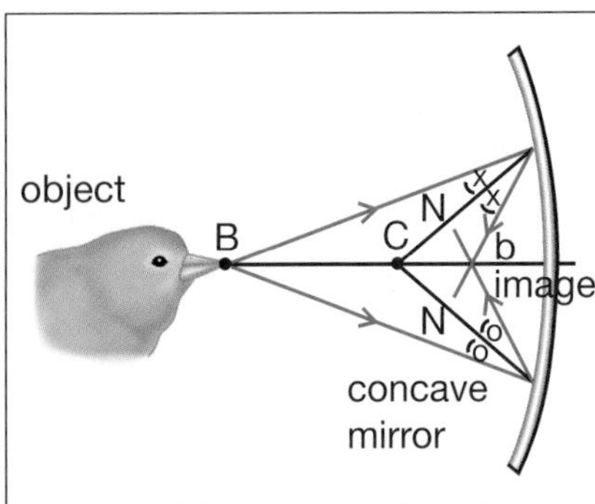

3. Where each incident ray meets the mirror, carefully measure and draw the reflected ray. Remember that the angle of reflection must be equal to the angle of incidence. Use arrows to show in which direction the reflected rays are travelling.
   - **(a)** Find out where the two reflected rays meet. This is where the image of the tip of the bird's beak will be found. Label that point with a small **b**. (The point should be on the line. However, getting close to the line is good enough.)
   - **(b)** A **real image** is one that is found where the reflected rays actually meet. A **virtual image** is one that is found where the reflected rays only seem to meet. Decide which type of image has been formed and record your decision.

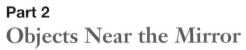

Part 2
Objects Near the Mirror

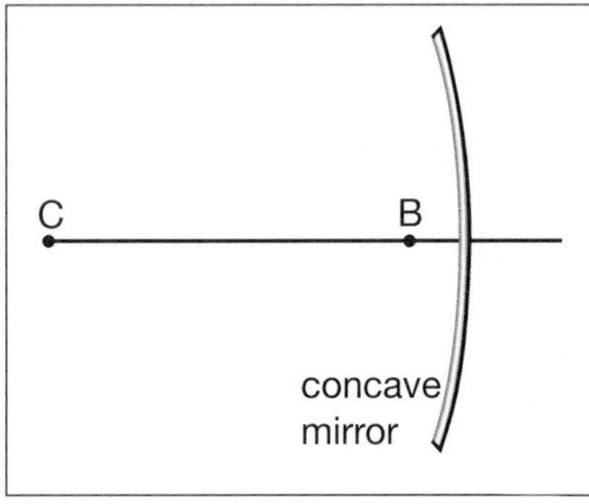

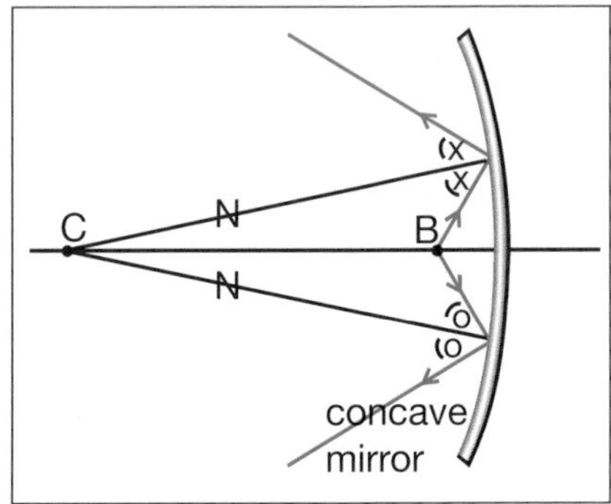

1. For your second diagram, draw a line across the second sheet of paper and locate point **C** near the left-hand side of the page. Draw your concave mirror using a radius of 5 cm. Place point **B** on the line 1 cm to the left of the mirror. This point will represent the bird's beak.

2. Draw two incident rays from point **B** to the mirror, as you did in step 2 of Part 1. Use arrows to show the direction in which the rays are travelling.
   - **(a)** Draw radii from the centre (**C**) to the points where the rays meet the mirror, as you did in step 2 of Part 1. Label these radii with an **N** for normal.
   - **(b)** Where each incident ray meets the mirror, carefully draw the reflected ray, as you did in step 3 of Part 1. Remember that the angle of reflection must be equal to the angle of incidence. Use arrows to show the direction in which the reflected rays are travelling.

It is possible to draw two angles that are equal to each other using just a mathematical compass and a ruler. Try to figure out how you could do this without even having to read the markings on the ruler.

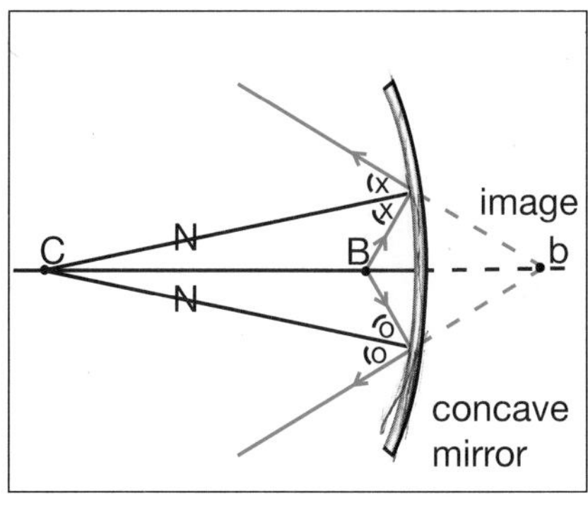

❸ In this case, the reflected rays spread out as they travel away from the mirror. However, if you were looking into the mirror, you would think that the rays were coming from point **b**. To find this point, draw the reflected rays backwards, as shown. Because these light rays are not actually coming from this image, what type of image is it?

Analyze

1. **(a)** What kind of image is formed when the object is far from a concave mirror?

(b) Where is the image located?

2. **(a)** What kind of image is formed when the object is near a concave mirror?

(b) Where is the image located?

Conclude and Apply

3. Why can a concave mirror project an image in front of itself?

4. Where must the object be located in order to form a virtual image behind the mirror?

5. Which of your diagrams would represent a concave mirror being used as

(a) a cosmetic or shaving mirror?

(b) a mirror for a telescope?

(c) the mirror behind the filament in the headlight of a car? (Remember that the beam from the headlight must spread out.)

Images Formed by Curved Mirrors

A concave mirror can produce two types of images. If an object is close to the mirror, the image seen is upright and larger than the real object. As in a plane mirror, the image looks as though it is located behind the mirror. Mirrors for shaving are often concave in shape because they are meant to be used at close range. On the other hand, when the object is far from the concave mirror, the image is inverted and smaller than the real object.

Convex mirrors are more common than concave mirrors. Convex mirrors always produce small, upright images, as shown in Figure 8.3. Because the image of each object is smaller in a convex mirror, you see more of the overall scene. This characteristic is applied in security mirrors in stores (see Figure 8.4). These mirrors allow one person at the counter to see most of the customers in the store. Parking garages sometimes have convex mirrors to allow drivers to see if an oncoming car is just around the corner.

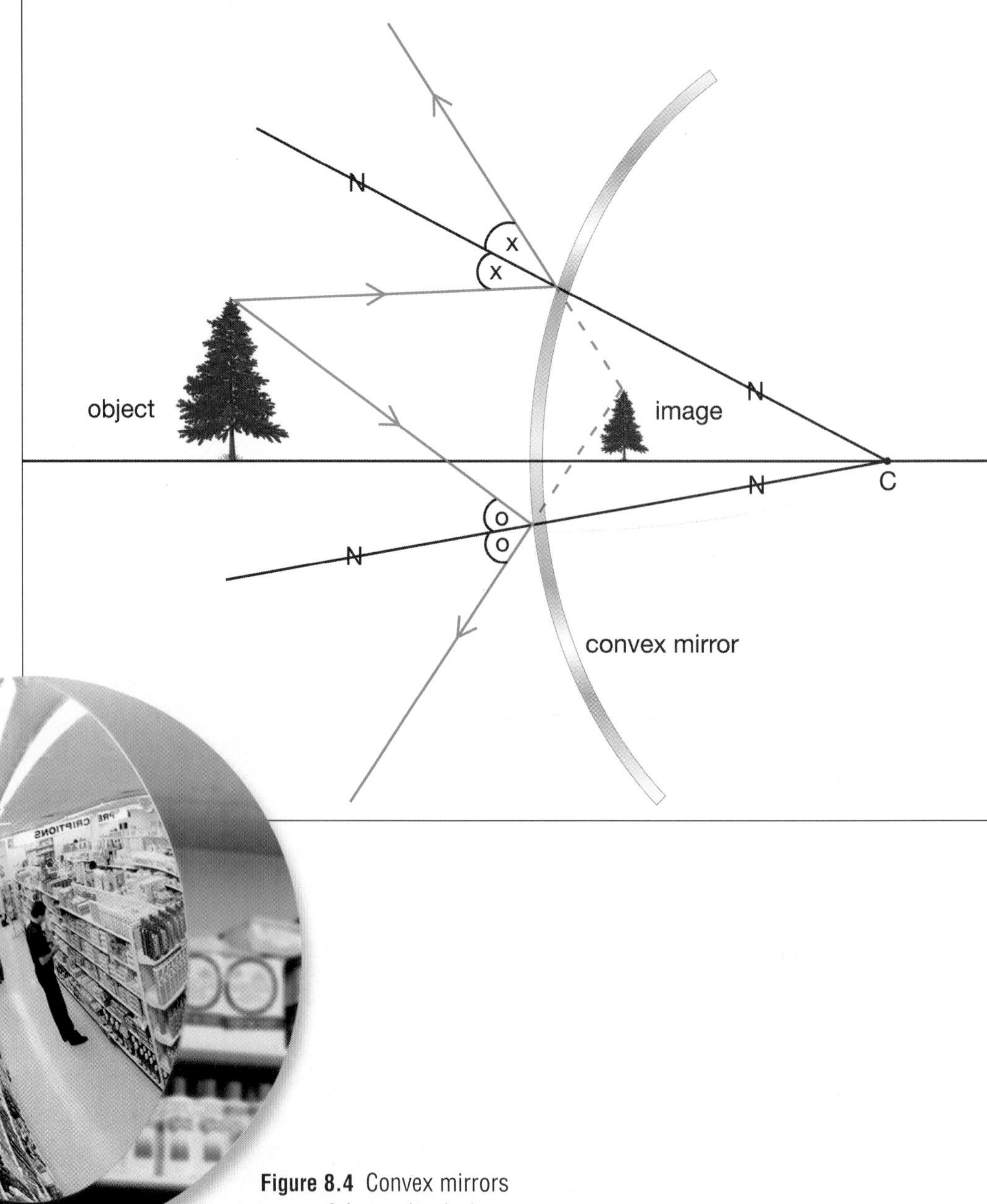

Figure 8.3 The image formed by a convex mirror is always small and upright.

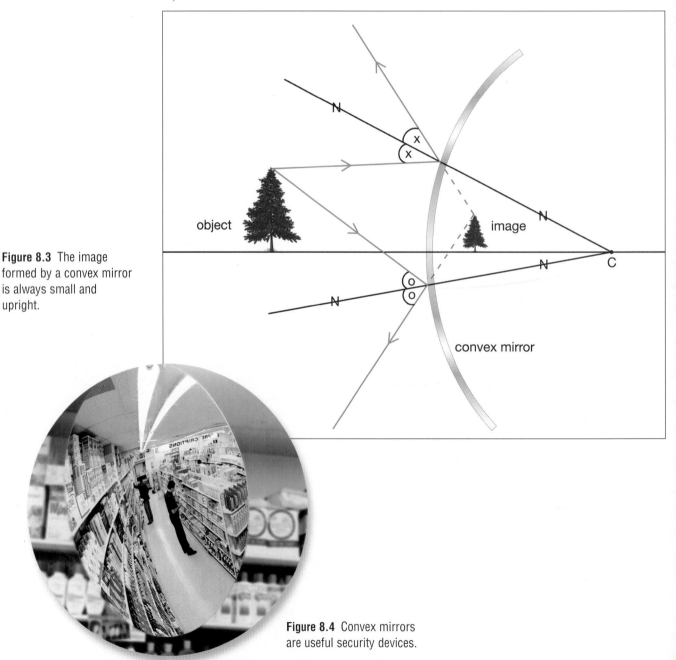

Figure 8.4 Convex mirrors are useful security devices.

Convex mirrors are also used as sideview mirrors on cars (see Figure 8.5). With a convex mirror, the driver can see several lanes at once. However, the image is small. This gives the impression that the cars seen in the mirror are far away when they are in fact much closer than they appear.

You can see that the shape of the mirror determines its properties and its uses. The next time you see a curved mirror, try to picture how the light rays are reflecting in order to form the image that you are seeing.

Figure 8.5 Why do these mirrors have labels saying "Objects may be closer than they appear"?

Check Your Understanding

1. Name the type or types of mirror that form(s)
   (a) a real image
   (b) a virtual image
   (c) a virtual image that is larger than the object
   (d) a virtual image that is the same size as the object

2. Sketch a diagram to show how a mirror inside a flashlight spreads out the flashlight beam. (Hint: Recall the diagrams you produced in Conduct an Investigation 8-A.)

3. Make a diagram similar to the one in Conduct an Investigation 8-A, Part 2, step 1. This time, use a radius of 6 cm. Place point **B** halfway between the mirror and **C**. Using the method that you learned in the investigation, find out what happens to the reflected rays. Why is this arrangement used in searchlights?

4. Make a diagram similar to Figure 8.3. Use a radius of 6 cm for the mirror. Place object **B** on the line 4 cm in front of the mirror. Locate the image of **B**.

5. Imagine you have sprayed a soup bowl with shiny, reflective paint.
   (a) Which part of the bowl serves as a convex mirror?
   (b) Which part of the bowl serves as a concave mirror?
   (c) How should you hold the bowl to see an inverted image of your face?
   (d) How should you hold the bowl to see a larger, upright image of your face?
   (e) How should you hold the bowl to see a smaller, upright image of your face?

6. **Apply** In a funhouse, what kind of mirror would you use to
   (a) make yourself look taller?
   (b) make yourself look shorter?

8.2 Lenses and Vision

Knowing how curved mirrors reflect light can help you understand how lenses affect light rays. A **lens** is a curved piece of transparent material, such as glass or plastic. Light refracts as it passes through a lens, causing the rays to bend.

A **concave lens** is thinner and flatter in the middle than around the edges (see Figure 8.6). Light passing through the thicker, more curved areas of the lens will bend more than light passing through flatter areas. This causes rays of light to spread out, or diverge, after passing through the lens.

A **convex lens**, also shown in Figure 8.6, is thicker in the middle than around the edges. This causes the refracting light rays to come together, or converge.

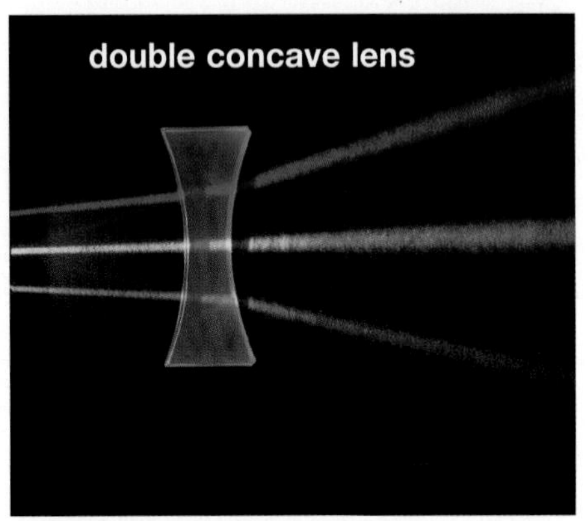

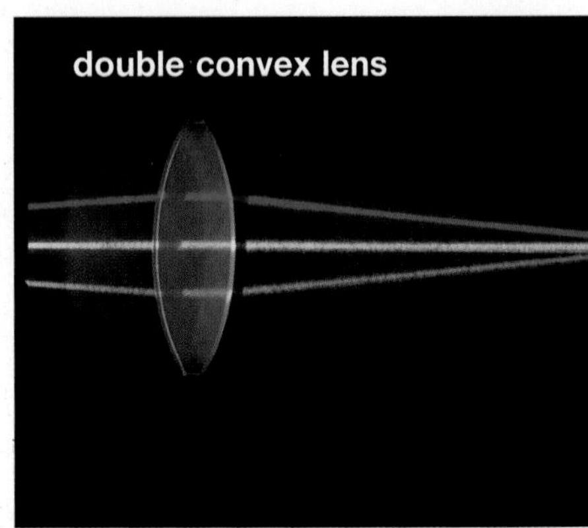

Figure 8.6 Light refracts when it travels through lenses. The double concave lens shown on the left causes light rays to spread out. The double convex lens shown on the right brings light rays close together.

Find Out **ACTIVITY**

See for Yourself!

Using simple materials, you can observe how concave and convex lenses work.

What You Need

flashlight
comb
concave lens
convex lens
piece of plastic wrap (5 cm x 5 cm)
water

What to Do

1. In a dark room, shine the flashlight through the teeth of the comb. Observe the shadows of the teeth on a table.
2. Now place a concave lens just beyond the comb and repeat step 1. What difference in the shadows do you observe?
3. Replace the concave lens with a convex lens and repeat step 1. What difference in the shadows do you observe?
4. Place a single drop of water on the piece of plastic wrap and hold the wrap over some words in your textbook. The water drop forms a simple convex lens. How does the appearance of the words change?

What Did You Find Out?

1. Based on your observations, predict what kind of lens is used to make a magnifying glass.
2. **Apply** List some possible uses for (a) a concave lens and a light beam, and (b) a convex lens and a light beam.

Lenses and Images

Lenses are probably the most useful and important of all optical devices. Eyeglasses, for example, were made from lenses as early as the thirteenth century. The type of lens used in an optical device determines whether the image is real or virtual.

An image forms where light rays from an object converge. The light rays spread out from points on the object. A convex lens refracts these rays so that they come back together (see Figure 8.7).

However, the lens directs light from the left portion of the object to the right portion of the image. Similarly, light from the top of the object is directed to the bottom of the image (see Figure 8.8). Thus, an image formed by a convex lens is sometimes inverted rather than upright. Because your eye sees the image of the flower by light rays that are actually coming from the image, the image is a real image. An image on a movie screen is a real image since light actually travels from the screen to your eye. Any image that can form on a screen is a real image. Both film projectors and overhead projectors use a convex lens to create real images.

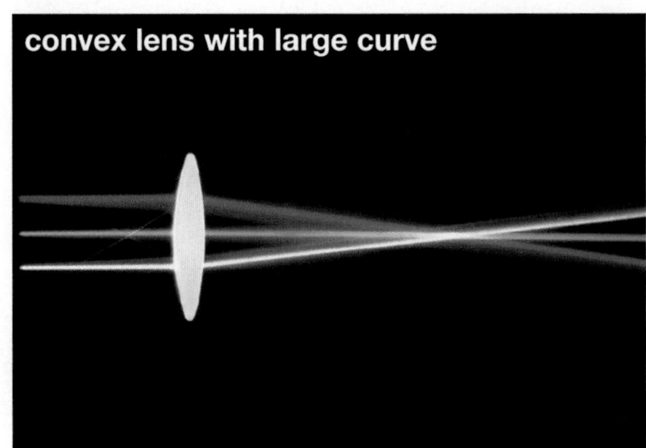

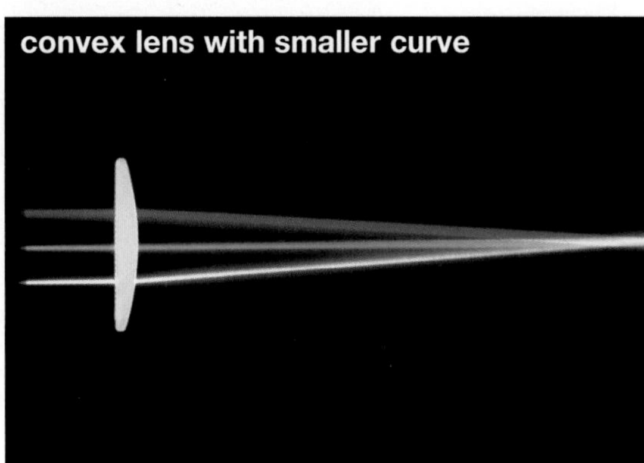

Figure 8.7 A convex lens causes parallel rays to converge. Why do the rays from the lens on the right converge farther from the lens compared to the rays on the left?

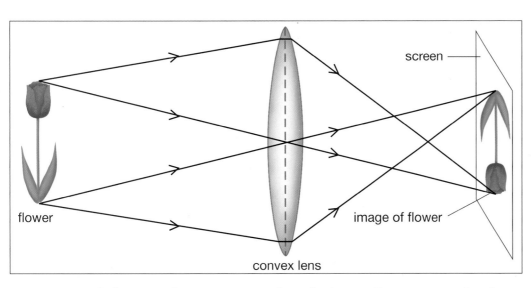

Figure 8.8 Images formed by a convex lens are inverted, or upside down.

In some cases, light rays only seem to come from the image. Do you remember the bird and the mirror in Chapter 7? The bird is convinced that its "friend" is on the other side of the mirror, but there is no other bird. Light rays are coming from the friend *as if* they had started from behind the mirror. In fact, the rays are just bouncing off the mirror, so a virtual image is produced. Light rays do not actually come

Figure 8.9 Computers are sometimes used to create effects called *virtual reality*. This digital composite image shows a biochemist using a virtual reality system to investigate molecular interactions. In what way are the computer effects virtual?

from a virtual image (see Figure 8.9). Plane mirrors always produce virtual images. Remember that if you cannot form an image on a screen, the image must be virtual. Sometimes a convex lens can be used to produce a virtual image. For example, a magnifying glass is a convex lens used to produce an enlarged, virtual image.

Finally, concave lenses produce virtual images from real objects. Concave lenses are used in combination with convex lenses in cameras and telescopes. As the following section explains, they are also used in eyeglasses to correct a vision problem known as "near-sightedness."

Eye Spy

The convex lens is the type of lens found in the human eye. This lens takes light rays that are spreading out from an object and, by refraction, **focusses** them, or brings them back to a point. This focussing of light rays allows us to see objects. In a normal eye, light refracts through the lens onto a light-sensitive area at the back of the eye called the **retina**. The image you see is formed on the retina (see Figure 8.10A).

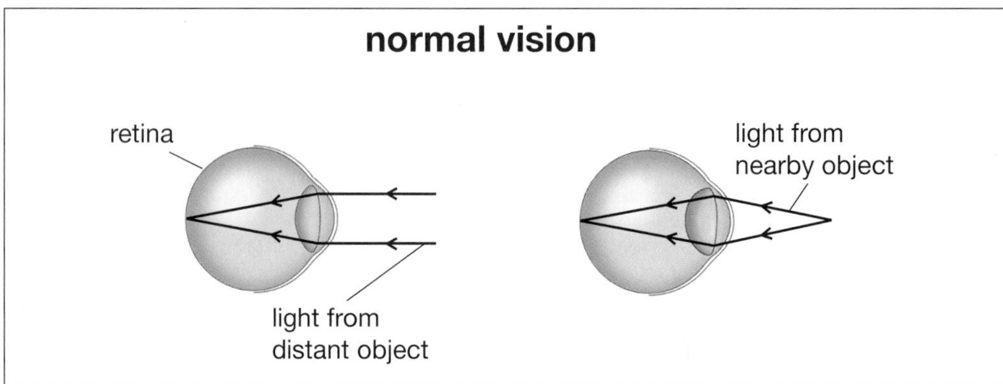

Figure 8.10A These diagrams show how the lens in a normal human eye focusses light rays onto the retina.

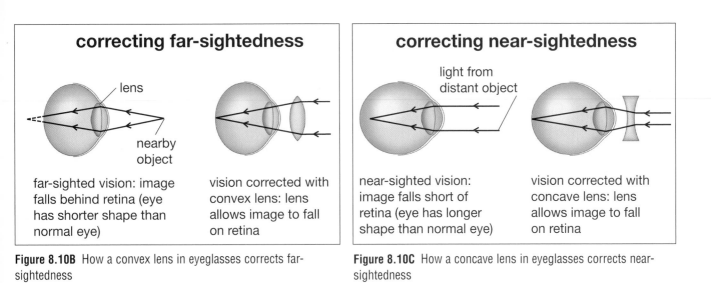

correcting far-sightedness

lens

nearby object

far-sighted vision: image falls behind retina (eye has shorter shape than normal eye)

vision corrected with convex lens: lens allows image to fall on retina

Figure 8.10B How a convex lens in eyeglasses corrects far-sightedness

correcting near-sightedness

light from distant object

near-sighted vision: image falls short of retina (eye has longer shape than normal eye)

vision corrected with concave lens: lens allows image to fall on retina

Figure 8.10C How a concave lens in eyeglasses corrects near-sightedness

However, for some people the eye has a longer shape. This means the image forms *in front of* the retina. These people are **near-sighted** — they have trouble seeing distant objects. For other people, the eye has a shorter shape and the image has not formed by the time the light reaches the retina. These people are **far-sighted** — they have trouble seeing objects that are close to them.

Knowledge of how light behaves when it travels through lenses helps eye specialists to correct vision problems. As Figure 8.10B shows, a convex lens placed in front of a far-sighted eye helps bend the light rays so that the image appears on the retina. As Figure 8.10C shows, a concave lens placed in front of a near-sighted eye moves the image back so that, again, it forms on the retina.

Word CONNECT

The word "optics" has not been defined in this unit, even though nearly everything you have learned so far has been related to optics. Think about what you have been learning, then write a definition for optics.

Comparing the Eye and the Camera

There are many similarities between the human eye and the camera (see Figure 8.11). You already know that you see objects when light is reflected to your eye, refracted through your lens, and focussed on your retina. In a camera, the lens refracts the light and the film senses the light. Explore how a camera works in the next investigation. Then read the pages that follow for a more detailed comparison of the eye and the camera.

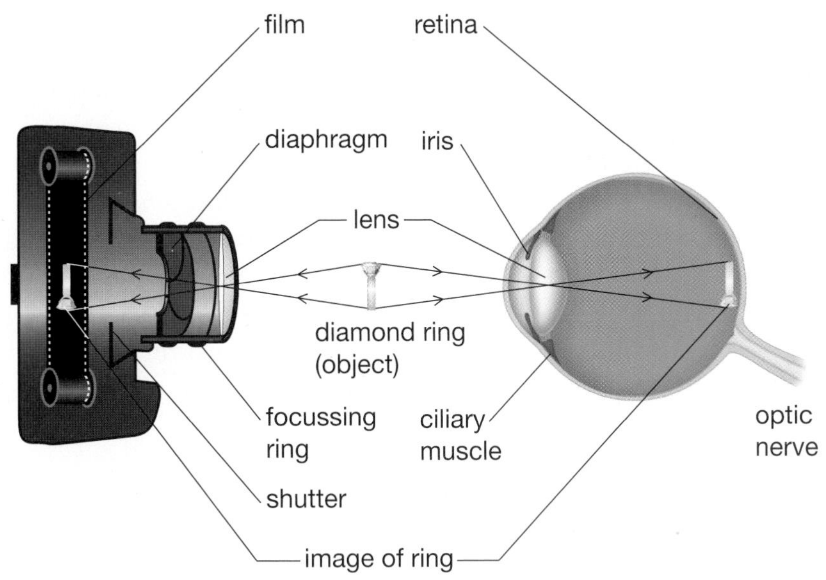

Figure 8.11 A comparison of the camera and the human eye. Study this diagram and try to infer the function of each part of the camera and the eye.

The Camera

A basic camera is a rather simple device. To make a camera, all you need is a convex lens, film, a box to keep the light out, and a shutter to let the light in when you wish. Why, then, do we sometimes take photos that are too dark or out of focus?

To answer this question, you will examine two processes in this investigation:

1. what happens when you focus a camera
2. how to adjust the brightness of the photograph

Problem

How is the image produced in a camera?

Safety Precautions

- Exercise extreme caution when dealing with open flames.
- Keep skin, clothing, and hair well away from the flames.
- Long hair should be tied back so that it does not swing into the flame.
- If a fire does start, dowse it with water or smother it with a coat or fire blanket. Alert your teacher immediately.
- Be careful when using sharp objects such as scissors.

Apparatus

convex lens
candle and matches
dish large enough for the candle to sit on
mathematical compass
book
scissors
metre stick

Materials

masking tape
3 pieces of black Bristol board (about 20 cm × 15 cm)
piece of white paper (about 15 cm × 13 cm)

Procedure

Part 1

Getting Ready

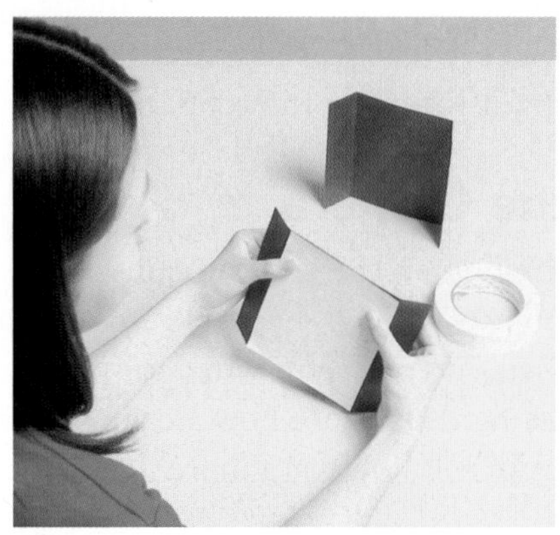

1. Fold two pieces of black Bristol board as shown, to make two optics stands. Cover the front of one of the stands with white paper to make a screen. In this investigation, the white paper will represent the film in the camera.

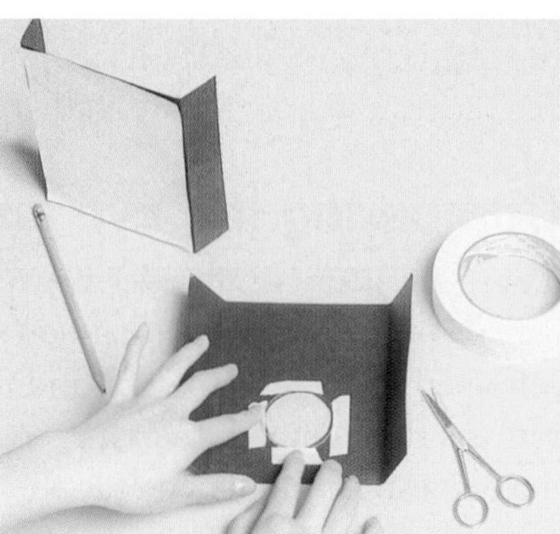

2. Draw a circle 10 cm from the bottom of the second stand. The circle should be slightly smaller than the diameter of the convex lens. Cut out the circle. Tape the lens securely over the back of the hole. Make sure that the tape covers just the edge of the lens. This lens will represent the camera lens.

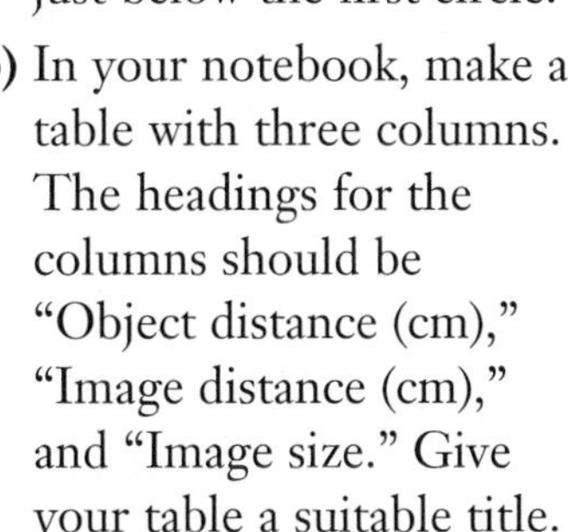

3 Take the third piece of Bristol board and mark a point about 6 cm from the top of the piece and midway between the sides. Using that mark as the centre, draw a circle with a diameter of 1 cm. Cut out the circle.

(a) Cut out a second circle with a diameter of 2 cm about 6 cm from the bottom of the Bristol board, just below the first circle.

(b) In your notebook, make a table with three columns. The headings for the columns should be "Object distance (cm)," "Image distance (cm)," and "Image size." Give your table a suitable title.

(c) Your teacher will light the candle, drip some wax onto the plate, and set the candle upright in the melted wax, holding it there until the wax has solidified.

Part 2

Setting Up the Camera

When you take a photograph, you usually hold the back of the camera firmly against your face. Because the film is found just inside the back of the camera, the film is also held firmly in place.

1 Place the book near one edge of the table. It will represent your face. Then place the screen against the book. The black Bristol board represents the back of the camera. The white paper represents the film. Do not move the screen during this investigation.

2 Your teacher will place the lighted candle 1 m from the screen. The candle flame will be the object upon which the camera is being focussed.

Now place the lens (of the camera) between the candle and the white film. Darken the room. Move the lens back and forth until you see a sharp image on the "film."

(a) Measure the distance from the lens to the object (object distance). Then measure the distance from the lens to the film (image distance). Record these values in your table. Also record how the size of the image compares to the size of the object. Is it larger, smaller, or the same size?

(b) Repeat steps 2 and 2(a), but move the candle 10 cm closer to the film. Record your observations.

(c) Continue moving the candle closer to the film, 10 cm at a time, until it is impossible to form a sharp image on the film.

Skill POWER

For tips on making tables, turn to page 546.

CONTINUED

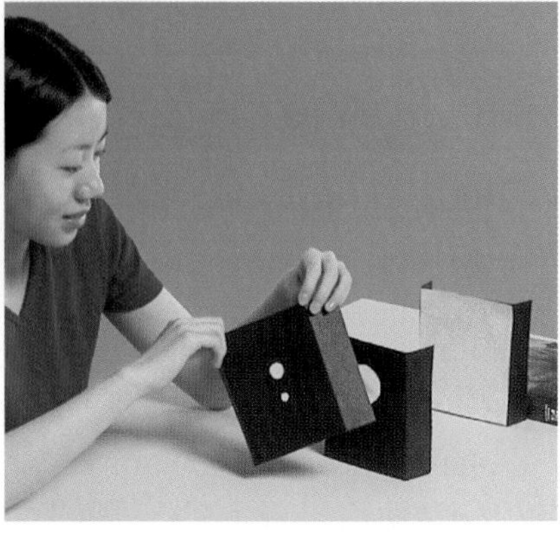

3 Move the candle back to a distance of 1 m from the film. Move the lens until you have focussed a sharp image on the film. Place the 2 cm hole in front of the lens and notice the brightness of the image.

(a) Now place the 1 cm hole in front of the lens. How does the brightness change? Is it as bright as with the 2 cm opening? Is it less bright? Or more bright?

(b) Do you see only part of the image? Record your observations.

***Computer* CONNECT**

Enter the data from Conduct an Investigation 8-B into a spreadsheet. Make a graph of the data using a graphing program.

***Math* CONNECT**

In Conduct an Investigation 8-B, you changed the brightness of the image using two openings. Light comes in through the area of the opening. Try calculating the area of each opening. Then divide the larger area by the smaller area. According to your calculations, how should the brightness of the images compare?

Analyze

1. Make a line graph of your observations. Place the object distance on the horizontal axis (x-axis) and the image distance on the vertical axis (y-axis). Choose a suitable title for your graph. On your graph, mark the following:
   (a) the region in which the image is smaller than the object
   (b) the region in which the image is larger than the object
   (c) the point where the object and image are closest to being the same size

Conclude and Apply

2. As the object moves toward the "camera," which way must the lens move to keep the image focussed on the film?
3. What happens to the image distance as the object distance decreases?
4. How do the distances compare when the object and the image are the same size?
5. What happens to the brightness of the image when you reduce the diameter of the opening into the camera by half? Do you see less of the image?

Putting It in Focus

In a camera, if an object moves closer to the film, the lens must move farther from the film to keep the image in focus. This is what you do when you focus the camera. In the human eye, you cannot move the lens farther away from the retina. Instead, the **ciliary muscles** change the shape of the lens (refer back to Figure 8.11 on page 247). If the object you are looking at comes closer to you, these muscles make the lens bulge in the middle. This keeps the object in focus on the retina without having to stretch the eyeball.

The process of changing the shape of the lens to adjust for different object distances is called **accommodation**. As people become older, the lens stiffens and loses its ability to change shape. It can no longer become thick enough to focus on close objects. As a result, many people wear convex lenses as reading glasses.

The shortest distance at which an object is in focus is called the **near point** of the eye. The longest distance is the **far point** of the eye. For the average adult human eye, the near point is about 25 cm away (see Figure 8.12A). However, babies can focus on objects only about 7 cm away (see Figure 8.12B). The far point is said to be infinity. After all, a person with normal vision can see the stars and they can be many thousands of light years away.

Figure 8.12A Most adults can focus on objects that are about 25 cm away.

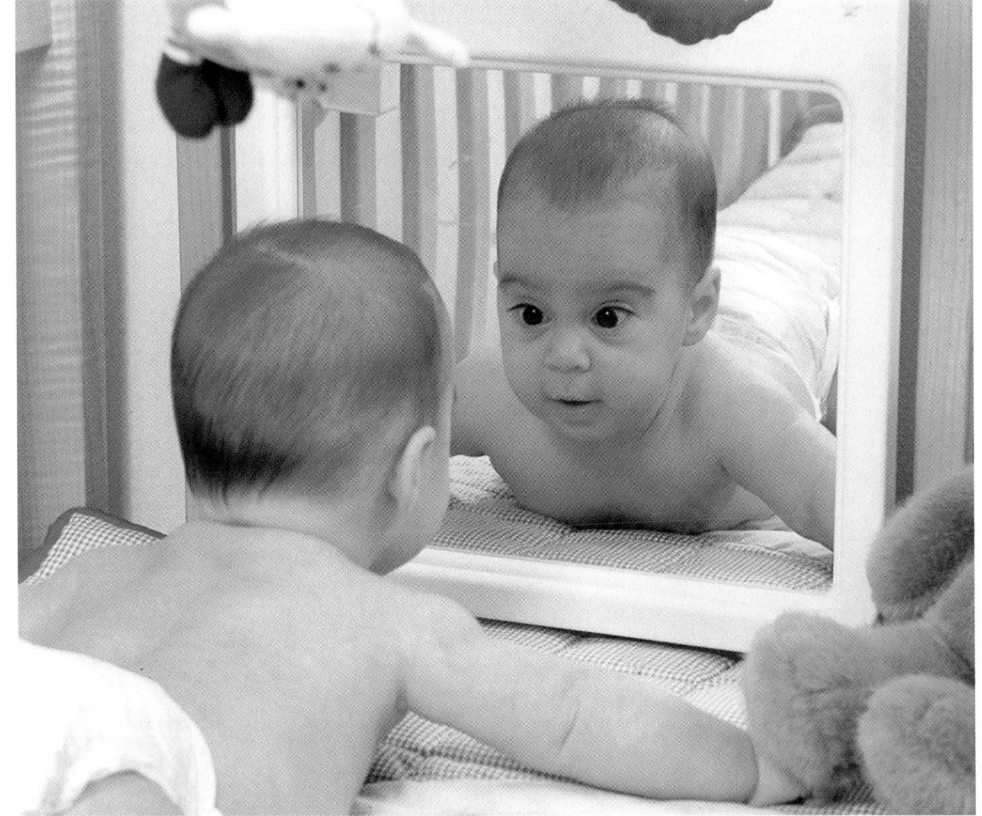

Figure 8.12B Most babies can focus easily on objects that are very close.

Bringing in the Light

If you are photographing a scene and a cloud suddenly covers the Sun, the brightness of the scene decreases. As a result, the amount of light reaching the film decreases. If the light is too dim, the film will not receive enough light to record the image clearly. In this case, the camera's diaphragm and shutter can be adjusted to allow the correct amount of light to reach the film. The **diaphragm** is a device that controls the **aperture** (the opening) of a lens or optical system. The **shutter** is a device that limits the passage of light. The aperture in the diaphragm can let more light into the camera by being opened wider, as shown in Figure 8.13. The shutter can let in more light by staying open longer.

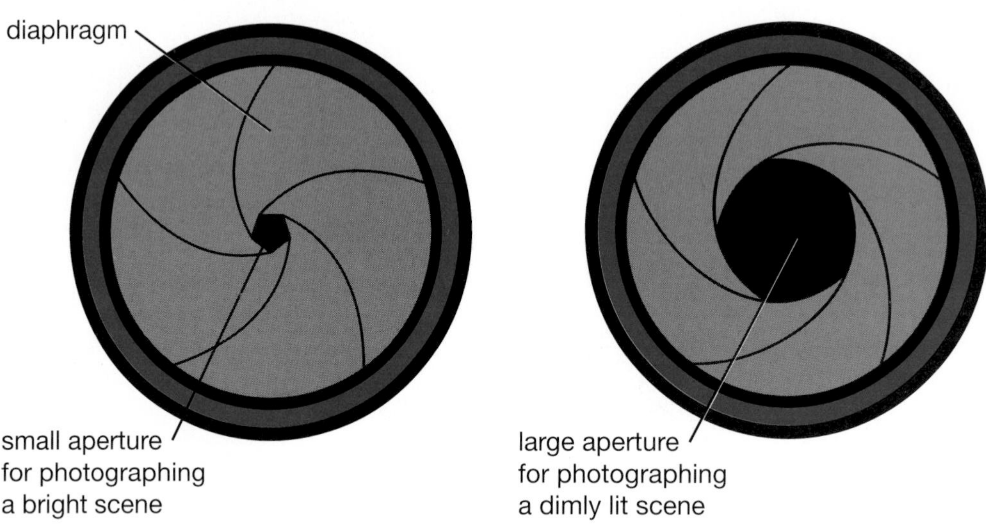

Figure 8.13 The diaphragm controls the amount of light that enters a camera. The opening in the diaphragm is called the aperture.

In the human eye, the **iris**, which is the coloured ring, functions like the diaphragm of a camera (see Figure 8.14). If the light is dim, the iris increases the size of the eye's opening to let in more light. This opening is called the **pupil**, which appears as the dark centre of your eye.

DidYouKnow?

Frogs' eyes are eight times more sensitive to light than human eyes.

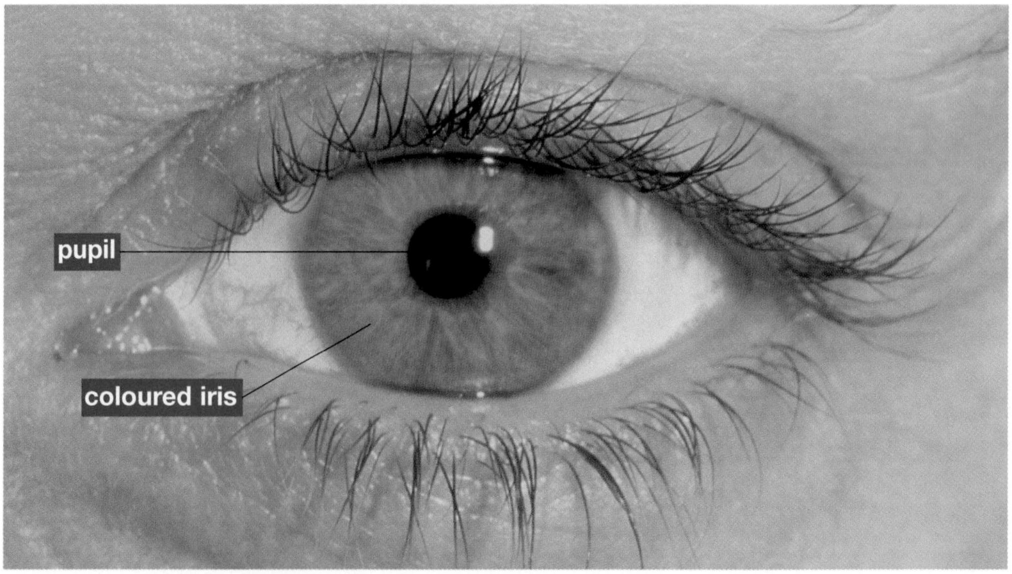

pupil
coloured iris

Figure 8.14 The iris controls how much light enters the human eye.

The natural adjustment in the size of the pupils is called the **iris reflex**. You are generally unaware of this reflex, which is extremely rapid. What happens when you first walk into a dark movie theatre on a bright, sunny day? You are probably unable to see anything at first. The split-second iris reflex is the first in a series of adjustments the eye makes to enable you to see the interior of the theatre.

When you leave the theatre and walk into bright sunlight, the opposite reaction occurs. The daylight glare makes your eyes feel uncomfortable. However, the iris reflex quickly shrinks the size of the pupils, less light enters the eyes, and other adjustments take place so that your eyes feel comfortable again.

Seeing the Image

At the back of the camera, you will find the film. The image is focussed onto this light-sensitive material. The light energy causes chemical changes in the film to record the image.

In the human eye, the retina senses the light. When the cells in the retina detect light, they create small electrical impulses that travel from the retina to the brain through the **optic nerve**. The point where the optic nerve enters the retina does not have any light-sensing cells. This point is known as the **blind spot.** You can easily demonstrate the presence of your blind spot by following the steps outlined in Figure 8.15. Note that each eye sees what the other misses because the blind spots are not in the same place.

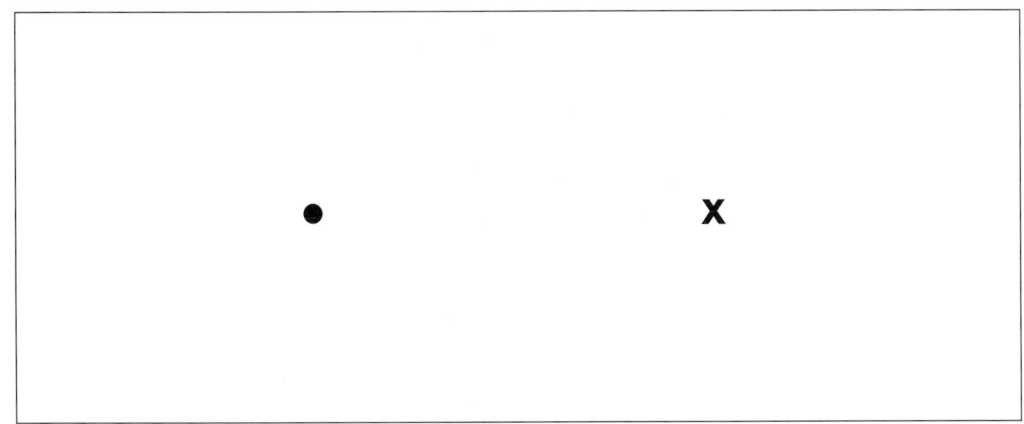

Figure 8.15 Locating your blind spot. Hold this book at arm's length. Cover your right eye with your hand. Stare at the X while you move the book slowly toward yourself. The dot should disappear and then reappear as its image moves onto your blind spot and then off again.

The parts of a camera are contained in a rigid, lightproof box (see Figure 8.16). In the eyes, layers of tissue hold the different parts together. In order to keep the eye from collapsing, the eyeball is filled with fluids called **humours** (see Figure 8.17). In addition to keeping the eye rigid, humours help refract the light that enters the eye. The next investigation will reinforce what you have learned about the parts of the eye and the camera.

Word CONNECT

It may seem strange that the fluids in the eye are called "humours." However, *humour* is an old word for bodily fluids. At one time it was thought that a person's moods were governed by these fluids. You could be in good humour, out of humour, or in foul humour, and it was all thought to be due to imbalances in these bodily fluids. Explain how you think today's common meaning of humour was derived from these original meanings.

Figure 8.16 Cameras differ in size and complexity, but the parts of all cameras are housed in a rigid, lightproof box.

DidYouKnow?

Contact lenses are very thin lenses placed directly on the cornea of the eye. A thin layer of tears between the cornea and the contact lens keeps the lens in place. Most of the refraction occurs at the air-lens surface.

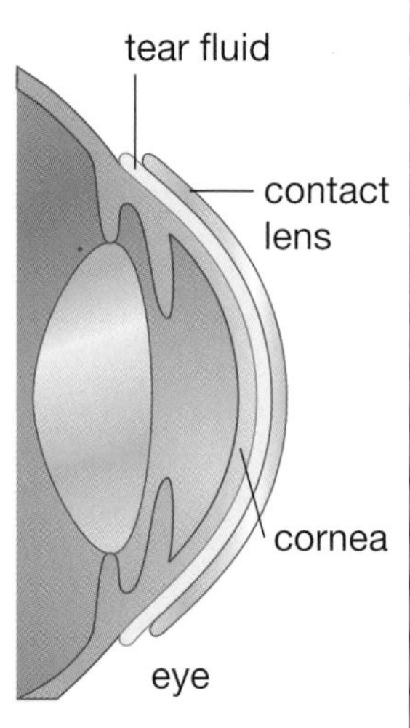

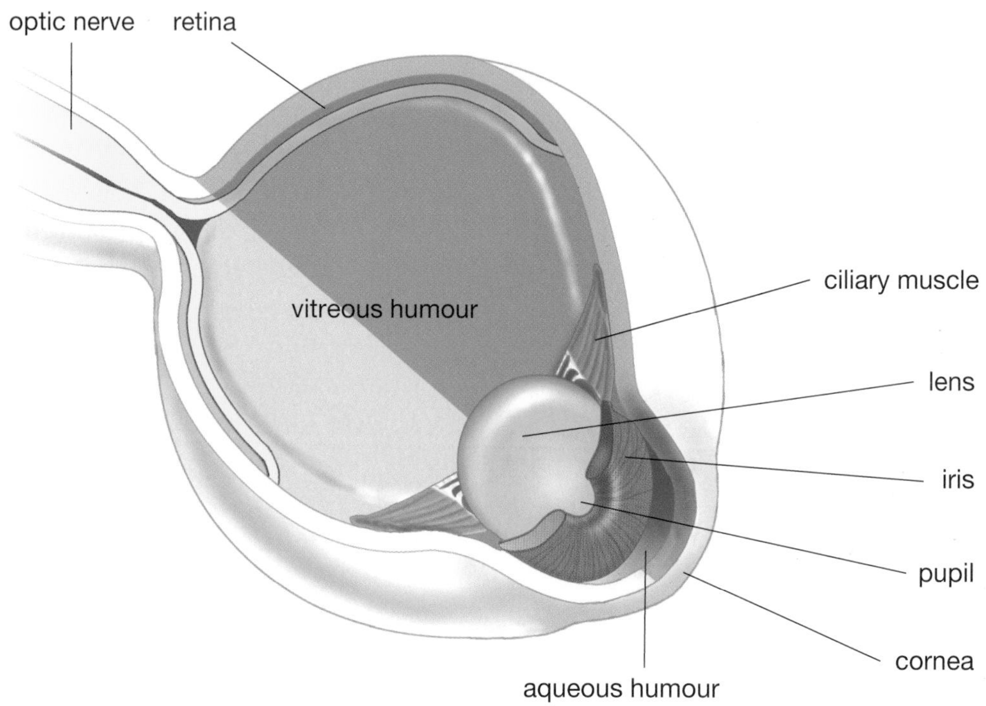

Figure 8.17 The eye is an almost spherical object that is filled with fluids.

INVESTIGATION 8-C

Looking at Systems

Think About It

A **system** is a set of connected parts whose action is controlled in specific ways. The parts work together in such a way that a change in one part can result in a change in another part. Both the eye and the camera are examples of a system.

What to Do

1. Work in a group. Decide which system you intend to model — the eye or the camera. Members of the group will take on different roles. For example, if the group is portraying a camera, students will play the following roles:
   - the *object* that is being photographed
   - the *lens* of the camera, which moves in response to a change in location of the object
   - the *motor* to move the lens
   - the *rangefinder* to tell the motor which way to move the lens
   - the *diaphragm* to let in the correct amount of light
   - the *motor* to adjust the diaphragm
   - the *light meter* that signals to the diaphragm motor to make the opening larger or smaller
   - the *film* on which the lens focusses the image

   If your group is portraying an eye, the roles are similar. For example, muscles replace motors; the iris replaces the diaphragm; the retina performs the functions of the light meter and the film; and nerve cells replace the rangefinder.

2. Assume that you are photographing (or viewing) the flame of a candle that moves toward and away from the camera (eye). As the object comes closer, the amount of light entering your camera or eye would increase. Decide in advance how the lens and the diaphragm (or the iris) would have to respond to the new location of the candle flame.

3. The lens must signal to its motor when to stop or start. This process of telling the motor when to stop is called feedback. In other words, **feedback** is the return of information about the result of a process or activity. The diaphragm must also give feedback to its motor to communicate how large or small to make the aperture.

4. During the course of this investigation, the candle flame could grow brighter or dimmer while standing in one location. The group should decide in advance how to handle such a situation.

Analyze

1. After the role-play, discuss with your group whether you could have improved on your system. Did your system function as well as possible? Did your system have effective feedback?
2. Give several more examples of systems with which you are familiar. Which parts of the systems are involved in feedback?
3. In earlier science studies, you learned that sources of *input* into a system can include people, materials, and energy. A system's *output* is the actual result obtained from a system, for example, the light that comes on when a light switch (the input) is pressed. Identify the input and the output in (a) a camera, and (b) the human eye.

Career CONNECT

An Eye on Education

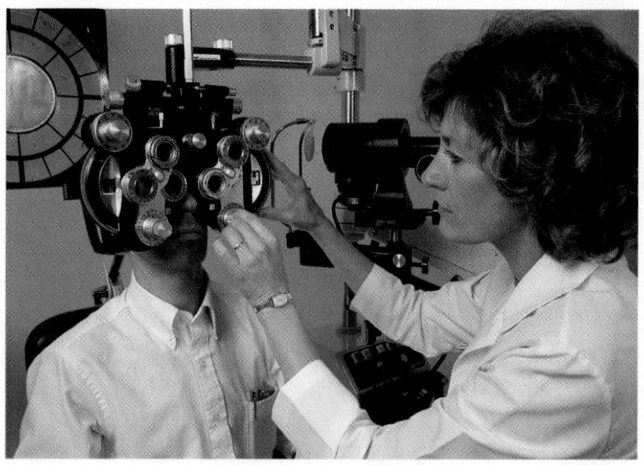

If you are having trouble with your vision, you visit an optometrist, a specialist trained to examine patients' eyes for defects and then prescribe lenses or exercises to correct the problem. How might you become an optometrist?

First, you would need to complete two or three years of university courses in many different sciences. After completing these courses, you would apply to one of the two schools of optometry in Canada, at either the University of Waterloo or l'Université de Montréal. If accepted, you would spend the next four years learning about the physics of light and lenses and the anatomy of the human eye. You would learn how to set up your own practice and spend time dealing with patients in clinics. If you studied hard, after four years, you could be one of the 100 optometrists in all of Canada to graduate each year. You must then pass a provincial examination before getting your optometrist's licence and opening an office.

Six or seven years of preparing for a career might seem like a long time. However, these years of study would lead to a very rewarding career. Deciding what amount of education is ideal for you is an important step in choosing a career path. Would you like to enter the work force right after high school? Are you interested in receiving a one-, two-, or three-year college diploma, or a three- or four-year university degree? Do you plan to invest more than four years in education and training for a career? Do some research to identify careers that require the amount of education you think is right for you.

Check Your Understanding

1. Describe a test to determine whether an image is real or virtual.

2. As an object comes closer to a convex lens, what happens to
   (a) the size of the image?
   (b) the attitude of the image?
   (c) the location of the image?

3. Draw a diagram to show what happens when light passes through
   (a) a concave lens
   (b) a convex lens

4. State which part of the human eye corresponds to each of the following parts of the camera, and explain why:
   (a) the film
   (b) the diaphragm
   (c) the aperture
   (d) the lens

5. **Thinking Critically** In this section, you have seen a number of ways in which the camera and the eye are similar. Describe three ways in which the camera and the eye are dissimilar.

6. **Thinking Critically** The stability of the feedback system in either the camera or the human eye depends on a number of factors. For example, too much light will damage both the film in a camera and the retina in the eye. In a group, brainstorm other factors that could damage these parts of the system, thus disrupting the system's stability.

8.3 Extending Our Vision

Human knowledge about our planet and the universe was very limited until we developed tools to extend our vision. We now have the ability to peer into the tiny world of micro-organisms and out into the vast reaches of outer space. The tools we use for these inquiries seem quite different from each other, but they are based on the same understandings of light, mirrors, and lenses.

Telescopes

Telescopes like the one shown in Figure 8.18 help us see distant objects more clearly. In a **refracting telescope**, light from a distant object is collected and focussed by a convex lens called the **objective lens**. A second lens, called the **eyepiece lens**, works as a magnifying glass to enlarge the image (see Figure 8.19A). A **reflecting telescope** uses a concave mirror to collect rays of light from a distant object. The mirror is called the primary or **objective mirror**. It forms a real image, which is then magnified by the eyepiece lens (see Figure 8.19B on the next page).

Did You Know?

The first telescope may have been invented by Hans Lippershey, a Dutch optician. In 1608, Lippershey was testing some lenses and discovered their magnifying power by accident. The following year, the Italian scientist Galileo Galilei developed a telescope that could see mountains on the Moon, sunspots, stars in the Milky Way, and moons around Jupiter.

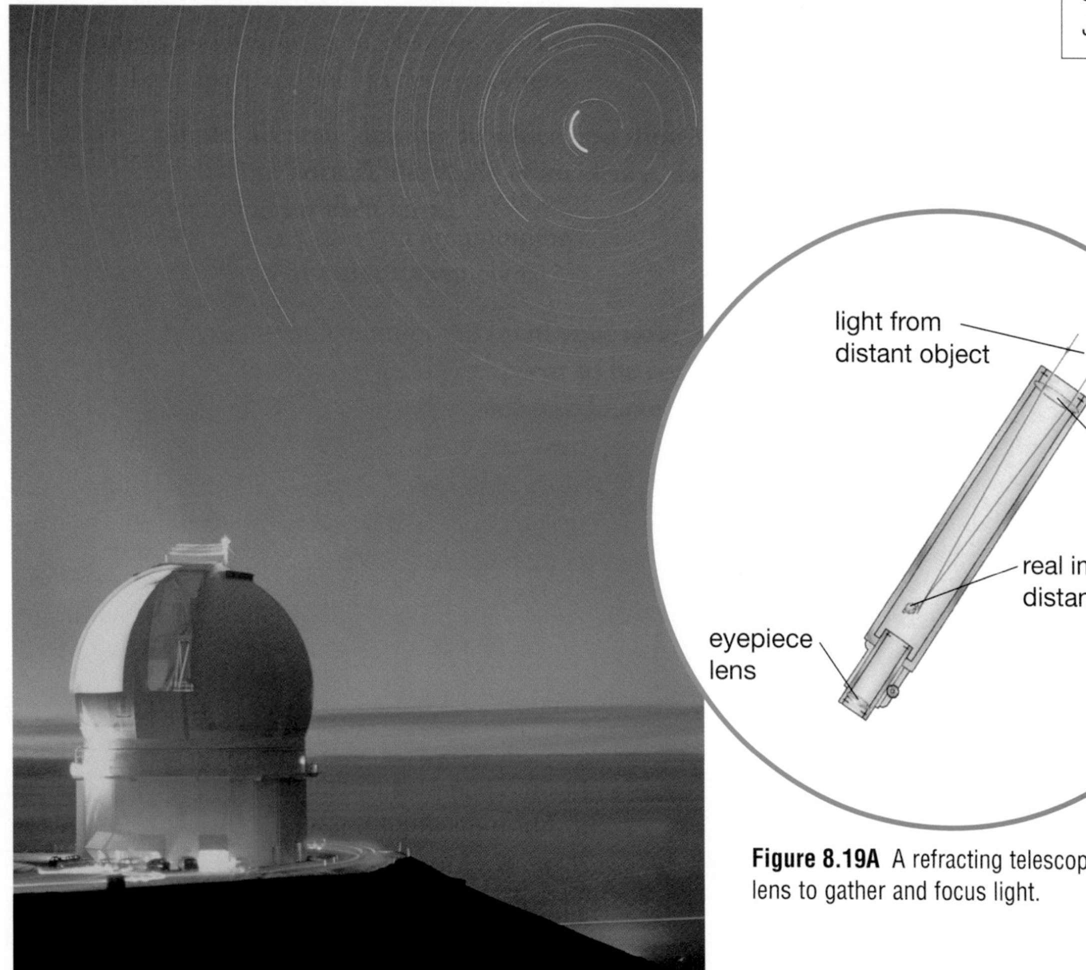

Figure 8.18 The Mauna Kea telescope, jointly sponsored by Canada, France, and Hawaii, is located near the summit of Mauna Kea, an extinct volcano in Hawaii. It uses a concave mirror that measures 3.6 m in diameter. What does a concave mirror do to light?

Figure 8.19A A refracting telescope uses a convex lens to gather and focus light.

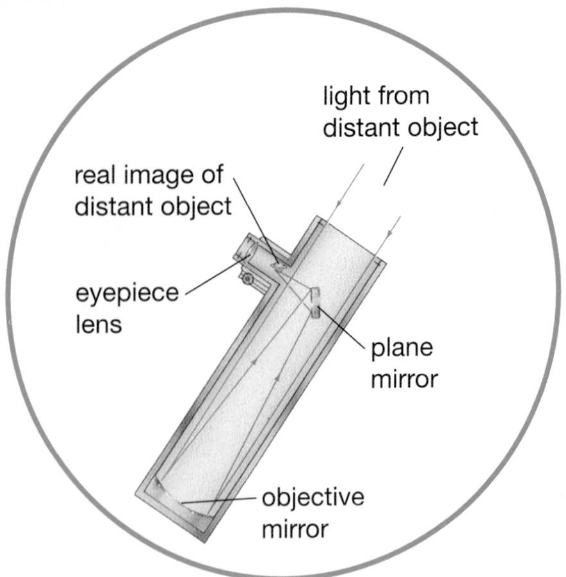

Figure 8.19B A reflecting telescope uses a concave mirror to gather and focus light.

The lens in a refracting telescope and the mirror in a reflecting telescope both act as a collector of light. Reflecting and refracting telescopes must have a large collector (either a lens or a mirror) in order to gather as much light as possible from the distant object. The collector then focusses the light into an image. As you learned in Conduct an Investigation 8-B, the farther the image is from the lens, the greater the magnification. Similarly, the farther the image is from the mirror in a reflecting telescope, the greater the magnification. For the greatest magnification, the telescope needs to have as large a distance as possible between the object being viewed — a star or a planet, for example — and its image. This explains why some telescopes are so enormous.

Find Out **ACTIVITY**

Make Your Own Refracting Telescope

In this activity, you will make a simple refracting telescope using two lenses.

What You Need

2 mounted convex lenses of the same diameter (one having a smaller curve than the other)

What to Do

1. Place the mounted lenses on a table and look in a straight line through both lenses.
2. Find out which combination of lenses will let you see a magnified image of a distant object. This object could be a poster on the far wall of the classroom, for example. You may have to move the objective lens back and forth until you can see a clear image.
3. Choose a specific area of the image. Estimate how much larger that part of the image is when viewed through the lens, compared with its size when viewed with the unaided eye.
4. Make a diagram showing the arrangement of the far object, the two lenses, and your eye. Label the objective and eyepiece lenses.

What Did You Find Out?

1. How many times larger does the image appear to be?
2. What is the attitude of the image?
3. In a refracting telescope, which lens has the greater curve?
4. **Design Your Own** How could you use a piece of waxed paper to find out whether the image between the lenses is real or virtual? Design your own experiment to find out.

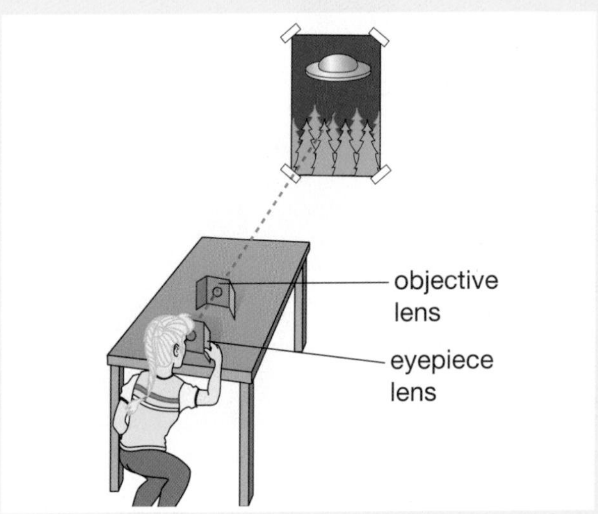

Binoculars

Binoculars are actually two reflecting telescopes mounted side by side. You can imagine how difficult it would be to hold up two long telescopes. In binoculars, the telescopes are shortened by placing glass blocks inside. These glass blocks, called **prisms**, serve as plane mirrors. Rather than travelling down the long tube of a telescope, light in binoculars is reflected back and forth inside a short tube, as you can see in Figure 8.20.

prism binoculars

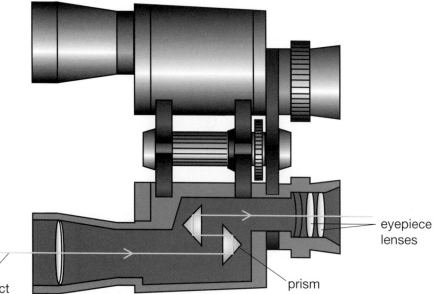

eyepiece lenses

light from distant object

prism

Figure 8.20 Binoculars use prisms to reflect light.

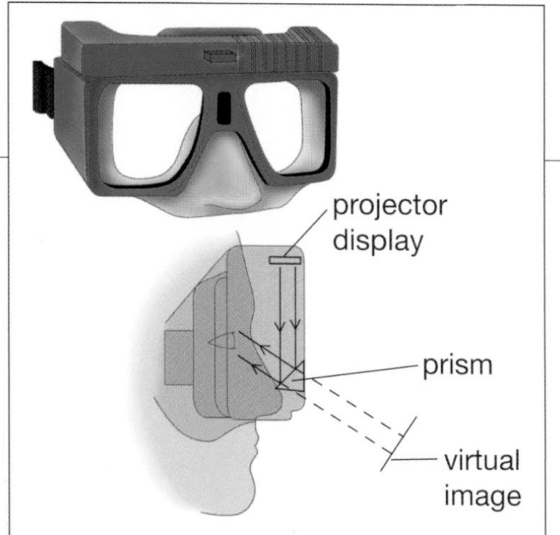

Prisms are used to reflect light in a device known as a "Heads Up Display" or HUD. Light reflects through prisms to make an image appear in front of a person's eyes. This device helps airline pilots, who need to check air speed and altitude while at the same time looking out the plane's windscreen. Pilots can monitor the glowing displays projected in front of them even as they watch the approaching runway.

HUDs have also been developed for scuba divers. The device is positioned inside the diver's mask. The diver can check depth, water pressure, and air consumption by examining an image of this data projected directly in front of the diver. This spares the person the inconvenience of having to check the readings on separate gauges that indicate depth, pressure, and the amount of air left in the diver's tanks. Where else do you think HUDs could be useful?

INVESTIGATION 8-D

Just Having a Look Around

In earlier science studies, you may have seen how a periscope works. Now you know how a telescope works. There are situations in which the two might be combined. A submarine commander needs to stay underwater and still see in all directions above the surface of the water. Technicians working with explosive substances might wish to watch the results of their work from behind the safety of a shield. In both cases, a combination of telescope and periscope would be useful.

Challenge

Design and construct a device that will allow you to see a magnified view of objects that are behind you.

Materials 

convex lens (with a large curve)	2 plane mirrors
convex lens (with a small curve)	scissors
stiff cardboard or Bristol board	tape
	ruler

Design Criteria

A. The mirrors and lenses should be safely mounted and positioned on a flat surface.

B. All the items listed under "Materials" must be used.

Plan and Construct 

1. In your group, plan how the mirrors and lenses should be arranged to allow you to see a magnified view of objects behind you.

2. Before arranging the items, decide
   (a) which lens is to be the objective lens
   (b) which lens is to be the eyepiece lens
   (c) how far apart the lenses should be to act as a telescope. You may have to experiment with the lenses.

3. Show your plan to your teacher for approval. This plan must include the distance between the lenses and the location of the mirrors.

4. Construct any mounts (optics stands) that you need.

5. Arrange the items on the table according to your plan. Then look into the eyepiece to see if your arrangement works well.

Evaluate

1. Did your arrangement work as you had planned?

2. Suggest a way to improve the arrangement to give you a better magnification or a clearer view.

INTERNET CONNECT

www.school.mcgrawhill.ca/resources/
Search the Internet to find images from the Hubble Space Telescope. Visit the above web site. Go to **Science Resources**, then to **SCIENCEPOWER 8** to find out where to go next. Download images of Saturn, Mars, the Andromeda Galaxy (M31), and two other images that you find interesting. Find out how far each of the objects is from Earth and how big the mirror is in the Hubble Space Telescope. Present the images in a poster describing the achievements of the Hubble Telescope.

Microscopes

As you learned in Chapter 1, a magnifying glass is a simple kind of microscope. Effective magnifying glasses can magnify objects up to ten times their normal size. A compound light microscope can magnify objects up to 2000× (see Figure 8.21). Like telescopes, compound microscopes have an objective lens that forms a real image of the object and an eyepiece lens that magnifies the image further. As you also learned in Chapter 1, compound microscopes use more than one lens in the objective and eyepiece to improve the sharpness of the image (see Figure 8.22).

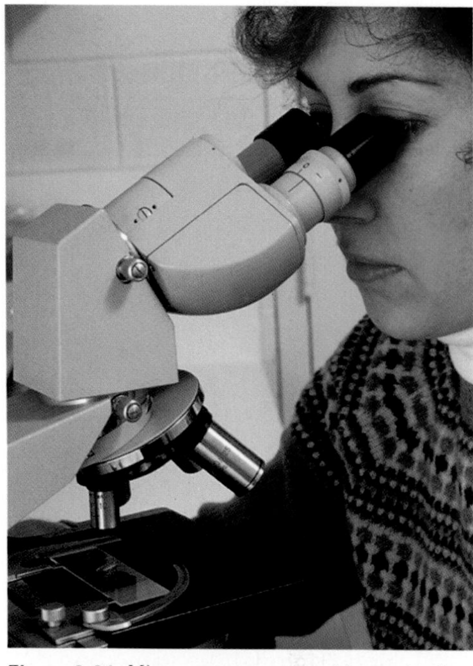

Figure 8.21 Microscopes are used to probe the world of cells, bacteria, and other living things too small to be seen with the unaided eye. This laboratory technician, who carries out fish hatchery research for the Ontario Ministry of Natural Resources, is scanning a tissue sample for signs of disease in fish.

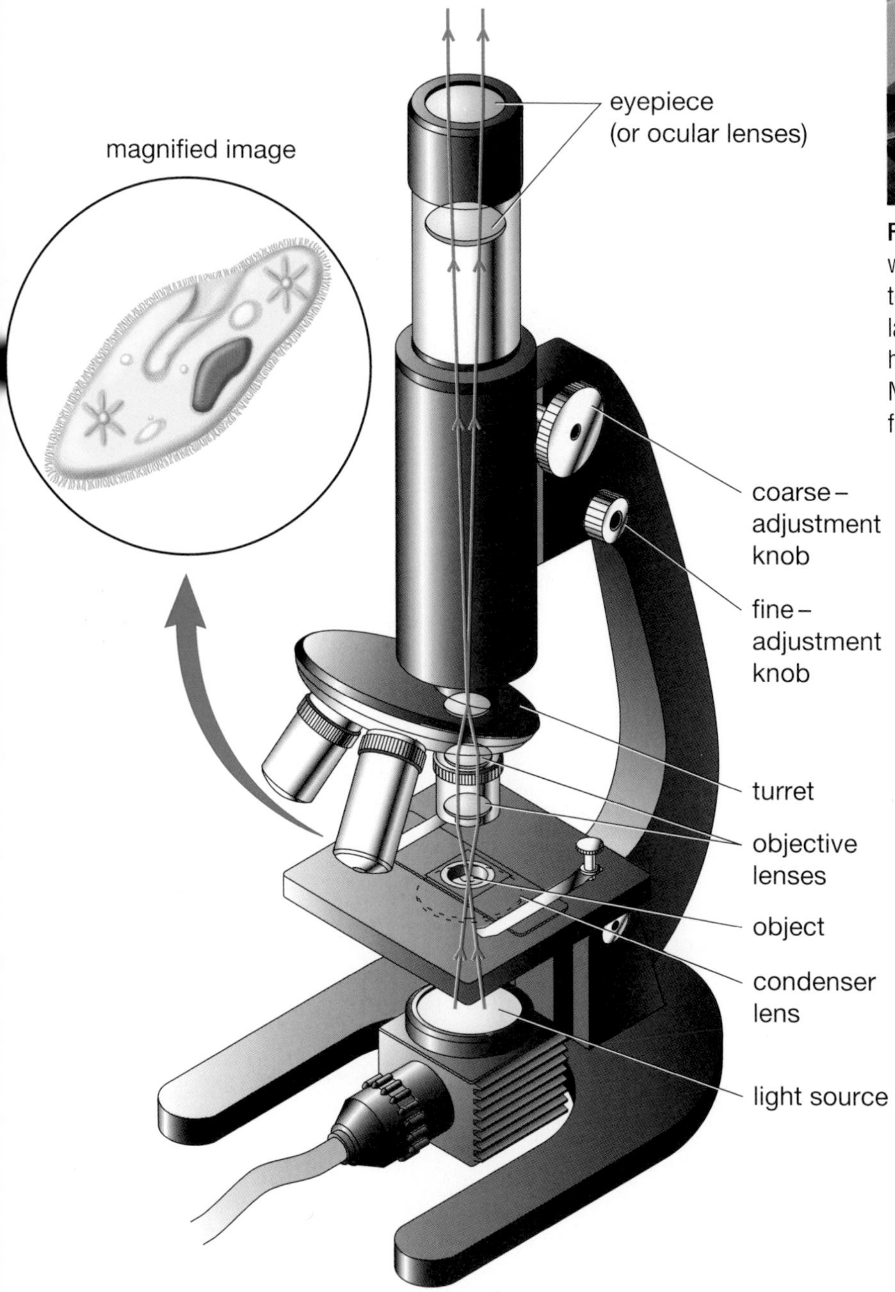

Figure 8.22 A compound light microscope uses several lenses to magnify an object.

Find Out **ACTIVITY**

Making a Microscope

Because a microscope, like a telescope, uses two lenses to magnify an object, you should be able to make a simple microscope out of the same two lenses used in the previous telescope activity.

What to Do

1. Find out which combination of lenses will let you see a magnified image of a near object. You should not place the object too close to the objective lens or you will not get any image at all!
2. Estimate how much larger the image is than the object. To do this, keep both eyes open and compare the same parts of the image and the object.
3. Draw a diagram showing the arrangement of the near object, the two lenses, and your eye. Label the lenses.

What Did You Find Out?

1. How many times larger than the object is the image?
2. What is the attitude of the image?
3. In a microscope, which lens has the greater curve?

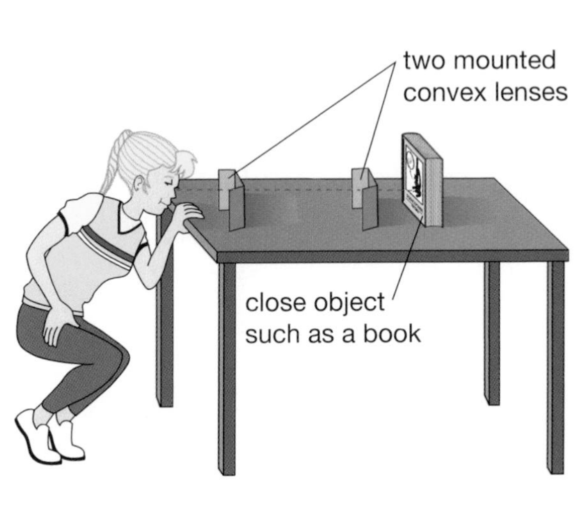

Check Your Understanding

1. State the function of a telescope's
   (a) objective lens or mirror
   (b) eyepiece lens
2. Name the lens that has the greater curve in
   (a) a refracting telescope
   (b) a compound microscope
3. What type of image is formed by the objective lens of a telescope or a microscope? What is the attitude of this image?
4. What role do prisms play in a pair of binoculars?
5. **Apply** In general, describe the attitude of
   (a) a real image
   (b) a virtual image

CHAPTER at a glance

Now that you have completed this chapter, try to do the following. If you cannot, go back to the sections indicated.

Explain how the appearance of an image depends on the shape of the reflecting surface. (8.1)

Explain how a concave mirror forms different kinds of images depending on how close the object is to its reflecting surface. (8.1)

Describe how a convex mirror always produces a smaller, upright image of the object. (8.1)

Give some uses for curved mirrors. (8.1)

Explain the difference between a real image and a virtual image. (8.1)

Describe how the distance between the object and the convex lens affects the characteristics of the image. (8.2)

Describe how the camera and the human eye are examples of systems and explain the role of input, output, and feedback. (8.2)

Describe how convex lenses and concave mirrors are used in telescopes. (8.3)

Describe how convex lenses are used in microscopes. (8.3)

Look at the diagram below and explain how binoculars work. (8.3)

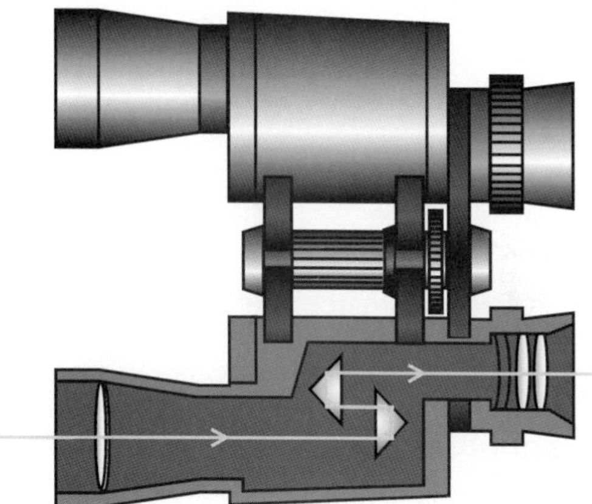

Prepare Your Own Summary

Summarize this chapter by doing one of the following. Use a graphic organizer (such as a concept map), produce a poster, or write the summary to include the key chapter ideas. Here are a few ideas to use as a guide:

- Describe some applications of concave and convex mirrors.
- Copy the table below and complete it.
- Describe some applications of concave and convex lenses.
- Explain how lenses can be used to correct vision.
- Compare and contrast how optics systems such as the camera and the human eye work.
- Compare and contrast a reflecting telescope and a refracting telescope.

Position of object	Mirror	Image type (real or virtual)	Image size (smaller, same, or larger)	Image attitude (upright or inverted)
far from mirror	concave			
far from mirror	convex			
close to mirror	concave			
close to mirror	convex			
far from mirror	plane			
close to mirror	plane			

CHAPTER 8 Review

Key Terms

concave
convex
attitude
real image
virtual image
lens
concave lens
convex lens
focusses
retina
near-sighted
far-sighted
ciliary muscles
accommodation
near point
far point
diaphragm
aperture
shutter
iris
pupil
iris reflex
optic nerve
blind spot
humours
system
feedback
refracting telescope
objective lens
eyepiece lens
reflecting telescope
objective mirror
prisms

Reviewing Key Terms

If you need to review, the section numbers show you where the terms were introduced.

1. In your notebook, match the description in column A with the correct term in column B.

A	B
• the point where the optic nerve enters the retina	• concave mirror (8.1)
• bulges outward	• inverted (8.1)
• curves inward	• objective lens (8.3)
• an image from which light comes	• virtual image (8.1)
• an image from which no light comes	• optic nerve (8.2)
• attitude of the image in a microscope	• iris (8.2)
• controls the size of the pupil	• retina (8.2)
• lens in a refracting telescope	• accommodation (8.2)
• path from the retina to the brain	• convex mirror (8.1)
• corresponds to the film in a camera	• real image (8.1)
	• prism (8.3)
	• blind spot (8.2)

Understanding Key Ideas

Section numbers are provided if you need to review.

2. Draw the surface of a mirror that is
   (a) plane
   (b) concave
   (c) convex (8.1)

3. Draw the cross-section of a glass lens that has
   (a) one plane surface and one convex surface
   (b) two concave surfaces
   (c) two convex surfaces (8.2)

4. **(a)** What are some things that a lens can do to light? (8.2)
   (b) What are some uses of lenses? (8.3)

5. Why does your eye have a blind spot? (8.2)

6. You are looking at a friend. If the friend starts to move away, what must the lens in your eye do to keep the friend in sharp focus? (8.2)

7. If you step out of a lighted house into a dark backyard, what sudden change occurs in your eye? What is this process called? (8.2)

8. State three ways in which the compound microscope is similar to the refracting telescope. (8.3)

Developing Skills

9. Draw a concave mirror as you did in Conduct an Investigation 8-A, but place it near the right side of the page. Use a radius of 6 cm. Place point **B** 4 cm to the left of the mirror and 1 cm above the line.
   (a) Locate the image of point **B** by drawing the incident and reflected rays.
   (b) Why would you expect an object that is located at **B** to have a larger, inverted, real image?

10. Draw a convex mirror near the middle of the page. Use a radius of 6 cm. Place point **B** 2 cm to the left of the mirror and 1 cm above the line.
   (a) Locate the image of point **B** by drawing the incident and reflected rays.
   (b) Why would you expect an object that is located at **B** to have a smaller, upright, virtual image?

11. (a) Diagram A below shows a side view of a convex lens used as a contact lens to help someone read. Why is it designed this way?
   (b) Diagram B shows a side view of a concave lens. How might you change its shape so it could be used as a contact lens? Remember, it must remain a concave lens. What kind of vision problem do you think lens B is meant to correct?

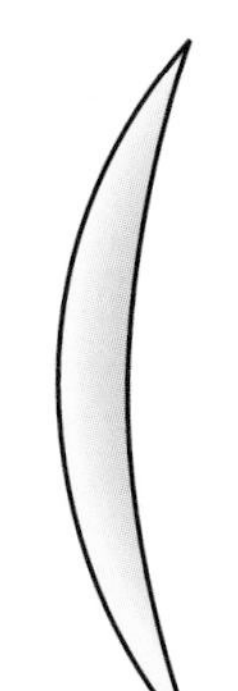

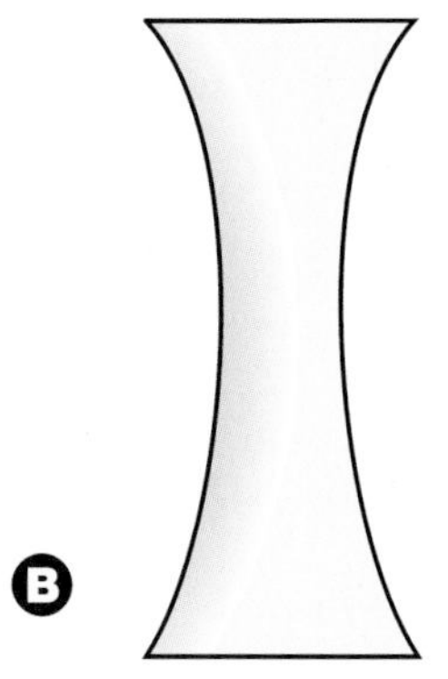

12. In question 3 on page 264, which lenses could focus light to a bright point? What are some uses for lenses shaped like this?

13. Make a diagram showing the similarities and the differences between the human eye and the camera. Label your diagram.

Problem Solving/Applying

14. A magnifying lens is a convex lens used to examine tiny objects. Explain how a magnifying glass creates an enlarged image. What kind of image is this? Use a diagram in your explanation.

15. You are designing a telescope to study distant galaxies. What would be the main feature of your design if you wanted your telescope to gather as much light as possible? What might be some limitations on your design?

Critical Thinking

16. What happens to the size and location of the image formed by a convex lens as the object comes closer to the lens?

17. How could you make a large image of your face seem to hover above ground in a dark, foggy street?

18. How could you make ray diagrams much easier to draw when the object is a point on the line?

Pause& Reflect

Go back to the beginning of this chapter on page 236 and check your original answers to the Getting Ready questions. How has your thinking changed? How would you answer those questions now that you have investigated the topics in this chapter?

CHAPTER

9 Light, Colour, and

Getting Ready...

- How do we see colour?
- What makes blue light different from red light?
- Why do people who work with X rays need to protect themselves?

Science Log

On a traffic light, the colours red, amber, and green represent stop, caution, and go. In your Science Log, list several other objects or devices that use colours for practical purposes. Then write answers to the questions above in your Science Log.

What do members of the "green" movement believe in? What does it mean if you are feeling "blue"? Have you ever heard a successful enterprise described as being "in the red"? Colours are so important to how people perceive the world that they are used to describe much more than how things look. Colours are often associated with moods, emotions, events, symbols, and attitudes.

What exactly *is* colour? Why do some objects appear red while others appear blue? What do you think is causing the bands of colour in the photographs shown here?

By studying this chapter, you will deepen your understanding of colour and the link between colour and visible light. You will then discover how light relates to a range of invisible radiation, from radio waves to gamma rays. The chapter concludes by examining some of the many applications of invisible radiation.

Radiant Energy

Spotlight On Key Ideas

In this chapter, you will discover

- the source of colours
- the difference between additive primary colours and subtractive primary colours
- applications of primary additive colours, such as colour television
- how the human eye perceives colours
- how to explain the behaviour of light using a scientific model
- how we use electromagnetic radiation, of which light is only a small part

Spotlight On Key Skills

In this chapter, you will

- combine filters and lights to produce desired colour effects
- use a diffraction grating to analyze colour spectra
- design your own controlled experiment
- use scientific notation as a way to write large numbers
- record data in a table

Starting Point Activity

What Is Colour?

Does light contain colour, or is colour something that light picks up when it strikes a coloured object? You can begin to answer this question by doing this activity.

What You Need

overhead projector

pieces of red, green, and blue cellophane or colour filters

pieces of coloured construction paper, including red, green, blue, and other colours

What to Do

1. Darken the room. Hold a piece of red construction paper in front of the screen in the light of the overhead projector.
2. Place two layers of red cellophane on the glass plate (stage) of the overhead projector. What colour does the construction paper appear to be under red light? Record your observation.
3. Repeat step 2 using green cellophane, then blue cellophane.
4. Repeat the activity with the remaining colours of construction paper.

What Did You Find Out?

1. Record your results in a table. Write the three colours of light as the column headings and all the colours of construction paper as the row headings. Give your table a suitable title.
2. Which combinations of cellophane and construction paper resulted in the construction paper looking black?
3. **Apply** For the paper to appear blue, what colour of light must shine on the paper? What colour must the paper be?

9.1 The Source of Colours

At one time, people believed that colour was something added to light. When white light struck a green leaf and green light was reflected, people believed the leaf was adding green to the light. However, the Starting Point Activity suggests that a leaf will only appear green if there is green light already in the sunlight, before it strikes the leaf. Where, though, is the colour in sunlight?

Figure 9.1 The gorgeous colours of a fall day blaze through forests because the colours are already contained in the light striking the leaves.

Find Out **ACTIVITY**

A Shower of Colour

How is colour related to light?

What You Need

sharp pencil

overhead projector

piece of cardboard large enough to cover the stage of the overhead projector

equilateral prism

sheet of white paper

What to Do

1. Use a sharp pencil to punch a hole in the middle of the cardboard. **CAUTION:** Be careful when punching the hole.
2. Place the cardboard on the stage of the overhead projector so that a beam of light passes through the hole.
3. Shine the light beam from the overhead projector onto one side of the prism.
4. Using the sheet of white paper, locate the area where the light exits the prism. List the sequence of colours you see. Then compare your list with the lists your classmates made.

What Did You Find Out?

1. You allowed white light to refract through a prism, and you observed colours emerge. Where did the colours come from?
2. Do all colours refract the same amount? Give a reason for your answer.
3. What do you think you would see if you were able to put all the colours back together again?

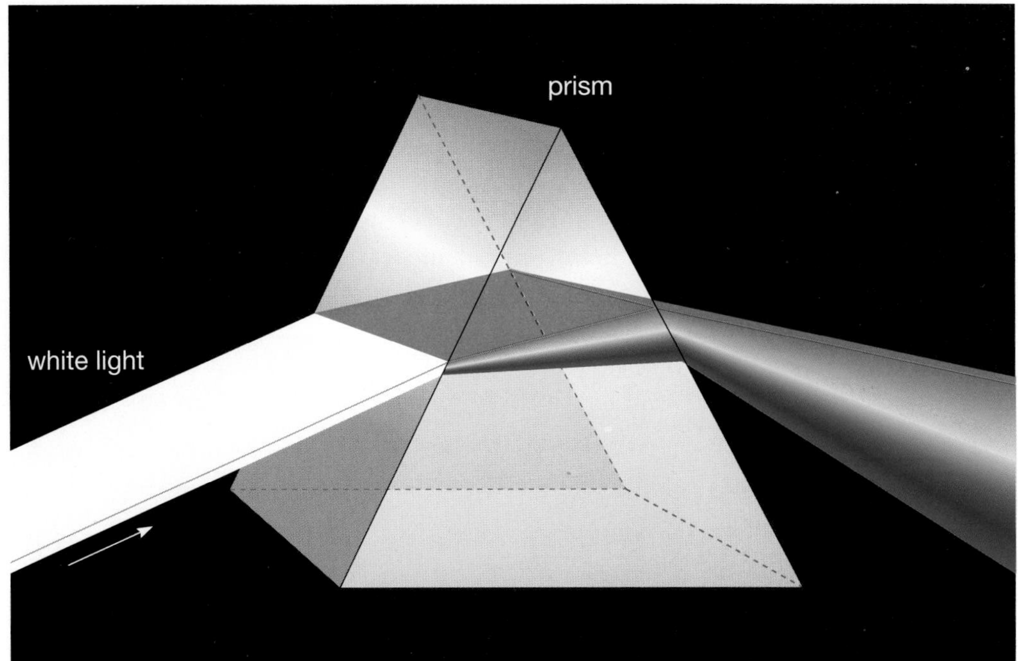

Figure 9.2A Newton observed that a prism refracted light into different colours.

In the seventeenth century, English scientist Sir Isaac Newton conducted a famous experiment. He placed a prism so that a thin beam of white light could pass through it. When white light travelled through the prism, Newton saw bands of colour emerge. He observed that each band of colour was refracted at a different angle, producing a rainbow effect (see Figure 9.2A). Newton realized that the different colours must have been present already in the white light since the prism was not the source of the colours.

Next, Newton passed these colours through a reversed prism. This time, only white light emerged, as shown in Figure 9.2B. In this way, Newton showed that colour was a property of visible light. He proposed that white light such as sunlight is the result of mixing together all the different colours of light.

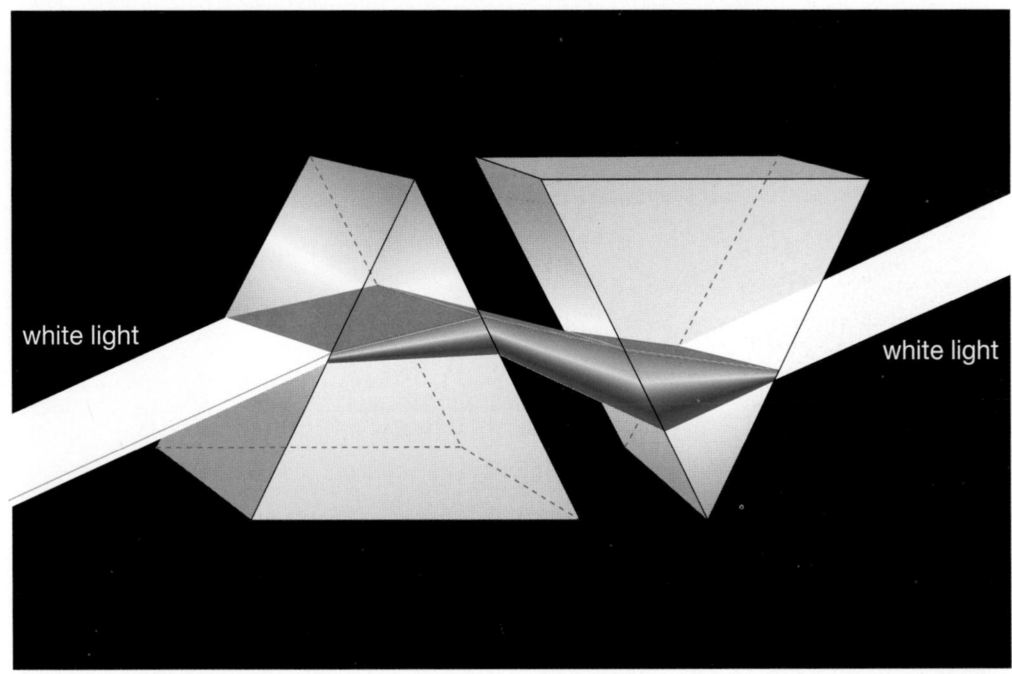

Figure 9.2B When passed through a second prism, the refracted light combined to form white light once again.

The Spectrum

When white light is refracted into different colours, the resulting pattern is called a **spectrum**. For sunlight, the colours range from red through orange, yellow, green, blue, indigo, and violet. This pattern is called the **solar spectrum**. You can remember the order of the seven colours in the solar spectrum by means of a memory aid: the name ROY G. BIV. The "R" stands for red, "O" for orange, and so on. Make up a memory aid of your own to help you remember the order of the colours in the spectrum. A rainbow is an example of a colour spectrum (see Figure 9.3).

Word **CONNECT**

The word *spectrum* comes from Latin and means "spectre" or "apparition." These words are also synonyms for the word "ghost." The plural of spectrum is spectra. In a dictionary, look up another word with the root "spectre" and use it in a sentence.

Figure 9.3 This rainbow was photographed in southeastern Ontario. Notice the pattern of colours ranging from red at the top, to violet at the bottom.

When light strikes an object, the light may be reflected off the object, absorbed by the object, or transmitted through the object. When white light passes through a blue bottle, the glass absorbs all the colours of light except the blue (see Figure 9.4). Only the blue light is reflected or transmitted. This explains why the bottle appears blue.

Pause& Reflect

You have seen that a prism can break up light into the colours of the spectrum. What do you think is serving as a prism in the case of a rainbow? Answer this question in your Science Log. Use the word "refract" or "refraction" in your answer.

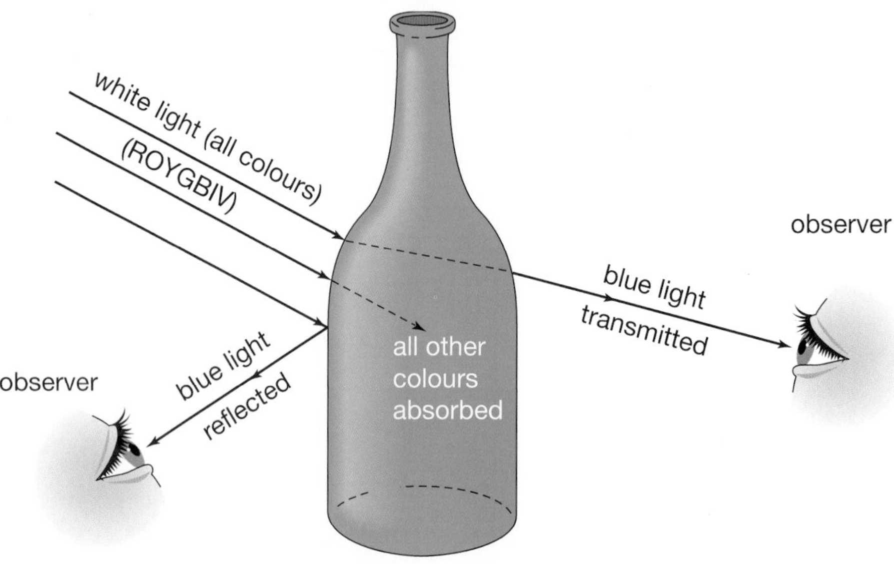

Figure 9.4 All the colours of light except blue are absorbed by a blue bottle. Only the blue light is reflected or transmitted.

Seeing Red

The colour you see when light strikes an opaque object depends on which colours are reflected and which ones are absorbed. The paper on this page reflects all colours, so you see it as white. The print on the page absorbs all colours, so you see it as black. Any colours you see in coloured photographs depend on the colour of light reflected from the photographs.

Examine Figure 9.5. A tomato under white light looks red because it is reflecting only red light to your eyes, as shown in photograph A. What happens to all the other colours if only red light is reflected? The other colours have been absorbed by the skin of the tomato. In photograph B, you see the tomato under red light. The tomato still appears red since red light is not absorbed by the tomato. However, in photograph C, the tomato is illuminated by only blue light. Because blue light is almost completely absorbed by the tomato, very little light can reflect back. Thus, the tomato looks black, with just a hint of blue.

DidYouKnow?

Many animals other than humans, for example, cows, deer, and dogs, cannot see colours. They see the world in only black and white.

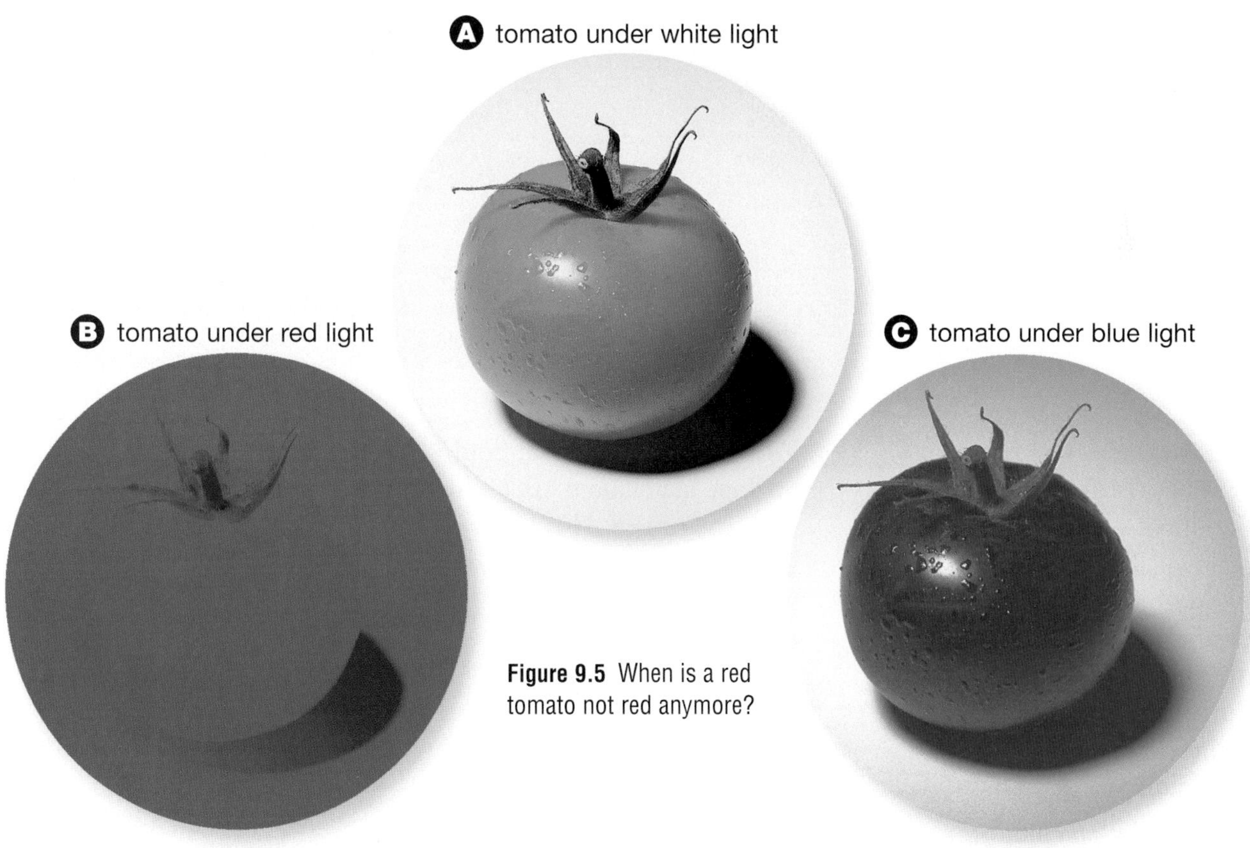

Figure 9.5 When is a red tomato not red anymore?

Pause& Reflect

Colours are reflected light that we can see. What happens when we see no reflected light? Close your eyes and cover them with your hands. You should see only black. Now look at something black, such as a pair of shoes. The shoes appear black because they have absorbed all the colours in the light. Black objects do not reflect any light. In your Science Log, write a definition of the colour "black."

CONDUCT AN

INVESTIGATION 9-A

Spotlight on Colour

You have seen that white light is a mixture of red, orange, yellow, green, blue, indigo, and violet light. However, do you need all these colours to make white light? Is it possible to trick the human eye into thinking that all the colours are present when some are missing? You will find out in this investigation.

Problem

How can you produce the effect of white light using fewer than the seven colours of the spectrum?

Apparatus

3 floodlights (1 red, 1 green, 1 blue) mounted on separate stands, all facing a screen or a light-coloured wall

Procedure

1. Darken the room as much as possible. Turn on all three floodlights. Can you produce a glowing patch of white light? You can change the brightness of each colour by moving the floodlights closer to the screen or farther away from it.
   - **(a)** Turn off the red floodlight. What colour do you see?
   - **(b)** Turn on the red floodlight and turn off the green. What colour do you see?
   - **(c)** Turn on the green floodlight and turn off the blue. What colour do you see?
2. With all floodlights on, hold up your hand about 10-15 cm from the screen. How many shadows do you see? Try to name their colours.

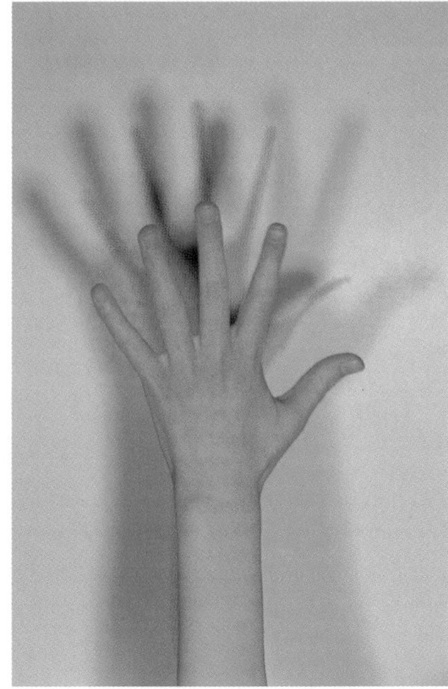

Analyze

1. How close did you get to producing white light using just red, green, and blue light?
2. Name the colour that is produced by combining
   - **(a)** green and blue
   - **(b)** red and blue
   - **(c)** red and green

Conclude and Apply

3. Try to explain the colours that you saw in the shadows of your hand.
4. How might a stage lighting technician use the information that you just gained?

Additive Primary Colours

The three colours red, green, and blue are called the **additive primary colours**. They are called additive colours because adding all three together in the proper amounts will make white light, as shown in Figure 9.6. The light of two additive primary colours will produce a secondary colour. The three **secondary colours** are yellow, cyan, and magenta.

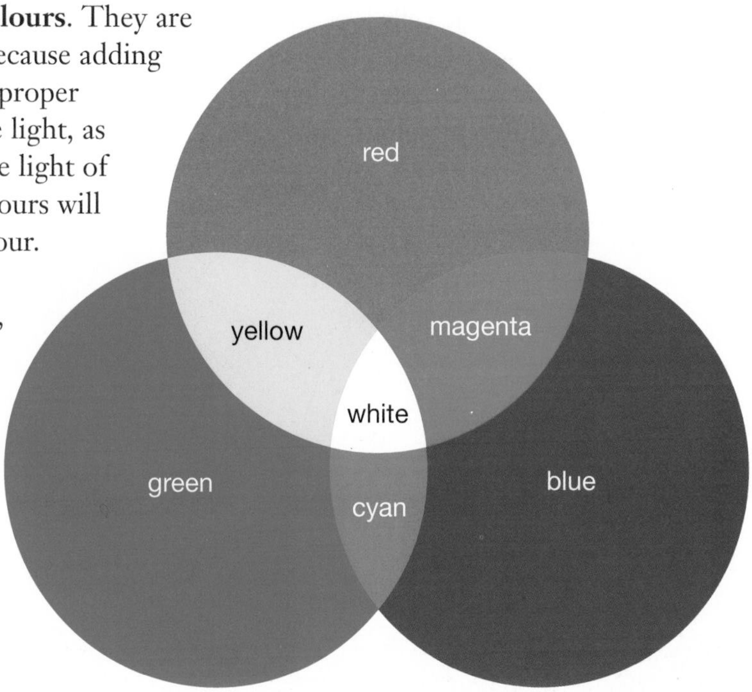

Figure 9.6 Combining the additive primary colours red, green, and blue produces white light.

DidYouKnow?

An *electron* is a negatively charged particle orbiting the nucleus of an atom. You will learn more about atoms and electrons in your grade 9 studies.

Television screens use additive primary colours. The screen contains many groups of three tiny phosphor dots (see Figure 9.7). Each dot glows with a different colour when it receives energy from the electrons inside the picture tube of the television set. In each group, one dot glows red, a second glows green, and the third glows blue. If all three dots glow brightly, you see white light.

The phosphor dots on a television screen are too tiny to be seen from a distance. Thus, you see sections of colour rather than individual dots. A section on the screen will look yellow if the red and green dots in the section glow. If the red and blue dots glow, you will see magenta. What combination of glowing dots produces cyan? All other colours can be created by varying the intensities of the three primary colours.

Figure 9.7 All colours on a television screen are composed of various combinations and intensities of red, green, and blue light.

How We See Colour

The retina of the human eye contains two types of cells that respond to light (see Figure 9.8). Some cells look like tiny cylinders and are called **rods**. They detect the presence of light. The other cells are called **cones**, again because of their shape. The cones detect colour. There are three types of cones, each of which responds to a different colour.

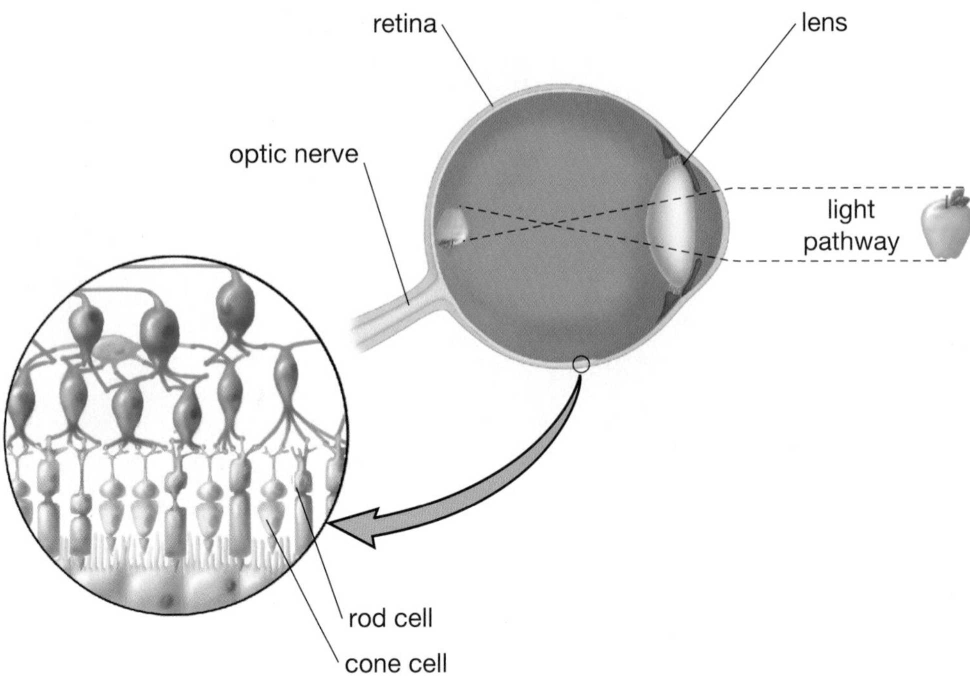

Figure 9.8 Rods and cones, named for their shapes, are the two types of light-detecting nerve cells in the retinas of your eyes.

As you saw in the spectrum produced by a prism, white light contains red, orange, yellow, green, blue, indigo, and violet light. However, cones in the human eye respond mainly to red, green, and blue light. This is why the eye can be tricked into thinking that a beam of light is white when it contains only those three colours. This is also why all other colours are seen by the eye in terms of the relative amounts of red, green, and blue light sensed by the cones.

Signals from all three types of cone cells and the rod cells travel along the optic nerve to the brain. The brain then interprets the shape and colour of the object that you see.

Almost no mammals except primates – a category that includes apes, monkeys, and humans – possess colour vision. In those mammals that do have colour vision, it is very basic. However, many other animals possess an excellent sense of colour. It is highly developed in birds and fish, and insects such as bees see a broader spectrum and probably a greater variety of colours than humans can. To a bee, vegetation we see as basic green would be transformed into a range of hues. Birds have the best developed colour sense of any class of animal. They see an enormous variety of colours compared to humans!

Some people's eyes have defective cone cells. Because of this defect, they have difficulty detecting some colours. This condition is known as **colour blindness**. A simple test for colour blindness is shown in Figure 9.9.

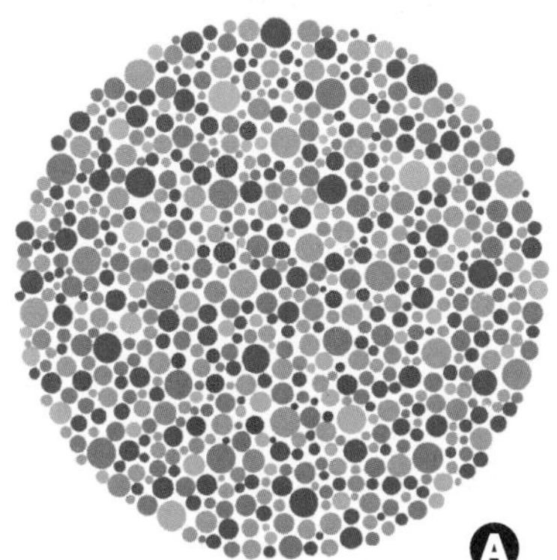

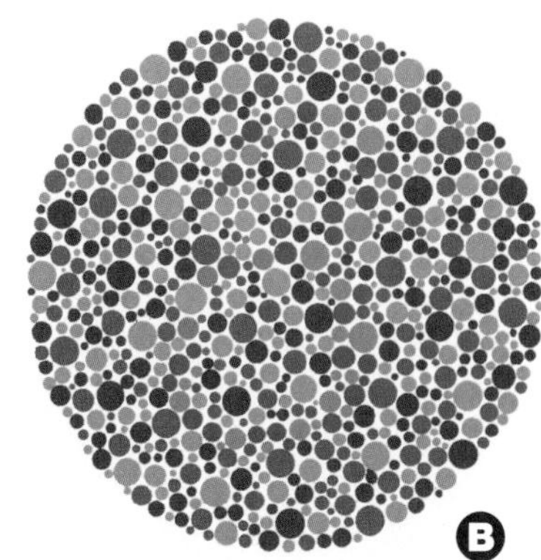

Figure 9.9 People who cannot detect the difference between green and yellow cannot see the "5" in picture A. Those who cannot detect the difference between red and green see the "8" as a "3" in picture B. People with total colour blindness cannot read either number.

Find Out **ACTIVITY**

Making a Colour Wheel

You can see how the additive primary colours blend together in this simple activity.

What You Need

circular piece of cardboard (about the size of the lid of a yogurt or margarine container)

white paper the same size and shape as the cardboard circle

red, green, and blue markers or paints

sharp pencil

tape or glue

What to Do

1. Colour the paper using red, green, and blue markers. Your paper circle should look like the one below.

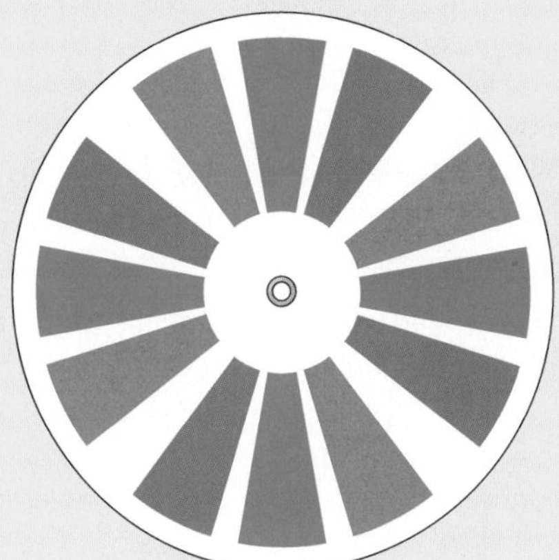

2. Attach the paper to the cardboard.
3. Using a sharp pencil, punch a hole through the middle of the disk. **CAUTION:** Use care when punching the hole. If you have difficulty, ask your teacher for help. Push the pencil through the hole and spin the disk rapidly. (For best results, do this outdoors in bright sunlight.)

What Did You Find Out?

1. What colour did you see when you spun the disk? The result should have been a fairly clear white. If your mixture looked too red, make up a second paper disk and adjust the size of the colour strips so that the red ones are smaller than the others. Repeat the activity.
2. Sometimes the white will appear as dark grey. How could you brighten the mixture?

Pause& Reflect

In your Science Log, list some reasons why people use colour in their clothing and homes. What are some ways in which animals use colour?

Subtractive Primary Colours

Now take a closer look at the secondary colours — cyan, magenta, and yellow. These colours are called **subtractive primary colours** because some portion of white light has been removed in order to produce each colour.

Because cyan is missing only red light, if you shine cyan light and red light onto the same screen, you should see white light reflected. For this reason, cyan and red are called **complementary colours**. They are "complementary" because together they form white light (see Figure 9.10). In the same way, magenta and green are complementary colours. So, too, are yellow and blue. Table 9.1 summarizes the colours that are present in subtractive primary colours, and the colour that is missing.

Table 9.1 Subtractive Primary Colours

Light colour	Includes	Missing
cyan	blue + green (B + G)	red (−R)
magenta	red + blue (R + B)	green (−G)
yellow	red + green (R + G)	blue (−B)

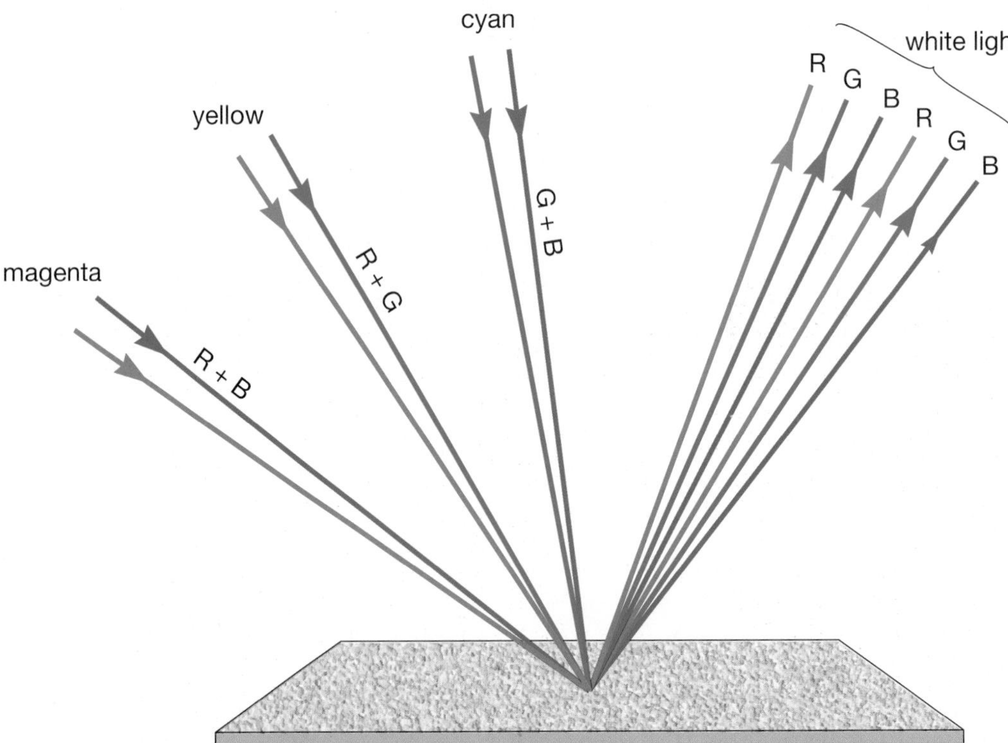

Figure 9.10 What will you see if you shine cyan, magenta, and yellow light onto the same spot on a screen? You would end up with a combination of red, green, and blue. Your eye would see the result as white light.

Using Colour Filters

Colour filters subtract colours from the light they transmit. Remember the blue bottle and why it appears blue? Suppose you combine two filters of secondary colours. For example, place a magenta filter on the stage of an overhead projector. Next, place a cyan filter so that it partially overlaps the magenta filter and turn on the projector. What colour do you see where they overlap? Assume that white light consists of red, green, and blue. The magenta will subtract the green from the light. The cyan will subtract the red. This leaves only the blue to shine through, as shown in Figure 9.11.

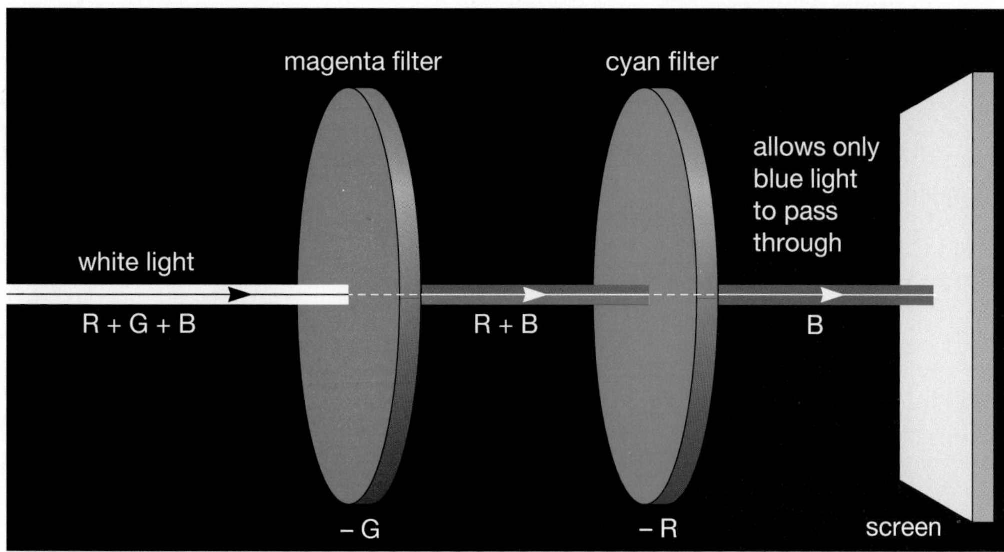

Figure 9.11 Combining magenta and cyan filters

Now suppose you used one secondary colour and one primary colour. For example, place a red filter on top of a yellow filter. What do you see? If you follow the steps in Figure 9.12, you will observe that only red can emerge from this combination. You will experiment with other combinations of filters in the next investigation.

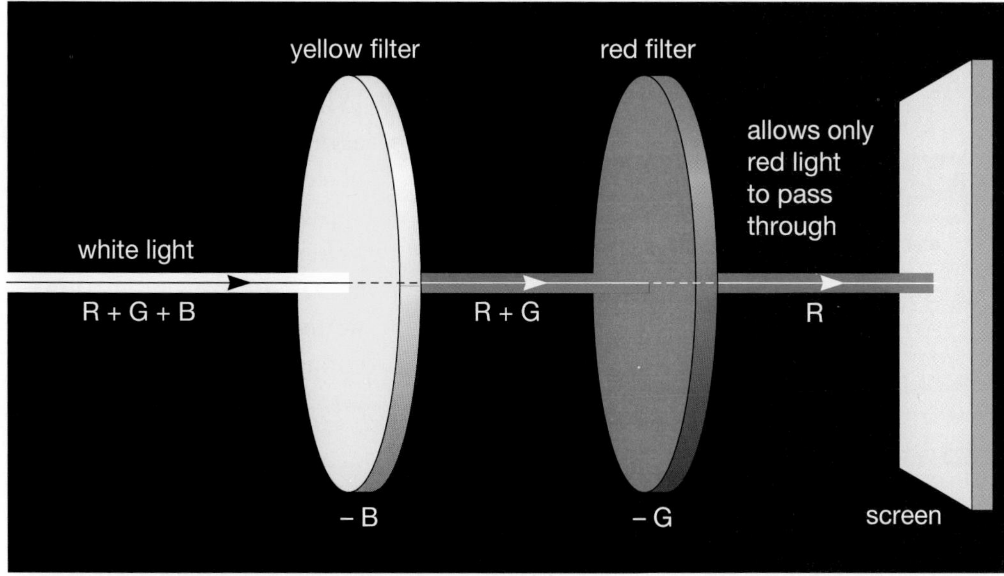

Figure 9.12 Combining yellow and red filters

Working with Filters

Staging plays and live shows such as rock concerts involves using filters placed over stage lights in order to create interesting effects. The stage manager and the lighting technician must be able to predict exactly what effects will be produced by different combinations of filters. This type of information is what you will be looking for in this investigation.

Problem

How accurately can you predict the effects produced by shining white light through different combinations of colour filters?

Materials

overhead projector or flashlight

selection of coloured cellophane or filters (red, green, blue, magenta, cyan, and yellow)

Procedure

1. Copy the table below into your notebook. You will be shining white light through two filters placed one on top of the other. There are 30 possible combinations.
2. For each pair of filters, predict what colour of light will pass through the filters. Enter your prediction in the corresponding box in the table.
3. Darken the room.
4. Turn on the overhead projector and test out each pair of filters by placing the cellophane layers on top of each other. If you are using a flashlight, layer the cellophane over the flashlight.
5. When the colour you see agrees with your prediction, place a check mark in the appropriate box in the table. When the colour does not agree with your prediction, write in the name of the colour that you do see.

Filter Predictions and Results

Colour of top filter	Colour of bottom filter					
	red	green	blue	cyan	magenta	yellow
red	XXX					
green		XXX				
blue			XXX			
cyan				XXX		
magenta					XXX	
yellow						XXX

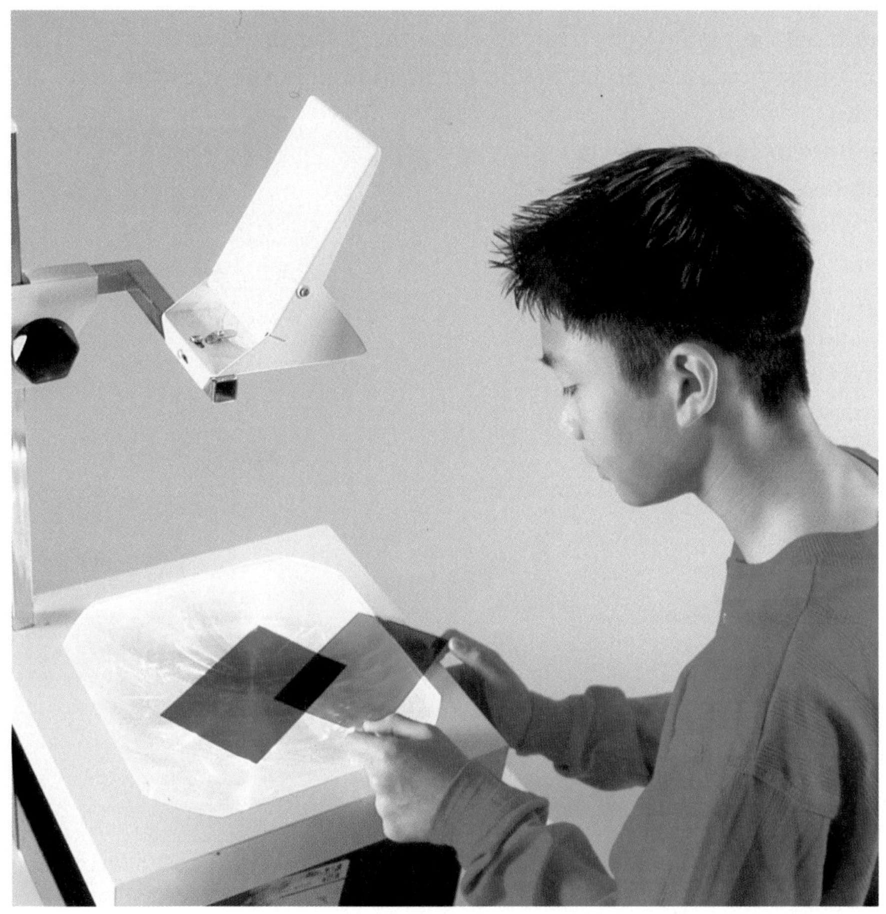

Analyze

1. Count the total number of trials and the total number that you were able to predict correctly. What percentage of the combinations did you predict correctly?

2. For which filters did you seem to have difficulty making predictions?

Conclude and Apply

3. How might you change this investigation to produce better results?

4. An actor is wearing a green costume. How will the costume look if the actor steps onto the stage when it is lit by blue light?

5. Another actor is wearing a magenta costume. How will the costume look if she steps onto the stage to join the other actor under the blue light?

Math CONNECT

Go to the mathematics books you are using to find out how to calculate percentages.

Applications of Colour Filters

Colour filters have many uses. Each time you take a colour photograph, for example, you are using filters. Photographic film is made of layers of materials that are sensitive to light. The top layer responds to blue light. Beneath this layer is a yellow filter that blocks any blue light from reaching the lower levels of the film. The bottom layers of the film are exposed by the green and red portions of the light from the scene.

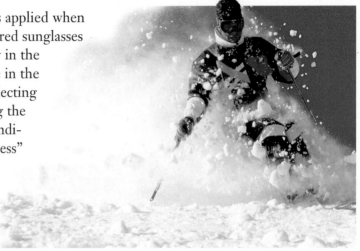

This same principle is applied when skiers wear amber-coloured sunglasses while skiing. The yellow in the glasses prevents the blue in the bright sunlight from reflecting off the snow and striking the eyes. This prevents a condition called "snow blindness" — temporary blindness caused by the glare of light reflected by large expanses of snow.

Check Your Understanding

1. (a) What is a spectrum?
   (b) List the colours of the solar spectrum, beginning with violet.
2. What common assumption about white light did Newton show to be wrong?
3. What determines the colour of an object?
4. (a) Name the three additive primary colours.
   (b) Name the three subtractive primary colours and indicate which colour is absorbed by or missing in each one.
5. Predict the effect of passing white light through the following combinations of filters. Explain your reasoning in each case by referring to additive or subtractive primary colours and their combinations.
   (a) a blue filter followed by a magenta filter
   (b) a yellow filter followed by a green filter
   (c) a cyan filter followed by a magenta filter
   (d) a yellow filter followed by a blue filter
6. **Apply** You have been given six filters — red, green, blue, cyan, magenta, and yellow. By using them individually or in combinations, how many ways can you produce blue light?
7. **Thinking Critically** Does the order in which the light passes through the filter combination affect the resulting colour? Explain your answer.

Did You Know?

Cyan, yellow, and magenta inks are used to produce the colours you see in magazines. If you look at a magazine photograph with a magnifying glass, you should be able to see tiny dots of these subtractive primary colours. These inks also provide the colour for computer printers.

9.2 A New Way of Thinking About Light

You have seen many situations in which light travels in straight lines. For example, light that shines through a rectangular window will produce a sharp, bright rectangle of light inside the room. Sir Isaac Newton tried to explain why this happens. He proposed that light beams are made of streams of extremely tiny, fast-moving particles. These tiny particles of light, he suggested, could only travel in straight lines. Newton believed that the particles could not travel around objects. Explore Newton's idea further in the activity below.

Find Out **ACTIVITY**

An Unexpected Behaviour of Light

You already know what happens to light when it passes through a regular-sized window. However, would the same result occur if the window is very small?

What You Need

showcase lamp
microscope slide
piece of aluminum foil slightly larger than the microscope slide
sharp knife such as a hobby knife
red filter or cellophane; blue filter or cellophane
ruler
coloured pencils or labels

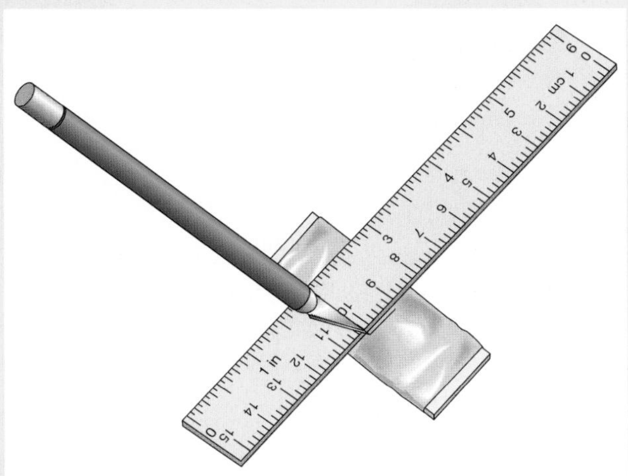

What to Do

1. Place the aluminum foil on one side of the microscope slide. Smooth the foil out and wrap it slightly around the edges of the slide.
2. Using a ruler to guide the knife, cut a slit in the foil partway across the width of the slide. **CAUTION:** Be careful when using the knife. If you cut yourself, tell your teacher immediately.
3. Make the room as dark as possible. Then turn on the showcase lamp.
4. Look through the slit at the filament of the showcase lamp. Draw a diagram showing how the filament appears. If you see colours, show them on your diagram. You may use coloured pencils or labels.
5. Hold a red filter over the slit and look at the glowing filament again. Now use a blue filter in place of the red. Draw the two patterns that you see.

What Did You Find Out?

1. Describe what you saw when you looked through the single slit without a filter.
2. What difference did it make when you replaced the red filter with a blue filter?
3. In what ways is the light doing something that you would not have expected?

Looking at Waves

Based on what you observed in the previous Find Out Activity, light does not seem to consist of speedy little particles that travel in straight lines. When light passes through a small opening, it spreads out around each side of the opening. To explain this behaviour, Dutch scientist Christiaan Huygens (1629–1695) proposed that light travels as a wave, not as a stream of particles.

Examine Figure 9.13. The high parts of the wave are called **crests**. The low parts between the crests are called **troughs**. As the wave moves to the right, the duck rises up as the crest passes. Then the duck drops down as the trough passes.

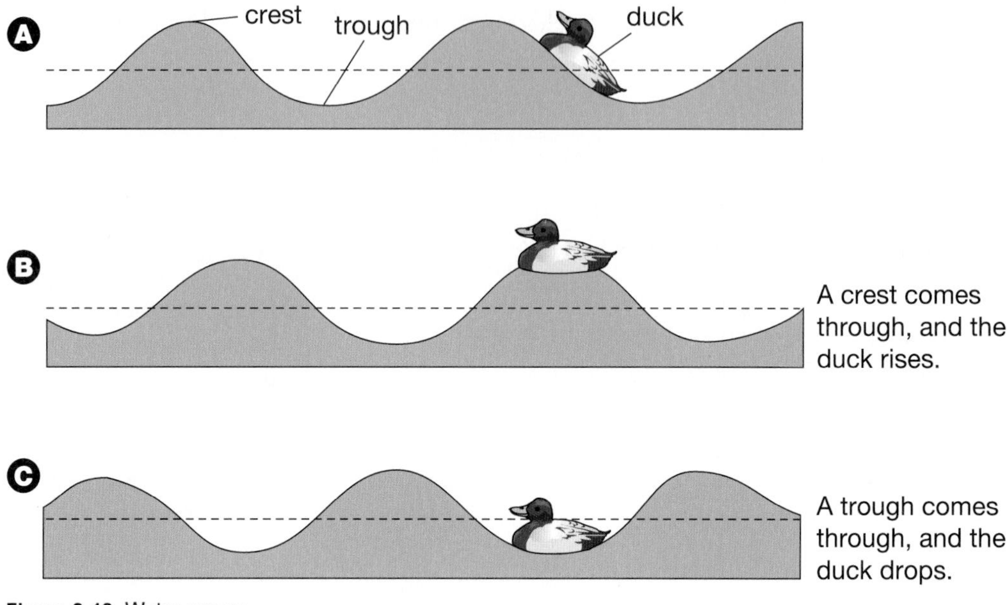

Figure 9.13 Water waves

The duck and the water around it do not move forward. They just move up and down as the wave passes. The distance from crest to crest, or from trough to trough, is called the **wavelength**. Alternatively, you can think of a wavelength as the distance covered by one complete crest plus one complete trough (see Figure 9.14).

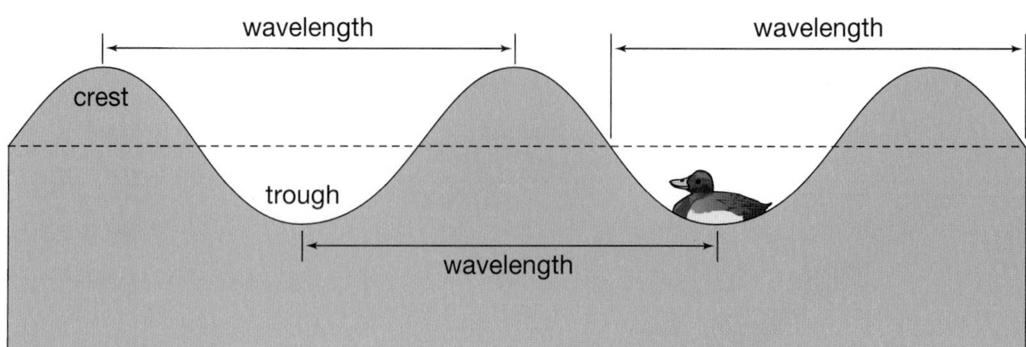

Figure 9.14 A wavelength can be measured in three different ways: the distance from crest to crest; the distance from trough to trough; or the distance covered by one complete crest plus one complete trough.

If the wind starts to blow, the water will become rougher. Soon, higher crests and deeper troughs will form. The height of the crest or the depth of the trough from the rest position is called the **amplitude**, as shown in Figure 9.15. The amplitude is a good indication of the amount of energy transmitted by the wave.

The rate at which the duck and the water move up and down is called the frequency, *f*. **Frequency** is the number of cycles completed by a vibrating object in a unit of time. Frequency is usually measured in **hertz**, or cycles per second. If something vibrates 20 times in a second, its frequency is said to be 20 hertz (20 Hz).

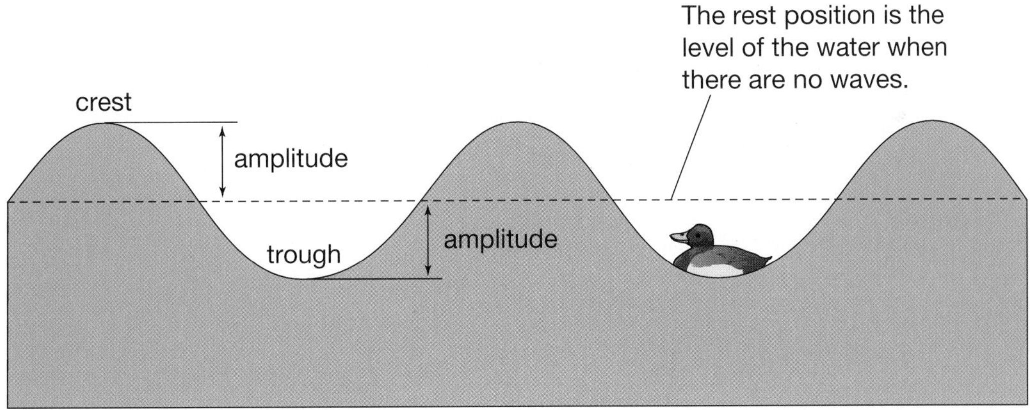

Figure 9.15 The amplitude of a wave indicates the height of the crests or the depth of the troughs from the rest position.

The Wave Model of Light

Scientists have developed a model of light by looking at how light behaves, and then trying to explain what they see. As you learned in earlier science studies, a model is a way of representing something in order to try to understand it better and to make predictions. One of the explanations of light behaviour is the **wave model of light**, which pictures light travelling as a wave. This model does not explain everything about how light behaves, but it helps explain a great deal. Thinking about light in terms of waves helps explain otherwise unpredictable behaviour, such as why light curves around an opening. Just as for water waves, the distance between the crests or the troughs of a wave of light as it travels through space is called a wavelength.

When light passes through a small opening, the waves spread out. How far the waves spread out depends on the wavelength of the wave. Waves with short wavelengths spread out very little. Waves with long wavelengths spread out more. Explore the relationship between frequency and wavelength in the next investigation. Then discover the relationship between wavelength and colour in Conduct an Investigation 9-D on page 286.

CONDUCT AN

INVESTIGATION 9-C

Exploring Frequency and Wavelength

If you observe a duck floating on the water, you might notice that when the crests of the waves are far apart, it takes longer for the duck to bob up and down. However, when the crests are close together, the duck bobs up and down much more frequently. This suggests that there is a relationship between frequency and wavelength. You will examine this relationship in this investigation.

Problem

What is the relationship between the frequency and the wavelength of a wave?

Apparatus

slinky-like coil

overhead projector and screen

transparent container with a flat bottom, such as a casserole dish

pencil

Materials

piece of tape or string

water

Procedure

Part 1

Waves on a Coil

1. Work with a partner. Attach a piece of tape or string at about the halfway mark of the coil.
2. Hold each end of the coil and stretch it out along the floor. Be careful not to stretch it too much. These coils can be damaged easily. **CAUTION:** Handle the coil carefully to prevent it from "knotting up."
3. Hold one end of the coil firmly in place as your partner moves the other end of the coil slowly from side to side. Draw a diagram of the appearance of the wave that results. Label the diagram "low-frequency wave." Indicate the wavelength on your diagram. Use arrows to show the directions in which the marked coil moves.

4. Repeat step 3 but have your partner move the end of the coil quickly from side to side to provide a higher frequency. Draw a sketch of the appearance of the resulting wave. Indicate the wavelength. How might you label this diagram?

Part 2

Waves in Water

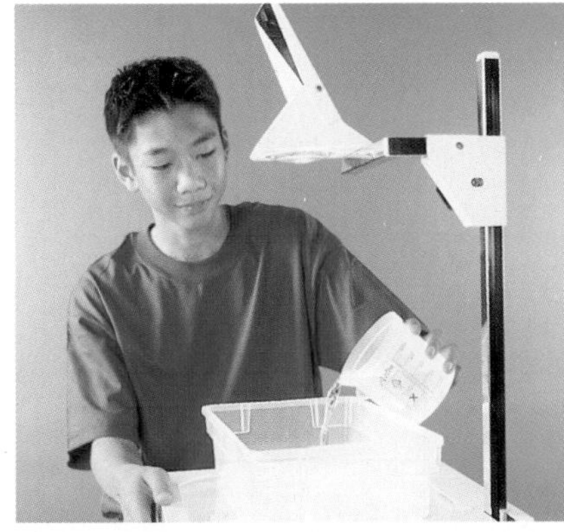

1. Carefully place the container on the stage of the overhead projector. Fill the container with water to a depth of about 1 cm.

2. Dip the eraser end of the pencil into the water to create ripples. Focus the projector until the ripples are sharp.

3. Move the pencil slowly up and down in the water. The ripples should appear as light and dark rings on the screen. The bright rings are crests and the dark rings are troughs. You have probably noticed that the troughs are easier to detect than the crests. Draw a diagram of the pattern you see. Label your diagram "low-frequency waves."

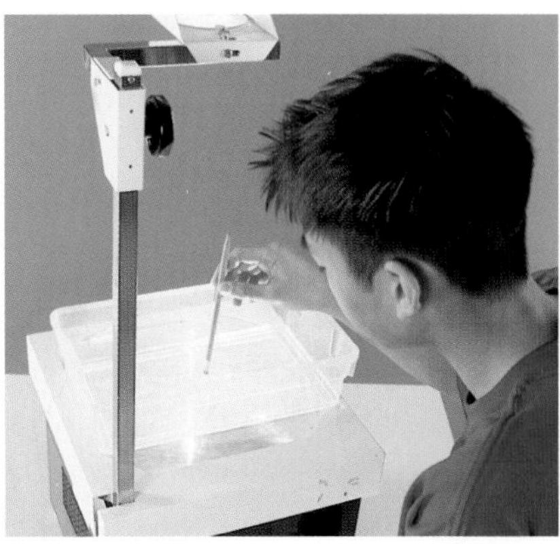

4. Now move the pencil up and down at a faster rate. Draw a diagram of the dark rings you observe. How might you label this diagram?

Pause& Reflect

A female soprano sings a higher frequency (higher pitch) note than a male baritone. Which singer is producing waves of longer wavelength? Write your answer in your Science Log.

Analyze

1. What happened to the wavelength of the coil when you moved the coil more quickly from side to side?
2. As the coil wave travelled from one student to the other, in which direction did the marked coil move?
3. What do you call the distance between the dark rings that you saw on the overhead projector?
4. What happened to the wavelength of the water waves when you moved the pencil up and down at a faster rate?

Conclude and Apply

5. What happens to the size of the wavelength when the frequency increases?
6. What happens to the size of the wavelength when the frequency decreases?
7. In general, how is frequency related to wavelength?

Why Are Colours Different?

The wave model of light provides an answer to one of the original questions that began this chapter. What makes blue light different from red light? You will discover the answer in this investigation.

A diffraction grating consists of a piece of glass or plastic made up of many parallel slits. Diffraction gratings commonly have as many as 12 000 slits per centimetre. If light travels as a wave, this grating will intensify the effects you saw previously when you observed light through a single slit in aluminum foil.

Problem

If light behaves like a wave, how is the wave for one colour different from the wave for another colour?

Apparatus

diffraction grating (or rainbow glass)
showcase lamp

Safety Precautions

Never look directly at the Sun. Damage to your eyes could result.

Procedure

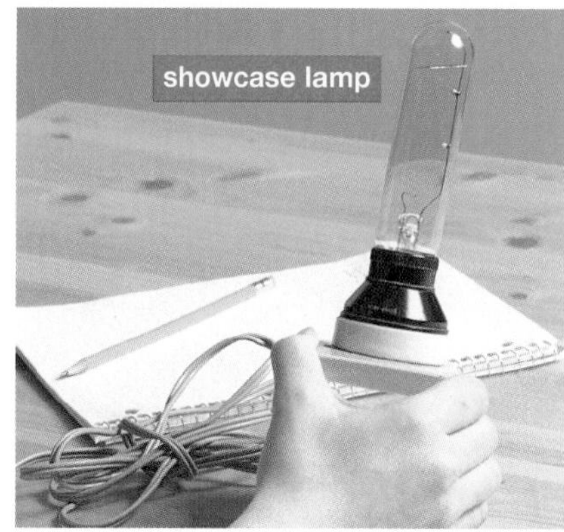

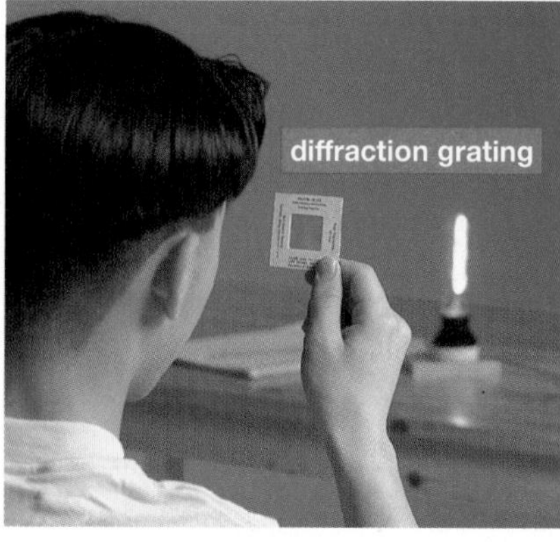

1. Set up a showcase bulb so that its filament is vertical.
2. Darken the room and turn on the lamp.
3. Position yourself a few metres away from the showcase lamp.
4. Hold the diffraction grating so that the slits are vertical. Look at the lamp through the diffraction grating. You may notice a pattern that repeats itself. Just concentrate on the patterns immediately to the left and right of the bright filament. Record your observations as a diagram. Label any colours you see.

The colour with which a heated object glows can indicate its temperature. Astronomers use this knowledge to study stars. Some stars glow red or yellow, while others glow white. Conduct research at your library or search the Internet to discover how astronomers use colour to determine the temperature and the age of stars.

Analyze

1. In what way is light behaving like a wave in this investigation?
2. You learned that waves with longer wavelengths spread out more than waves with shorter wavelengths. Based on that information and your observations in this investigation, which colour of light has the longest wavelength? Which colour has the shortest wavelength?

Conclude and Apply

3. Copy the following diagram into your notebook. Try to place the colours that you saw onto the diagram in the same way they appeared to you.

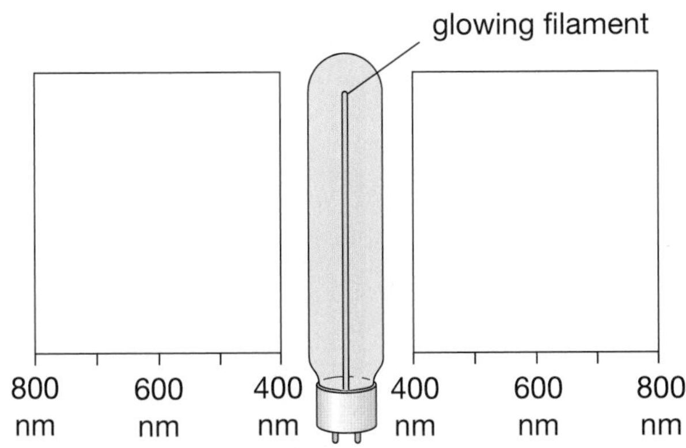

Note: The unit of length used here is the *nanometre* (nm). 1 nm = 0.000 000 001 m.

4. Based on your observations, state the wavelengths for
   - (a) red light
   - (b) orange light
   - (c) yellow light
   - (d) green light
   - (e) blue light

Extend Your Knowledge

5. Use a sharp pencil to punch a hole in the middle of a large piece of cardboard. **CAUTION:** Be careful when punching the hole. If necessary, ask your teacher for help. Allow sunlight to pass through the hole and fall on a piece of white paper. Look at the bright spot through the diffraction grating. **CAUTION:** Do not look directly at the Sun.
   - (a) List the range of colours that you observe in the solar spectrum.
   - (b) Which colour is closest to the bright spot?
   - (c) Which colour is farthest from the bright spot?
   - (d) Which colour or colours seem to be the brightest?
   - (e) Are there any major differences between the spectrum produced from the showcase bulb and the solar spectrum?
6. Look through the diffraction grating at a glowing fluorescent tube.
   - (a) List the range of colours that you observe in the fluorescent spectrum.
   - (b) Which colour is closest to the glowing tube?
   - (c) Which colour is farthest from the glowing tube?
   - (d) Which colour or colours seem to be the brightest?
   - (e) Are there any major differences between this spectrum and the others that you have observed?

Light Waves in Action

We can explain sunsets such as the one in Figure 9.16 using the wave model of light. As light waves from the Sun travel through Earth's atmosphere, they strike particles of different sizes, including dust and grit. The longer wavelengths of the reds and oranges tend to pass around these particles. The shorter wavelengths, especially blue and violet, strike the particles and reflect and scatter off them, as shown in Figure 9.17.

Figure 9.16 A sunset near Sooke, British Columbia. As the golden disk of the Sun sinks below the horizon, low-lying clouds light up with shades of orange, red, and purple. Where do these colours come from?

At sunset, the sunlight we see passes through about 700 km of Earth's atmosphere. At this time of the day, the sunlight passes through many more particles than earlier in the day. This allows plenty of opportunity for many of the blue and violet waves to be reflected away. Red and orange waves remain to colour the clouds of the sunset.

Some of the waves with shorter wavelengths are reflected into space. However, if you look straight overhead at sunset, you will see the blue light that has been reflected toward you.

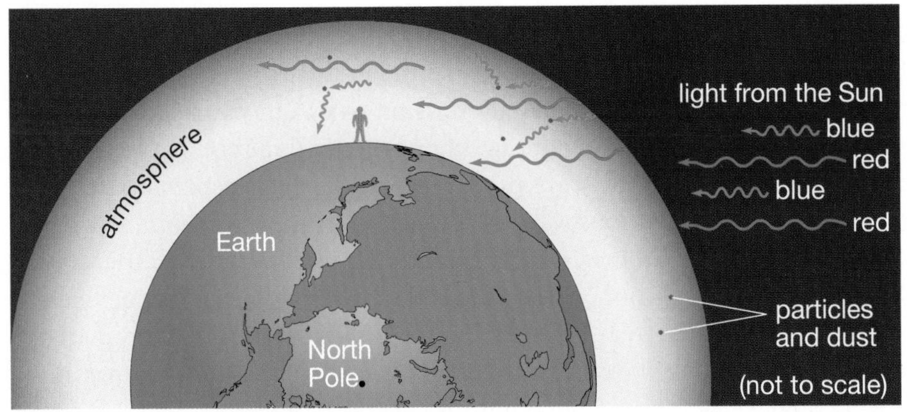

Figure 9.17 Some wavelengths of light can pass around particles in Earth's atmosphere; other wavelengths are reflected and scattered by them.

DidYouKnow?

Volcanic eruptions throw tonnes of dust and grit into the atmosphere. This often results in spectacular sunsets for months after an eruption.

Making Sunsets in the Classroom

If you can put an appropriate number of particles in the path of light, you can mimic the wave effects of blue skies and sunsets. In this activity, you will pass light through water that contains a small amount of powdered coffee whitener.

What You Need

powdered coffee whitener
spoon
large beaker or tall glass
overhead projector
screen
cardboard large enough to cover the stage of the overhead projector
scissors
water

What to Do

1. Cut a hole in the cardboard just slightly smaller than the base of the beaker or glass. **CAUTION:** Be careful when using sharp objects such as scissors.
2. Place the cardboard on the stage of the overhead projector.
3. Almost completely fill the beaker with water, and place the beaker over the hole in the cardboard.
4. Turn on the projector and focus the light onto a screen. Observe the appearance of the water as light passes up through it. Also observe the colour of the light on the screen.
5. Add a tiny amount of the whitener to the water and stir. Describe any changes in the appearance of the water and in the light on the screen.
6. Try adding different amounts of whitener until you obtain the best effect.

What Did You Find Out?

1. Were you successful in separating the reds and the blues?
2. How does this demonstration relate to blue skies and red sunsets?
3. **Apply** Sunrises do not generally show as much colour as sunsets. What does that tell you about the air quality at night compared to daytime? Keep in mind that the light at sunrise has passed through early morning air. The light at sunset has passed through daytime air.

Laser Light

In 1966, scientist Theodore H. Maiman became the first person to use a process called **l**ight **a**mplification by the **s**timulated **e**mission of **r**adiation, or **laser** light. Laser light is quite different from the other types of light you have studied. However, an understanding of the wave model of light helped scientists develop laser light and its many applications.

As you have already seen, an incandescent light bulb gives off many different colours. In terms of the wave model, this means that many different frequencies and wavelengths are produced, as shown in Figure 9.18A. Moreover, the waves are all jumbled. Crests from one wave might overlap a trough from another, making the waves work against each other. Such light is called **incoherent**.

A laser, on the other hand, emits waves with only one frequency or wavelength, as shown in Figure 9.18B. In addition, all the waves line up to work together. Such light is called **coherent**.

Figure 9.18A An incandescent bulb gives off many frequencies of incoherent light.

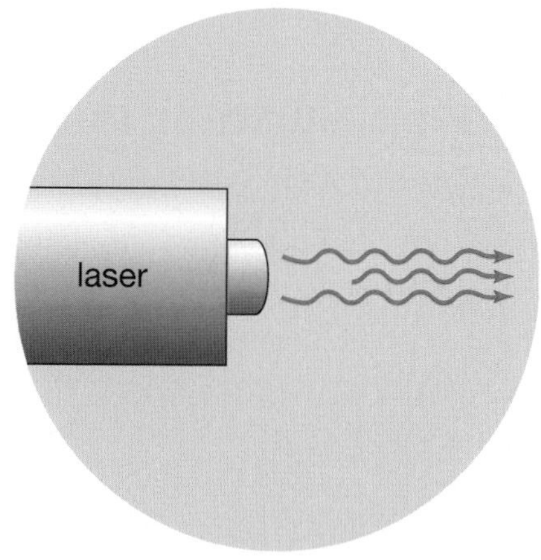

Figure 9.18B A laser gives off a single wavelength (frequency) of coherent light.

Scientists are using laser light to trap and cool tiny particles of matter. Why? Many properties of matter change at temperatures near the coldest temperature possible, which is absolute zero (–273.15°C). Physicists are trying to bring small samples of matter to the coldest possible temperature, in order to better understand magnetism and the behaviour of particles in matter. They can bring matter to extremely low temperatures by using special lasers to decrease the vibrations of the particles that make up the material.

Lasers are used for many purposes. They help read bar codes at supermarket checkout counters. They sense tiny pits on a compact disc to play music. You may have seen police officers using lasers to measure the speed of cars. Doctors use lasers as scalpels when performing surgery on delicate tissues such as the eye. On a larger scale, infrared (heat) lasers are used in industry to make precise cuts through metal. Figures 9.19A and B show how lasers are used in fibre optics.

DidYouKnow?

Just one optical fibre can carry thousands of phone conversations at the same time.

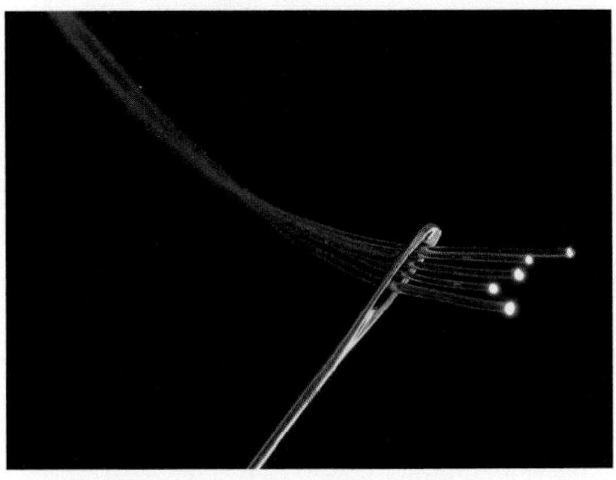

Figure 9.19A Because of its strength and coherence, laser light can be used to carry information over long distances through fibre optic cables. Such cables use the laws of reflection to carry light signals.

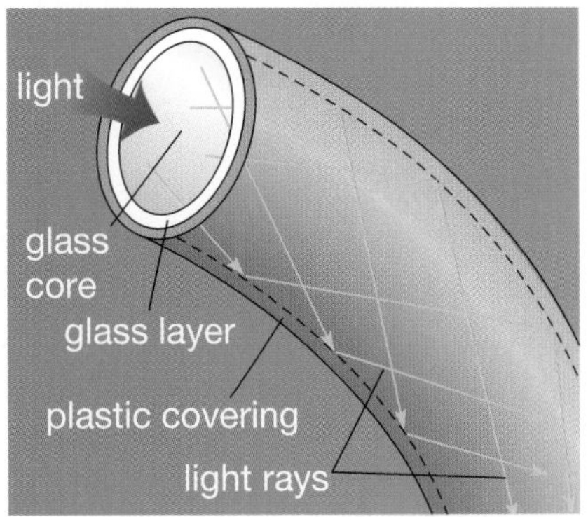

Figure 9.19B An optical fibre is designed to reflect light so that it is piped through the length of the fibre without escaping, except at the ends.

Career CONNECT

Does This Job Appeal to You?

When Ron Kumabe splices two pieces of glass fibre optic cable together, he does it very carefully. First, he has to join the two plastic-coated cylinders precisely or not enough light will bounce through the cable to carry the telephone company's communication signal. Second, if he accidentally bends the cable, it will break and be useless. In addition to these difficulties, Ron must do this tricky work outdoors, at the location where the newly joined cables will be buried underground.

As a fibre optics splicer, Ron does the following:

- works with his hands
- is outdoors much of the time
- uses specialized tools
- works as part of a team
- provides a service to a customer
- must be careful and precise in his work
- is not required to dress in business clothes
- works in a new location with each new customer

Based on the statements above, does this sound like the job for you? Do some of the aspects of this job match your own career expectations? For each job feature listed, decide if you consider that aspect of Ron's job desirable or not. If it is not desirable, decide what you would prefer. Write your own career wish list based on your responses, beginning "I would enjoy a job in which I…."

Across Canada

When John Polanyi was young, he was more interested in politics and poetry than in science. Following the path of his father, who was a professor of chemistry and philosophy, Polanyi discovered he had a talent for science and quickly became fascinated by the energy given off by chemical reactions.

When Dr. Polanyi began working at the University of Toronto in 1956, he measured the invisible light produced by chemical reactions. His discoveries eventually led to the development of lasers. In 1965, Polanyi made a new kind of laser that used energy from chemicals to emit infrared rays (you will learn about infrared radiation in the next section). Today, as you have already learned, lasers are used in many applications, such as surgery, bar codes, and metal cutting.

In 1986, Dr. Polanyi was awarded the Nobel prize in chemistry for his research into the behaviour of particles during chemical reactions. In a 1996 speech, he said, "Discoveries don't come from well-oiled machines, but from scientists." He stresses that imagination plays a key role in scientific discoveries. He also places value on asking questions and searching constantly for answers.

John Polanyi

Check Your Understanding

1. Draw a wave with a wavelength of 4 cm and an amplitude of 1 cm. Label a crest, a trough, the amplitude, and the wavelength.
2. (a) A buzzer vibrates 900 times in 1 s. What is its frequency?
   (b) A guitar string vibrates 880 times in 2 s. What is its frequency?
   (c) A ball bounces on the floor 10 times in 50 s. What is its frequency?
3. If the crest of a wave stretches over 3 cm, how long is the wavelength?
4. (a) Describe one way in which light behaves like a wave instead of a particle.
   (b) Why do scientists prefer to talk about a wave model of light instead of saying that light *is* a wave?
5. **Apply** A woman went into a store to buy purple shoes to match her purple purse. Inside the store she found shoes that matched perfectly. However, once she got outside, she found that they no longer matched well. Why might this have happened?

9.3 Beyond Light

Earlier in this unit, you learned that the Sun is the most abundant source of light on Earth. However, there is far more to sunshine than meets the eye! In addition to the visible energy that we call light, the Sun also radiates invisible energy. The light we can see is just a tiny band of a much broader spectrum of visible and invisible energy.

You have seen how water waves can be used to represent how light moves through space. However, light is a very different kind of wave from those that travel through water. In a water wave, water particles vibrate up and down as the wave passes through the water. In a light wave, electrical and magnetic fields vibrate. As a result, light is classified as electromagnetic radiation. Visible light energy and all the invisible forms of radiant energy exist on the **electromagnetic spectrum**, as shown in Figure 9.20.

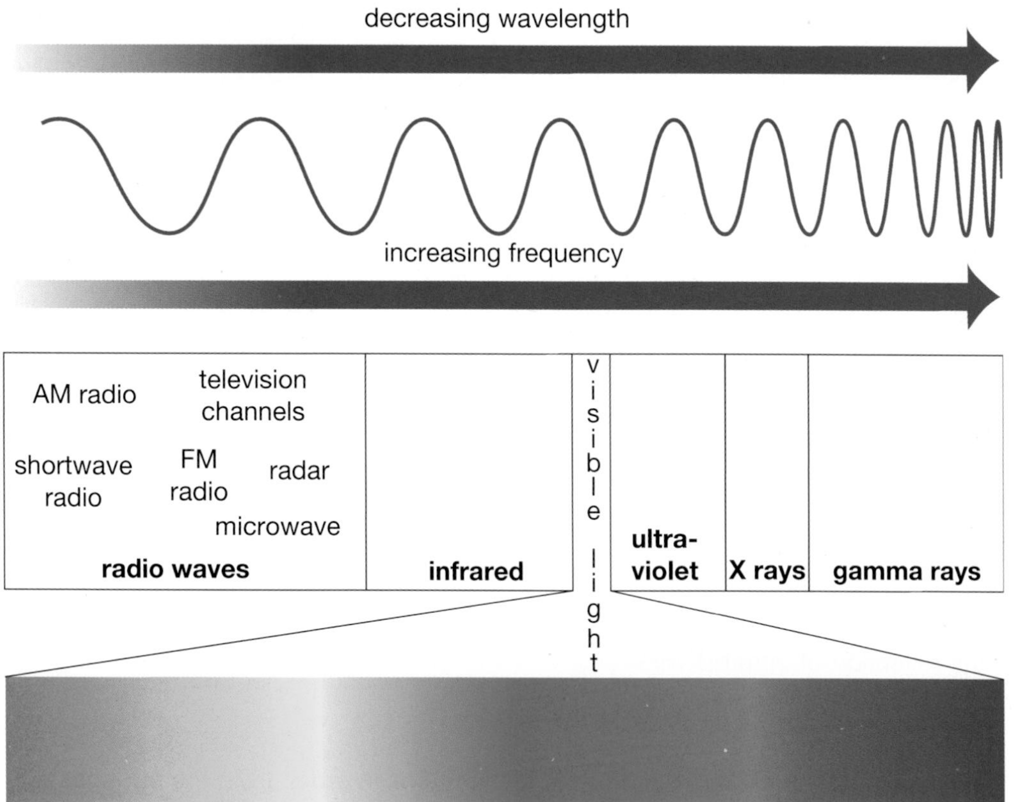

Figure 9.20 The electromagnetic spectrum

Different colours of light represent different frequencies and wavelengths of light. Table 9.2 shows the wavelength and frequency of the colours in the visible spectrum. In the rest of this section, you will examine more closely the bands of invisible energy beyond the red and violet ends of the visible light spectrum.

Table 9.2 Wavelength and Frequency of Colours in the Visible Spectrum

Colour	Wavelength in vacuum (nm)	Frequency (Hz)
red	700	4.3×10^{14}
orange	600	5.0×10^{14}
yellow	580	5.2×10^{14}
green-blue	500	6.0×10^{14}
violet	400	7.5×10^{14}

Math CONNECT

The frequency of red light is given as 4.3×10^{14} Hz. This method of expressing a large number is called *scientific notation.* The number 10^{14} means $10 \times 10 \times 10 \times 10 \times 10 \times 10 \times 10 \times 10 \times 10 \times 10 \times 10 \times 10 \times 10 \times 10$. How would you express the number 5 000 000 000 in scientific notation?

Skill POWER

To learn more about scientific notation, turn to page 537.

DidYouKnow?

A nanometre (nm) is equal to 0.000 000 001 m. Expressed in scientific notation, this is 1.0×10^{-9} m. Red light has a wavelength of 700 nm or 0.000 000 7 m (7.0×10^{-7} m). The nanometre is so tiny that about 500 000 of them would fit across the thickness of a sheet of paper.

Figure 9.21 Infrared lamps called "brood lights" are used to keep baby chicks warm.

Infrared Radiation

If you could take red light, which has a wavelength of about 700 nm, and somehow stretch it out to 1000 nm, it would no longer be light. It would be heat radiation or **infrared radiation**. Your eyes would no longer see it, but your skin would sense it. You can feel infrared radiation when you bring your hand close to the side of a cup of hot chocolate.

There are many uses for infrared radiation. Because infrared radiation is heat radiation, anything that is warmer than its surroundings emits infrared rays. Based on this understanding, motion sensors and burglar alarms have been developed to detect a change in the infrared levels in an area if a person is present. Some restaurants use infrared heat lamps to keep food warm. Figure 9.21 shows an infrared lamp being used to warm a baby chick.

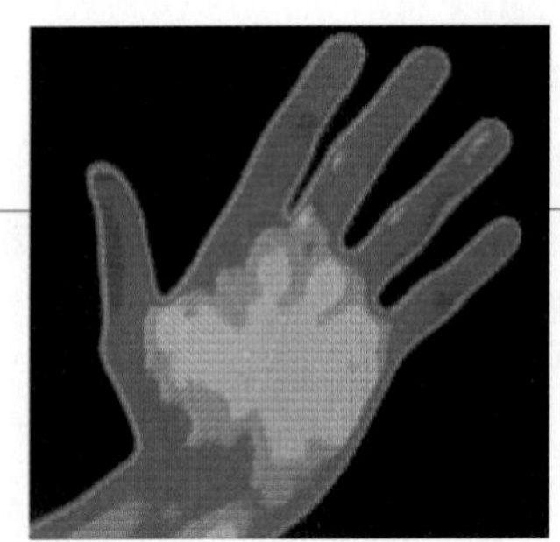

A thermogram of a hand, like the one shown here, can help diagnose circulation problems by detecting infrared radiation. Warmer areas show up as yellow, orange, and red. Cooler areas are green, blue, and black. Would the problem area show up as warmer or cooler than the other parts of the hand?

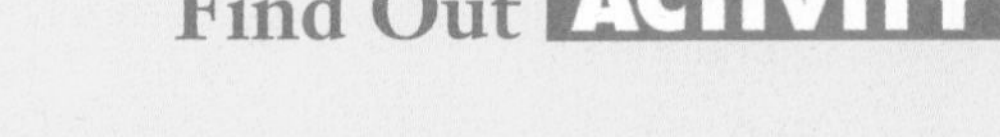

Infrared Reflections

You have just learned that infrared radiation is like light. As a result, you would expect that infrared rays should behave like light. Modern remote control units use infrared radiation to send a signal to the receiver in a television or VCR. In this activity, you will test the reflection of infrared rays off various surfaces.

What You Need

television set or VCR with a remote control unit

several different types of surfaces (for example, a smooth surface, a rough surface, a metallic surface, and a non-metallic surface)

What to Do

1. Use each surface to try to reflect radiation from the control unit toward the television or VCR.
2. Aim the control unit away from the television or VCR and hold one of the surfaces in front of it. Be sure to stand slightly to one side so that you do not block the reflected radiation.
3. Repeat step 2 for each type of surface.
4. Observe which kind of surface provides the strongest reflection.

What Did You Find Out?

1. Does infrared radiation reflect from all the surfaces?
2. List the surfaces in order, from the strongest reflector to the weakest reflector.
3. **Thinking Critically** Do the results of this activity support the prediction that infrared radiation should behave like light?
4. **Design Your Own** You have seen that light spreads out slightly when it passes through a small opening. Will infrared radiation behave in a similar way? Design an experiment to test this possibility. Decide on a control and which variable(s) you will investigate. Show your experimental design to your teacher.

Radio Waves

If you could stretch the infrared wave out again even farther so that the wavelength becomes a few millimetres long, you would begin to obtain **radio waves**. Radio waves have a longer wavelength and a lower frequency than visible light. Different types of radio waves have different uses.

Microwaves have the shortest wavelength and the highest frequency of all the radio waves. Microwave ovens use a specific wavelength or frequency of microwave that is strongly absorbed by water particles. When the water particles in the food absorb microwaves, they begin to vibrate quickly and become hot. Only foods that contain water particles can be heated using microwaves.

Microwave frequencies are also used in telecommunications (see Figure 9.22). Microwaves can be transmitted to telecommunications satellites that orbit Earth. The satellites receive microwave signals, amplify them, and retransmit them to a new location.

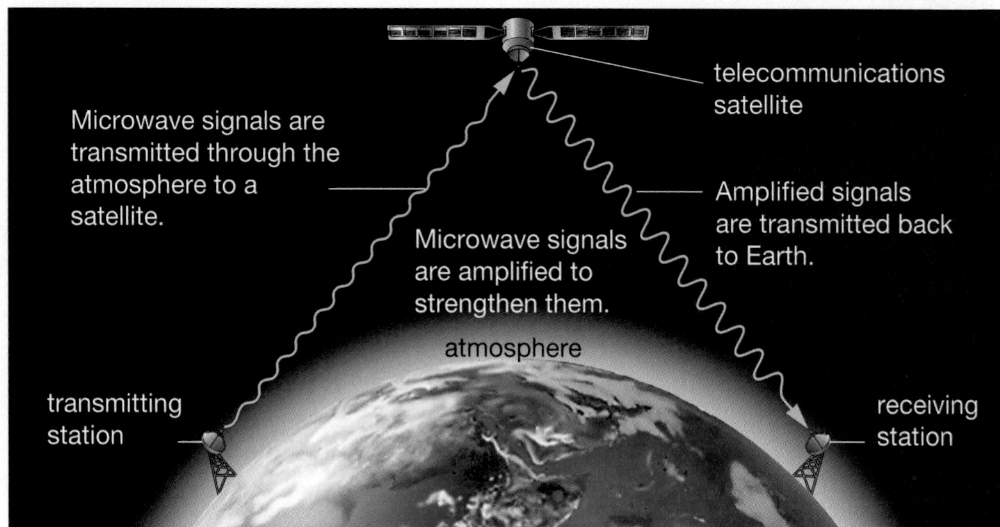

telecommunications satellite
Microwave signals are transmitted through the atmosphere to a satellite.
Amplified signals are transmitted back to Earth.
Microwave signals are amplified to strengthen them.
atmosphere
transmitting station
receiving station

Figure 9.22 Signals sent by satellites can travel vast distances. One satellite can replace many ground-based relay stations. (This illustration is not drawn to scale.)

Microwaves are also used in **radar**. In this case, microwaves are beamed out through the air. The waves that reflect from an object tell the radar operator the location and speed of the object. As Figure 9.23 shows, air traffic controllers depend on radar to guide aircraft.

Radio waves with longer wavelengths than microwaves are used to broadcast radio and television programs. Most radio waves are harmless. In fact, radio waves are streaming through you right now. However, it is not known for certain whether long-term exposure to sources of radio waves could be harmful to people (see Figure 9.24).

Figure 9.23 Air traffic controllers use radar to guide airplanes safely during take-offs and landings.

Figure 9.24 Scientists are studying whether there is any danger from long-term exposure to radio waves, especially those emitted by electronic equipment such as computers, microwave ovens, and televisions. So far, studies have not yielded clearcut evidence of harm.

Ultraviolet Radiation

At the violet end of the electromagnetic spectrum, wavelengths of 200 nm are known as **ultraviolet (UV) radiation**. This radiation is extremely energetic. It causes tanning, which is the skin's way of trying to protect itself from the ultraviolet waves. Ultraviolet radiation can damage the cornea, the front surface of the eye. This can lead to a fogging of the cornea, causing a slow loss of vision. As Figure 9.25 shows, sunglasses that block ultraviolet radiation will protect the cornea from damage.

In recent years, more and more ultraviolet radiation from the Sun has been reaching the surface of Earth. This is due to a decrease in the Earth's **ozone layer**. Ozone, a form of oxygen, is located in Earth's atmosphere about 20–25 km above the ground (see Figure 9.26). When UV radiation from the Sun reaches the atmosphere, ozone absorbs much of the UV radiation, preventing it from reaching Earth's surface.

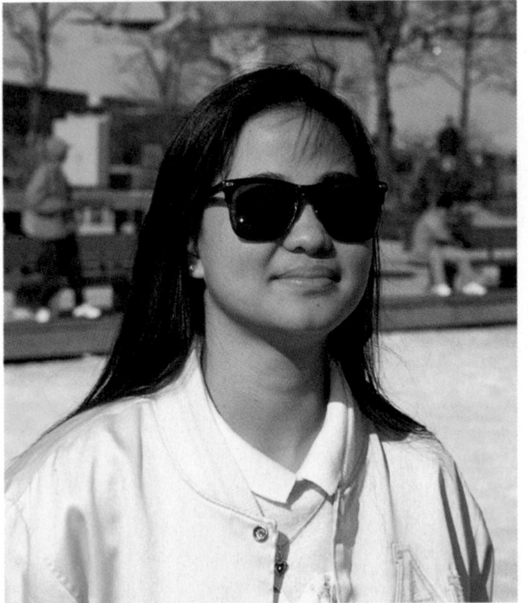

Figure 9.25 You can prevent damage to your skin from sunlight by wearing light clothing that covers your arms and legs and by using sunscreens. Sunglasses that block ultraviolet radiation can help protect your eyes.

However, Earth's ozone layer appears to be thinning. Some chemicals used in aerosol spray cans are able to rise through the atmosphere and break down the ozone particles. Freon, a gas once frequently used in refrigerators and air conditioners, can also destroy ozone particles. Fortunately, many spray products such as deodorants and hair sprays are now manufactured without ozone-destroying chemicals. Also, the use of Freon has been banned in many countries.

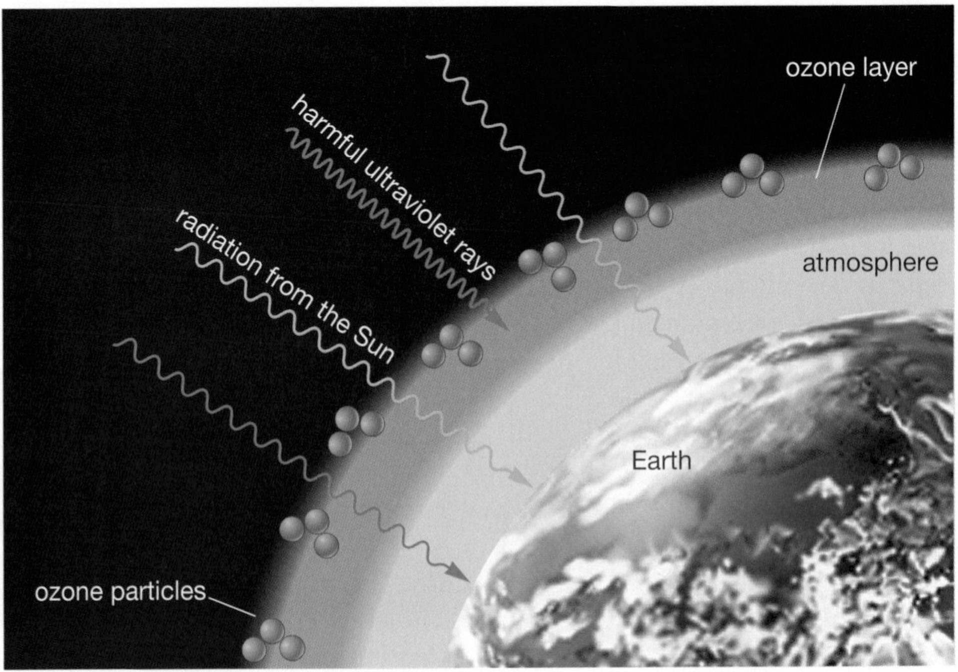

Figure 9.26 Earth's ozone layer (not to scale)

Did You Know?

Bumblebees can see ultraviolet radiation, which helps them detect patterns on flowers.

X rays

At even shorter wavelengths and higher frequencies on the radiant energy spectrum are **X rays**. These waves are very penetrating and extremely energetic. X rays pass easily through tissue such as skin and muscle. However, X rays are absorbed by bone (see Figure 9.27).

People who work with X rays protect themselves from harmful radiation by leaving the room while the equipment is being used. When a dentist takes an X ray of your teeth, he or she places a shielding pad on your body to protect you, and then leaves the room before operating the X-ray machine.

INTERNET CONNECT

www.school.mcgrawhill.ca/resources/

Search the Internet to explore how infrared waves, ultraviolet waves, and X rays help investigators solve crimes. Visit the above web site. Go to **Science Resources**. Then go to **SCIENCEPOWER 8** to find out where to go next.

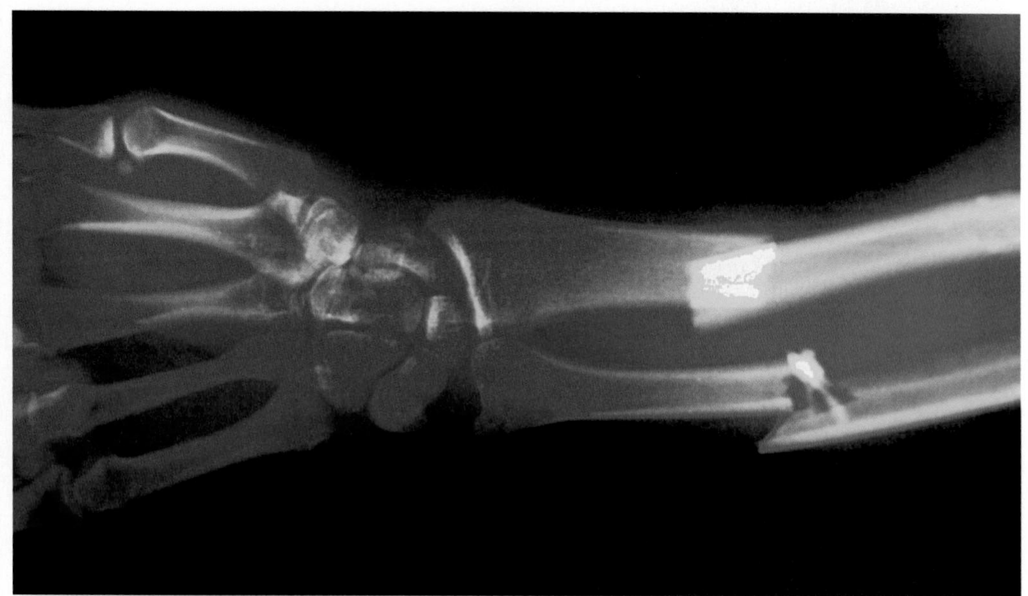

Figure 9.27 X rays are commonly used to locate a break in a bone, such as this forearm fracture.

Across Canada

When astronomers spot a new comet with their telescopes, University of New Brunswick physicist Dr. Li-Hong Xu knows her telephone may start ringing. Colleagues around the world will be asking her if the comet's infrared light shows the telltale signature of her favourite substance, methanol. In outer space, methanol can be found in clouds between stars, in the birthplaces of new stars, and in comets. The vibrations of the particles in methanol produce infrared light patterns, which Dr. Xu studies with spectroscopes (instruments used to observe a spectrum of light). The results of her studies help astronomers determine how cold and thin the outer space material is.

Methanol is important not only in space, but also on Earth; for example, as a race car fuel and as a raw material for making plastics. Thus, Dr. Xu also receives calls from various industries seeking her advice about applications for methanol.

Dr. Li-Hong Xu

Born in Suzhou, China, a city famous for its gardens and scholars, Li-Hong Xu grew up in the turmoil of the Cultural Revolution of the late 1960s. Because of the social and political upheaval, there were often no regular classes in schools. However, Li-Hong studied on her own by borrowing textbooks from her friends' older sisters and brothers. Her advice to Canadian students is, "Take advantage of the stable environment in Canada to learn all you can; you never know whether you will get a second chance. Knowledge is a valuable possession that will accompany you throughout life."

Gamma Rays

Gamma rays, as shown in Figure 9.20 on page 293, have the shortest frequency and the highest energy of all the radiant waves in the electromagnetic spectrum. Gamma rays result from nuclear reactions and can kill cells. However, sometimes the ability to destroy cells is useful. When doctors discover a cancerous tumour in a patient's body, they may decide to destroy the tumour with gamma rays. This is known as **radiation therapy**.

Pause& Reflect

In your Science Log, make a list of all the forms of invisible radiant energy that you have learned about, in order of increasing frequency. List one application for each form of radiant energy.

As you have learned, visible light represents only a small part of the electromagnetic spectrum. We are surrounded by energy waves. However, many of these waves are invisible. They exist in frequencies we cannot see. Nonetheless, these invisible waves serve important roles, from transmitting music to our homes to saving lives in hospitals.

Figure 9.28 People who work with gamma rays must use shielding to protect themselves from the radiation. In addition, they wear a badge, like the one shown here, that registers how much exposure they have had to gamma rays.

Check Your Understanding

1. List five types of radiation other than visible light.
2. (a) Which part of the electromagnetic spectrum has the highest frequency?
   (b) Which part of the electromagnetic spectrum has the lowest frequency?
3. Identify a use for
   (a) X rays
   (b) microwaves
4. (a) Which type of radiation from the Sun causes tanning?
   (b) Why is the ozone layer important to us?
5. Convert the following to metres.
   (a) 100 nm
   (b) 900 nm
6. **Apply** What can you do to reduce the damage to the ozone layer?
7. **Thinking Critically** Express the numbers in question 5 above in scientific notation.

CHAPTER at a glance

Now that you have completed this chapter, try to do the following. If you cannot, go back to the sections indicated.

Explain why objects appear to be coloured. (9.1)

Identify the three additive primary colours, the three subtractive primary colours, and complementary colours. (9.1)

Explain how additive primary colours are used in stage lighting and in colour television. (9.1)

Explain how the human eye perceives colours. (9.1)

Identify the colours in the spectrum of white light. (9.1, 9.2)

If the walls of this experiment box were lit by red light instead of white light, what colours would the walls appear to be? (9.1)

State how the wavelength changes when the frequency of a wave changes. (9.2)

Describe evidence that shows that light behaves as a wave. (9.2)

Explain the difference between colours in terms of frequencies and wavelengths. (9.2)

Explain applications of light waves, such as blue skies, red sunsets, and laser light. (9.2)

Use the prefix "nano" and scientific notation. (9.2)

Define the electromagnetic spectrum. (9.3)

Indicate how frequency and wavelength change across the spectrum. (9.3)

Describe some applications and dangers associated with the different regions of the electromagnetic spectrum. (9.3)

Discuss problems related to the depletion of the ozone layer. (9.3)

Prepare Your Own Summary

Summarize this chapter by doing one of the following. Use a graphic organizer (such as a concept map), produce a poster, or write the summary to include the key chapter ideas. Here are a few ideas to use as a guide:

- Why does a blue hat appear blue? When might it look black?
- List the additive and subtractive primary colours and explain how they are related.
- Outline how to determine what colour of light will pass through a combination of filters (cyan, yellow, magenta).
- Draw a labelled diagram of a wave.
- Explain the role of a scientific model.
- Give evidence to support
  - **(a)** a wave model for light
  - **(b)** the idea that red light has a longer wavelength than blue light.
- Describe the difference between light from an incandescent light bulb and light from a laser.
- List the different regions of the electromagnetic spectrum. Describe each wave frequency as "low," "moderate," "high," or "extremely high."

CHAPTER

9 Review

Key Terms

spectrum
solar spectrum
additive primary colours
secondary colours
rods
cones
colour blindness
subtractive primary colours
complementary colours
crests
troughs
wavelength
amplitude
frequency
hertz
wave model of light
laser
incoherent
coherent
electromagnetic spectrum
infrared radiation
radio waves
microwaves
radar
ultraviolet (UV) radiation
ozone layer
X rays
gamma rays
radiation therapy

Reviewing Key Terms

If you need to review, the section numbers show you where these terms were introduced.

1. In your notebook, match the description in column A with the correct term in column B.

A

- unit of frequency
- causes tanning
- red plus blue
- longest wavelength of the electromagnetic spectrum
- produces coherent light
- a pattern of colours in light
- longest wavelength of visible light
- produces spectra by refraction
- absorbs ultraviolet radiation
- distance from crest to crest

B

- spectrum (9.1, 9.2)
- prism (9.1)
- radio waves (9.3)
- ultraviolet radiation (9.3)
- ozone layer (9.3)
- laser (9.2)
- hertz (9.2)
- magenta (9.1)
- wavelength (9.2)
- red (9.1)
- gamma ray (9.3)
- cones (9.1)

Understanding Key Ideas

Section numbers are provided if you need to review.

2. List the colours in the visible spectrum in order, beginning with red. (9.1)

3. Name the three additive primary colours. (9.1)

4. Name the three subtractive primary colours and indicate which primary colour is missing in each case. (9.1)

5. What colour will you see if you shine green light and blue light onto the same spot on a white screen? (9.1)

6. **(a)** State two uses for the additive primary colours. (9.1)

   (b) State two uses for the subtractive primary colours. (9.1)

7. Which cells in the retina are responsible for detecting colour? (9.1)

8. A man is wearing a cyan tie. How will it look if he steps into a room in which the light is red? Explain your answer. (9.1)

9. If a crest has a length of 2 m and a height of 50 cm, determine
   (a) the wavelength of the wave (9.2)
   (b) the amplitude of the wave (9.2)

10. How does the wavelength of a wave change if its frequency decreases? (9.2)

11. Radio waves have a lower frequency than light waves. Which of the two has the shorter wavelength? (9.3)

12. Why is the thinning of the ozone layer a problem? (9.3)

13. State one application for each of the following:
    (a) radio waves (9.3)
    (b) infrared radiation (9.3)
    (c) ultraviolet radiation (9.3)
    (d) X rays (9.3)
    (e) gamma radiation (9.3)

Developing Skills

14. White light is shone upon two filters — a cyan filter placed on top of a yellow filter. What will emerge from the filters? Explain your reasoning.

15. What evidence exists that light travels like a wave?

16. Using diagrams, compare and contrast
    (a) rod cells and cone cells
    (b) infrared radiation and ultraviolet radiation

17. **(a)** A singer's vocal cords vibrate back and forth 1320 times in 4 s. What is the frequency of the singer's note in hertz?
    (b) A yo-yo goes up and down 10 times in 20 s. What is its frequency in hertz?

Problem Solving/Applying

18. Imagine aliens have landed on Earth. You find out that their visual organs have only two types of colour-sensing cells. One type responds mainly to green. The other responds to blue and violet. Which colours would they see as well as we do? How could you write a secret message on paper in such a way that the aliens would not notice the message?

19. Imagine that you are the only lighting technician for a show. You have red, green, blue, magenta, cyan, and yellow filters. In the middle of the performance, you misplace your red filter. What could you do to replace it with the filters you have on hand?

20. Find a mathematical relationship between the values for frequency and wavelength in Table 9.2 on page 293.

Critical Thinking

21. In Figure 9.2A on page 269, identify the colour that **(a)** refracts the most when it passes through the prism; and **(b)** refracts the least when it passes through the prism.

22. The word SCIENCE has been painted on a black wall. The "S" is red, the "E"s are blue, and the rest of the letters are green. What will you see if you shine magenta light on the word? Explain your answer.

Pause& Reflect

Go back to the beginning of this chapter on page 266 and check your original answers to the Getting Ready questions. How has your thinking changed? How would you answer these questions now that you have investigated the topics in this chapter?

Ask an Expert

Solar cells convert light from the Sun and other sources into electrical energy. Have you ever seen a solar cell? Judy Kitto has seen thousands of them. Her company, Solar Power Systems, designs and sells a wide range of solar systems to customers in central Ontario. Her own house is an example of a solar home system.

Q What exactly is a solar power system?

A It's a system of solar cells built into panels connected to a battery and then to some other equipment. The solar cells provide the energy to charge the battery with electrical power, and the battery provides the electrical power to run the other equipment.

Q What kinds of systems do you design and sell?

A I've designed all kinds of systems. Farmers often want a system to power an electric fence or to pump water from a pond or river to their fields. People with cottages or homes in remote areas that have no hydro-electric service sometimes ask me to design whole-home power systems. They may be using propane or kerosene lamps for light and they may have no electric appliances, such as toasters or hair dryers. Solar power can be more convenient for them, and cheaper in the long run.

Q Can you run everything in a house on solar power?

A Well, solar power generates electricity, so technically you can do anything with it that you could do with electricity. Solar power is not always practical, though. It would take a very large, very expensive system to heat my home in the winter, for example. Our daylight hours are shortest when the weather is coldest. In other words, the time when you need the most power is the time when the least power is available. I use wood to heat my home, but solar power runs my lights and many other appliances.

Q Are your clients mostly farmers and home-owners then?

A No, I also sell systems to boat-owners to power their lights and their fishfinders. Recreational vehicle owners buy solar systems, too, because these systems allow them to park anywhere and still have electricity to cook or to turn on the lights. Recently, I devised a small solar system to churn up and aerate the water in a fish pond to keep the fish healthy.

Q Does everything stop working at night when there is no Sun?

A That's where the batteries come in. During the day, the solar panels send a lot of electrical energy to the batteries all the time. Then, when I turn on a light or make toast, I use some of the electricity stored in the battery. When I design a system, I am careful to find out how much electricity the client needs and to recommend a battery large enough to store more power than the client will probably need. Many home systems also have a backup generator that runs on gasoline. If the batteries get low, the generator comes on automatically to charge them up again.

Q If it rains for six days straight, wouldn't the generator have to run the whole time?

A Grey, rainy days still provide some solar energy.

These 18 solar panels produce 900 W of power. The panels power Judy's indoor lights, outside sensor lights, water softener, water pressure pump, washing machine, iron, microwave oven, toaster, electric garage door, satellite dish, two TVs, two VCRs, two offices, two computers, a photocopier, and a fax machine.

You'd be surprised by how little light it takes to power the panels. Once, on a very clear, still night, our panels generated one amp of electricity just from the light of the full Moon!

Q How did you learn everything you need to know to design and sell a wide range of solar systems?

A Mostly, I learned about solar power from reading the product manuals. My husband and I started this business together, and at first he was the one with the electronics know-how. Then he got sick and I had to take over. I read everything I could, including all of the material written by the manufacturers of the solar products. I also had the knowledge I'd gained from living in our solar house for seven years or so. I learned more with each new system I developed. Some of my clients were sceptical at first, but they seem to be convinced, now, that I know what I'm talking about!

EXPLORING Further

Battery Bonus

Compare two solar-powered calculators, one that has a backup battery and one that does not. (Most calculators that have a backup battery will say so on the calculator itself or on its packaging. It may be called a "dual-powered" calculator.)

Try some simple equations on each of the calculators, in a brightly lit room, in a dimly lit room, and while moving from a bright area to a dim area. How much light is necessary to power the calculators? What is the advantage of having a backup battery in a calculator?

An *Issue* to Analyze

A SIMULATION

Bright Lights in the Big City

Think About It

Picture any large Canadian city. At night, neon signs, traffic lights, illuminated billboards, electronic notice boards, streetlights, and spotlights compete for your attention. In the downtown core, glowing skyscrapers tower above you. On a clear evening, the city sends a dome of light up into the sky that can be seen 100 km away.

Recently, an editorial appeared in a city's local paper that struck a chord with readers. The journalist pointed out the waste involved in this nightly display of lights, spurring hundreds of readers to write letters in response. City Council responded to citizens' concerns by creating an Illumination Task Force, of which you will be a member. The task force will assess the problem of excessive use of lights and will report back to City Council, submitting a set of recommendations. Read the editorial below, as well as some of the letters sent in by readers to get background information about the issue.

Background Information

Editorial: June 10

Just Say "No" to the Nightly Light Show

Want to know how Canada wastes $250 million a year, while needlessly harming the environment and reducing the quality of life for Canadians? Light pollution. In the United States, the tab is even higher. Members of the International Dark-Sky Association, a group of astronomers against light pollution, estimate that the United States wastes up to $1 billion a year because of light pollution. Other estimates reach as high as $2 billion.

Many North Americans don't even know that light pollution is a problem. Oil and chemical spills in rivers, lakes, and coastal waters are cause for outrage. However, few people complain about the millions of dollars' worth of wasted light and heat "spilling" into the sky every night.

Light pollution wastes energy and money and blocks our view of the Milky Way, one of nature's greatest marvels. Each year, up to 10 000 birds die in one large city alone by crashing into lit office towers. In addition, all this light doesn't necessarily help us see better. Inefficient lights cause glare that actually decreases our night vision.

Our city is especially guilty. Buildings downtown are lit 24 hours a day, wasting electrical energy on empty offices. Most of our city's 120 000 streetlights are inefficient mercury-vapour and high-pressure sodium lights. Light is often directed up and to the sides, not down toward the ground where we need it. Like many cities in North America, our lighting use is ten times the lighting standard set by the Illuminating Engineering Society of North America.

The truly shocking aspect of all this waste is that new technology makes it avoidable. Energy-efficient lights greatly reduce waste. Simply turning off unused lights would save millions of dollars. Don't let City Council ignore the problem any longer. Help save taxpayers' money and help protect the environment by letting city councillors know how you feel.

Re: Just Say "No" to the Nightly Light Show (June 10)

I couldn't agree more with the editorial about our city's pollution of the night skies. As a member of the International Dark-Sky Association, I heartily support reduction of light pollution. Did you know that on a clear night in the countryside, 1500 to 2000 stars are visible to the naked eye? In many Canadian cities, people are lucky to see 50. Urban observatories no longer receive funding for important astronomical research because light pollution obliterates most of the stars. Funding now goes to remote observatories in Hawaii and Chile. Help reduce light pollution so that we can all see the night skies again.

Re: Just Say "No" to the Nightly Light Show (June 10)

One reason cities use so many lights is to prevent crime. Businesses use security lights to discourage theft and many parks have lights to permit safe use of public areas at night. I'm all for saving money and seeing stars at night, but not at the expense of public safety. We must preserve the option of using lights to discourage crimes against both people and property. People should feel safe when using parking lots at night, for example. Public security demands that lighting be used to cut down on crime. We must not overlook this important function of lights in the downtown core.

Plan and Act

1. The following people have been invited to participate in the task force:
   - a member of an environmental group
   - a representative from a lighting manufacturer
   - a member of City Council
   - a member of the police force

2. What do you think each person's point of view might be before the task force meeting? What are some concerns and considerations each person might want to discuss? What facts or arguments are likely to cause these people to change their points of view?

3. Your teacher will give your group the role of one of these people, along with additional information to help you plan your presentation. As a group, research your role, and then present a strong case at the task force meeting. (Your teacher will provide you with a blackline master outlining the correct *Procedure for a Public Hearing*.)

4. After presentations and questions, members of the task force will vote on proposed recommendations. If accepted by the majority, proposals will be adopted as recommendations for City Council.

5. Your task, as a class, will be to assess all the presentations made at the task force meeting, along with the final recommendations drawn up for presentation to City Council.

Analyze

1. Did the task force have trouble agreeing on final recommendations? Which issues, if any, were the most difficult to resolve?

(a) Which of the recommendations relied upon societal changes, such as shifting people's attitudes toward wasted light or changing laws to influence people's behaviour?

(b) Which recommendations relied upon scientific advances, such as adopting new, light-efficient technologies?

2. Which proposed changes should be implemented first? Why?

3. Did you have all the information you needed at the meeting or were there some questions for which no group could formulate answers? Suggest resources where you might find answers to any remaining questions.

Protect It With Light!

WANTED

Young, creative company to design and install a laser security system

The local museum has recently acquired a rare and priceless dinosaur fossil and wishes to put it on display. The museum wants to install the latest in laser security systems to protect the fossil. Innovative and creative companies who understand the properties and principles of light and optics are invited to submit designs for their own state-of-the-art systems.

Challenge

Design a simulation of a security system that uses reflection, refraction, absorption, and transmission of light to surround and protect the museum's priceless fossil with a single beam of light. If the beam is broken, an alarm will sound.

Safety Precautions

- Be careful when using sharp objects such as scissors.
- The edges of mirrors can be sharp; be careful when handling them.
- A glue gun is hot and the glue remains hot for several minutes.

Materials

floor plan of the room in the museum where the fossil will be displayed (to be supplied by your teacher)

mirrors (concave, convex, and plane)

reflective material (e.g., aluminum foil, metal, glass)

…ashlight or ray box

…modelling clay, tape, glue or glue gun,
…y other construction materials of your

…or four, design and
…l model of your (simu-
…tem, based on the floor
…r teacher.

B. The system should consist of a single beam of light that creates a perimeter around the fossil by means of refraction, reflection, and transmission of light.

C. There must be a fixed starting point for the beam, as well as a specific end point where a sensor will be located.

D. You must place one mirror (plane, convex, or concave) somewhere in the room. This mirror will permit the guard in the security booth to see the entrance to the room, as outlined in the floor plan.

E. With your team, prepare a written and oral presentation of your working model. In your written presentation, explain the path of the light beam – where it goes, what happens to it, and why. Also include the suggested cost of your system. In your oral presentation, demonstrate how your model works by using a light source to show the path of the beam of light. Point out the selling features of your design and outline why you think the curator of the museum should buy it.

Plan and Construct

1 Prepare a design brief that includes the following:

- the name of your company and an appropriate logo
- what you are making and the materials you will be using

- a one-paragraph "sales pitch" stating the credentials of your company. Be sure to mention what you have learned from this unit that qualifies you to design a museum security system that uses light.

2. Prepare a design proposal. Your proposal should feature an overhead scale drawing of the museum's display room (based on the floor plan) with the additional equipment required by your company; for example, types of mirrors and their locations. The drawing should have a solid red line to show the path the light beam is expected to take. In addition, make a three-dimensional model of your design, including mirrors and objects in the room, based on your scale drawing of the room.

3. As you build your system, keep a learning log of the steps you have taken. Record any difficulties in creating your design and what you did to overcome them.

Evaluate

1. How well does your model demonstrate the properties of light in a useful way?
2. How well did your team members translate their design into a three-dimensional model?
3. How well does your model "protect" the rare fossil?
4. Were your company's written and oral presentations clear, creative, and informative?
5. After seeing the competition, how do you think you might improve your design, your model, or your presentation?

- Watch a video in which your favourite singer or group performs. Describe in writing the lighting effects used in the performance. Use your knowledge of colour and filters to explain the lighting effects.
- Research advances in diagnosing and treating eye problems that have been made in recent years. What new kinds of contact lenses and eyeglasses are available? What kinds of materials are used in corrective eyewear? What are the advantages of having glass lenses? Plastic lenses? Will there be a time when most common eyesight problems are corrected surgically rather than with eyeglasses? Present your findings in the form of a report a poster.

UNIT 4

Water Systems on Earth

A vast amount of water plunges over the rim of Niagara Falls. On a summer's day, about 3000 m^3 of water thunders down onto the rocks below every second. How long do you think it would take for this mighty flow to equal the total volume of water used by all Canadians in one day? Did you guess 11 h? Where does all that water come from? Will it ever run out?

Like Niagara Falls, water is always moving. However, the total amount of water in the world remains the same. Most water on Earth is in the oceans. These huge bodies of water influence weather patterns, affecting areas hundreds of kilometres inland. Water is also home to thousands of species of plants and animals, many of which we eat as food.

In this unit, you will discover how water is distributed around the world. You will also learn how water affects our lives in many ways. As people use water — for drinking, farming, manufacturing, recreation, and other uses — they alter water systems. Some changes are obvious, such as the oil spills at sea that injure and kill animals. Other changes are less obvious. These changes may lead to water shortages or floods. By studying Earth's water systems, you will better understand the importance of water in our lives and to all life on the planet. You will also better understand the importance of sustainable development of Earth's water resources. Sustainable development meets the needs of the present generation without hindering the ability of future generations to meet their needs as well.

Unit Contents

CHAPTER

10 A World of Water

Getting Ready...

- What happens to rain after it falls?
- Where does the water in your taps come from?
- Will Earth run out of water some day?

Science Log

Try to answer the questions above in your Science Log, using words and illustrations. Fill in further details as you read this chapter.

Skill POWER

For tips on how to make and use a Science Log, turn to page 534.

Earth has been called "the blue planet." From space, its surface appears to be mostly oceans of liquid water. Water vapour is found in Earth's atmosphere. Frozen water occurs on mountaintops and at the North and South poles. The bodies of animals and plants consist mainly of water. For example, sixty-five percent of your body mass is water.

Water is not only plentiful, it is also always on the move. It moves in and out of our bodies. It flows from ice cubes melting in a drinking glass. It evaporates into the air from wet clothes and falls from clouds as rain or snow. The best way to understand Earth's water supply is to study it as a system — a closed system in which water constantly moves around between sea, sky, land, and life.

Spotlight On Key Ideas

In this chapter, you will discover

- how water exists in various states on Earth's surface
- how water is distributed around the planet
- how water circulates between land, ocean, and atmosphere
- how rivers and lakes are formed
- why flooding occurs
- how people have tried to prevent or control flooding

Spotlight On Key Skills

In this chapter, you will

- design and conduct an investigation to determine what happens when water falls on different surfaces
- make a model showing how river valleys are formed
- study and interpret maps and tables of water systems
- make a simple model of a wetland
- analyze data showing what happens to precipitation falling in Canada
- measure the amount of water vapour released into the air by plants
- compare the acidity of rainwater, tap water, and other samples

Starting Point 

The Restless Liquid

Water occurs on Earth not only as a liquid but also as a solid (ice) and a gas (water vapour). What conditions make water change from one of these states to another? You can actually see the changes taking place by observing a jar of water for a few minutes.

What You Need

glass jar with twist-off lid (1 L)
ice cubes
water at room temperature

What to Do

1. Pour water into the jar to a depth of about 1 cm.
2. Turn the jar lid upside down and set it over the mouth of the jar. Put 3 to 4 ice cubes in the lid.
3. Observe what happens in the jar over the next 10 min. Sketch your observations.

What Did You Find Out?

1. Name the process that changes
   (a) liquid water into water vapour
   (b) water vapour into liquid water
   (c) liquid water into ice
   (d) ice into liquid water
2. Explain how this activity might be used as a model for weather systems on Earth.

10.1 Water on the Move

Figure 10.1 Where is the water in this river going?

Pause& Reflect

Have you noticed earthworms on the surface of the ground after a heavy rainfall? In your Science Log, suggest a reason for this observation. Why might you see earthworms in one area but not in a nearby area that receives the same amount of rain?

Have you ever gone for a walk in your neighbourhood just after a heavy rainstorm? What do you notice? Water drips from leaves and runs along gutters or drainage ditches. There are puddles in a parking lot, and footpaths across a playing field are muddy. A few hours later, the puddles are gone and the sidewalks are dry. What has happened to the water? Every drop of rain that falls to Earth must go somewhere. To follow the trail of the vanishing water, you can start by considering some of the different places where rain first falls. The type of surface on which rain falls and the slope of the surface affect what happens to rainwater, as you will see in the next investigation.

INTERNET CONNECT

www.school.mcgrawhill.ca/resources/

Did you know that a stalactite in a cave begins as a single drop of water clinging to the roof of the cave, but may take thousands of years to form? If you are interested in fascinating water facts, go to the above web site. Go to **Science Resources**, then to **SCIENCEPOWER 8** to know where to go next. Start a scrapbook or a mini-encyclopedia containing amazing water facts.

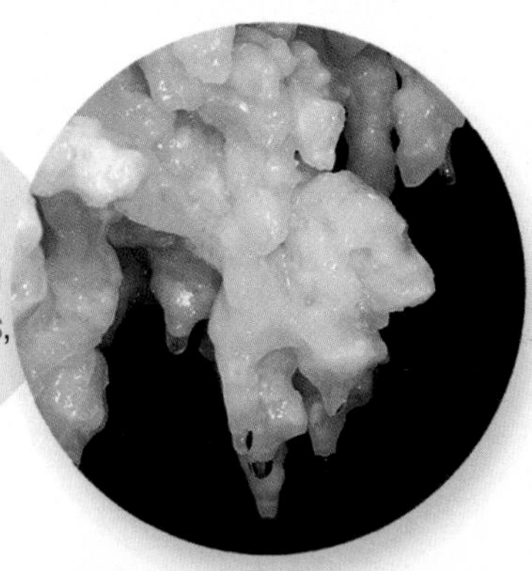

CONDUCT AN

INVESTIGATION 10-A

What Happens to Rainwater? 

Work in a group to study how Earth's surface affects what happens to rainwater falling on it. Consider the types of solid surfaces on which rain falls and the slope of the surface. Make a hypothesis and then test it by designing your own experiment.

Problem

What happens to water falling on different surfaces?

Safety Precautions

Apparatus
graduated cylinder
sink or basin
protractor
rigid piece of acrylic plastic
sponges
small blocks of wood
watering can

Materials
paper towels
water

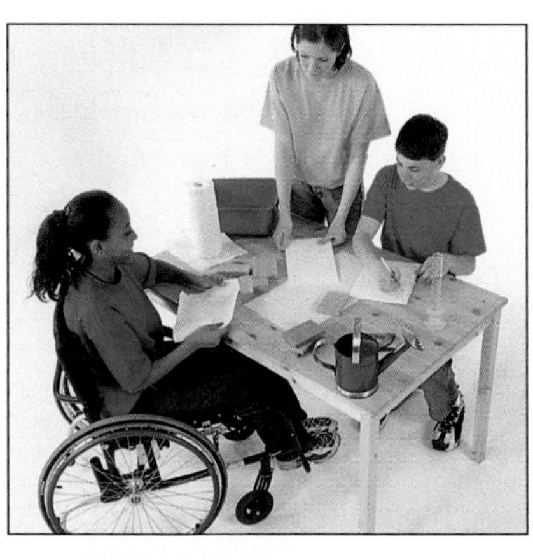

Procedure

Note: To review how to set up a controlled experiment using different variables, see pages IS-8–IS-13.

1 Before you begin to plan your group's investigation, consider these questions:

(a) Your class is comparing three different surfaces: plastic, paper towels, and a sponge. What is different about each surface?

(b) Which surface represents a parking lot? Sandy soil? A grassy area?

2 Before carrying out your investigation, design a procedure and have your teacher approve it. Then prepare a table for recording your observations. What are you measuring? What are the variables? Which variable(s) stay(s) the same? Which variable(s) change(s)? How will you measure the effect of changing the angle of slope of each surface?

(a) Wipe up any spills as wet floors are slippery.

(b) Wash your hands after this investigation.

Skill POWER

For tips on working in groups, turn to page 536.

Analyze

1. **(a)** Assess whether your procedure allowed you to make clear comparisons. What criteria did you use to make your assessment?

   (b) How might you improve the experimental design?

2. **(a)** What other variables could you have investigated?

   (b) What additional materials or procedures would you need to do this?

Conclude and Apply

3. This investigation is a model to show what happens to rainwater falling on land. Based on your observations, write three different outcomes for the following: A raindrop fell on ________ and it ________ because ________.

Figure 10.2 Water quickly runs off smooth, non-porous surfaces.

Run-off

In the previous investigation, you observed how water behaves on different surfaces. On smooth, hard surfaces, water quickly runs off — these surfaces are **impermeable**. That is why, as Figure 10.2 shows, an umbrella or a nylon jacket with a hood can keep you dry in a rain shower. On surfaces with many pores, however, water soaks in — these surfaces are **porous**. You have probably experienced this if you have been caught in a rain shower while wearing a cotton sweatshirt. Rain penetrates the cotton and soaks right through to your skin. The same results occur when raindrops hit Earth. Rainwater runs off impermeable surfaces such as bare rock and asphalt, and soaks into porous surfaces such as gravel and soil.

Rainwater that flows off the land surface is called **run-off**. The amount of run-off depends on the type of surface and the steepness of the slope. Run-off also depends on the amount of precipitation and the time span over which it falls. Heavy rainfall may run off even a porous surface, because the water does not have time to soak in. Pulled by gravity, the flowing water runs downhill. Over bare ground, flowing water may carry away loose soil and particles of rock. Over areas covered with vegetation, run-off is slowed down. Eventually, run-off may enter a stream or river, and continue a journey that can take it hundreds of kilometres from where it first fell.

A River's Journey

You may have stood beside a large river, watching the water flow. Perhaps you have jumped across a small, trickling stream. Like highways, rivers and streams connect one place with another. Unlike highways, however, water courses always move in only one direction — downhill! Notice the direction of flow the next time you look at a stream or a river. The area **upstream** (from which the water is coming) is at a higher elevation than the area **downstream** (to which the water is going).

Most rivers begin their journey on the steep slopes of a mountain range, as shown in Figure 10.3. The young river is usually narrow, straight, and fast-moving. It tumbles down waterfalls and forms white-water rapids. The water carries along any loose rocks or boulders. The tumbling rocks wear away the stream bottom and carve a steep channel of land known as a **valley**, as shown in Figure 10.4.

On gentler slopes, rivers move more slowly. As one stream joins another, the volume of water grows larger. The ceaseless movement of the river removes rocks and soil along the river's banks. This process widens the valley as well as deepens it.

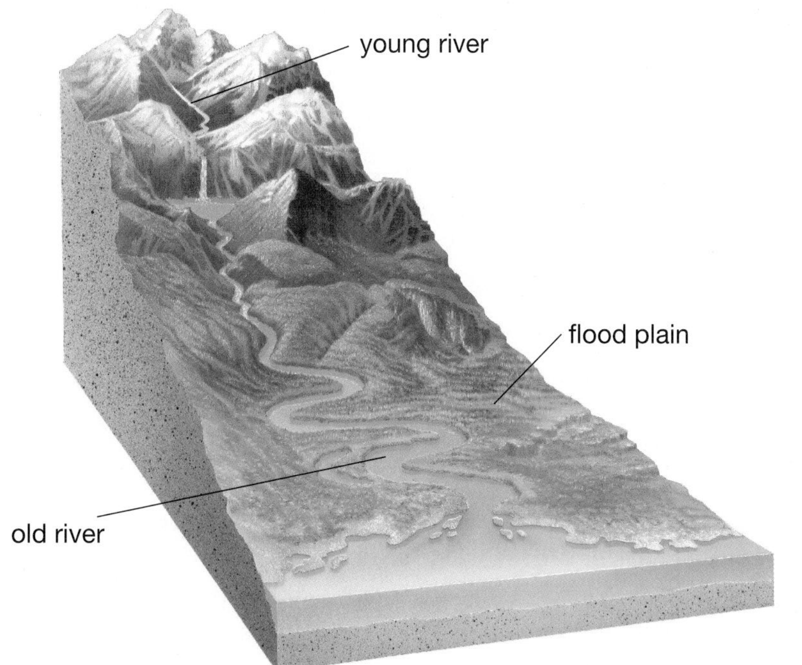

Figure 10.3 The stages of river development. A river system includes narrow, rapid streams at its beginning and a broad, slow-flowing river at its last stage, just before it flows into the sea.

Pause& Reflect

In your Science Log, explain why water flowing in a stream or a river will eventually end up in a lake or an ocean.

Figure 10.4 Fast-moving water has created a V-shaped valley.

Figure 10.5 This broad, wide river winds across a level flood plain.

The final destination of a river is a **lake** or **ocean** — large bodies of water filling basinlike depressions in Earth's surface. Before reaching the ocean, many rivers cross **plains** — wide, gently-sloping areas of land near the coast. On these level plains, the river is easily deflected sideways, and begins to wind in loops across the broad valley floor (see Figure 10.5). The river slows here, depositing some of the fine rock particles or sediment carried by the flowing water.

Word CONNECT

Water imagery is often featured in literature and music. Find and bring to class a poem, short story, or song in which water is a central image. Write your own poem, story, or song lyrics based on water imagery.

CONDUCT AN

INVESTIGATION 10-B

Making a Model of a River

Streams and rivers flow in channels of land, known as valleys. Streams and rivers also help create these channels. This investigation will model how water shapes river valleys.

Problem

How does flowing water create a valley?

Safety Precautions

Apparatus

stream table or large rectangular pan
plastic tubing (attached to water tap)
bucket (for overflow)

Materials

rock fragments of different sizes:
clay (very fine)
sand (small)
gravel (medium)
pebbles (large)

Procedure

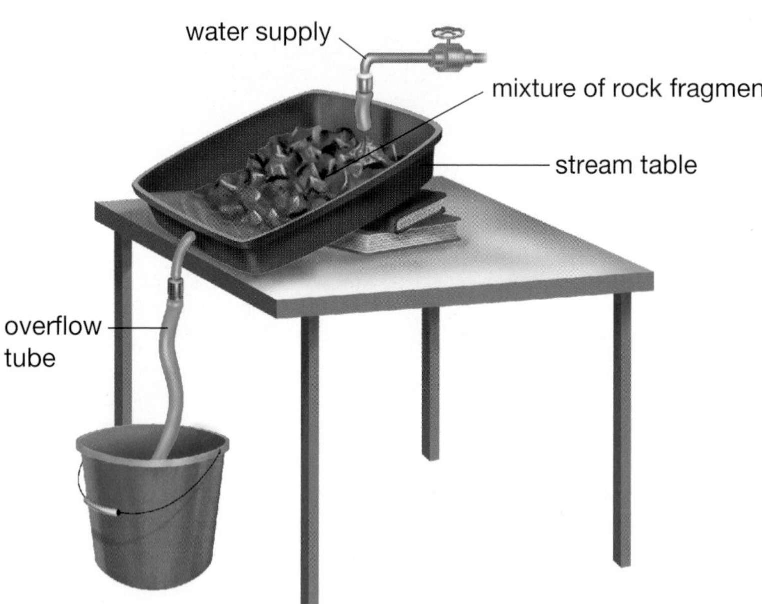

1. With your teacher's assistance, set up a stream table similar to the one shown in the diagram.

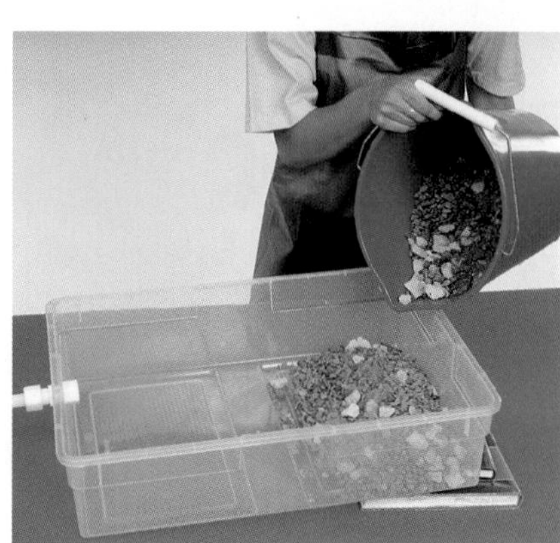

2. Pour the mixed rock fragments into the top half of the stream table to a depth of about 5 cm.

3. Add water to the lower half of the stream table to represent a lake.

4. With the tubing at the top of the stream table, slowly turn on the tap to produce a steady drip of water for about 1 min.

5. Gradually increase the water flow to a gentle, steady stream. Observe how the water flows through the rock materials and how it affects different rock fragments. Record your observations.

6. Predict what will happen to the model river if you
   (a) make the water flow faster
   (b) tilt the stream table to a steeper angle
   (c) reduce the angle of slope of the stream table

7.  Working in groups, test your predictions by carrying out the experiments described in step 6. Sketch what you observe in each case.
   (a) Wipe up any spills as wet floors are slippery.
   (b) Wash your hands after this investigation.

Skill POWER

For tips on scientific drawing, turn to page 557.

Analyze

1. What happened to each size of rock fragment when the water was flowing slowly?
2. What happened to each size of rock fragment when the water was flowing quickly?
3. Describe what happens to the stream of water where it joins the "lake" at the lower end of the stream table.
4. Under what conditions does the model river produce a deep, narrow valley?
5. Under what conditions does the model river produce a shallow, wide valley?

Conclude and Apply

6. What kind of water flow do you think created the Grand Canyon or the Niagara River Gorge?

Flood Plains

Imagine you must cross a room while holding a glass full of juice in one hand and a plate full of juice in the other. It would be much harder to keep juice from spilling off the plate because of the plate's shallow shape. A broad, shallow river channel is equally poor at holding in water. In periods of heavy rain, a river may overflow its banks and spill out over a wide area of the valley floor. This part of the river valley is called the **flood plain** (see Figure 10.6).

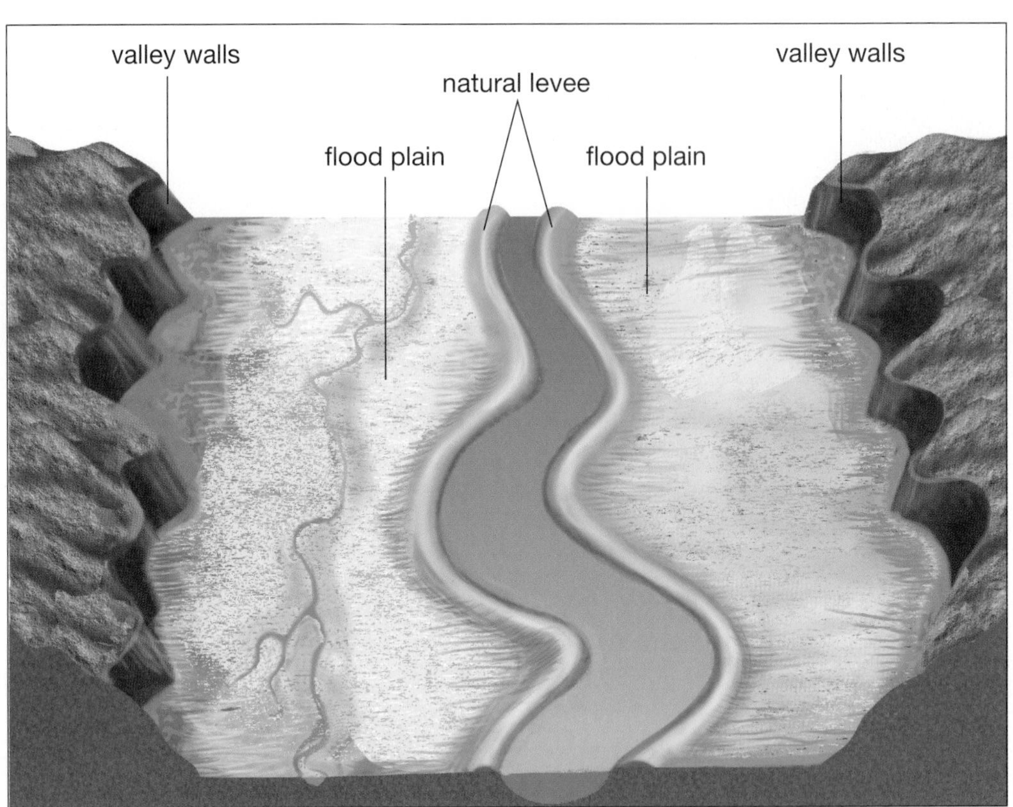

Figure 10.6 When rainfall is heavy, a river may overflow its banks and spill onto a flood plain on either side of the river.

For people, a river flood is both good and bad. When a river overflows onto its flood plain, it spreads sediments in a thick layer over the valley floor. The fine particles of sediment, including minerals and soil nutrients, make flood plains the most fertile farming areas in the world. In dry parts of the world, such as Egypt and parts of India, flood plains are very important in producing food.

The benefit of floods occurs after the flood waters have gone, having drained back into the river, soaked into the ground, or evaporated. The rich soil left behind attracts large numbers of people to live on or near the flood plain.

The destructive effects of floods come when the waters are rising. Imagine you are on a farm near the Red River in Manitoba after heavy spring rains. You watch as the rising river water appears on the horizon and advances across the surrounding fields. It creeps steadily closer to your farmhouse, and now you can do nothing to stop it. It washes over the doorways, cascades down stairs, and rises up past walls and windows. Figure 10.7 shows damage caused by the Red River flood of 1997. Figure 10.8 shows people struggling to build walls of sandbags to try to hold back the rising water.

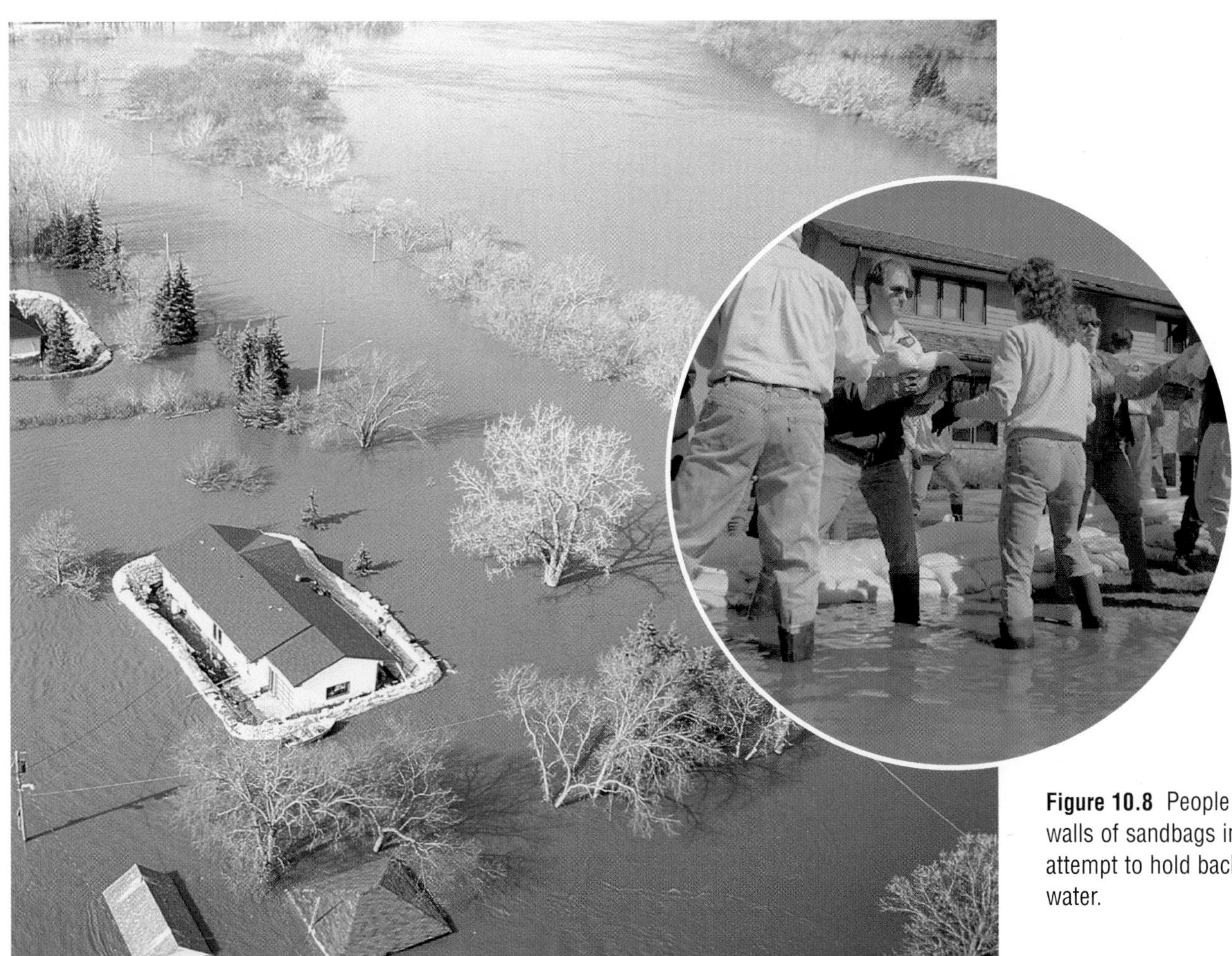

Figure 10.8 People build walls of sandbags in an attempt to hold back flood water.

Figure 10.7 In 1997, the Red River flood in Manitoba caused great social and economic damage. The rising flood waters forced more than 28 000 people to evacuate their homes.

Floods wash away houses, cause billions of dollars' worth of damage every year, and kill people. Heavy rainfall is not the only cause of floods. In Canada, a warm spring after a cold winter is a dangerous time for people who live in river valleys. A quick thaw increases the run-off from melting snow, called **meltwater**. If the ground is still frozen, and is therefore impermeable, the meltwater cannot soak in and thus it flows into the river valley. This spring run-off is much more than the river can hold. As a result, the water level rises beyond the banks of the river.

Human activities can both increase and decrease the risk of floods. For example, removing natural vegetation such as trees and shrubs increases run-off, especially on steep slopes. Replanting trees and grass can help reduce this risk. On large river systems, people build dams to hold back water and control its flow. The sides of a river valley can be heightened by building elevated ridges called **levees**. Drainage channels can be dug to divert water around towns and villages. Floods are likely to be an increasing problem as long as people clear land in river valleys, replacing vegetation with buildings and pavement. All precipitation must go somewhere. If it cannot soak into the ground, it will flow into a river system and raise the water level.

Pause& Reflect

Can a town be flooded without having any rain? Recall that river systems can extend over hundreds of kilometres. A change in a stream in one area can affect a river a long distance away. In your Science Log, describe a situation in which a rainfall or a thaw in area A causes a flood in area B. Make a sketch of the areas and use the terms "upstream," "downstream," and "flood plain."

Drainage Basins

Do rivers ever run out of water? In past centuries, people were puzzled by this question. The amount of rain and snow that fell in an area did not seem large enough to account for all the water in rivers and lakes.

Figure 10.9 This simplified drawing represents a drainage basin.

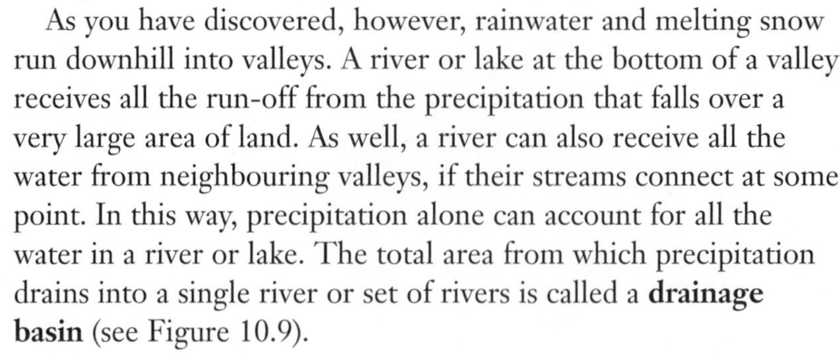

As you have discovered, however, rainwater and melting snow run downhill into valleys. A river or lake at the bottom of a valley receives all the run-off from the precipitation that falls over a very large area of land. As well, a river can also receive all the water from neighbouring valleys, if their streams connect at some point. In this way, precipitation alone can account for all the water in a river or lake. The total area from which precipitation drains into a single river or set of rivers is called a **drainage basin** (see Figure 10.9).

Every stream is part of a large drainage basin. A small stream in your neighbourhood will flow until it meets other small streams. The streams join larger rivers, and large rivers merge into major waterways such as the St. Lawrence River.

The boundary of a drainage basin is formed by the crest of a hill or mountain. The mountain crest forms a **drainage divide**. In North America, a continuous ridge of mountain ranges divides the continent into two main drainage areas, as shown in Figure 10.10. On one side, water flows northeast to Hudson Bay or southeast to the Gulf of Mexico. On the other side, water flows west to the Pacific Ocean. This long ridge of mountains dividing North America is called the **Continental Divide**.

From what provinces does water drain into the Gulf of Mexico? Why do you think this is so?

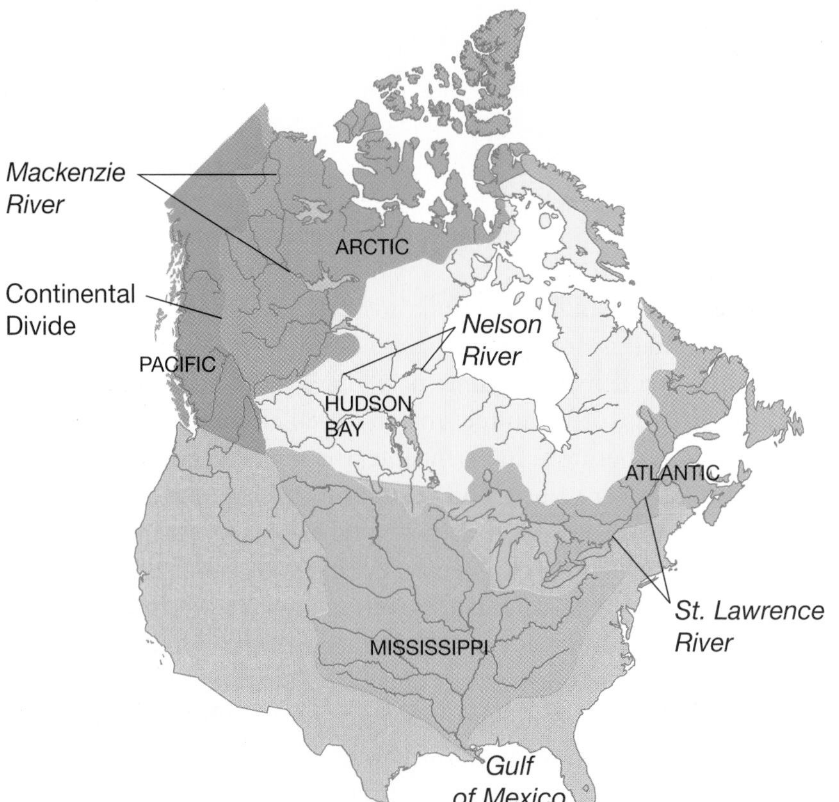

Figure 10.10 Canada has a total of five major drainage basins. Most precipitation falling on Canada drains into the Pacific, Atlantic, and Arctic Oceans, and Hudson Bay. Interestingly, some of Canada's river water also ends up in the Gulf of Mexico.

INVESTIGATION 10-C

How Much Water?

Think About It

Study the table below. It lists the sizes of Canada's five major drainage basins and the volumes of water that flow from each river into the oceans. The volume of water flowing from a river into an ocean is called the **mean discharge**.

Canada's Major Drainage Basins

Drainage basin	Area (km^2)	Mean discharge (m^3/s)
Pacific	1 009 064	24 100
Arctic	3 583 265	16 400
Hudson Bay	3 860 136	30 900
Gulf of Mexico	26 677	25
Atlantic	1 570 071	33 400

What to Do

Rank the drainage basins in order of

(a) size (area in km^2), from 1 (largest area) to 5 (smallest area)

(b) mean discharge, from 1 (largest volume) to 5 (smallest volume)

Analyze

1. Explain the difference in rankings.
2. To which coastline does most of Canada's water flow — north, east, or west?
3. What percentage of the total flow goes in this direction?
4. **Apply** Study the map of Canada's drainage basins in Figure 10.10. Suppose you wanted to canoe "down" the Mackenzie River. Plot a course of your journey that starts near the beginning of the Mackenzie River and follows the river's path to the sea. In which direction will you travel? How will the river change? In which sea will your journey end?

Computer CONNECT

Use the values in the table to create a spreadsheet. Make a bar graph of the data, with area on the horizontal axis (*x*-axis) and mean discharge on the vertical axis (*y*-axis).

Skill POWER

For tips on how to make a bar graph, turn to page 546.

Streams of Pollution

Within a drainage basin there is much more than rivers and valleys. There are farmer's fields, forests, factories, mines, mills, highways, towns, and cities. The rain and snow that fall on all these places also eventually end up in streams and rivers. Study Figure 10.11. How might different human activities in a drainage basin affect the river systems?

Figure 10.11 Many different activities on land add pollution to a river system. How many polluting activities can you identify in these photographs?

Run-off water carries fertilizers and pesticides from fields. These can harm plants and animals that live in water systems. When trees are removed by logging, this removal of vegetation increases the amount of soil washed into streams. The water becomes muddy, making it difficult for plants and animals to survive. Waste chemicals may get into rivers from mines, mills, factories, and households.

If you live in a large city, you may not realize that the water running down your sink or toilet eventually ends up in a river or lake. What about the rain and snow that fall on cities? Rain and snow carry oil from the roads and dissolved salt from melted slush in winter. Look along the gutters on every city street (see Figure 10.12). Where does the run-off go after it trickles down those storm sewers and out of sight? What other substances might be carried in this water? In rural areas, where does the wastewater go?

Figure 10.12 Where do storm sewers take rainwater?

Later in this unit, you will consider some impacts of pollution on plants and animals living in rivers, lakes, and oceans. You will also learn about ways of reducing water pollution. For now, think about the idea that water is always on the move. In the past, people dumped wastes into rivers because it was an inexpensive and easy way to dispose of unwanted materials. The ever-flowing rivers carried the wastes away. Where did the wastes go? The wastes did not really disappear. They simply moved downstream, perhaps to another community where people took their drinking water from the river. As populations grew, however, people realized there is no "away." Because water systems connect each community with other communities, we all share our drinking water.

Check Your Understanding

1. What two characteristics of land produce run-off?
2. Give two reasons why a puddle might disappear.
3. Compare a river at the start of its journey with the river at the end of its journey.
4. In what way does flooding benefit people?
5. Name two ways of controlling a river flood.
6. What is a drainage basin?
7. Rainwater in southern Alberta can end up in the Gulf of Mexico. How are these two places connected?
8. **Thinking Critically** Formulate your own question related to flowing water and explore possible answers to this question.

10.2 Water Above and Below Ground

DidYouKnow?

The amount of fresh water underground is about 37 times the amount on the surface in rivers and lakes. About half this underground water saturates the rock and soil to a depth of nearly 1 km. The remaining half of the water is even deeper — trapped 1–5 km below Earth's surface.

Rivers and lakes are the most visible form of fresh water, but most of the precipitation that falls on land sinks out of sight into the ground. Below the surface, this water continues moving downwards through connected pores and cracks in the soil and rock. Eventually, it reaches a layer of non-porous, impermeable rock, such as granite, which forms a barrier. When the water, now called **groundwater**, cannot move any deeper, it begins to back up and fill the pores in the material above the impermeable layer.

Some types of rock, such as sandstone, are full of pores and can hold a lot of water. As every space in the rock becomes filled, the groundwater forms a sort of underground river system, connected by pores instead of stream channels. This system of water flowing through porous rock is called an **aquifer**. Like rivers on the surface, the water in aquifers moves, although its flow is very slow. It might take 80 years or more to travel through 1 km of sand. The layer of porous rock in which all pores are full of water is called a **zone of saturation**. The top of this saturated layer forms a **water table**, as shown in Figure 10.13. The level of a water table may change, rising closer to the surface in wet seasons and sinking lower in dry seasons.

Pause& Reflect

What effects might plant roots and earthworms have on pore spaces in soil? In your Science Log, make a hypothesis and then describe how you could test it.

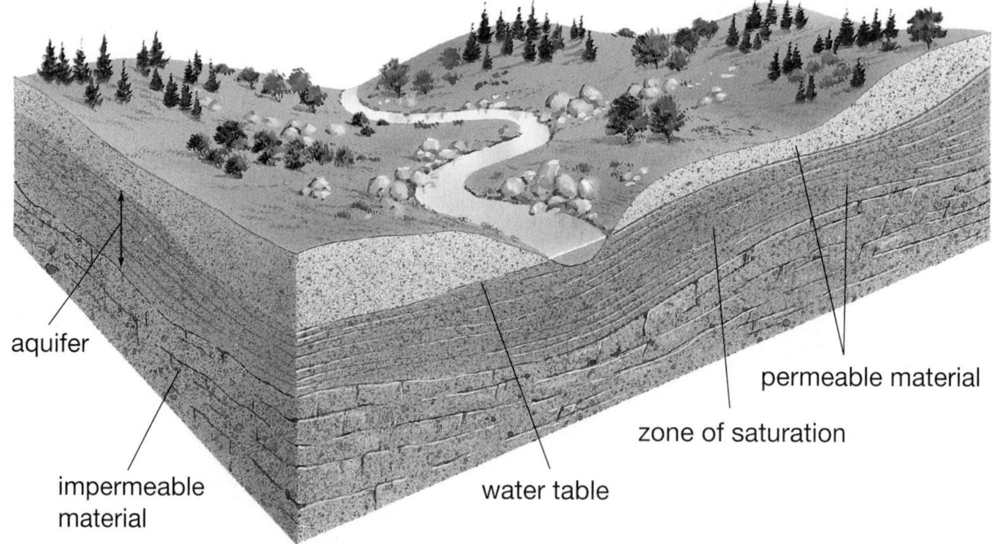

Figure 10.13 Cross-section of a groundwater system

Wells and Springs

Have you ever been to a beach and scooped out a hole in the sand? As if from nowhere, water appears in the hole. You have dug below the water table, and water moves into the hole from the saturated sand surrounding it. In almost any part of the world, even in deserts, a hole dug deep enough will reach water from the reservoir under the ground. Wells are made in this way, by drilling a hole through the soil and rock to a point below the water table (see Figure 10.14). From there, water is pumped up to the surface. If too many wells are dug in one area, they may remove groundwater faster than it is replaced by precipitation. If that happens, the water table will drop deeper and the wells may run dry.

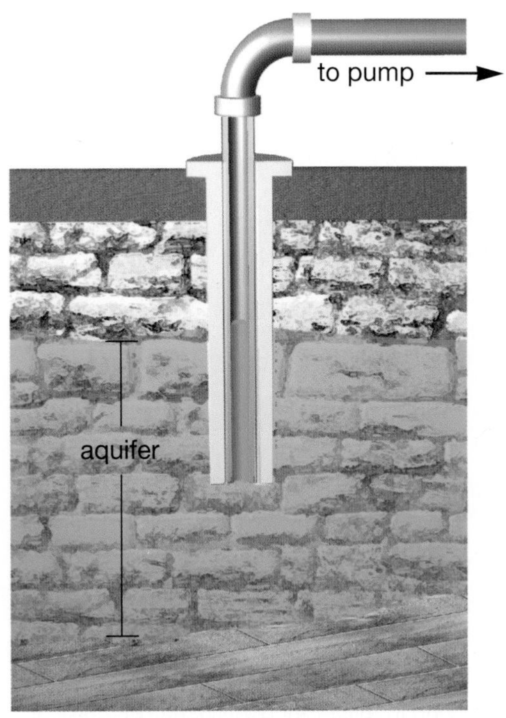

Figure 10.14 Wells are dug deep enough to reach a point below the water table.

DidYouKnow?

In the twelfth century, in the French region of Artois, people began digging wells from which water flowed without any need of a pump. Such wells today are called *artesian wells*, after the name of the region in France.

Skill POWER

To review how to measure liquids using a graduated cylinder, turn to page 540.

Find Out ACTIVITY

How Much Pore Space?

Some materials have a greater volume of pore spaces in them than other materials. How can you measure pore space? Find out in this activity.

What You Need

2 identical beakers
100 mL sand
100 mL gravel
water
graduated cylinder

What to Do

1. Take two identical beakers. Put 100 mL of sand in one and 100 mL of gravel in the other.
2. Slowly pour 100 mL of the water from the graduated cylinder into the beaker of sand until the water just covers the sand. Record the amount of water used.
3. Repeat the procedure with the beaker of gravel.

What Did You Find Out?

1. Which material has more pore space? Why?
2. How might the total volume of pore space change if you crushed the gravel into smaller pieces? Draw a sketch to explain your answer.
3. What variables did you need to keep constant in this activity?

Extension

4. Is soil from your school grounds more porous or less porous than sand? Make a prediction, then repeat the activity using 100 mL of soil.
5. What criteria did you use to assess your solution in step 4?

In some areas, groundwater flows naturally out onto the surface, forming a **spring**. This usually happens on hillsides or in gullies, where the water table is exposed by a dip in the land. Some springs produce steaming hot water. Hot springs occur where groundwater is heated by rocks that come in contact with molten material under Earth's surface. A well-known hot springs area in Canada is Banff, Alberta, as shown in Figure 10.15.

Figure 10.15 Banff Hot Springs, Alberta

Water Above Ground

On its journey down slopes or through the ground, water may arrive at a basin or depression in Earth's surface. Such basins fill with water to form lakes or wetlands, such as swamps, marshes, and bogs. Lakes may be large or small, deep or shallow. They may have streams flowing into and out of them, or they may be completely surrounded by dry land (see Figure 10.16).

Word CONNECT

Find the precise meaning of each of the following words: marsh, bog, swamp, fen, slough, prairie pothole.

Figure 10.16 The Great Lakes contain a total of about 25 000 km^3 of water — one fifth of all the world's fresh surface water.

By studying the contours of the land and the structure of rock and soil, scientists know that lakes continually form and then gradually disappear throughout the history of Earth. Lake Ontario, for example, was once much deeper than it is today. The evidence can be seen in the slope of the old lake bed that rises from the edge of today's lake. Another example is Death Valley, California, which is a flat, hot, dry desert today. Twenty thousand years ago it was covered by a lake 55 m deep. The desert floor is made of sediment left behind after the lake water evaporated.

How do wetlands develop? Many lakes fill in from the sediment brought to them by rivers and then deposited in the still waters. As sand and gravel settle on the bottom, the lake gets shallower. Water plants grow and turn the shallow lake into a swamp, bog, or marsh (see Figure 10.17). In time, the open water disappears altogether.

Figure 10.17 This marsh was once a lake.

In the past, wetlands in many parts of Canada were considered wastelands. Wetlands have neither large supplies of drinking water as do lakes, nor can they be built on or farmed, like dry land. As a result, wetlands were commonly drained, filled in, and used for agriculture. Wetlands, however, are important habitats for many species of animals and plants. They also filter pollutants out of water, prevent soil erosion, and reduce flood damage. You can discover how wetlands function by making a model of one in the next investigation.

Career CONNECT

Wetland Watchers

Marshes are home to many different creatures that require very specific water levels to thrive. If the land is drained or if a dam or other structure does not allow rainwater to run off naturally, the animals and plants in the area may leave, get sick, or die. Marsh managers or wildlife technicians who look after marshes monitor water levels and often adjust the levels using a system of water channels, pumps, and dykes. Consider some of a marsh manager's responsibilities:

- monitoring and adjusting water levels
- maintaining the marsh property, including a system of dykes
- conducting plant and animal surveys
- assisting scientists with wildlife and environmental research
- managing people to ensure hunters, birders, and other visitors to the marsh do not interfere with nesting, feeding, and other wildlife needs.

The first two duties would likely require science skills and math skills. The second duty might also require some practical knowledge such as landscaping or carpentry. The remaining duties involve science, or more specifically, zoology or biology skills and knowledge. The duties involving science would also require communication skills because the marsh manager reports his or her findings and interacts with research scientists and visitors to the marsh.

Make a list of the job responsibilities of a career that interests you. Beside each, mark the letter M, S, or C if you think the duty requires math, science, or communication skills.

INVESTIGATION  10-D

Making a Model of a Wetland

Find out how marshes, bogs, and swamps affect the water flowing through them by making a model of a wetland.

Problem

How does a wetland filter pollutants and control flooding?

Safety Precautions

Dispose of the muddy water as instructed by your teacher.

Apparatus

graduated cylinder

timer or watch with a second hand

roasting pan (about 30 cm x 25 cm x 4 cm works well)

modelling clay

plastic bucket or container

Materials

piece of indoor/outdoor carpeting

water

beaker or jar of muddy water

Procedure

1. Spread modelling clay in one half of the roasting pan to represent land. Shape the clay to slope gradually downward from the end of the pan to the centre. Press the clay to the sides of the pan to seal the edges.

2. Cut a strip of indoor/outdoor carpeting about 5 cm wide and long enough to span the width of the roasting pan. Lay the carpet strip tightly against the lower edge of the clay. This strip of carpet represents a wetland between dry land and an area of open water.

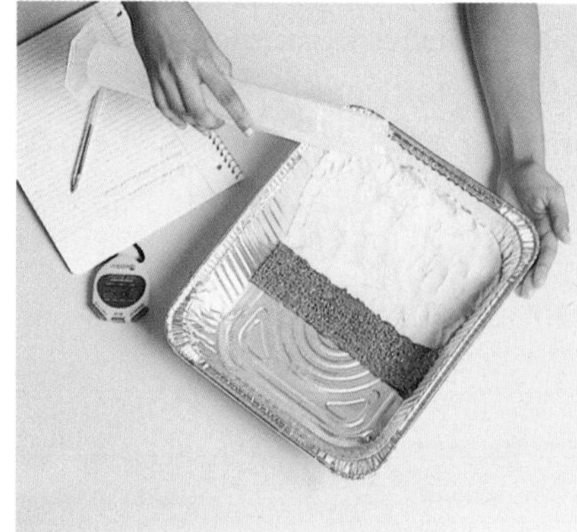

3. Measure 250 mL of water into a graduated cylinder. Start the timer (or note the time) and pour the water onto the top of the clay slope. Record how long it takes for the water to drain completely into the other half of the pan.

4. Pour out the water and remove the strip of carpet. Repeat step 3.

5. Pour out the water and replace the strip of carpet. Make sure you press the carpet tightly to the lower edge of the clay slope.

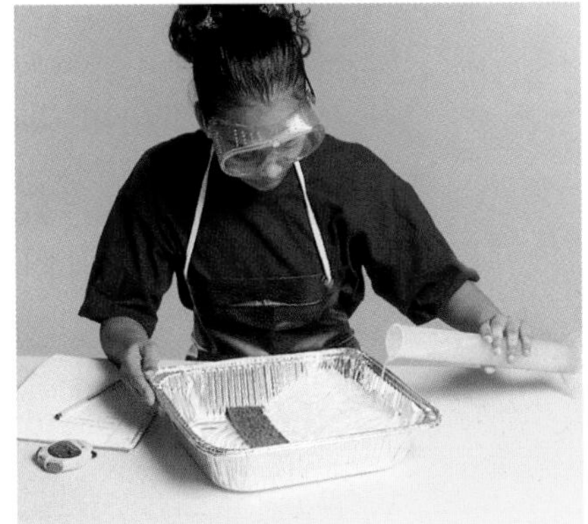

6. Measure 250 mL of muddy water into a graduated cylinder. Pour the water onto the top of the clay slope. Record your observations.
   - **(a)** Wipe up any spills as wet floors are slippery.
   - **(b)** Wash your hands after this investigation.

Analyze

1. How did the carpet (wetland) affect the rate of run-off of water from the land (clay)? Explain your results.
2. Compare the appearance of the muddy water in the jar with the water that ends up in the open part of the roasting pan. Explain your observations.

Conclude and Apply

3. In many coastal areas, wetlands have been drained to allow the building of cottages near the seashore. Suggest why developed areas such as these are often subject to flooding and erosion during storms.
4. Suggest how plant roots in a swamp might help prevent the buildup of sediment and pollution in a lake next to the swamp. What might happen to the lake if the wetland were drained and filled in?

Groundwater and Surface Water

Do you know where the tap water in your home and school comes from? Some communities get their water from lakes and rivers, while others are supplied by groundwater. Does the source of the water matter?

Usually, the supply of water depends on the location. Many towns and cities are built near large rivers and lakes (see Figure 10.18). These open waters have many uses besides supplying tap water. For instance, they may be used for activities such as fishing, recreation, transporting goods, producing hydro-electricity, and carrying away treated (non-hazardous) wastes. Groundwater does not have these other uses, but it is a much more plentiful source of fresh water. Table 10.1 compares some of the other advantages and disadvantages of groundwater vs. surface water.

Figure 10.18 Many towns and cities are built near sources of water. Regina, Saskatchewan overlooks Wascana Lake.

Table 10.1 Comparing Groundwater and Surface Fresh Water

Surface water	Groundwater
• may contain disease-causing micro-organisms	• usually free of disease-causing micro-organisms
• temperature fluctuates	• temperature is nearly constant (this is useful for industries that use water for cooling)
• may carry sediment	• usually does not carry sediment
• chemical composition may fluctuate	• chemical composition is usually constant
• supply is affected by droughts	• supply is not usually affected by short droughts
• easily polluted	• less easily polluted
• supply is limited to areas with rivers and lakes	• supply is available over wide areas that lack dependable surface water
• usually contains fewer dissolved solids	• usually contains more dissolved solids
• usually less costly to develop	• usually more costly to develop

Check Your Understanding

1. What two types of rock are needed to produce an aquifer?
2. What is the relationship between a water table and a well?
3. Someone tells you it is a waste of time to drill for water in a desert. Explain why you agree or disagree.
4. What conditions produce a hot spring?
5. Name an example of a wetland and describe how it is formed.
6. Describe two ways in which wetlands can benefit people.
7. Name an advantage and a disadvantage of using groundwater as a source of drinking water.

10.3 Water, Land, and Atmosphere

You have learned that some precipitation runs off the land into streams and rivers, and then pours into the oceans. Some precipitation goes underground. Some goes into wetlands and lakes. No matter where falling rain and melted snow travel, however, the journey does not end. Recall the Starting Point Activity on page 311. You began with water in the bottom of a jar and then observed water drops appear near the top of the jar and under the lid. The ability of water to disappear from view and to reappear somewhere else is not magic but the result of evaporation and condensation.

DidYouKnow?

If there were no water cycle, the current rate of evaporation of water from the oceans, lakes, rivers, and land would completely dry out Earth in about 3000 years.

The Water Cycle

Evaporation converts liquid water from the surface of oceans, lakes, rivers, and land into water vapour. This process explains why puddles disappear and streets dry out after a rain shower has ended. There is always some water vapour in the atmosphere, even if you cannot always see it. Water vapour remains in the atmosphere until it cools, forms clouds, and condenses. Liquid water then falls from the clouds as precipitation, taking us back to the beginning of the journey. To be more precise, there is no real beginning. Nearly all the water on Earth moves continually between the oceans, land, and atmosphere in a **water cycle** without an end or a beginning (see Figure 10.19).

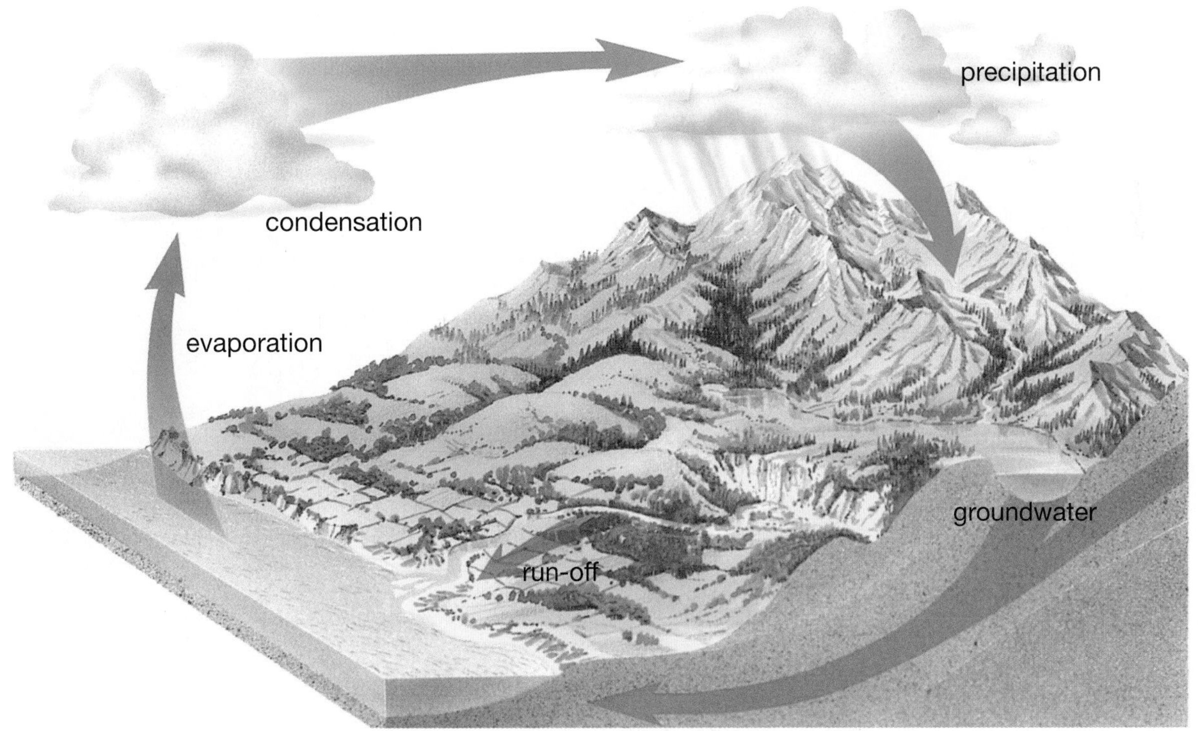

Figure 10.19 Water moves from Earth to the atmosphere and back to Earth again in the endless water cycle.

Measuring Rates of Transpiration

You learned in Chapter 3 that plants absorb water from the soil and release it through their leaves in the process of transpiration. This process is an important part of the water cycle. In fact, transpiration from an area of forest or grassland can release more water vapour into the air than evaporation from soil, lakes, and rivers in the area. In this investigation, you will measure the rate of transpiration from an area of grass.

Problem

How can you measure and compare rates of transpiration?

Safety Precautions

Be careful when using the stapler and the mallet.

Apparatus

- graduated cylinder
- 2 plastic sheets (1 m^2 each)
- 2 small plastic dishes
- 8 wooden stakes (20 cm each)
- tape measure or metre stick
- mallet
- stapler
- 2 marbles or pebbles

Procedure

1. Locate an area of short grass and a nearby area of bare soil where you can set up your experiment. This investigation is best done in the morning on a sunny day.

2. Measure out an area of 1 m^2 on the grass.

3. Use the mallet to place a stake at each corner of the square, leaving a height of 10 cm above the ground.

4 Place a small plastic dish (such as an empty yogurt container) at the centre of the square.

5 Stretch the plastic sheet over the square. Fix the corners of the sheet to the stakes with staples so the sheet is suspended 10 cm above the ground.

6 Place a marble or pebble at the centre of the sheet directly over the plastic dish.

(a) Repeat steps 2 to 6 on a nearby area of bare ground.

(b) Leave the set-up in place for at least 1 h.

7 Using a graduated cylinder, measure the volume of water collected in each dish. Record your results.

Skill POWER

To review how to measure the volume of a liquid, turn to page 540.

Analyze

1. Why would you expect water to collect in the plastic dishes?

2. Why did you set up a plastic sheet over a bare area?

3. Explain any differences you observed in the volume of water collected in the two locations.

Conclude and Apply

4. From your results, what inference can you make about the role of transpiration in the water cycle?

A glass of water that you drink today may include some of the same water a dinosaur bathed in millions of years ago!

Heat is needed to evaporate water. Scientists estimate that almost one quarter of the Sun's energy reaching Earth is used in evaporating water. Each year, about 520 000 km^3 of water from Earth's surface are converted to water vapour and rise into the air. Water vapour does not remain in the atmosphere for long. On average, there is enough water vapour in the atmosphere for only a ten-day supply of precipitation worldwide. Most precipitation falls on the ocean surface, and most of the water vapour in the atmosphere evaporates from the oceans.

In the late seventeenth century, the English scientist Edmund Halley was the first to demonstrate that water on Earth keeps moving in a complete cycle. He calculated the rate of evaporation of water from a dish and used this rate to estimate the amount of water evaporated from the Mediterranean Sea. He compared this quantity with the amount of water flowing into the Mediterranean from rivers. The two amounts were equal, showing that evaporation and precipitation alone can explain why rain keeps raining and rivers keep flowing.

Distribution of Water

The water cycle balances evaporation with precipitation over Earth as a whole. However, the distribution of water varies greatly from one part of the planet to another. Massive bodies of salt water — the oceans — cover more than two thirds of Earth's surface (see Figure 10.20). About 97 percent of all the water on the planet is found in the oceans, as shown in Figure 10.21. They fill wide, deep depressions called **ocean basins**. If you drop a rock from a boat while sailing on the open ocean far from land, the rock may sink as far as 4 km beneath the waves before touching the ocean floor. This depth is four times as great as the average height of the continents above sea level.

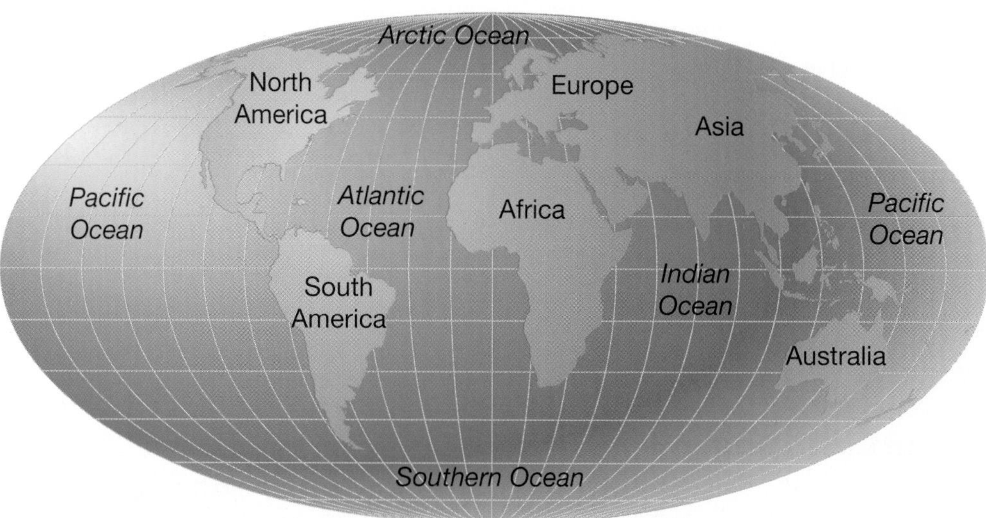

Figure 10.20 The oceans cover about 70 percent of Earth's surface, or 360 million square kilometres. The total volume of the oceans is about 1.35 billion cubic kilometres.

After the ocean water, the next largest amount of water is frozen in large masses of ice, forming ice sheets at the North and South poles and glaciers in high mountaintops (see Figure 10.22). These large masses of compressed snow and ice cover about 10 percent of Earth's land area and lock up about three quarters of the

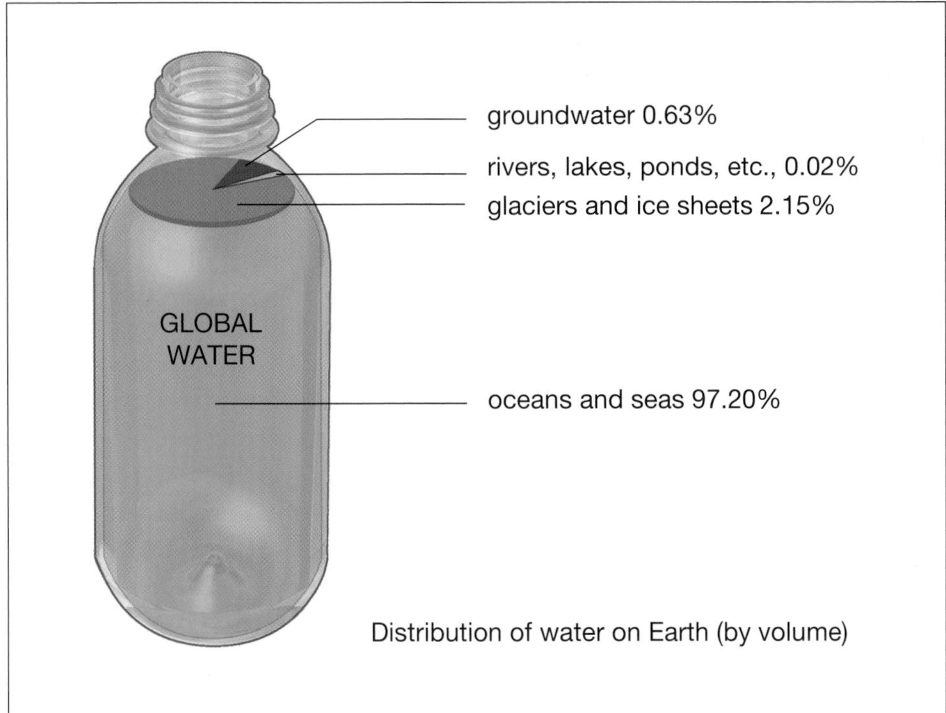

Figure 10.21 If all the water on Earth were contained in a 1 L bottle, only about half a teaspoonful would be fresh water.

Math CONNECT

Look at the percentages of water distribution on Earth in Figure 10.21. Draw a circle graph to represent these percentages.

Skill POWER

To review how to draw a circle graph, turn to page 546.

planet's fresh water supply in its solid state. Indirect evidence from rock erosion and fossil remains indicates that glaciers have been more extensive in the past. During periods called ice ages, when the global climate was much colder, glaciers spread over as much as 28 percent of the land. You will investigate the links between water and climate in Chapter 12.

Little more than 0.5 percent of the water on the planet is found as liquid fresh water — the form most easily available for use by people. Most fresh water (about 97 percent) occurs below the surface of the land as groundwater. The tiny remainder accounts for all the water on the land surface in rivers, streams, ponds, lakes, and fresh-water wetlands.

Figure 10.22 A glacier consists of frozen fresh water.

Air Pollution and Water Systems

Pause& Reflect

Seawater is a solution of salts in water. When seawater evaporates, what happens to the water part of the solution (the solute)? What happens to the salt part (the solvent)? Write your ideas in your Science Log, with an explanation.

The water cycle moves water back and forth between the atmosphere, land, and oceans. This process also purifies water as it recycles it. For example, most of the water vapour in the atmosphere comes from the oceans, but rainwater is fresh water, not salt water. Evaporation followed by condensation distils seawater, separating pure water from salts and other dissolved substances.

Many human activities affect processes in the water cycle, creating problems for people, plants, and animals. Although the water cycle purifies water, pollutants can be added into the cycle at any stage. For example, you have already learned how pollutants can wash into rivers from the land. Pollutants can also enter water vapour in the atmosphere. One example is acid rain, which affects large areas of Canada. **Acid rain** is caused by dissolved sulfur dioxide and nitrogen oxides in the atmosphere. These are waste gases released by coal-burning industries, metal smelters, and automobiles. These gases combine with water vapour in the atmosphere to form sulfuric acids and nitric acids. These chemicals return to Earth in precipitation that can be more acidic than vinegar (see Figures 10.23A and B).

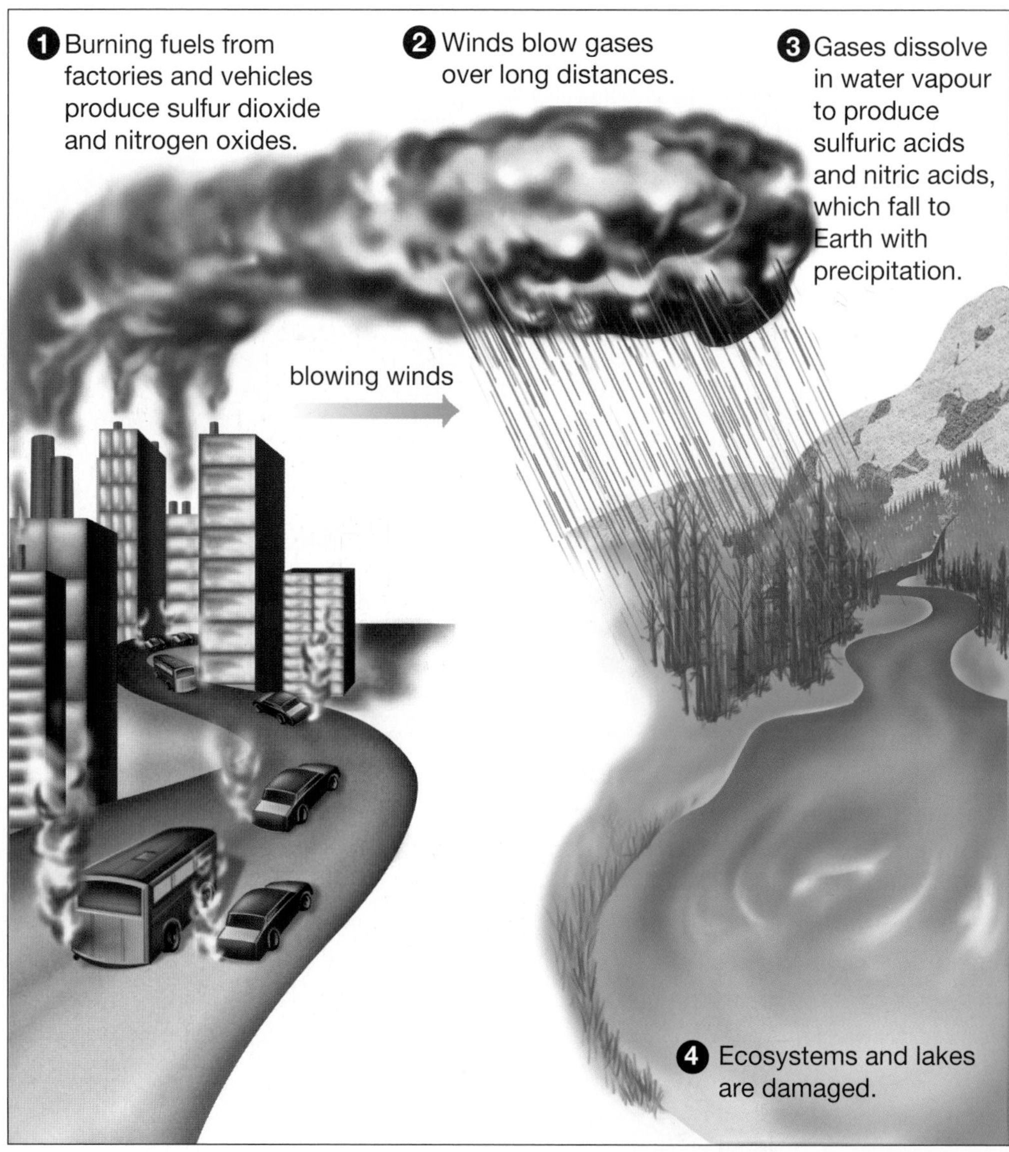
1 Burning fuels from factories and vehicles produce sulfur dioxide and nitrogen oxides.
2 Winds blow gases over long distances.
3 Gases dissolve in water vapour to produce sulfuric acids and nitric acids, which fall to Earth with precipitation.
blowing winds
4 Ecosystems and lakes are damaged.

Figure 10.23A Polluting gases become part of the water cycle when acid rain is produced.

Acid rain ←

Most acidic rain recorded 2.0

Vinegar 2.8

Normal rain 5.6

Human blood 7.4

Average pH of Great Lakes 8.3

0 1 2 3 4 5 6 7 8 9 10 11 12 13 14

ACIDIC

NEUTRAL

ALKALINE (Basic)

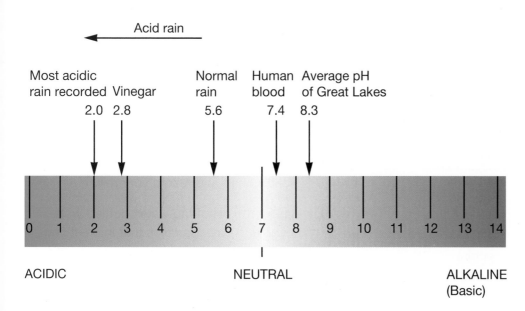

Figure 10.23B The pH scale is used to describe how acidic a solution is. A pH of 7 is neutral (neither acid nor base). The lower the pH number, the greater the acidity. Normal rain has a pH of 5.6. Acid rain has a pH below 5.6.

Figure 10.24 Acid rain can slowly dissolve stone, causing damage to structures.

Winds carry atmospheric pollution from industrial areas to areas of lakes and forests, where acid rain can slowly kill or damage plants and animals. When acid rain sinks into the ground, it can dissolve toxic metals in the soil and rock, such as aluminum, mercury, and lead. These dissolved toxic metals might eventually work their way through the ground into streams and rivers, causing more damage. Acid rain also causes damage to the human-made environment. The acids dissolve certain building materials, such as limestone and marble. Old monuments and statues in Europe, and the Houses of Parliament in Ottawa, for example, have all been damaged by acid rain (see Figure 10.24).

Career CONNECT

Testing the Waters

Water quality is a serious issue. If our drinking water becomes polluted, it can make us sick. Even swimming in a contaminated lake can cause illness. That is why Gareth Gumbs' job is so important. He is an environmental chemist who works in a laboratory. Part of his job involves testing water samples to see if the water is safe. His experiments help people figure out if a landfill site is contaminating a town's water supply, for example, or if a beach is unsafe for swimming.

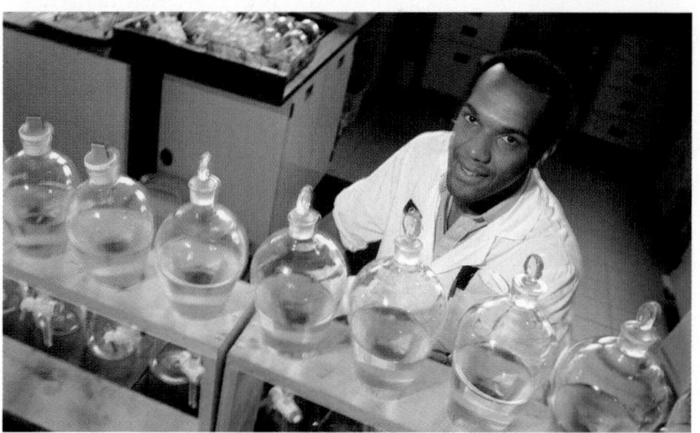

Gareth is also hired to do experiments to determine how human activity and industry are affecting water quality. For example, some lakes have high levels of acid rain. The acid can dissolve some of the surrounding rock and release toxic substances such as mercury into the water. In one study, Gareth and his colleagues analyzed many fish samples over the course of a year. They needed to find out how much mercury was in the body tissues of the fish because of the water they were inhabiting.

Gareth's job requires good problem-solving skills. Not all of the tests he runs are straightforward. In many cases, he must consider similar tests that other researchers have done, choose the one most suitable for the study he is working on, and then adapt the test to suit his needs. Slowly and precisely, he does the experiments and analyzes the results. Then, he fits the pieces of the puzzle together to provide a clear analysis for the client who has requested the environmental study.

Based on what you have learned about the career of environmental chemist, make two lists. In one, list the aspects of Gareth's job that appeal to you. In the other, list the aspects of his job that do not appeal to you. Give a brief explanation for the item in each list.

Find Out **ACTIVITY**

How Pure Is Your Rainwater?

You may live in a city or in a rural area. Is one place more likely to have acidic rainwater than another? If so, why? In this activity, you can perform a simple pH test to compare the acidity of your rainwater with that of several other liquids.

What You Need

pH indicator (for swimming pools)
tap water
bottled water
vinegar
test tubes
beaker

What to Do

1. Collect a sample of rainwater in a clean beaker.
2. Place about 2 mL of the rainwater into a test tube.
3. Take a very small amount of the pH indicator and add it to your sample. Shake the test tube gently until the indicator dissolves.
4. Match the colour of your sample to the indicator chart and record your results.
5. Measure the pH of tap water, bottled water, vinegar, and other samples that interest you. Record your results in a list of increasing acidity.
6. Wipe up any spills as wet floors are slippery.
7. Wash your hands after this activity.

What Did You Find Out?

1. How does the acidity of your rainwater compare with that of normal rain (5.6)?
2. If your rainwater varies from normal pH, suggest why.

To decrease the amount of acid rain, efforts have been made to reduce pollution from industries and automobiles. Governments around the world have passed laws to cut down pollution entering the atmosphere from factory smokestacks and car exhausts. Although these laws have helped, acid rain continues to be a problem. For one thing, some countries' laws are not strict enough. Other solutions include using different forms of fuel that do not produce polluting gases. These include natural gas, solar power, and wind power.

Check Your Understanding

1. From where does most water vapour in the atmosphere evaporate?
2. In what form is most fresh water found?
3. Where is most of the liquid fresh water on land located?
4. How does the water cycle help purify water?
5. What process involving plants is an important part of the water cycle? Briefly describe this process.
6. How can air pollution in one place produce water pollution in another place some distance away? Give an example.

CHAPTER at a glance

Now that you have completed this chapter, try to do the following. If you cannot, go back to the sections indicated.

Name the three states in which water occurs and describe where you would find a naturally-occurring example of each state. (10.1)

Explain, using diagrams or words, how precipitation falling on a mountainside ends up in the ocean. (10.1)

Name Canada's five main drainage basins. (10.1)

Explain what causes river flooding and describe how flood damage can be controlled. (10.1)

Give a reason why some precipitation sinks underground and explain what determines its maximum depth. (10.1)

Name two ways in which people obtain underground water. (10.2)

Describe the processes by which a lake may change into a wetland. (10.2)

Copy the sketch of the water cycle shown here and add the missing labels. (10.3)

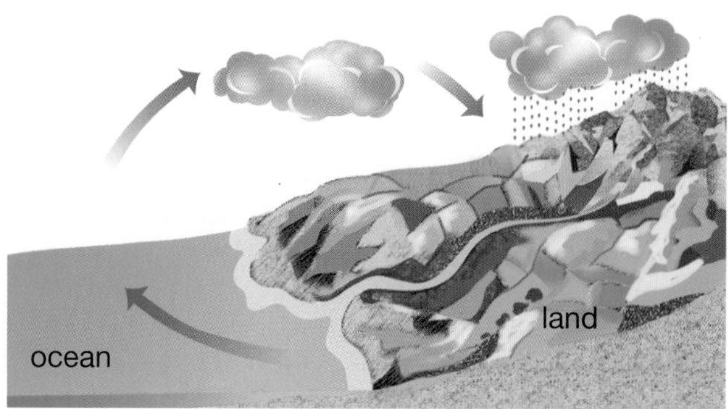

Give an example of how a human activity causes pollution to enter the water cycle and describe an effect of the pollution. (10.3)

List four locations in which water is distributed on Earth, in decreasing order of abundance (that is, from most plentiful to least plentiful). (10.3)

Prepare Your Own Summary

Summarize this chapter by doing one of the following. Use a graphic organizer (such as a concept map), produce a poster, or write the summary to include the key chapter ideas. Here are a few ideas to use as a guide:

- Where does water vapour in the atmosphere come from?
- Why do rivers flood?
- Where does well water come from?
- How is acid rain formed?
- How might the logging of trees pollute a river?
- Copy the diagram of a water table shown here. Add labels to identify permeable and impermeable layers of rock and an aquifer. Use these terms to explain the origin of spring water.
- Where is most water on Earth found?
- Why do rivers flow into the sea or into a lake?

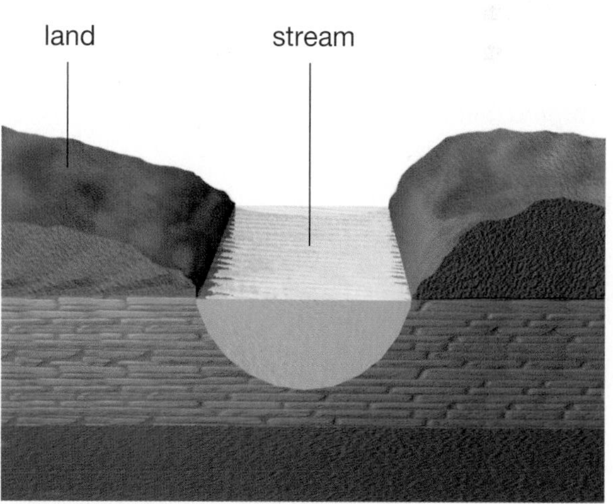

- How could you make a simple model of the water cycle?
- Why can activities in one place pollute water in a different place?

CHAPTER 10 Review

Key Terms

impermeable
porous
run-off
upstream
downstream
valley
lake
ocean
plains
flood plain
meltwater
levees
drainage basin
drainage divide
Continental Divide
mean discharge
groundwater
aquifer
zone of saturation
water table
spring
water cycle
ocean basins
acid rain

Reviewing Key Terms

If you need to review, the section numbers show you where these terms were introduced.

1. Write a sentence to describe the relationship among the terms in each group.
   (a) drainage basin, run-off (10.1)
   (b) groundwater, water table (10.2)
   (c) zone of saturation, water table (10.2)
   (d) water cycle, evaporation, condensation (10.3)
2. Complete each sentence by choosing the correct word from the pair given.
   (a) The direction from which a river flows is upstream/downstream. (10.1)
   (b) The boundary between drainage basins is a drainage divide/Continental Divide. (10.1)
   (c) The top of a saturated layer of rock is a water table/well. (10.2)
   (d) Water flows from the ground in a lake/spring. (10.2)
   (e) Water enters the atmosphere by condensation/evaporation. (10.3)
   (f) Water leaves the atmosphere by precipitation/condensation. (10.3)

Understanding Key Ideas

Section numbers are provided if you need to review.

3. What factors determine how much of precipitation becomes run-off? (10.1)
4. Compare and contrast two characteristics of a young stream and an old stream. (10.1)
5. What factors determine the level of a water table? (10.2)
6. What two main processes are involved in the water cycle? (10.3)
7. What supplies the energy for the water cycle? (10.3)
8. Make a flow chart to illustrate the events in the water cycle in sequence, beginning with precipitation falling to Earth. (10.3)

Developing Skills

9. Copy and complete the spider map of precipitation shown below.

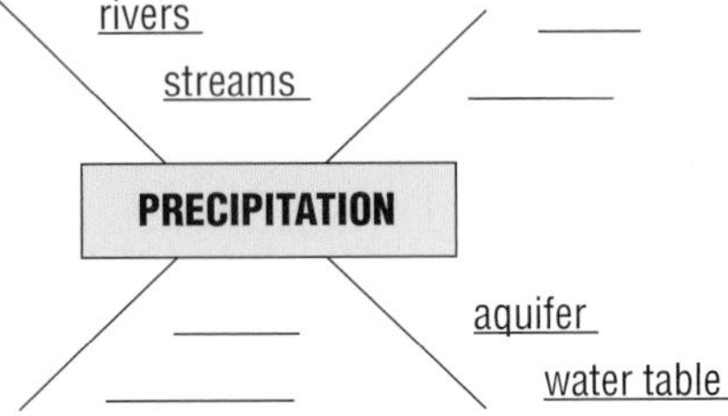

10. The data shown on page 341 were obtained by students using a stream table to compare the run-off time of 200 mL of water from a surface of sand and clay. The experiment was repeated with the stream table placed at three different angles of slope.
    (a) Make a graph of the data, with angle of slope on the horizontal axis (x-axis) and run-off time on the vertical axis (y-axis). Use one coloured line to show run-off times on clay, and another coloured line for run-off times on sand.

(b) What is the relationship between angle of slope and run-off time?

(c) From which surface, and at which angle, would the greater volume of water run in 1 min?

Effect of Varying Slope on Run-off Time

	slope (°)	run-off time (s)
Clay	15	4.58
	30	3.97
	60	2.40
Sand	15	8.57
	30	8.00
	60	7.30

Problem Solving/Applying

11. A farming family had been getting plenty of water from their well for many years. One winter, a developer completed a new housing development on land close to the farm. The family did not notice any change in their water supply until the end of the following summer, when their well ran dry. Explain what happened. How could the family avoid this problem next year?

12. What sort of rocks would you look for if you were trying to find an aquifer? Explain.

13. How can the volume of run-off from an area be reduced? Why would you want to control run-off?

14. Design a method you could use to condense water vapour in the atmosphere. Draw a labelled sketch of your design.

15. Consult library resources or search the Internet for information about world ice distribution patterns over several geological time periods. Draw a map showing these changes in ice distribution on Earth.

16. Formulate your own question about wetlands, flooding, water pollution, or a topic of your choice, and explore possible answers.

Critical Thinking

17. Explain how groundwater, aquifers, and the water table are related to each other.

18. Why does the volume of water flowing in a drainage basin vary throughout the year?

19. The peak season of water flow in a drainage basin is different in northern Ontario than in southern Ontario. Explain how the peak seasons are likely to differ and why.

Pause& Reflect

1. Whether you get your drinking water from a tap, from a well, or from bottled spring water, it is all recycled. In your Science Log, write a paragraph or poem, or make a drawing that expresses the idea of water moving endlessly through the phases of the water cycle.
2. Go back to the beginning of this chapter on page 310 and check your original answers to the Getting Ready questions. How has your thinking changed? How would you answer these questions now that you have investigated the topics in this chapter?

CHAPTER

11 The Oceans

Getting Ready...

- How does salt get into ocean water?
- Why are some beaches sandy, while others are rocky?
- How deep are the deepest parts of the ocean?

In your Science Log, try to answer the questions above. Expand and revise your answers after completing this chapter.

The ocean is never still. You can see it moving in the waves that crash ashore. Sit on a beach for a few hours, and you will see the water move gradually in or out as the ocean slowly rises or falls. On the open water, sailors witness broad, slow-flowing "rivers" at the ocean's surface. These movements carry enormous amounts of water over hundreds of kilometres from one part of the ocean to another. The ocean is in constant motion below the surface, too. Huge masses of cold water creep along the ocean floor, then rise from the depths to the surface.

What effect do these moving waters have on our lives? No matter where you live, you cannot escape the influence of the ocean. The ocean shapes the land along the shoreline. Its waters provide us with food, and its floor holds resources of oil, gas, and minerals. Many of the items you buy in a store were transported across oceans from other continents. Even the weather is affected each day by the oceans. How has the ocean affected your life today?

Spotlight On Key Ideas

In this chapter, you will discover

- what determines the shape, the height, and the wavelength of an ocean wave
- how beaches, bays, and headlands are formed
- what makes ocean tides rise and fall
- what the ocean floor looks like
- what causes ocean currents

Spotlight On Key Skills

In this chapter, you will

- make a model to show how waves affect a beach
- use data to draw a cross-section of the sea floor
- design your own investigation of how to protect beaches from erosion
- investigate the buoyancy of fresh water compared to salt water
- change the density of water to illustrate how currents are formed

Starting Point ACTIVITY

The Mysterious Ocean Environment

Because we live on dry land, we may not always realize the importance of the oceans to life on Earth. When you look at a map of the world, you can see that the oceans dominate the planet. Oceans cover two thirds of Earth's surface. The volume of water in the oceans is eleven times the volume of the land above the oceans. The oceans are a vast and difficult environment to explore, and much about them remains unknown.

What to Do

1. Before studying this chapter, think about what you already know about the ocean. You will use this knowledge to create an ocean environment in your classroom.
2. As a class, brainstorm everything you associate with the word "ocean." List this information on a large chart with the word "ocean" in the middle.
3. Organize the information into categories. For example, "Forms of Life in the Ocean," "Physical Conditions in the Ocean," and "Resources from the Ocean."
4. Work in groups to produce classroom displays based on each category. For example, you might paint murals; make mobiles; produce a cassette tape of ocean sounds; write a poem; or make models of beaches, ships, or sea organisms.
5. As you study this chapter, look for new ideas to add to your displays.

11.1 The Coastline: Where Sea Meets Land

Figure 11.1 Some of the largest waves occur along the coasts of California and Hawaii. The waves in these places have been blown by wind over thousands of kilometres of open ocean.

Surfers ride ocean waves, using the motion of the water to carry them to the shore. What causes waves? You can find the answer in a bowl of hot soup! If you blow on the soup to cool it, your breath makes small ripples on the surface of the liquid. Ocean **waves** are just large ripples, set in motion by steady winds.

Waves begin on the open sea. Their height depends on how fast, how long, and how far the wind blows over the water. An increase in any one of these variables can cause an increase in wave height. Normal winds produce waves of 2–5 m in height. Hurricane winds can create waves 30 m high — two thirds the height of Niagara Falls! Even on a calm day, there is usually a steady movement of smooth waves near the shore. These smooth waves are called **swells**. They are caused by winds and storms far out at sea.

Whether large or small, waves on the water have features in common with all the other types of waves studied by scientists — such as sound waves, light waves, and radio waves. First, waves have height, as shown in Figure 11.2. As you learned in Chapter 9, section 9.2, a wave's height is measured from its crest (the highest part of the wave) to its trough (the lowest part of the wave). Second, waves also have a wavelength, which is the distance from one crest to the next. Third, waves have a speed of motion, which is measured by the time required for one wave to pass a given point.

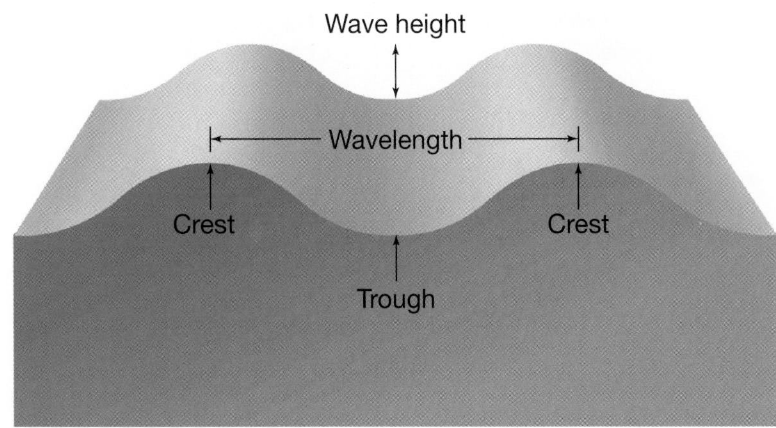

Figure 11.2 Features of a wave

Waves of Energy

A small boat drifting offshore bobs up and down with the waves, moving slightly back and forth in the same place. The boat does not move along with the waves that roll under it toward the beach. You can discover why by making waves with a rope in the Find Out Activity below. Waves travel along the rope away from your hand. However, each part of the rope remains the same distance from you, moving only up and down.

Find Out **ACTIVITY**

Making Waves

If you stand on a beach and watch the waves roll onto the shore, the water appears to be flowing across the surface of the ocean. In this activity, you can study wave motion.

What You Need

piece of rope (about 1.5 m long)
felt tip pen

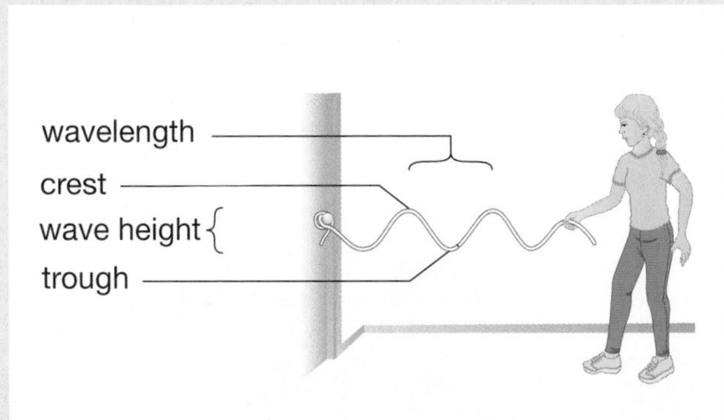

What to Do

1. Tie one end of the rope to a doorknob (or the back of a heavy chair).
2. With the felt tip pen, make a mark 1 cm long about halfway along the rope.
3. Hold the other end of the rope in your hand and stand away from the door.
4. Move your hand up and down once to make a wave.
5. Experiment with ways of making taller waves, faster waves, or more waves in the rope at one time. Note what hand movements produce each pattern.

What Did You Find Out?

1. In which direction does a wave move along the rope?
2. Describe the motion of the ink mark on the rope. Did the ink mark move along the rope?
3. What is the relationship between the amount of energy you use to move your hand and the wave patterns you produce?

DidYouKnow?

Some of the highest ocean waves are not caused by winds but by earthquakes, volcanic eruptions, or landslides on the sea floor. These events give the ocean water a push, just as you can make a wave in a bathtub by pushing with your hands underwater. Sea-floor events sometimes produce giant waves called *tsunamis.* The wavelength of a tsunami may be 150 km, and it may travel over the open ocean at a speed of up to 800 km/h. When a tsunami approaches land, the speeding water is pushed up into a towering, destructive wave with a height of 15 m or more.

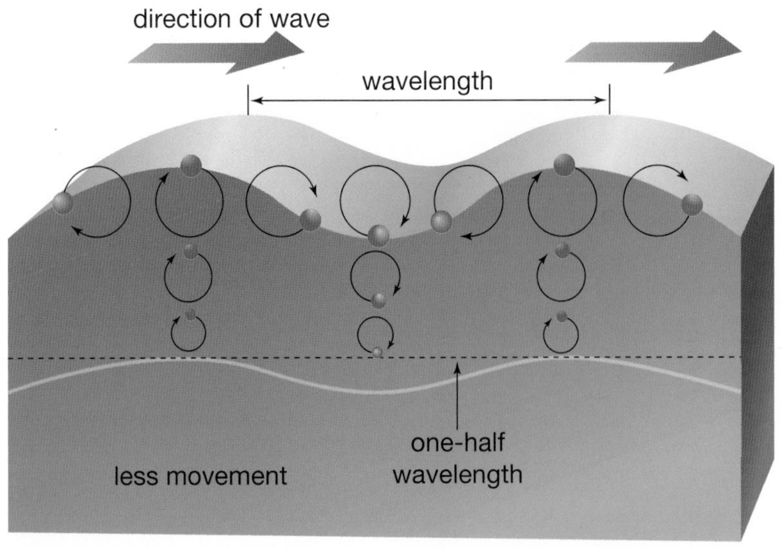

Figure 11.3 Individual particles of water move in circles as a wave passes through the water. The circles are smaller as depth increases. At a depth below about half the wavelength, motion ceases.

Near the ocean surface, water particles move in a circular motion as a wave passes. The particles end up in nearly the same position as where they started, as shown in Figure 11.3. As each water particle moves, it bumps into the next particle and passes its energy along.

When a wave reaches shore, it changes shape. As the trough of the wave touches the beach, it is slowed down by friction. The crest of the wave, however, continues moving at the same speed. The wavelength shortens, and the wave height increases. The crest of the wave eventually outruns the trough and topples forward. The wave collapses onshore in a tumble of water called a **breaker** (see Figure 11.4). A surfer can use the forward motion of the crest to ride a wave to shore.

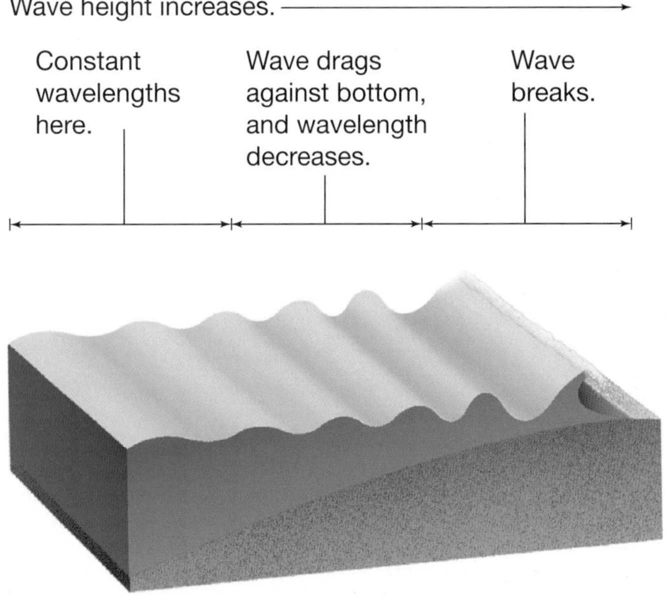

Figure 11.4 As a wave approaches the shore, its wavelength decreases and its height increases. It collapses onshore as a breaker.

How Waves Change Shorelines

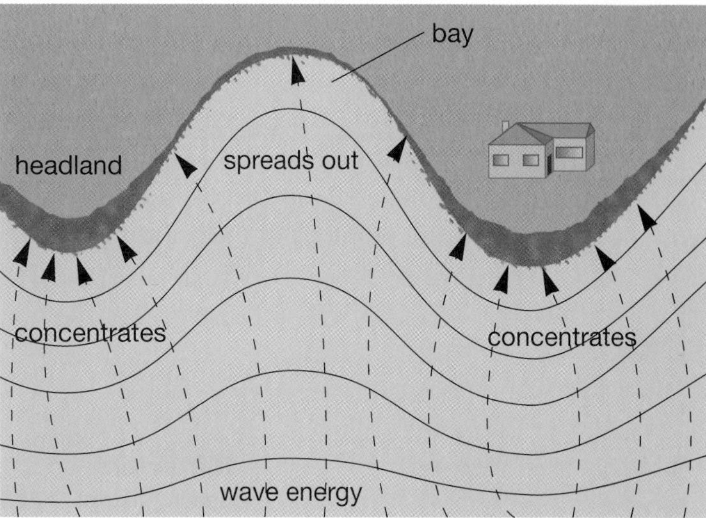

Figure 11.5 The energy of the waves is concentrated on the headlands. In bays, waves reach the shore with much less force than on headlands.

The energy in waves shapes and reshapes coastlines around the world. Along some coastlines, churning waters swirl rocks and pebbles on narrow strips of land between steep sea cliffs and the ocean. On other coastlines, gentle waves lap broad, gently-sloping beaches of fine sand. Why does one coastline have a sandy beach while another does not? You can find answers in the type of rock in each area and in the shape of the coastline.

Some coastlines have a zigzag shape that forms a series of **bays** (parts of the ocean that reach into the land) and **headlands** (areas of the land that stick out into the sea). Waves approaching shore strike headlands first (see Figure 11.5). They slam into the rocks with a force of up to thousands of kilograms per square metre (see Figure 11.6A). This impact breaks off chunks of rock from the headland, producing vertical sea cliffs that extend into deep water. After hitting headlands, waves lose much of their energy. They slow down and spread out into nearby bays and coves. In these places, the water arrives on shore with much less force compared with the water striking headlands.

At the base of a sea cliff, masses of loose rock and pebbles swirl around in the breaking waves. Softer rock near the base of a cliff is removed by this action and then washed away. Over time, this process of erosion may hollow out the rock to form sea caves and arches in the cliff, as shown in Figure 11.6B. Seawater can also dissolve certain minerals in rock, increasing erosion by chemical action. The combination of all these processes can erode areas of rocky shoreline by as much as 1 m in a year.

DidYouKnow?

In a single day, about 14 000 waves crash onto a shore.

Figure 11.6A Waves can change the shape of shorelines.

Figure 11.6B Sea caves and arches are formed by erosion by waves and wind.

CONDUCT AN

INVESTIGATION 11-A

Waves and Beaches

What happens to the material on a beach when waves strike it? Depending on the type of rock and the slope of the coastline, waves can either erode the land or build it up.

Problem

How do waves affect beaches?

Safety Precautions

Apparatus

beaker or measuring cup (500 mL)
clear plastic or glass pan or small aquarium
ruler
small block of wood
clock or watch
plastic pail or container

Materials

beach mixture 1 (450 mL sand + 150 mL gravel)
beach mixture 2 (450 mL gravel + 150 mL sand)
water

Procedure

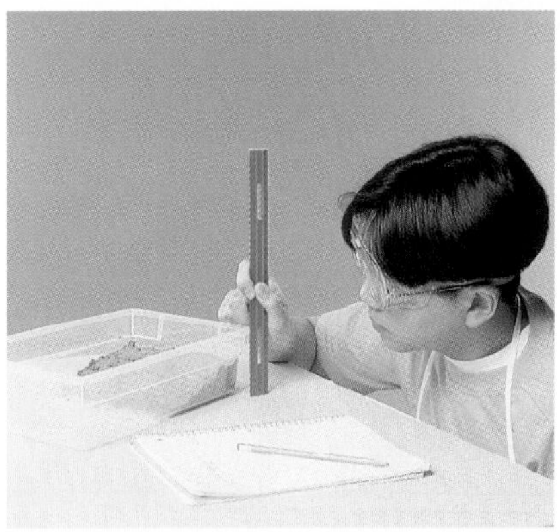

1. Using beach mixture 1, build a small beach at one end of the pan or aquarium. Use your ruler to measure the height of the beach at the top of the slope.

2. Then measure the width of the beach (from the top of the slope to the bottom of the slope). Draw a side view of the beach to scale.

3. Carefully pour water into the other end of the pan or aquarium until the water level reaches about one third of the way up the beach.

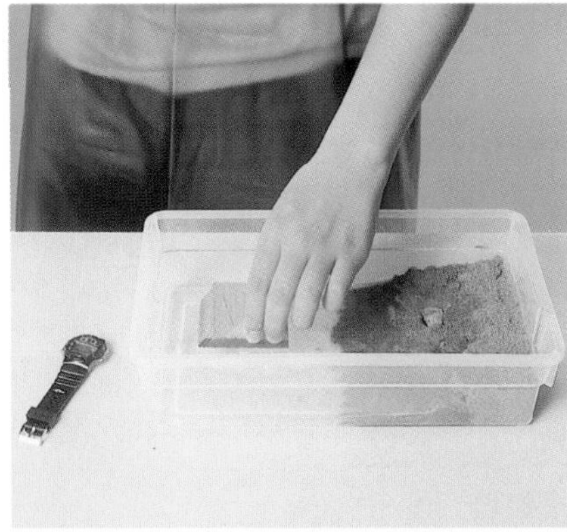

4. Make waves by holding the block of wood in the water and quickly moving it up and down for 2 min. Try to keep the speed and size of the waves constant.

5. After 2 min, draw a side view of the beach to scale. (Use your ruler to measure its dimensions.)
   (a) Label the position of the sand and the gravel.
   (b) Carefully pour off the water into a container provided by your teacher.

6. Rebuild a beach with a slope twice as steep as the first. Measure the dimensions of the beach and make a sketch of it.

7. Repeat steps 3 to 5. Then empty the water and beach materials into a container provided by your teacher.

8. Repeat steps 1 to 7, using beach mixture 2.
   (a) Wipe up any spills as wet floors are slippery.
   (b) Wash your hands after this investigation.

Skill POWER

For tips on estimating and measuring, turn to page 540.

Analyze

1. Based on your models, describe the effect that wave action has on a beach made mostly of sand compared with a beach made mostly of gravel.
2. How does the slope of a beach affect erosion by waves?
3. How do the materials on a beach affect erosion by waves? Explain your observations by referring to the difference in the mass of grains of sand and pieces of gravel.

Conclude and Apply

4. Based on the results of this investigation, what effect do you think a large storm at sea might have on a sandy beach?
5. Beach erosion is a problem for many seaside communities. Suggest what might be done to prevent a beach from eroding.

How Beaches Are Formed

What happens to all the fragments of rock nibbled from the coast by crashing waves? As they rub against each other in the surging water, rock fragments are smoothed and ground down into smaller pebbles and grains of sand. Along steeply-sloping shorelines, these rock fragments wash back into the sea. This leaves a shoreline of only bare rock, with scattered boulders and larger stones. Where the shoreline has a gentler slope and calmer waters, smaller rock fragments can settle and build up, forming a broad beach (see Figure 11.7).

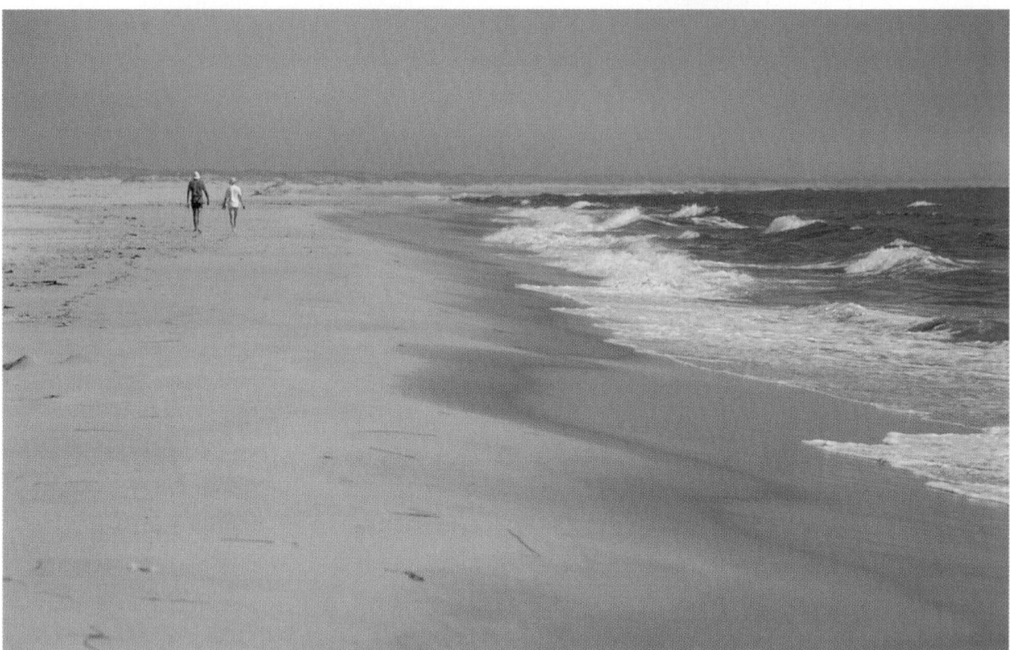

Figure 11.7 Ocean waves are a powerful and constant agent of change along coastlines. Like the fingers of a sculptor, they poke and probe the land, taking away bits here and adding pieces there. The result can be a sandy beach like this one.

Beaches are deposits of sediment that run along the shoreline. The materials that form a beach range in size from fine grains of sand less than 2 mm in diameter to pebbles and small boulders. Most beach sediments are fragments of hard minerals such as quartz. Beaches can also include other minerals of various colours, or fragments of seashells and coral.

Due to the continuous action of waves, beaches are in a constant state of change. In winter, strong winds bring larger waves that remove more sediment from the beach than they deposit. The beach erodes and becomes narrower. Winter waves may damage boardwalks and beachside dwellings. Calmer summer weather produces low, gentle waves that deposit sediments on shore, rebuilding the beach.

Word CONNECT

A poem about the sea written by Canadian poet E.J. Pratt (1882–1964) has these lines:

It took the sea a thousand years,
A thousand years to trace
The granite features of this cliff,
In crag and scarp and base.

Explain what the poet is describing. Write your own short poem or haiku describing ocean waves.

Figure 11.8 Barriers are used along some shorelines to protect beaches from erosion.

Many seaside communities spend a great deal of money to preserve their beaches. They may build seawalls and breakwaters, as shown in Figure 11.8. These barriers of concrete and steel help slow down the speed and force of waves. Jetties are walls that extend from the land into the ocean. They are built to prevent sand from blocking the mouth of a river or harbour. Along some sandy beaches, rows of short walls called groins are built at right angles to the beach. They help reduce the movement of sand along the shoreline. However, most of these structures are only temporary solutions. Efforts to protect beaches are costly and never-ending. Some large vacation resorts even truck in sand to replace the sand removed by waves.

Career CONNECT

From Sea to Sea to Sea

Canada is a country bordered by oceans on three sides. For that reason, the Canadian Coast Guard has an important job to do. It employs many people on its boats as well as in its onshore offices. Some of its services include:

- setting up and expanding satellite navigational systems and aids for ships, such as buoys and lighthouses
- keeping northern channels clear of ice, freeing ice-trapped ships, and escorting ships through frozen passages
- providing a ship-to-shore radio communication link
- monitoring and assisting ships in distress
- promoting boating safety and implementing safety standards
- protecting the marine environment, by cleaning up oil and chemical spills, for example

The Canadian Coast Guard College in Sydney, Nova Scotia, accepts between 20 and 30 new students each year into its four-year course. Applicants are secondary school graduates. Imagine that you are applying to be accepted into the Canadian Coast Guard College. In addition to a transcript (official record) of your high school grades, you would need to send a covering letter highlighting your experience or interests that relate most closely to the Coast Guard's objectives. Make a list of your recreational interests; clubs you belong to; part-time, summer, or volunteer jobs you may have had; or any other relevant experience. Decide which of these experiences you think will be most important to those judging your application. Write a covering letter highlighting these points.

Tides

If you stand on a beach in one place for long enough, your feet will get wet as the sea level rises. Beaches are sometimes covered by water, and sometimes they are not. They are covered and uncovered in a regular daily cycle by the slow rise and fall of the ocean, called **tides**. The upper and lower edges of a beach are determined by the high-tide mark and the low-tide mark.

Centuries ago, people realized that the cycle of tidal movement is linked to the motion of the Moon. The largest tidal movements (called **spring tides**) occur when Earth, Moon, and Sun are in a line (see Figure 11.9A). At these times, tides are extra high and extra low. The smallest tidal movements (called **neap tides**) occur when the Sun and Moon are at right angles to each other (see Figure 11.9B). On these days, there is little difference in depth between high and low tides. The difference in level between a high tide and a low tide is called the **tidal range**.

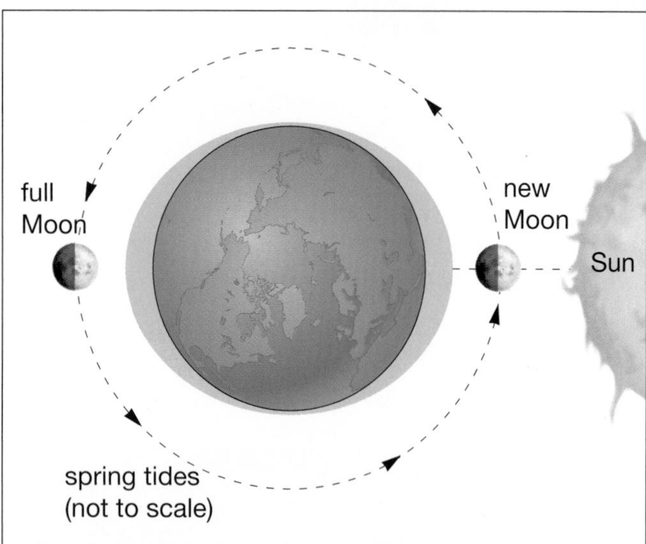

Figure 11.9A Spring tides occur twice per month, at full Moon (when Earth is between the Moon and the Sun) and at new Moon (when the Moon is between Earth and the Sun).

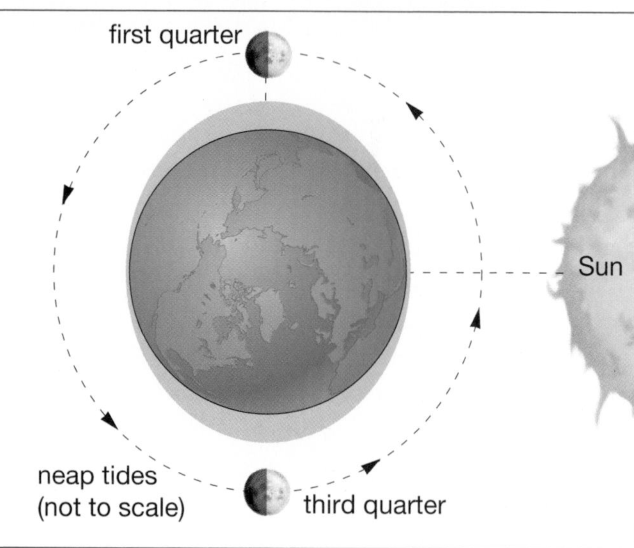

Figure 11.9B Neap tides occur twice per month, during the first-quarter phase and third-quarter phase of the Moon.

The link between Earth, the Moon, the Sun, and tides is gravity. As you learned in previous studies, gravity is the force of attraction between two masses. Tidal movements result mainly from the pull of the Moon's gravity on the ocean. The Sun is much farther from Earth than the Moon is. Thus, the Sun has less than half as much influence on the tides as the Moon does, despite the Sun's much greater size.

Can you explain the difference between spring tides and neap tides by studying Figures 11.9A and B? During spring tides, the Sun adds its gravitational pull to the Moon's, producing a large tidal range. During neap tides, the Sun's pull works against the Moon's. This produces high tides that are not very high and low tides that are not very low.

If you look at the water moving up a beach as the tide rises, you may think that the volume of the ocean is increasing. However, the bulge of water that produces a high tide along one coastline draws water away from the other side of the ocean.

This causes a low tide along the opposite coastline (see Figure 11.10). As Earth turns on its axis, different locations on Earth's surface face the Moon. The result is a sequence of high and low tides that follow each other around the world. On many of Earth's shorelines, tides rise and fall about twice a day (see Table 11.1).

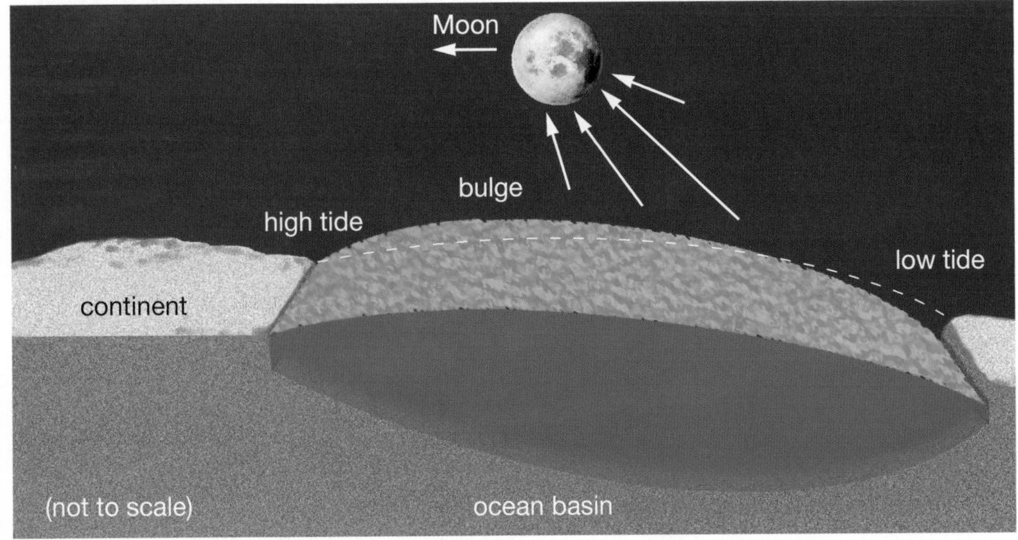

Figure 11.10 The gravitational pull of the Moon shifts water from one side of the ocean to the other.

If you live on Canada's east coast and telephone a friend on the Atlantic coast in Europe, you will find that the time of day there is 4 h later than where you are. If you and your friend phone each other from a beach, you will also notice that when you see a rising tide your friend sees a falling tide.

Look at Figure 11.11. The Moon's pull causes the ocean to bulge on the side of Earth facing the Moon. At the same time, there is a second bulge of water on the side of Earth that faces away from the Moon. This occurs partly because the Moon pulls more strongly on Earth itself than on the ocean on the far side. This leaves a bulge of ocean behind. The two areas of high tide on opposite sides of the planet draw water away from the areas in between them. In these areas, the ocean's surface falls.

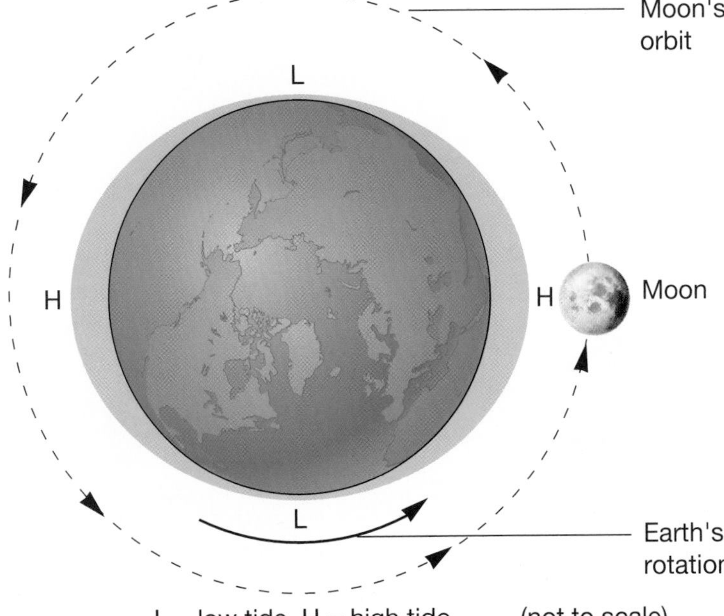

Figure 11.11 High tides occur on the side of Earth closest to the Moon and on the side farthest away.

The times of each tide have a pattern that follows the cycle of the Moon. Study Table 11.1. What is the time interval between one tide and the next? Predict the time of the next low tide on July 2.

Table 11.1
Sample Tide Table

Date	Tide	Time
July 1	High	1:00 a.m.
	Low	7:13 a.m.
	High	1:25 p.m.
	Low	7:38 p.m.
July 2	High	1:50 a.m.
	Low	

DidYouKnow?

Lakes also have tides, but the change in water level is usually too small to be noticeable. In a very large lake, such as Lake Superior, the tidal range is only a few centimetres.

In mid-ocean, the rise and fall of the ocean averages less than 1 m. Along shorelines, the tidal movement is more noticeable. The shape of a shoreline can have a great influence on the size of the tidal range. For example, in the Gulf of Mexico the tidal range is only about 0.5 m. The Gulf has a narrow passage, or mouth, to the open ocean, and a long, curved coastline (see Figure 11.12). A rising tide that enters the narrow mouth spreads out around the bay, giving a small tidal range. In the Bay of Fundy, on Canada's east coast, the opposite occurs. The bay there is long and V-shaped. Tides enter the wide mouth of the V and pile up as they are funnelled down to the narrow end of the bay. The tidal range in the Bay of Fundy can be as great as 20 m (see Figures 11.13A, B, and C).

Figure 11.12 The Gulf of Mexico has a small tidal range. Tides that enter the small mouth of the gulf spread around the long coastline of the bay.

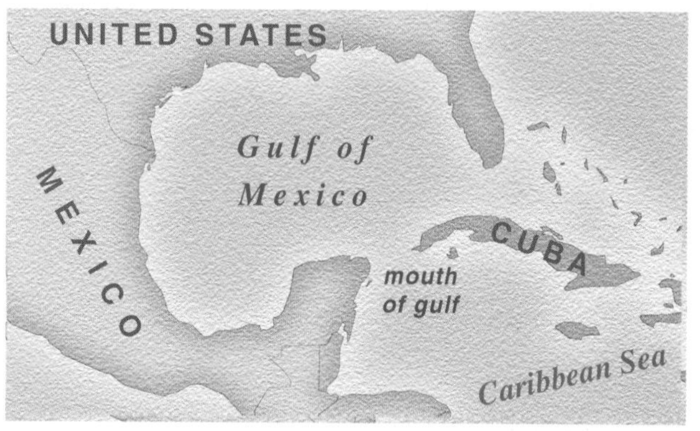

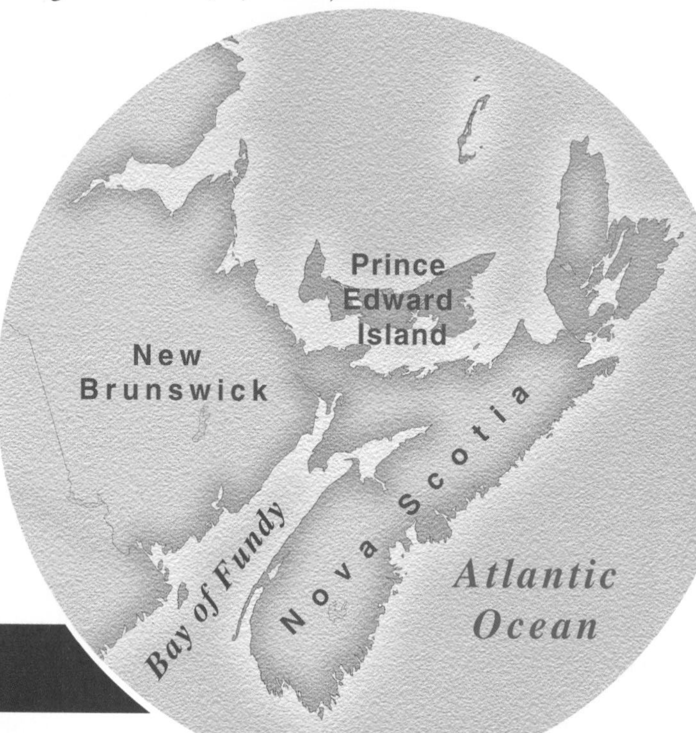

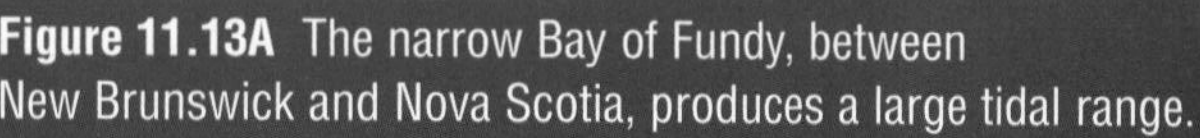
Figure 11.13A The narrow Bay of Fundy, between New Brunswick and Nova Scotia, produces a large tidal range.

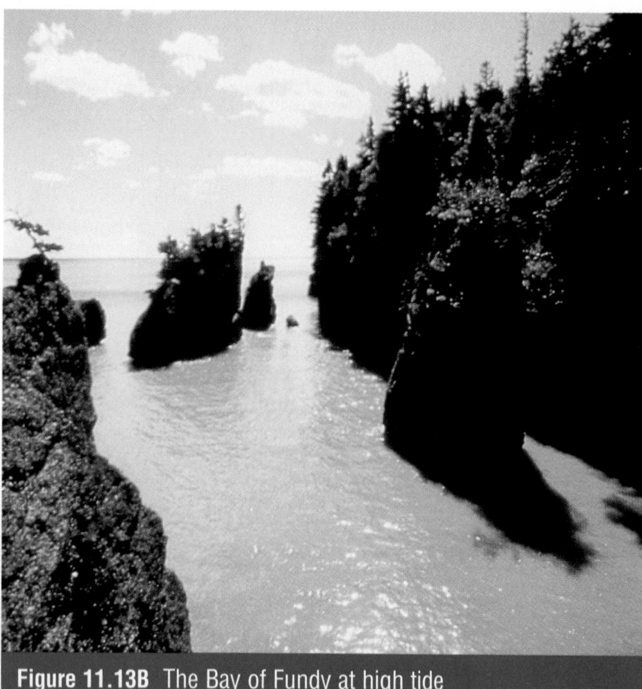
Figure 11.13B The Bay of Fundy at high tide

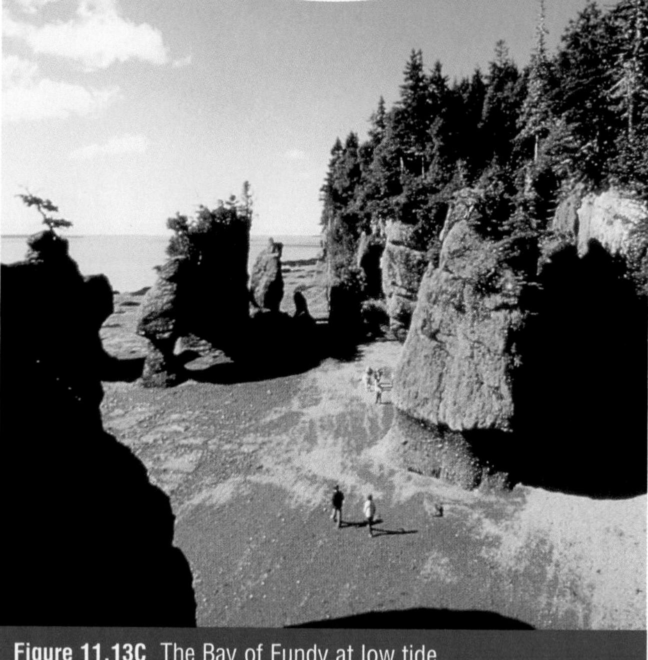
Figure 11.13C The Bay of Fundy at low tide

Find Out **ACTIVITY**

Shorelines and Tidal Range

Ocean basins are containers that hold the sea. The size and shape of these containers affect the height of tides along the shorelines. You can observe this effect for yourself in this activity.

What You Need

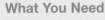

containers of varying width, depth, and shape (each with a capacity of at least 500 mL)
ruler
water
beaker (500 mL)

What to Do

1. You will be pouring the same volume of water into each container.
   (a) Draw a small sketch of each container, showing its size and shape. Use the sketches as headings for a data table.
   (b) Predict which container will exhibit the greatest depth of water and which will exhibit the smallest depth of water.
2. Fill the beaker to the 500 mL mark with water. Pour all the water into one container.
3. Using the ruler, measure the depth of water in the container. Record your measurements in the table.
4. Repeat steps 2 and 3 for all your containers.
5. Wipe up any spills as wet floors are slippery.
6. Wash your hands after this activity.

What Did You Find Out?

1. What features of a container produce the greatest depth in a given volume of water?
2. What features of a container produce the smallest depth in a given volume of water?

Check Your Understanding

1. How does a breaker form?
2. What processes produce bays and headlands?
3. How is a sandy beach formed?
4. Name three factors that determine the height of a wave.
5. Explain the difference between spring tides and neap tides.
6. Explain why the Bay of Fundy has a large tidal range.
7. **Apply** Why is it important for sailors to keep track of high tides and low tides?
8. **Thinking Critically** Why do beach sands from different locations vary in sediment size, colour, and texture?
9. **Design Your Own** Using materials and procedural steps similar to those in Conduct an Investigation 11-A, design your own investigation to determine the most effective types of barriers in protecting beaches from erosion. In a group, brainstorm how to construct models to represent various types of jetties and sea walls. Have your teacher approve your experimental design before you proceed.

11.2 Ocean Basins

The oceans form the largest ecosystem on Earth. Until the twentieth century, we knew very little about them. Oceans are vast, and much remains to be discovered about what lies below the ocean surface. Beneath the waves, most of the ocean is pitch-black. Visible light does not penetrate the water beyond a depth of about 100 m. Starting with simple techniques, and later using advanced technology, scientists have come to learn more and more about the amazing undersea world.

A Journey on the Ocean Floor

Imagine you could empty the oceans of water and take a journey along the sea floor. What would you see? For one thing, features in the ocean basins are much bigger than on land. There are mountain ranges taller than the Himalayas, steep valleys deeper than the Grand Canyon, and vast plains wider than the Canadian Prairies (see Figure 11.14). In most cases, the forces that shaped these features are different from those that shaped rivers and lakes. There are no winds, rivers, rain, or ice on the seabed to erode and carve the rock. Instead, the origin and formation of the ocean basins are due mainly to the movements of Earth's tectonic plates.

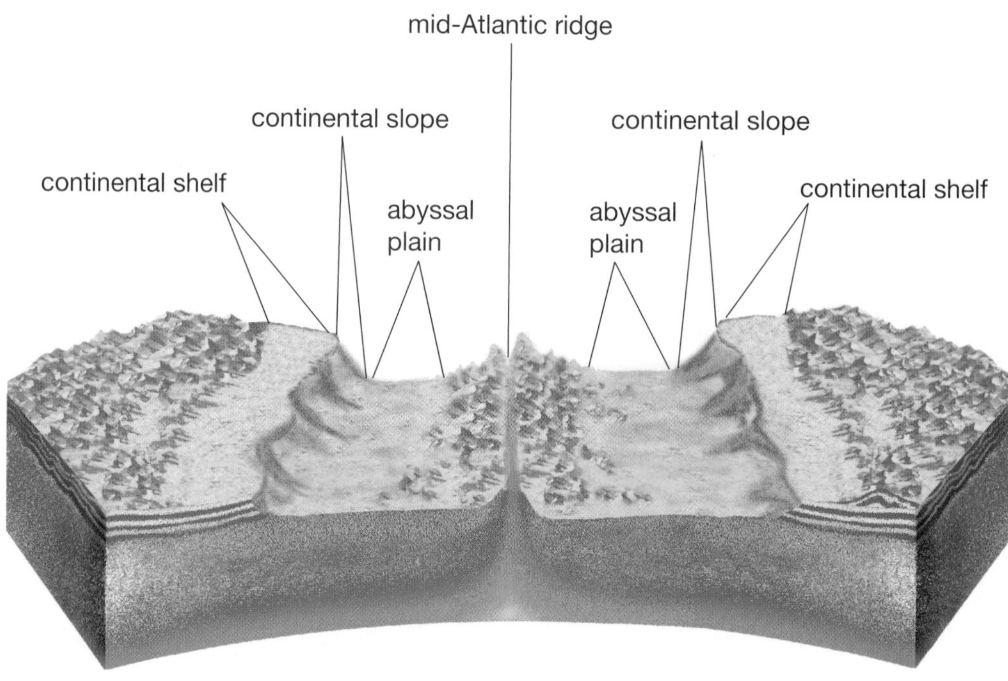

Figure 11.14 The ocean basins contain mountain ranges, deep valleys, and wide plains.

Recall from your earlier studies that Earth's crust is made of large plates of rock that move slowly over time. One result of this movement is that the oceans have not always been shaped as they are today. As tectonic plates move apart or together, some oceans expand, while others shrink. For example, precise measurements have shown that the Atlantic Ocean is expanding, carrying North America and Europe farther apart at a rate of about 3 cm per year (see Figure 11.15).

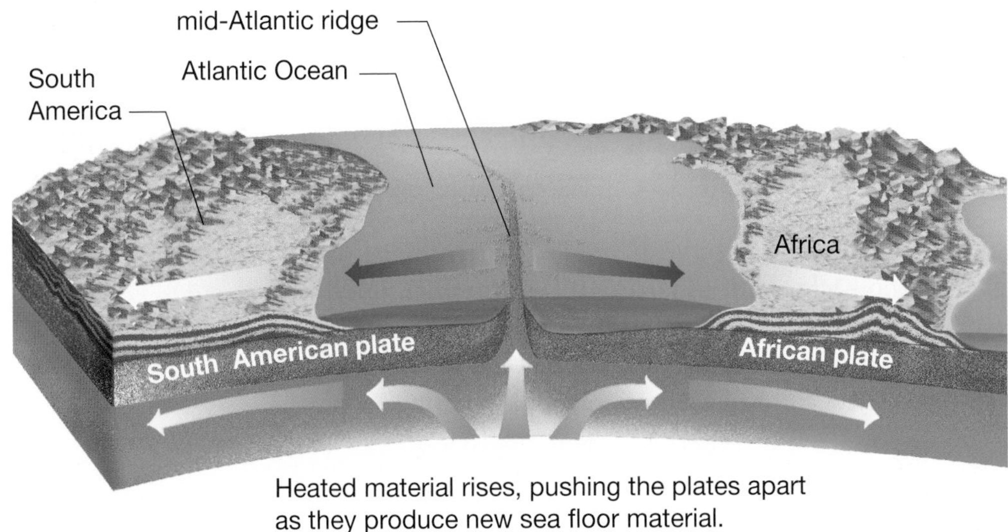

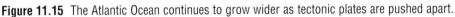

Figure 11.15 The Atlantic Ocean continues to grow wider as tectonic plates are pushed apart.

Pause& Reflect

Put your hands on a globe of the world, placing one hand on either side of the Atlantic Ocean. Imagine your hands are tectonic plates, and begin to move them slowly apart. What would happen at either side of the Pacific Ocean as you do this? In your Science Log, write a short paragraph or draw a sketch to explain why trenches are located around the rim of the Pacific Ocean.

What features of the sea floor are related to plate movements? Long undersea mountain chains called **mid-ocean ridges** run along the centre of the oceans. These ridges are the youngest areas of the sea floor and are still being formed by volcanic eruptions. Molten lava flows from these ridges, quickly hardening into new plate material that pushes tectonic plates farther apart. Mid-ocean ridges are more than 1000 km wide and rise 1000–3000 m above the sea floor.

Large submarine mountain peaks are called **seamounts**. They are usually older, inactive volcanoes at the edges of mid-ocean ridges. These peaks were carried farther from the centre of the active ridge as the plates moved apart. Seamounts are often found in clusters and are most common in the Pacific Ocean.

Along some margins of the sea floor, narrow, steep-sided canyons, called **trenches**, are formed where the edge of an ocean plate pushes against the edge of a continental plate. As the plates move together, the ocean plate is forced to bend steeply down beneath the heavier continental plate. Most trenches occur around the margin of the Pacific Ocean. The deepest trench, called the Marianas Trench, extends 11 km below sea level. This distance is nearly seven times deeper than the Grand Canyon — deep enough to submerge an object as tall as Mount Everest (see Figure 11.16).

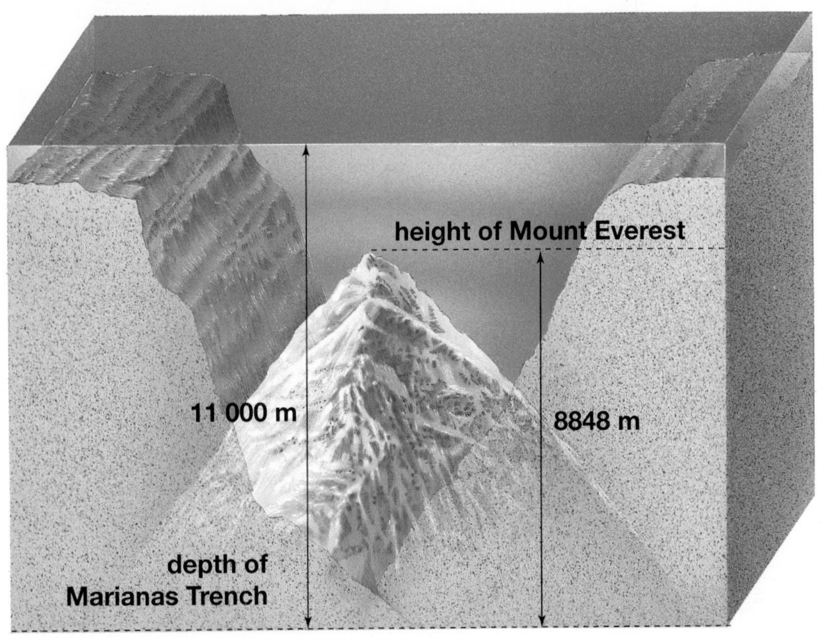

Figure 11.16 The top of the tallest mountain on land, Mount Everest, would be under more than 2000 m of water if its base were set in the Marianas Trench.

What's Down There?

How can you study something that you cannot observe directly? That is the challenge for oceanographers — people who study the oceans. In this investigation, you will simulate a simple method used to map the sea floor. Work with a partner. First, you will make a model of a sea floor. Then you will use a method of discovering its shape. Finally, you will record data and make a cross-sectional drawing.

Problem

How can you find out what the sea floor looks like without seeing it?

Safety Precautions

Be careful when using sharp objects such as scissors.

Apparatus

wooden blocks or other objects of various sizes
scissors
ruler
pen

Materials

shoebox
drinking straw
tape
graph paper

Procedure

1. To make the model sea floor, place different objects in the shoebox. It will be more interesting if you include items of different sizes. Do not let your partner see what you are doing.

2. To discover the shape of the model sea floor, you will measure its depth at regular intervals. Start by drawing a line along the centre of the shoebox lid. Cut a slot from this line a little wider than the width of a drinking straw.

3. Using a ruler, mark off intervals 2 cm apart along the length of the slot.

   (a) Make a table to record the distance from the lid to the sea floor at each interval.

   (b) Give your table a title.

Skill POWER

For tips on how to make a table, turn to page 546.

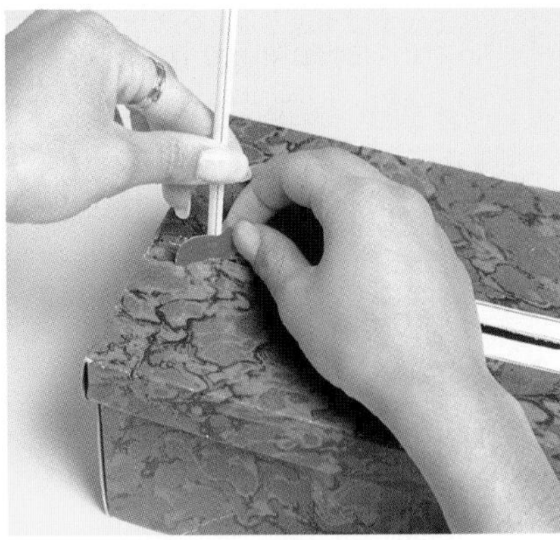

4. Push the straw through the slot at the first interval, holding it vertically. Lower the straw until it touches the model sea floor. Carefully attach a small piece of tape to the straw at the level of the box lid.

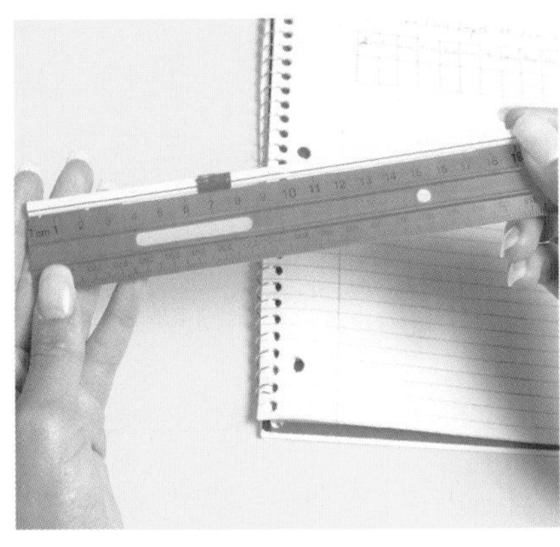

5. Withdraw the straw and measure the distance between the bottom of the straw and the tape. Record this distance in your table.
   - **(a)** Remove the tape.
   - **(b)** Repeat the depth measurements at each interval.

6. On a piece of graph paper, use your data to draw a cross-section of the model sea-floor shape to scale.
   - **(a)** Your drawing (called a profile) will show the terrain of the model sea floor as it would appear if you sliced through the ocean basin along a straight line.
   - **(b)** Wash your hands after this investigation.

Analyze

1. Visually compare the model sea floor and your profile.
2. Did the measuring method you used provide an accurate cross-section of your model sea floor?
3. Did you miss recording any features in your model? Why do you think this is so?

Conclude and Apply

4. You made measurements along a single line. What are the limitations of using this approach?
5. How could you map the entire model of your sea floor?
6. Do you think that a method similar to this one would be useful to map the real ocean floor?

Between the high mountain ranges at their centre and the deep trenches at their edges, the ocean floors are remarkably flat. These wide, open features of the deep sea are called **abyssal plains**. They are formed of thick deposits of sediment, up to 1 km deep in places.

The sediments come from the continents, brought to the ocean edge by rivers. How do they reach the deep sea floor? They are carried there in great underwater landslides. These landslides are started by earthquakes, or simply by the force of gravity. From time to time, massive volumes of mud and sand slump down the slopes at the edge of the continents. Dropping to deeper waters, mud-filled currents spread their deposits evenly over the abyssal plains.

Continental Shelves

Ocean basins do not begin at the coastline. Instead they begin many kilometres out at sea. The area between the coast and the edge of the ocean basin is actually a submerged part of the continent, called the **continental shelf** (see Figure 11.17).

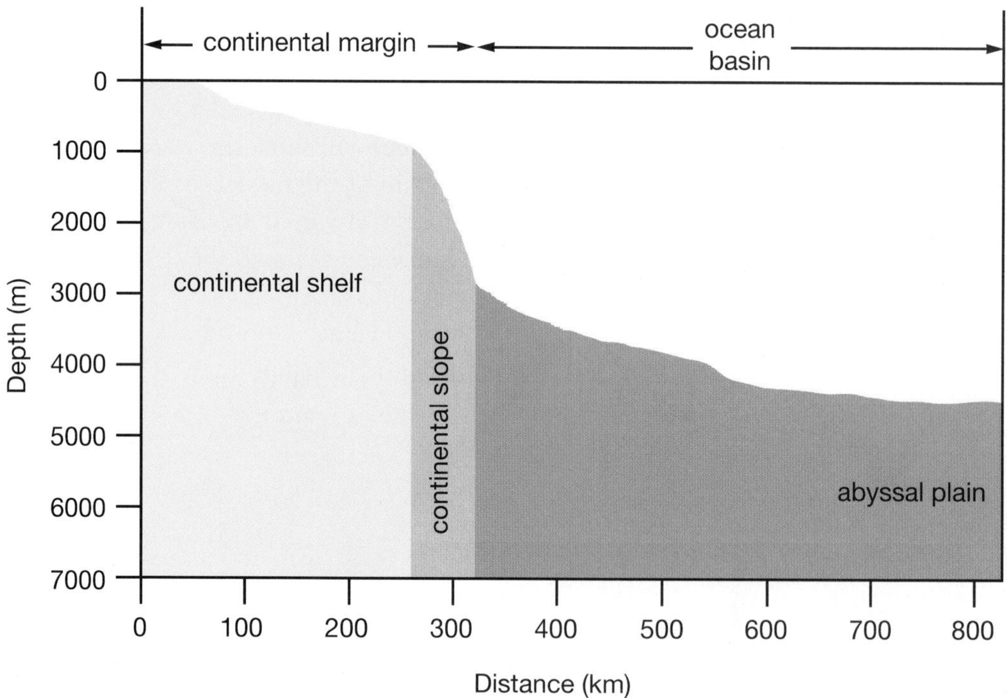

Figure 11.17 The ocean floor is divided into two main regions — the continental margin (which is part of the continental rock) and the ocean basin.

Continental shelves slope gradually away from the land before dropping steeply downward at the shelf edge. The narrowest continental shelves occur along coastlines with high mountains, where plates are moving together. The width of these shelves is usually less than about 30 km. Along other coastlines, the continental shelf is much wider. It may extend more than 300 km out into the sea.

From the edge of the shelf, the **continental slope** plunges at a steep angle to the sea floor. Continental slopes are usually less than 200 km wide and descend to about 3 km. Beyond the base of the continental slope lies the floor of the ocean basin.

INVESTIGATION 11-C

Mapping the Ocean Floor

Think About It

What does the ocean floor look like? In Conduct an Investigation 11-B, you drew a profile of a sea floor you made yourself. Now you can use this skill to draw a profile of the sea floor under the Atlantic Ocean. The data you will use are actual measurements taken from a ship during a research voyage between New Jersey and Portugal. Identify these two places on a map before you begin.

What to Do

1. Make a graph like the one shown below. Give your graph a title.
2. Study the data in the table shown on the right. It lists a series of depths to the ocean floor. The measurements were taken at 29 different points along a line.
3. Plot each data point on your graph. Connect the points with a smooth line.

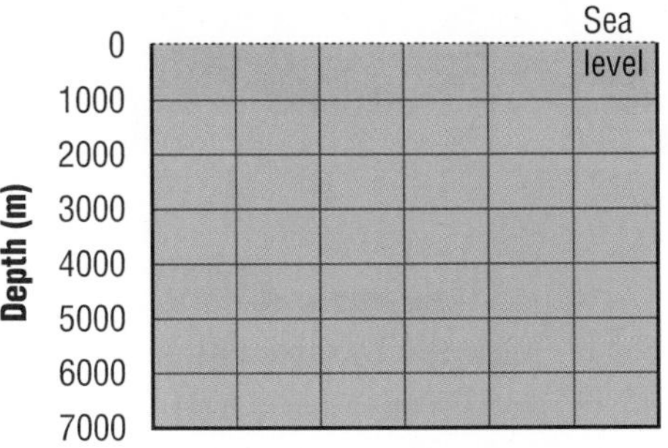

Analyze

1. List the sea floor features that occur east of New Jersey
   (a) between 160 and 1050 km
   (b) between 2000 and 4500 km
   (c) between 5300 and 5600 km
2. Does your profile give an accurate picture of the sea floor? Explain. (Hint: Consider the distance between the two locations and the water depth shown on your graph.)
3. Why do you think the depth of the ocean floor is 0 for station 20? What feature do you think this might be?

Data for Sea-Floor Profile

Station number	Distance east of New Jersey (km)	Depth to ocean floor (m)
1	0	0
2	160	165
3	200	1800
4	500	3500
5	800	4600
6	1050	5450
7	1450	5100
8	1800	5300
9	2000	5600
10	2300	4750
11	2400	3500
12	2600	3100
13	3000	4300
14	3200	3900
15	3450	3400
16	3550	2100
17	3600	1330
18	3700	1275
19	3950	1000
20	4000	0
21	4100	1800
22	4350	3650
23	4500	5100
24	5000	5000
25	5300	4200
26	5450	1800
27	5500	920
28	5600	180
29	5650	0

Exploring the Oceans

The first map of the sea floor was produced in the 1870s. Scientists on board the expedition ship *HMS Challenger* used wire lines lowered at intervals to the ocean floor. By using this method, they discovered the long undersea mountain range in the middle of the Atlantic Ocean.

An improvement on this method — called *sonar mapping* — uses sound waves to probe the seabed. The depth of water is found by sending sound waves directly down from a ship and measuring the time it takes for the signals to hit the sea floor and bounce back to the surface (see Figure 11.18). Sound travels at a speed of about 1500 m/s underwater, and can be transmitted through several thousand kilometres of water.

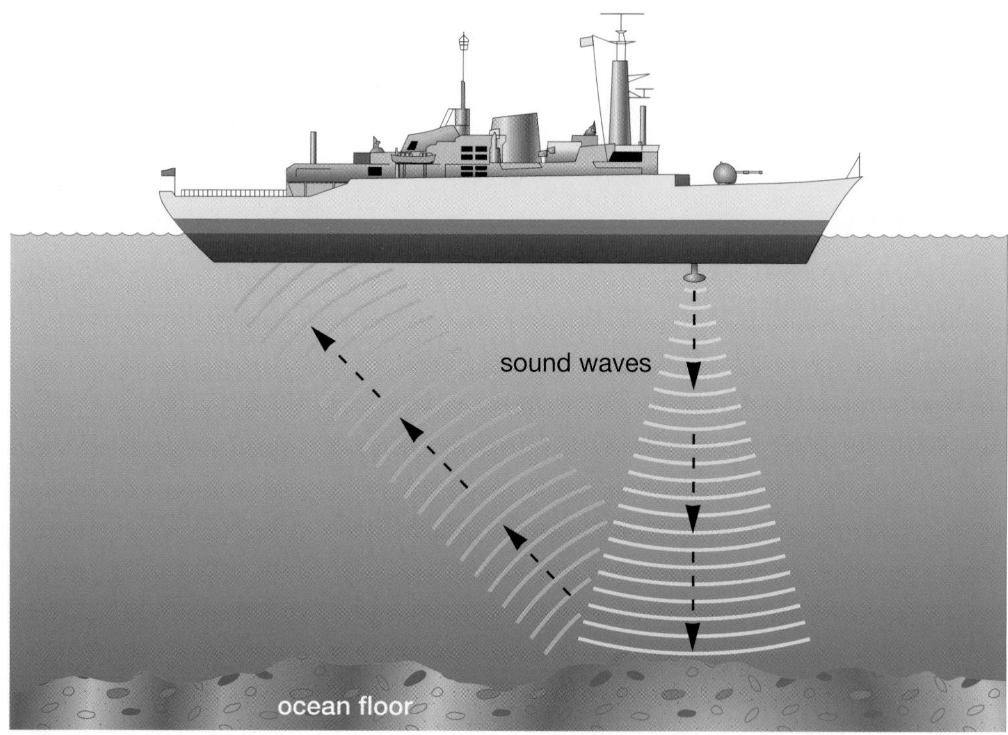

Figure 11.18 Sonar mapping of the sea floor

Skill POWER

For tips on using the Internet, turn to page 559.

Today, we have detailed pictures of the oceans produced by satellites in orbit far above Earth (see Figure 11.19). Spacecraft can automatically record data using radar, infrared light, or other means of measuring features on Earth. A great advantage of satellites over ships is that satellites can survey very large areas of ocean in a relatively short time. Satellites are also able to record and transmit data in all kinds of weather, both day and night.

Lines made of new, strong materials, such as Kevlar™, allow scientists to anchor buoys at fixed points on the ocean floor without having the buoys swept away by strong currents. Instruments attached to these buoys transmit air and sea temperatures to satellites. These data help scientists predict weather changes. Satellites also help oceanographers monitor water movements, by tracking the path of drifting buoys.

INTERNET CONNECT

www.school.mcgrawhill.ca/resources/

To learn more about satellite imagery and to view images of the ocean floor, visit the above web site. Go to **Science Resources**, then to **SCIENCEPOWER 8** to find out where to go next.

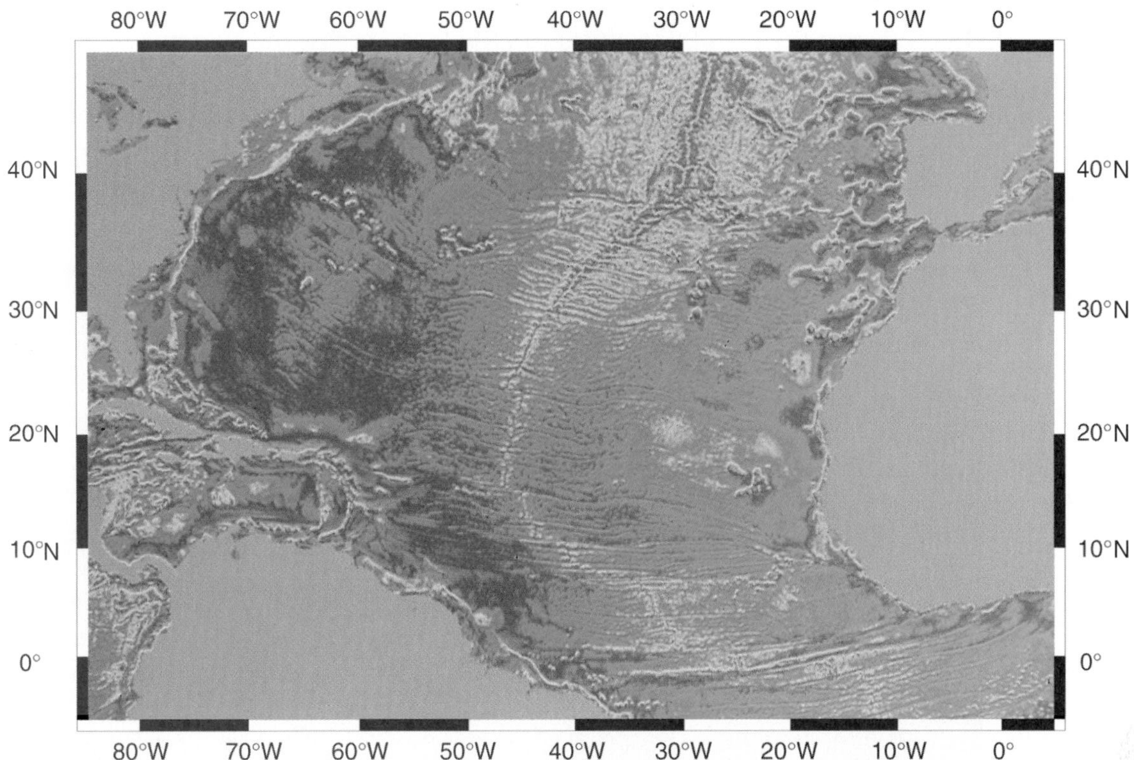

Figure 11.19 This image of the Atlantic sea floor was produced by instruments on a satellite.

The most dramatic information we have about the deepest reaches of the ocean comes from submersibles and deep-towed camera sleds (see Figure 11.20). Cameras towed from ships can take thousands of high-resolution photographs a day. When overlapped, the photographs build up a picture of the sea floor with details measured in centimetres. New technology, including computer analysis, has opened an exciting era of ocean exploration that will continue for many years into the future.

Figure 11.20 A deep-towed camera sled is used to photograph the sea floor.

By sending and receiving signals, satellites can measure the distance to a point on the ocean surface within 0.1 m. Such amazing accuracy shows that the ocean surface has bumps and dips in it. These are caused by the gravitational effects of contours on the sea floor. For example, an undersea volcano 2000 m high causes the ocean surface to bulge upward by about 2 m. In contrast, deep sea trenches cause the ocean surface above them to dip down.

If the total amount of salt in the oceans were dried and spread over the whole Earth, it would produce a layer 45 m thick. This is deep enough to bury a 15-storey apartment building.

Salt Water

Rivers around the world flow down from hills and across the land before finally pouring their waters into the ocean basins, as shown in Figure 11.21. Traces of muddy water from the largest rivers, such as the Amazon River in South America, can be detected as far as 1000 km out from the coastline. Ocean water, however, is not the same as river water. If you have ever swum in the sea and accidentally got some water in your mouth, you know the main difference. Ocean water is salty.

Figure 11.21 Brown, silt-laden water from a river meets the blue water of the Atlantic Ocean.

On average, 1000 g of seawater contain 35 g of dissolved salts. This is usually expressed as 35 parts per thousand (ppt). By far the most common material in this solution is sodium chloride. This is the same chemical substance as the table salt you use to season food. The next most plentiful salts are composed of sulfates, magnesium, calcium, and potassium (see Figure 11.22). The measure of the amount of salts dissolved in a liquid is known as **salinity.**

Where do the salts in the ocean come from? Most started out in rocks on land. Rivers and groundwater flowing over the rocks pick up the salts and carry them to the ocean. Fresh water, in fact, contains many of the same dissolved substances as seawater, but in far smaller proportions. For example, the concentration of sodium in seawater is about 16 000 times greater than in fresh water.

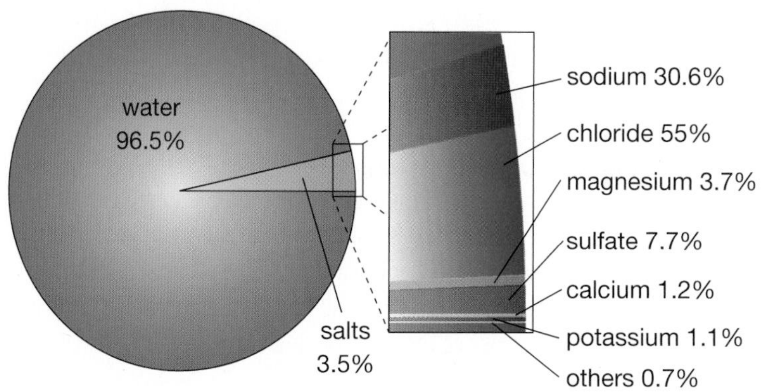

Figure 11.22 Ocean water contains about 3.5 percent salts. Traces of almost every chemical substance on Earth can be found in seawater, including gold and silver. Most of these substances, however, occur in extremely tiny quantities.

Although rivers from different areas of land may carry different salts into the seas, the motion of sea water mixes them together. Thus, samples of water from any part of the ocean generally contain similar substances.

Another source of chemicals in seawater is volcanoes. Molten lava and gases from volcanic eruptions on the sea floor add materials such as chlorine and sulfur to seawater. Gases from volcanic eruptions on land circulate first in the atmosphere. From there, they may dissolve directly in the surface waters of the ocean, or be carried from the atmosphere in rainfall.

Find Out **ACTIVITY**

How the Ocean Gets Its Salt

Why is the ocean saltier than the rivers that flow into it? Where does the salt come from? Why does the ocean level not rise as rivers continually pour water into ocean basins? This activity provides clues that help answer these questions.

What You Need

graduated cylinder
2 paper cups
paper coffee filter
15 mL salt
15 mL soil
sharp pencil
sheet of black construction paper
modelling clay
large plate or tray
water

What to Do

1. Use the point of the pencil to punch five holes in the bottom of the paper cup. **CAUTION:** Be careful as you punch the holes.
2. Place the coffee filter inside the cup.
3. Make short legs of modelling clay for the cup. Set up your materials as shown.
4. In the empty cup, stir together the soil and the salt.
5. Pour the soil/salt mixture into the coffee filter.
6. Pour 20 mL of water onto the soil/salt mixture. Allow the liquid to drain through the holes in the cup onto the construction paper.
7. Allow the paper to dry. Record your observations.
8. Wipe up any spills and wash your hands after this activity.

What Did You Find Out?

1. Why does salt move from the soil mixture onto the construction paper?
2. Why is salt left behind on the paper after it dries?
3. What process causes water to leave the ocean? (This process balances the water entering the ocean, and explains why the ocean level does not rise.)
4. Why is river water not salty?

A Steady State

Did You Know?

The ocean basins are no older than 200 million years! By determining the age of fossils found on the ocean floor, scientists can estimate the time at which the ocean floor was formed.

Ocean basins were once believed to be like large storage bowls into which rivers continually dump their loads of salts. If this were true, the oceans would be getting steadily saltier over time. However, a little calculation shows this cannot be true. By estimating the amount of dissolved substances rivers carry to the oceans each year, scientists calculated it would take about 12 million years for all the world's rivers to transport the total mass of salts now in the oceans. Since the oceans are much older than 12 million years, they cannot be getting saltier.

Measurements indicate that the salinity of the oceans varies slightly from place to place — between 30 and 40 ppt. You will discover why in the next section. Overall, however, the proportions and amounts of dissolved salts in the oceans appear to have remained about the same for the past 100 million years or more. The salinity of the oceans is in a steady state. Therefore, salts must leave seawater as well as enter it.

How do salts leave seawater? As one example, chlorine leaves the ocean and enters the atmosphere in salt spray produced by waves. Some dissolved salts react with suspended solids in the water and fall to the ocean floor as solid sediments. Other chemicals are removed from the water by organisms that use silica and calcium to make bones or shells (see Figures 11.23A and B). When the organisms die, these materials sink to the sea floor.

Figure 11.23A Star-like radiolarians and diatoms (microscopic algae) have shell-like coverings made of silica.

Figure 11.23B Foraminifera are tiny marine organisms with shells made from calcium materials. Both diatoms and foraminifera are among the most abundant organisms in the ocean. A large part of the thick sediments that cover the seabed are made from their remains.

CONDUCT AN

INVESTIGATION 11-D

Comparing Buoyancy

What is the difference in the buoyant force exerted by fresh water and salt water? You can measure the difference precisely using the following method. Prepare a table to record your results. Give your table a title.

Problem

How can you compare the buoyant force of fresh water and salt water?

Safety Precautions

Be careful not to drop the 500 g mass onto your foot.

Apparatus

2 beakers (1 L each)
spring scale (5 N)
mass (500 g)

Materials

tap water
salt solution (35 g salt in 1000 mL water)

Procedure

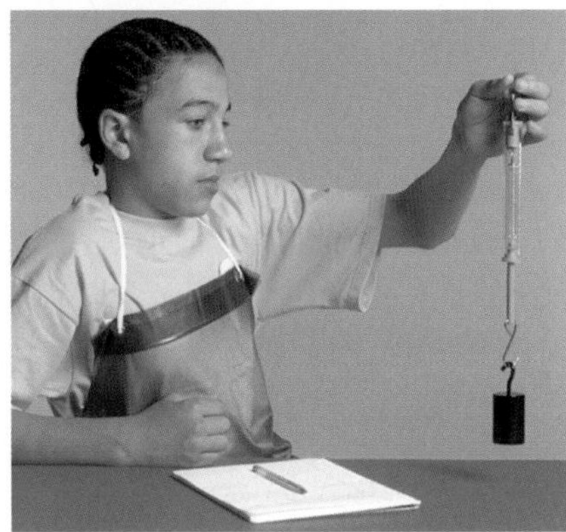

1. Measure and record the weight of the 500 g mass suspended in air.
2. Fill a clean beaker to the 400 mL mark with tap water.
3. Lower the 500 g mass on the spring scale into the water until it is just submerged. Do not let the mass touch the beaker. Record the weight in your table.
4. Dry the 500 g mass thoroughly.
5. Repeat steps 2 and 3, using salt solution instead of tap water.
   (a) Wipe up any spills as wet floors are slippery.
   (b) Wash your hands after this investigation.

Skill POWER

To review how to make a data table, turn to page 546.

Analyze

1. When the 500 g mass was immersed in liquid, did its weight appear to increase or decrease?
2. By how much did the weight of the 500 g mass appear to change when you immersed it in
   (a) tap water?
   (b) salt solution?

Conclude and Apply

3. The buoyant force exerted on an object by a liquid is expressed in the following equation:

   Buoyant force = Weight in air – Weight in liquid

   Calculate the buoyant force exerted on the 500 g mass by tap water and by salt solution. Which liquid has the greater buoyant force?

Recall from Chapter 5 that the salinity of seawater gives seawater different physical properties compared with fresh water. For example, objects (including people) float more easily in seawater than in fresh water (see Figure 11.24). They appear to weigh less. This occurs because the upward force exerted by seawater (the buoyant force) is greater than that of fresh water.

Figure 11.24 Water near the surface of the Dead Sea between Israel and Jordan has a salinity of about 300 ppt. This is about nine times greater than the average ocean salinity.

Check Your Understanding

1. Name a substance found in seawater that comes from
   (a) land
   (b) the atmosphere
2. Explain why oceans are not getting saltier.
3. Why are large areas of sea floor covered with thick layers of sediment containing calcium?
4. Name two twentieth-century technologies that have helped scientists explore the deep ocean. What information has come from each technology?
5. Describe how the following features are formed:
   (a) sea mounts
   (b) trenches
6. What are continental shelves?
7. Why does seawater have a greater buoyant force than fresh water?
8. **Thinking Critically** Why is it useful to know what the sea floor looks like?

11.3 Ocean Currents

Strange things sometimes wash up onto beaches with the waves. A message in a bottle is not very common, but many objects lost or thrown from ships far out at sea eventually make their way to land. Floating objects are carried over thousands of kilometres of ocean by broad, continuous movements of ocean water called **currents**. An ocean current is like a massive river within the ocean. Like a river, a current flows in one direction and connects one place with another.

Find Out **ACTIVITY**

Winds and Currents

How do winds affect surface currents? If you turn a skateboard upside down and run your hand across a wheel, friction between your hand and the wheel starts the wheel moving. The direction in which you move your hand determines the direction in which the wheel spins. In the same way, winds blowing over the surface of the ocean cause the surface waters to move. The surface currents flow in the same direction as the wind. What happens when a moving ocean current reaches land?

What You Need

rectangular pan
10 small circles of paper from a hole punch
water
drinking straw

What To Do

1. Fill the pan with water. Float the paper circles along one long edge of the pan.
2. Hold the straw with one end just above the paper circles and blow slowly and steadily through the other end. Observe and record what happens.
3. Place an object in the centre of the pan, such as a stone, to simulate an island. Repeat the activity.
4. Wipe up any spills as wet floors are slippery.
5. Wash your hands after this activity.

What Did You Find Out?

1. How does moving air affect the surface water in the pan?
2. How does the shape of the pan affect your results?
3. How does the "island" affect your results?

Surface Currents

Currents of water at the ocean surface are driven by winds. Most surface currents flow in the top 100–200 m of water. The steady flow of currents results from major wind patterns. These wind patterns blow in fairly constant directions around the world (see Figure 11.25).

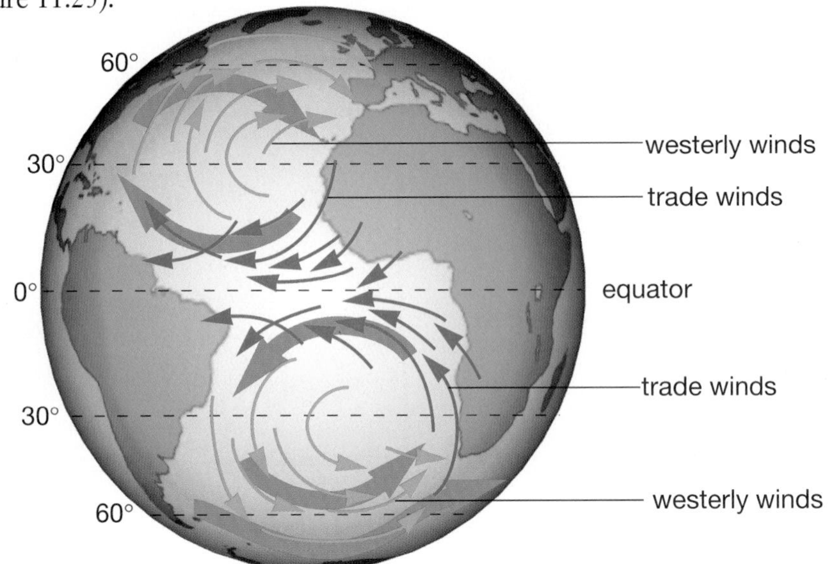

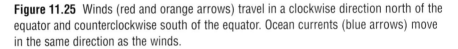

Figure 11.25 Winds (red and orange arrows) travel in a clockwise direction north of the equator and counterclockwise south of the equator. Ocean currents (blue arrows) move in the same direction as the winds.

Three factors influence the direction of winds and surface currents:

- uneven heating of the atmosphere
- rotation of Earth
- continents

All winds begin as a result of uneven heating of the atmosphere. Warm air expands upward and outward as its particles move farther apart. This produces an area of low air pressure. Cool air, with a higher pressure, moves into the area of low pressure. The moving masses of air create winds (see Figure 11.26).

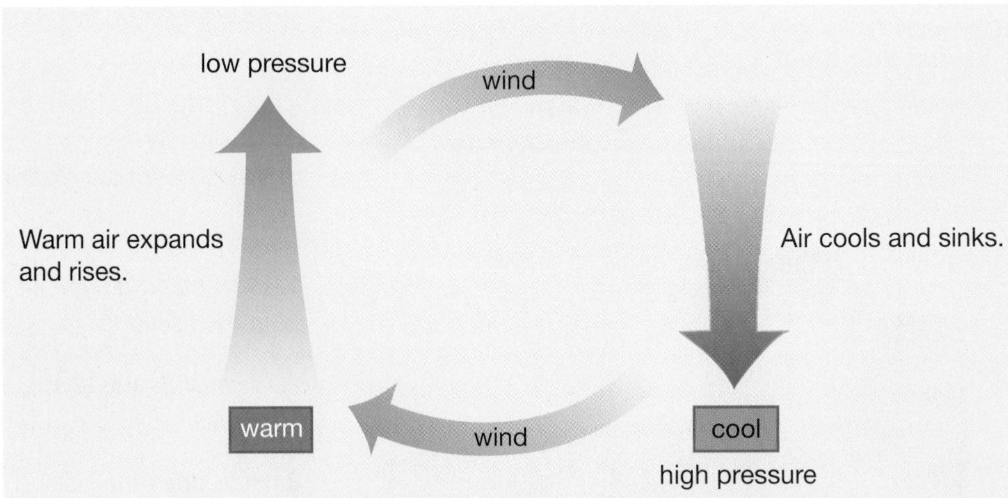

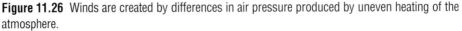

Figure 11.26 Winds are created by differences in air pressure produced by uneven heating of the atmosphere.

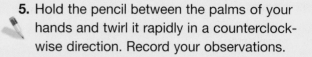

At Home **ACTIVITY**

Winds on a Rotating Earth

How does Earth's rotation affect wind direction?

What You Need

construction paper	mathematical compass
scissors	medicine dropper
sharp pencil	ink

CAUTION: Be careful when using scissors and other sharp objects.

What to Do

1. Use the compass to draw a circle with a 20 cm diameter on the sheet of paper.
2. Using scissors, cut out the circle.
3. Push the point of the pencil through the centre of the circle.
4. Place one drop of ink on the paper circle next to the pencil.
5. Hold the pencil between the palms of your hands and twirl it rapidly in a counterclockwise direction. Record your observations.

What Did You Find Out?

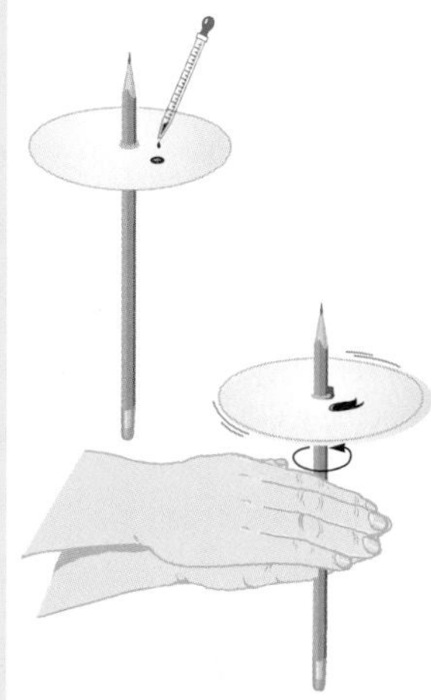

1. What does the ink drop represent in this model?
2. Does the surface of the paper disk represent Earth's northern or southern hemisphere? Explain your answer.
3. Does the ink move in a straight line or a curved line?
4. How would the result differ if you twirled the paper disk in the opposite direction? Try it and test your prediction.

The rotation of Earth produces a deflection (bending) of moving currents called the **Coriolis effect**. As wind and water currents flow over Earth's surface, the planet turns beneath them from west to east. This motion causes currents in the northern hemisphere to turn to their right relative to the Earth's surface. In the southern hemisphere, the effect causes currents to turn to their left (see Figure 11.27). The overall result of the Coriolis effect is that winds along the equator blow from the east. These winds, called **trade winds**, push ocean currents toward the west. Toward the polar regions, **westerly winds** drive currents the opposite way, from west to east, as shown in Figure 11.25.

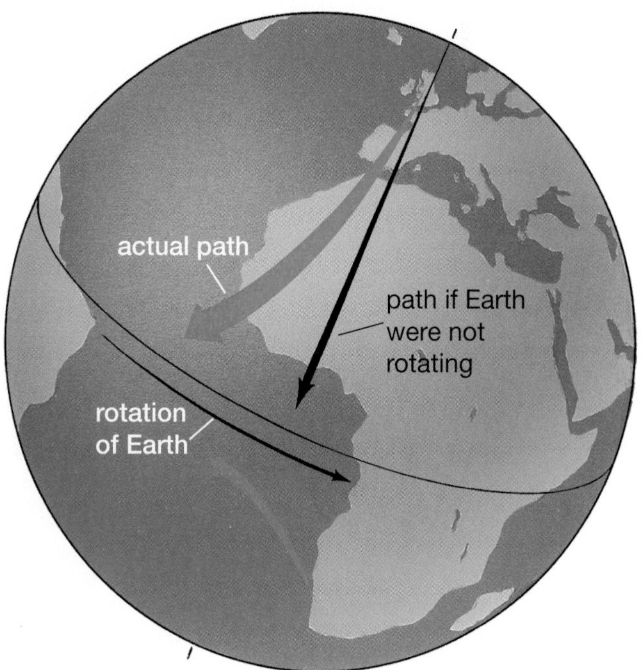

Figure 11.27 Air currents are deflected by the rotation of Earth.

Pause& Reflect

As water gets colder, its density increases. However, this is true only up to a certain temperature. Water achieves its maximum density at a temperature of 4°C. When water is cooled to 0°C, its density decreases. Therefore, bodies of water freeze from the top down rather than from the bottom up. Imagine if these differences in water density did not exist. How would life on Earth be affected if water bodies froze from the bottom up?

The third influence on ocean currents is the continents. As you observed in the Find Out Activity on page 371, moving currents are forced to turn when they meet a solid surface. Continents deflect east–west currents either north or south. The combined influence of winds, continents, and the Coriolis effect keeps ocean currents circulating clockwise in the northern hemisphere and counterclockwise in the southern hemisphere.

The Temperature of Ocean Water

Suppose you are on a boat in the middle of the ocean near the equator. You lower a thermometer over the side of the boat and record the water temperature at different depths until the thermometer reaches the sea floor. How would you expect the temperature to change, and why?

Figure 11.28 shows the results of such measurements. You can see that water temperature does not decrease steadily with depth. It changes sharply to form three distinct layers. As you might expect, water is warmest near the surface.

Almost all the heat in the ocean comes from the Sun. (Volcanic activity adds small amounts of heat on parts of the sea floor.) Winds and waves mix the heat evenly through the surface waters, forming a **mixed layer**. Because the Sun's energy does not penetrate very far, however, the water temperature begins to drop rapidly below a depth of about 200 m. Between 200 m and 1000 m, the temperature may fall from 20°C to a chilly 5°C. This region of rapid temperature decline is called the **thermocline**. Below the thermocline, ocean water remains very cold down to the ocean floor, where its temperature is close to the freezing point.

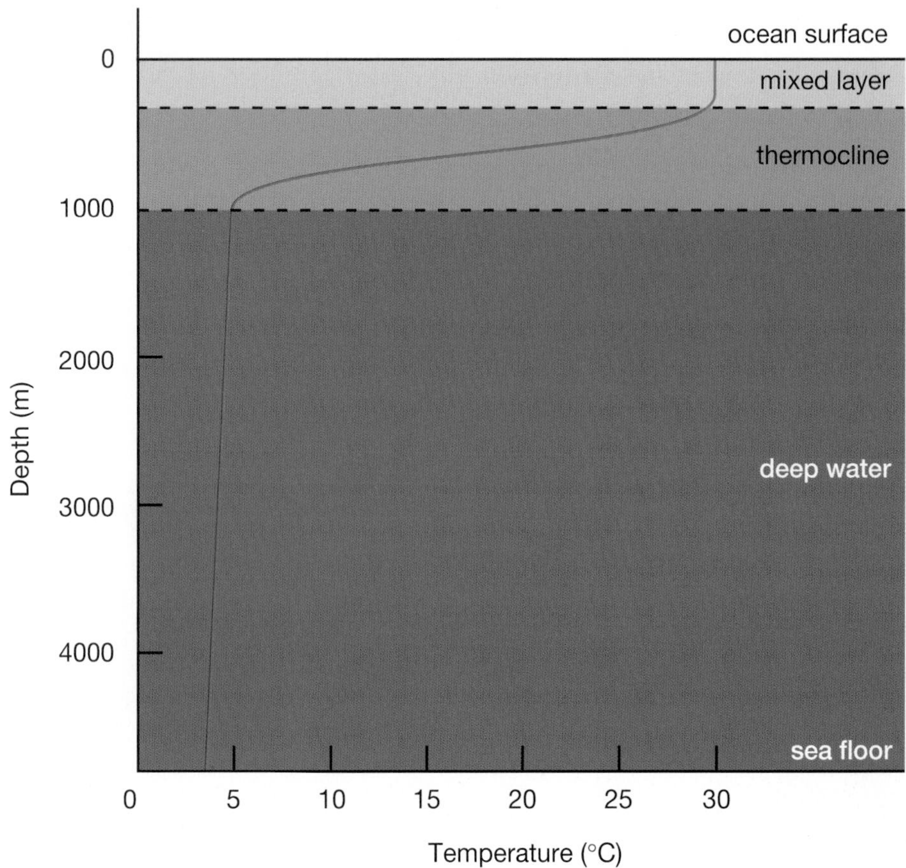

Figure 11.28 Temperature zones in the ocean

Density Currents

Water at different temperatures has different densities. Cold water is more dense than warm water, and tends to sink. In the oceans, sinking masses of cold water flow downward and move along the ocean floor. These masses of cold water produce **density currents** that flow beneath the surface waters (see Figure 11.29).

Density currents are also produced by differences in the salinity of seawater. The greater the salinity, the denser the water. How might salinity in certain parts of the ocean be increased or decreased? One way to lower salinity is to add more fresh water. For example, seawater is less salty at the mouth of large rivers. Fresh water also enters the ocean where icebergs and glaciers melt, and in regions with high levels of precipitation.

In contrast, regions with hot, dry conditions and a high rate of evaporation produce seawater with a higher-than-average level of salinity. For example, the deepest waters of the Mediterranean (a nearly-enclosed sea) have a salinity of about 3.8 percent. These deep, salty waters form a density current that moves along the sea floor and spills out of the Mediterranean into the Atlantic Ocean. Above this current, the less salty waters of the Atlantic flow in the opposite direction, entering the Mediterranean in a surface current (see Figure 11.30).

Just as evaporation increases salinity, so does freezing. Evaporation turns water into a gas (water vapour). Freezing turns water into a solid (ice). In both cases, salt is left behind in solution in the remaining water, increasing its saltiness. Dense, salty water is produced in oceans cold enough to form ice on the sea surface.

DidYouKnow?

Cold water flowing along the bottom of the ocean takes about 1000 years to travel from Antarctica to the equator!

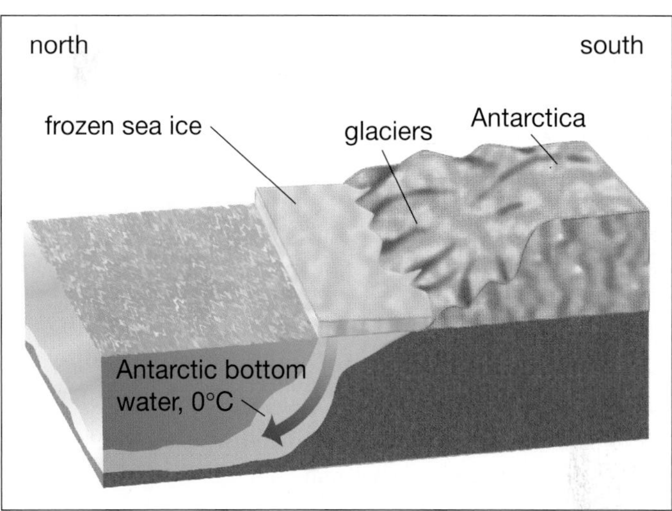

Figure 11.29 Density currents are formed by sinking, dense water that flows along the sea floor.

Upwelling

When dense water sinks, it forms a vertical, downward-moving current. Another type of vertical current flows upward, from the sea floor to the ocean surface. This current is called an **upwelling**. Upwellings are most common along coastlines where strong winds blow offshore. These winds push the surface water away from the land. Cold, deep water then rises from below to replace the surface water that has been moved out to sea. Figure 11.31 on page 374 shows the process of upwelling.

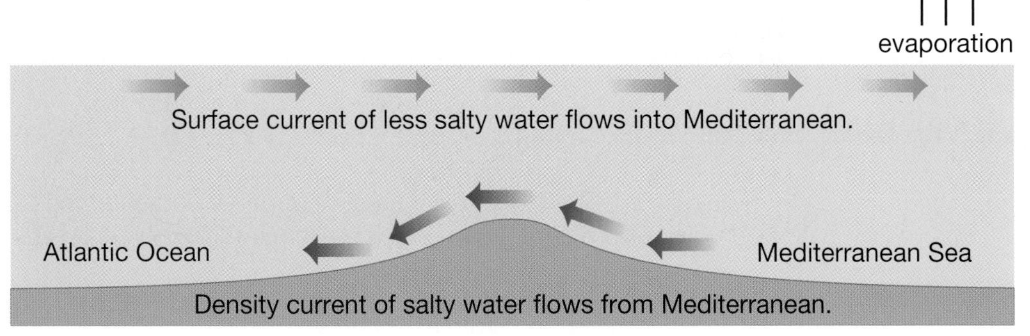

Figure 11.30 Flow of seawater into and from the Mediterranean Sea

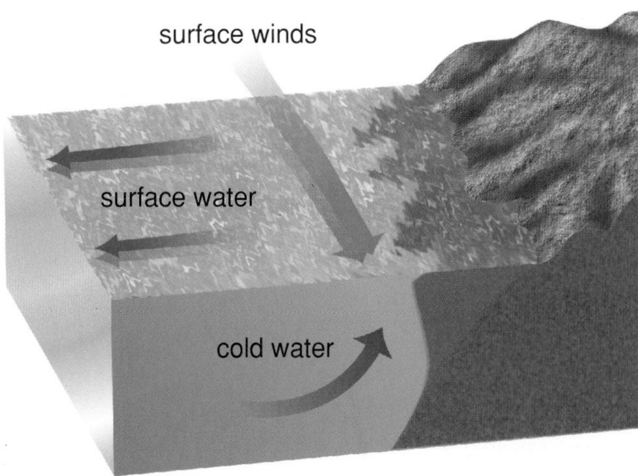

Figure 11.31 Upwelling brings cold water from the depths to the surface along the shoreline.

The coldest ocean waters are found near Greenland and off the coast of Antarctica. In these places, glaciers and cold winds keep surface water temperatures cool. Extremely cold, dense water from Antarctica sinks and creeps slowly northward along the ocean floor.

Upwellings have an important effect on the ecology of the sea and on human fisheries. The upwelling water contains large amounts of nutrients from the sea floor, such as phosphates and nitrates. Plants living in the surface waters use these nutrients to grow. The plants, in turn, attract fish to areas of upwelling. Major fishing industries are based on upwellings along the coasts of Oregon, Washington, and Peru.

Check Your Understanding

1. What is an ocean current?
2. How is a surface current produced?
3. How is a density current produced?
4. Why are some currents warm and others cold?
5. Name three factors that affect the movement of ocean currents.
6. How does upwelling affect the fishing industry?

Find Out **ACTIVITY**

Make a Density Parfait

What You Need

5 test tubes
teaspoon
hot water
cold water
4 colours of food colouring (red, green, blue, yellow)
funnel

What to Do

1. Label four test tubes A, B, C, and D.
2. Fill test tubes A and B halfway with hot water.
3. Fill test tubes C and D halfway with cold water.
4. Pour about one-eighth teaspoon of salt into test tubes A and C. Cover the top of each test tube with your thumb and shake it gently to mix in the salt.
5. Place one drop of food colouring in each test tube as follows: A – red, B – green, C – blue, D – yellow.

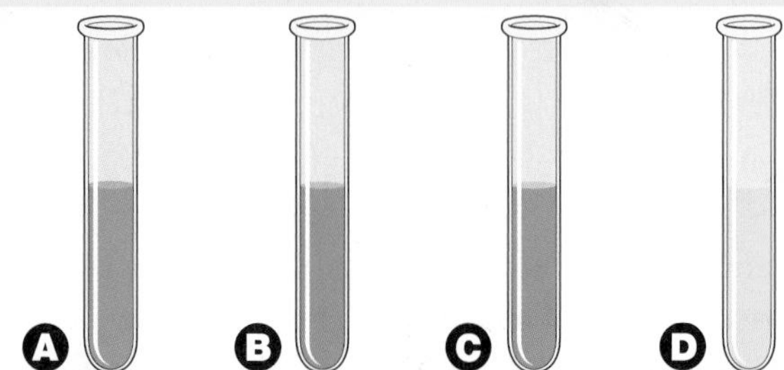

6. Pour a small amount of water from each test tube into the fifth test tube in the following order: C, D, A, B.

What Did You Find Out?

1. What happened? Why do you think this happened?
2. Predict what would happen if you changed the order of pouring. Test your prediction.
3. How does this activity relate to ocean currents?

CHAPTER at a glance

Now that you have completed this chapter, try to do the following. If you cannot, go back to the sections indicated.

Copy the diagram below and add labels to identify the main characteristics of waves. (11.1)

Describe how breakers are formed. (11.1)

Explain why some shorelines are rocky while others have sandy beaches. (11.1)

Use diagrams to illustrate the relationship between tides and the orbit of the Moon. (11.1)

Name the main features on the sea floor and describe how each is formed. (11.2)

Discuss ways in which technology has aided exploration of the ocean basins during the twentieth century. (11.2)

Explain why ocean water is salty. (11.2)

Describe the causes of different types of currents in the ocean. (11.2)

Make a graph showing how ocean temperature varies with depth. (11.3)

Explain why water may have a different density in different parts of the ocean. (11.3)

Give an example of how ocean currents affect a human activity. (11.3)

Prepare Your Own Summary

Summarize this chapter by doing one of the following. Use a graphic organizer (such as a concept map), produce a poster, or write the summary to include the key chapter ideas. Here are a few ideas to use as a guide:

- What causes waves?
- How do waves affect shorelines?
- What causes tides?
- How does the shape of a shoreline affect tidal range?
- What process produces most features on the sea floor?
- Why are most trenches on the sea floor found around the rim of the Pacific Ocean?
- What is a continental shelf?
- How are satellites used to explore oceans?
- What is meant by salinity?
- What causes surface currents?
- What causes density currents?
- Why does ocean temperature vary?
- Copy the following diagram and add labels to identify how air currents are deflected by the rotation of Earth. What is the name of the process shown?

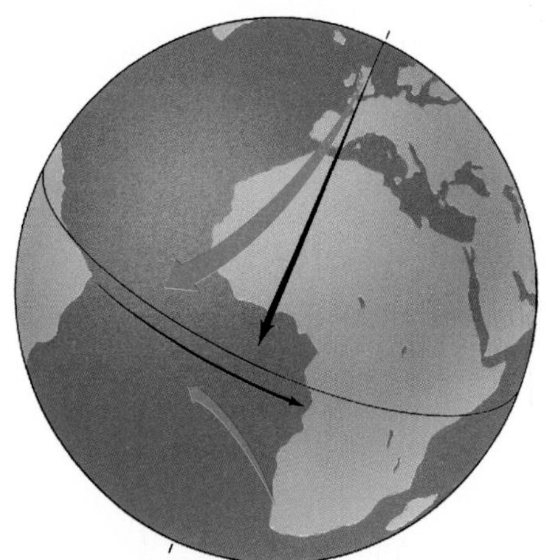

CHAPTER

11 Review

Key Terms

waves
swells
breaker
bays
headlands
tides
spring tides
neap tides
tidal range
mid-ocean ridges
seamounts
trenches
abyssal plains
continental shelf
continental slope
salinity
currents
Coriolis effect
trade winds
westerly winds
mixed layer
thermocline
density currents
upwelling

Reviewing Key Terms

If you need to review, the section numbers show you where these terms were introduced.

1. Copy the following sentences, adding the correct words from the list of key terms above.

(a) ▬▬▬ and ▬▬▬ are both produced by winds. (11.1, 11.3)

(b) ▬▬▬ are produced by the Moon's gravitational effect. (11.1)

(c) ▬▬▬ and ▬▬▬ are produced by undersea volcanic activity. (11.2)

(d) ▬▬▬ are produced where one tectonic plate moves under another. (11.2)

2. Explain the difference between

(a) spring tides and neap tides (11.1)

(b) a continental shelf and a continental slope (11.2)

(c) surface currents and density currents (11.3)

Understanding Key Ideas

Section numbers are provided if you need to review.

3. Describe the motion of water particles in a wave. (11.1)

4. What causes breakers to form? (11.1)

5. What causes high tides? (11.1)

6. Where does the sediment on abyssal plains come from? (11.2)

7. How are seamounts formed? (11.2)

8. Describe the general circulation pattern of surface currents

(a) north of the equator

(b) south of the equator (11.3)

9. Describe two ways in which the density of seawater can increase. (11.3)

10. What is upwelling? (11.3)

Developing Skills

11. The data below show how the depth of the ocean varies. The surface area is the area of ocean surface that covers each particular range of depth. Make a bar graph using these data and answer the questions that follow.

Range of depth (m)	Surface area (million km^2)	Percentage of surface area (%)
0–200	2.74	7.5
200–1000	15.5	4.3
1000–2000	15.2	4.2
2000–3000	24.5	6.8
3000–4000	70.8	19.5
4000–5000	119.1	33.0
5000–6000	84.1	23.5
over 6000	4.4	1.2

(a) The most productive ocean waters are over the continental shelves, which have a maximum water depth of about 200 m. What percentage of the world's oceans covers the continental shelves?

(b) What percentage of the oceans lies below 3000 m?

(c) What is the most common range of depth shown on your graph?

(d) What is the least common range of depth?

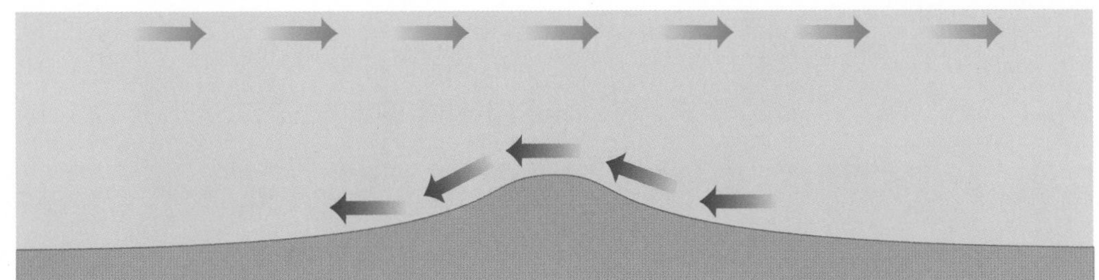

12. Copy the diagram above showing currents of water moving between the Atlantic Ocean and the Mediterranean Sea. Add the following labels in the correct positions on the diagram: more dense, less dense, surface current, density current, evaporation. Write a brief caption for the diagram, explaining the process it shows. Use the word "salinity" in your explanation.
13. Using a computer, make a spreadsheet that includes data about ocean surface currents and density currents, evaporation. Use characteristics such as wind, temperature, density, depth, salinity, and anything else you think describes each current. Make graphs relating these characteristics with the ocean currents. Use your graphs to show the differences between the two types of current.

Problem Solving/Applying

14. You are standing on a beach watching a large beach ball floating about 200 m offshore. Waves are rolling in and breaking on the beach, but the ball seems to remain in the same place. Explain this observation.
15. The data below were collected from samples of water in the Atlantic Ocean. One sample came from near the surface, one from a depth of 750 m, and the third from near the ocean floor.

Sample	Temperature (°C)	Density (g/mL)
1	6	1.02716
2	3	1.02781
3	14	1.02630

Which sample do you think came from near the ocean floor? How do you know?

Critical Thinking

16. Explain why a sandy beach cannot be considered a permanent feature of the shoreline.
17. Why are V-shaped valleys and rounded hills commonly found on land but not on the ocean floor?
18. From time to time, trade winds near the equator reverse direction and blow from west to east. This event is called *El Niño*. Explain how *El Niño* could affect upwelling off the coast of Peru.
19. The Gulf Stream is an ocean current that flows across the Atlantic Ocean. This current carries warm water from the Straits of Florida north to Newfoundland and then east to northern Europe. How does the Gulf Stream affect the climate of northwestern Europe?

Pause& Reflect

1. Now that you know how geological features on the ocean floor are formed, compare this with the formation of lakes and rivers described in Chapter 10.
2. Go back to the beginning of this chapter on page 342 and check your original answers to the Getting Ready questions. How has your thinking changed? How would you answer those questions now that you have investigated the topics in this chapter?

CHAPTER

12 Water, Climate, and

Getting **Ready...**

- Is Earth heating up or cooling down?
- What is life like in the ocean?
- Will there always be enough water for people to use?

Try to answer the questions above in your Science Log. You will discover more detailed answers as you read this chapter.

The amount of water on Earth has not changed in billions of years; however, the distribution of water has changed. Changes in global climate patterns affect the timing and amount of precipitation falling in different areas. As a result, some rivers and lakes may dry up, while others flood. In turn, changes in the distribution of water have dramatic impacts on life. Years of drought turn forests into deserts, and grasslands into dustbowls. Over centuries, changing climates have even raised and lowered sea levels, and formed new lakes and river valleys on the land.

In this chapter, you will learn how climates have changed in the past and are still changing today. You will also learn how organisms that live in water are affected by changing climates and human activities. People depend on water for many uses and must manage supplies to meet human demands. For example, dams create reservoirs that store water, helping to even out seasonal changes in a river's flow. However, these reservoirs also flood many square kilometres of land and destroy habitat for plants, wildlife, and people. How can we sustain our water resources? How can we continue to obtain all the water we need now and for future generations? Will our efforts to obtain water endanger the life that depends on Earth's water systems?

Human Activity

Spotlight On Key Ideas

In this chapter, you will discover

- how the oceans and lakes affect climate
- what causes an ice age
- why cold oceans produce more food than warm oceans
- how human use of water affects the environment

Spotlight On Key Skills

In this chapter, you will

- analyze the effect of bodies of water on climate
- evaluate the evidence for future climate changes
- design a system for filtering water
- measure how much water your family uses in a day
- design your own experiment to compare different methods of cleaning oil spills from beaches

Starting Point

Water and You

How is water important in your life?

What to Do

1. As a class, start brainstorming and writing responses to the following questions.
   (a) How do you use water directly? How do you use water indirectly?
   (b) How do you depend on other organisms that use water?
   (c) What different forms of water do you use?
   (d) Explain how energy can be produced by water.
   (e) Name the different items that you use that are manufactured using water.
   (f) How are you affected by precipitation?
2. When you have listed ways in which water has an impact on your life, think about ways in which you have an impact on water.
   (a) What kinds of wastes produced by your household go into water?
   (b) Give an example of how people use more water than they really need.
   (c) How might any items you buy or use pollute water when they are manufactured, used, or discarded?
   (d) How do you think your activities might affect organisms living in water?
3. List all the forms of water you are aware of in your community.
4. From your notes and ideas, produce a poster or display on the theme of "Water and Daily Life." Include as many ideas in your poster as you can.

12.1 Water and Climate

Pause& Reflect

In your earlier science studies, you learned to distinguish between "climate" and "weather." In your Science Log, write clear definitions for each of these terms. Check your definitions against those in the Glossary at the back of this book.

To warm a bath of cool water, you must turn on the hot-water tap and stir the bath water with your hand. As the bath water circulates, you can feel the heat move from one end of the bathtub to the other. Similarly, surface currents in the ocean carry heat from one place to another. Warm currents begin near the equator, where the Sun's heat is most intense. As these warm currents circulate, they affect the climate and sea life of the regions to which they move.

Hot and Cold Currents

If you study a map of the world, you will see that Britain is as far north as northern Ontario and Hudson Bay. However, Britain's climate is much milder. In southwest England, the winter seasons are mild enough to allow subtropical palm trees to grow! This mild climate is mainly due to a current of warm water called the Gulf Stream. The Gulf Stream starts in the Caribbean Sea and flows north along the east coast of North America. Then it turns northeast and crosses the Atlantic Ocean. The Gulf Stream carries warm water to Iceland and the British Isles (see Figure 12.1).

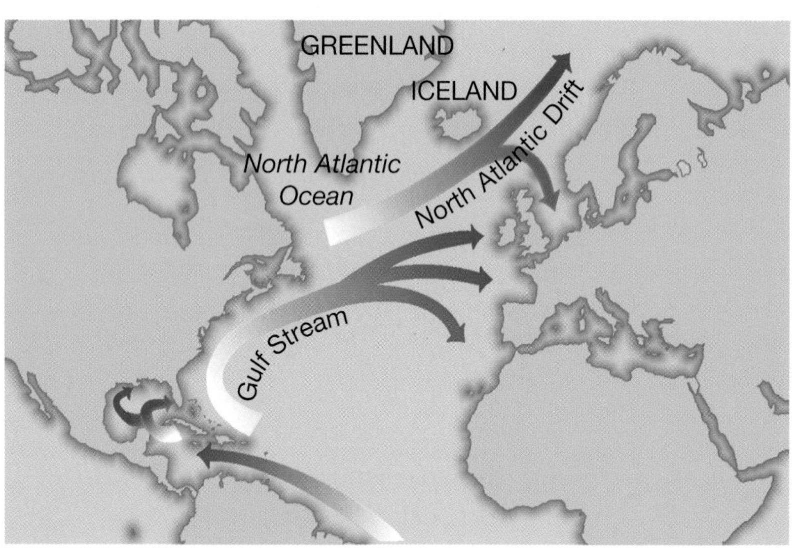

Figure 12.1 Map of the North Atlantic Ocean showing the Gulf Stream (not to scale)

Warm ocean currents affect climate by transferring their heat to the atmosphere. Water has a very high **heat capacity**, which means that it takes a relatively long time to heat up or cool down. As a result, large bodies of water act as heat reservoirs in the winter, remaining warmer than the land nearby. The difference in temperature between the water and land affects the weather systems near the shoreline. These systems produce breezes that alter the processes of evaporation and condensation near the shoreline.

While warm currents flow from the equator, cold currents flow from the Arctic and Antarctic regions. Cold currents also affect the climate, by drawing heat from the air. For example, the Labrador Current on Canada's east coast flows south from Baffin Bay along the east coast of Newfoundland (see Figure 12.2). The icy cold water carries icebergs from the far north into the North Atlantic Ocean. In these waters, the icebergs create a hazard on busy shipping routes. It was in this part of the Atlantic Ocean, southeast of Newfoundland, that the *Titanic* collided with a large iceberg and sank in 1912.

Not only icebergs but also thick mists plague this region. The mists are produced when the cold Labrador Current meets the warm Gulf Stream. As the warm, moist air above the Gulf Stream blows over the colder water of the Labrador Current, it condenses, producing some of the thickest fogs in the world.

DidYouKnow?

The volume of water in ocean currents makes even the mightiest rivers appear tiny by comparison. For example, the Gulf Stream moves about 26 000 000 m^3 of water per second, or about 1000 times more than the Mississippi River!

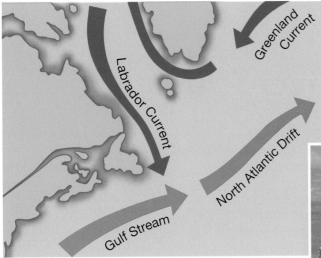

Figure 12.2 The cold Labrador Current meets the warm Gulf Stream southeast of Newfoundland. Icebergs are brought to this part of the North Atlantic Ocean by the cold waters from the north.

Find Out **ACTIVITY**

Water and Air

How does the temperature of the ocean affect the climate? This activity demonstrates one way.

What You Need

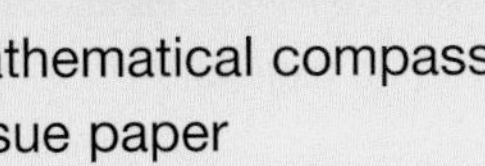

mathematical compass	masking tape
tissue paper	fine thread (30 cm)
scissors	cup of hot tap water

What to Do

1. Use the compass to draw a circle with an 8 cm diameter on the tissue paper.
2. Starting at the centre of the circle, draw a spiral to the edge of the circle, as shown in the diagram.
3. Carefully cut along the line to make a paper spiral. **CAUTION**: Be careful when using sharp objects such as scissors.

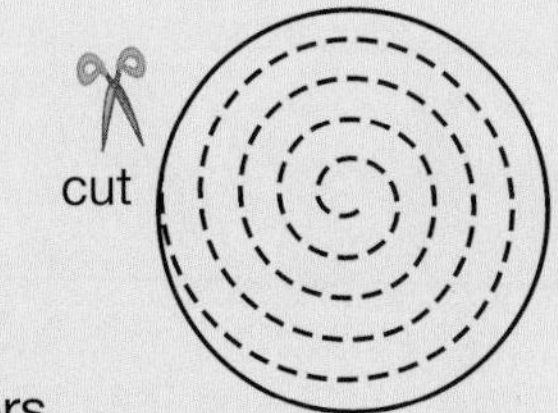

4. Attach one end of the thread to the centre of the paper spiral, using a small piece of tape.

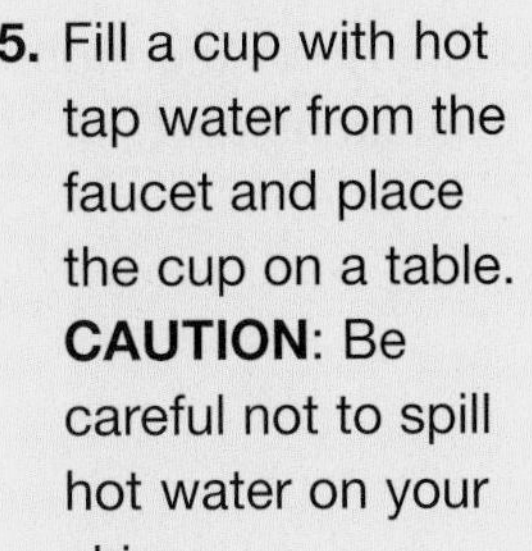

5. Fill a cup with hot tap water from the faucet and place the cup on a table. **CAUTION**: Be careful not to spill hot water on your skin.
6. Holding the free end of the thread, suspend the paper spiral about 10 cm above the surface of the hot water. Observe and record what happens.
7. Wipe up any spills and wash your hands after this activity.

What Did You Find Out?

From your observation, suggest what happens to the air above the hot water. Use the words "energy" and "currents" in your answer.

INVESTIGATION 12-A

How Does a Lake Affect Climate?

Think About It

You have learned that warm and cold ocean currents affect the climate of coastal areas. Do you think a large lake also has an effect on the area surrounding it? You can find out by studying climate data.

What to Do

1. Recall what you know about the water cycle and about how ocean currents affect the air above them.
2. Use your knowledge to analyze the data in the table below. These figures were collected from four cities located at different distances from Lake Erie. *Average monthly temperature range* is the average difference between the maximum and minimum temperature for the month. *Annual precipitation* is the total amount of precipitation per year.

Analyze

1. Make line graphs of these data. Plot distance from the lake on the x-axis (horizontal axis) and plot the other variables on the y-axis (vertical axis). If you decide to plot all of the data on one graph, use different colours to draw each line.
2. How does distance from the lake affect temperature, frost-free days, and precipitation?
3. Using your knowledge of heat capacity and the water cycle, suggest explanations for the relationships between the variables.
4. Why are these data important? Suggest several possible uses for this information. For example, if you were a farmer, how could these data help you decide what crops you should grow on your farm?

Skill POWER

For tips on making line graphs, turn to page 546.

Climate Patterns for Four Cities near Lake Erie

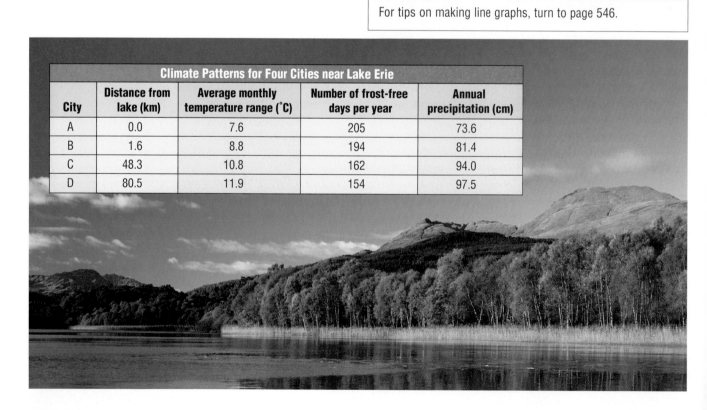

City	Distance from lake (km)	Average monthly temperature range (°C)	Number of frost-free days per year	Annual precipitation (cm)
A	0.0	7.6	205	73.6
B	1.6	8.8	194	81.4
C	48.3	10.8	162	94.0
D	80.5	11.9	154	97.5

Changing Climates

Is the climate in your region similar to the climate a hundred years ago — or a million years ago? You know from your own experience that the weather changes from day to day and alters with the seasons. Weather records and other evidence show that climates also change over years and centuries.

For example, every few years a change in water currents in the Pacific Ocean produces a short-term warming of the climate called ***El Niño***. Normally, the trade winds in the Pacific Ocean blow warm currents west along the equator. This creates very warm surface water temperatures in the western Pacific Ocean and upwelling in the eastern Pacific Ocean. During an *El Niño* year, however, the wind direction is reversed and warm water is blown eastward across the ocean (see Figure 12.3).

Wacky Weather

El Niño takes blame for strange events

The world has witnessed some weird, often devastating weather in the past year [1998].

Heavy rains and floods in normally parched Southern California. Out-of-control forest fires during what should have been the rainy season in Borneo. Tornados in Minnesota in March, an almost unheard-of event.

Closer to home, many Canadians warmed to an unusually mild winter. But in eastern Ontario and Quebec the relatively balmy temperatures came with a price – early thaws and heavy spring flooding, not to mention January's devastating ice storm.

Although individual storms are hard to predict, the strange weather came as no surprise to forecasters and climatologists. They had seen it all before as a result of the Pacific Ocean phenomenon called El Niño – a periodic buildup of warm water off the coast of South America.

Not every weather calamity can be blamed on El Niño. For instance, experts aren't sure whether it led to the January ice storm.

But past El Niños have caused dramatic changes to global weather patterns.

The 1997-98 *El Niño* was the strongest and most widespread on record. The photograph above shows houses in danger of sliding into the ocean as a result of a severe storm, caused by *El Niño*, that has resulted in a landslide.

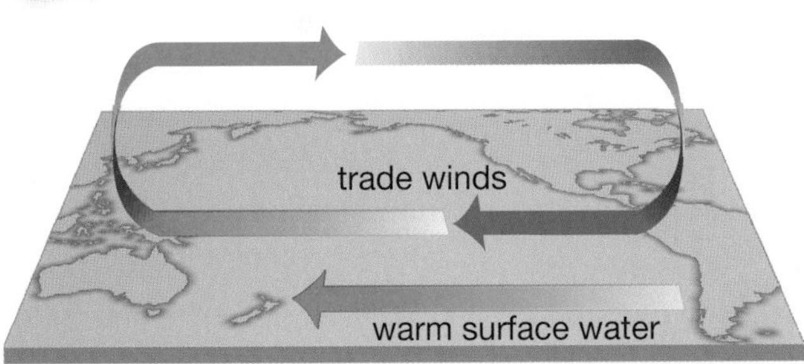

normal year

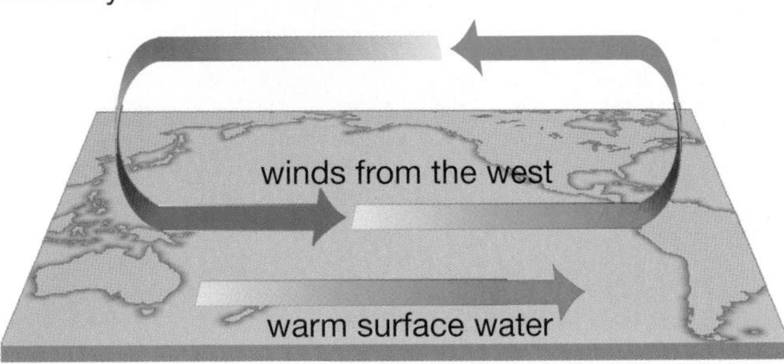

El Niño year

Figure 12.3 Wind and water current patterns in the Pacific Ocean during a normal year and an *El Niño* year

INTERNET CONNECT

www.school.mcgrawhill.ca/resources/

You can see satellite photos of weather features on your computer. Track the path of hurricanes and tornadoes as they happen and watch clouds circulate over the oceans. To follow current weather events, visit the above web site. Go to **Science Resources**, then to **SCIENCEPOWER 8** to find out where to go next.

As wind and water currents change direction in the Pacific Ocean, they force changes in the patterns of wind and water currents elsewhere. Because the oceans and atmosphere affect each other, *El Niño* can disrupt normal weather patterns around the world. Wind currents are pushed off their usual paths, delivering their energy and weather to different areas than usual. During an *El Niño* year, winter in Canada and the northern United States is milder than usual. The south-eastern United States gets more precipitation, while Australia and Africa experience droughts.

Changing ocean currents can have a major effect on sea life. *El Niño* may increase ocean temperatures off the coast of Peru by as much as 7°C. This change causes some populations of fish, birds, and other marine animals to move farther north. Other plants and animals may die because they cannot survive very long in the warmer waters.

Skill POWER

For tips on using the Internet, see page 559.

Ice Ages

El Niño is an example of a short-term change in Earth's climate. There have also been long-term changes in Earth's climate. Long-term changes alter weather conditions around the world for centuries. For example, over the last three million years, Earth has had at least four major periods of cooling called **ice ages**. The most recent ice age began about 120 000 years ago and ended only about 11 000 years ago. During this period, the climate was very different from what it is today.

At the peak of the last ice age, scientists estimate that average air temperatures around the world were about 5°C lower than they are now. That may not seem like much, but imagine how a difference of a few degrees in temperature might be enough to start an ice age. A cooler-than-average winter blankets the land with more snow than usual. A cooler-than-average summer does not produce enough heat to melt all the snow by summer's end. The next winter adds more snow. This makes it even harder for the following summer's heat to melt all the snow. When more of the land is covered with snow and ice, a greater amount of sunlight is reflected back into the atmosphere. This adds to the cooling. Over hundreds and thousands of years, the snow piles up to heights of 1 km or more. The tremendous weight of the snow causes the lower layers to turn to ice, forming **continental glaciers** (see Figure 12.4).

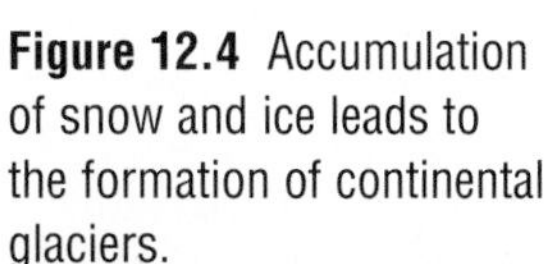

Figure 12.4 Accumulation of snow and ice leads to the formation of continental glaciers.

Ice and the Water Cycle

During the last ice age, much of North America was as cold as Greenland is today. Average temperatures were slightly above 0°C in the summer! Glaciers covered the land from the Arctic to as far south as the Great Lakes (see Figure 12.5). How do we know this? Glaciers create distinct features in the land. For example, boulders and stones frozen in the bottom of a moving glacier leave long, parallel scratches in the underlying rock. Glaciers also erode the sides of valleys, producing gently-curving U-shaped valleys. When glaciers melt at the end of an ice age, the run-off carries away tiny rock fragments and sediment. These rock fragments are deposited in ridges that remain long after the glaciers have melted. From evidence such as this, scientists can determine where glaciers have previously existed.

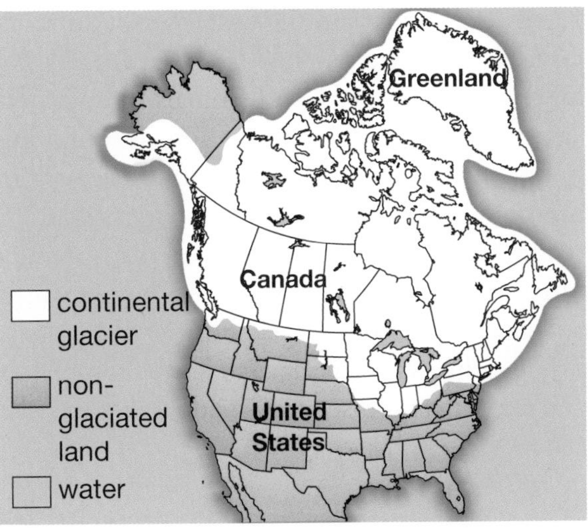

Figure 12.5 During the last ice age, glaciers in North America covered an area three times as large as they do today.

Most glaciers move less than 1 m per day. However, a glacier in Alaska holds the record for being the fastest glacier on Earth. In 1937, the Black Rapids Glacier advanced at the incredible rate of 30 m per day! Why do you think this glacier was moving so swiftly?

At Home **ACTIVITY**

How Do Glaciers Move?

Glaciers are massive mountains of ice. Yet observation reveals that they can move. Some glaciers move a few centimetres each day, while others may move many metres in a single day. This activity demonstrates how this process may happen.

What You Need

cake pan or plastic tub	brick
refrigerator freezer	wire rack

What to Do

1. Get permission from an adult at home to use the cake pan, wire rack, and freezer.
2. Fill the cake pan with water.
3. Place the pan and wire rack in the freezer until the water is frozen solid.
4. Remove the ice block from the pan.
5. Place the ice block on the wire rack in the freezer.
6. Set the brick on top of the ice block. **CAUTION:** Be careful not to let the brick slide off the ice and onto your foot. Leave the brick in the freezer for 24 h.
7. Observe the bottom of the ice block and record your observations.
8. Wipe up any spills and wash your hands after this activity.

What Did You Find Out?

1. Did you notice any changes in the block of ice? Suggest an explanation for any change that you observed.
2. How might your observation help explain the movement of glaciers during an ice age?

By studying the previous extent of glaciers, scientists estimate that the total volume of ice in the world during the last ice age was about three times as great as it is today. What effect would this increased volume of ice have on the water cycle? With more water locked solid in ice, and with no flowing rivers to return water to the ocean, sea levels fell by as much as 100 m. Lower seas exposed more of the continental shelf. This increased the total area of continental land relative to the ocean.

Lower sea levels in the past created different coastlines. Offshore islands, such as Great Britain, were joined to the mainland. A **land bridge** connected Alaska with Siberia. This allowed land animals to migrate between North America and Asia. The evidence for these migrations is found in the distribution of fossil remains on both continents. The new coastlines also altered ocean currents, affecting the movements of fish and other sea organisms.

With the end of an ice age and the melting of the glaciers, sea levels rose once more. New lakes and river valleys were formed as melting ice flooded into basins and channels eroded by ice. Forests and grasslands spread north, planting their roots in the sediments that glaciers left behind. Much of the landscape of hills, rivers, and lakes that you see in Canada today was shaped by the events of the last ice age.

Scientists are drilling into a glacier in Greenland with a tool that is literally "cool." The project uses a drill about 13 cm across to extract ice cores from up to 3 km below the surface of the glacier. Ice holds trapped bubbles of gas that tell scientists about the composition of the atmosphere at the time the ice was formed. By taking samples from deep within the glacier, scientists can gather data on climate changes going back 200 000 years!

Find Out

Canada in the Year 22 000

Will there be another ice age? Some scientists think we are living in a warm period between ice ages. Based on past cycles, they estimate another ice age may begin within about 20 000 years from now. Others think Earth's climate is becoming warmer. Atmospheric pollution and clearing of trees for lumber and farming are among the human activities that may help produce this warming trend. Either way, a climate change will have a major impact on Earth's water systems and, therefore, on all life on the planet.

What to Do

1. What kind of evidence would indicate that Earth may be getting cooler or warmer over the long term? Direct measurements of average temperatures are one indication. What indirect factors might also suggest that the climate is changing? When you have decided what evidence to investigate, use the Internet and library resources to collect information.

What Did You Find Out?

2. Based on your findings, decide whether there is better evidence for a cooling or a warming trend in the global climate.
3. List some of the consequences of a future climate change (either cooler or warmer).
4. Describe how your life might be different in this changed climate.
5. Draw a map of Canada in the future, after the climate has changed. Show how coastlines, rivers, lakes, and vegetation might differ from the way they are today.

Causes of Climate Change

You have discovered how a small change in average temperature might start a chain of events that produces an ice age. Why might the temperature of the atmosphere fall in the first place? You can probably think of some ideas of your own. Here are some hypotheses scientists have suggested to explain why ice ages happen:

- There may be occasional reductions in the thermal energy given off by the Sun.
- An increase in volcanic activity may add large volumes of dust to the atmosphere. This might reduce the amount of the Sun's energy reaching Earth.
- Periods of mountain formation would increase the area of high mountain ranges on Earth. The extra snow remaining on these cold peaks through the summer reflects sunlight and may reduce the temperature (see Figure 12.6).
- The movement of tectonic plates alters the shape of the oceans. This change affects the flow of ocean currents. With less mixing between hot and cold waters, some regions might become cold enough to start an ice age.
- Changes in the tilt of Earth's axis, or in its orbit around the Sun, may produce colder climates.

Scientists are still gathering data from many different areas of research to test which of these hypotheses about climate change are most likely. What is certain is that Earth's climate has changed often in the past and will continue to change in the future.

STRETCH Your Mind

Scientists have formed several hypotheses to explain how ice ages begin. What might cause ice ages to end? Write one or two hypotheses of your own, then conduct research at your local library or use the Internet to find some answers to this question.

Figure 12.6 Valley glaciers are found around the world today in mountain ranges at very high elevations.

Check Your Understanding

1. What changes occur in the Pacific Ocean during *El Niño*?
2. Why is the weather close to a lake different from the weather farther away from the lake?
3. How is a continental glacier formed?
4. Why do sea levels fall during an ice age?
5. How many major ice ages have there been during the past three million years?
6. **Thinking Critically** Why are fossils of ocean fish found in places geographically removed from present-day oceans?
7. Formulate your own question about water and climate, and use print or electronic resources to explore possible answers.

12.2 Life in the Water

What kinds of animals live in the oceans? You probably think of fish and whales. What about lobsters, shrimps, oysters, sea worms, squid, sea stars, sponges, jellyfish, anemones, corals? A far greater variety of organisms live in water than on land. In oceans, as on land, different environments with different conditions provide varied habitats for many types of organisms (see Figure 12.7).

You have already learned about some of the different ocean environments. There are warm waters and cold waters. There are differences in salinity and differences in the amount of light. Some organisms live near the surface in open water, while others live on the sea floor or along the sea coast. How might each of these conditions affect the types of organisms living there?

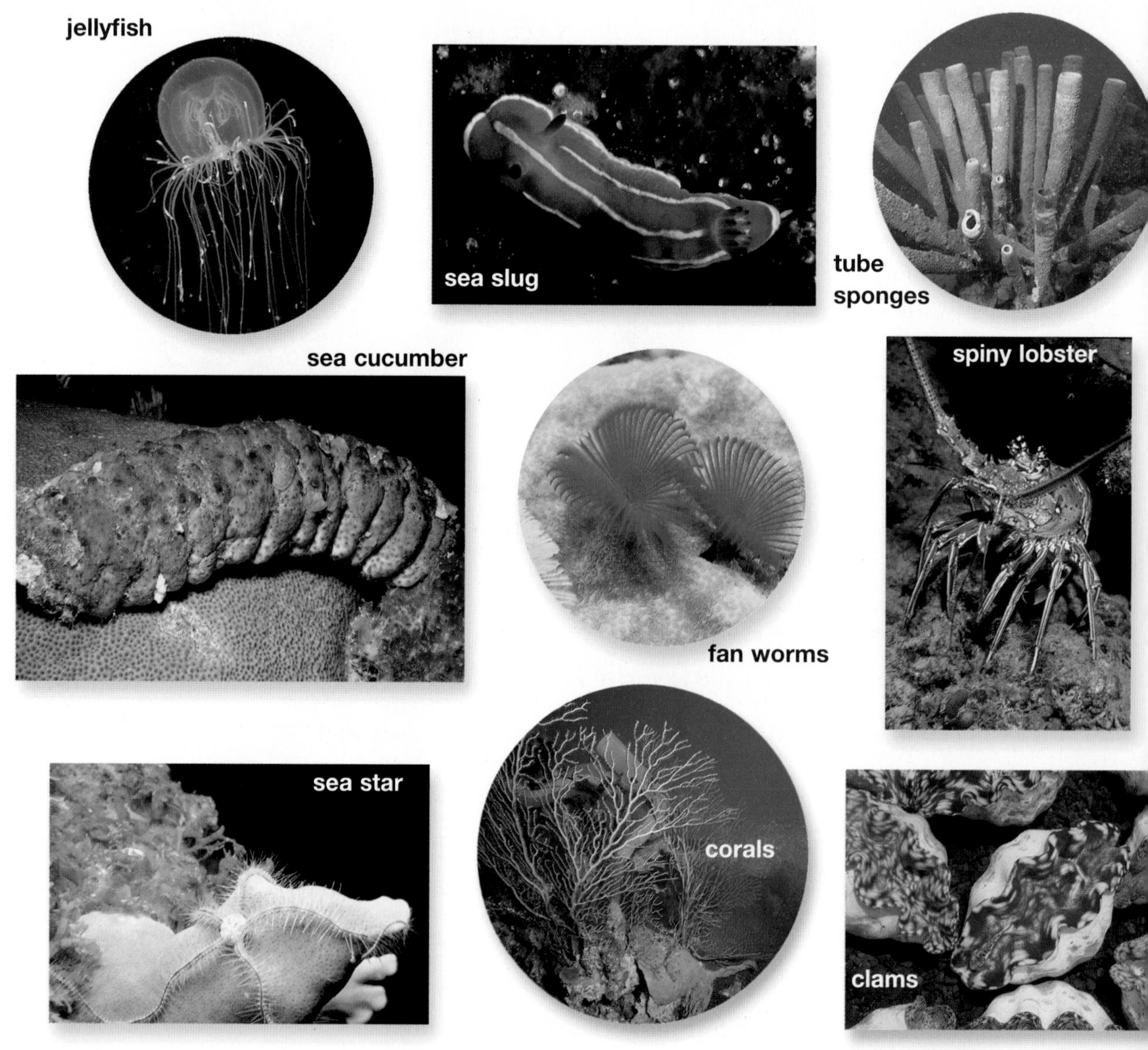

Figure 12.7 The oceans contain a spectacular variety of marine organisms.

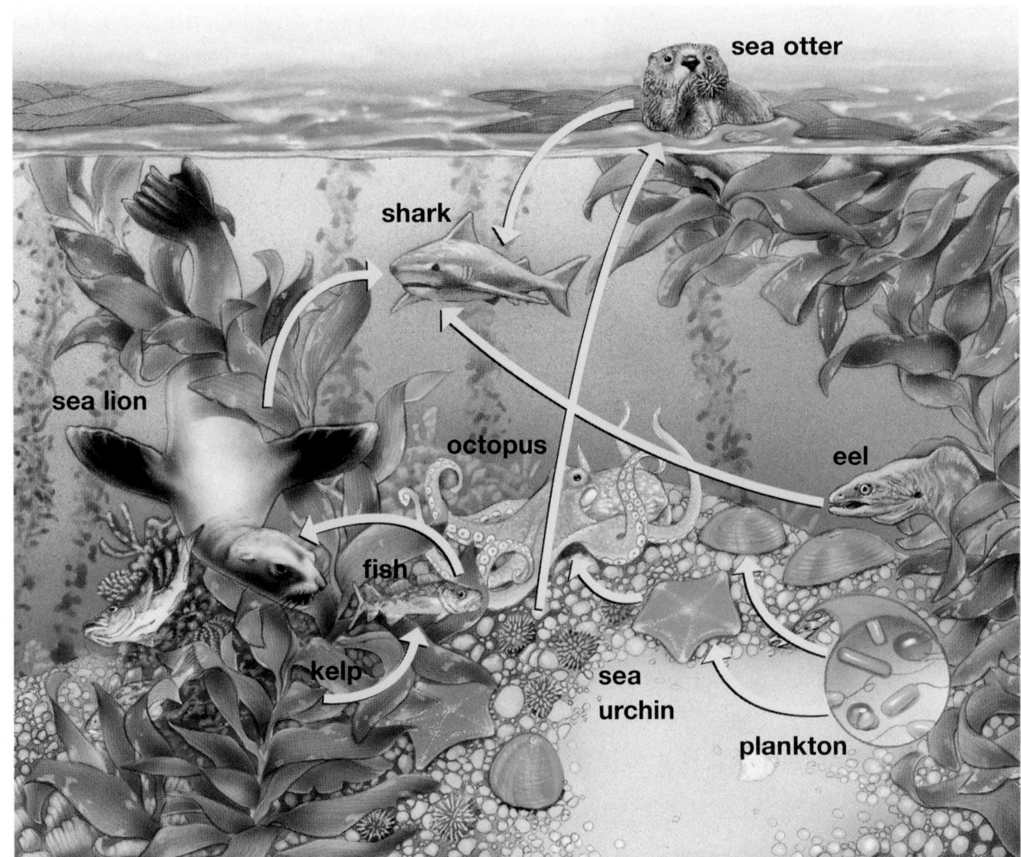

Figure 12.8 The basis of ocean food chains and food webs is plankton, which is made up of microscopic organisms. Smaller animals are food for larger animals through the food web. Arrows show the flow of energy in a food web.

As on land and in fresh water, organisms in the ocean ecosystems are linked together by food chains — sequences of feeding relationships between organisms. The basis of food chains and food webs — series of overlapping food chains — is plants and many unicellular organisms, which make their own food by photosynthesis. Because there is not enough light for photosynthesis below a depth of about 80 m, most plants in the ocean float near the surface. The majority of these plants are microscopic algae, ranging in size from 0.002 to 2 mm. They drift in vast numbers in the currents, forming the "grass" of the sea. Nearly all other marine life depends on these tiny plants (see Figure 12.8).

The surface waters of the ocean are also home to numerous tiny, drifting animals, as well as plants. Together, these drifting organisms are called **plankton**. (Plant plankton is called **phytoplankton**, and animal plankton is called **zooplankton**.) Plankton is the source of food for small fish, shrimps, and crabs. In turn, these organisms are food for larger fish. At the top of the food webs are animals such as seals, whales, and large fish such as sharks (see Figure 12.9 on page 390).

Where would you expect to find the greatest abundance of organisms in the sea? Although the oceans are vast, more than 90 percent of marine life occurs in the top layer of water, to a depth of about 180 m.

DidYouKnow?

Single-celled organisms called *diatoms* make up more than half the plankton in the ocean. Phytoplankton may increase by as much as 300 percent in one day. A cubic metre of ocean water may contain 700 000 organisms!

Figure 12.9 Some species of whales, such as this grey whale, feed directly on plankton at the bottom of the food chain. They strain the tiny organisms from the water on rows of comblike filters in their huge mouths.

The growth of plants depends on nutrients as well as light. Land plants obtain nutrients from the soil through their roots. Marine plants absorb nutrients from the surrounding water. In the oceans, nutrients slowly sink to the sea floor. How do marine plants obtain them? Nutrients are brought back to surface waters by upwelling currents. The most productive areas of the ocean are where cold, deep currents rise and mix with surface currents.

At Home **ACTIVITY**

Dissolved Gases

As well as using light and nutrients, plants need carbon dioxide for photosynthesis and oxygen for respiration. Land plants obtain these gases from the atmosphere. In the ocean, plants use gases dissolved in the water. How might water temperature affect the amount of gases available?

What You Need

2 new, unopened containers of carbonated soft drink

refrigerator

What to Do

1. Make a hypothesis about the effect of temperature on dissolved gases. (In other words, does a liquid dissolve more gas when it is warm or when it is cool?)
2. Describe a method of comparing the amounts of dissolved carbon dioxide in carbonated soft drinks.
3. Place one container of soft drink in a refrigerator and leave the other at room temperature for at least 3 h.
4. Open both containers and compare their amounts of dissolved carbon dioxide gas using the method you described in step 2. Record your observations.
5. Wipe up any spills and wash your hands after this activity.

What Did You Find Out?

1. Was there more dissolved gas in the cold liquid or the warm liquid?
2. Based on your observations, would you expect warm oceans or cold oceans to hold more dissolved gases?
3. Based on your answer to question 2, would you expect warm oceans or cold oceans to have a greater number of organisms? Why?
4. **Critical Thinking** Describe the relationship between temperature and plant growth
   (a) on land
   (b) in water
   If your answers to (a) and (b) are different, explain why.

Life on the Sea Floor

When deep-sea explorers first took underwater research vessels to the sea floor in the 1970s, they were amazed at what they saw. The explorers discovered new varieties of animals living 2.5 km below the ocean surface! These organisms do not use the Sun's energy or phytoplankton for food. Their source of energy is chemicals that flow from cracks, or **sea-floor vents**, in the ocean floor (see Figure 12.10).

Sea-floor vents are found along mid-ocean ridges. Hot water flows from these vents, carrying dissolved minerals and gases from below the ocean floor. The most important chemicals are sulfur compounds such as hydrogen sulfide gas. Certain bacteria can use these chemicals to produce food and oxygen by the process of **chemosynthesis**. The bacteria form the base of food chains around the vents. They are eaten by tiny larvae and other organisms, which, in turn, support populations of giant clams, tube worms, crabs, barnacles, and other unique creatures of the sea floor.

Figure 12.10 Sea-floor vents called black smokers form plumes of black iron sulfide particles. The temperature of the water here may be over 300°C when the vent erupts.

Fisheries

People have been catching fish in rivers, lakes, and oceans since prehistoric times. In oceans, the most productive areas for fish are the shallow waters above continental shelves. The largest populations of fish live in these waters, and most of the world's commercial fisheries capture fish there. One of the richest fisheries in history is in an area off Canada's east coast called the Grand Banks. The Grand Banks lie southeast of Newfoundland. The continental shelf in this region forms a large rectangular ridge, thousands of square kilometres in area. Water depth over the Banks averages 55 m. The cold waters of the Labrador Current meet the warm Gulf Stream here. This produces good conditions for the plankton on which larger fish depend for their food.

The Grand Banks were reported by the explorer John Cabot in 1498. Cabot had sailed to North America from England in search of a trade route to Asia and discovered huge numbers of fish on the Banks. For hundreds of years following their discovery, the Grand Banks were a major international fishing ground for the capture of cod, haddock, herring, and other species (see Figure 12.11 on page 392). In recent years, however, there has been a massive decline in fish populations, putting many people in Newfoundland out of work. The decline may be due to any number of causes, including overfishing and changes in water temperature or nutrient levels.

Figure 12.11 Fisheries were plentiful on the Grand Banks of Newfoundland before the collapse of the fish populations.

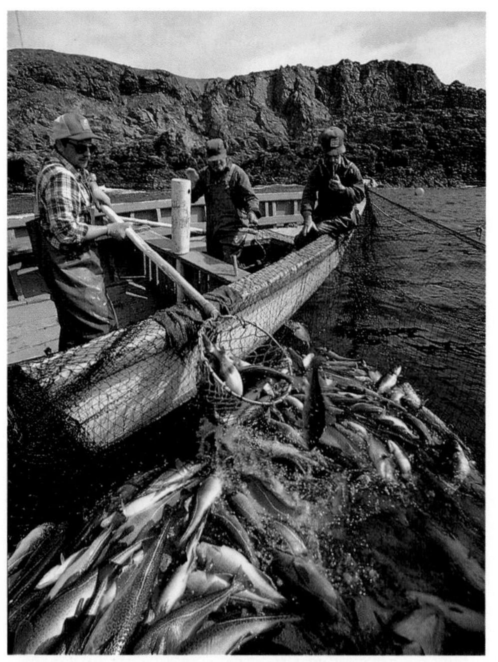

In the Great Lakes, the abundance and mix of fish populations have been altered over the past century. This change has been caused by fishing, introduction of new species by people, and pollution. How might fishing affect the balance of fish populations in a lake? Commercial fisheries tend to capture species of larger fish and fish with the highest market price. As the populations of these fish decline, smaller, less commercial fish species may increase in numbers. In this way, fisheries alter food chains, competition among species, and use of the habitat. The total abundance of fish may be similar, but the types and sizes of fish may be very different.

Did You Know?

The Red Sea gets its name from a species of microscopic algae that floats near the surface of the water. These algae have a reddish pigment for photosynthesis, instead of the more common green pigment.

Cities and farms along the shores of lakes produce a variety of pollutants that get into the water. Improved laws and regulations have reduced the amounts of toxic chemicals entering lakes. However, even small quantities of toxic chemicals can enter food chains and create health hazards over the long term. Non-toxic chemicals also produce problems. For example, rivers and run-off from farmers' fields bring extra nutrients into a lake, allowing increased growth of algae in the water. The result is a population explosion of algae called an **algal bloom**. The algae cover the water surface with a film of green slime (see Figure 12.12).

The natural ecosystem of a lake can be dramatically altered by the arrival of new species of plants or animals not originally found there. New organisms may be deliberately introduced by people. Species also find their way into a new habitat as a result of other forms of human activity.

Figure 12.12 Too many nutrients in fresh-water lakes and ponds can cause populations of algae to increase rapidly. The algae reduce the oxygen supply available to other organisms in the water.

For example, lampreys are parasitic fish that normally breed in rivers and streams (see Figure 12.13). In the last century, lampreys lived along the St. Lawrence River and in Lake Ontario. Niagara Falls created a barrier that prevented them from moving into Lake Erie and the other Great Lakes. When the Welland Canal was built, however, it connected Lakes Ontario and Erie. Lamprey swam through the canal and invaded the remaining Great Lakes. Lamprey have had a major effect on commercially valuable fish such as lake trout.

Figure 12.13 Sea lamprey feeding on a trout

Check Your Understanding

1. Give three reasons why plants are more abundant in cold seas than warm seas.
2. Draw a simple food chain found in the ocean.
3. Why is phytoplankton called the "grass" of the sea?
4. What is chemosynthesis?
5. What human activity might cause an algal bloom in a lake? Explain.
6. **Thinking Critically** Why does adding or removing species from an ecosystem often create unexpected problems?
7. **Thinking Critically** Here is a typical food web in the Antarctic Ocean and a fresh-water pond food web. Use these two food webs as a guide to create two food webs of your own: one an Arctic Ocean food web, and the other a fresh-water food web showing organisms' relationships in the ponds and lakes in your own province. Compare your two food webs.

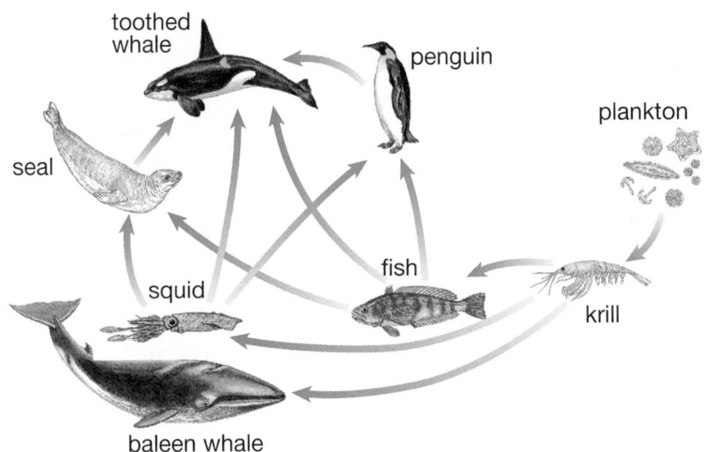

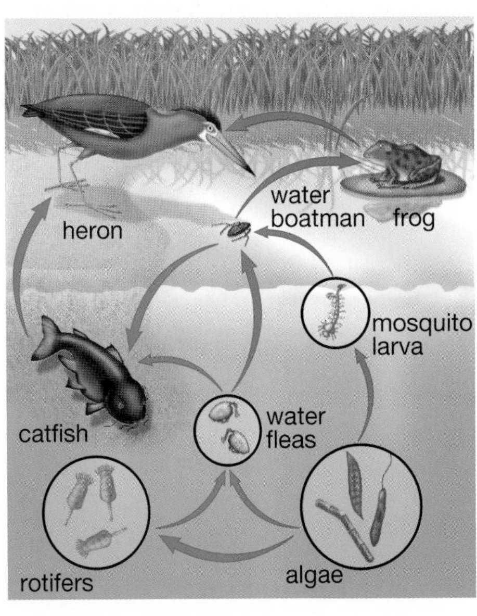

12.3 People and Water

People are part of the water cycle. We take fresh water from rivers, lakes, and underground. We use it for drinking, cooking, washing, farming, manufacturing, and other activities. The water we use for drinking must be treated to make it **potable**, or safe to drink, as shown in Figure 12.14. Once it has been used, the dirty water may be treated again to remove harmful wastes. The water then goes back into rivers, lakes, groundwater, and the sea.

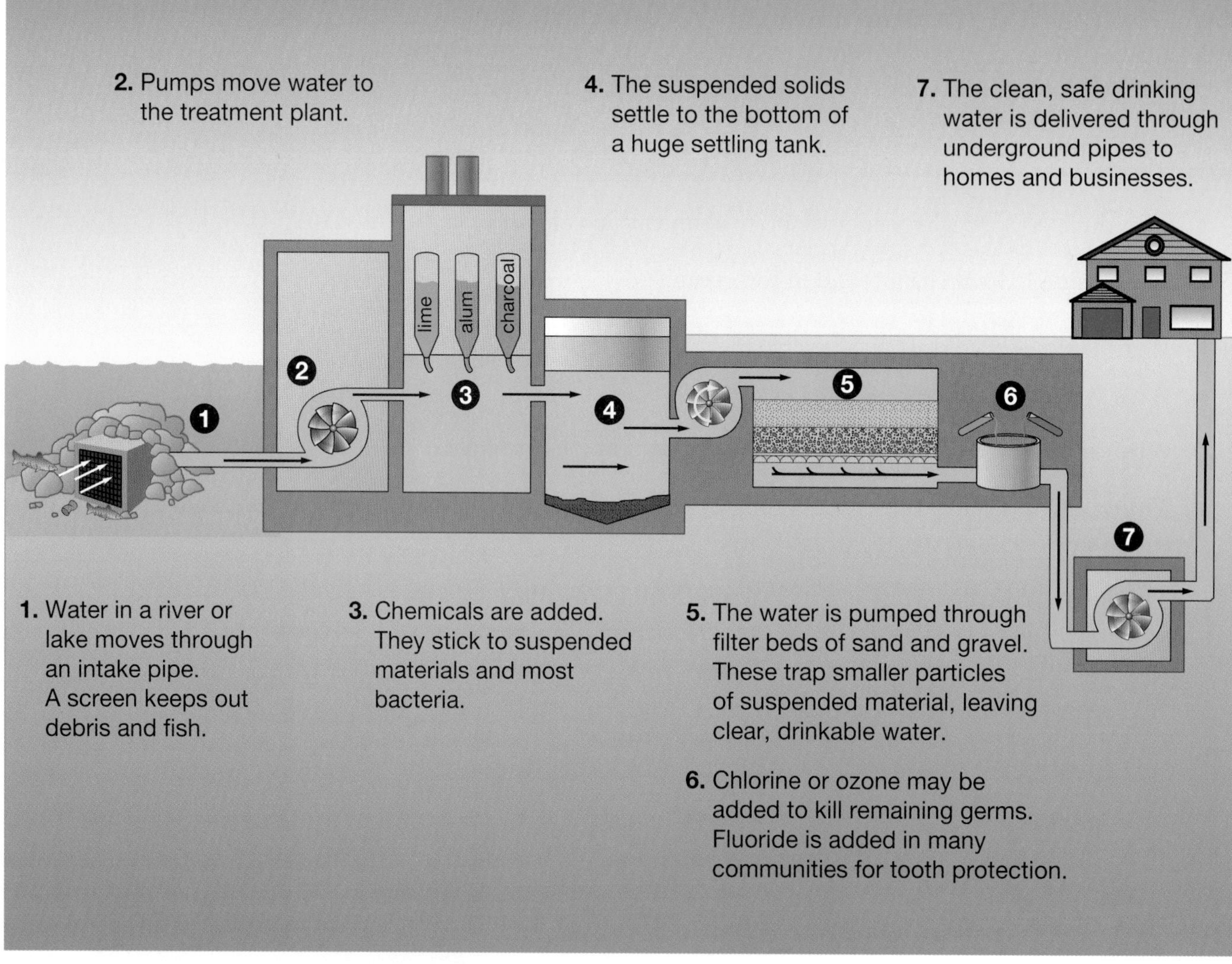

2. Pumps move water to the treatment plant.
4. The suspended solids settle to the bottom of a huge settling tank.
7. The clean, safe drinking water is delivered through underground pipes to homes and businesses.
lime
alum
charcoal
1
2
3
4
5
6
7
1. Water in a river or lake moves through an intake pipe. A screen keeps out debris and fish.
3. Chemicals are added. They stick to suspended materials and most bacteria.
5. The water is pumped through filter beds of sand and gravel. These trap smaller particles of suspended material, leaving clear, drinkable water.
6. Chlorine or ozone may be added to kill remaining germs. Fluoride is added in many communities for tooth protection.

Figure 12.14 Stages in the treatment of drinking water

Did You Know?

In developed countries, such as Canada, the average person uses up to 500 L of water per day. In the world as a whole, the average person uses about 5 L of water per day. In most countries in Africa, Asia, and South America, 80 percent of the people have no access to clean drinking water. They do not have faucets in their home, but must walk long distances to a river or well and carry the water back.

DESIGN & DO

INVESTIGATION 12-B

How to Clean Water

Imagine you are part of a team of engineers hired to provide water to a community. The only water source nearby is a muddy river. Design a system that will remove all the mud and suspended material from the river water, leaving it clear.

Challenge

Design and construct a model of a water-filtering system.

Materials

plastic containers, plastic tubing, filtering materials, funnel

Safety Precautions

- Do **not** test your water by drinking it. The goal of this investigation is to produce a filtration system that will remove mud and suspended materials from the water, but not necessarily any dissolved materials.
- Wipe up any spills and wash your hands after this investigation.

Design Criteria

A. Your filtration system should make the water as clear as possible.

B. The filtration system should be efficient and process water in a reasonable amount of time.

C. The filtration system should be easy to operate.

D. Availability of materials and cost factors might be included in your choice of materials.

Plan and Construct

1. In your group, select the materials from which to build your filtering system.
2. Make a sketch showing how your system will filter water.
3. Build your system and test it.

Evaluate

1. How successful was your system in cleaning water?
2. What improvements could you make to your system?
3. Would your water filtering system create other problems in the environment?
4. Would your system affect living organisms in the river?
5. Would your system be able to process the large volumes of water needed by a community? Explain.
6. If possible, have your samples tested by a water-testing laboratory.

Extend Your Knowledge

7. How does Earth filter water? Is your design similar to Earth's water-filtering process? Explain.

Altering the Water Cycle

Every city depends on water. If you study a map of Canada, you will see that most cities are built near a lake or river. Since ancient times, engineers have developed ways of carrying water from its source to where it is needed. This may include diverting rivers, laying underground pipes, and building water storage tanks. At the same time, waste water, sewage, and run-off have to be separated and drained safely away from the city.

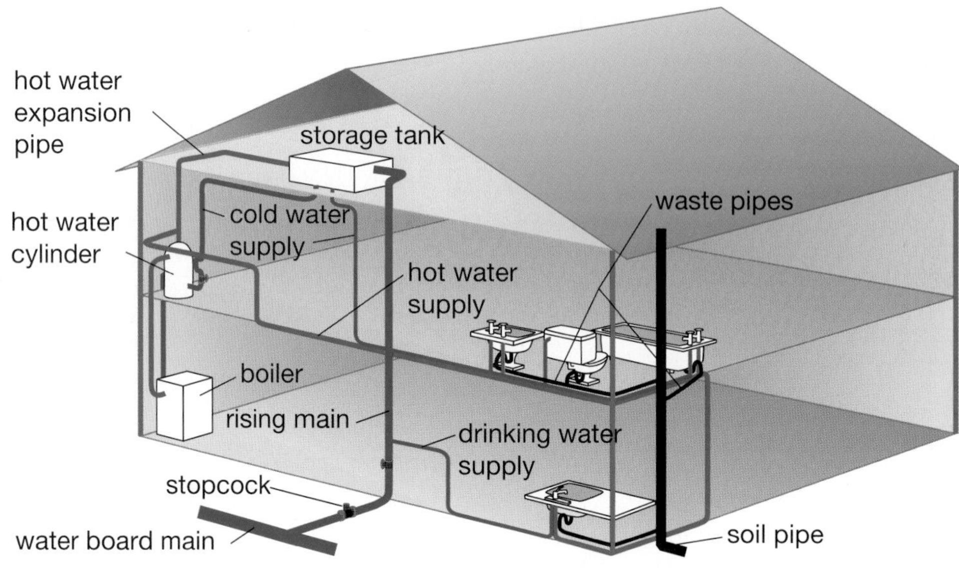

Figure 12.15 Water is distributed throughout a building by a series of pipes and drains.

We still distribute water and dispose of liquid wastes in much the same ways as people did thousands of years ago. The taps in your house can be traced back through a maze of pipes to your community's reservoir. Your toilets and drains are connected to a sewage treatment plant (see Figure 12.15). Maintaining a reliable water supply is an important issue for city planners and politicians.

Variations in the weather from year to year can make it difficult to balance the supply of water with the demand. Some areas get too little rainfall, while others get too much. Rain may fall in one season but not another. One way of making water supplies more reliable is to build a dam across a river. A dam holds back the water flow. This causes a river to flood land upstream, forming a lake or reservoir. As well as creating a year-round source of water for the community, reservoirs can be used for recreation (see Figure 12.16). Another reason for building dams is to produce hydro-electric power. The electricity is generated from the energy of falling water, which spins turbines in the dam structure (see Figure 12.17).

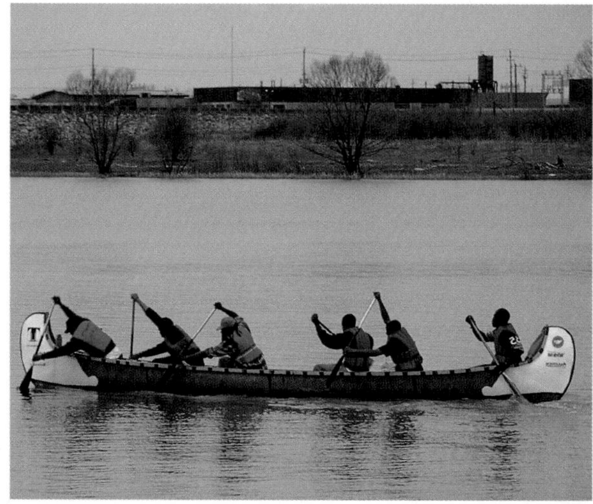

Figure 12.16 Reservoirs used to store water can also be used for recreation.

Dam building dates back thousands of years. The Egyptians built dams to control floods and help irrigate crops nearly 5000 years ago. In this century, more and bigger dams have been built than ever before, diverting and controlling the flow of almost every major river system on Earth. While dams have produced short-term benefits in many regions, they also bring many negative impacts. The investigation on the next page illustrates some of the issues in dam building.

Figure 12.17 In 1950, there were about 5000 large dams in the world. Today, there are over 38 000! The large cylinders shown here house a dam's turbines.

Career CONNECT

People Power

This is a picture of Debbie McGurk at work. She is an electrical operator at a dam that provides hydro-electric power. Her responsibilities include

- monitoring the amount of water passing through the dam at all times
- opening or closing the dam to match the community's need for electricity
- making sure the water level of the lakes and rivers of the system does not drop below acceptable limits
- cutting off electricity to an area of the plant that needs maintenance, so crews working there are safe from electrocution

To prepare for the job of electrical operator, Debbie took a three-year course at the power company. Considering her responsibilities and what you see in the photo, what sorts of questions do you think the people who interviewed Debbie asked in order to make sure she was the right person for this type of work?

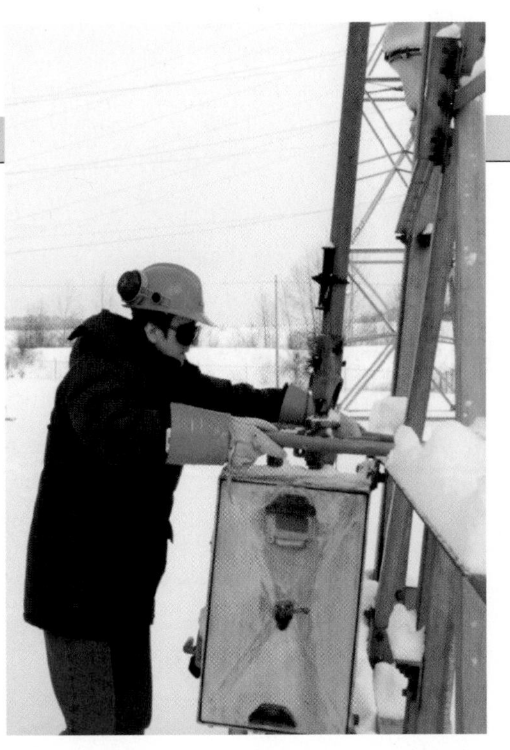

INVESTIGATION 12-C

The James Bay Hydro-Electric Project

In this investigation, you will read an account of a major hydro-electric dam project built in Canada. After this introduction, you will carry out further research on the project. You will then organize the material into a summary and discuss some of your findings and ideas as a class.

1. The Area

La Grande Rivière flows 800 km across northern Québec and empties into James Bay. Three nearby rivers (the Eastmain, Opinaca, and Caniapiscau) were diverted into La Grande by a series of dams and dykes built during the 1970s and 1980s.

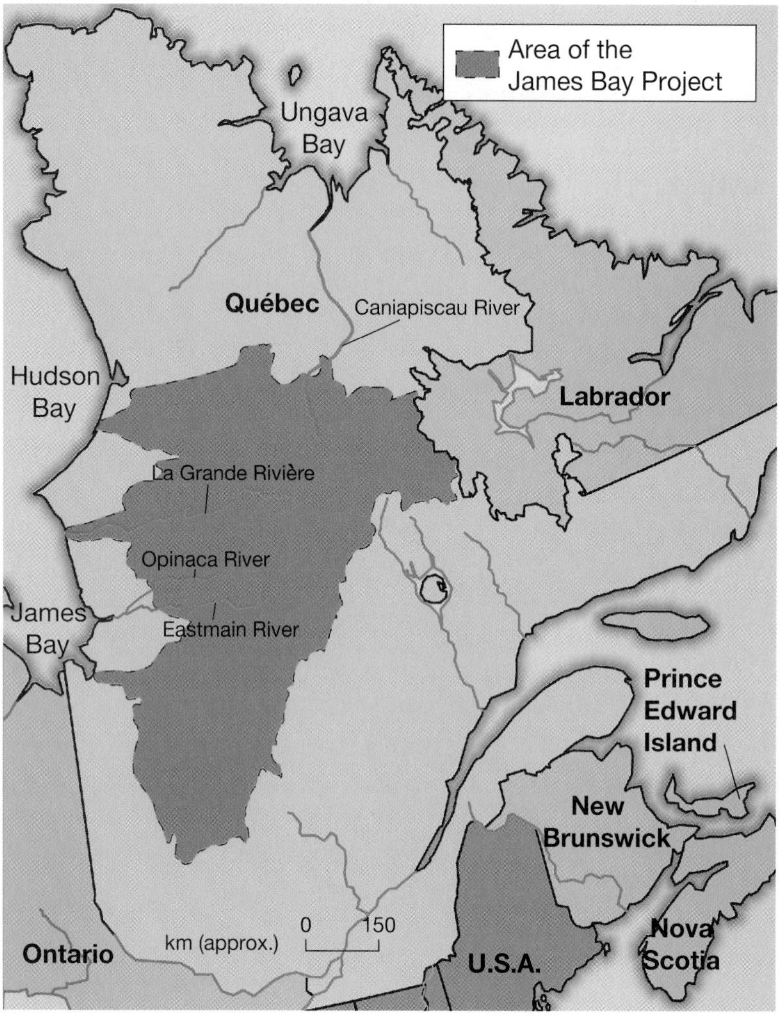

2. The Project

Construction of the dams consumed more than 1.6 million tonnes of fuel, and moved enough rock to build the Great Pyramid 80 times over. To allow supplies for construction to be flown in, five small airports were built in remote areas of wilderness. Over 8000 km of transmission lines were strung through forest and wetland to carry electricity from the dam site.

The damming of La Grande cost more than \$21 billion. There are plans to build 38 more major dams along other rivers flowing into James Bay. These dams will create new reservoirs having a total surface area as large as Lake Erie. The total area to be covered by the planned project is over 380 000 km^2! These plans may change, however, due to changing political and economic conditions.

3. Goals of the Project

The government of Québec developed the James Bay project to produce electricity for the province and for export to the northeast United States. The electricity would also be used to develop industries such as aluminum and magnesium smelters. These industries consume large amounts of energy.

The dams first generated power in 1979. By 1989, most homes in northern Québec had converted from oil or gas to electricity. The percentage of homes using only electricity in the province increased from 30 percent to 60 percent.

If completed, the James Bay project will produce 28 million kilowatts of energy. This is equal to about 25 percent of Canada's present electricity consumption.

4. **Effects of the Project on the Environment and Wildlife**

The river diversions doubled the mean annual flow of water in La Grande Rivière. The Eastmain River was reduced to only a trickle. More than 2500 km^2 of land were flooded behind the dams. Flooded soil and decaying vegetation released toxic methyl mercury into the water.

Reservoirs created by the dams destroyed the original wildlife habitat. For example, the feeding areas used by caribou were no longer available to migrating herds. Other areas of wilderness were disturbed through the building of roads and power lines.

The James Bay project could affect food chains in James Bay and Hudson Bay. At the base of these food chains are algae and plankton. These organisms depend on the flow of nutrients brought by rivers and on the mixing of fresh and salt waters. Damming La Grande increased the normal winter flow of the river by eight times. Changes in water temperature or salinity in the bays could alter current flows and local climate. These changes could affect the lives of plants, animals, and people across thousands of kilometres.

5. **Effects of the Project on People**

Areas flooded by the project are part of the traditional home of Aboriginal Cree and Inuit peoples. Hundreds of people had to move after construction began. Toxic chemicals from flooded soil made their way through food chains into pike and lake trout, creating health concerns among Aboriginal people who eat these fish. Flooded lands forced caribou to alter their migration routes, also affecting Aboriginal people who depend on caribou for food. After several years of legal dispute, the government of Québec compensated the northern residents with $225 million.

Some of Canada's Native people hunt caribou for their meat and hides.

La Grande Rivière dam, commonly referred to as the LG-1 dam

What to Do

1. As a class, prepare a chart listing the positive and negative outcomes of the James Bay project, based on the information given on these pages.

2. Use your library and computer resources to gather more information about the current state of the project and its effects on the environment and people of the area. Add your findings to the chart.

3. Form smaller groups representing different points of view about the project. They might be the views of a politician, a person employed by the project, a resident of the province, a member of an Aboriginal Cree or Inuit community, an ecologist, or some other interested person. In your group, discuss how the project affects the person you are representing. Prepare a short statement expressing the person's point of view to the class.

4. If you were asked whether a project such as this should be carried out today, would you answer "Yes" or "No"? Explain why.

Energy from the Seas

Oceans contain abundant energy in the form of heat, currents, tides, and waves. However, it is costly and difficult to harness these forms of energy with present-day technology. Small tidal-power plants have been built in France and Russia. Norway has an electric plant powered by ocean waves. Inventing efficient ways to use ocean energy is a challenge for future engineers.

Other forms of energy such as fossil fuels are found in the ocean floor. Oil and natural gas were formed millions of years ago. These resources were formed from the decomposed material of marine organisms buried under thick layers of sediment. To locate these oil fields, research ships send down sound pulses that penetrate the sea floor. The pattern of reflected sound waves tells geologists which areas are most likely to hold oil.

Most offshore oil and gas is extracted from continental shelves. In these areas, the sea is not too deep for drilling. To obtain the fuel, drills are lowered from huge floating platforms. The platforms are towed out to sea and anchored by scaffolding to the sea floor (see Figure 12.18). About 20 percent of the world's oil supply is now obtained by undersea drilling.

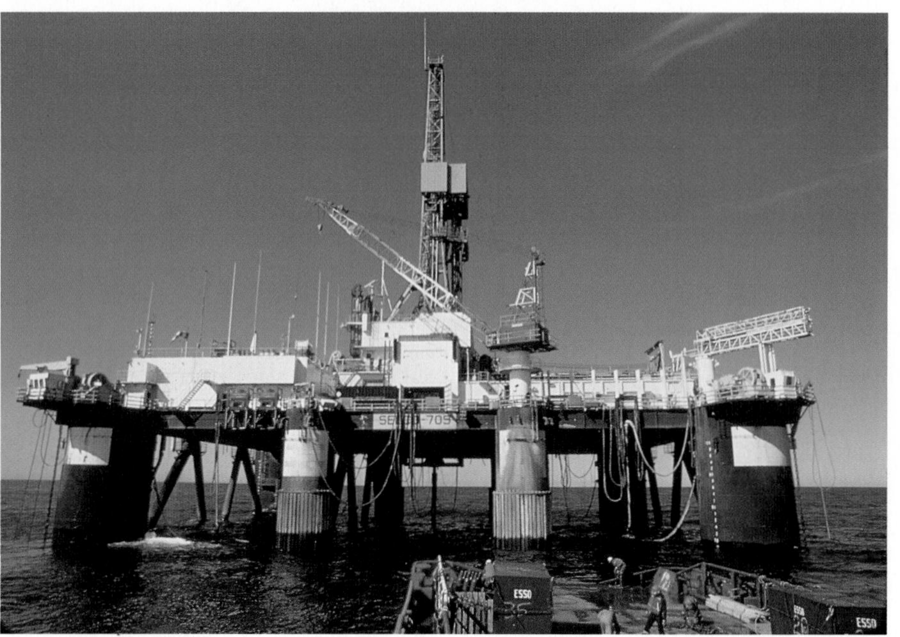

Figure 12.18 An offshore drilling platform extracts oil from the ocean floor.

If the offshore oil field is close to land, the oil may be pumped onshore through pipelines laid along the sea floor. If drilling is farther from land, the oil is stored in huge tanks on the sea floor. The oil is then pumped into oil tankers that carry it to a nearby port.

What are the risks of extracting oil and gas from the ocean floor? You may have read news stories about oil tankers that leak their oil. Spilled oil pollutes large areas of sea and shoreline and kills marine organisms. Workers on offshore drilling platforms risk their lives in sea storms with strong winds and high waves. In areas with earthquake activity, such as the coast of California, there is a chance that an earthquake could topple oil rigs or fracture undersea pipelines and storage tanks.

Pollution of the Seas

You have learned why areas of ocean close to coastlines have the most marine life. These areas are also at greatest risk from pollution. Coastlines receive pollutants from land that is washed into the sea. The pollutants are carried in rivers and in run-off from cities and farms. The coastlines also receive pollutants discarded by ships at sea or from drilling platforms or broken pipelines (see Figure 12.19).

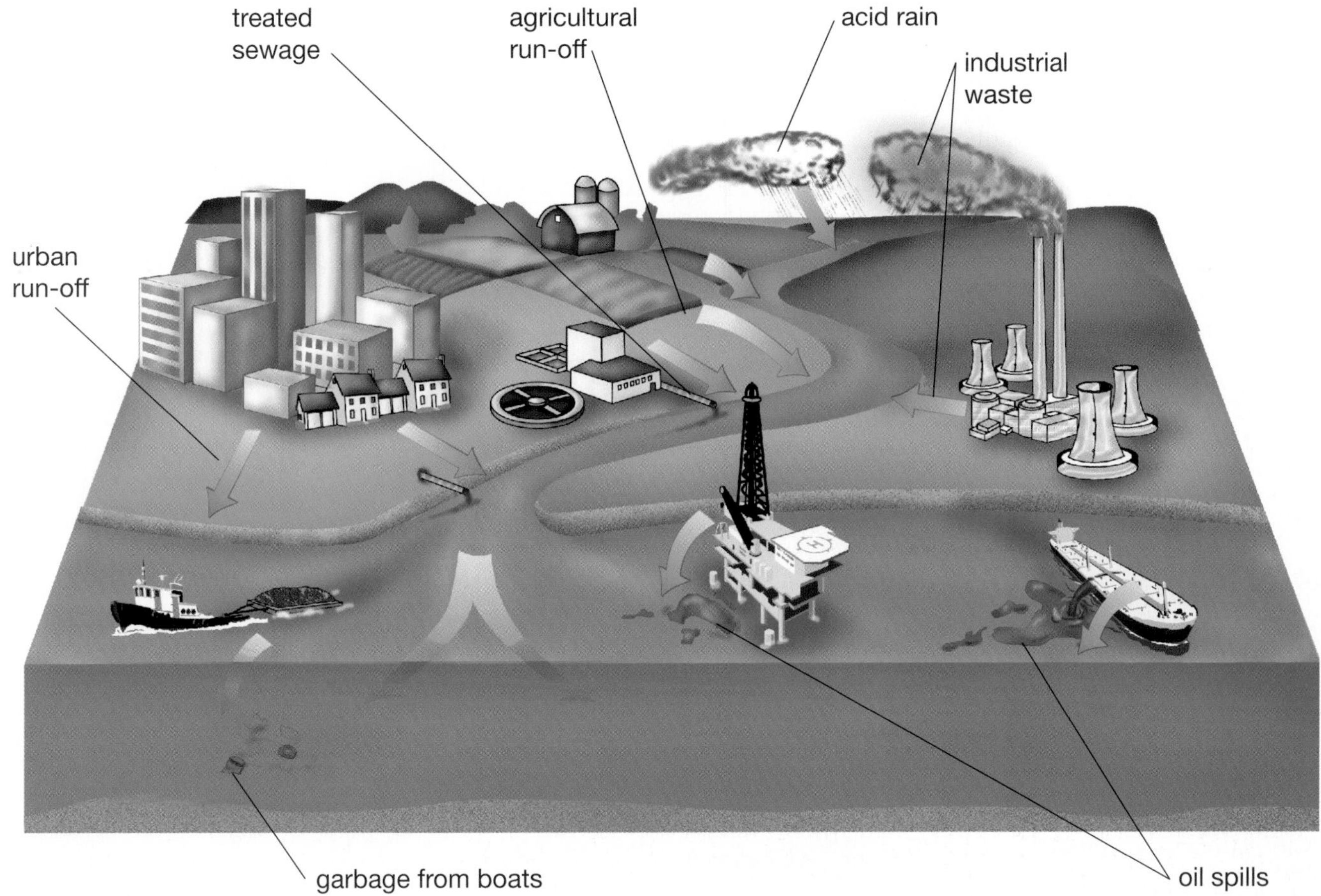

Figure 12.19 Sources of pollution in the seas

You may think that the vast volume of water in the ocean would dilute polluting chemicals and make them less harmful. However, pollutants may enter the bodies of living organisms and be stored in their tissues. As they pass from organism to organism along the food chain, the pollutants become more concentrated. The feeding method of some animals also helps concentrate pollutants. For example, shellfish filter huge volumes of water through their bodies in order to trap microscopic food particles. In this way, shellfish also trap and concentrate dispersed chemicals. People are at the top of marine food chains. As a result, our health is at risk as the oceans become more polluted.

Chemical pollution is not the only threat to marine wildlife. Sea birds, turtles, whales, dolphins, seals, and other marine animals may be injured or killed by abandoned fishing lines, nets, plastic bags, needles, and other objects carelessly disposed of (see Figure 12.20). Circulating ocean currents carry garbage, chemicals, and oil spills around the world. Careless habits in one place eventually affect all life on the planet.

The fate of the oceans may finally affect life everywhere on Earth. Scientists estimate that phytoplankton produce more than three quarters of the oxygen in Earth's atmosphere. If populations of these microscopic plants are harmed by pollution, the global atmosphere may become less suitable for many animals and plants.

Figure 12.20 Many beaches show evidence of sea pollution.

CONDUCT AN

INVESTIGATION 12-D

Water Use in Your Home

Have you ever wondered how much water your family uses? There are many household uses for water, from washing the dishes to taking a shower. In this investigation, you will estimate the amount of water used in your home in one day. You will also use class data to estimate the amount of water used in your community. Finally, you will suggest ways in which you can conserve water.

Problem

How much water does your family use in a day?

Procedure

1. Prepare a table with the headings listed below. Give your table a title. Record how much water your family uses for different purposes during a day.

2. Make a list of all the places in your house where water is used. You may want to place record sheets at each location and have family members help you keep records.

3. Enter your data into the table and calculate totals used by each activity. The table on the right gives some average figures of water consumption for particular uses. (Your appliances and water fixtures may have higher or lower rates of flow.)

Average Quantity of Water Used for Various Activities

clothes washer	230 L per use
dishwasher	65 L per use
toilet flush	20 L per use
bath	130 L per min
shower	25 L per min
faucet	12 L per min

Activity	Total time used	Rate of water use (see the table)	Total volume of water

Skill POWER

To learn how to make a stem-and-leaf plot, turn to page 546.

Analyze

1. Calculate the total amount of water used in your household in one day. Divide the total by the number of people in your household to get an average use per person per day.
2. Sort the class data on water use in the home using a stem-and-leaf plot, to show how the range of numbers is distributed from the smallest to the largest.
3. As a class, find the average and the maximum and minimum quantities of water used.
4. Use the average to calculate the total water consumption by the population in your town or city.

Conclude and Apply

5. Are there any dripping faucets in your home? Measure and calculate how much water might go down the drain each day from a dripping tap.
6. List some ways you could conserve water in your home.

Sustaining Water Resources

Is water a sustainable resource? Recall what you have learned about the distribution of water on Earth (Chapter 10). Most fresh water is underground or frozen. In addition, water is not evenly distributed around the world because of differences in average annual rainfall. Some places have much precipitation, while others have very little.

The first problem in managing water is quantity. Does the supply of water in a particular area match people's demand for water? The second problem in managing water is quality. Is the water suitable for human use, or is it polluted?

As the human population grows, demand for water increases. Increasing consumption has reduced the quantity of water in reservoirs and underground aquifers throughout the world. When water becomes scarce, people must look for alternative sources, limit their use of water, and wait for more rain. At the same time, human activities, such as agriculture, mining, and pulp and paper manufacturing, are adding huge amounts of waste materials to Earth's water systems. Water pollution has made many sources of fresh water unusable, or even dangerous to use without costly treatment to remove harmful chemicals or microbes.

Many countries now have to import water from outside their borders. Neighbouring countries sometimes come into conflict over water supplies. For example, one country may build a dam that reduces the flow of water to a neighbouring country. Furthermore, pollution may flow down a river and across a border into the next country. Shortages of water have forced some countries to develop costly **desalination plants**, which produce fresh water by removing salts from seawater (see Figure 12.21). Other countries have proposed towing icebergs from Earth's poles to supply fresh water as they melt.

Math CONNECT

Suppose an average shower uses 25 L of water per minute. If you shower 5 min each day, how much water will you use in a year? How much water will you save by reducing your shower time to only 4 min? How else could you reduce water use in the shower?

Figure 12.21 A desalination plant in Mexico makes fresh water from salt water.

Across Canada

Host of CBC TV's "The Nature of Things," David Suzuki is a household name in Canada. Dr. Suzuki is chair of a non-profit organization named after him, The David Suzuki Foundation. His foundation works to find solutions to environmental and social problems. The foundation also works to inform people about the problems we face and possible sustainable solutions so that the public can make the government pay attention and change laws.

Dr. Suzuki began his career as a geneticist. He has been a professor at the University of British Columbia since 1969. In the past 25 years, he has won numerous awards for his contributions to science, environmental awareness, television, and radio. Today, most of his time and energy are directed toward seeking solutions to environmental issues.

The need to take greater care of our life support systems, especially air, soil, water, and biological diversity, is something David Suzuki strongly promotes. He often speaks publicly to encourage others to change the habits of our consumer culture. He tries to walk or bicycle instead of driving and dislikes anything disposable. One issue he is especially concerned about is that of dioxins leaking into the Great Lakes; another is the decreasing numbers of wild salmon on the west coast of Canada.

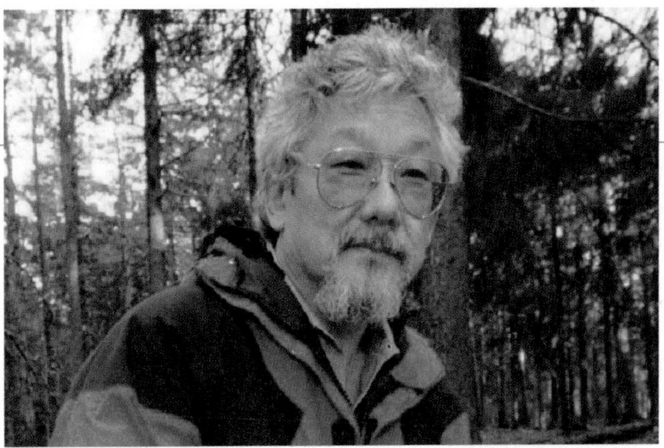

David Suzuki

David Suzuki is the author of many books, including *Metamorphosis* (1987) and *The Sacred Balance* (1997, reprinted 1998). He lives in Vancouver, British Columbia.

Figure 12.22 By looking after the waterways in their neighbourhood, people can help protect the larger system of which they are a part.

For long-term water management to succeed, water managers and engineers must study water as part of a larger ecological system. Rivers, lakes, and wetlands carry out important functions as well as supply water. Rivers help distribute fertile silt during floods. Wetlands act like sponges, holding and purifying water and preventing flooding. Bodies of water are important habitats for many species, including those that people use for food. As wetlands are drained, rivers diverted, and water habitats polluted, we lose much more than water alone.

Much of the damage to water systems can be reversed. In many parts of the world, people are taking actions to restore rivers and lakes to better health (see Figure 12.22). The first step is to stop pollution. Industrial and domestic wastes that were once dumped into rivers and lakes must now be disposed of in other ways, or treated to remove harmful materials. Run-off into waterways can be reduced by planting protective buffer zones of trees, shrubs, and grasses along the edges of rivers and lakes. Plants also prevent erosion and help restore fish habitats.

The second step to make water supplies sustainable is conservation. It is far less expensive to avoid wasting water than to develop methods to obtain more. The prices now charged to supply water to homes and industries in Canada are relatively low. As a result, people often do not appreciate its true value and do not use it efficiently.

The health of our water systems depends on everyone. The choices we make about how we use water will affect all people and all other living organisms. In order to sustain our water resources, it is important that we keep lakes, rivers, and oceans clean for today and for the future.

DidYouKnow?

Millions of people in the world now die each year from lack of clean drinking water. It is predicted that up to one third of the world's countries may suffer from permanent water shortages during the early part of the twenty-first century.

Word CONNECT

The novelist and poet Hermann Hesse wrote:

"Love this river, stay by it, learn from it."

Yes, he wanted to learn from it, he wanted to listen to it. It seemed to him that whoever understood this river and its secrets, would understand much more, many secrets, all secrets.

How is a river or lake special to you? Write a poem or short story about what you have learned from a lake or river.

Check Your Understanding

1. List five purposes for which people use water.
2. Give two reasons why water is scarcer in some countries today than in the past.
3. Give three reasons why dams are built.
4. Why are pollutants in the sea not made harmless by dilution?
5. What two steps can help sustain water resources for future generations?
6. **Thinking Critically** In the 1970s in northern Ontario, hundreds of residents along the English-Wabigoon river system lost their jobs as commercial fishers and suffered health problems because of mercury poisoning. The mercury was dumped into the river as waste material by a local paper mill before 1970. Using print or electronic resources, find out why this pollution of the river will remain a danger in the water and fish for another century.

CHAPTER at a glance

Now that you have completed this chapter, try to do the following. If you cannot, go back to the sections indicated.

Compare the ocean with the land in terms of temperature changes and availability of light. (12.1)

Relate a change in conditions in the ocean to a short-term change in the climate. (12.1)

Describe the chain of events leading to the formation of continental glaciers. (12.1)

Draw a flow chart to demonstrate how ice age temperatures produce a fall in sea level. (12.1)

Explain why most ocean food chains and food webs begin in surface waters. (12.2)

Give three reasons why plankton drifts with the currents. (12.2)

Contrast the source of energy for food chains on land with the source of energy for food chains based around sea-floor vents. (12.2)

Discuss some issues related to water sustainability. (12.3)

List three benefits and three negative effects of dam building. (12.3)

Explain how pollutants can harm organisms such as fish even after the pollutants are diluted in the ocean. (12.3)

Identify ways in which individuals in your community can reduce water pollution and avoid wasting water. (12.3)

Prepare Your Own Summary

Summarize this chapter by doing one of the following. Use a graphic organizer (such as a concept map), produce a poster, or write the summary to include the key chapter ideas. Here are a few ideas to use as a guide:

- Copy the following diagram and add labels to explain what it is illustrating.

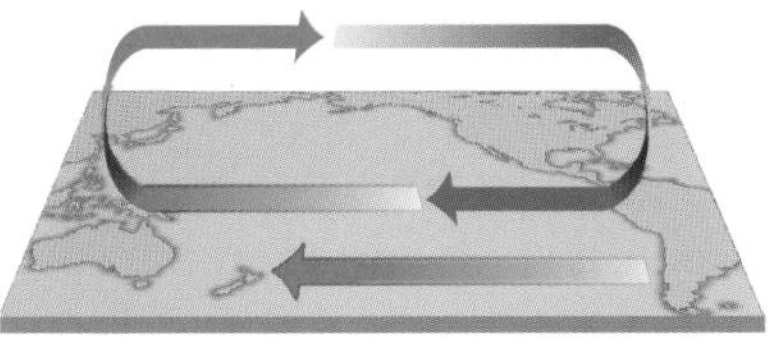

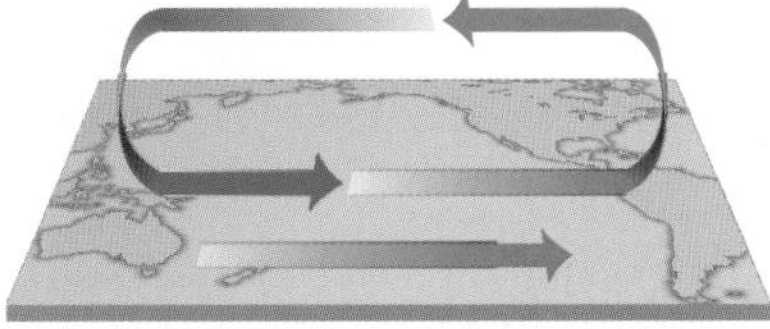

- How does an ice age affect the distribution of animals both on land and in the oceans?
- Why do some scientists think we are living in a period between ice ages?
- How do plants in the ocean obtain the sunlight and nutrients they need for photosynthesis?
- Explain why some large whales are at the top of long food chains and why others are at the top of short food chains.
- List or sketch the steps used to treat water taken from a lake or river in order to make it suitable for household use.
- Why are the waters along a coastline especially at risk from pollution?
- Describe some ways in which people can restore polluted streams and rivers so that they are suitable for fishing and swimming.

CHAPTER
12 Review

Key Terms

heat capacity
El Niño
ice ages
continental glaciers
land bridge
plankton
phytoplankton
zooplankton
sea-floor vents
chemosynthesis
algal bloom
potable
desalination plants

Reviewing Key Terms

If you need to review, the section numbers show you where these terms were introduced.

1. In your notebook, write the key terms missing from each of the following sentences:
   (a) ________ consists of small floating plants. (12.1)
   (b) ________ consists of small floating animals. (12.1)
   (c) Bacteria on the sea floor use ________ to produce food from chemicals. (12.1)
   (d) Chemicals seep into the ocean through ________ on the sea floor. (12.1)

2. What is the difference between chemosynthesis and photosynthesis? (12.2)

3. Write a definition of a land bridge, giving an example of where one might occur. (12.2)

Understanding Key Ideas

Section numbers are provided if you need to review.

4. Why is phytoplankton important to life in the sea? (12.1)

5. How do food chains at sea-floor vents differ from food chains in other parts of the ocean? (12.1)

6. Describe the extent of the continental glaciers in North America during the last ice age. (12.2)

7. Give two scientific hypotheses that could explain the origin of an ice age. (12.2)

8. List five human activities that pollute water. Suggest how to reduce the risk of each. (12.3)

Developing Skills

9. The graph below shows changes in global temperatures over the past 200 000 years. Ice ages occurred when temperatures were coolest. Study the graph and answer the questions that follow.

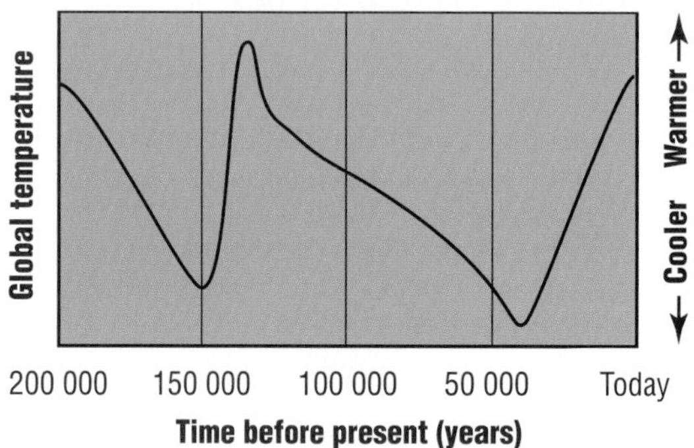

   (a) How many ice ages occurred over the past 200 000 years?
   (b) When did they occur?
   (c) How do temperatures today compare with temperatures 150 000 years ago?
   (d) About how long did it take for the last ice age to get to its coldest from the warmest conditions that came before it?
   (e) About how long has it been since the coldest temperature during the last ice age?

10. **Design Your Own** Design an experiment to compare different methods of cleaning oil spills from beaches. List the materials you will need and the procedure you would use. Include a data table to record results. To help you get started, here are some materials you could use to simulate a beach, an oil spill, and cleaning materials: pan, cooking oil, gravel, water, paper, liquid soap.
   (a) What variable(s) must be held constant in your investigation to ensure a fair test?
   (b) What criteria would you apply to assess your solution to the problem posed in the investigation?

11. Describe how crop fertilizers on a farmer's fields can affect fish in a nearby lake.

12. Copy and complete the following concept map of sustaining water resources.

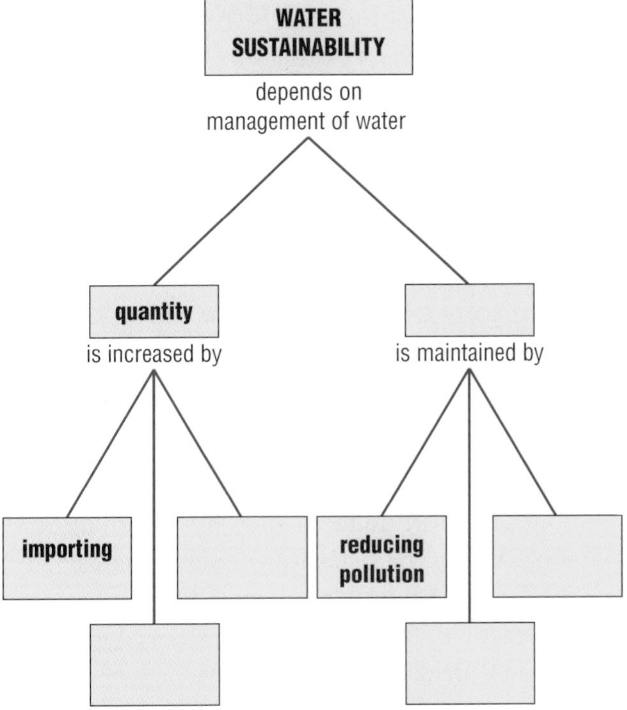

Problem Solving/Applying

13. Some icebergs are several kilometres long. If melted, they could provide billions of litres of fresh water.
    - (a) What areas of the world would benefit most from using icebergs as a source of fresh water?
    - (b) What practical problems would arise in moving icebergs from the polar regions to other parts of the world?
    - (c) For each problem in (b), suggest a solution.

Critical Thinking

14. Would you expect to find more marine organisms in waters near the coast or in deeper waters far out at sea? Explain your answer.

15. Why is water sustainability an international problem?

16. If you could compare a map of North America before the last ice age with one after the last ice age, you would see that lakes and rivers have different locations. Why?

17. A younger student says to you: "I don't eat fish, so it won't affect me if the sea is polluted." Explain why you would agree or disagree with the student's statement.

18. Before the James Bay hydro-electric project began, Premier Robert Bourassa of Québec described the large rivers flowing into James Bay as "a waste." Explain why you agree or disagree with his description.

INTERNET CONNECT

www.school.mcgrawhill.ca/resources/
Research the history of lampreys in the Great Lakes and produce a report for your class. Include the following in your report: how lampreys feed, how they affect fish populations in the Great Lakes, and how lamprey populations in the Great Lakes are controlled. Visit the above web site. Go to **Science Resources**, then to **SCIENCEPOWER 8** to find out where to go next.

Pause& Reflect

1. One reason Canadians have not valued water more is that it has always been relatively cheap and easy to obtain. How might it affect you if you had to pay each time you used water, or if you had to travel many kilometres to obtain water?
2. Go back to the beginning of this chapter on page 378 and check your original answers to the Getting Ready questions. How has your thinking changed? How would you answer those questions now that you have investigated the topics in this chapter?

UNIT 4 Ask an Expert

"What goes 'round, comes 'round." That saying has a different meaning for Carole Mills than it does for the rest of us. Carole is a biochemist and a contaminants researcher. She is the manager of the Contaminants Division of the Department of Indian and Northern Affairs in Yellowknife, Northwest Territories. She knows that whatever harmful chemicals humans throw into the environment will probably end up in the Arctic. These chemicals affect not only Canada's Aboriginal people and wildlife, but also people and wildlife in arctic regions around the globe. Because of her work, Carole was asked to represent Canada's indigenous people on the United Nations' committee to reduce the use of harmful contaminants throughout the world.

Q What exactly is a contaminant?

A A contaminant is something that ends up where it doesn't belong and where it can cause problems. The contaminants that I study are chemicals called organo-chlorines, heavy metals such as mercury, and radioactivity. These are materials that people use in industries or spray as pesticides, and so on. Unfortunately, these contaminants can make people and other animals very sick.

Q Why are there contaminant problems in the Arctic? There isn't very much industry up there, is there?

A No. In fact, most of the worst contaminants have been banned in Canada and many other countries for years. These contaminants are still being used in some countries, however, and when contaminants enter the environment, they don't stay in one spot. Rainwater can wash a chemical or other contaminant into a river, which flows into an ocean. Ocean currents could take it anywhere.

Some of these chemicals also evaporate, just as water does, and end up in the air. Once in the air, they drift wherever the wind takes them. They can drift quite far in a short amount of time. When the Chernobyl nuclear reactor in the former Soviet Union exploded, for example, radioactivity from the disaster showed up in Arctic Canada in just a few days.

Q Do these drifting contaminants create health problems for people all over the world?

A Yes, they can cause health problems. The people in the Arctic are more at risk, however, for two reasons. First, northern animals, such as fish and walruses, are traditional foods of people living in the Arctic. When the food and water become contaminated, so do these animals. When the animals become contaminated, so do the people who eat them.

Q What is the second reason?

A The second reason is that more contaminants end up in the Arctic than in most other places on Earth because of something called "global distillation" or the "grasshopper effect." Since the contaminants can evaporate, they go through a cycle much like the water cycle. As the chemicals evaporate, they rise and may drift for a distance, then they cool and condense on the ground. When the temperature rises again, the chemicals evaporate again, and so on. If the vapour drifts north into arctic climates, however, the colder temperatures prevent the chemicals from evaporating, and so they are trapped here.

Q **What led you to study contaminants in the Arctic? Have you always lived in Yellowknife?**

A No, I grew up in Ottawa, but spent my summers in the Arctic with my Dene relatives. It was my job that brought me north full time. While I was getting my university degree in biochemistry, I did volunteer work in the environment department of the Assembly of First Nations. As soon as I graduated, the department hired me to work on the Eagle Project, a study of contaminants in the Great Lakes and their effect on local Aboriginal people.

After that, I worked with the Northwest Territories health department to study the effects of contaminants on the territories' people, and then with the Dene nation and Indian and Northern Affairs on similar projects. I moved up to Yellowknife in 1993 and I've been here ever since. It's wonderful to be closer to so many members of my family.

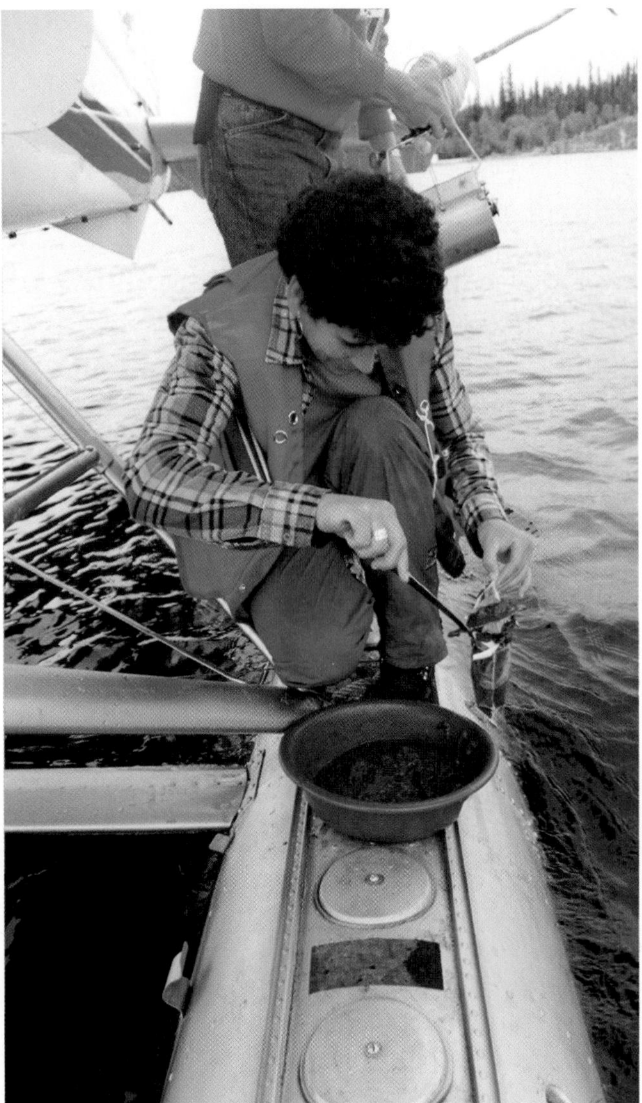

This researcher is testing water for contaminants.

EXPLORING Further

Dry Up

Carole says colder temperatures mean less evaporation. Try this activity to see for yourself. You will need some water, paper towels, scissors, a refrigerator, and a nearby countertop or other surface.

1. Cut two equal squares of paper towel.
2. Wet each piece of paper towel. Hold the pieces vertically over a sink until they stop dripping.
3. Lay one piece of paper towel flat inside the refrigerator. Lay the other piece flat on a nearby surface. Be sure the surface is free of any drafts. (If the refrigerator has open shelves, lay the other piece on a baking rack on the counter so it is exposed to similar air circulation.)
4. Check your paper towels periodically. Which appears to be drying faster? Is Carole's observation accurate?

UNIT 4

An *Issue* to Analyze

A SIMULATION

There's More to a Lake Than Water

Think About It

Lake Treadwell is the source of water for residents of the nearby town of Clairville. People from the surrounding area also use the lake for swimming, fishing, and boating. A small summer camp for students has been situated on the shore of the lake for many years.

Recently, the lake has become more popular in summer with tourists from other parts of the province. Town officials plan to double the size of the camp and to add attractions such as a lakeside restaurant and a launching ramp and mooring for motorboats. Increased tourism will benefit merchants in the town and provide jobs at the lake site. Some residents, however, are worried that the developments might pollute the lake. Will the water still be safe for drinking and swimming? What will happen to the animal and plant life in the area? Will larger numbers of people at the lake create conflicts among swimmers, boaters, and fishers?

This situation is an example of the kinds of challenges that face towns and cities across Canada. Making decisions about the best use of water resources is a task that combines personal and social values with scientific facts and logical thinking. In this simulation, you will prepare a presentation giving a particular point of view about the proposed development at the lake.

Background Information

- Construction of the restaurant, boat ramp, and expanded camp facilities will provide short-term jobs for labourers, tradespeople, truckers, and merchants.
- Construction plans include the widening of the road to the lake. This will involve the removal of some roadside trees, which may increase erosion.
- The merchants and owners in the completed development will pay taxes, which will increase the money available for community services such as roads and schools.
- Sewage and wastewater from the new development will be treated on-site in septic tanks.
- Boats may dump wastes or leak oil or gasoline into the lake. Increased numbers of vehicles and campers on the lakeshore might add pollutants to the runoff, such as detergent, shampoo, insecticides, windshield washing fluid, and cooking oil.
- Tourists might toss away garbage such as bottles, cans, wrappers, and fishing line.
- The restaurant and camp will provide jobs for local youth.
- The development will add to the recreation opportunities for both tourists and residents of Clairville.

- Merchants in Clairville will benefit from the increased numbers of people visiting the town and buying their products.
- People attracted to the lake for swimming, birdwatching, or fishing may have their recreation disturbed by increased traffic, motorboats, and pollution.
- Larger crowds and conflicting uses will increase the chance of accidents.
- Animal and plant life at the lake will be affected in different ways by the development. Some organisms will have their breeding sites disturbed, while other organisms may be harmed by pollution or garbage.
- The ecological balance of the area may be upset by overpopulation of certain species. For example, plants such as algae may thrive on increased nutrients; animals such as crows, jays, or squirrels may be attracted by waste food.
- Leaking sewage or growth of bacteria may create health problems.
- Added amounts of chemicals, silt, and garbage in the lake increase the cost of filtering and purifying the water before it is supplied to households in town.

Plan and Act

1. Plan to attend a town meeting to make a presentation about the development. The following people will be making a formal presentation at the meeting:
   - a guide who leads birdwatching tours of the lake
   - an unemployed resident of Clairville
   - a city councillor
   - an executive of a construction company
2. What do you think each person's point of view will be before the meeting? What are some concerns and considerations each person might want to discuss? What facts or arguments, if any, are likely to cause these people to change their points of view?
3. Your teacher will give your group the role of one of these people, along with additional information to help you plan your presentation. As a group, research your role, and then present a strong case at the town meeting. (Your teacher will provide you with a blackline master outlining the correct *Procedure for a Public Hearing*.)

4. At the end of the town meeting, your task, as a class, will be to approve or reject the proposed development. Base your decision on the overall balance of advantages and disadvantages that you record during the discussion.

Analyze

1. What was your final decision?
2. Were all points of view well researched and well presented? If not, how do you think they could be improved?
3. Were all participants fairly treated? If not, how could you change the procedure to ensure fair treatment?
4. How did your understanding of science and technology help you make your decision?

UNIT 4 PROJECT

The "Blood" of the Earth

Earth's water cycle is similar to blood circulation within the human body. Just like blood, water circulates nutrients and minerals and regulates the temperature of the planet. Also like blood, water can become "diseased" and therefore unable to carry out its functions. Humans and nature can make Earth's water supply unusable in a variety of ways. For example, water that is trapped in glaciers or that contains large amounts of dissolved salt must be treated before people can use it. In addition, acid rain, oil spills, and damage to lakes and rivers caused by dumping waste chemicals into them can damage or destroy our water supply.

Challenge

Prepare a research presentation on an environmental topic that involves Earth's water systems.

Materials

newspapers and magazines

other print resources

electronic information resources

computer and access to the Internet (optional)

multimedia equipment (optional)

Research Criteria

A. In your group, choose a specific environmental topic related to water (see the list below). You might select an event or a disaster that has occurred in the past or one that might occur if human activities continue to cause damage to water systems.

B. Find several sources of information related to your topic. Include both print and electronic media.

C. Consider the following questions as you do your research:

- What are the causes of the problem or disaster? Is there one specific source of the problem or are there several sources? Which cause or source is the most important?
- What are the effects of the problem or disaster? Consider the immediate and long-term effects on humans and on the environment. Also, take into account both local and global effects.
- If you chose to research a problem or disaster that has already occurred, how did people deal with it? Was the response to the problem or disaster successful? Why or why not?
- If you chose a potential problem or disaster, how might people deal with it? What technology exists to help humans handle the situation, in terms of both prevention and response?
- What alternatives can you think of in response to the problem?

D. Include your own questions and answer these, as well.

Plan and Present

1. Here are some suggested topics from which you can choose. You can also select a topic of your own and have your teacher approve it.
   - Oil spills: Why do they occur? How are they cleaned up? Are they preventable? If so, how?
   - Desalination: In what parts of the world do people rely on desalination for fresh drinking water? How cost-effective is desalination?
   - Acid rain: How is Canada affected by acid rain? What are Canada and other countries doing to prevent or reduce it?
   - Sale of Canada's fresh water supply: Does Canada sell its fresh water? If so, what effects might this have on the Canadian environment and on Canada's water systems?
   - Water treatment: What chemicals are used to make water safe to drink? Are there alternatives to certain chemicals such as chlorine, for example?

A rainforest before clearcutting

A rainforest after clearcutting

- Deforestation and destruction of rainforests: Where and why are rainforests being cut down? What are the short-term and long-term effects of this deforestation?
- Industrial pollution: Who is polluting and where, and what are some ways to prevent or reduce water pollution?

2. Your presentation can be in the form of a web page, a multimedia presentation, or a written report. You might include the following when designing your presentation:
   - a map or series of maps related to your topic
   - answers to the questions listed in step 1, plus additional information that you think your audience will need to know
   - pictures that demonstrate the environmental impact of your chosen topic
   - a bibliography, including all text resources, web sites, and newspaper or magazine articles used in your research
   - a special section showing the connections between this textbook and your topic. Consider all aspects of the water cycle and how your topic relates to it when you present this component.

Evaluate

1. Did all the members of your team contribute equally to the project? If not, how might you ensure the sharing of tasks in the future?
2. Did team members research the topic thoroughly and carefully?
3. Did the presentation satisfy the Research Criteria?
4. Was your presentation clear, interesting, and complete? Did it hold the audience's attention?
5. If you could improve your research strategies or your presentation in any way, what would you do?

Plan and write an illustrated *Emergency Response Handbook* outlining how city officials, rescue workers, citizens, and others could respond most effectively to a natural disaster such as a flood, a severe hail storm, or an ice storm.

UNIT 5

Mechanical Advantage and Efficiency

A thundering herd of rampaging elephants would be a fearful sight if you were anywhere near it. The strength of those giant legs, tusks, and trunks could do a lot of damage. However, elephants can be trained to use their strength to help humans with difficult tasks. Elephants in India and Asia have been trained to carry logs as large as tree trunks. If an elephant can lift a large tree, it could easily lift you. Could you lift an elephant?

This may sound like a strange question at first, but think for a moment. Before humans had electric motors and gasoline engines to power cranes, they managed to move huge stones to build immense structures such as pyramids and cathedrals. The scientific principles that made their human-powered machines work are still applied today.

In this unit, you will learn how some small, human-powered tools work. You will see that tools as simple as a pair of scissors function on the same principles as massive equipment powered by fluid pressure and heat engines. You will discover the main factors in the efficient operation of mechanical systems. You will also design and build your own mechanical devices — including some powered by hydraulics and pneumatics — and investigate their efficiency. Finally, this unit examines factors affecting the manufacturing of a product, including the needs of the consumer.

Unit Contents

CHAPTER

13 Mechanical Systems:

Getting Ready...

- How are levers used in outer space?
- Why does shifting gears make a bicycle easier or harder to pedal?
- How is heavy cargo lifted onto a ship?

Science Log

What is the largest object you have ever tried to lift? At the time, did you think, "There must be an easier way to do this"? There probably was. Based on what you already know, try to answer the Getting Ready questions above in your Science Log. As you study this chapter, you will discover some "better ways" to lift and move large loads.

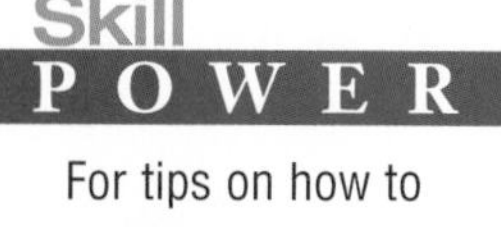

Skill POWER

For tips on how to make and use a Science Log, turn to page 534.

Logging has been an important resource industry in Canada for more than 300 years. Early loggers did not have elephants to carry logs for them, nor did they have gasoline-powered heavy equipment. Instead, they had to develop and use human-powered machines to transport logs. Logging is only one of many industries in which massive objects must be lifted and moved. Think about some everyday mechanical devices used to move large objects. Perhaps you have helped change a flat tire on a car. If so, you probably used a jack to lift the car.

In this chapter, you will review what you learned in previous studies about simple, human-powered machines. These mechanical devices sometimes make exerting a force easier. Think of a bottle opener, for example. A machine can also increase the speed at which you exert a force (for example, when you ride a bicycle). You will see that no machine is completely efficient, and that you can calculate how much of the work that you or a motor puts in is wasted due to friction.

Design and Function

Spotlight On Key Ideas

In this chapter, you will discover

- how work and efficiency are related in a machine
- that all machines have a mechanical advantage
- why speed and velocity are not the same
- how to find the velocity ratio of some kinds of machines
- how to boost efficiency

Spotlight On Key Skills

In this chapter, you will

- learn how to calculate a machine's mechanical advantage and efficiency
- design and build your own mechanical device using levers
- learn how to increase the mechanical advantage of a pulley
- predict and test the mechanical advantage of a compound pulley
- design and build a prototype of a crane, using a pulley and a wheel and axle

Starting Point ACTIVITY

How Hard Must You Press?

To move a boulder using a large stick, how should the stick and the boulder be arranged to allow the person to use the least amount of force? Test the arrangement with smaller objects to find out.

What You Need 

rock (about 2–3 kg) or object of similar mass
table knife (or rigid, flat metal bar)
dowel (or pencil)

What to Do

1. Slide about 1 cm of the tip of the knife under the rock. **CAUTION:** Be very careful when handling the table knife. Do not point it at anyone.
2. Push the dowel under the knife so that it is about 3 cm from the knife tip.
3. Press on the end of the knife handle. Notice the amount of force you must exert to lift the edge of the rock.
4. Repeat step 3 twice more, once with the dowel about 5 cm from the knife tip and then with the dowel about 7 cm from the knife tip. Each time, notice the amount of force you must exert.

What Did You Find Out?

1. Did the amount of force you had to exert to lift the rock depend on the position of the dowel? If so, which position required the least amount of force?
2. What should the person in the illustration above do to make her task easier?
3. Suggest how you could make a quantitative measurement of the force needed to lift the rock.

13.1 Machines, Work, and Efficiency

Figure 13.1 How are the screwdriver and the teeter-totter like the knife in the Starting Point Activity?

When you exerted a force on the handle of the knife in the Starting Point Activity, the knife then exerted a force on the rock. If you were to exert a force on a screwdriver, the screwdriver would exert a force on something else, as shown in Figure 13.1. Both these objects — as well as the teeter-totter shown in Figure 13.1 — act as levers.

As you learned in previous studies, a **lever** is a simple machine that changes the amount of force you must exert in order to move an object. It consists of a bar that is free to rotate around a fixed point. This fixed point, the **fulcrum**, supports the lever (see Figure 13.2). The fulcrum is the lever's point of rotation. The force that you exert on a lever to make it move is called the **effort force**. This term is used to describe the force supplied to any machine in order to produce an action. The **load** is the mass of an object that is moved or lifted by a machine such as a lever. In other words, the load is the resistance to movement that a machine must overcome. The distance between the fulcrum and the effort force is called the **effort arm**. The distance between the fulcrum and the load is called the **load arm**.

Word CONNECT

In a dictionary, find the origin of the word "lever." Then look up the meaning of the word "leverage" and use it in a sentence.

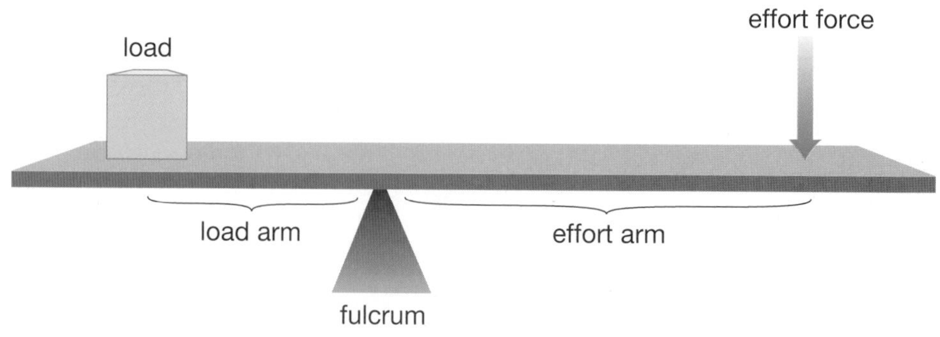

Figure 13.2 A lever is a simple machine consisting of a bar that rotates around a fixed point, the fulcrum.

You can discover levers in many different situations. However, all levers are sorted into one of three classes, depending on the position of the effort force, the load, and the fulcrum, as shown in Figures 13.3A, B, and C. As the photographs show, different classes of levers are used for different purposes.

In a **Class 1 lever**, the fulcrum is between the effort and the load. A pair of scissors is an example of a Class 1 lever. This class of lever can be used either for power or for precision.

A **Class 2 lever**, such as a wheelbarrow, always exerts a greater force on the load than the effort force you exert on the lever. In this type of lever, the load is between the effort and the fulcrum.

In a **Class 3 lever**, — a hockey stick, for example — the effort is exerted between the fulcrum and the load. When using a Class 3 lever, you must exert a greater force on the lever than the lever exerts on the load. However, the load can be moved very quickly.

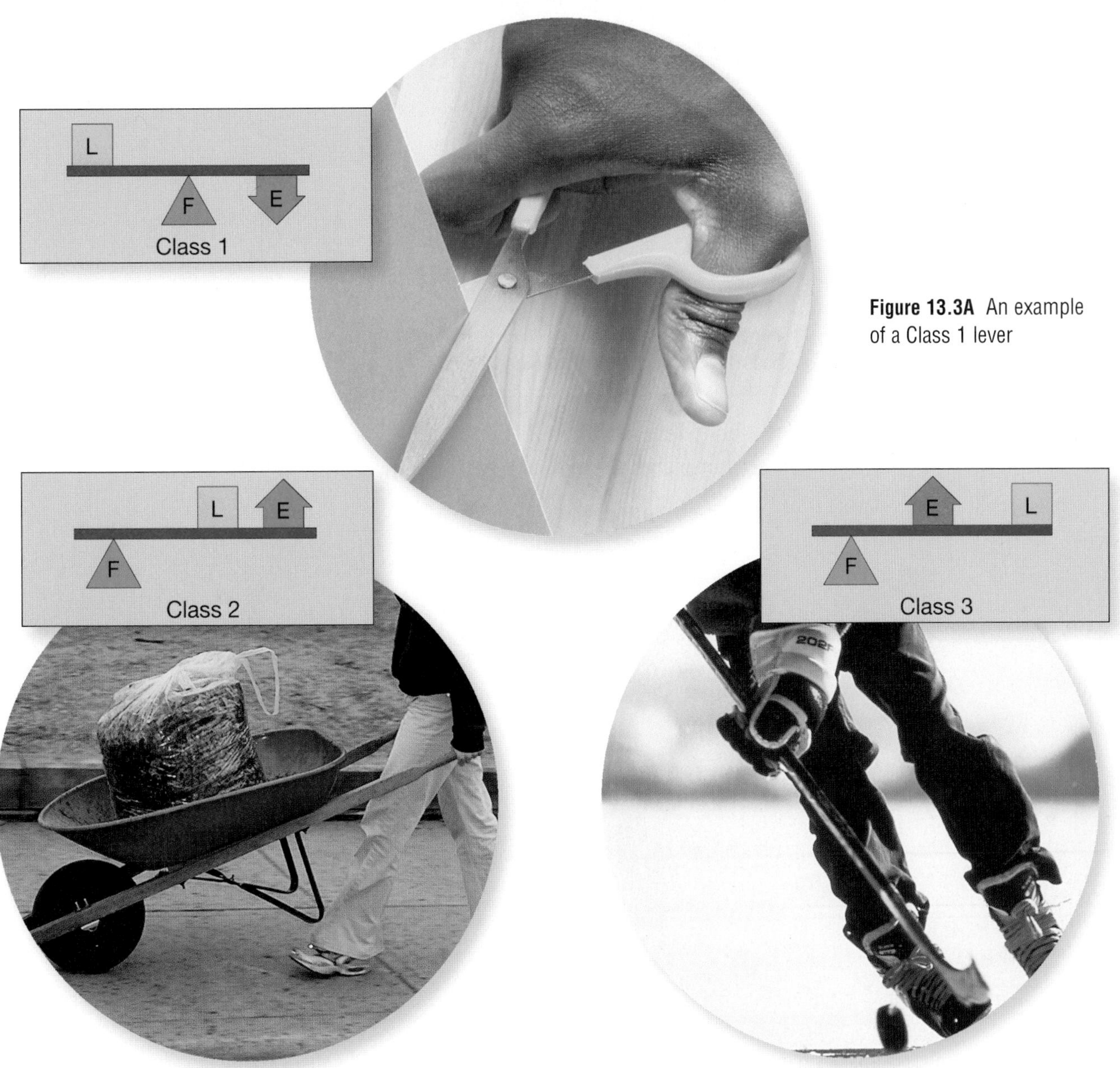

Figure 13.3A An example of a Class 1 lever

Figure 13.3B An example of a Class 2 lever

Figure 13.3C An example of a Class 3 lever

DidYouKnow?

Have you ever rowed or sailed a boat? The oars in a rowboat and the rudder of a sailboat are both Class 1 levers. What class of lever do you think a canoe paddle is?

Bones and Muscles: Built-in Levers

Every time you move a finger, arm, or toe, you are using a lever. Your bones act as levers and each of your joints acts as a fulcrum. Tendons attach muscles to your bones. When a muscle contracts, the tendon exerts an effort force on the bone. The load might be something that you are lifting or pulling. The load could also be your own body, for example, when you do a knee bend.

As Figure 13.4 shows, most of the levers in your body are Class 3, but you can find Class 1 and Class 2 levers as well.

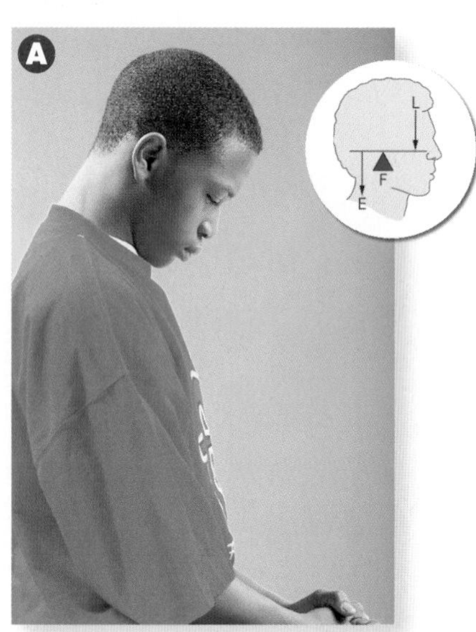

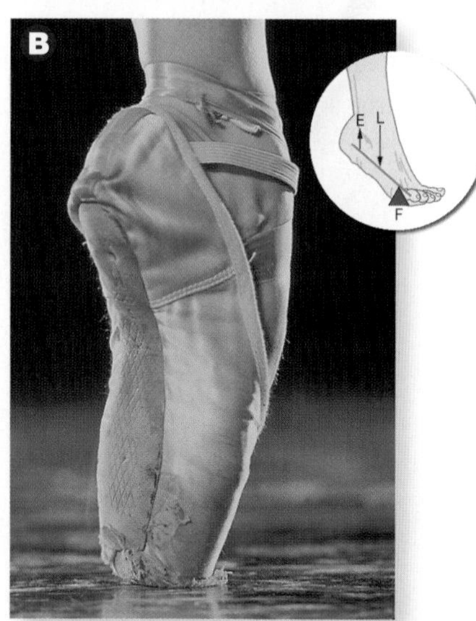

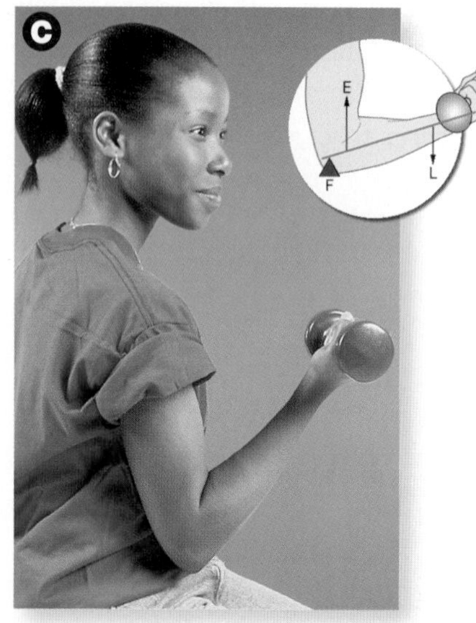

Figure 13.4 Your body's system of muscles and bones contains natural examples of levers, including Class 1 (A), Class 2 (B), and Class 3 (C).

Look at the body levers shown in Figure 13.5, and decide the class of each lever. (Remember that weight, which is a force, is measured in units called newtons, N.)

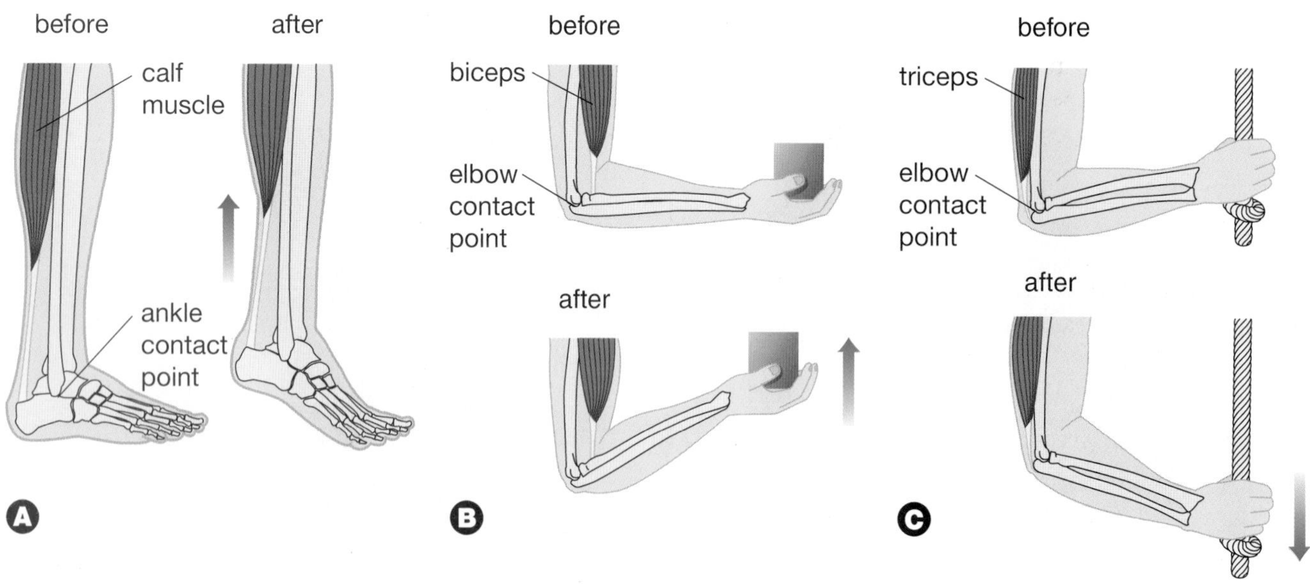

Figure 13.5A The calf muscle provides the effort force. Assume that a body weight of 600 N is the load.

Figure 13.5B The biceps muscle provides the effort force. The hand is lifting a 15 N object.

Figure 13.5C The triceps muscle provides the effort force. The hand is pulling the rope down with a force of 30 N.

An Arm in Space

One of the most exciting technological applications of levers is the Space Shuttle Remote Manipulator System, shown in Figure 13.6A. This system is usually called the Canadarm because it functions much like a human arm, and it was designed and built in Canada. The "joints" are moved by gears. As the gears turn, they move the "arms" that resemble levers.

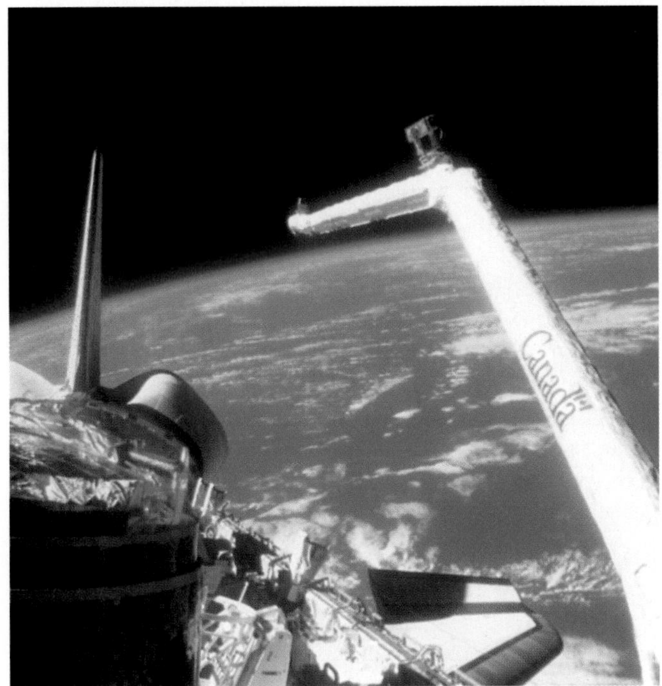

Figure 13.6A The Canadarm is an amazing application of gears and levers in outer space.

Figure 13.6B The Space Station Mobile Servicing System will be equipped with a smaller two-armed robot – the SPDM – to do complex repair jobs in space.

The Canadarm is a valuable addition to the space shuttle program because it helps launch and recover satellites from the shuttle's cargo bay. One of the Canadarm's most important missions was the repair of the Hubble Space Telescope. This orbiting telescope can see farther and more clearly than any ground-based optical telescope. (You may have learned about the Hubble Space Telescope in Chapter 8.)

A larger version of the Canadarm — the Space Station Mobile Servicing System — is shown in Figure 13.6B. This system will assist in building and maintaining the International Space Station. The base of the system will move along rails spanning the entire length of the space station. When stretched out straight, the arm will be more than 17 m long. It will be equipped with a smaller two-armed robot that can do delicate repair jobs that astronauts themselves have done on space walks until now.

Sixteen countries, including Canada, Russia, Japan, and the United States, are co-operating in the planning and building of the International Space Station. If everything goes as planned, the space station will be completed by the end of the year 2004.

INTERNET CONNECT

www.school.mcgrawhill.ca/resources/

Learn more about the Canadarm and the International Space Station by going to the above web site. Go to **Science Resources**, then to **SCIENCEPOWER 8** to find out where to go next. Present your findings in an oral or written in-class report.

The Inclined Plane

In previous studies, you learned about the **inclined plane** — a ramp or a slope that reduces the force you need to exert to lift something.

Suppose you must lift a 100 kg box of camping gear a distance of 1 m, from the ground up into the back of a truck. Lifting the box straight up and into the truck would be difficult. However, if you use a board to make a ramp, as the girl in Figures 13.7A and B is doing, you could probably push the box up, even though you would have to exert force to overcome the friction between the box and the board.

Figure 13.7A Olivia used an inclined plane to help her load boxes of camping gear into the truck. The inclined plane decreases the effort force Olivia needed by increasing the distance through which her effort force was applied. How could Olivia have made her job even easier?

Figure 13.7B After loading the first box, Olivia replaced the short ramp with a longer one. She increased the distance she had to push the boxes, but she decreased the effort force she had to exert.

Find Out **ACTIVITY**

Easy Does It

Why is it easier to climb a gentle hill than a steep mountain trail or a cliff face? You can answer this question using a ramp, a spring scale, a toy car, a support stand, a stack of books, and some string.

What to Do

1. Hook the spring scale to a laboratory stand.
2. Tie the string to the front of the toy car, then hook the free end of the string over the spring scale.
3. Pull down gently so that the car is pulled straight up into the air. Record the reading on the spring scale.
4. Now fix the ramp in place. Put the car at the bottom of the ramp and pull it steadily up the ramp, keeping the string hooked over the spring scale. Is the force recorded on the spring scale larger or smaller than before?

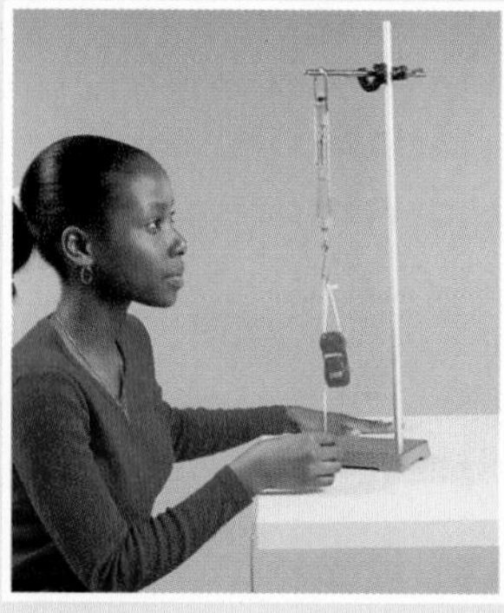

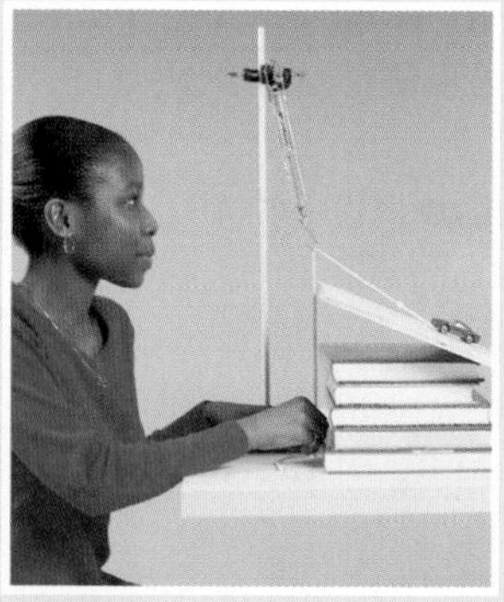

A Fair Trade

If someone offered you two five-dollar bills in exchange for a twenty-dollar bill, would you hand over the twenty? Of course not — it would not be fair. Perhaps you have been wondering how a machine can exert more force than it receives.

When you use a lever or a ramp (or any other kind of machine), you are putting in work. Stated simply, **work** is the transfer of energy through motion. *Work is done only when force produces motion in the direction of the force.* In the Find Out Activity, you saw that pulling a toy car up a slope takes less effort force than lifting it straight up. However, there is a trade-off — you have to pull the car the entire length of the ramp, not just the shorter distance straight up. You can calculate the work you did by multiplying your effort force by the distance through which you applied the force:

$$\text{Work} = \text{Force} \bullet \text{Distance}$$

Try multiplying force and distance for steps 3 and 4 of the Find Out Activity. You should get roughly the same answer each time. When you used the ramp, you exerted less force but over a longer distance. Thus, in the end you did the same amount of work.

Work is energy in action. Like energy, work is measured in units called **joules**. The joule is named after English scientist James Prescott Joule (1818–1889). In Chapter 6, you learned that 1 N is approximately the weight of a 100 g mass. When you lift a 1 N weight a distance of 1 m, you do 1 J (joule) of work.

DidYouKnow?

How big is a joule? A sandwich made of chicken and lettuce on a bun, for example, weighs about 1 N. If you lift the sandwich 1 m, you have done about 1 J of work. The energy you get from eating the sandwich, however, is about 2 000 000 J!

What Is Mechanical Advantage?

Mechanical advantage is the comparison of the force produced *by* a machine to the force applied *to* the machine. In other words, mechanical advantage is the comparison of the size of the load to the size of the effort force. The smaller the effort force compared to the load, the greater the mechanical advantage. You can use the following formula to calculate mechanical advantage:

$$\text{Mechanical Advantage } (MA) = \frac{\text{Load force } (F_L)}{\text{Effort force } (F_E)}$$

Suppose you are a passenger in a truck that gets stuck in mud. You and the driver use a tree branch as a lever to lift the truck out of the mud, as shown in Figure 13.8 on the next page. If you apply an effort force of 500 N to the branch, and the back of the truck weighs 2500 N, then the mechanical advantage of the branch-lever is 5. Note that no units are used to express mechanical advantage:

$$\begin{aligned}\text{Mechanical Advantage } (MA) &= \frac{\text{Load force } (F_L)}{\text{Effort force } (F_E)}\\ &= \frac{2500\text{ N}}{500\text{ N}}\\ &= 5\end{aligned}$$

The branch-lever has exerted a force 5 times greater than the force you exerted on it. This means the branch-lever made the job of lifting the truck 5 times easier. Any machine with a mechanical advantage greater than 1 allows the user to move a large load with a relatively small effort force.

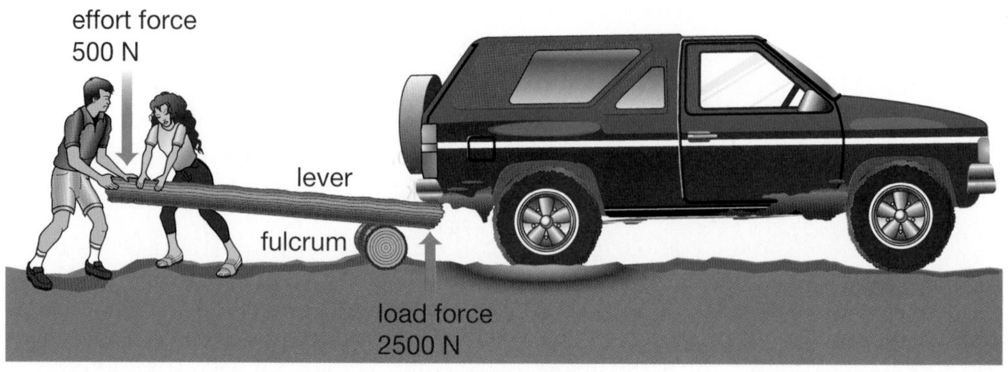

Figure 13.8 The mechanical advantage of this branch-lever is 5.

A machine can also have a mechanical advantage that is less than 1. Imagine you are riding your bicycle. You exert an effort force of, say, 736 N downward as you push on the pedal. The resulting load force that causes the bicycle to move forward is 81 N. The mechanical advantage of the bicycle is calculated as follows:

$$\begin{aligned}\text{Mechanical Advantage } (MA) &= \frac{\text{Load force } (F_L)}{\text{Effort force } (F_E)}\\ &= \frac{81\text{ N}}{736\text{ N}}\\ &= 0.11\end{aligned}$$

Finally, a machine may have a mechanical advantage equal to 1. For example, suppose the effort force needed to raise a flag up a flagpole is 120 N. The load force — the flag plus the rope — is also 120 N. Therefore, the mechanical advantage of the pulley on the flagpole is 1:

$$\begin{aligned}\text{Mechanical Advantage } (MA) &= \frac{\text{Load force } (F_L)}{\text{Effort force } (F_E)}\\ &= \frac{120\text{ N}}{120\text{ N}}\\ &= 1\end{aligned}$$

Pause& Reflect

Some machines do not have any effect on the effort force that you exert. They simply change the direction of the effort force. For example, when you pull down on the cord of window blinds, the blinds go up. Only the direction of the force changes. The effort force and the load are equal, so the mechanical advantage is 1. Try to think of other mechanical devices that have a mechanical advantage of 1. Write your ideas in your Science Log.

Math CONNECT

A crafty coyote is trying to use a catapult to launch a heavy rock. The rock, with a mass of 1000 kg, sits on one end of a plank. The coyote figures that if he jumps on the other end of the plank, his 25 kg mass will be enough to launch the rock into the air. Calculate the mechanical advantage the catapult must have for the coyote's plan to work.

Find Out ACTIVITY

Sharpen Up with Scissors

Is one way of using scissors easier than another? Make a hypothesis about whether it takes less effort force to cut cardboard with the tip of the blades or with the base of the blades near the hinge.

What You Need

scissors
piece of heavy cardboard or folded paper

What to Do

1. Test your hypothesis. Try to cut the cardboard with the tip of the scissors.

CAUTION: When using scissors, always cut away from your body.

2. Open the scissors wide, put the cardboard close to the hinge of the scissors, and again, make a cut.

What Did You Find Out?

1. Does one method make cutting the cardboard easier than the other method?
2. Explain your observations based on what you have learned about levers, effort force, and mechanical advantage.

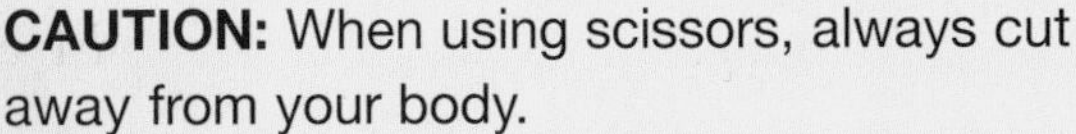

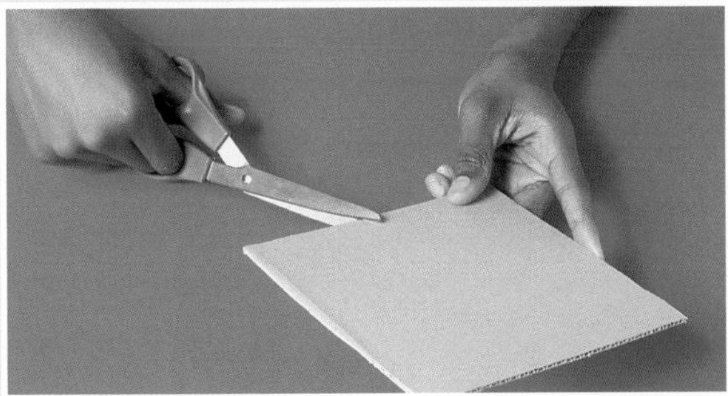

Levers in Action

You can calculate the mechanical advantage for any machine, including levers. Do you think the relative positions of the fulcrum, the effort force, and the load will affect a lever's mechanical advantage? Make a hypothesis and then test your prediction.

Problem

Is the mechanical advantage of the three classes of lever greater than 1, equal to 1, or less than 1?

Apparatus

sturdy board
brick (or similar heavy mass)
strong string

Procedure

1. Place the board on a desk or work surface, with half its length extending over the edge.
2. Place the brick on the desk on top of the end of the board. This makes a Class 1 lever, with the edge of the desk acting as the fulcrum. **CAUTION:** Handle the brick carefully. Do not allow it to fall on your foot.
3. Try to lift the brick by pushing down on the free end of the board.

4. Repeat step 3 with most of the board's length on the desk surface.

5. Repeat step 3 with most of the board extending over the edge of the desk.

6. Compare the amount of effort force you must exert in each position in steps 3 to 5. Record your observations.

7. Now tie the brick to the board so that the brick hangs underneath it. Put one end of the board on the desk and hold the other end. This makes a Class 2 lever. Try to lift the brick while it is hanging at two or three different places along the board.

8. Finally, tie the brick to the far end of the board. This makes a Class 3 lever. Try lifting it while holding the board in two or three different places. You will need to make sure the end of the board stays in place on the desk.

Analyze

1. (a) Which class or classes of lever exert(s) a load force greater than your effort force?

   (b) Which class or classes of lever exert(s) a smaller force on the load than you exert on it?

2. Does a Class 1 lever always exert a load force that is greater than your effort force?

Conclude and Apply

3. Write a statement comparing the mechanical advantages of Class 1, Class 2, and Class 3 levers.

Another Way to Calculate Mechanical Advantage

Levers can exert a force on a load that is either greater than or less than the effort force you exert. If the load is less than the effort force, the lever's mechanical advantage is less than 1. For example, a mechanical advantage of $\frac{1}{2}$ shows that your effort goes only half as far compared to a lever with a mechanical advantage of 1.

The concepts of mechanical advantage and work can be linked. In the last investigation, you may have noticed that the longer the effort arm (the distance between the fulcrum and the effort), the less effort force it took to lift the brick. Thus, you gained a mechanical advantage. Recall that Work = Force • Distance. You traded distance for force — you had to move the board farther, but moving it was easier. *However, the amount of work you did was the same.* This suggests another way to calculate the mechanical advantage of levers:

$$\text{Mechanical Advantage } (MA) = \frac{\text{Load force } (F_L)}{\text{Effort force } (F_E)} = \frac{\text{Effort arm}}{\text{Load arm}}$$

If the effort arm of the branch-lever mentioned earlier were 3 m, and the load arm were 0.3 m, then the mechanical advantage would be calculated as follows:

$$(MA) = \frac{\text{Effort arm}}{\text{Load arm}} = \frac{3\text{ m}}{0.3\text{ m}} = 10$$

By using the branch as a Class 1 lever, you have increased your effort force by 10 times.

Although it might seem strange, there are situations in which you might want to *reduce* the force you exert. For example, you might need a machine to perform a delicate precision task. Think about how tweezers work as you study the photograph below. Which class of lever do they use? Infer whether the mechanical advantage of tweezers will be greater than 1, equal to 1, or less than 1.

Surgeons use special tools in a type of microsurgery sometimes called "keyhole surgery" because only a small incision is needed. A long tube is pushed through the incision to the part of the patient's body requiring surgery. Fine wires running through the tube operate tiny levers to cut and sew as needed. The surgeon watches the operation on a television screen connected to a tiny camera at the end of the tube.

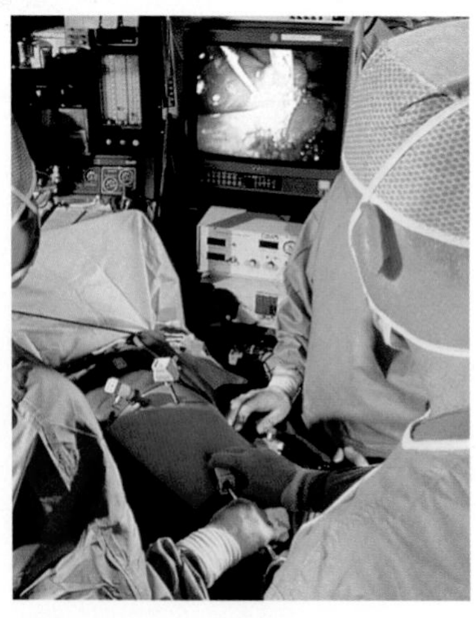

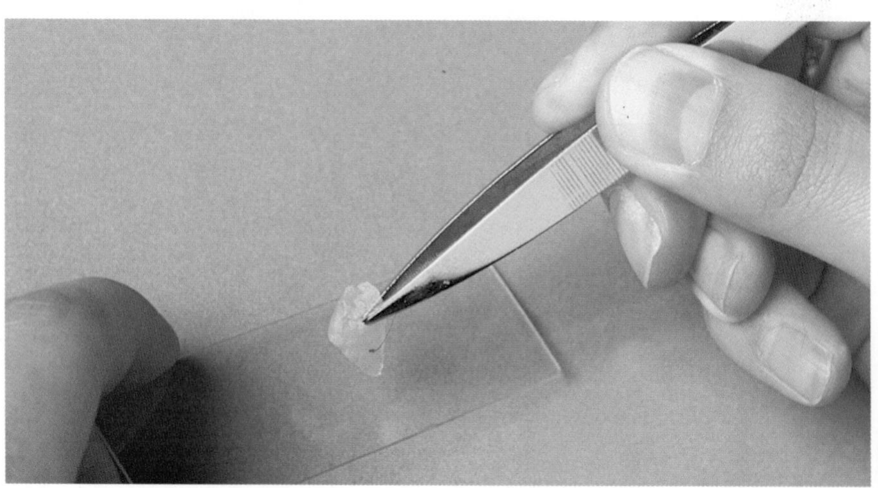

No Machine Is 100 Percent Efficient

An ideal machine would transfer all of the energy it received to a load or to another machine. However, real machines do not work this efficiently. Some of the energy is always lost. Another way of saying this is that the work put out by a machine is always less than the work put into a machine.

No lever, or any other machine, is perfect, but some machines come closer than others. The **efficiency** of a lever tells you how much of the energy you gave to the machine was transferred to the load by the machine. Efficiency is a comparison of the useful work provided *by* a machine or a system with the work supplied *to* the machine or system. Efficiency is usually stated as a percentage. You can calculate the efficiency of a lever by using this formula:

$$\text{Efficiency} = \frac{\text{Work done by lever on load}}{\text{Work done on lever by effort force}} \times 100\%$$

The higher the efficiency, the better the lever is at transferring energy. A "perfect" machine would transfer all the work done by the effort and would be 100 percent efficient. However, the efficiency of real machines is always less than 100 percent. Why? Every time a machine does work, some energy is lost because of friction. Think about a pair of hedge trimmers. As you close the handles, the blades rub against each other. If the blades are rusty, they will tend to stick even more. You could summarize this situation by means of the following word equation:

$$\text{Work done } \textit{on} \text{ a machine} = \text{Work done } \textit{by} \text{ the machine} + \text{energy lost as heat due to friction}$$

Many machines can be made more efficient by reducing friction. You can usually do this by adding a lubricant (such as oil or grease) to surfaces that rub together, as shown in Figure 13.9. After a time, dirt will build up on the grease or oil, and the lubricant will lose its effectiveness. The dirty lubricant should be wiped off and replaced with clean grease or oil. Explore efficiency further in the next investigation.

Many car engines are only about 20 percent efficient. Where does all the "lost" energy go? The fastest car can brake much faster than it can accelerate. Why do you think the brakes in a car are more efficient than the engine?

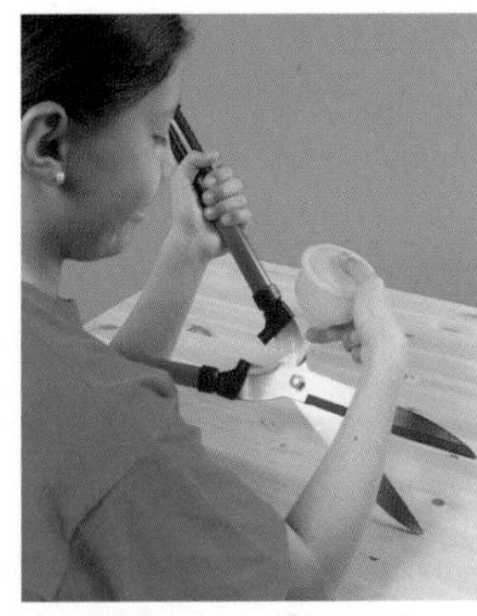

Figure 13.9 You can improve a machine's efficiency by oiling parts of it to reduce friction.

Pause& Reflect

Does a lever always do as much work on the load as you do on the lever? Suppose you have a summer job trimming hedges. Someone leaves the hedge-trimming shears out in the rain, causing the bolt at the joint to get rusty. The next time you trim the hedge, you discover that you have to exert a much greater force on the handles than you did before. You are doing more work on the shears, but the shears are doing the same amount of work on the hedge that they did before the joint rusted. In your Science Log, explain why this happens.

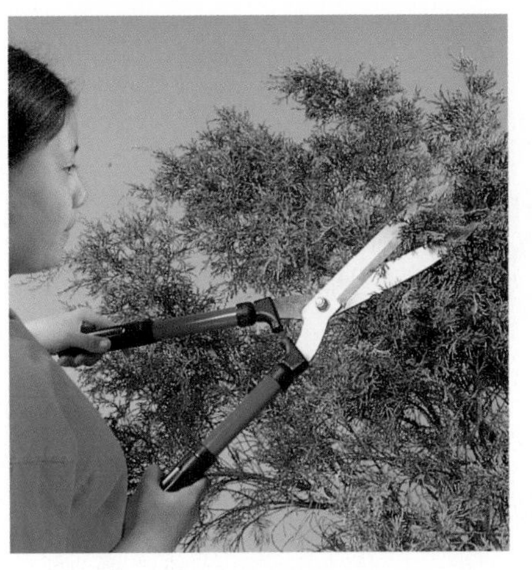

INVESTIGATION 13-B

Staying Green at Sea

Imagine you are working with a group of environmentalists who are taking a survey of garbage found in the Atlantic Ocean using the Global Positioning Satellite system. For the long sea voyage, the team decided to take along several cases of pop. However, this posed a problem: How will you dispose of the empty cans? You cannot throw them overboard. Since space is limited on the sailboat, your team has decided to make a device that will crush the cans manually in a neat and efficient way.

Challenge

Design and build a can crusher for use in a limited space.

Materials

Styrofoam™ cups, dowels (for joints), wood strips (for lever arms), short wood plank (for base), handsaw, hand drill, spring scale or bathroom scale, ruler

Safety Precautions

- Use a saw or hand drill only under adult supervision.
- Fasten the piece of wood in a vice before you start, and always keep your hands well clear of the saw or drill.

Design Criteria

A. Your device must be safe, simple, compact, and it must crush the cans as uniformly as possible. It can be wall-mounted or free-standing.

B. Your device must use as little manual effort and materials as possible.

C. Test your device on Styrofoam™ cups. Keep in mind that if your device were made of steel, it would be strong enough to crush aluminum cans.

Plan and Construct

1. With your team members, decide what sort of lever or levers your can crusher will use. Remember that your device must crush the cup neatly and completely flat.
2. Using materials from the list to represent metal forms, design a working prototype of a can crusher strong enough to crush the Styrofoam™ cup, which represents a model can.
3. Make a technical drawing of your design and show it to your teacher.
4. Decide how you will measure the effort force and the load for your device.
5. When your can crusher is finished, make scale drawings of its top and side views, including measurements.

Evaluate

1. Calculate the mechanical advantage and the efficiency of your can crusher. (To calculate the efficiency, you will need to measure and record the load distance and the effort distance.)
2. Which class of lever does your can crusher use? Does it use more than one lever?
3. If you were redesigning your can crusher, what would you change?
4. Take a class poll on which can crusher worked best.

Speedy Levers

You saw earlier that Class 3 levers (and sometimes Class 1 levers) exert a force on the load that is smaller than the effort force. In return, they move the load a greater distance. However, increasing distance also increases speed. That is why you hit a hockey puck with the end of a metre-long stick. **Speed** is the rate of motion, or the rate at which an object changes position.

Figure 13.10A How can a small contraction (shortening) of the biceps muscle produce the long, fast movement of the pitcher's hand? Most of the levers inside your body have a mechanical advantage smaller than 1. Therefore, your muscles usually have to exert a greater force on the lever (bone) than the lever (bone) can exert on the load.

Figure 13.10B Why does the spinning pizza dough remain more or less in the same place?

Look at the baseball pitcher and the pizza chef in Figures 13.10A and B. In both cases, the biceps muscle moves only a small amount to produce the effort force needed to make the hand move rapidly through a relatively large distance. The structure of the levers in the human body makes it possible to perform delicate tasks with precision, as well as major tasks requiring tremendous speed and flexibility.

Speed in Different Directions

Look again at the baseball pitcher and the pizza chef. They are both generating a great deal of speed, but there is a difference. The pizza dough, unlike the baseball, stays more or less in the same place. Why does this happen? The **velocity** of a moving object includes direction as well as speed. The chef can keep changing the direction of the dough's motion to stop it from flying across the kitchen.

Suppose the pitcher throws a pitch that the batter hits, as shown in Figure 13.11. The speed of the ball may not appear to change very much. Where did all the batter's force go? Force changes not only speed but also velocity. Force was used to switch the *direction* of the speeding ball so that it flies over the outfield fence and into the crowd.

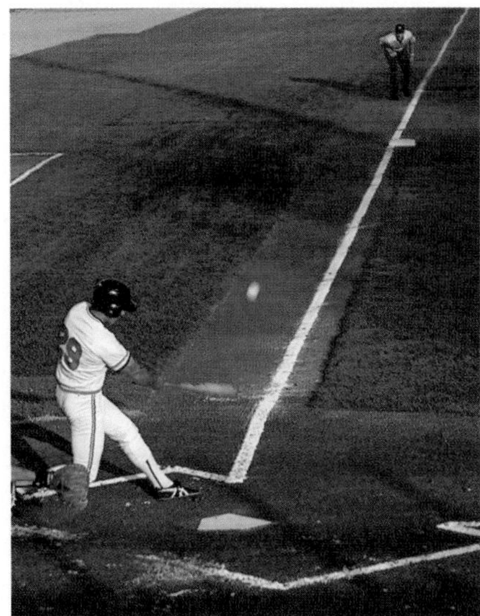

Figure 13.11 This baseball player sends the ball flying by using the bat to do work on it. The ball's velocity is a measure of speed in a given direction.

Machines Made to Measure

Industrial designers study the dimensions of the human body in great detail to make sure that every part of a machine or a product — such as the ones shown in Figure 13.12 — will fit the person using it. Body weight, height, size, age, and sometimes gender are factors taken into account when designing products. These products can range from cars to office furniture to light switches. The science of designing machines to suit people is called **ergonomics** (from the Greek words *ergon*, meaning "work," and *nomos*, meaning "natural laws").

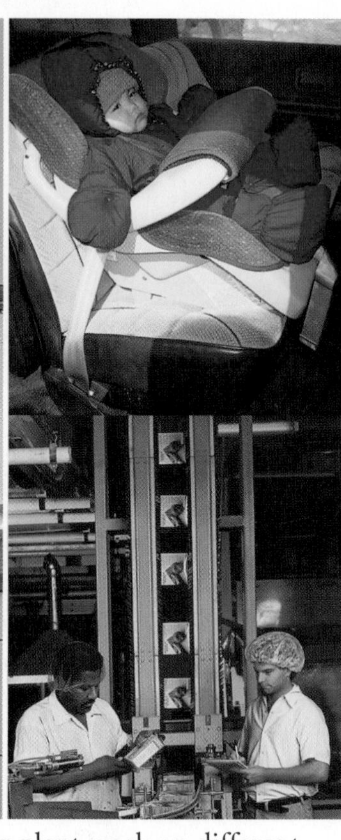

Figure 13.12 This space suit, kitchen, child's car seat, and assembly line in a factory have all been designed to ensure that they are easy, comfortable, and safe for people to use.

To increase efficiency, employees in a manufacturing plant work on different lines. A *line* is an area that has all the equipment needed to complete one product or process, for example, putting tablets in plastic bottles, labelling the bottles, or shrink-wrapping a set of bottles for shipping. Engineers improve the machines on a line to make them easier and more comfortable for people to use.

Ergonomics is especially important in the design of work environments where occupational safety is an issue. For example, a common workplace disorder known as **carpal tunnel syndrome** causes numbness and pain in the thumb and first three fingers. Study Figure 13.13 to find out what the "carpal tunnel" is. Carpal tunnel syndrome results from repetitive movements of the fingers, such as working at a computer keyboard. If the tendons that attach muscles to bones in the wrist become irritated, they swell and start to squeeze the nerve inside the carpal tunnel. If the condition is not treated soon after the symptoms appear, severe pain as far up as the shoulder can result. The damage could become permanent.

The most common treatment for carpal tunnel syndrome is a brace that holds the wrist straight and prevents irritation of the tissues near the carpal tunnel.

INTERNET CONNECT

www.school.mcgrawhill.ca/resources/

To learn more about ergonomics, visit the above web site. Go to **Science Resources**, then to **SCIENCEPOWER 8** to find out where to go next. Based on what you discover, make a sketch of a well-designed computer work station and present your design in class.

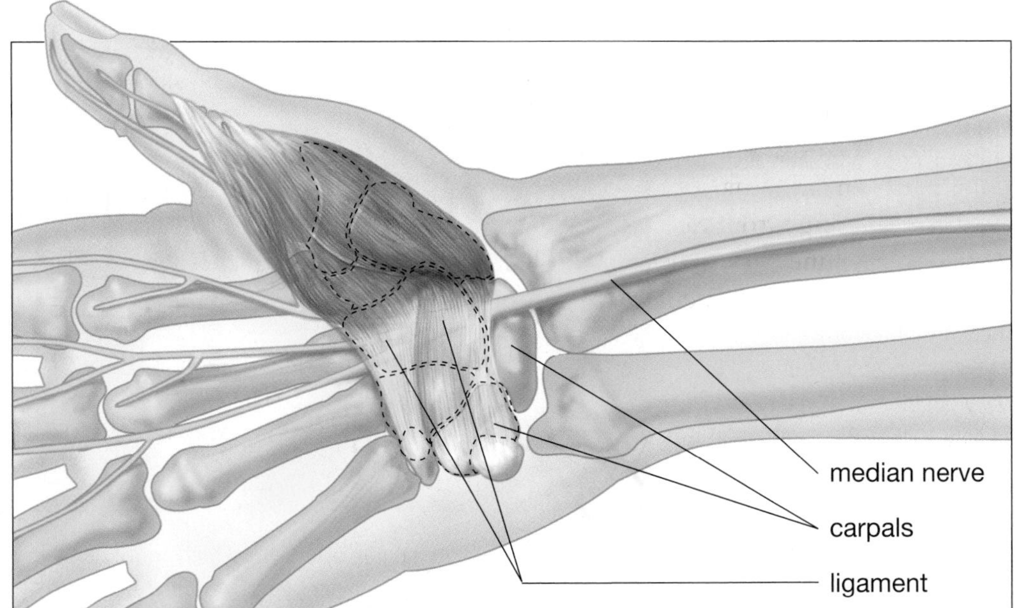

Figure 13.13 The carpals are small bones in the wrist. A large nerve, called the median nerve, runs through a tunnel formed by the bones and a large ligament crossing over these bones. The median nerve then branches out to the thumb and first three fingers.

Pause& Reflect

Imagine you are an ergonomist (an ergonomics designer) working on the International Space Station program. What sorts of problems might you have to solve? Remember that the astronauts will be working in cramped positions as well as in zero gravity. Write your ideas in your Science Log.

Another way to avoid carpal tunnel syndrome is to get rid of the keyboard altogether and to operate the computer using a special pen-like device, which keeps the wrist flat. "Palm pilots" (hand-held computers) are already pioneering this approach (see Figure 13.14A). Perhaps one day home computers will be able to "read" your handwriting — however messy it is! Voice-activated computer programs are also reducing the incidence of carpal tunnel syndrome (see Figure 13.14B).

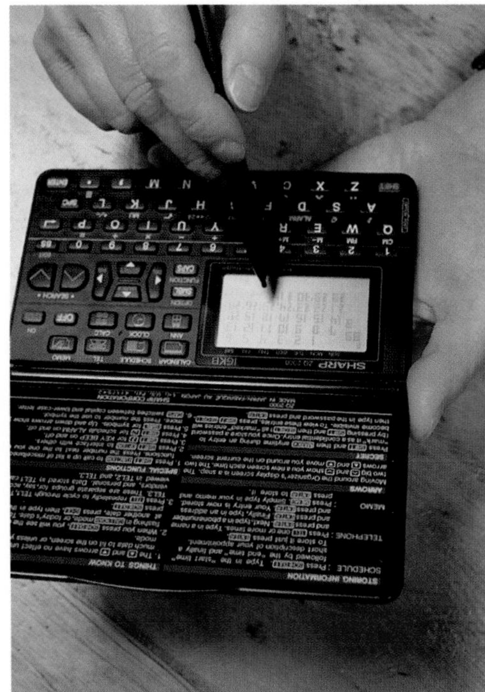

Figure 13.14A Use of palm pilots such as this one reduce repetitive strain injuries.

Figure 13.14B Voice-activated computers may one day eliminate the need for keyboards altogether.

The simplest activities can be complicated for astronauts who have to stay in space for a long time. For example, how would you brush your teeth in zero gravity? Free water floats around in large bubbles, so you cannot use a sink or taps. Therefore, astronauts use a specially invented foamless toothpaste that does not require water and is easily swallowed.

Career CONNECT

Working Naturally

Imagine that you could draw a line from the middle of the top of your head, down over your forehead, nose, and chin, along the length of your body, all the way to the floor. This imaginary line would divide your body in half. This division is called *symmetry*, and it is a very important feature of how our bodies function. Your body works best when it is more or less centred, or aligned symmetrically. If you reach out or twist around too far to one side, you will lose your balance or suffer a sprain. If you hold something heavy out in front of you, you will eventually get sore arms or a sore back.

Ergonomists carefully study the symmetry of the human body. They specialize in making the job, tools, and equipment fit the people who use them. Also called "human factors engineers," ergonomists ensure that the design of all kinds of equipment — from forklifts to desk chairs, from automobile instrument panels to toothbrushes — makes them safe, efficient, and comfortable for people to use. They know how each part of the human body moves naturally, as well as what movements are unnatural and could cause strain or injury. Ergonomists match their designs to natural, symmetrical body movements that maximize efficiency and minimize discomfort at the worksite. The symmetrical "split" computer keyboard shown here allows a more natural position for the hands than a one-piece keyboard. Using this type of keyboard can help prevent carpal tunnel syndrome.

Check Your Understanding

1. Classify the levers in the illustrations as Class 1, Class 2, or Class 3.
2. How much work, in joules, must you do to lift an elephant weighing 60 000 N up 1.5 m onto the back of a truck?
3. You have found a ramp to lead up to the back of the truck. If the slope of the ramp is 18 m long, how much force (in newtons) will you and your team need to exert to move the elephant up the ramp? (Use your answer to question 2.)
4. If you exert a force of 100 N on a hockey stick, and the stick exerts a force of 20 N on the puck, what is the mechanical advantage of the stick?
5. If the "effort arm" distance for the hockey stick in question 4 (between your "fulcrum" hand and your pushing hand) is 25 cm, how long is the stick? (Use your answer to question 4.) If your hand is pushing at a speed of 20 km/h, how fast will the puck move?
6. Explain the difference between speed and velocity.
7. **Thinking Critically** Think of a practical use for a lever with a mechanical advantage of 1. Draw a sketch of this lever in action.

13.2 Machines That Turn

Earlier, you discovered that you can lift a heavy load as long as you can find a lever that is long enough and strong enough to do the job. Sometimes, however, levers are not practical, as shown in Figure 13.15. Fortunately, there are many other kinds of machines that can give you a mechanical advantage great enough to move a heavy load with a much smaller effort force. Think about this question: How could you modify a lever to make it shorter, but still able to move a load over a longer distance? Look for clues in Figure 13.16, which shows a person loading a motor boat onto a boat-trailer.

Figure 13.15 No one would ever try to lift an elephant like this!

A Lever that Keeps on Lifting

The device the person is using to move the boat is called a winch. A **winch** consists of a small cylinder and a crank or handle. Study Figure 13.16 to see how a winch works. Notice that the axle of the winch is held in place and acts like a fulcrum. The handle is like the effort arm of a lever. Exerting a force on the handle turns the wheel. This motion is much like the effort force on a lever. However, you do not reach "bottom" with the handle. You just keep turning.

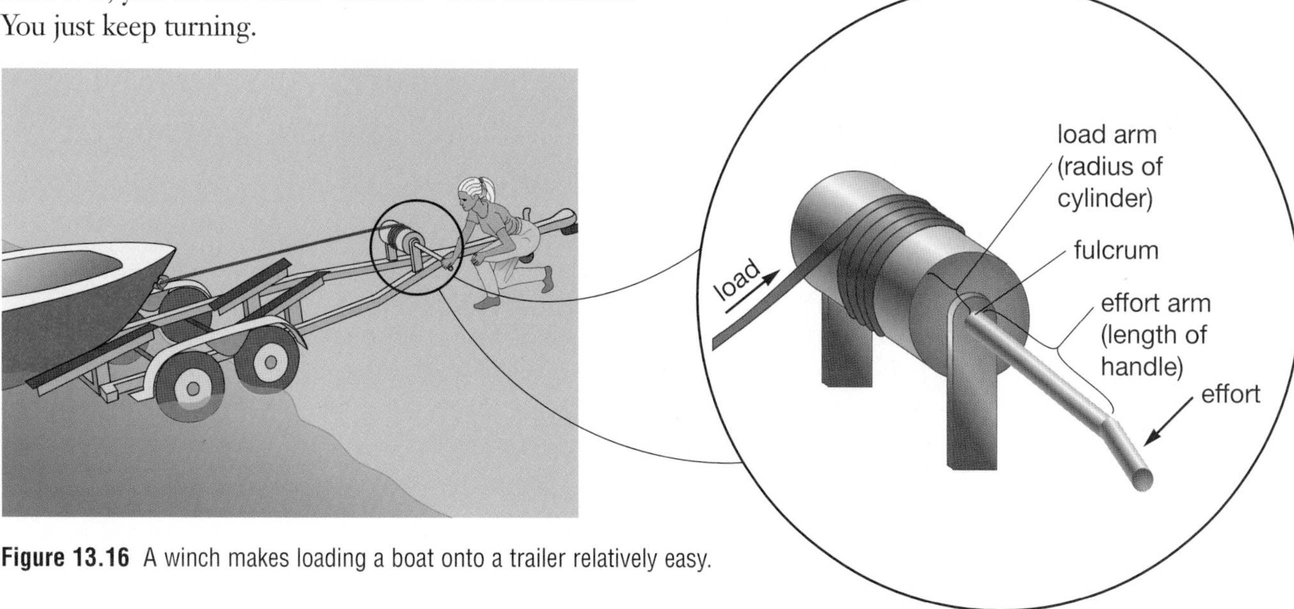

Figure 13.16 A winch makes loading a boat onto a trailer relatively easy.

Notice that the **radius** of the wheel — the distance from the centre of the wheel to the circumference — is like the load arm of a lever. The force that the cable exerts on the wheel is like the load on a lever. Since the handle is much longer than the radius of the wheel, the effort force is smaller than the load.

Using a winch is like using a short lever over and over again. The handle of a manual pencil sharpener and the reel on a fishing rod are examples of winches.

Machines in a Spin

A winch is just one example of a wheel-and-axle device. As you can see in Figure 13.17, wheel-and-axle combinations come in a variety of shapes and sizes. The "wheel" does not even have to be round. As long as two turning objects are attached to each other at their centres, and one causes the other to turn, you can call the device a **wheel and axle**.

You can hardly open your eyes without seeing a wheel-and-axle machine of some sort. Study Figure 13.17 and identify the wheel-and-axle devices. Remember that some instruments or machines have more than one wheel-and-axle combination. The wheel and axle is more convenient than a lever for some tasks, and, like a lever, it provides a mechanical advantage.

Figure 13.17 Each of these objects contains a wheel and axle.

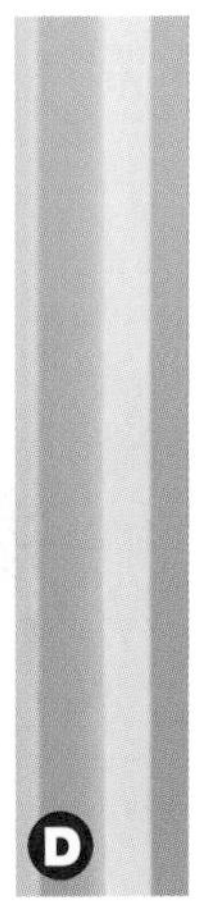

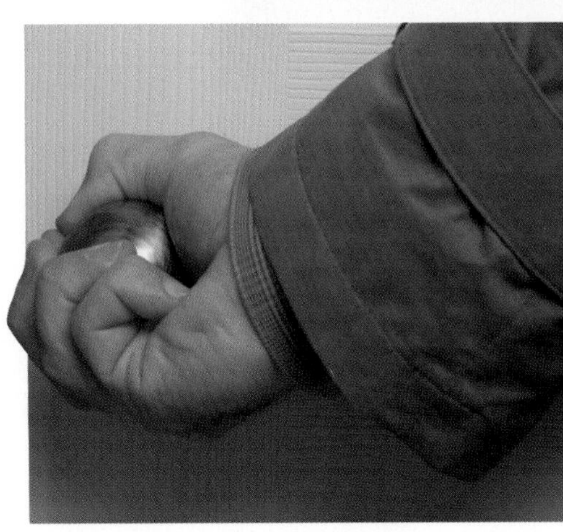

Speed and Action

Gaining a mechanical advantage is one benefit of using a wheel-and-axle device. Just like a lever, a wheel-and-axle device can also generate speed, as shown in Figures 13.18A and B. In return, however, these machines require a large effort force and produce a smaller force on the load.

Figure 13.18A Look at the pedals and the front wheel on this tricycle. Is the effort force exerted on the wheel or the axle? What does the clown get in return for the effort put into the machine?

Figure 13.18B What are the possible benefits of the huge wheel on this old-fashioned bicycle?

Gearing Up

A wheel-and-axle device provides speed for a race car zooming around a track. However, the wheel and the axle are attached to each other, so each makes the same number of rotations every second. Suppose you wanted to make one wheel rotate faster than another wheel. For example, a clock has a second hand, a minute hand, and an hour hand, each rotating at different speeds from the same point.

As you learned in previous studies, a **gear** is a rotating wheel-like object with teeth around its rim. A group of two or more gears is called a **gear train**. Two different gear trains (A and B) are shown in Figure 13.19. The teeth of one gear fit into the teeth of another. When the first gear turns, its teeth push on the teeth of the second gear, causing the second gear to turn. The first gear, or **driving gear** (often called the **driver**), may turn because someone is turning a handle or because it is attached to a motor. The second gear is called the **driven gear** (often called the **follower**). Figures 13.20A and B on page 438 illustrate two different applications of gears. Find out what gears can help you to accomplish by doing the next activity.

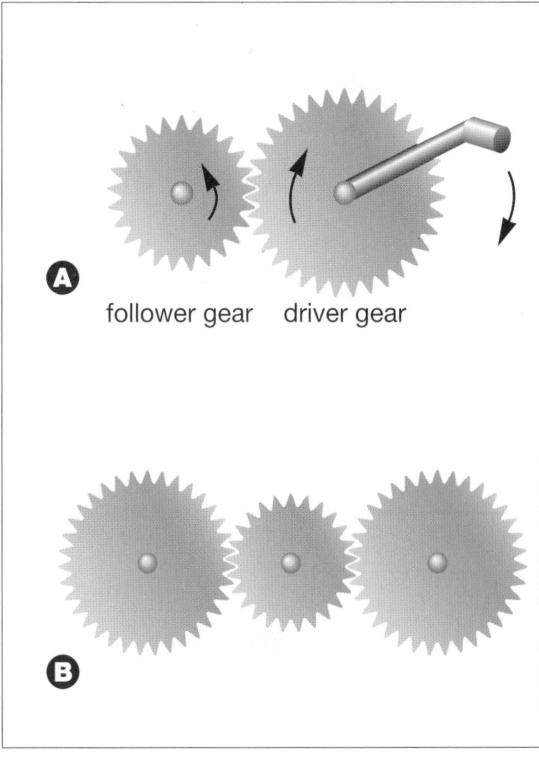

Figure 13.19 A gear train consists of two or more gears in contact with each other.

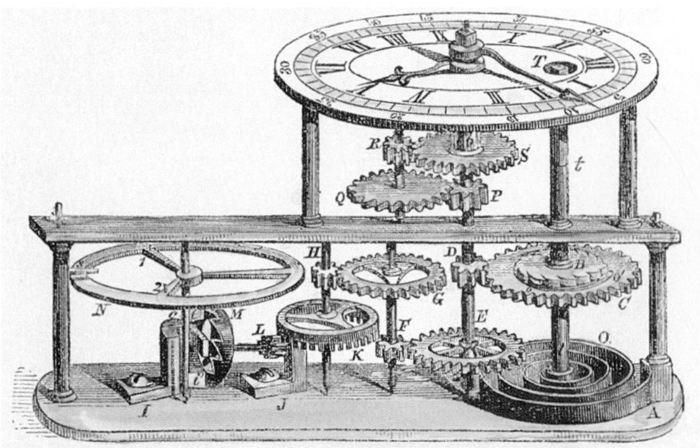

Figure 13.20A This diagram shows how the gears inside an old-fashioned clock ensure that the minute hand makes exactly 60 full rotations when the hour hand makes one full rotation.

Figure 3.20B The gears inside a large telescope are designed so that the telescope can track the constant slow motion of stars across the sky with incredible precision.

Find Out **ACTIVITY**

Turnaround Time

How many times does the follower gear turn when the driver gear makes one full turn? Does the number of rotations depend on how much larger the driver gear is?

What You Need

set of gears of different sizes, for example, from a Spirograph™ or Lego Technik™ set
felt tip pen and ruler

What to Do

1. With a felt tip pen, make a mark on one tooth of each of the two gears, at the spot where they touch.
2. Turn one gear and count the number of times the smaller gear turns when the larger gear makes one full turn. Record this number.
3. Measure and record the diameters of each of the gears.

4. Divide the diameter of the larger gear by the diameter of the smaller gear. Record your answer. Compare this number with the number you recorded in step 2.
5. Count the number of teeth on each gear. Divide the number of teeth on the larger gear by the number of teeth on the smaller gear. Record your answer. Compare this number with the numbers you calculated in steps 2 and 4.

What Did You Find Out?

1. Why does the smaller gear complete one full rotation before the larger gear does? (Look at the felt tip marks as the gears go around.)
2. If the larger gear had three times as many teeth, how many rotations would the smaller gear make in one rotation of the larger gear? How much bigger would the larger gear be in this case?
3. Explain two different measurements that you could use to predict the numbers of turns a small gear will make every time the large gear makes one full turn. What would you predict about the mechanical advantage of this gear combination? Write a statement that summarizes your conclusion about gears.

Going the Distance

Can one gear turn another gear without touching it? Does this sound impossible? Think about the gears on your bicycle. One set of gears is attached to the pedals and another to the rear wheel. A chain connecting the gears allows the front gear to turn the gear on the rear wheel, some distance away. A gear with teeth that fit into the links of a chain is called a **sprocket**. Figure 13.21 compares gears in contact with each other and gears in a sprocket.

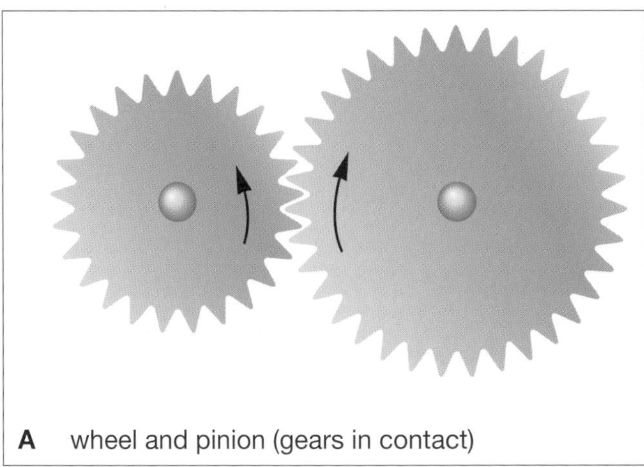

A wheel and pinion (gears in contact)

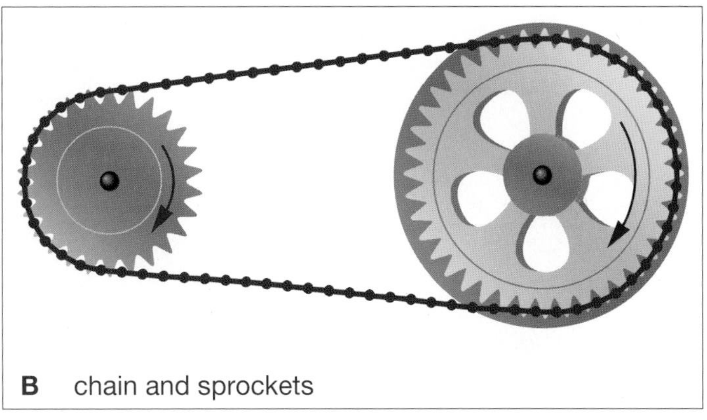

B chain and sprockets

Figure 13.21 Take a look at this comparison of gears in contact (A) and gears, or sprockets, connected by a chain (B). While the gears in contact turn in opposite directions, the gears connected by a chain turn in the same direction.

Each link of a bicycle chain moves the same distance in the same period of time. Thus, if the front sprocket moves the chain a distance equal to 45 teeth, the back sprocket will also move through a distance of 45 teeth. However, the back sprocket may have only 15 teeth and the front sprocket may have 45 teeth. As a result, the back sprocket would make three full turns for every one complete turn of the front sprocket. The relationship between the speed of rotations of a smaller gear and a larger gear is called the **velocity ratio**. In this example, the bicycle has a velocity ratio of 3. Here is the formula for calculating velocity ratio:

$$\text{Velocity ratio} = \frac{\text{Number of driver gear teeth}}{\text{Number of follower gear teeth}}$$

In the next investigation, examine the velocity ratio of gears in a bicycle.

CONDUCT AN

INVESTIGATION 13-C

Downshift for Power!

How do bicycle gears help your bicycle go faster, or help you pedal up a hill? What is the difference between high gear and low gear? Why would you want more than one gear on your bicycle, anyway? This investigation will demonstrate how gears on a bicycle can give you a mechanical advantage.

Problem

How does the velocity ratio change as you switch between different gears on a bicycle, and how does this affect the force you need to pedal the bicycle?

Apparatus

bicycle with double set of racing gears

Procedure

1. Make a data table like the one shown below. Give your table a suitable title. You may have to change the number of rows and columns, depending on the number of sprockets on the bicycle you are using.
2. Count the number of teeth on each of the front sprockets. Record these numbers in the row of your table to the right of the heading, "Number of teeth." Make sprocket number 1 the largest sprocket.
3. Count the number of teeth on each of the back sprockets. Record these numbers in the column below the heading, "Number of teeth." Again, make sprocket number 1 the largest sprocket.

		Front sprockets		
		1	2	3
	Number of teeth			
Back sprockets	1			
	2			
	3			
	4			
	5			
	6			

4 For each box in the rest of the table, divide the number of teeth in the front sprocket (at the top of the column) by the number of teeth in the back sprocket (in the first column of the table). This gives you the velocity ratio of each gear combination.

Analyze

1. What do the data indicate about the number of times the back sprocket and the wheels turn when the front sprocket and the pedals make one full turn?
2. Explain what you think "high gear" and "low gear" mean.
3. If the velocity ratio increases when you change gear, will the mechanical advantage of the bicycle increase or decrease? (Hint: Remember about trading force for distance or speed.)

Conclude and Apply

4. Why do you need to pedal faster to go at the same speed when your bicycle is in a lower gear?
5. Which gear helps you go faster on level ground? Why?
6. Why do you use low gear when going up hills?

Boosting Efficiency

You have seen that gears are modified wheel-and-axle machines. A gear is simply a wheel with teeth along its circumference. Effort exerted on one gear causes another gear to turn. The mechanical advantage of a pair of gears is found by dividing the radius of the effort gear by the radius of the load gear.

As you have learned, some of the effort force put into any machine must overcome friction. For example, some of the effort force you exert when you pedal a bike must overcome the friction of the pedal gear rubbing against the bicycle chain. This reduces the efficiency of the bicycle. Low-efficiency machines lose much of the work put into them because of friction; high-efficiency machines do not.

You can boost the efficiency of a machine such as a bicycle. You have seen that you can increase efficiency by adding a lubricant such as oil or grease to the surfaces that rub together. If a bicycle's chain, gears, and other moving parts are cleaned and lubricated periodically, the bike will operate more efficiently. Also, keeping the tires properly inflated will reduce friction between the road and the tires. Similarly, keeping car tires properly inflated and changing the engine oil to keep it clean will increase the efficiency of a car. A more efficiently running car gives better gas mileage and saves both money and energy.

Check Your Understanding

1. If you wanted a winch to have a mechanical advantage of 4 and the radius of the axle was 5 cm, how long would the handle have to be?
2. Explain how a screwdriver could be used
   (a) as a lever **(b)** as a wheel-and-axle device
3. If the chain of your bicycle is running from the largest front sprocket to the smallest back sprocket, are you in high gear or low gear? Would you use this gear while pedalling up a steep hill or cruising fast on a level road? Explain your answer.
4. A gear with a 9 cm radius is turning another gear that has a radius of 3 cm. If the large gear is turning at a rate of 10 revolutions per second, what is the rotation rate of the small gear? What is the velocity ratio of this gear system?
5. Define efficiency and give the formula for calculating the efficiency of a mechanical system.
6. **Thinking Critically** If a bicycle has two sprockets on the front and four sprockets on the back, how many different gear combinations would it have?
7. **Design Your Own** Formulate your own question about the efficient operation of mechanical systems and design your own investigation to explore possible answers.

13.3 Pulley Power

Figure 13.22 How does this weight machine allow the woman to lift weights safely and comfortably?

You learned in previous studies that a **pulley** is a grooved wheel with a rope or a chain running along the groove. You can see an example of pulleys in action in Figure 13.22. A pulley is similar to a Class 1 lever. Instead of a bar, a pulley has a rope. The axle of the pulley acts like a fulcrum. The two sides of the pulley are the effort arm and the load arm.

Pulleys can be fixed or movable, as shown in Figure 13.23. A **fixed pulley** is attached to something that does not move, such as a ceiling, a wall, or a tree. A fixed pulley, such as the one used at the top of a flagpole, can change the direction of an effort force. When you pull down on the effort arm with the rope, the pulley raises the object attached to the load arm. Thus, a single fixed pulley simply changes the direction of the motion and makes certain movements more convenient. Once the flagpole pulley is attached, you can raise and lower the flag without ever climbing to the top of the flagpole!

A **movable pulley** is attached to something else, often by a rope that goes around the pulley itself. If a rope is fixed to the ceiling and then comes down around the pulley and back up, you can lift and lower the pulley itself by pulling on the rope. The load may be attached to the centre of the pulley.

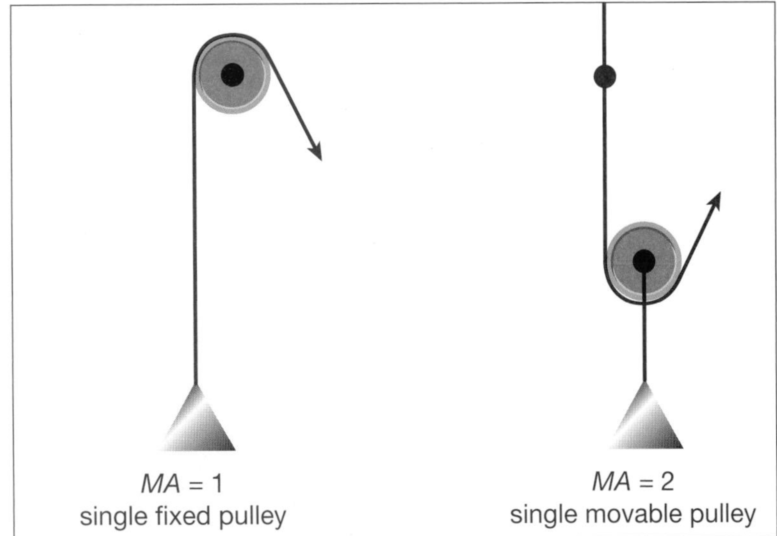

Figure 13.23 Pulleys can be fixed or movable.

INVESTIGATION 13-D

How Silly Can It Be?

Think About It

Sometimes machines featuring pulleys are designed just for the fun of it. In the twentieth century, the cartoonist Rube Goldberg drew many pictures of ridiculously elaborate machines for doing everyday tasks, often with unexpected parts like old boots or broomsticks. His cartoons became so popular that "overdesigned" and accident-prone machines in real life are still called "Rube Goldberg devices."

Look at the Rube Goldberg device in this picture. Could a machine like this work if you built it?

What to Do

1. Try to figure out, step by step, how the device in the picture works. Which step do you think is the most likely not to work?

2. Now design your own Rube Goldberg device on paper. It might open a door when someone rings the bell, stir a cooking pot, dress someone in a hat and scarf, or even do several tasks at once. Make sure your design has at least four distinct steps, and try to use as many different types of machines — levers, winches, pulleys, ramps, wheels and axles — as you can.

Analyze

1. List the different kinds of machines in your device, in order of use.

2. Describe in writing how your device works, step by step.

3. Exchange your written description with a partner, and see if your partner can follow the operation of your device.

4. **Apply** After receiving your teacher's approval, try to build a simple version of your device, and see if you can get it to work.

Supercharging Pulleys

You saw that a wheel-and-axle combination can be compared to a lever. It would seem logical to analyze a pulley in the same way. However, if you imagine a pulley as a lever, you will discover that the "effort" arm and the "load" arm are the same. How do pulleys help you lift heavy loads, therefore?

You have seen that a single pulley can make lifting a load more convenient. Combinations of pulleys are required to lift extremely heavy or awkward loads, such as cargo on a ship (see Figure 13.24). The complex pulley system shown in Figure 13.25 is a combination of fixed and movable pulleys, called a **block and tackle**. Depending on the number of pulleys used, a block and tackle can have a large mechanical advantage. You have probably noticed that pulley systems designed to lift extremely heavy loads have long cables running around several pulleys. How can you determine the mechanical advantage of a **compound pulley** — one made up of several pulleys working together? To find out, perform the investigation on the next page. As a warm-up, you can also do the activity below.

Figure 13.24 In the shipyards of port cities such as Vancouver and Halifax, pulley systems lifting huge loads onto ships are a familiar sight.

Figure 13.25 A block and tackle

Tug of War

How can you increase the mechanical advantage of a pulley?

What You Need

2 broom handles or similar smooth poles
rope or twine (about 4 m)

What to Do

1. Two students hold the upright broom handles between them, side by side.
2. Tie one end of the rope to one broom handle, and pass it once around the other handle.
3. A third student should try to pull the handles together using the rope, while the other two try to hold them apart.
4. Now wind the rope a couple more times around the handles, and try again. Experiment with different numbers of rope windings.

What Did You Find Out?

1. Does increasing the number of rope windings make it easier for the student pulling the rope to move the handles together?
2. What forces do the two students holding the handles experience?

Extension

3. Is there any change in how far the student has to pull the rope as the number of windings increases?

CONDUCT AN

INVESTIGATION 13-E

Easy Lifting

Industrial pulley systems are usually made up of many pulleys working together. As you have just learned, such a combination of pulleys is called a compound pulley. Why do compound pulleys make lifting easier?

Problem

How can you predict the mechanical advantage of a compound pulley? How can you test the efficiency of a pulley system?

Safety Precautions

- Be very careful not to drop any heavy weights.
- Have your teacher check your apparatus before you make any measurements.

Apparatus

10 N spring scale
1 kg mass
support stand, held firmly in place
2 single pulleys
2 double pulleys
1 triple pulley

Materials

rope (at least 6 m)

Procedure

1. Make a data table like this one. Give your table a title.

Trial	Load	Effort	Number of ropes
A			
B			
C			
D			
E			

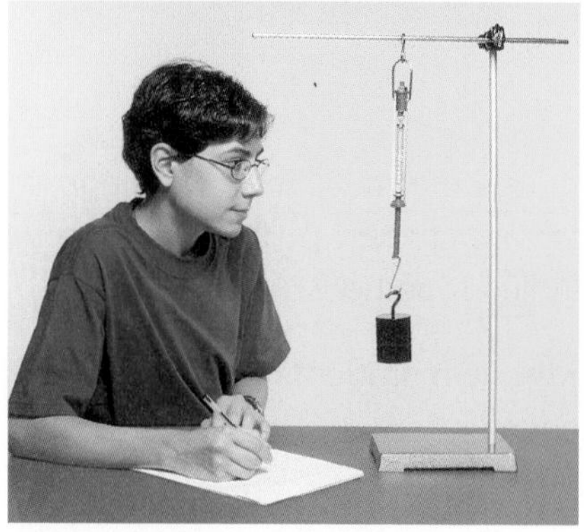

2. Suspend the mass on the spring scale and note the weight. Record this value in every row of your table, in the column labelled "Load."

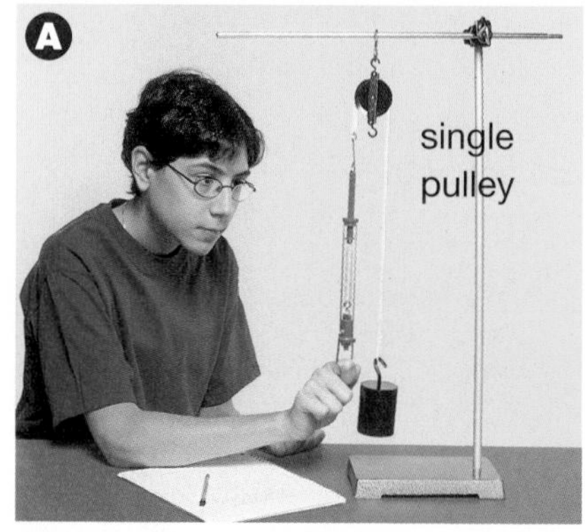

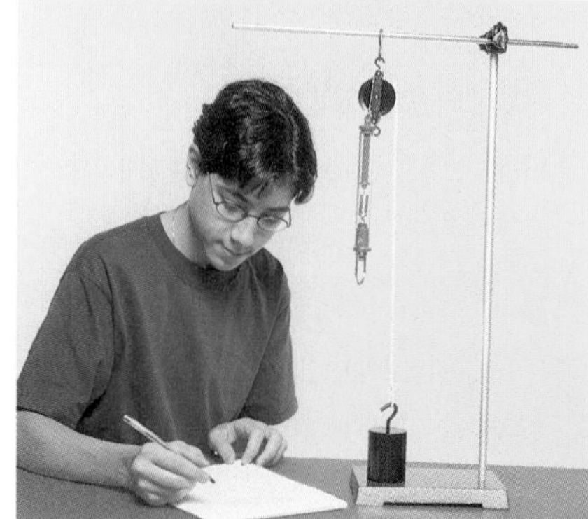

3. Assemble the apparatus as shown for Trial A above. While supporting the load with the spring scale, read the amount of force shown on the scale. Record this number in the column of your table labelled "Effort."

4. Count the number of ropes that are supporting the load and record the number in the column labelled "Number of ropes."

5. Repeat steps 3 and 4 for trials B through E, shown below.

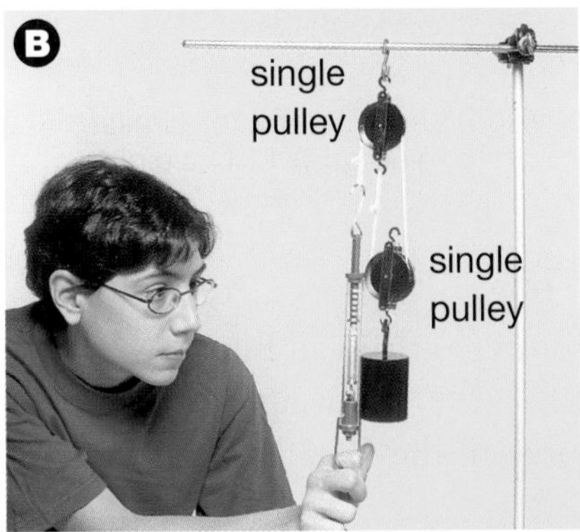

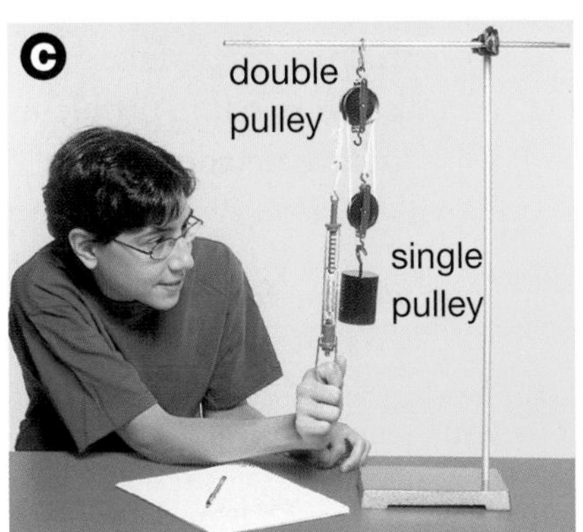

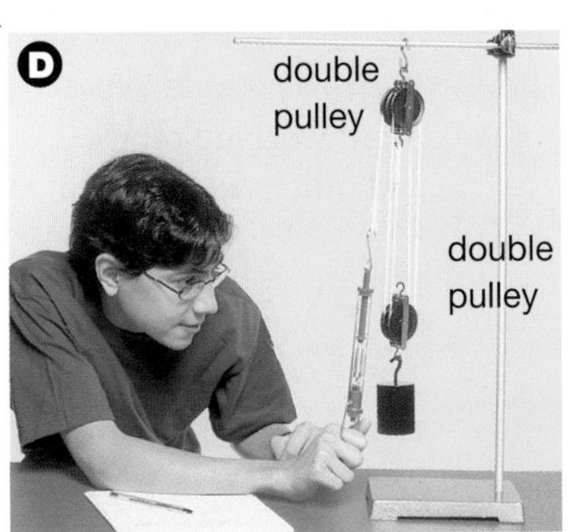

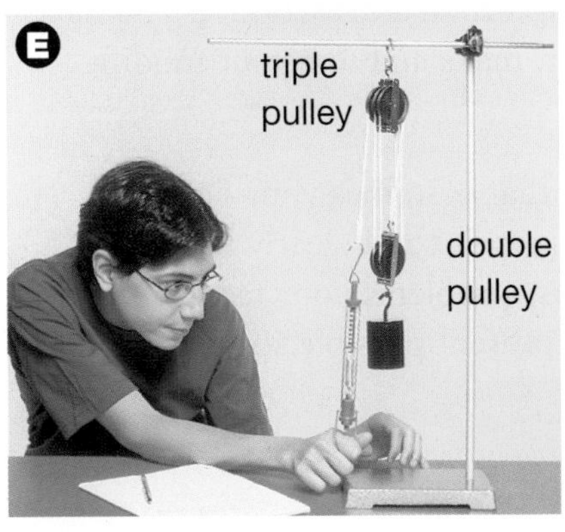

Analyze

1. Make an analysis table with the following headings to record the results of your calculations. Give your table a title.

Trial	Number of ropes	Mechanical advantage
A		
B		
C		
D		
E		

2. For each trial, divide the weight of the load by the effort force. Record this number in your table in the column labelled "Mechanical advantage."

3. For each trial, copy "Number of ropes" from your data table to your analysis table.

Conclude and Apply

4. Compare the mechanical advantage with the number of ropes for every trial. What conclusion can you draw from this comparison?

5. Devise a method for determining the efficiency of the compound pulley in Trial E. You can calculate efficiency using the same formula you used previously for levers:

$$\text{Efficiency} = \frac{\text{Work done on load}}{\text{Work done by effort force}} \times 100\%$$

Remember that work equals force multiplied by distance. Thus, you will need to know how far the effort and the load moved as well. What is the efficiency of your pulley system? Why is it less than 100 percent?

Computer CONNECT

If you have access to a computer spreadsheet program, you may want to use it for your data tables.

INVESTIGATION 13-F

Pick It Up

Imagine you and your group are a team of engineers working for the Ace Crane Company. You are in the process of developing a new crane to be used in the construction industry.

Challenge

Design and build a prototype of a crane with a wheel-and-axle system that can lift weights of up to 12 000 N. The motor that must be used to turn the crane's wheel-and-axle system can generate a force of 4000 N on the rim of the axle.

Materials

wood, cardboard, dowelling, Lego™ parts (or similar construction kit parts), string, glue gun, 12 N weight

Design Criteria

Build a model (a prototype) that can lift a load of 12 N (to represent the 12 000 N weight) with an effort of 4 N (to represent the 4000 N force).

Plan and Construct

1. As a group, discuss potential designs. Make technical drawings and discuss possible problems with each design until you have decided on a design that you think will work.
2. Select the materials you will use for the prototype.
3. Show your plan to your teacher for approval.
4. Collect your materials and draw a blueprint. Assign the tasks among your team members.
5. Construct your prototype. You should have some members of your team working on the wheel-and-axle system and others working on the body of the crane.
6. Test your crane using a 12 N load.

Evaluate

1. Does your mechanical device satisfy all the conditions in the Challenge? If not, how could you modify the design to make it work? If you have the opportunity, make and test your modifications.
2. Write a report that describes your device. Include your blueprint and clearly label each part. Discuss any problems your team had with the device and present possible solutions.

Pulleys in Action

Since 1985, more than 500 students from Canada, the United States, and other countries have enjoyed a unique educational experience offered by a school called Class Afloat™. The school is based in Montréal, Québec and in Calgary, Alberta. Every year, 48 students board a sailing school vessel, the *S.V. Concordia*, that offers shipboard studies at the grade 11, grade 12, and junior college levels. Students spend up to 300 days studying, travelling, and working — and gaining lots of hands-on experience with pulleys. The pulleys are part of the rigging of the sailing school on which the students live and study. The ship's 15 huge sails are hoisted aloft by pulleys to a maximum height of 35 m above the deck. Figure 13.26A shows students adjusting the sails by pulling on ropes wound around pulleys. Figure 13.26B shows a close-up of one of the ship's pulley systems.

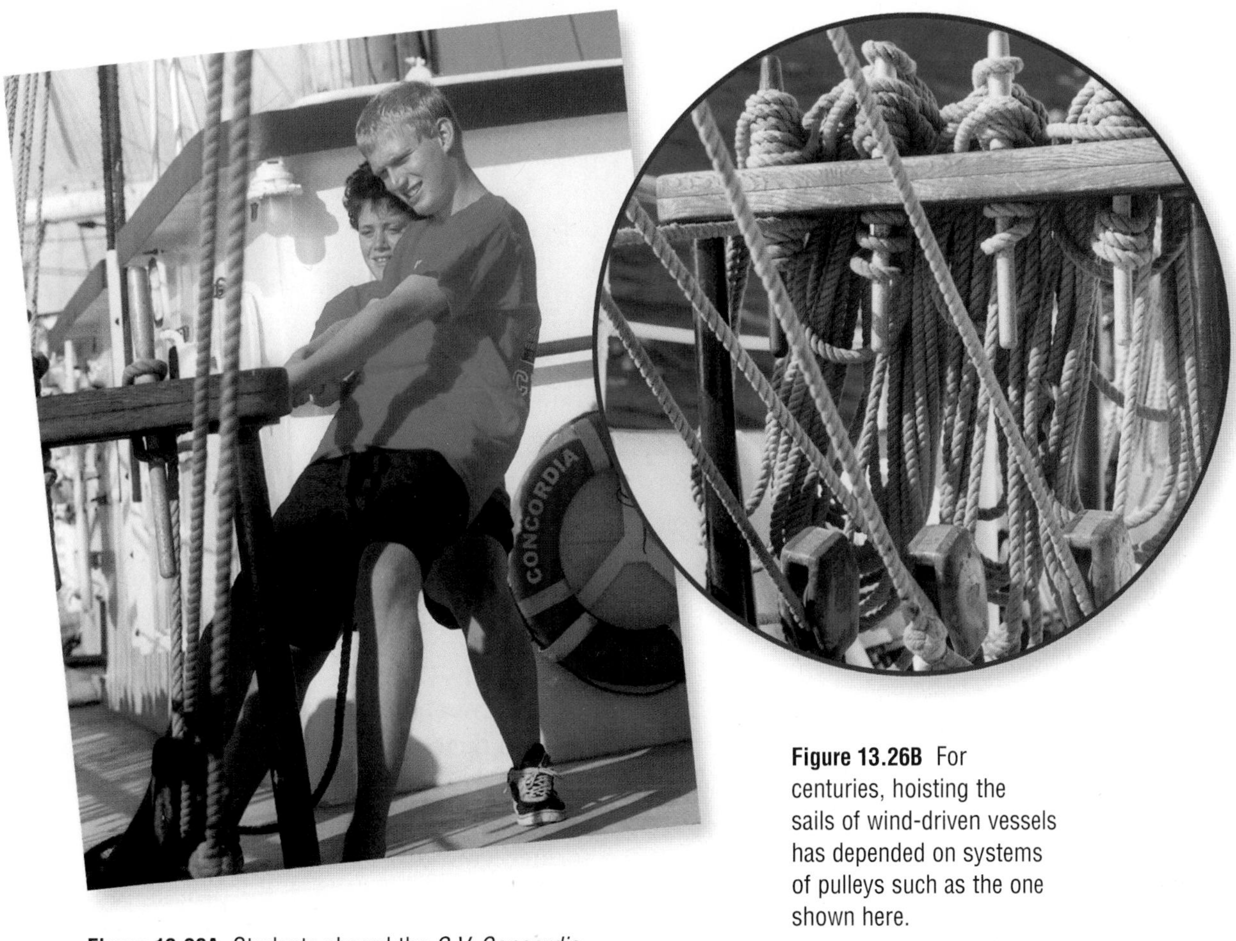

Figure 13.26A Students aboard the *S.V. Concordia* tug hard on a rope that winds through a pulley.

Figure 13.26B For centuries, hoisting the sails of wind-driven vessels has depended on systems of pulleys such as the one shown here.

The students sail this 58 m yacht to 32 ports of call on five continents. Lessons taught in the shipboard classroom are combined with practical field experience. Studies in marine biology, for example, include visits to specific research stations. By the end of their course of study, students have learned a great deal, particularly in science and social science. They also know first-hand the advantages of lifting a weight by means of a pulley system!

Check Your Understanding

1. Draw a sketch of a single pulley in an arrangement that gives a mechanical advantage of 1. Then draw a sketch of a single pulley in an arrangement that gives a mechanical advantage of 2.

2. Suppose you weigh 450 N and you can lift 150 N. Which pulley system in the investigation on pages 446–447 would you need to use to lift yourself?

3. What is the mechanical advantage of the pulley system shown in Diagram A?

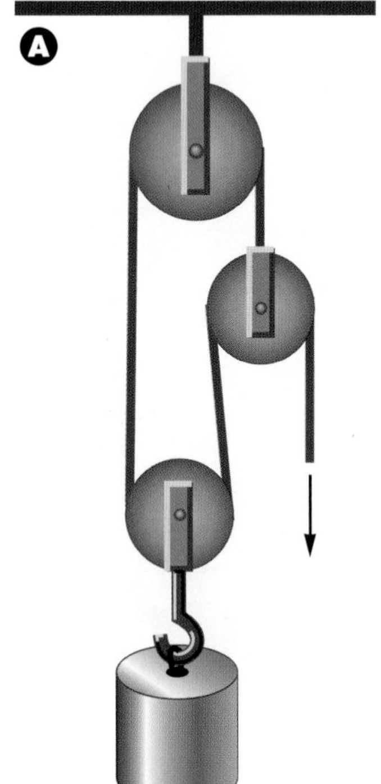

4. Find the overall mechanical advantage of the pulley system shown in Diagram B.

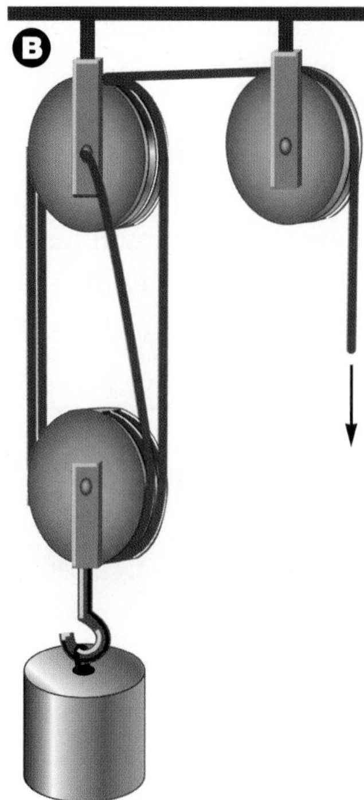

5. Distinguish between a single fixed pulley and a single movable pulley, and describe the advantages of each.

6. Make a series of drawings to show, in sequence, how some of the simple machines in a bicycle work to make the bicycle move. Start with the rider's feet applying force to the pedals.

7. **Design Your Own** Design your own experiment about an aspect of mechanical systems that you wish to explore. Be sure to identify dependent and independent variables, and to specify a control. After you have performed your investigation, list criteria for assessing your solution to the problem.

CHAPTER at a glance

Now that you have completed this chapter, try to do the following. If you cannot, go back to the sections indicated.

Define the terms "effort force" and "load," as they apply to levers. (13.1)

Explain how bones can act as levers. Include the function of muscles and joints in your explanation. (13.1)

Define the term "work" and express it as a mathematical formula. (13.1)

Define the term "mechanical advantage" and express it as a mathematical formula. (13.1)

Explain why machines, including levers, are not 100 percent efficient. Use the definition of efficiency in your explanation. (13.1)

State the difference between speed and velocity. (13.1)

Give an example of a lever that has a mechanical advantage smaller than 1. Describe its purpose and tell to which class it belongs. (13.1)

Describe some ways to boost efficiency. (13.2)

Describe a wheel-and-axle device and explain how it functions much like a lever. (13.2)

Define the terms "driver gear" and "follower gear." (13.2)

Define "velocity ratio" and express it as a mathematical formula. (13.2)

Explain the difference between high gear and low gear on a bicycle. (13.2)

Determine the mechanical advantage of the compound pulley shown here. (13.3)

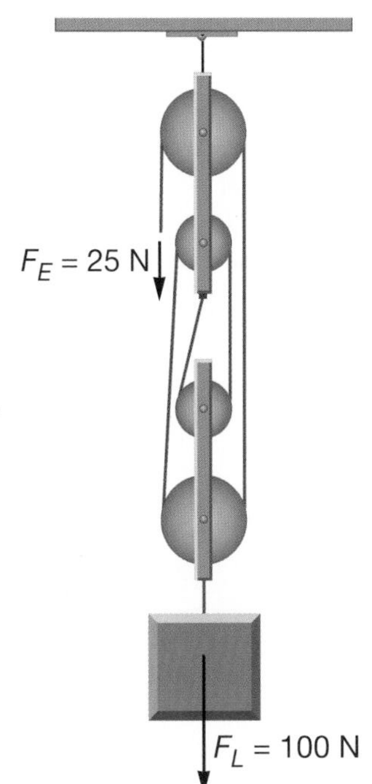

Prepare Your Own Summary

Summarize this chapter by doing one of the following. Use a graphic organizer (such as a concept map), produce a poster, or write the summary to include the key chapter ideas. Here are a few ideas to use as a guide:

- How can a teeter-totter be adjusted so that a large child and a small child can play on it together?
- Look at the garden hoe on the right.
  - **(a)** What sort of lever is it?
  - **(b)** What does the longer arrow show?
  - **(c)** Which is the fulcrum, A, B, or C?
  - **(d)** What are the lengths AC and BC called?
  - **(e)** Calculate the mechanical advantage of the hoe.

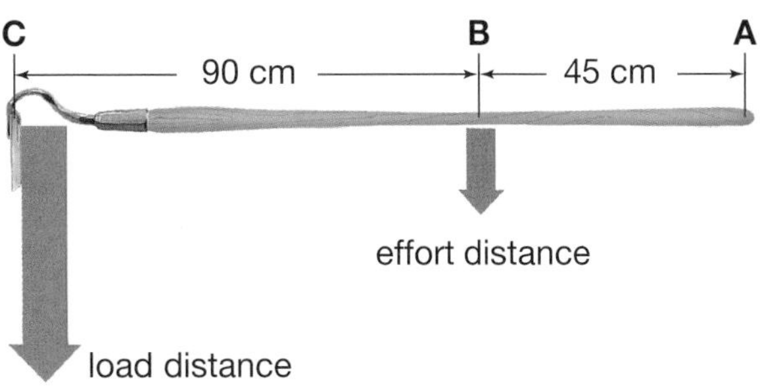

- Explain what an ergonomist does and give some examples of industries in which an ergonomist might work.
- How can a winch make it easy to load a boat onto a boat-trailer?
- Explain a situation in which you would use a pulley instead of a lever.

CHAPTER

13 Review

Key Terms

- lever
- fulcrum
- effort force
- load
- effort arm
- load arm
- Class 1 lever
- Class 2 lever
- Class 3 lever
- inclined plane
- work
- joules
- mechanical advantage
- efficiency
- speed
- velocity
- ergonomics
- carpal tunnel syndrome
- winch
- radius
- wheel and axle
- gear
- gear train
- driving gear (driver)
- driven gear (follower)
- sprocket
- velocity ratio
- pulley
- fixed pulley
- movable pulley
- block and tackle
- compound pulley

Reviewing Key Terms

If you need to review, the section numbers show you where these terms were introduced.

Copy the crossword puzzle into your notebook and complete it using some key terms listed above.

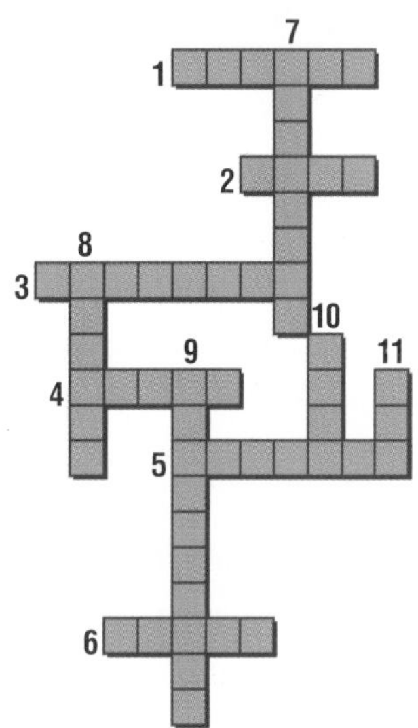

Across

1. This gear turns another gear. (13.2)
2. A lever exerts a force on this and lifts it. (13.1)
3. Its teeth fit into the links of a chain. (13.2)
4. A bottle opener is this kind of tool. (13.1)
5. The point where a lever does not move. (13.1)
6. Turn the handle of this machine and it will pull on a cable and wrap it around a cylinder. (13.2)

Down

7. This tells how fast and in what direction an object moves. (13.3)
8. A wheel that turns and allows a rope to move over it easily. (13.3)
9. The percentage of the work done on a machine that the machine then does on a load. (13.1)
10. A wheel with teeth. (13.2)
11. The distance from the fulcrum to the load is the load ______. (13.1)

Understanding Key Ideas

Section numbers are provided if you need to review.

12. Which class of lever always has a mechanical advantage that is less than 1? Give an example of this type of lever. (13.1)
13. State at least three situations in which it would be more practical to use a wheel and axle rather than a lever. (13.2)
14. Explain how a winch is like a lever. (13.2)
15. Sketch a diagram showing how you could use one single pulley and one double pulley to gain a mechanical advantage of 3. (13.3)
16. Which of these does *not* describe what a machine does? (13.1, 13.3)
    (a) changes the effort force
    (b) transforms energy
    (c) changes the direction of a force
    (d) does work

17. Describe as many ways as you can in which a simple machine can make work easier. (13.1, 13.2, 13.3)

18. Many machines, including levers, wheel-and-axle devices, and pulleys, exert a greater force on a load than you exert on the machine. What do you have to do in return for a mechanical advantage that is greater than 1? (13.1, 13.2, 13.3)

Developing Skills

19. Complete the following concept map of simple machines, using the following words and phrases: lever, wheel and axle, effort force, work, pulley, mechanical advantage, load, efficiency.

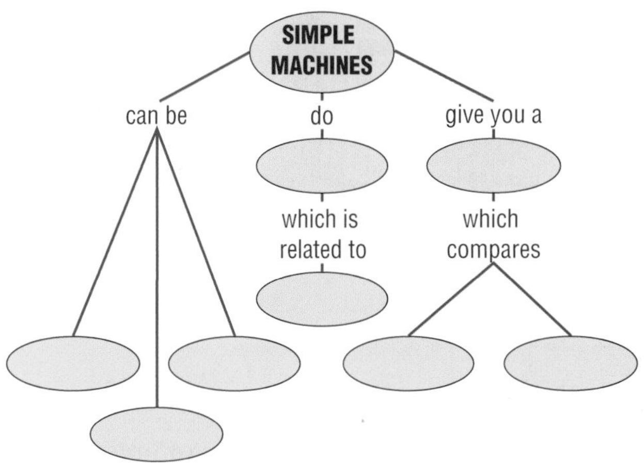

20. Look at the photograph of the man chopping a log with an axe. If the man exerts an effort force of 80 N and the load force of the wood is 320 N, what is the mechanical advantage of the axe?

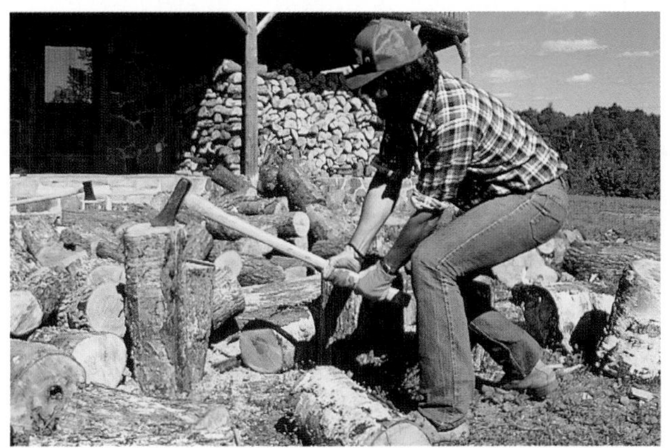

Problem Solving/Applying

21. Design a pulley system that is similar to the platform of a window washer. You should be able to stand on the platform and pull on a rope that lifts the platform with your own weight on it. Sketch your pulley system.

22. Why might you choose a gear that would make you pedal extremely fast while your bicycle was travelling slowly?

Critical Thinking

23. Design a manually operated machine that will load a 5 t elephant into a truck.

24. The engine of a car turns a shaft that runs down the centre of the underside of the car. The shaft is perpendicular to the rear axle. Explain how gears can be used to allow the rotating shaft to make the rear axle turn.

25. Machines make work easier, but you always have to do more work than the machine does on the load. Explain this statement.

Pause& Reflect

1. A good ice skater can glide quickly across the ice with only a little effort, because the small area and smooth surface of the blades mean there is very little friction between the ice and the skates. The pressure of the blades on the ice melts the ice a bit. When the ice melts, it leaves a thin film of water between the skate blades and the ice. The water layer acts as a lubricant and helps the blades slide smoothly across the ice without sticking. Think of other ways in which friction can be reduced in machines. Write your ideas in your Science Log.
2. Go back to the beginning of this chapter on page 416 and check your original answers to the Getting Ready questions. How has your thinking changed? How would you answer those questions now that you have investigated the topics in this chapter?

CHAPTER

14 Fluid Pressure and

Getting Ready...

- How are hydraulics used to land an airplane?
- How can air pressure be used to save lives?
- What makes a hovercraft hover?

Science Log

Try to show in a drawing how air, oil, or another fluid could generate a force large enough to power a ride such as the one in the photograph. Then, write answers to the questions above in your Science Log. As you study this chapter, you can revise your answers or add more details.

If you were to go on a ride like the one shown here, you would feel the tremendous power of this machine as it lifts and whirls its riders at dizzying speeds. Machines such as this use fluids, under high pressure, to generate forces large enough to lift dozens of people at a time. The smaller inset photograph shows the lifting mechanism that operates the ride hydraulically.

Chapter 14 reinforces the relationship between force and fluid pressure that you learned about in Chapter 6. You will review **hydraulics** (the study of pressure in liquids) and **pneumatics** (the study of pressure in gases). This chapter also introduces a scientific principle known as Pascal's law. This law explains how fluid pressure in a closed system can be used to operate mechanical devices such as amusement park rides. Find out how air, oil, or another fluid can generate a force large enough to lift a heavy load. You can then apply what you have learned about hydraulics and pneumatics to build your own mechanical system.

Mechanical Systems

Starting Point ACTIVITY

Squeeze Effect

What happens to a liquid and a gas under pressure?

What You Need

plastic squeeze bottle with a tiny opening in the cap
water
string (20–30 cm)
sink or basin

What to Do

1. Fill a plastic squeeze bottle with water. (A shampoo bottle with a small opening in the cap is ideal.)
2. Hold the bottle and direct the opening into a sink or a basin. Squeeze the bottle rhythmically. Try different pressures.
3. Empty the squeeze bottle and remove the cap. Poke a string through the hole in the cap. Tie a knot in both ends of the string. Put the cap on the bottle with the string dangling inside the bottle.
4. Predict what you think will happen when you squeeze the bottle. Then squeeze it.
5. Wipe up any spilled water after this activity.

What Did You Find Out?

1. Describe what happened to the water-filled bottle when you squeezed it.
2. How did the stream of water respond to different pressures?
3. Describe what happened to the bottle with the string inside when you squeezed it.
4. **Apply** Describe a real-life situation in which water pressure or air pressure must be controlled.

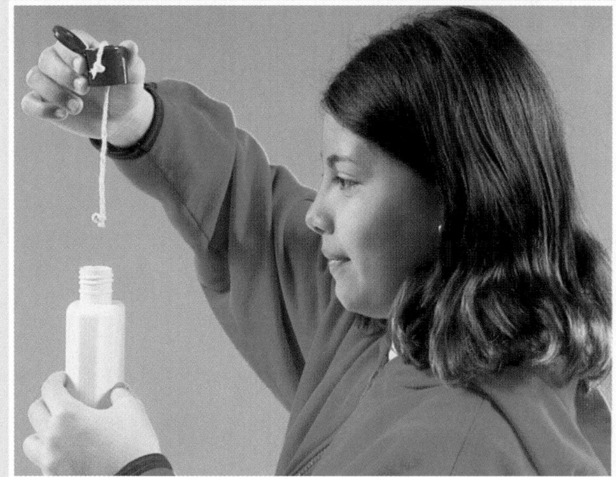

Spotlight On Key Ideas

In this chapter, you will discover

- how the diameter of a tube affects a fluid's flow rate
- how gases and liquids respond to pressure
- how Pascal's law is applied in hydraulic and pneumatic systems
- how a hydraulic lift can be used to gain a mechanical advantage
- how mechanical systems are made up of subsystems

Spotlight On Key Skills

In this chapter, you will

- make a line graph showing the relationship between the diameter of a tube and a fluid's flow rate
- calculate the mechanical advantage of a hydraulic lift
- build your own model hydraulic lift
- design your own controlled experiments
- design and construct a mechanical system made up of several subsystems

14.1 Hydraulic and Pneumatic Systems

Word CONNECT

In a dictionary, look up the origin of the word "pneumonia."

If you lived for 100 years and breathed normally, how much air would you breathe in and out of your lungs?

Each time you squeezed the bottle in the Starting Point Activity, you applied a force that pushed a fluid — either water or air — out of the bottle. This activity demonstrated first a simple hydraulic system, and then a simple pneumatic system.

Did you know that your life depends on a pneumatic system? Your body's respiratory system is more complex than any pneumatic machinery. As you learned in Chapter 3, this system is made up of lungs; tubes that allow air to enter and leave the lungs; and muscles that cause your lungs to expand and contract. Breathing depends on changes in air pressure. When you are breathing normally, your muscles make your lungs expand and draw in about 500 mL of air. Simply relaxing will push the air back out. You breathe in and out about 12 times per minute. When you are active, like the girls in the photograph, you breathe more quickly and more deeply.

Figure 14.1 During strenuous activity, your breathing quickens and deepens.

Find Out 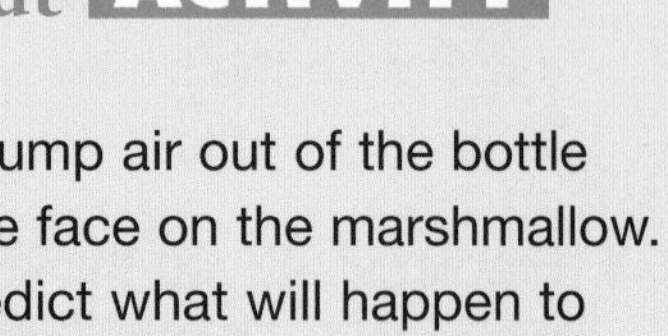

Making Faces

Try this simple activity to observe the effects of changes in air pressure.

What You Need

marshmallow

glass bottle (the mouth should be slightly larger than the marshmallow)

modelling clay

hand-held suction pump

pen or marker

What to Do

1. Draw a face on a marshmallow and put the marshmallow in a glass bottle.
2. Wrap modelling clay around a suction pump. Insert the end of the pump into the bottle.
3. Seal the mouth of the bottle with modelling clay.
4. You are going to pump air out of the bottle while observing the face on the marshmallow. Before you do, predict what will happen to the marshmallow. Now pump air out of the bottle. (Make sure there are no leaks!) What happens to the marshmallow?
5. Stop working the suction pump. What happens to the marshmallow?

What Did You Find Out?

Explain your observations based on your knowledge of air pressure.

Your existence also depends on a complex hydraulic system. As you learned in Chapter 3, blood must be kept under pressure so that it can be pumped to all parts of your body. Your heart is the pump that moves blood through the blood vessels with pressure that rises and falls. Like the rhythmic squeezing of the water bottle, each time your heart beats, it exerts a force on your blood and pushes it along. In Chapter 6, you learned that blood pressure is exerted by blood against the inner walls of arteries and, to a lesser extent, capillaries.

Your heart is an amazing hydraulic device. Over the course of a lifetime, it can pump nearly 4 billion times without stopping, and it can circulate a total of nearly 500 million litres of fluid without wearing out. Throughout your lifetime, your heart will pump enough blood to fill 13 supertankers, each holding one million barrels!

When blood leaves the left ventricle, it travels first through arteries, then capillaries, and finally veins, before it returns to your heart. Figure 14.2 shows how the arteries become smaller and smaller until they are tiny, thin-walled capillaries through which oxygen and waste products pass easily.

Did You Know?

If the blood vessels in your body were connected end to end, they would extend almost 100 000 km. That is two and a half times around the planet Earth!

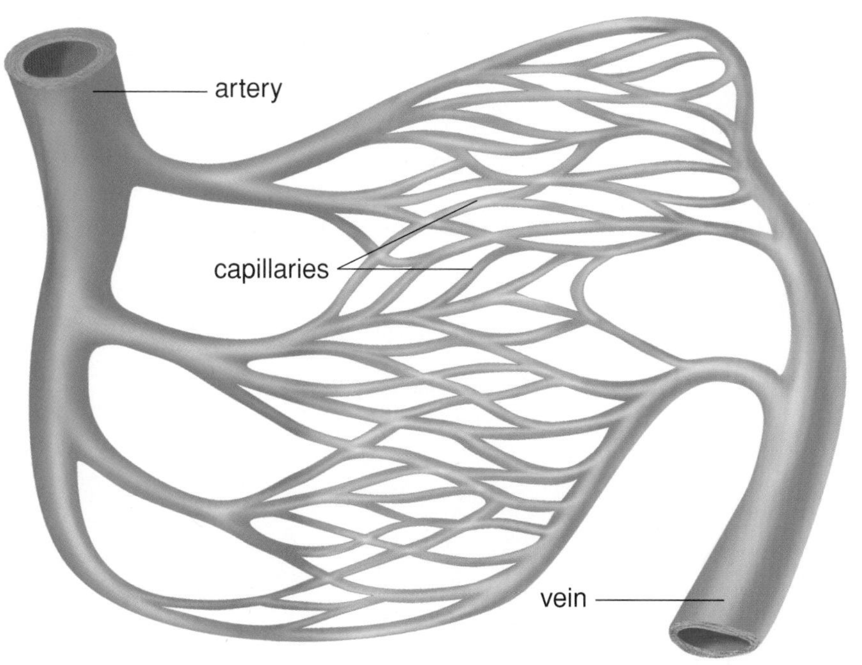

Figure 14.2 The aorta, the largest artery, branches into smaller and smaller arteries that lead to capillaries, which are tiny vessels that carry blood to every part of the body.

The diameter of the largest blood vessel in your body is just under 2 cm. The smallest vessels, the capillaries, are less than 0.0001 cm in diameter. It takes a great deal of pressure to push a fluid through tubes with small diameters, such as capillaries. You can observe this for yourself in the next investigation.

CONDUCT AN

INVESTIGATION 14-A

Fluid Friction and Flow Rate

When you think of the relationship between fluids and friction, you probably think of fluids as lubricants that reduce friction. For example, you know that oil is used to lubricate the moving parts of machines to keep them functioning smoothly. However, fluids also experience friction as they flow through a tube or a pipe. In this investigation, you will observe one of the features of hydraulic systems that affects fluid friction.

Problem

How does the diameter of a tube or a pipe affect friction, and therefore the flow rate, of the fluid that passes through it?

Safety Precautions

Apparatus

50 mL modified syringe (the plunger of the syringe must slide freely)

4 pieces of plastic tubing of varying diameter (20 cm long and made of the same material)

stopcock or clamp (to close off the tubing)

500 g mass

support stand

clamp (to attach the syringe to the support stand)

stopwatch or watch with a second hand

wood square (about 5 cm x 5 cm)

glue gun

plastic dishpan

felt tip pen

ruler

Materials

water

Note: Your teacher will glue wood squares to the tops of the syringe plunger ahead of time so the glue has time to dry.

Procedure

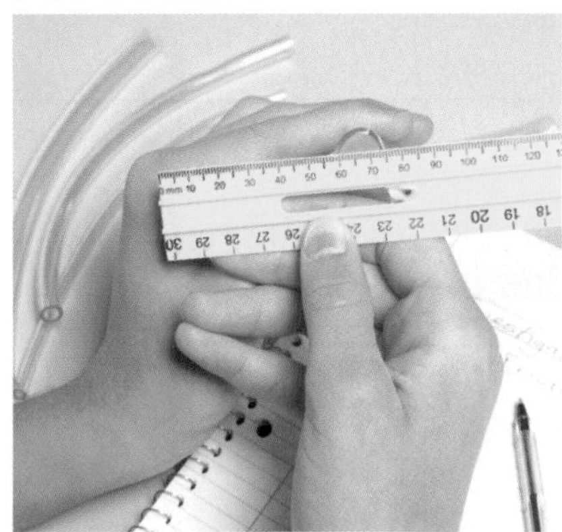

1. Measure and record the inner diameter of each of the four pieces of tubing.

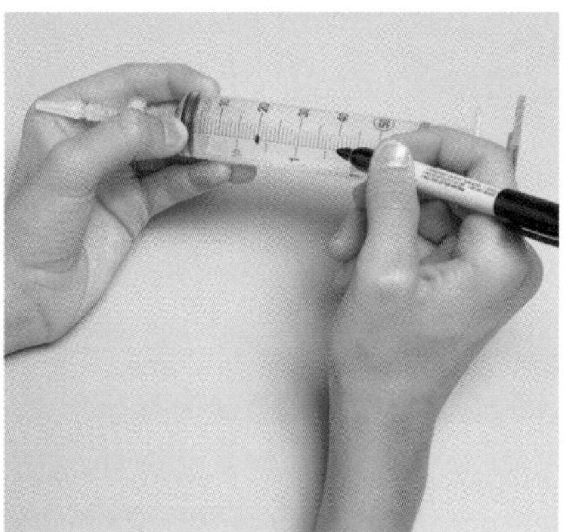

2. With the felt tip pen, make two marks on the syringe, so that the volume of fluid between the marks is about 20 mL.

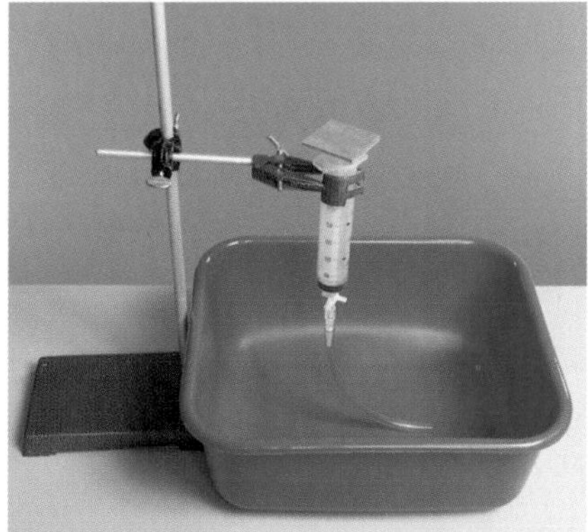

3. Assemble the apparatus with one of the lengths of tubing, as shown.

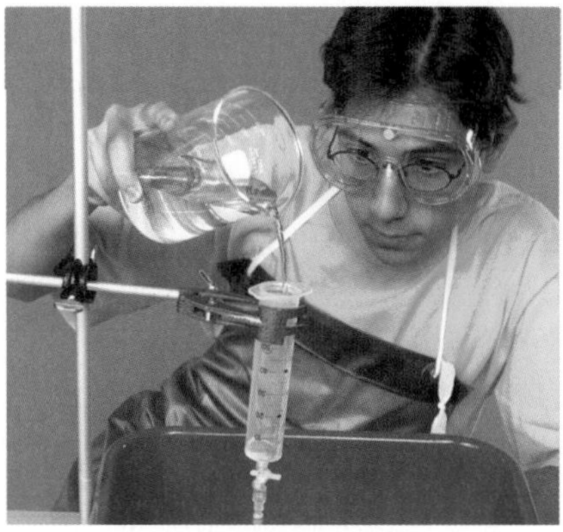

4 Close the stopcock and fill the syringe with water to a level above the marks. Briefly open the stopcock until water fills the tubing and begins to leak out the end. Then close the stopcock again. Make sure there are no air bubbles in the syringe or in the tubing.

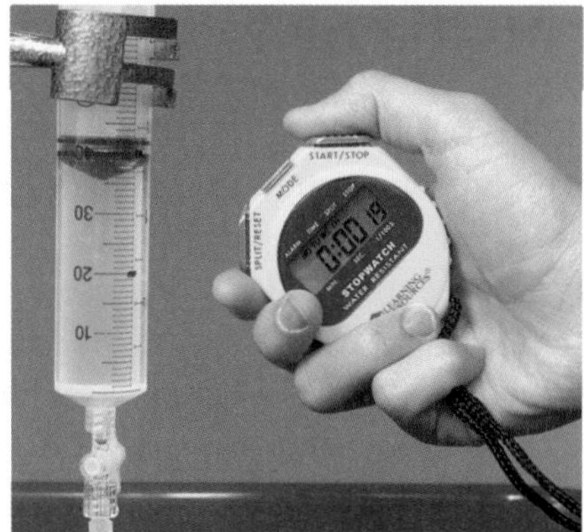

6 When the plunger reaches the first mark on the syringe, start the stopwatch.

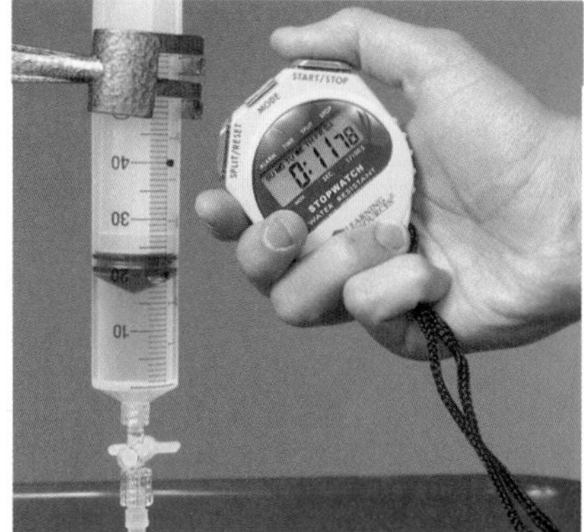

7 When the plunger reaches the second mark on the syringe, stop the stopwatch. Record the time it took for the water to flow through the tubing.

8 Repeat steps 3 to 7 for each of the other three lengths of tubing. Wipe up any spilled water after this investigation.

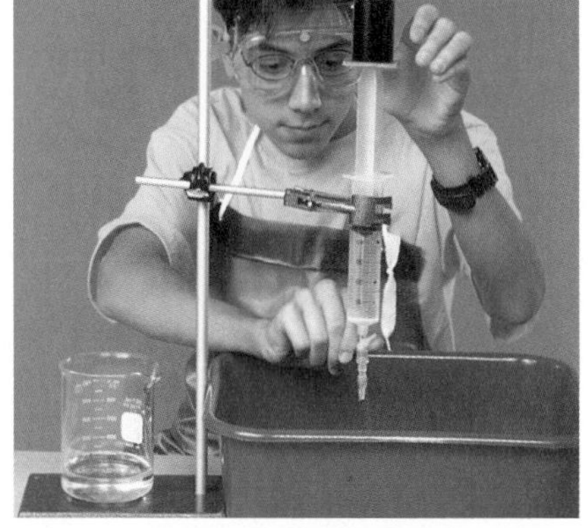

5 Place the 500 g mass on the wood platform on the plunger of the syringe. Open the stopcock. **CAUTION:** Place the mass carefully on the wood platform so that it does not fall off the work surface onto your foot.

Skill POWER

To review how to make a line graph, turn to page 546.

Analyze

1. Make a line graph. Put time in seconds on the horizontal axis (x-axis). Put the diameter of the tubing in centimetres on the vertical axis (y-axis).
2. Plot a point on the graph for each of the four lengths of tubing you used.
3. Connect the points with a smooth curve.

Conclude and Apply

4. Why was it important to use the same syringe, the same mass, the same distance of plunger movement, and the same length of tubing in each step of the investigation?
5. Examine your graph. Write a statement describing the effect of the diameter of a tube on the rate at which a fluid flows through the tube.
6. Imagine that you are designing some hydraulic equipment. How would your observations in this investigation influence your design?

Closed Systems

Your body's circulatory system is an example of a **closed system**, a self-contained collection of parts. Your heart pumps blood to all the cells of the body and back to the heart through a continuous network of blood vessels that are directly connected to one another.

Like your circulatory system, the human-made hydraulic systems that you will study in this chapter are examples of closed systems. As you learned in Chapter 6, hydraulic systems are mechanisms that use the force of a liquid in a confined space, such as an oil pipeline. Hydraulic systems apply two essential characteristics of fluids — their incompressibility and their ability to transmit pressure. Do the Find Out Activity below to see an important principle of fluid pressure in action.

Find Out ACTIVITY

How Far Up?

If water is forced out of holes of different sizes in a tube, predict which stream of water will rise the highest.

What You Need

plastic dishpan
50 mL modified syringe (the plunger of the syringe must slide freely)
plastic tubing punctured with four holes of different sizes
stopper or clamp to close off the tubing
water-resistant tape
water

What to Do

1. Insert the stopper securely into one end of the tubing (or clamp the tubing).
2. Tape the tubing to the bottom of the dishpan. Make sure the holes in the tubing point up.
3. Fill the syringe with water. Attach it to the open end of the tubing, as shown in the photograph.
4. Gently press the plunger of the syringe. Observe the streams of water coming through the holes.
5. Wipe up any spilled water after this activity.

What Did You Find Out?

1. Did the streams of water coming from the different-sized holes rise to nearly the same level or to different levels? Did your observations support your prediction?
2. Discuss the results of this activity with your group. Try to agree on an explanation for your observations.
3. Where might you see something like the apparatus in this activity in your daily life?

Skill POWER

For tips on working in groups, turn to page 536.

Pascal's Law

Have you ever squeezed a water-filled balloon? Did you notice how the walls of the balloon bulged out in all directions? Squeezing a water-filled balloon demonstrates Pascal's law. **Pascal's law** states that pressure exerted on a contained fluid is transmitted undiminished in all directions throughout the fluid and perpendicular to the walls of the container. French physician and scientist Blaise Pascal (1623–1662) first observed that the shape of a container has no effect on the pressure at any given depth (see Figure 14.3).

Many hydraulic and pneumatic systems make use of Pascal's law. Figure 14.4 shows one such system: the hydraulic lift. A **hydraulic lift** is a mechanical system that raises heavy objects, such as a vehicle on a service station lift. Just like your circulatory system, a hydraulic lift uses a fluid under pressure in a closed system.

As the illustration shows, a hydraulic lift consists of a small cylinder and a large cylinder. The cylinders are connected by a pipe. Each cylinder is filled with a hydraulic fluid, usually oil. (Water is not used in a hydraulic lift for two reasons — it is not a good lubricant, and it can cause parts of a system to rust.) Note that each cylinder also has a type of platform, or piston, that rests on the surface of the oil.

Pause& Reflect

Write answers to these review questions in your Science Log: What is the unit for pressure? What type of fluid is compressible and what type is incompressible? What happens when you exert a force on an incompressible fluid?

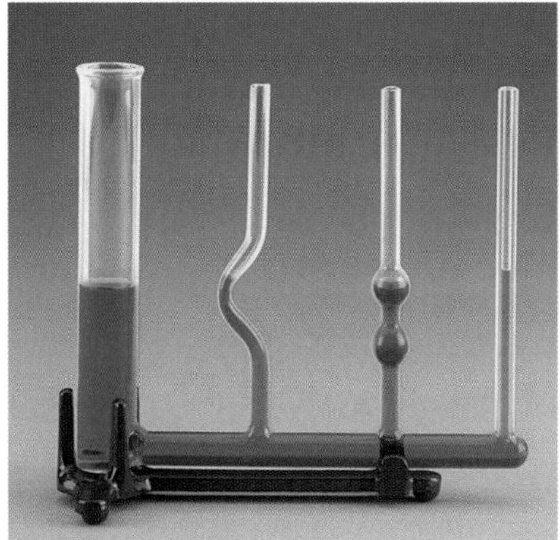

Figure 14.3 Pascal's vases show that a container's shape has no effect on pressure.

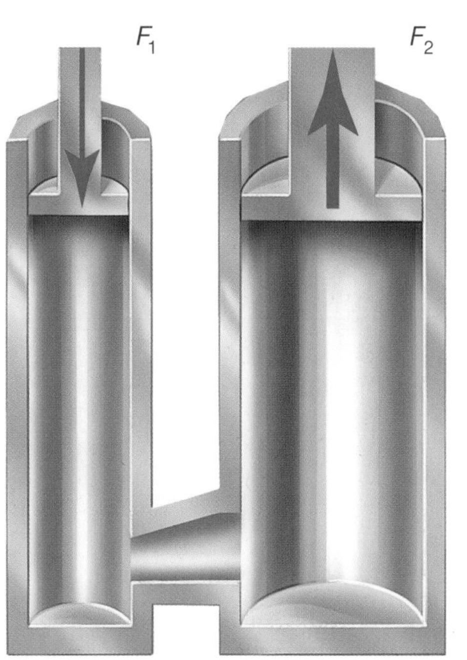

Figure 14.4 In this hydraulic lift, pressure applied to a small piston is transmitted to a large piston by means of a hydraulic fluid.

Every time you squeeze out some toothpaste, you are applying Pascal's law!

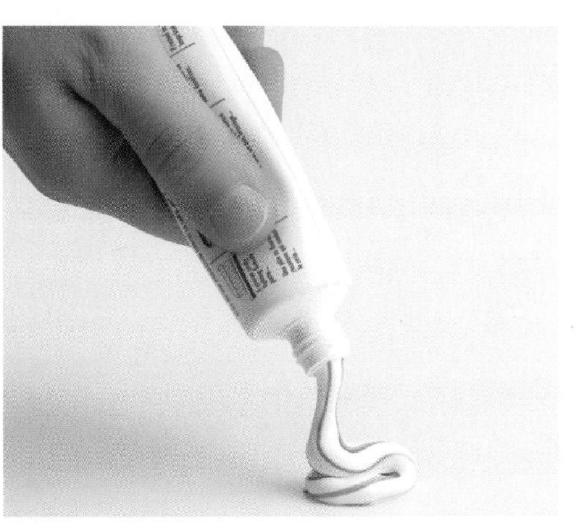

Suppose you apply 500 N of force to the small piston with an area of 5cm^2. The pressure on the small piston is expressed in the following equation. Recall that **pressure** (P) is **force** (F) divided by the **area** (A) over which the force is acting.

$$P = \frac{F}{A}$$
$$= \frac{500 \text{ N}}{5 \text{ cm}^2}$$
$$= 100 \text{ N/cm}^2$$

Pascal's law states that this pressure is transmitted unchanged throughout the liquid. Therefore, the large piston will also have a pressure of 100 N/cm^2 applied to it. However, the total area of the large piston is greater than the area of the small piston. The large piston's area is 50 cm^2. Thus, the total force on the large piston is 100 N/cm^2 × 50 cm^2 = 5000 N. This is *ten times* the force applied to the small piston. Using this hydraulic machine, you could use your own weight to lift something ten times as heavy as you are!

In Chapter 6, you learned that the **pascal** (Pa) is the standard unit of pressure. One pascal of pressure is a force of one newton per square metre. This is a small pressure unit, so most pressures are given in kilopascals (kPa).

To sum up, a hydraulic lift uses a liquid to produce a large force on a load when a small effort force is exerted on the liquid. The fact that a small effort force can produce a large force on a load means that a hydraulic lift can provide a mechanical advantage.

Pascal's Law and Mechanical Advantage

Study Figure 14.5. Note that the area of the small piston of the hydraulic lift is 1 unit. If you push down on the piston with a force of 10 N, you will increase the pressure in the fluid by 10 N per unit of area. Now examine the area of the large piston. There are nine squares. Each square has the same area as the small piston. According to Pascal's law, the pressure on every unit of area on that piston will have increased by 10 N. Since there are 9 units of area, the total increase in force on the large piston will be:

$$\frac{10 \text{ N}}{\text{unit area}} \times 9 \text{ unit areas} = 90 \text{ N}$$

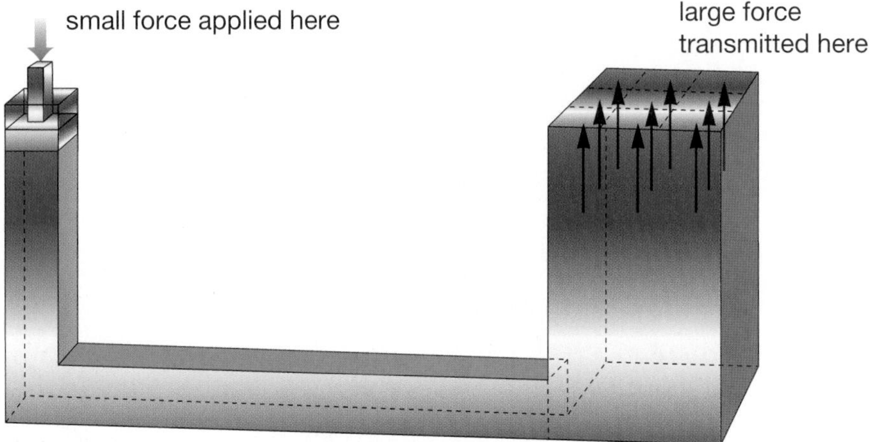

Figure 14.5 This simplified diagram of a hydraulic lift shows how a small effort force can produce a large force on a load.

By exerting 10 N of effort force, you could cause the large piston to exert 90 N of force on a load. Thus, the hydraulic lift provides a mechanical advantage. As you learned in Chapter 13, mechanical advantage is the load divided by the effort force:

$$MA = \frac{\text{Load}}{\text{Effort force}}$$

$$= \frac{90\text{ N}}{10\text{ N}}$$

$$= 9$$

This hydraulic lift has a mechanical advantage of 9. However, remember that you would have to push the piston nine times farther than the distance you could lift the load. You would have to increase the effort distance because the work done on the small piston must be at least as great as the work done on the load. (Recall that work equals force multiplied by distance.)

For example, suppose you wanted to lift a 90 N load a distance of 2 m using the hydraulic lift in Figure 14.5. Approximately how far would you have to push the piston as you exert your effort force? You could find out by doing the following calculations (note the formula for work):

$$W = F \bullet d$$

$$W(\text{effort}) = 10\text{ N} \bullet d\ (\text{effort})$$

$$W(\text{load}) = 90\text{ N} \bullet 2\text{ m}$$

$$W(\text{effort}) = W(\text{load})$$

Therefore,

$$10\text{ N} \bullet d\ (\text{effort}) = 180\text{ J}$$

$$d\ (\text{effort}) = \frac{180\text{ J}}{10\text{ N}}$$

$$= 18\text{ m}$$

Pause& Reflect

How does Pascal's law explain your observations in the Find Out Activity on page 460? Write a response to this question in your Science Log.

To lift a 90 N load a distance of 2 m using the hydraulic lift, you would have to push the piston 18 m. This effort distance is nine times the load distance.

INVESTIGATION 14-B

What a Lift!

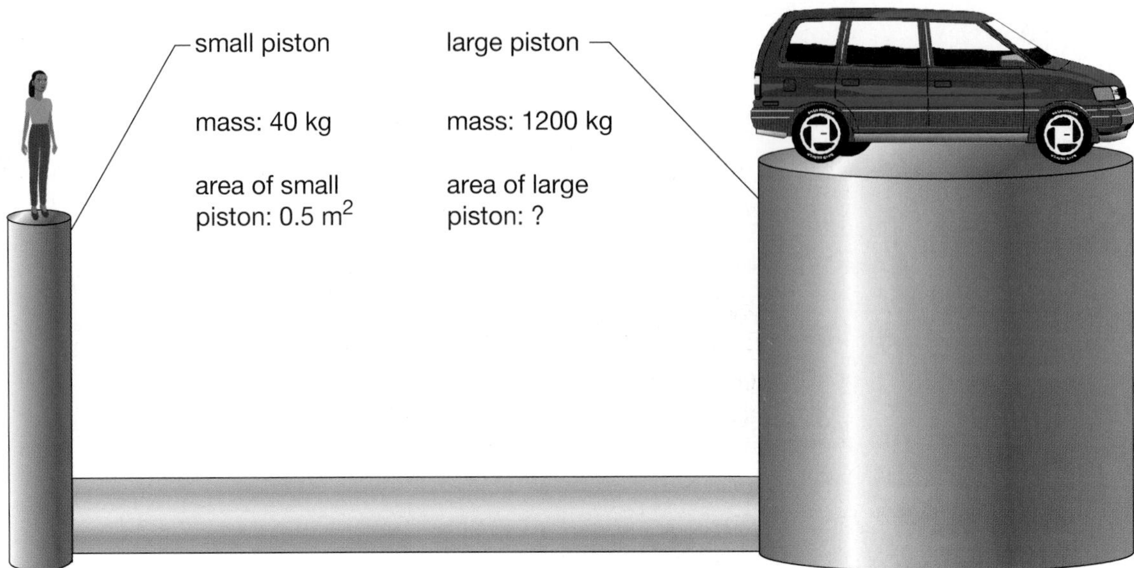

Think About It

In a hydraulic lift, how large a piston would you need to lift a minivan? Imagine you are standing on one piston of a hydraulic lift and a minivan is on the other piston. The area of your piston is 0.5 m^2. Suppose you have a mass of 40 kg and the minivan has a mass of 1200 kg. How large must the other piston be to lift the minivan?

Recall that mass is measured in grams (g) and kilograms (kg). Weight, which is a force, is measured in newtons (N). A kilogram of mass on Earth's surface weighs 9.8 N.

What to Do

1. Estimate the size of the large piston in the hydraulic lift. Think of an area that is about the same size as the large piston. What is the area of your kitchen table? Of your bedroom floor? Of your living room? Of your classroom? Which area do you estimate is closest to the area of the large piston in the hydraulic lift?
2. Study the diagram, then calculate the area of the large piston. (Hint: Remember that the ratio of the two masses is the same as the ratio of the two areas of the pistons.)

Analyze

1. What result did you get when you calculated the area of the piston supporting the minivan? How close was your estimate to this result?
2. Do you think that this design for a hydraulic lift is practical? Explain your answer.

Skill POWER

To review estimating, turn to page 540.

DESIGN & DO

INVESTIGATION 14-C

Build Your Own Hydraulic Lift

Now that you understand how a hydraulic lift operates, design and build your own working model of one.

Challenge

Design and build a model of a hydraulic lift that will exert a large force on a load when you exert a small force on the lift.

Materials

10 mL modified syringe
50 mL modified syringe (the plungers of both syringes must slide freely)
narrow plastic tubing
1 kg mass
250 g mass
variety of smaller masses
2 wood squares (5 cm x 5 cm)
2 support stands
4 stopcocks or clamps
glue gun
water

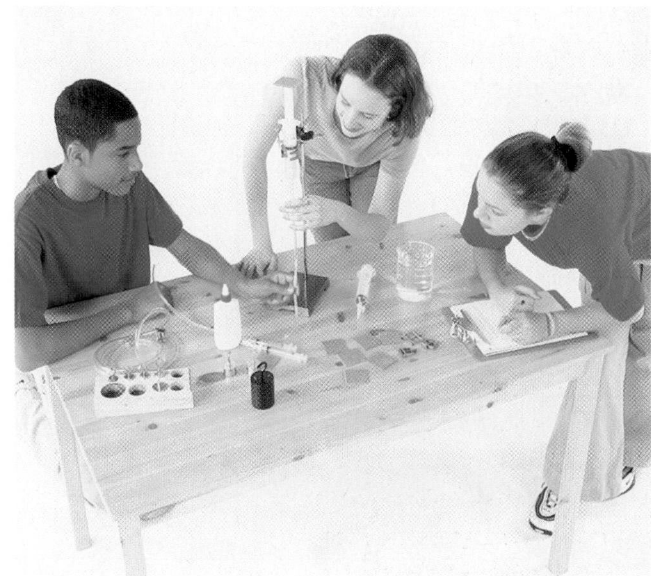

Design Criteria

A. Your model hydraulic lift must exert a force in one place when you exert a force in a different place.

B. There must be no air bubbles in the tubing and in the syringes.

C. Your model hydraulic lift must provide an observable mechanical advantage.

Plan and Construct

1. With your group, predict what arrangement might allow you to balance the 250 g and 1 kg masses on the two modified syringes. Test your prediction. **CAUTION:** Place the masses carefully on the wood platform each time, so that they do not fall off the work surface onto your foot. Also, the glue gun is hot and the glue remains hot for several minutes.
2. Predict what arrangement might allow you to raise the 1 kg mass using the least amount of force. Test this prediction as well.
3. Wipe up any spilled water after this investigation.

Evaluate

1. Did your model hydraulic lift produce a mechanical advantage? How could you tell?
2. Suppose you need to raise a 1 kg mass using an even smaller force. How could you modify your model hydraulic lift to achieve this?

Extend Your Skills

1. How could you calculate the work done by your effort force?
2. How could you find out the pressure exerted by the water in the modified syringes?

Hydraulics at Work

Have you ever seen a bulldozer clearing an area to build new homes? You may have seen a backhoe digging a trench for a new water line or a sewer pipe. Have you ever watched a "cherrypicker" in action? Perhaps you've seen a farmer driving a tractor in a field. In all these cases, you were watching hydraulic equipment at work.

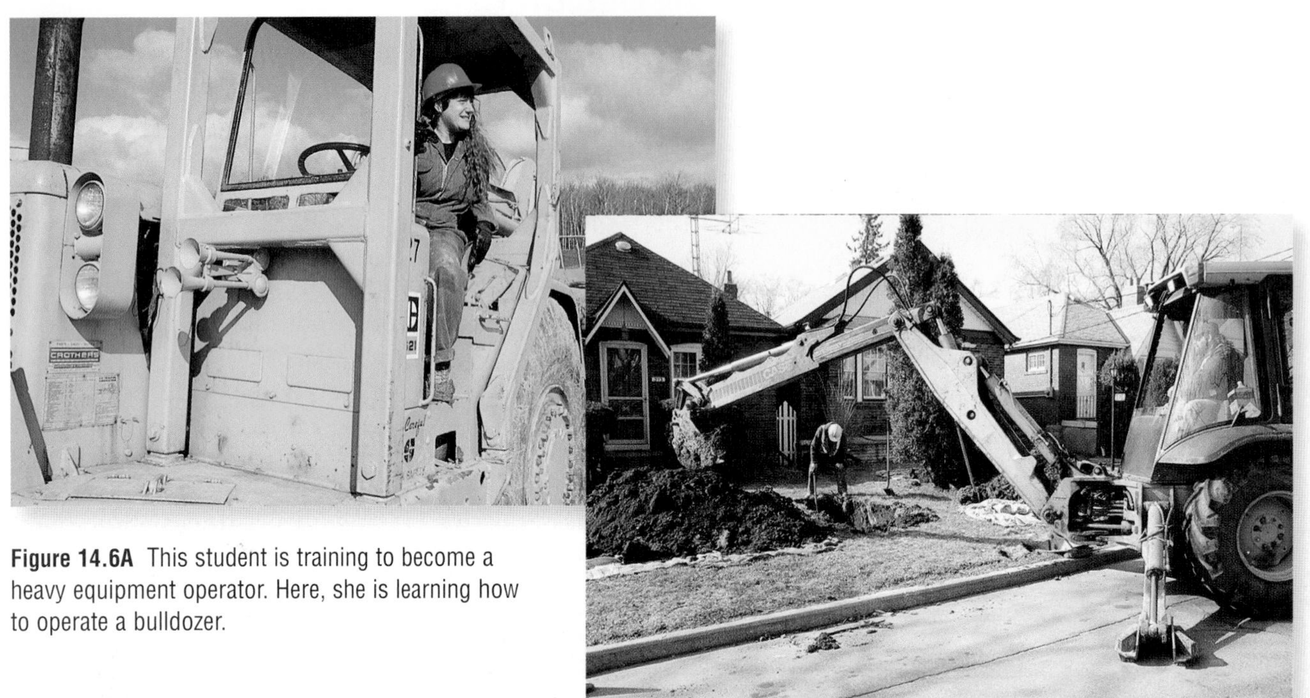

Figure 14.6A This student is training to become a heavy equipment operator. Here, she is learning how to operate a bulldozer.

Figure 14.6B This backhoe is digging up a lawn to install a gas pipe. (For a diagram showing how a backhoe works, turn to page 479.)

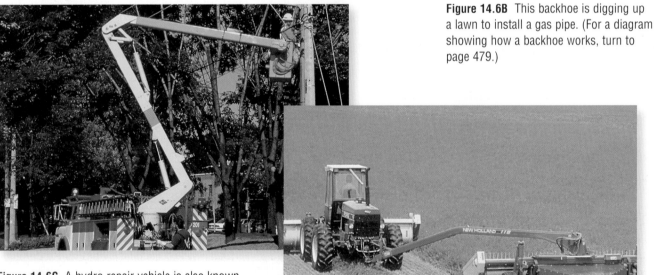

Figure 14.6C A hydro repair vehicle is also known as a "cherrypicker."

Figure 14.6D Why do you think this farm equipment is called a "bi-directional tractor"?

Unlike the simple hydraulic systems you have explored so far, the huge machines shown in Figure 14.6 are not operated by plungers that workers push manually! Instead, the machines contain tanks filled with hydraulic fluid and pumps that generate pressure. In most hydraulic equipment, the energy for pumping is supplied by a gasoline engine or by an electric motor. Valves direct the high-pressure fluid through steel pipes to the parts of the machine where the pressure of the fluid is needed to generate large forces to lift or to dig. Often the steering and braking systems in large machines are powered by the high-pressure hydraulic fluid as well.

INTERNET CONNECT

www.school.mcgrawhill.ca/resources/

As a class project, start an Internet search for as many types of hydraulic equipment as you can find. Visit the above web site. Go to **Science Resources**, then to **SCIENCEPOWER 8** to know where to go next. Decide how you want to keep track of all the machines and instruments that you find. You could create a bulletin board display, a poster, or a trade magazine entitled *Popular Hydraulics.* (Use a library if you do not have access to the Internet.)

Skill POWER

For tips on doing research using a library or the Internet, turn to page 559.

Career CONNECT

Heavy-Duty Work

Merritt Shilling is a heavy equipment operator. He knows how to operate everything from a huge excavator to a small bulldozer safely and efficiently. These specialized vehicles often have a separate control for each hand and as many as three foot pedals. Learning how to control several mechanical systems at once takes special training. Merritt got his training on the job and temporarily moved to a larger community for the experience he wanted. His career path looks something like this:

- completed high school
- cleared snow using a truck with an attached plow
- worked for a landscaper driving tractors and bulldozers
- worked for a large metal company driving huge forklifts and other equipment
- returned to his Native community and now works for local contractors driving whatever equipment each job requires

To become a heavy equipment operator, you could attend a privately-run heavy equipment school. You could also complete a college course in truck driving or in heavy equipment mechanics, or work with an experienced operator as an apprentice.

Locate someone in your community who works as a heavy equipment operator, or contact a spokesperson for your province's workers' compensation board (check the provincial government pages in the telephone book). Ask about the safety issues related to this type of work. What safety gear must be worn on the job? For example, is ear protection necessary? What are the most common risk situations that could arise, and how can these risks be reduced? How can an operator prevent accidents from occurring? Present answers to these or other questions in a brief oral or written in-class report.

Hydraulics in Flight

When an airplane such as the one shown in Figure 14.7A taxis along a runway, the pilot steers the plane by means of the nose wheel. During takeoff, the pilot may lower the flaps. To make a turn while airborne, the pilot moves the ailerons up or down and adjusts the rudder. To keep the plane level, the pilot adjusts the elevators. Landing a plane is a multi-system process. To slow these giant aircraft when they touch down, the pilot lowers the flaps and the slats and raises the spoilers, while at the same time applying the brakes.

Figure 14.7A These aircraft have three separate hydraulic systems, as well as an emergency backup system.

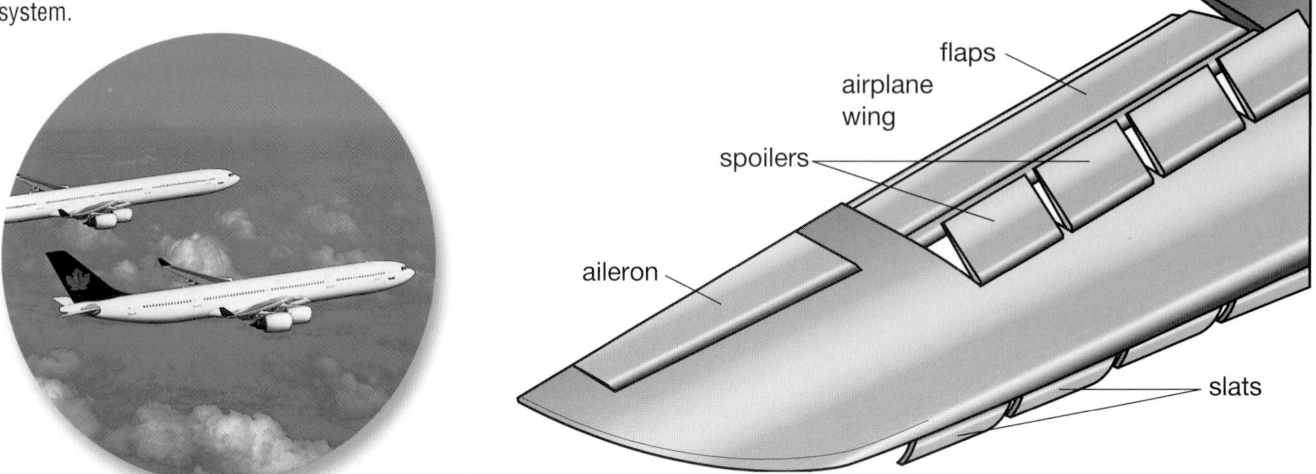

Figure 14.7B The various parts of an airplane wing are raised and lowered hydraulically when the pilot lands the plane.

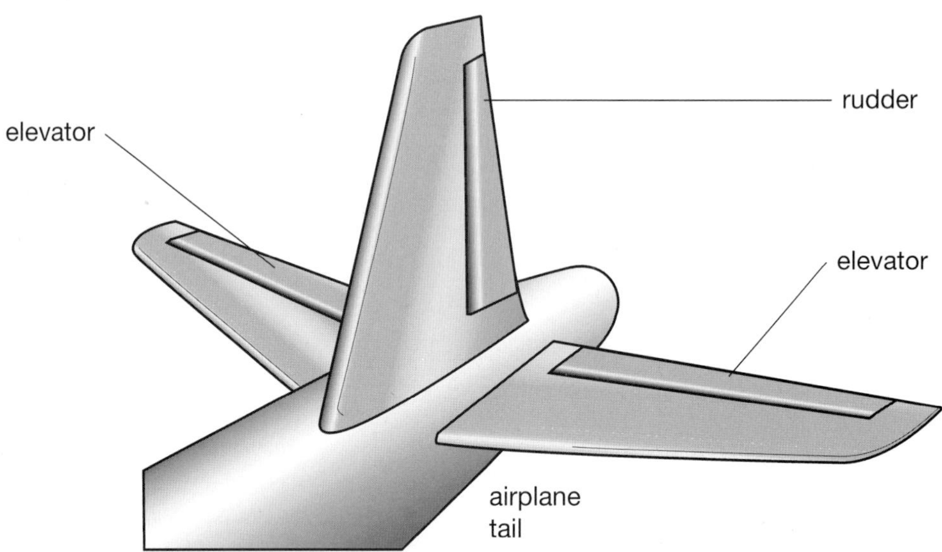

Figure 14.7C Hydraulics are responsible for tail adjustments that enable the pilot to turn the plane while airborne.

Every mechanical system mentioned in the paragraph above is powered by hydraulics (see Figures 14.7B and C). The precise designs of the hydraulic systems are different for different models of aircraft, but the basic principles are the same.

Each airplane in Figure 14.7A is an Airbus A340. It has three completely separate hydraulic systems. These are called the green system, the blue system, and the yellow system. If one system malfunctions, there are one or two other systems to back it up. For example, the green and yellow systems both control the flaps. The green and blue systems control the slats. All three systems control the ailerons.

The green system relies on fluid pressure generated by engines 1 and 4. The blue system relies on fluid pressure generated by engine 2, and the yellow system relies on pressure from engine 3. If an engine fails, additional backup motors provide pressure for the hydraulic systems. What happens if all the systems fail? Could the pilot still control the guidance systems? The answer is yes. An emergency air-driven generator drops out from a door in the bottom of the plane. As shown in Figure 14.8, this generator resembles a fan. It has a propeller that spins when outside air strikes it. Since jets travel at tremendous speeds, the air turns the propeller extremely rapidly. This rapid turning motion generates alternative power to supply the hydraulic systems.

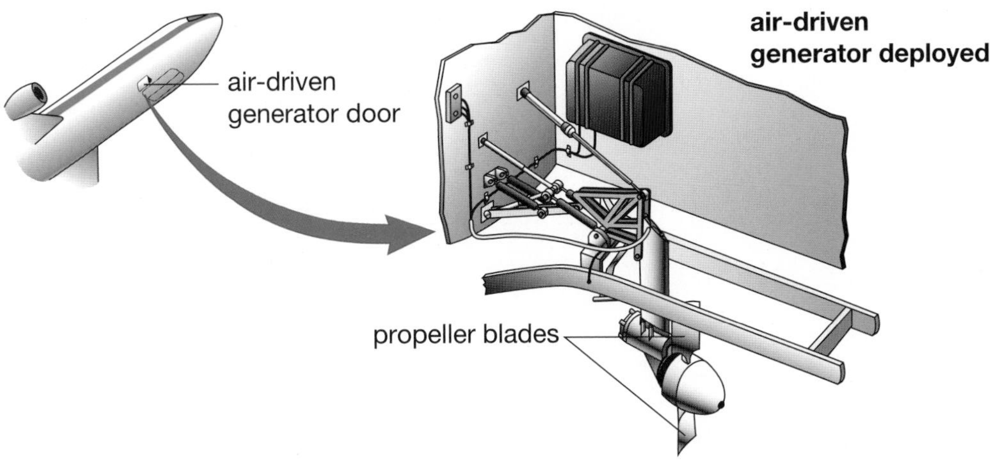

Figure 14.8 If hydraulic pressure in an Airbus A340 fails, an air-driven generator drops down from a door in the bottom of the plane. The air rushing past turns the blades of the generator, which produces both electricity and fluid pressure.

Check Your Understanding

1. Explain the difference between force and pressure.
2. Pressure is measured in pascals (Pa). What combination of units is the same as a pascal?
3. When you exert force on a fluid in a closed container, does the pressure increase, decrease, or remain constant?
4. State Pascal's law.
5. List four instruments or machines that use hydraulics.
6. In a drawing, show how to set up a model hydraulic system with a mechanical advantage of 4.
7. **Design Your Own** You discovered that a tube with a small diameter experiences more friction (slowing down a fluid's flow rate) than a tube with a larger diameter. What other characteristics of the tubing might increase or decrease flow rate? Make a hypothesis related to one of these characteristics and design an experiment to test your hypothesis. Identify which variable(s) will change in your experiment and which variable will remain constant. List criteria for assessing your solution.

14.2 Pressurized Air

Pause& Reflect

Based on what you learned in Chapter 6, record in your Science Log examples of machines and instruments that use pneumatic pressure. If you did not completely understand how these devices work, write questions in your Science Log. Look for answers as you learn more about pneumatics.

Moving air such as a summer breeze can have a pleasant effect (see Figure 14.9A). However, gases such as air sometimes exert extremely high pressures that can cause a great deal of damage, as shown in Figure 14.9B. You have seen news coverage of the wreckage caused by hurricanes and tornadoes. The tremendous energy of pressurized air can also be harnessed to perform important tasks. In this section, you will learn about some pneumatic systems that use pressurized air to do work.

Figure 14.9A On a sunny day, a gentle breeze feels refreshing.

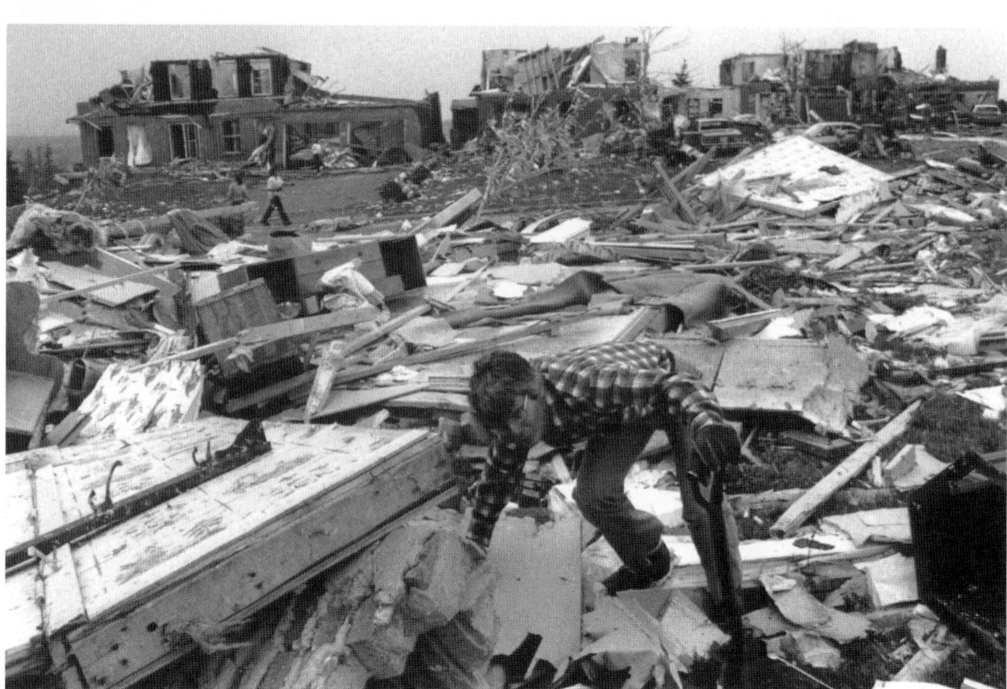

Figure 14.9B Air moving at great speeds can create pressures that demolish buildings and claim lives. This photograph shows the aftermath of a tornado that struck Barrie, Ontario, in 1985.

Pneumatics at Work

Pneumatic devices do not seal the gas — usually air — in a mechanical system in the same way that hydraulic machines seal in hydraulic fluid. Usually, the air passes through the pneumatic device under high pressure and then escapes outside the device. The high-pressure air may come from a machine that draws in outside air and compresses it. Hoses then carry the high-pressure air to the pneumatic device.

A common example of a pneumatic device is the jackhammer (see Figures 14.10A and B). You have probably heard the extremely loud noise of a jackhammer breaking up concrete when a sidewalk or a road is being repaired. Jackhammers are also used in the mining of coal, nickel, and gold. Bursts of air, under very high pressure, drive a part called a "chuck" in and out of the jackhammer at high speeds. Resembling a very large screwdriver, the chuck pounds the rocks or concrete into fragments.

DidYouKnow?

Using air pressure of about 620 000 Pa, more than 5.5 m^3 of air flow through a jackhammer every minute.

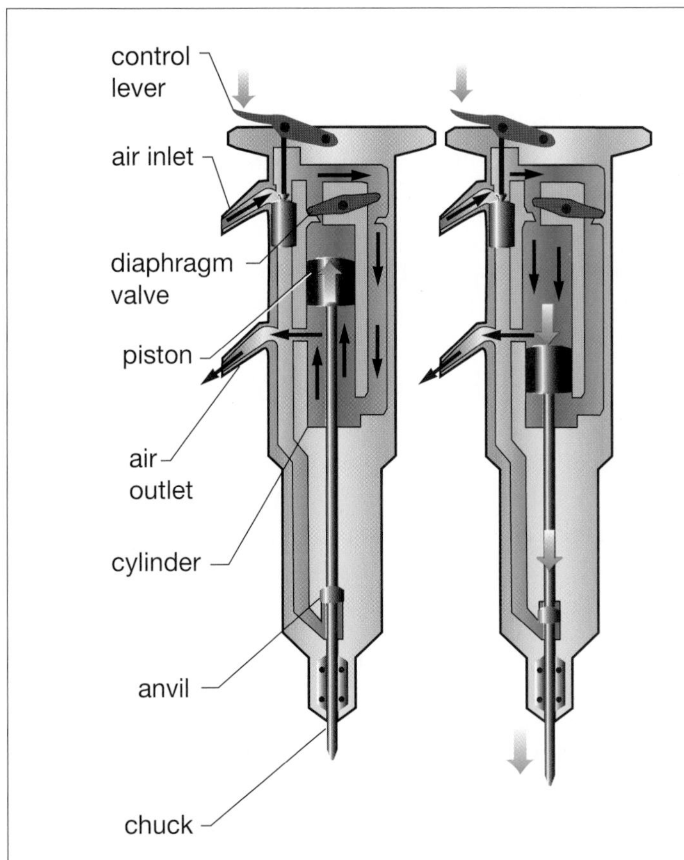

Figure 14.10A Cross-section of a jackhammer

Figure 14.10B Every time you hear the ear-splitting sound produced by a jackhammer, you are hearing compressed air at work.

Figure 14.11 Air pressure makes this staple gun work.

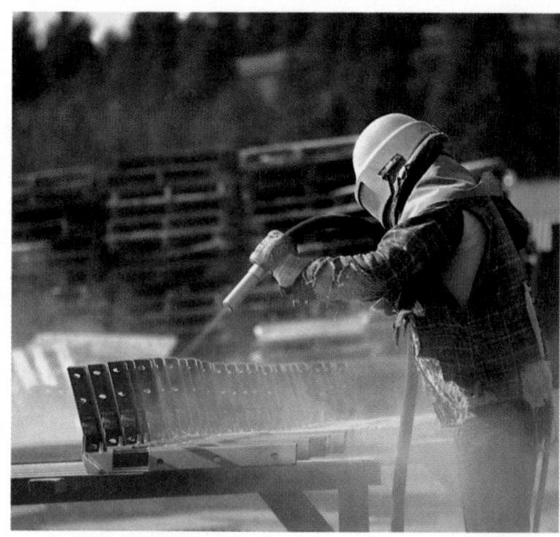

Figure 14.12 This worker is using a sandblaster to clean steel brackets.

Staple guns and pneumatic nailers use pulses of air pressure to drive staples or nails into solid objects. Staple guns are used in making furniture, woodworking, upholstering, and many other applications. Pneumatic nailers can even nail wood to concrete. A staple gun is shown in Figure 14.11.

Sandblasters do exactly what the name implies. High-pressure air blasts tiny sand particles out of a nozzle (see Figure 14.12). Sandblasting is an excellent way to remove dirt and paint from stone or brick. Old, dirty buildings or statues can be made to look new, as shown in Figures 14.13A and B. Can you imagine sanding a large stone building with sandpaper?

Besides improving appearances, sandblasting is also used for practical applications. For example, slippery granite or marble stairs can be made safer by being sandblasted. Sandblasting roughens the edges of the stairs to increase friction. The friction, in turn, prevents people from slipping on a step.

Figure 14.13A Before sandblasting

Figure 14.13B After sandblasting

Pause& Reflect

In your Science Log, describe how some modern sports shoes are similar to an air cast.

INTERNET CONNECT

www.school.mcgrawhill.ca/resources/

Continue to search the Internet for information about mechanical systems, but start adding pneumatic devices to your list. Visit the above web site. Go to **Science Resources**, then to **SCIENCEPOWER 8** to know where to go next. Each class member can look for at least one type of pneumatic equipment (other than the ones presented in this text) and present an oral or written report to the class. See what unusual devices you can find!

Figure 14.14 This "air cast" is used for both sprains and fractures.

The photograph above shows another application of pneumatics. Medical engineers have developed a type of cast filled with pressurized air. A solid frame with a balloonlike lining is fitted to the injured leg. Low-pressure air is pumped into the lining through a hose. Because the air pressure can be controlled precisely, the cast can be made to fit snugly and securely.

How do you think pneumatic pressure and hydraulic pressure compare? Find out in the following investigation.

The high-speed drill that dentists currently use is a pneumatic instrument that relies on pressurized air. The cutaway diagram on the right shows how a dentist's drill works. This technology has led to almost pain-free dentistry.

What does the future hold? A newly invented machine that drills teeth with a high-powered jet of water will make life easier for dentists — and for patients. The device, called the Millennium, works by pumping a jet of water at the teeth. The droplets of water are split by a laser into tiny particles. As these hit the enamel, they exert enough force to grind the tooth. This technology means there is no noise from a drill and no heat.

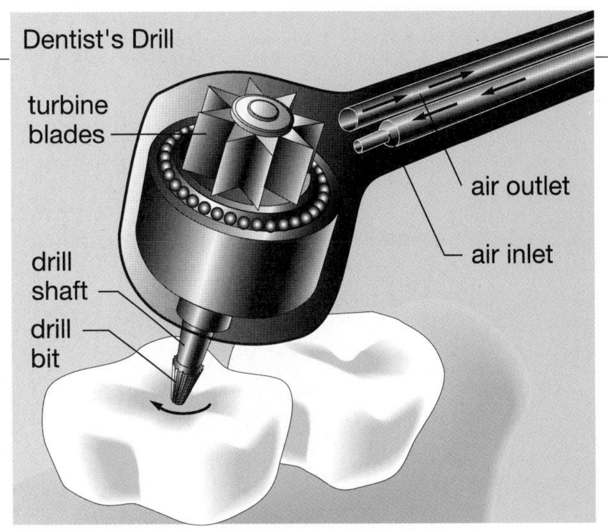

CONDUCT AN

INVESTIGATION 14-D

Comparing Pressure Exerted on a Gas and on a Liquid

What happens when you exert the same amount of pressure on a gas and on a liquid? Will the results be the same or different?

Safety Precautions

Apparatus

2 modified syringes (the plungers of both syringes must slide freely)

2 wood squares (5 cm x 5 cm)

2 masses (500 g each)

stopcocks

plastic dishpan

stopwatch or watch with a second hand

support stand

felt tip pen

glue gun

Materials

water

Procedure 

Note: Your teacher will glue the wood squares to the plungers ahead of time so the glue will have time to dry.

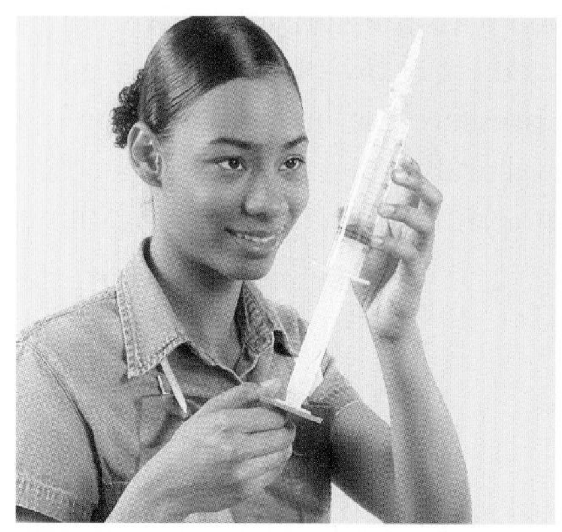

1. Fill one syringe with water. Turn it upside down over the dishpan and press the plunger until all of the air is gone.

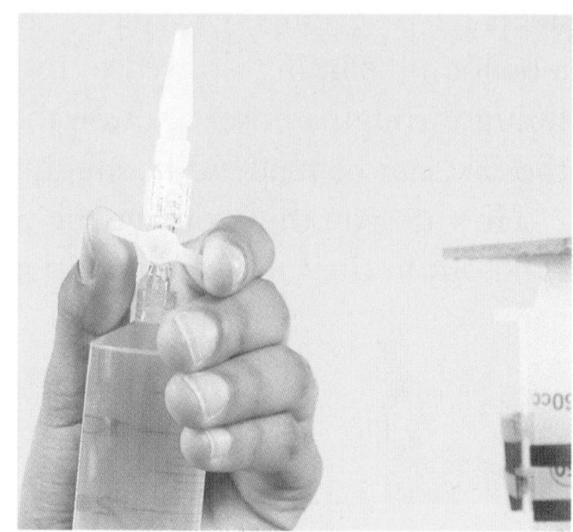

2. Close the stopcock.

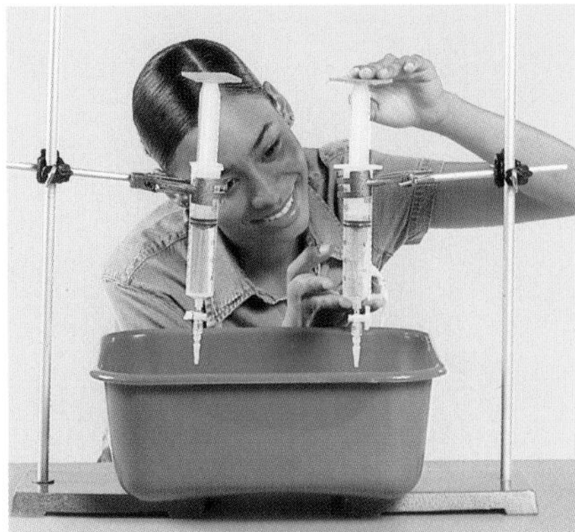

3 Assemble the apparatus as shown. Leave air in the second syringe. Adjust the plunger so that it is at the same position as the plunger in the water-filled syringe.

(a) Close the stopcock, making sure that all the connections are airtight.

(b) Wipe up any spilled water.

4 Observe the positions of the two plungers. Mark the positions of the plungers with a felt tip pen.

5 Place a 500 g mass on top of each syringe. Observe the new positions of the plungers. **CAUTION:** Place the masses carefully on the wood platforms so that they do not fall off the work surface onto your foot.

6 Before you open the stopcocks, make a prediction. Will the time it takes for each of the plungers to reach the bottom of the syringes be the same or different? Open the stopcocks one at a time and record the time it takes for each of the plungers to reach the bottom of the syringe. Wipe up any spilled water after this investigation.

Analyze

1. What happened when you put the mass on top of the water-filled syringe?
2. Was the result the same or different when you put the mass on the air-filled syringe?
3. Did one modified syringe empty faster than the other when you opened the stopcock? If so, which one emptied faster?

Conclude and Apply

4. What property of liquids did you demonstrate in this activity? What property of gases did you demonstrate? Use the term "viscosity," which you learned in Chapter 4, to explain your observation in question 3.
5. Write a statement that summarizes and compares how gases and liquids respond to pressure.

Riding on Air

Figure 14.15 shows a Canadian Coast Guard hovercraft used primarily for rescue operations. Not only can hovercraft float on air, but they can also transport people, cars, and equipment long distances over land or water.

Figure 14.15 Hovercraft are used not only for rescue operations but also for routine travel.

In a hovercraft, powerful pumps draw in outside air and pump it out through holes in the bottom of the hovercraft (see Figure 14.16). A "skirt" around the bottom holds in enough air to support the weight of the craft above water or land. Given enough air pressure, a hovercraft can support extremely heavy loads. Propellers drive the hovercraft forward, and rudders are used to steer it.

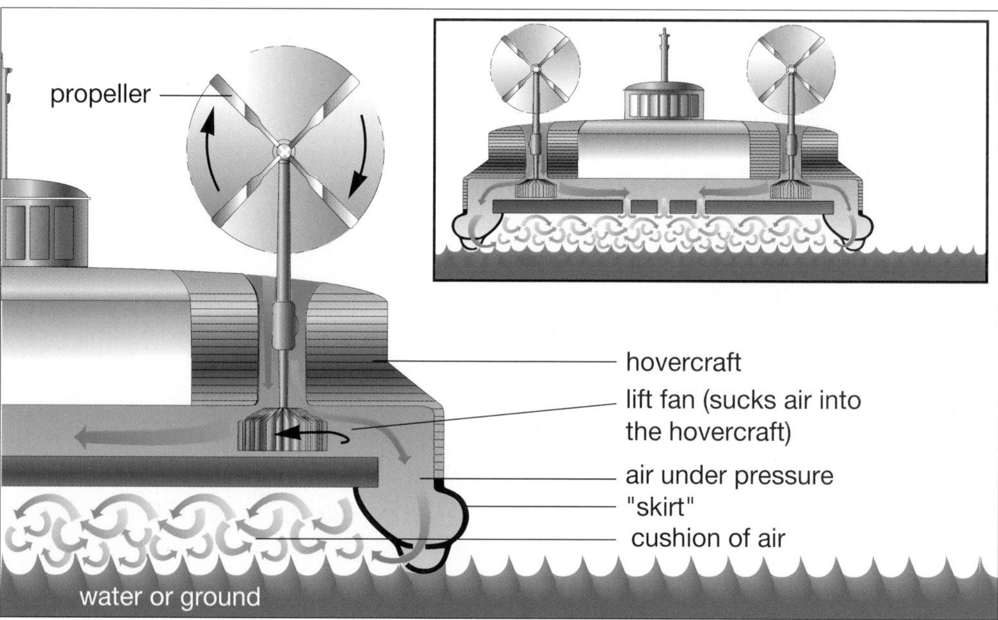

Figure 14.16 A hovercraft floats on a cushion of air.

Find Out **ACTIVITY**

Build a Model Hovercraft

See if you can send a miniature hovercraft skimming across a table.

What You Need

cardboard
empty thread spool
paper
scissors
pencil
glue gun
balloon

What to Do

1. Cut out a 10 cm square from the cardboard and use a pencil to punch a hole in the centre of the square. The hole should be the same size as the hole in the empty thread spool. **CAUTION:** Be careful when using sharp objects such as scissors and when punching the cardboard with a pencil. Also, the glue gun is hot and the glue remains hot for several minutes.
2. Glue the empty spool on top of the hole in the cardboard so that the holes line up.
3. Using the glue gun, seal the base of the thread spool so that no air can escape.
4. Cut out a circle of paper and glue it onto the top of the thread spool.
5. Using a pencil, punch a hole in the middle of the paper circle to line up with the hole of the spool.
6. Blow up the balloon and twist the neck. Stretch the mouth of the balloon over the top of the spool. Let the balloon go and give your hovercraft a nudge.

What Did You Find Out?

What are some ways that you could change your hovercraft design to make it go farther?

Check Your Understanding

1. Contrast the responses of gases and liquids to pressure.
2. Describe one important difference between the use of gases in pneumatic systems and the use of liquids in hydraulic systems.
3. Give three examples of pneumatic devices.
4. Describe how a hovercraft floats on air. You may use a drawing to illustrate your answer.
5. **Design Your Own** Formulate your own question about some aspect of pneumatic systems and design a controlled experiment to explore possible answers.

14.3 Combining Systems

In your library or Internet search for hydraulic and pneumatic machines and instruments, you have probably discovered that most machines use more than one system. In other words, most mechanical systems are made up of **subsystems**. For example, wheels and axles or levers might be used in combination with hydraulic or pneumatic systems. One common example of combined systems and subsystems is found in the brakes of a car.

The braking system shown in Figure 14.17 is known as "disc brakes." The brake pedal subsystem is a Class 2 lever. The force of the driver's foot on the brake pedal is the effort force. The load is the force on a piston that applies pressure on the brake fluid in the master cylinder. As the driver pushes down harder on the brake pedal, the effort force increases the pressure transmitted in the brake fluid. From the master cylinder, brake fluid flows through tubes that branch out to every wheel. The illustration shows the final action at each wheel. The brake fluid exerts pressure on brake pads that press on a disc. The friction between the brake pads and the disc slows and eventually stops the car.

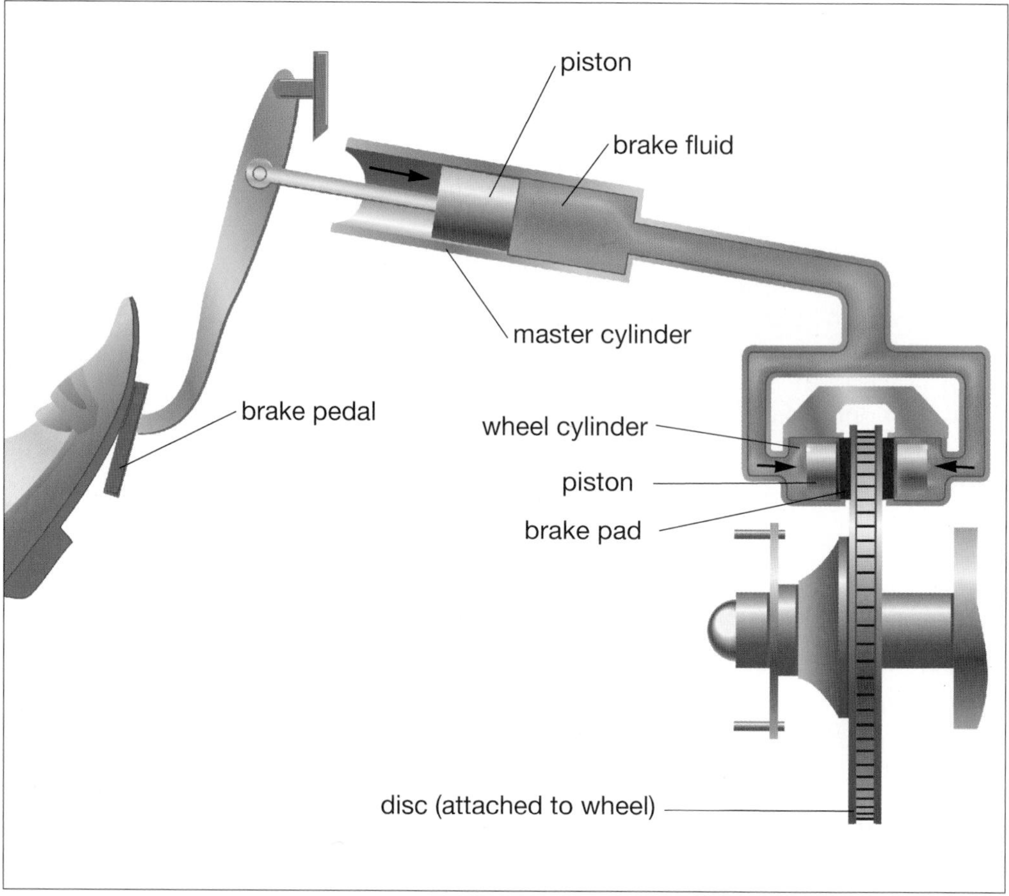

Figure 14.17 The pressure of the driver's foot on a brake pedal is transmitted by fluid pressure to the wheels of the car.

Figure 14.18 shows another example of a highly efficient combination of levers and hydraulics, the backhoe. Earlier in this chapter, you saw a photograph of a backhoe digging up a lawn. Also known as an excavator, a backhoe is a rotating combination of three levers. These three levers are called the boom, the dipper, and the bucket. As the diagram shows, this rotating assembly of levers is mounted on caterpillar tracks. The assembly swings around on a gear-like part called a slew ring. Powered by hydraulics, the three levers combine to place the bucket in any position. The boom is a Class 3 lever that raises or lowers the dipper. The dipper is a Class 1 lever that moves the bucket in and out. The bucket itself is a Class 1 lever that tilts to dig a hole and then empty its load of dirt or other material.

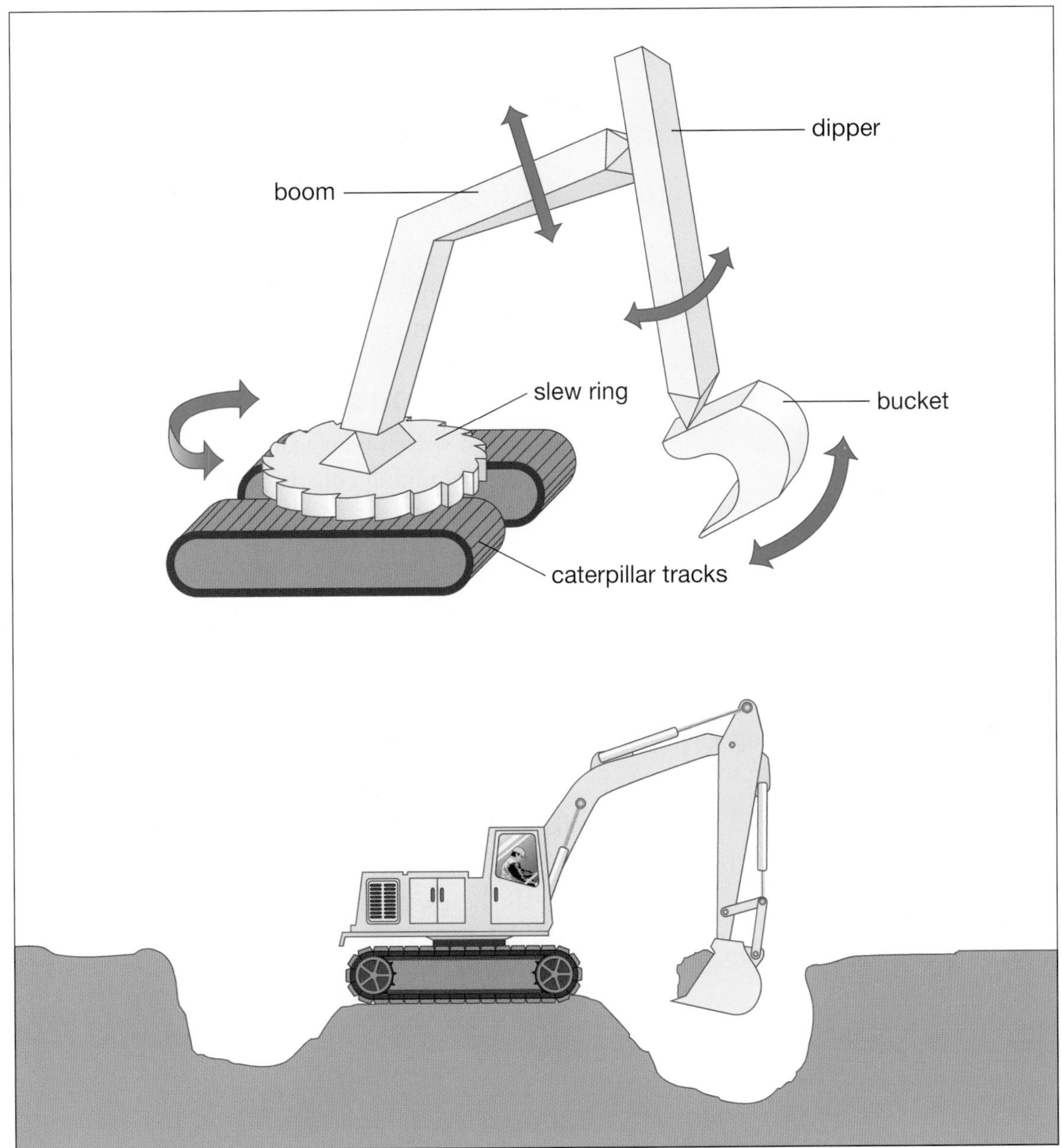

Figure 14.18 A backhoe is a rotating assembly of three levers combined with a hydraulic system.

INVESTIGATION 14-E

New, Improved Robots Required!

You are an engineer for a company specializing in the research and design field of robotics. Your company has been approached to design new and improved robots to handle hazardous wastes.

Challenge

Design and build a robotic arm that can transport hazardous waste in containers to a loading area.

Materials

jinx wood (1 cm × 1 cm)

dowelling (3 different diameters)

plywood platform (12 cm × 15 cm)

assorted wood screws, nuts and bolts, handles, gears, pulleys, winches, wheels, tubes, modified syringes, glue gun, saws, mitre box, small tools (e.g., a hand drill)

Safety Precautions

- A glue gun is hot and the glue remains hot for several minutes.
- Be careful when using tools such as saws and hand drills.

Design Criteria

A. Your robotic arm must be able to pick up a container and move it a minimum of 10 cm to the drop-off location.

B. The movement of the robotic arm described in A above must be completed in 1 min.

C. The robotic arm must be able to move up and down as well as side to side.

D. The robotic arm must have an operational jaw mechanism.

E. Three different mechanisms must be combined in the working prototype (model). These mechanisms must be chosen from the following list: gears, pulleys, cranks, wheels, hydraulics, and pneumatics.

F. Students may not touch the mechanism or the load directly at any time during the pickup, transport, and unloading.

Plan and Construct

1. Plan and sketch your team's solution on paper before beginning construction, and show it to your teacher.
2. How will the robotic arm manoeuvre and stop?
3. How will the simulated hazardous waste be picked up, transported, and unloaded?
4. Does the mechanism balance with and without the load?

Evaluate

1. How well did your team co-operate in arriving at the best solution using the design criteria as a guide?
2. How did preliminary sketches, planning, and experimentation contribute to a successful design?
3. Did your team make efficient use of materials and time, and did you follow safe and tidy work practices?
4. How well did your prototype demonstrate good design principles?

Life-Saving Fluid Pressure

Figure 14.19 shows a hydraulic walkway used in rescue operations. In Chapter 6, you learned about a rescue device known as the Jaws of Life. Recall that the spreaders and cutters in the Jaws of Life are operated by hydraulic pressure. Other types of rescue equipment, such as the cutting gun shown in Figure 14.20, are powered by pneumatic pressure. This cutting gun has four different-sized chisels that can be used interchangeably. These chisels can cut through deadbolt locks, hinges, doorposts, or even sheet metal buildings in just a few seconds to free someone trapped inside.

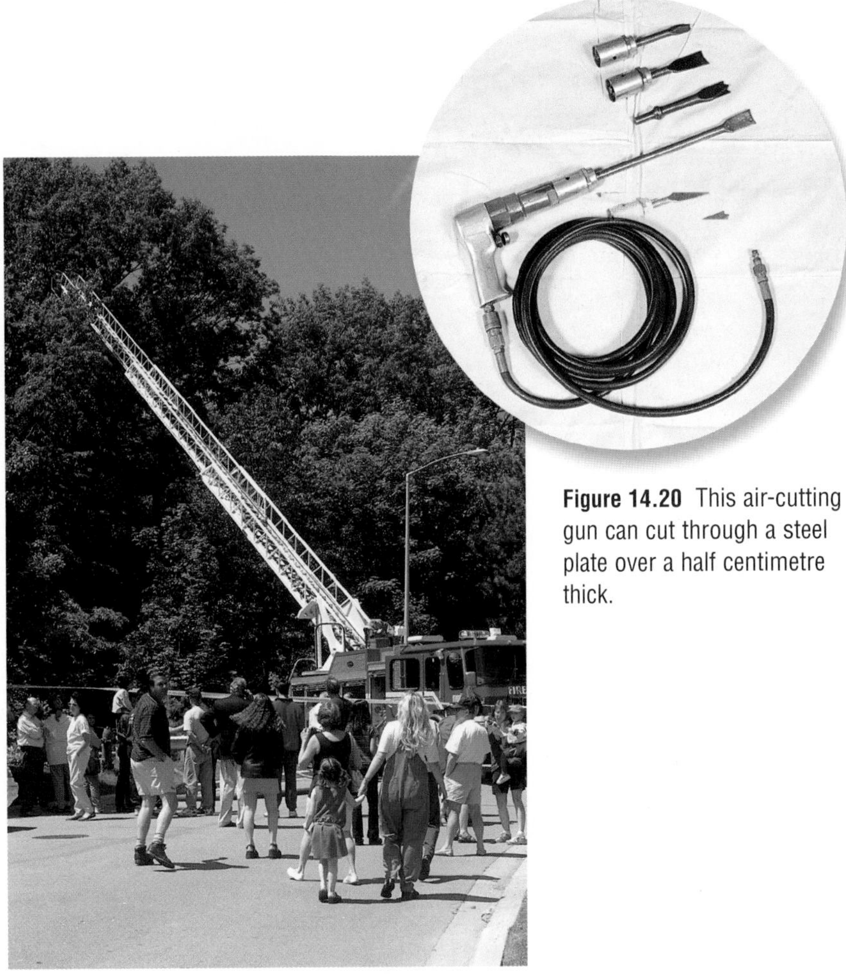

Figure 14.20 This air-cutting gun can cut through a steel plate over a half centimetre thick.

Figure 14.19 This hydraulic walkway was extended up into the trees to rescue a 69-year-old woman who had crashed her light aircraft in a wooded area in Don Mills, Ontario.

Across Canada

Inna Sharf

"My work involves simulation, modelling, and control of robotic systems for space applications, such as the Canadarm," explains Inna Sharf, professor of mechanical engineering at the University of Victoria in British Columbia. "The conditions in space are very different from those on Earth. With no gravity and a great deal of friction in a space environment, it is important to simulate the conditions of space when testing new machines and specialized equipment. Anyone who is developing new space technology depends on simulation tools to test whether a particular technological device will work effectively in space.

"I develop computer models to predict the motion of robotic systems in a space environment," Dr. Sharf adds. "We develop methods to ensure more precise control and manipulation of robotic arms in space," she says, describing the work of her research group at the Space & Subsea Robotics Lab. The focus of this group's research is primarily the needs of the Space Shuttle and International Space Station programs now underway. Many of Dr. Sharf's projects have been developed co-operatively with a Canadian aerospace company that makes advanced technology systems, including robotic machines for use in space.

Dr. Sharf says she enjoys "thinking about challenging problems and generating new results." She also enjoys interacting with her students.

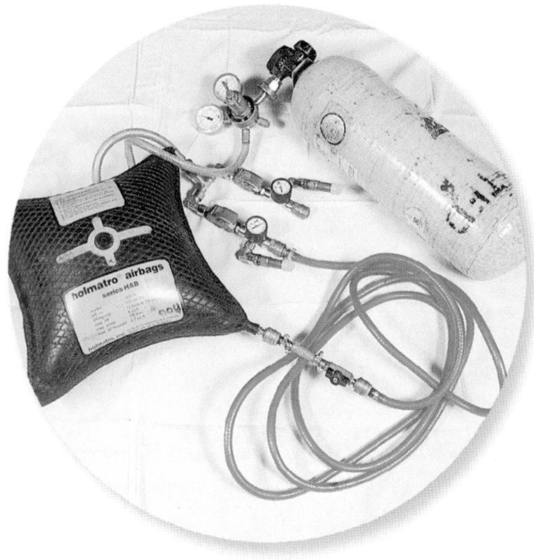

Figure 14.21 The canister on the right is filled with compressed air used to inflate the airbag shown on the left. Rescue workers use airbags to free people trapped under heavy objects.

Figure 14.22 Inflatable walkways help workers reach accident victims and carry them to safety. The walkways can also be used as rafts.

An earthquake or a tornado can cause a building to collapse, trapping people inside. To safely lift and support heavy beams or sections of fallen walls or concrete, airbags like the one shown in Figure 14.21 can be used. The rescue worker slides the bags into small spaces and then inflates them with air under extremely high pressure. The airbag shown in the photograph can lift a heavy object and hold it up while rescue workers free trapped people and carry them to safety. Sometimes a stack of airbags is needed to do the job.

INTERNET CONNECT

www.school.mcgrawhill.ca/resources/

As you complete your bulletin board display, poster, or magazine, search the Internet for machines that combine two or more subsystems. Visit the above web site, and go to **Science Resources**. Then go to **SCIENCEPOWER 8** to know where to go next. What machine combines a wheel and axle with a hydraulic system? Some pneumatic systems work in combination with levers. See how many combined subsystems you can find.

When an accident occurs on water, rescue workers may need to reach a damaged boat and bring an injured person to safety. The inflatable walkway shown in Figure 14.22 can help them do this. Air pressure not only makes these walkways float, but it also helps maintain their shape and stability.

Check Your Understanding

1. List three ways in which pneumatic or hydraulic equipment is used by rescue teams.
2. **Apply** Explain how brakes in an automobile work. Use a diagram in your answer.
3. **Apply** Design and sketch a mechanical system that uses a pulley or a lever in combination with a hydraulic or pneumatic device. Label the subsystems in the device.
4. **Thinking Critically** Formulate your own question related to how subsystems function in a mechanical device and explore possible answers.

CHAPTER at a glance

Now that you have completed this chapter, try to do the following. If you cannot, go back to the sections indicated.

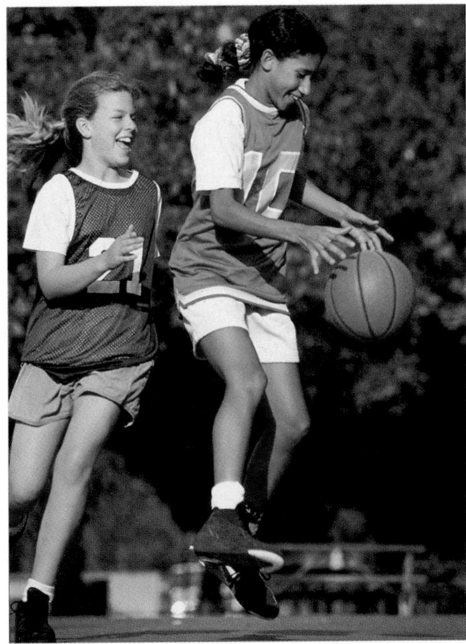

Describe the function of your body's pneumatic system. (14.1)

Describe the function of your body's hydraulic system. (14.1)

State the relationship between the diameter of a tube or a pipe and the flow rate of a fluid passing through it. (14.1)

Give an example of a closed system. (14.1)

State Pascal's law and describe some applications of this principle in hydraulic and pneumatic systems. (14.1)

Compare what happens when you exert the same amount of force on a gas and on a liquid. (14.2)

Give some examples of hydraulic and pneumatic devices and explain how they work. (14.1, 14.2)

Give an example of a mechanical system that is made up of several subsystems. (14.3)

Prepare Your Own Summary

Summarize this chapter by doing one of the following. Use a graphic organizer (such as a concept map), produce a poster, or write the summary to include the key chapter ideas. Here are a few ideas to use as a guide:

- What is the role of air pressure in breathing?
- How does the diameter of a tube or a pipe affect friction?
- What principle of fluid pressure is applied in a hydraulic lift?
- What happens when you exert the same amount of force on a gas and on a liquid?
- What is the mechanical advantage of the hydraulic lift shown here, if the effort force (F_1) is 100 N and the load raised (F_2) is 1000 N?

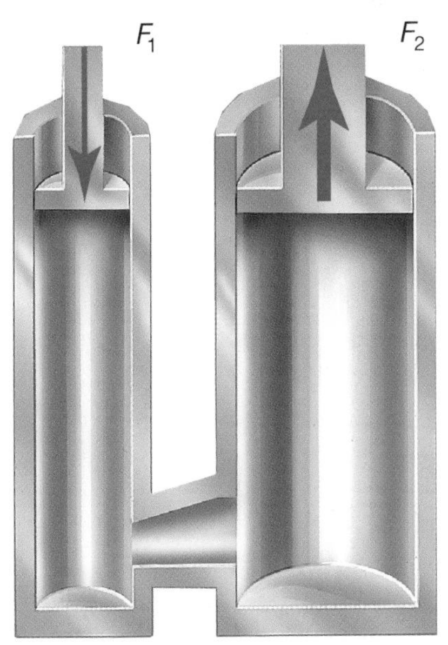

CHAPTER

14 Review

Key Terms

hydraulics
pneumatics
closed system
Pascal's law
hydraulic lift
pressure
force
area
pascal
subsystems

Reviewing Key Terms

If you need to review, the section numbers show you where these terms were introduced.

1. Which of the key terms best matches each of the following words or phrases?
   - **(a)** force per unit area (14.1)
   - **(b)** change of pressure is transmitted evenly throughout a fluid (14.1)
   - **(c)** circulatory system (14.1)
   - **(d)** unit of pressure (14.1)
   - **(e)** provides a mechanical advantage (14.2)
   - **(f)** brakes in a vehicle (14.3)

Understanding Key Ideas

Section numbers are provided if you need to review.

2. What hydraulic pump generates the pressure to force hydraulic fluid through 100 000 km of tubes? (14.1)
3. What is the main function of your body's pneumatic system? (14.1)
4. Restate Pascal's law in your own words. (14.1)
5. When you squeeze some toothpaste onto your toothbrush, you are applying Pascal's law. Give some other everyday applications of Pascal's law.
6. List four parts of an airplane that are controlled by hydraulic systems. (14.1)
7. List two differences between hydraulic and pneumatic systems. (14.2)
8. Describe four machines or instruments that use pneumatic systems. (14.2, 14.3)
9. Explain how a hovercraft works. (14.2)
10. Name several careers in which fluid pressure plays a role. (14.1, 14.2, 14.3)

Developing Skills

11. Copy the following flow chart into your notebook and fill in the blanks.

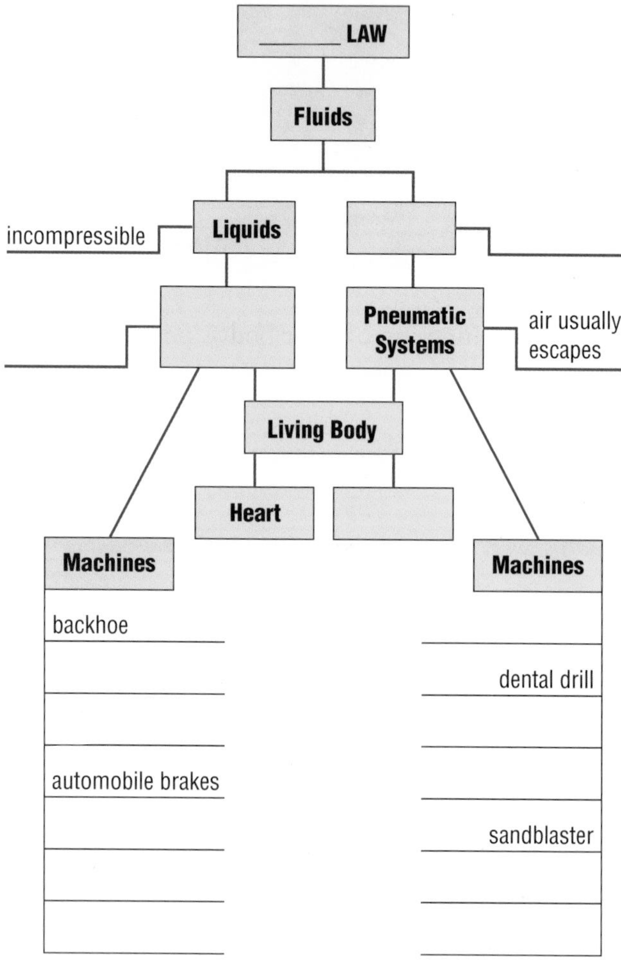

12. A typical high school student weighs 725 N and wears shoes that touch the ground over an area of 412 cm^2.
    - **(a)** What is the average pressure the student's shoes exert on the ground?
    - **(b)** How does the answer to (a) change if the student stands on one foot?

Problem Solving/Applying

13. In Conduct an Investigation 14-A, you investigated the relationship between fluid friction and flow rate using water. Choose any two liquids and design your own experiment comparing the fluid friction-flow rate relationship of the two liquids. Identify the variable(s) that will change in your experiment, as well as the variable(s) that will remain constant. Also, list criteria for assessing your solution to the problem you are investigating.

14. Assume that you are able to exert a force of 200 N on the piston of a hydraulic lift that has an area of 25 cm^2. What would the area of the other plunger have to be if you wanted to lift a load of 1000 N?

15. Sketch and describe a lever and a hydraulic lift that would both have a mechanical advantage of 4. Use numerical values to describe the length of the lever arms and the areas of the pistons in the hydraulic lift.

Critical Thinking

16. Hydraulic systems operate automobile brakes.
    (a) Describe one problem that could occur in a hydraulic system.
    (b) Think of one reason why a hydraulic system is more appropriate for brakes than a mechanical system made of levers, gears, or pulleys.

17. Think of one advantage and one disadvantage of using pneumatic systems to power rescue equipment such as the inflatable rescue walkway on page 482.

18. In tractors and other types of heavy equipment that feature hydraulic systems, oil is used as the hydraulic fluid. Often, the system includes an oil filter. The pump pushes the oil through the filter before it goes to the parts where it is used to exert pressure. Why do you think it is important to include oil filters in hydraulic systems?

19. The Heimlich manoeuvre is an emergency technique used to dislodge an object caught in the throat. How is this technique an application of Pascal's law?

20. An auto mechanic has four service bays in his shop for servicing cars or trucks. Only one bay has a hydraulic lift, but it can raise the heaviest truck that the mechanic will ever need to service. However, the shop is getting more and more jobs that require a lift, and the mechanic must invest in another one. Since hydraulic lifts are very expensive to install, the right choice will be critical. To help him make a decision, the mechanic reviews all of the jobs that were done in the shop over a typical two-week period. The list below represents the masses, in kilograms, of all the cars and trucks serviced by means of the lift during that period.

    Masses of cars and trucks (in kilograms): 2480, 4580, 1760, 4120, 2770, 5120, 2980, 1680, 5380, 1490, 4350, 2520, 2670, 1210, 3610, 2590, 2770, 1750, 4510, 2870, 1920, 3360, 1720, 2300, 2740, 1640, 2450, 2310, 4930, 1440, 2640, 2820, 3470

    Organize the data in a stem-and-leaf plot. Examine your plot. Use the organized data to help make a decision about the exact size of hydraulic lift that the owner of the shop should install. What is the greatest mass that the lift should be able to raise? Explain how you arrived at your decision.

Pause& Reflect

1. If you were to choose a job that involved the use of hydraulic or pneumatic equipment, what would interest you the most? Why? Write your response in your Science Log.
2. Go back to the beginning of this chapter on page 454 and check your original answers to the Getting Ready questions. How has your thinking changed? How would you answer those questions now that you have investigated the topics in this chapter?

CHAPTER

15 Temperature,

Getting Ready...

- How does burning fuel in a car make the car's wheels turn?
- How does steam power huge ocean liners?
- Why do manufacturers conduct product tests and consumer product surveys?

Science Log

All cars, trucks, and trains need a source of energy to make them move. In your Science Log, write down your ideas about how burning fuel can make wheels turn. As you study this chapter, go back and review what you wrote. Record your new learning in your Science Log.

A railway worker known as a "stoker" had to keep shovelling coal into the furnace of this old steam locomotive to keep the train moving. In a locomotive, burning coal heats water in the boiler. The water in the boiler turns to steam, which turns the gears, which move the wheels. Without the power of **steam** — the invisible gas into which water is changed by boiling — the train could not move.

You rarely see locomotives like this anymore, except perhaps in museums or on display. However, the motion of cars, trucks, ocean liners, and many other vehicles is based on the same scientific principles as the motion of a locomotive.

In this chapter, you will see how temperature, volume, and pressure interact to power engines in vehicles. You will learn how engines have become more efficient since the first steam engines, and what improvements are still needed. You will also see how the invention of the steam engine in the late eighteenth century led to large-scale manufacturing of goods in factories. As consumers of a wide array of goods and services, we face more choices and more responsibility for those choices than at any other time in human history.

Pressure, and Machines

Spotlight On **Key Ideas**

In this chapter, you will discover

- how a change in temperature affects the volume and pressure of a fluid
- how steam engines and internal combustion engines work
- why engines cannot be 100 percent efficient
- how the invention of the steam engine gave rise to the Industrial Revolution
- what features make machines and other consumer products "good value"
- how science and technology can improve products and find new uses for worn-out items

Spotlight On **Key Skills**

In this chapter, you will

- investigate the effect of temperature on the volume of a fluid
- determine the relationship between temperature and pressure
- graph data from a table
- build a mechanical device to transform one form of motion into another
- evaluate a product design in terms of consumer needs and desires
- evaluate the "user-friendliness" of written instructions
- conduct and analyze a consumer product survey

Starting Point 

Is Your Breath Hot or Cold?

While ice skating outdoors, you breathe on your hands to warm them. Later, indoors, you blow on a mug of hot chocolate to cool it. How can you warm your hands and cool your hot chocolate with the same air, your breath?

What to Do

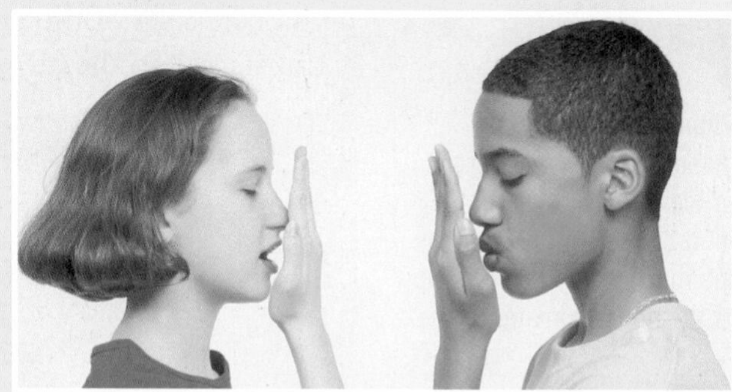

1. Hold your open hand in front of your face, as shown above. Take a deep breath, then open your mouth wide and breathe out.
2. Keep your hand in the same place and take another deep breath. Hold your lips tightly together, leaving only a tiny opening. Breathe out again.

What Did You Find Out?

1. Did your palm feel different depending on whether your mouth was wide open or nearly closed? Describe any difference you felt.
2. In both cases, do you think the temperature of the air inside your mouth was the same before breathing out? Explain.
3. In each case, was there a change in pressure of the air that was inside your mouth after you exhaled it? Explain.
4. Infer how the temperature and the pressure of a gas are related.

15.1 Pressure, Volume, and Temperature

What happens when a liquid is heated? Think again about the mug of hot chocolate mentioned in the Starting Point Activity. In previous studies, you learned that **temperature** is a measure of the thermal energy of the particles in a substance. **Thermal energy** is the total energy of the random motion of particles making up a substance. **Heat** is thermal energy transferred from one object or substance to another because of a difference in temperature. Like any other form of energy, heat is measured in joules.

According to the particle theory, particles move faster when they are heated because they gain more energy. Particles of hot chocolate have thermal energy that you can feel through the sides of the mug. You can also detect thermal energy if you take a sip of the hot drink (see Figure 15.1A). Some of the thermal energy is transferred to the air particles surrounding the mug.

Figure 15.1A You can detect the thermal energy of the hot chocolate in the mug by feeling the sides of the mug and by taking a sip of the hot chocolate.

When you blow on the hot chocolate, you cause high-pressure air to move over the surface of the liquid (see Figure 15.1B). As the air leaves your mouth, it begins to expand and cool. As it moves across the surface of the hot chocolate, the expanding air absorbs heat from the hot drink, and this helps to cool the liquid.

Figure 15.1B How is blowing on a mug of hot chocolate to cool it related to air pressure?

Find Out **ACTIVITY**

Double Bubble

You know that the boiling point of water is 100°C. When water is heated to its boiling point, its particles begin to move quickly enough to become a gas. The particles break free of the liquid and enter the air above it. What would happen if the air particles exerted less pressure on the water? This teacher's demonstration will help you find out.

What You Need

300 mL boiling water	ring stand
500 mL Florence flask	stopper for flask
funnel	cold tap water
large clamp	notebook; pen or pencil

What to Do

1. List all the equipment and materials your teacher sets out for this demonstration. Give the list a suitable heading.
2. Make a safety check list as your teacher describes the precautions being taken.
3. First, your teacher will attach a large clamp to the neck of a flask, and rest the flask's base on a ring stand. Why is the clamp needed?
4. Next, your teacher will pour 300 mL of boiling water into the flask. What safety precautions help avoid spills? Avoid burns?
5. After waiting several seconds, your teacher will plug the neck of the flask with a stopper. What is the purpose of waiting?
6. Now your teacher will tilt the stoppered flask, resting it on its side. Where is the air in the flask located at this point? Is the water still boiling?
7. Suppose the air in the flask is suddenly cooled. Predict what might happen to the water. Record your prediction.
8. Observe carefully as your teacher cools the air in the flask. What happens to the water in the flask?

What Did You Find Out?

1. Was your prediction in step 7 correct? If so, explain what previous knowledge helped you to make the correct prediction. If not, record now what you actually observed.
2. What effect would sudden cooling have on the air in the flask? Use the concept of air pressure to explain what happened to the water.
3. **Apply** Would water boil more slowly at the top of a tall mountain or at the bottom? Could you boil a pot of water on the Moon?
4. **Thinking Critically** Describe how your teacher cooled the air in the flask. Why did this method cool the air more rapidly than the hot water?

What do you see rising from a freshly poured cup of hot chocolate, coffee, or tea? We often use the word "steam" to describe what we observe when fast-moving particles of a hot liquid change into a gas and enter the air. In science, however, the word "steam" has a specific meaning. It refers to the invisible gas into which water is changed by boiling and which has the ability to make things move. Under certain conditions, this invisible gas can exert tremendous pressure. In this chapter, you will examine those conditions to see how steam and other hot gases can be used to operate machines. Begin your study of temperature, pressure, and machines by doing the next investigation.

Pause& Reflect

If you boil a pot of water without a lid for long enough, all the water will evaporate into the air. What clues would indicate that the water is turning into a gas? What would happen if you put a tight lid on the pot? If you were boiling a large pot of potatoes, why might you set the lid loosely on top of the pot? Write your responses in your Science Log.

CONDUCT AN

INVESTIGATION 15-A

Temperature and Fluid Volume

What effect does temperature have on the pressure and the volume of a fluid? How does this effect make machines run? This investigation will provide some clues to help you answer these questions. Make and record a hypothesis before you begin the experiment.

Problem

How does an increase in temperature affect the volume of air and of water?

Apparatus

500 mL Erlenmeyer flasks (2)

2 rubber stoppers with long, narrow glass tubes inserted into the holes in the stoppers

2 support stands

2 clamps

plastic dishpan

beaker or plastic cup

felt tip pens in different colours

thermometer

safety gloves

Materials

ice

water at room temperature

very hot tap water or water heated in a kettle

Safety Precautions

- Your teacher will insert the glass rods into the holes in the stoppers for you.
- Be careful when handling the glass rods and the thermometer. They are very fragile and break easily.
- Handle hot water with extreme care.

Procedure

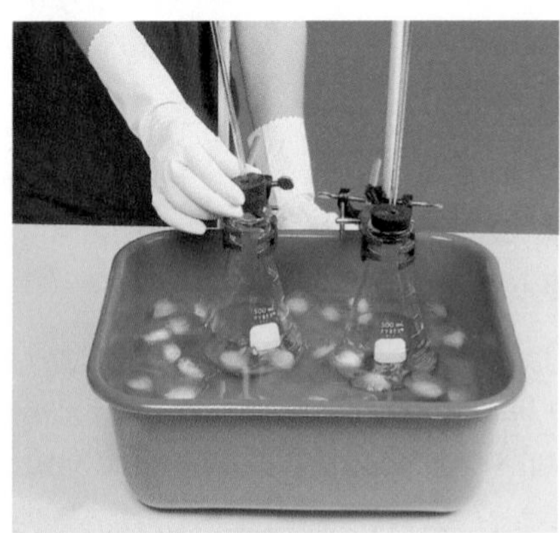

1. Fill the dishpan about half full of tap water and add a few cups of ice. Secure the flasks in the ice water with the clamps and support stands. Place the stoppers and the glass tubes in the top of the flasks.

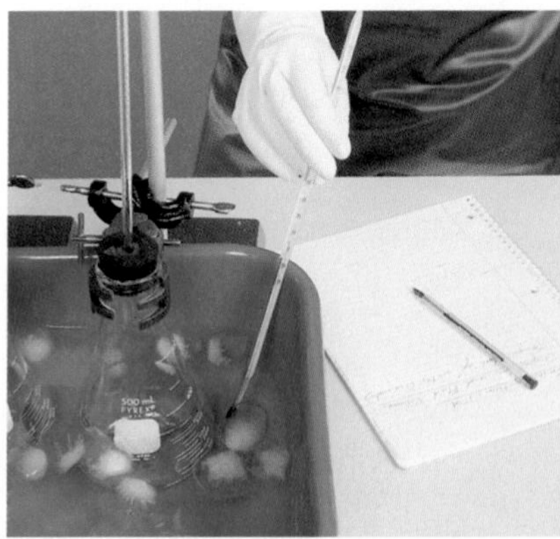

2. Let the flasks sit in the ice water for 3–4 min so the glass and the air can cool. Measure and record the temperature of the water in the dishpan.

Skill POWER

To review how to measure temperature using a thermometer, turn to page 540.

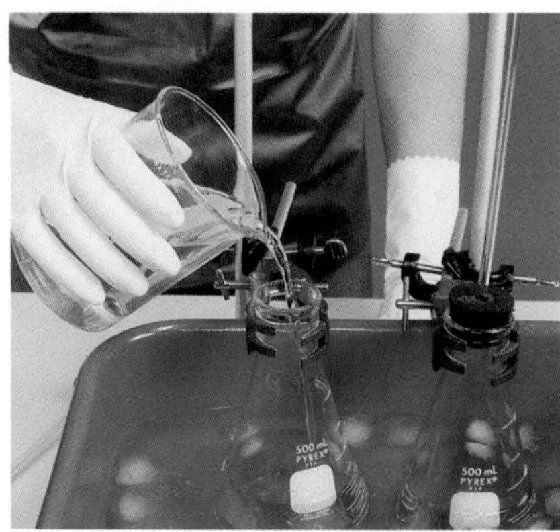

3. Remove one stopper and fill one of the flasks with the cold water. Be sure that no ice gets into the flask. When the water is all the way to the top of the flask, replace the stopper. Be sure that no air is left in the flask and that some water starts up the tube.

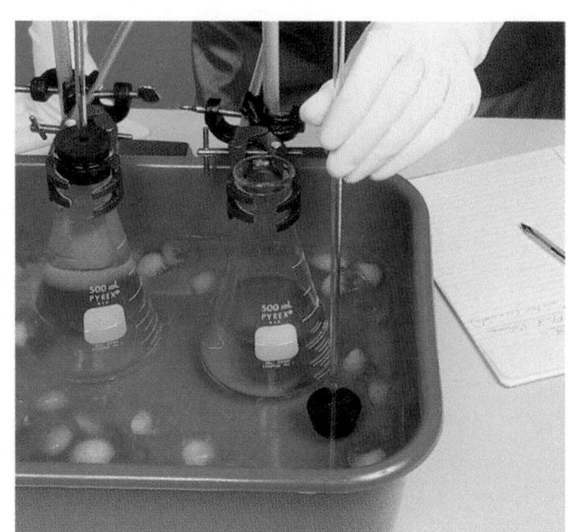

4. Remove the stopper from the empty flask. Touch the bottom of the glass tube to the water so that a bubble of water goes into the tube. Replace the stopper in the air-filled flask. The bubble of water will allow you to see any changes in the volume of air in the flask.

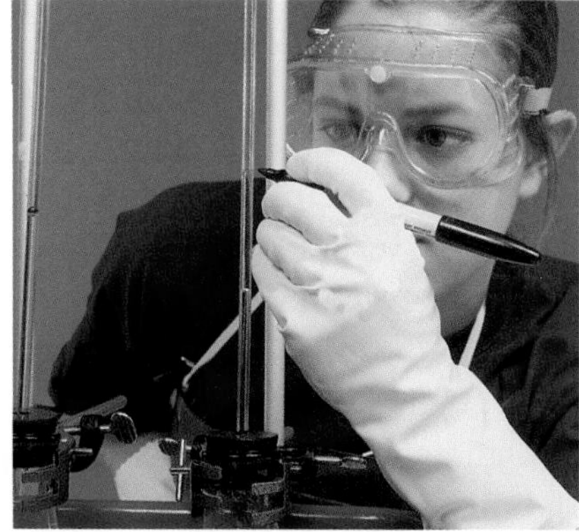

5. With the felt tip pens, mark each glass tube to show the level of the water in one tube, and the water bubble in the other.

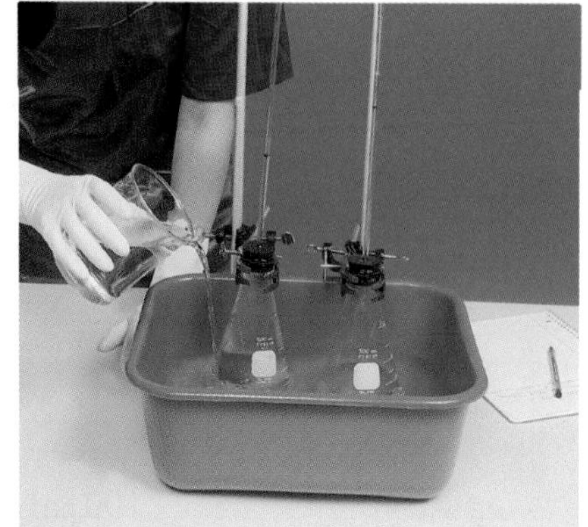

6. Using the beaker or cup, remove most of the ice water from the pan. Add room-temperature water. Leave the apparatus standing for about 5 min. Measure and record the temperature of the water in the pan.

7. Mark the level of the water and the air in the glass tubes using a felt tip pen in a different colour.

8. Repeat steps 6 and 7, substituting hot tap water or water heated in a kettle.

Analyze

1. Describe any changes you observed in (a) the water level, and (b) the air level in the flasks when the temperature changed.
2. What can you infer about the effect of temperature on the volume of a fluid?
3. Was the effect on the liquid (water) and the gas (air) similar, or was one change greater than the other?

Conclude and Apply

4. Write a statement that summarizes the relationship between temperature and the volume of a fluid.
5. Use the particle theory to explain your observations about the relationship between temperature and the volume of fluids.
6. How do you think this relationship between temperature and volume might be related to steam engines?

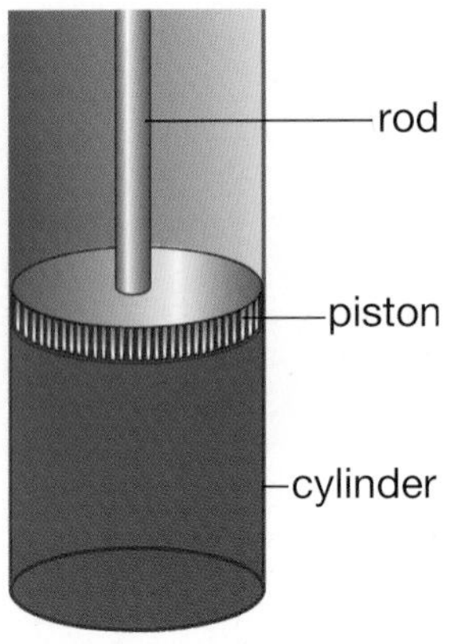

Figure 15.2 A simplified diagram of a piston. As you learned in Chapter 14, a piston is a movable disk or platform that fits inside a closed cylinder. When this piston moves, it causes an attached rod to move.

Working Fluids

In the previous investigation, you observed that liquids and gases expand when they are heated. As the temperature of a liquid or a gas rises, the volume of the liquid or gas increases. In other words, the volume of the fluid expands. The expansion of a fluid when it is heated can be used to make a piston move, and thus do work. This is how a **steam engine** functions. It burns fuel such as coal or wood to change water into steam in a boiler outside the engine. You might wonder, "Why use steam rather than simply liquid water?" (Hint: Which fluid, the water or the air, expanded more in Conduct an Investigation 15-A?)

If you poured 100 mL of water at 4°C into a measuring cup, then heated the water to 100°C, the volume of water would expand to 104 mL. An increase from 100 mL to 104 mL is quite small. What would happen if you continued to add heat to the water? It would boil. When water boils, it changes from a liquid to a gas.

If the entire 100 mL of water boiled into steam at 100°C at atmospheric pressure, it would expand to about 170 000 mL. In other words, when you convert liquid water into a gas, the volume increases to 1700 times its original volume! If you then heat the steam to 200°C, the volume would continue to increase to more than 200 000 mL, or 2000 times its original volume. There is a catch, however. These large volume increases for gases occur only if the air pressure exerted on the water stays constant, that is, the same, throughout the process. What happens to a gas if it is heated while held in a container that will not let it expand? Find out in the next investigation.

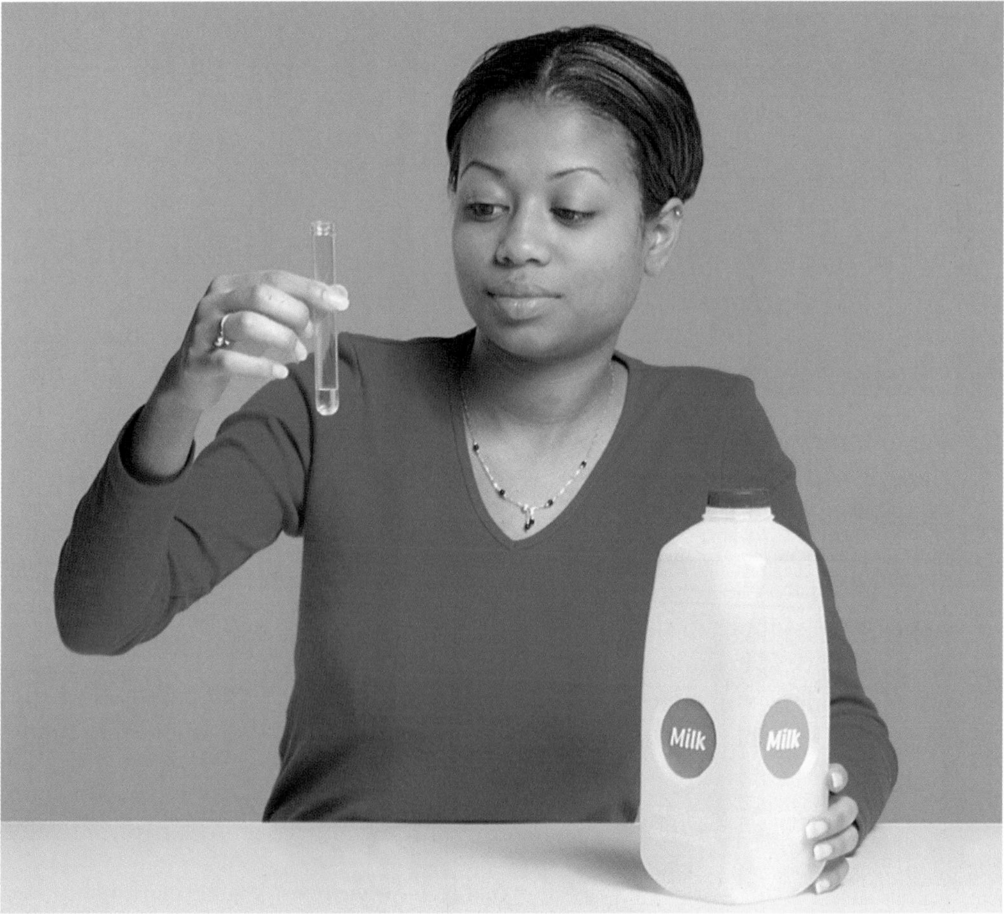

Figure 15.3 The test tube contains 1 mL of water. If this water were boiled and converted to steam at 200°C, the steam would fill the milk container.

INVESTIGATION 15-B

The Pressure Is Rising

Think About It

What would happen if you boiled 1 g of water into steam in a container that could not expand? What would happen if you continued to heat the steam to higher and higher temperatures? The table shows the results of an experiment in which 1 g of steam is heated from 100°C to 300°C inside a 1000 mL container.

Temperature vs. Pressure
of 1 g of Steam in a 1000 mL Container

Temperature (°C)	Pressure (kPa)
100	172
125	184
150	195
175	207
200	218
225	230
250	242
275	253
300	265

Note: Recall that "k" placed in front of a unit means 1000 of those units. For example, 1 km (kilometre) is equal to 1000 m (metres). In this table, 1 kPa is 1000 Pa of pressure.

What You Need

graph paper
pencil

Skill POWER

To review how to make a line graph, turn to page 546.

***Computer* CONNECT**

If you have access to a computer, you may want to prepare your graph electronically. Then compare your computer graph with your paper version.

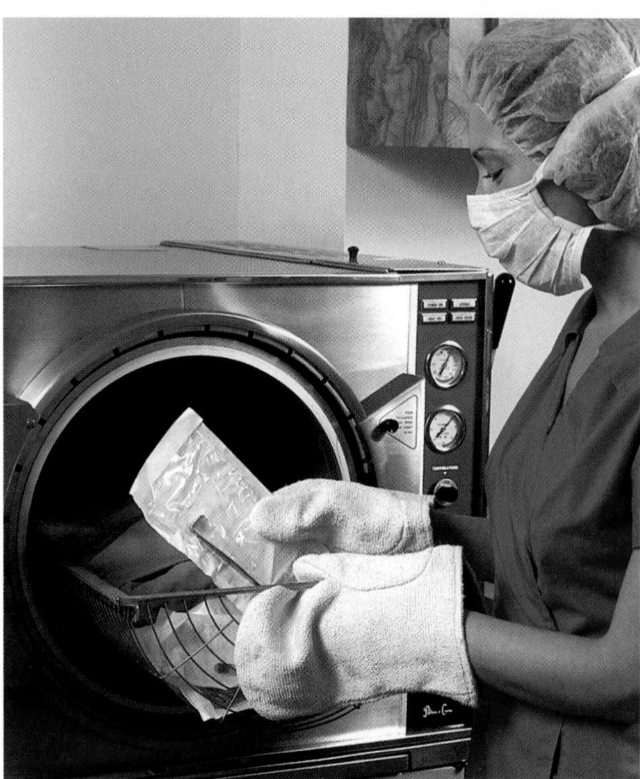

Steam at very high temperatures and a very high pressure is used in this autoclave to sterilize surgical instruments.

What to Do

1. Make a line graph showing pressure on the vertical axis (y-axis) and temperature on the horizontal axis (x-axis). Be sure to give your graph a title and to include labels and units on the axes.
2. Plot the points given in the table.
3. Draw a "line of best fit" on your graph.

Analyze

1. Write a statement that describes the relationship between the temperature and the pressure of a gas that is confined within a certain volume. Hint: What happens to the pressure when the temperature increases?
2. Use the particle theory to explain the relationship that you found between pressure and temperature when the volume of a gas is kept constant.

A kernel of corn is made of a tough shell called the pericarp and a starchy inside that contains a small amount of water. When you heat the kernel, the water inside evaporates into steam. The steam exerts so much pressure against the pericarp that it breaks open and the starch pops outward into a fluffy white snack — popcorn!

Summarizing Temperature, Volume, and Pressure

In the last two investigations, you learned two important principles involving pressure, volume, and temperature of gases. You will see how these principles apply to machines later in the chapter, so this is a good time to review them.

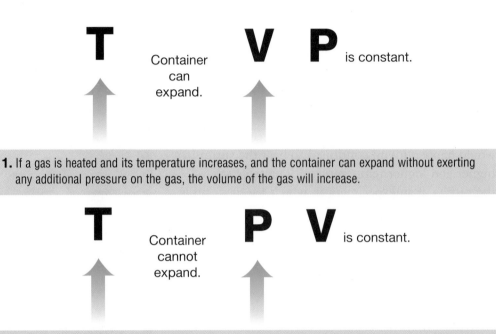

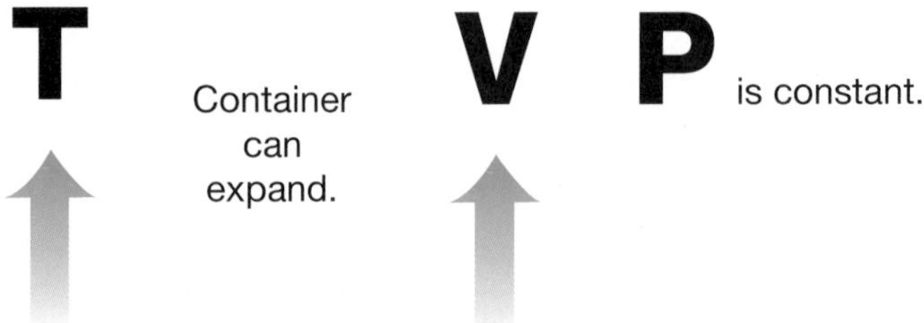

1. If a gas is heated and its temperature increases, and the container can expand without exerting any additional pressure on the gas, the volume of the gas will increase.

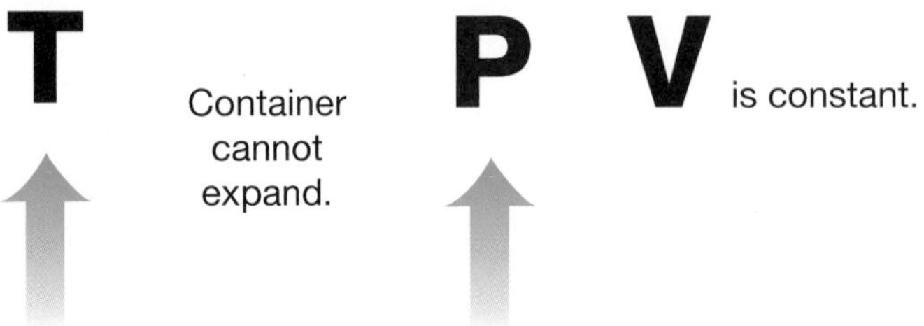

2. If a gas is heated and its temperature increases, and the gas is in a container with a fixed volume, the pressure of the gas will increase.

You may have realized by now that it is possible for both the pressure and the volume to increase when a gas is heated to a high temperature. This is exactly what happens in the cylinders and pistons of many machines. As the temperature of a gas in the cylinder increases, the pressure increases. Soon the pressure becomes great enough to move the piston. The expanding gas causes the piston to move and the motion of the piston does work, such as turning wheels on a steam locomotive.

Pause & Reflect

In your Science Log, describe in your own words the principles that relate pressure, volume, and temperature of a gas. Can you name any machines that have pistons that are moved by hot gases? If so, write down the names of these machines.

Check Your Understanding

1. What happens to the volume of water if it is heated in an open container?
2. If water is heated in an open container and the temperature reaches 100°C, what happens next?
3. If a gas is sealed in a container and the volume of the gas cannot change, what happens when the gas is heated?
4. **Apply** How can an expanding gas do work?
5. **Apply** Why do you need to pump more air into your bicycle tires in the fall when the temperature starts to drop?
6. **Thinking Critically** When you look at a whistling tea kettle that contains boiling water, you see a clear space above the spout. Above the clear space, you see a cloudy area. In which of these two areas is the steam located? What is in the clear space? What is the cloudy-looking substance?

15.2 Piston Power

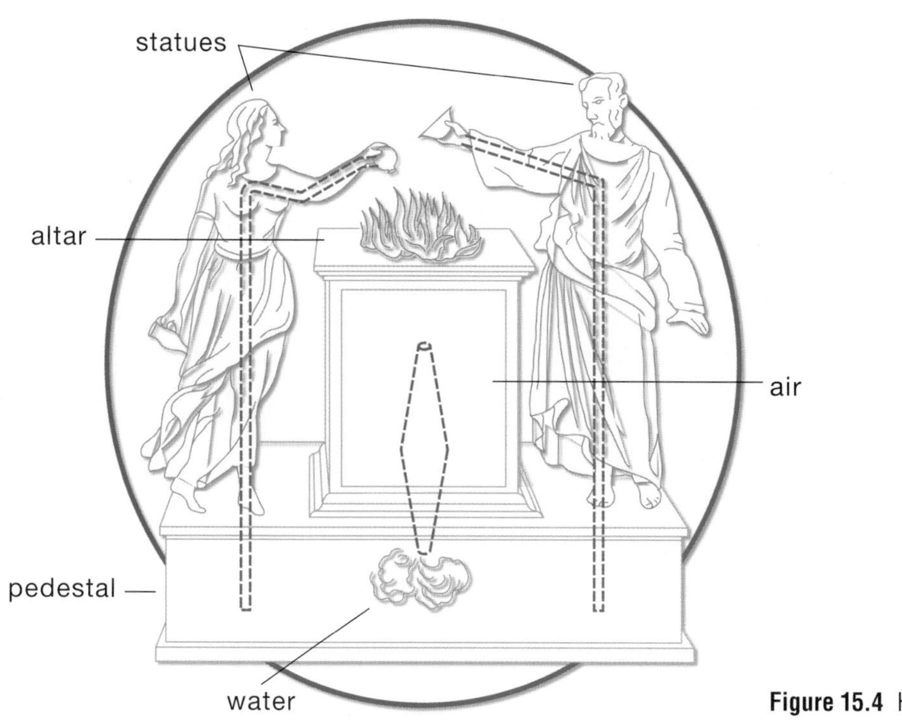

Figure 15.4 Hero's altar

Heat-operated mechanical devices have existed for a long time. About 150 B.C.E., for example, Hero of Alexandria in Egypt wrote a book describing many mechanical devices. No one knows for sure which ones Hero invented himself. These devices used gears, wheels and axles, pulleys, hydraulics, and pneumatics.

The device represented in Figure 15.4 appeared in a translation of Hero's writings. In this device, the pedestal at the bottom was filled with water and sealed. Pipes ran from near the bottom of the pedestal, up through the statues to the vessels in the statues' hands. The altar was filled with air. A tube connected the air in the altar to the water-filled pedestal below. What do you think would happen to the air pressure under the altar when a fire was lit on the altar? How would this affect the water in the pedestal? Where would the water go? What would this do to the fire? Imagine how this abrupt flow of water might strike an observer who did not understand the relationship between temperature, pressure, and volume of fluids! The water suddenly pouring out would seem like magic.

Steam Put to Work

Hero's writings showed that the early Egyptians understood many of the principles of science that are applied in machines today. For example, early Egyptians and Greeks used steam to run small devices. However, their "steam engines" could not power large equipment. Thomas Savery, an Englishman, built the first practical steam engine in 1699 to pump water out of flooded mines. This design was inefficient, though, and the pipes carrying the steam often burst. Thomas Newcomen improved upon the design in 1712, but it was James Watt (1736–1819) who greatly improved the efficiency of steam engines. His modifications made engines that wasted much less heat compared to previous versions.

Figure 15.5 At one time, steamboats were an important means of transportation in Canada. For example, from 1836–1957, more than 3000 steamboats travelled along the rivers and coasts of British Columbia and the Yukon, carrying gold seekers between the two regions. The *S.S Beaver,* shown above loading supplies from a log boom, was the first steamship to ply Canada's Pacific coast.

The invention of steam engines led to innovations in transportation (see Figure 15.5). In 1787, John Fitch demonstrated the first workable steamboat in the United States. When Robert Fulton built the first commercial steamboats in North America in the early nineteenth century, these "paddle-wheelers" were the fastest way to travel. In 1807, Fulton's steamboat *Clermont* steamed 241 km up the Hudson River from New York City to Albany in 32 hours.

Incredible as it may seem, for a time, a fleet of steamboats supplied the Canadian West. These steamships became a common sight in what many people assumed was a landlocked prairie. By 1879, seventeen ships travelled regularly on the North Saskatchewan, the South Saskatchewan, the Assiniboine, and the Red rivers. These steamboats transported the materials of the fur trade, as well as pioneers and farming equipment for the new society springing up on the Prairies.

Exactly how does steam cause a paddle wheel to turn? Under high pressure, steam flows into the top of the cylinder, as shown in Figure 15.6. The steam expands and pushes the piston down. At the same time, **exhaust valves** at the bottom of the cylinder open to allow steam to escape. Then steam enters from the bottom of the cylinder and pushes the piston back up. Once again, exhaust valves open to let steam escape from the top. As the process repeats itself again and again, the piston moves up and down. The rod of the piston is attached to gears and levers that do work. In a steamboat, the gears turn a paddle wheel that pushes against the water and propels the boat forward.

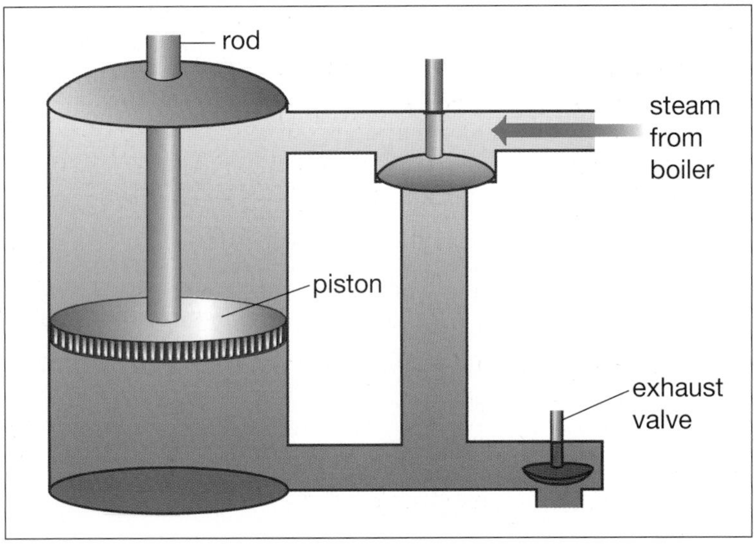

Figure 15.6 James Watt improved the steam engine by developing pipes and valves that allowed steam to push the piston both up and down.

The Industrial Revolution

The invention of the steam engine transformed society. No one can say for certain when the Industrial Revolution began. Simple machinery had been taking the place of hand labour since 1700. The water-driven spinning machine introduced in 1769, for example, could do the work of twelve workers. A combination of events in the late 1700s, however, transformed England and the world. First came James Watt's invention of an efficient steam engine in 1769 (see Figure 15.7A). A year later, Henry Cort developed a method of making iron using coal for fuel instead of wood. The iron to build machines and the engines to drive them led to the rapid development of mass-production industry. (**Mass production** is the manufacturing of large quantities of a standardized item by standardized mechanical processes. Modern examples include the manufacture of home appliances in a factory, the canning of foods in a food-processing plant, and the production of automobiles in an assembly plant.)

Within a few years, small towns such as Manchester and Birmingham in England became industrialized cities teeming with factories (see Figure 15.7B). Industrialization led to great social change. Unable to compete with the new factories, the spinners, weavers, and craftspeople from the villages flocked to the cities to find work. The transformation from a rural to an urban society had begun.

Many people feared the first steam-powered trains. In England in 1829, a locomotive called the *Rocket* won a race at a speed of 47 km/h — unbelievably fast for that time! In the nineteenth century, some people believed it was dangerous for humans to travel faster than 20 km/h.

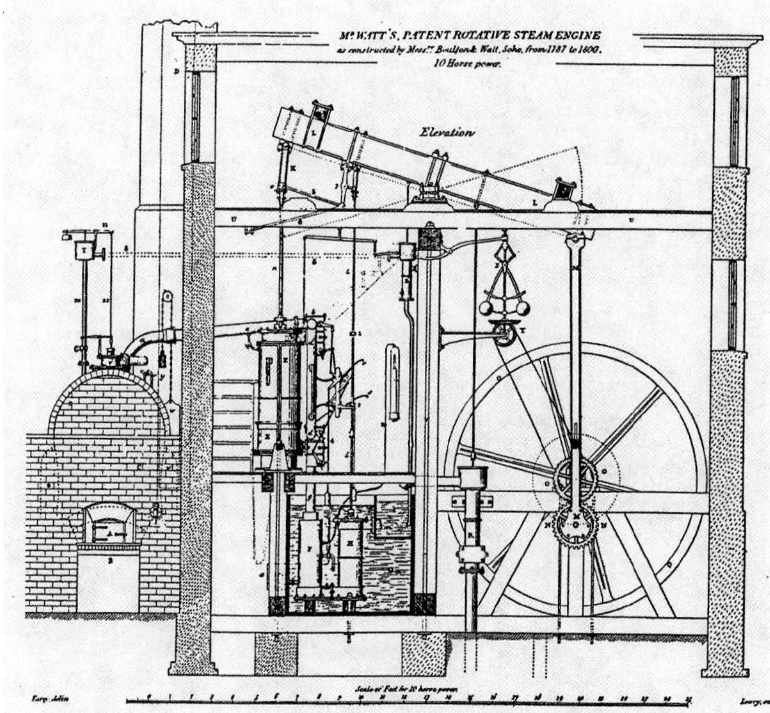

Figure 15.7A The plans for James Watt's steam engine. Watt's invention was one of the technological advances that gave rise to the Industrial Revolution in England.

Figure 15.7B In the late eighteenth and early nineteenth centuries, factories were built in Europe to mass-produce goods, and people moved from farms to cities to find work. Children as well as adults worked long, hard hours in these factories.

Turning Wheels

DidYouKnow?

The first record in history of a wheel turned by steam is found in the writings of Hero of Alexandria. Hero described what appeared to be a toy. The toy had a pot filled with water that was closed with a lid. A pipe ran from the lid to the inside of a hollow wheel. Bent pipes were attached to the edge of the wheel. When placed over a fire, the water in the pot began to boil. Steam was driven up the pipe into the hollow wheel. As the steam escaped through the bent pipes, the pressure of the steam caused the wheel to rotate.

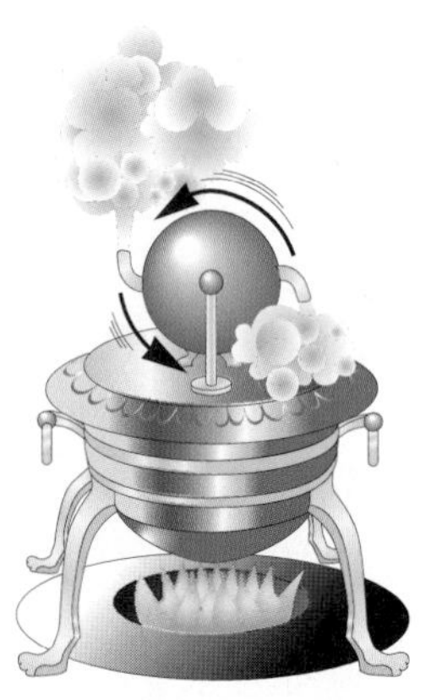

Figure 15.8 Steam powers huge ocean liners that can carry thousands of passengers at a time. This cruise ship is entering Burrard Inlet in British Columbia.

Paddle-wheeled riverboats are rarely seen today, but steam still propels most ocean liners, such as the one shown in Figure 15.8. In these huge ships, steam does not drive pistons up and down. Instead, the steam turns large turbines. A **turbine** is a rotary engine used to convert the motion of a fluid into mechanical energy. It consists of a number of fan blades attached to a central hub (see Figure 15.9). The blades rotate when steam moves past them at a high speed. The spinning turbine is attached to an axle that turns giant propellers. These propellers drive the ocean liner through the water.

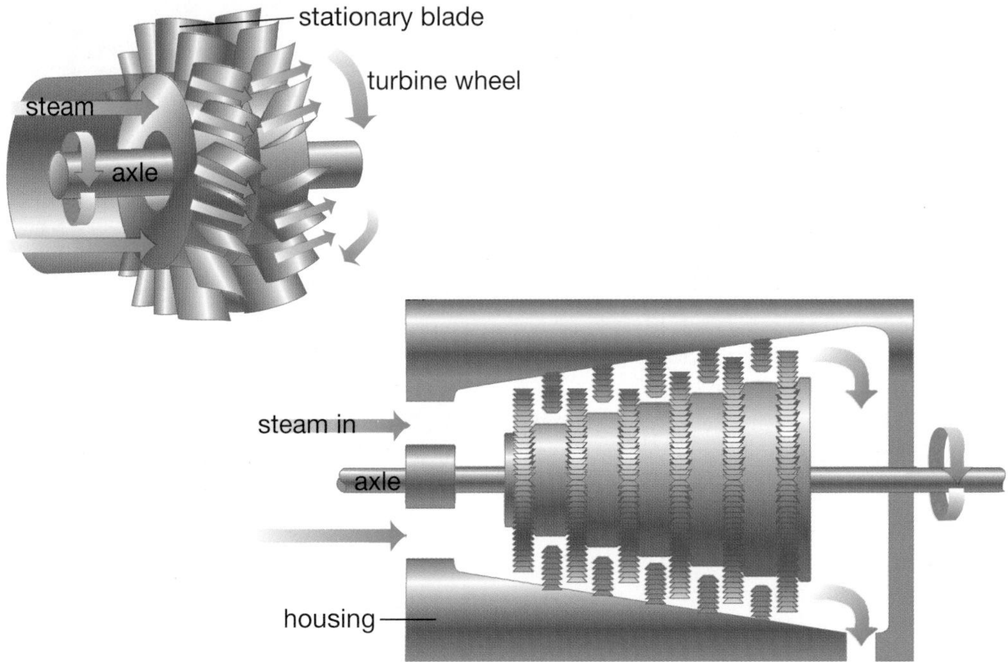

Figure 15.9 Stationary blades can increase a turbine's efficiency by carefully directing the angle at which the steam hits the spinning turbine wheel.

Turbines turn more than toys or propellers on ships. They are used in jet engines, and they turn shafts that operate many machines. Turbines also provide electricity. In thermo-electric generating stations, burning coal is used to heat water to steam. In other cases, nuclear reactors heat the water. In still other cases, turbines are powered by moving water to generate hydro-electric power.

Find Out **ACTIVITY**

Build a Model Steam Turbine

Watch the power of steam in action!

What You Need

3.6 L plastic bleach bottle
coat hanger wire
scissors
kettle with boiling water
drinking straw
glue gun or tape

What to Do

1. Remove the top from the bleach bottle. **CAUTION:** Make sure the bottle is very clean.

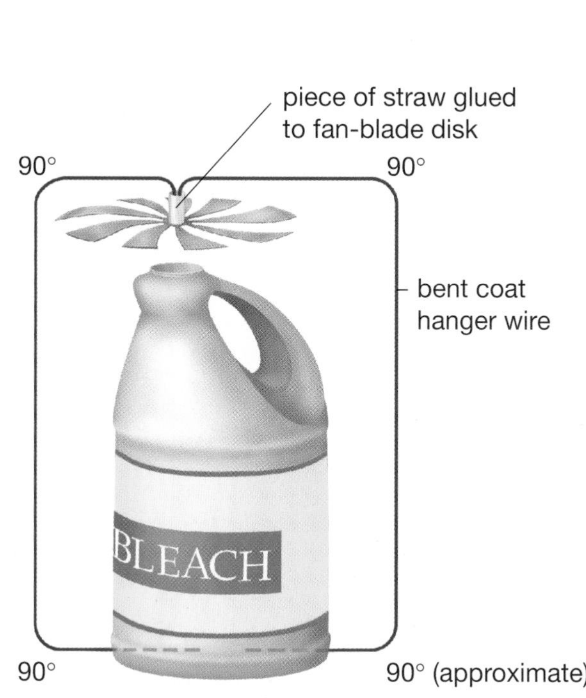

2. Cut out the bottom of the bottle. **CAUTION:** Be careful when using the scissors. Cut this bottom piece into fan-shaped blades, like a windmill, as shown.
3. Bend two pieces of coat hanger wire into a square-shaped frame. The frame should be large enough to allow the fan blades to turn within it.
4. Poke the end of each length of wire through the bottom of the bleach bottle. Then insert a piece of a drinking straw over the two top ends of the wire, as shown.
5. Glue or tape the fan-blade disk to the piece of straw (not to the wire), so the blades will not flop back and forth.
6. Using the handle of the bleach bottle, hold your turbine over a steaming kettle. **CAUTION:** Wear heat-resistant safety gloves and do this only under your teacher's supervision to avoid a severe burn. Alternatively, your teacher can hold the bottle carefully over the steam.

What Did You Find Out?

1. What did you observe when you held the bleach-bottle turbine over the steam?
2. **Apply** The water in the kettle was heated by means of a coil that conducts heat. What is the source of the heat that converts water into steam to turn the turbine in a thermo-electric generating station?

Burning Inside

During the Industrial Revolution, as the steam engine became more and more efficient, industries grew. Think for a moment about the size and the weight of the parts of a steam engine. A steam engine requires a furnace, coal or wood for fuel, a large boiler with a lot of water, and, finally, the actual engine and its pistons. If you could eliminate the furnace and the boiler, the engine would be much smaller and lighter. The desire to improve the steam engine's efficiency led to the development of the **internal combustion engine** in Germany in 1876. The term "internal combustion" describes the way the engine works. The combustion, or burning, of fuel occurs internally, that is, *inside* the engine. No external furnace, boiler, or water is needed. The fuel, gasoline, is burned right inside the cylinders.

Internal combustion engines usually have four, six, or eight pistons and cylinders. Each piston goes through the steps shown in Figure 15.10, but each piston does not carry out the same step at the same time. The diagram follows one piston through a cycle. The entire cycle is repeated many times each minute.

Most automobile engines have pistons that move either up and down or back and forth. A part called a **crankshaft** changes this up-and-down or back-and-forth motion to rotary motion, which turns the automobile's wheels. The power to move the pistons comes from the energy released by burning gasoline.

Figure 15.10 Automobiles move as a result of the transfer of thermal energy in their engines.

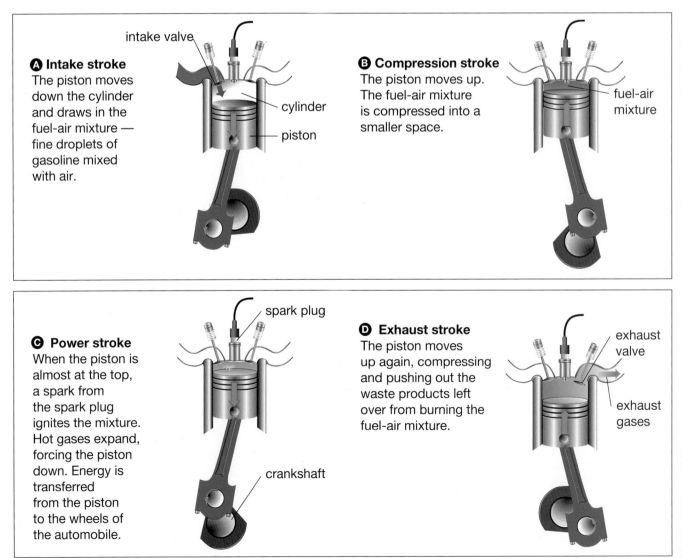

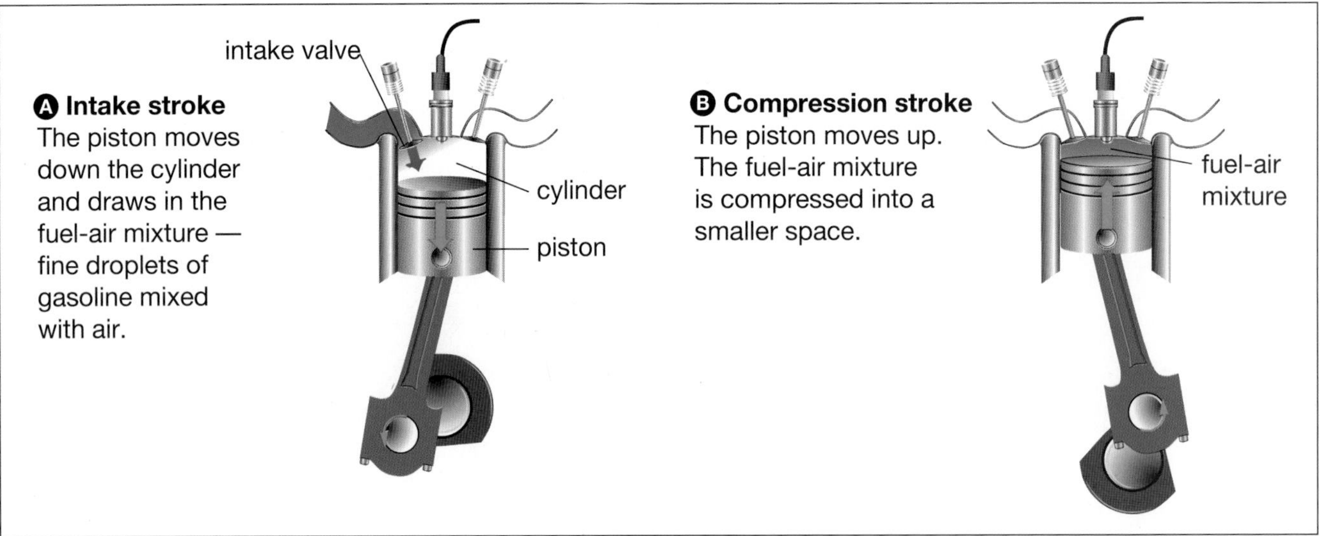

intake valve
A Intake stroke
The piston moves down the cylinder and draws in the fuel-air mixture — fine droplets of gasoline mixed with air.
cylinder
piston
B Compression stroke
The piston moves up. The fuel-air mixture is compressed into a smaller space.
fuel-air mixture

C Power stroke
When the piston is almost at the top, a spark from the spark plug ignites the mixture. Hot gases expand, forcing the piston down. Energy is transferred from the piston to the wheels of the automobile.

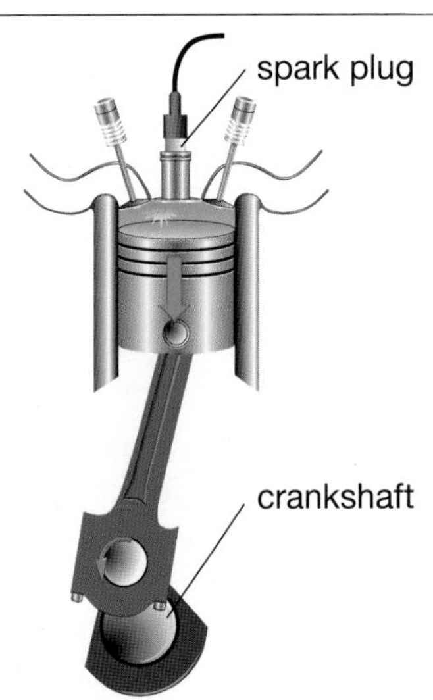

D Exhaust stroke
The piston moves up again, compressing and pushing out the waste products left over from burning the fuel-air mixture.

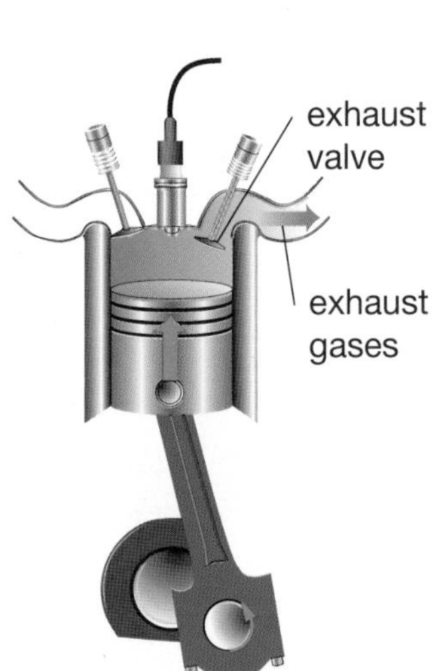

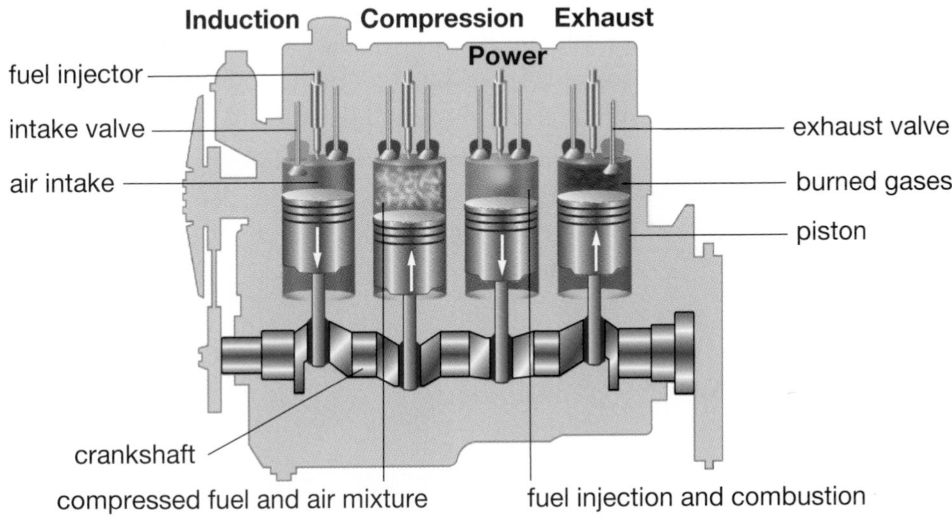

Figure 15.11 A diesel engine runs on diesel fuel rather than on gasoline.

Diesel engines are now used by almost all commercial vehicles and by an increasing number of cars. Diesel-powered vehicles are popular because their engines have lower running costs, partly because diesel fuel is less expensive than gasoline and partly because diesel engines are more efficient. In diesel engines, as Figure 15.11 shows, air is drawn into a cylinder and compressed with a piston. Compressing the air causes the temperature to increase. When diesel fuel is sprayed into the hot air, it ignites, pushing the piston down. As the piston rises, the exhaust is pushed out through a valve. Compare how the diesel engine works with how a gasoline engine works.

Pause& Reflect

In your Science Log, summarize the historical development of the use of fluid pressure. If you like, present your summary in the form of an illustrated timeline. The following list will help you get started.

- first record, in ancient literature, that steam pressure was used to operate a device
- first useful steam engine
- improvements in steam engines
- use of steam turbines
- first use of internal combustion engines

Escaping Energy

A **heat engine** is a device that converts thermal energy to mechanical energy continuously. An automobile engine is an example of a heat engine. Today's modern heat engines have something in common with the first steam engines. They are inefficient. You might wonder why engineers cannot design superefficient heat engines, with all the modern technology that is available. To understand why heat engines are not efficient, think about the meaning of efficiency. Recall that efficiency can be expressed by the following formula:

$$\text{Efficiency} = \frac{\text{Work done on load}}{\text{Work done by effort}} \times 100\%$$

Here is another way of expressing efficiency:

$$\text{Efficiency} = \frac{\text{Useful work a machine does}}{\text{Amount of energy put in}} \times 100\%$$

What is the form of the energy before it is used in a heat engine such as the engine in a car? It is chemical energy stored in fuel. As the fuel burns, the stored energy is converted into thermal energy. The hot gases inside the engine exert pressure on a piston or turbine and then escape from the engine. The gases are still hot when they escape. Therefore, much of the stored energy from the fuel is still in the gases — in the form of heat — after the gases have done all the work they were able to do in the engine. This heat escapes from engines into the atmosphere as pollution. This loss of energy limits the efficiency of all heat engines.

Although the overall use of fossil fuels has not increased much since 1982, and the amount of pollution given off by individual cars has been reduced, pollution levels remain high. More people now travel by automobile and they travel greater distances, making up for the other reductions (see Figure 15.12).

Figure 15.12 Hot exhaust gases carry away unused energy from vehicle engines. Exhaust also carries pollution in the form of carbon monoxide.

Thomas Savery, James Watt, and others are famous for designing and building steam engines. However, they knew very little about the scientific principles of heat engines. In the early nineteenth century, a French engineer named Sadi Carnot (1796–1832) formulated the theory of the "ideal heat engine." He showed that the maximum possible efficiency of a heat engine could be calculated from the temperatures of the intake gases and the exhaust gases.

For example, if the intake gases are at a temperature of 25°C and the exhaust gases are at 225°C, the maximum possible efficiency of the engine is about 40 percent. Use the equation and the conversion given in the box below to see how this efficiency is calculated. Remember, this is the maximum possible efficiency. This figure does not account for the friction between the moving parts of the engine or heat that escapes through the walls of the engine. A typical heat engine such as a car engine has an actual efficiency of about 20 percent.

INTERNET CONNECT

www.school.mcgrawhill.ca/resources/

Industrialization and the internal combustion engine have created a threat to the environment — smog. Use the Internet to do research on smog and the internal combustion engine. How does smog affect trees and other plants? What health problems are caused by smog? What can be done to reduce smog? Visit the above web site. Go to **Science Resources**. Then go to **SCIENCEPOWER 8** to know where to go next. (Use your library if you do not have access to the Internet.)

Carnot showed that the maximum efficiency of any heat engine can be calculated by using the following formula:

$$\text{Efficiency} = \left(1 - \frac{T_L \text{ (temperature of exhaust gases)}}{T_H \text{ (temperature of intake gases)}}\right) \times 100\%$$

However, the temperatures, T_L and T_H, must be given in degrees Kelvin, not Celsius. To convert degrees Celsius to degrees Kelvin, add 273.15 to the Celsius temperature. Use the formula to verify that the efficiency of the engine in the example above is approximately 40 percent. How might the work of scientists like Carnot help improve engine efficiency?

Keeping Cool with Thermal Energy

So far, the heat engines you have seen use fuel to do work. Pistons and turbines use thermal energy to generate mechanical energy. What about a refrigerator? If heat engines produce heat, how can they make things cold? For a clue to the answer to this question, do the Find Out Activity below.

Find Out ACTIVITY

That's Cool!

How can doing work cool something?

Safety Precaution

Do this activity only under teacher supervision.

What You Need

medicine dropper
ethyl alcohol
inner tube
air pump

What to Do

1. Fill a medicine dropper with ethyl alcohol and put a few drops on the inside of your wrist. Describe what happens. **CAUTION:** Be careful to wash off the ethyl alcohol after you have determined its effect.
2. Inflate an inner tube with an air pump. **CAUTION:** Be careful not to overinflate the inner tube. After the tube is full, feel the valve. Is it warm or cool? Have you done any work?
3. Let the tube stand for about 10 min. Open the valve and let the air out. Feel the valve again. Is it warm or cool?

What Did You Find Out?

1. When the ethyl alcohol was on your wrist, did the liquid become warmer or cooler? What happened to the ethyl alcohol and how did that make your wrist feel? Why?
2. Is the air in the inflated tube under more pressure or less pressure than the air outside the tube? How can you tell?
3. How would understanding what happens when fluids evaporate and condense help you understand how a refrigerator works?
4. **Apply** How do you think the expansion of gases is used in refrigeration?

A refrigerator uses two important properties of fluids. The first is that liquids evaporate at a faster rate as they get warmer. As liquids evaporate, they absorb heat. As they condense, they give off heat. You felt this happen when the ethyl alcohol evaporated from your wrist and left a cool spot.

The second property of fluids is that a gas becomes warmer when it is compressed and cooler when it expands. When you pumped air into the inner tube, you compressed and warmed the air. The valve felt warm. When you opened the valve, you allowed the high-pressure air to expand and cool.

Study Figure 15.13 on the following page to see how a refrigerator works. Mechanical refrigerators have four basic parts: an evaporator, a compressor, a condenser, and a refrigerant flow control (expansion valve). A **refrigerant** — a cooling liquid — circulates among the four parts, changing from liquid to gas and back to liquid.

DidYouKnow?

Evaporation of perspiration from the skin is an effective way of cooling your body. More than 2 million joules of thermal energy are carried away for each litre of liquid lost through evaporation of perspiration.

Figure 15.13 Liquids absorb heat when they evaporate. Refrigerators are designed to take advantage of this property of liquids. The cooling part of a refrigerator is a closed system of tubes in which a cooling liquid evaporates, condenses back into liquid, and then evaporates again. This is a continuous cycle.

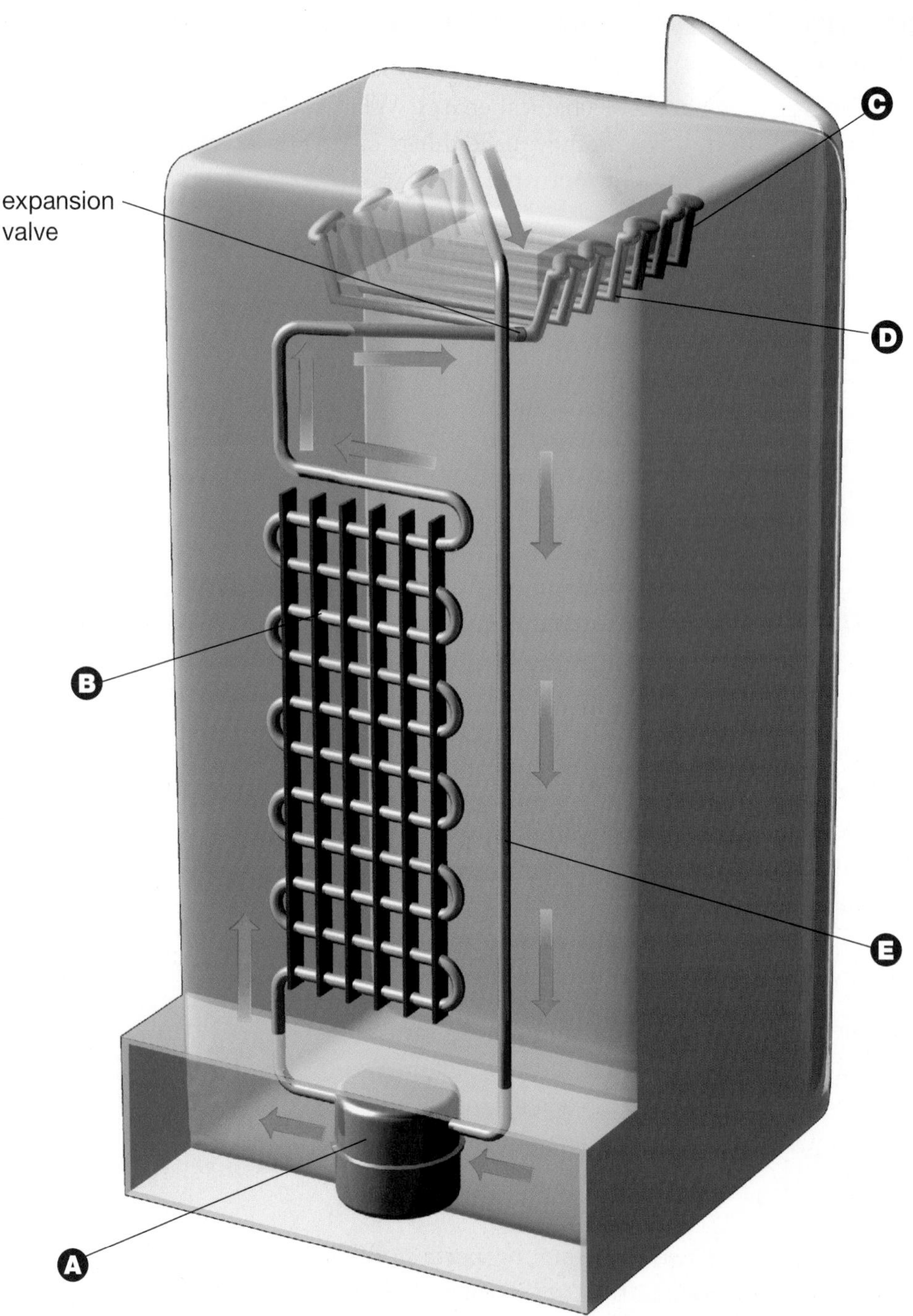

A Gas is compressed by the compressor, which makes the gas hot. The compressor is a pump that increases the pressure and then exhausts the gas at a higher pressure to the condenser.

B The hot gas moves to the condenser. In the condenser, heat from the gas is transferred to the room's air through the coils along the outside back of the refrigerator. The gas cools and condenses into a liquid.

C The liquid is pulled by suction through coils inside the refrigerator.

D In the evaporator, the liquid passes through an expansion valve to enter a region of lower pressure. The liquid absorbs heat from the food stored in the insulated refrigerator compartment and evaporates under the reduced pressure, turning back into a gas. The evaporator is where the refrigeration takes place.

E The gas flows back to the compressor. The cycle begins again.

If you opened the door of your refrigerator and left it open, would your kitchen eventually cool down? Think about where heat is produced and transferred as the refrigerator operates. The refrigerator takes heat out of the insulated portion of the compartment and gives off heat from the coils at the back. In addition, the heat engine of the compressor generates its own heat, which is also given off into the room. Believe it or not, leaving the refrigerator door open will eventually *increase* the room temperature!

Did You Know?

A chemical named Freon 12 was once the most common coolant used in refrigerators and car air conditioners. However, scientists discovered that Freon 12 contributed to the gases destroying Earth's ozone layer. Many countries have now banned the use of Freon in new appliances.

Check Your Understanding

1. Explain how a piston in a steam engine works. If you wish, use a labelled drawing.
2. How does a steam turbine differ from a steam engine?
3. Explain the meaning of each word in the term "internal combustion engine."
4. What two important properties of fluids does a refrigerator use?
5. **Apply** Sometimes steam engines are called "external combustion engines." Explain what this term means by comparing it to "internal combustion engines."
6. **Thinking Critically** The steam engine allowed factories to produce goods faster and more efficiently than production by hand. Why would the speed and efficiency of the engine encourage more people to open factories? Why did the growth in the number of factories change people's lives so dramatically?
7. **Thinking Critically** The illustration below shows how a gopher constructs its home so that one opening in the ground is higher than the other. How do you think this construction might serve to cool the gopher's home on hot days? (Hint: Think about the possible effects of differences in air pressure.)

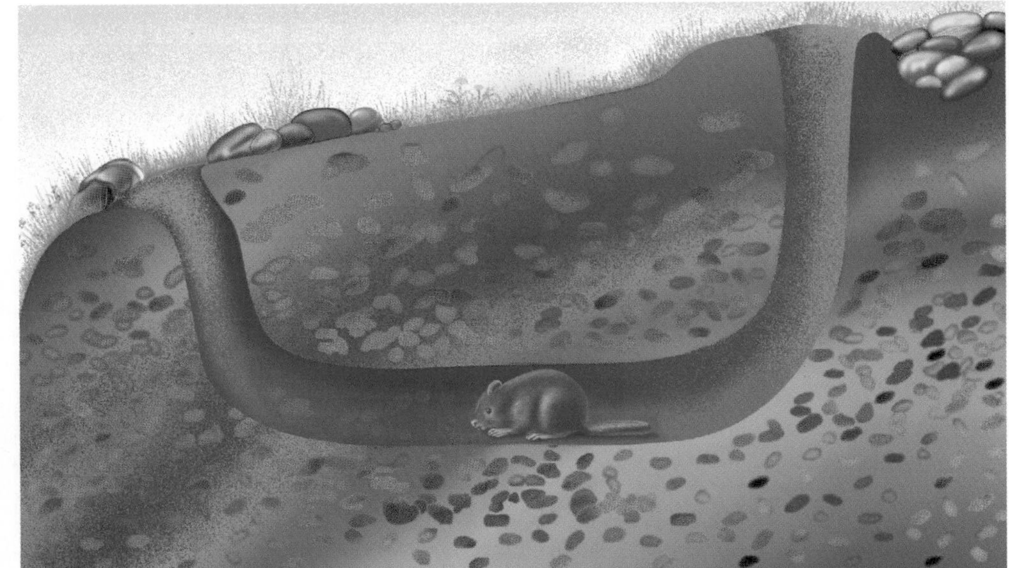

15.3 Consumer Needs, Manufacturing, and the Environment

DidYouKnow?

Earth's dwindling supply of fossil fuels and increasing pollution are spurring researchers to develop more efficient, cleaner ways of powering vehicles. Cars powered by solar cells, electricity, or gasohol (fuel made from wood, wheat, corn, sugar cane, potatoes, or animal wastes) may one day be common sights on city streets.

Imagine for a moment that you could buy a vehicle powered by an old steam-powered locomotive engine. Do you think many consumers would choose to buy this type of vehicle? It would be inconvenient, to say the least, to carry around a supply of coal to throw into the vehicle's furnace to keep the fire burning! What other problems might arise?

A vehicle with an internal combustion engine is a more efficient and practical choice. What other considerations might influence the person buying a vehicle? Price? Size? Safety? Would you think about how long the vehicle will last? How simple would it be to repair? Will it matter how efficiently the car uses fuel? How would you find this information? Could you trust advertisements produced by the manufacturer?

Organizations such as the Consumers' Association of Canada and the Canadian Standards Association (CSA) conduct independent tests comparing the efficiency, durability, strength, and safety of a wide assortment of products. These products range from seat belts and bicycle helmets to baby cribs and other furniture (see Figures 15.14A and B). Also, magazines such as *Carguide* publish "scorecards" — long-term road-test results rating ergonomics, steering, brakes, handling, safety, quality, and performance of vehicles.

Even before a product is manufactured, however, scientific tests are often conducted on control samples in order to observe, measure, and evaluate the product's performance. You have already learned about fair, unbiased testing in science. You know that every time you carry out a fair test, you must do the procedure in precisely the same way, changing only the variable you are measuring.

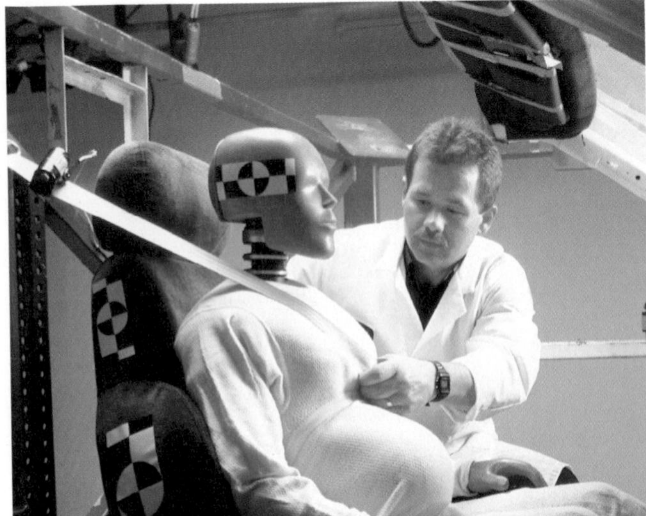

Figure 15.14A Government agencies and consumer protection groups often publish independent test results comparing and rating many consumer products. Here, a "pregnant" crash-test dummy is being used to test seat belts.

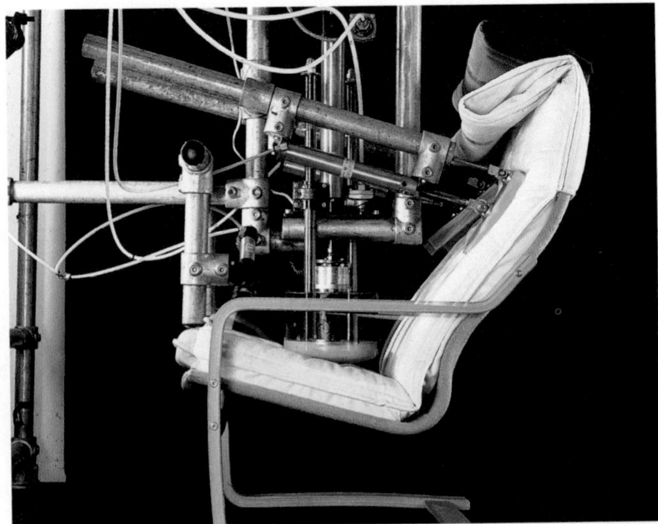

Figure 15.14B This chair is being product-tested to determine its strength and durability.

All other variables are controlled, that is, kept the same. By controlling all but the variable to be tested, you can conclude that any difference in results is due only to the one variable you have changed.

Suppose a manufacturer wants to find out which sample of hair conditioner developed in a research lab works best. In a fair test of these samples, the researchers would control variables such as the following: the volume of hair conditioner used; the time during which the hair conditioner is applied before rinsing; the volume of water used to rinse off the conditioner; the temperature and the pressure of the water used in the rinse; and the type of hair on which the conditioner is applied (the same person would participate in the test of the samples). Exactly the same procedures would be used in testing each sample. Careful repeat testing would be done, as well. The test results are evaluated to determine which sample of hair conditioner performed best. The sample that best demonstrates the characteristics of an effective hair conditioner would be selected for manufacture and distribution to consumers.

The Decision to Buy

Would environmental concerns be a factor in your decision to buy a product such as a car? Even efficiently running engines waste a great deal of energy and contribute to environmental problems such as smog and acid rain. Noise pollution can also be a nuisance, especially in large cities that have many vehicles. Would you consider buying a bicycle instead, or taking public transit?

The decision to buy a car or any other product can involve much more than personal preferences such as colour, brand, or price. For example, a consumer might ask the following questions:

- Is the product safe? Does it meet approved government or CSA standards?
- Was it manufactured in a country that allows unfair or dangerous employment conditions?
- Did making the product add to the world's pollution? Will using the product add to pollution?
- How much energy was used in making the product? Was the energy supplied by a non-renewable resource such as oil or gas, or a renewable resource, such as wind, water, or solar power?

Every time you spend money on a product, your choice supports how the product was made by sending money back to the manufacturer to make more!

INVESTIGATION 15-C

Sharpen Up!

Engineers face many challenges when designing new mechanical equipment. One challenge is simply getting the machine to work. For many machines, this can involve changing one form of motion into another. A second challenge can be to manufacture the machine as cheaply and efficiently as possible so the price for the consumer will remain low.

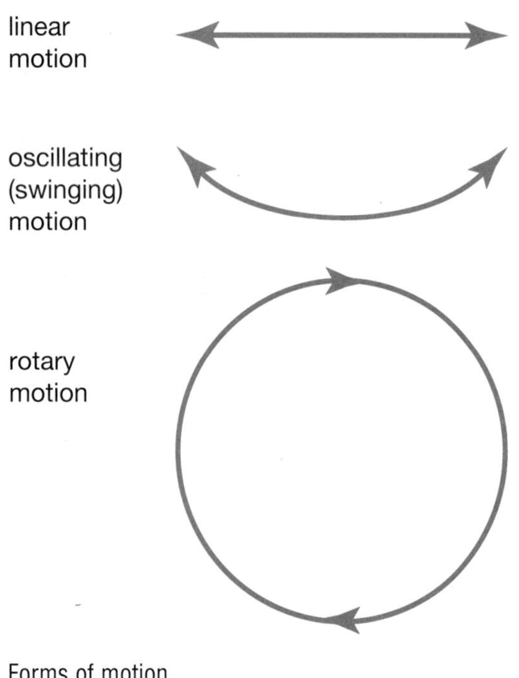

Forms of motion

Challenge

You work as part of a design team in the product development department of a company that manufactures pencil sharpeners. Your job is to create a modified design for a pencil sharpener to help the company increase its sales. Company managers do not want the new product to be too expensive, so you will have to keep costs as low as possible.

Materials

mechanical pencil sharpener ($6.00)
metre sticks ($1.00 each)
tape ($1.00/m or any portion of a metre)
coat hangers ($0.50 each)
chairs ($4.50 each)

Design Criteria

A. Your design cannot involve turning the sharpener's crank using your hands. However, you may use your hands to tip a chair back and forth to power the sharpener.

B. Your design must have a way of holding the pencil in place while it is sharpened.

C. Your design must change one form of motion into another.

D. Everyone on your design team must give themselves an hourly wage before starting. Time how long your design and construction take and add the total wages to your product's cost.

E. Each of the materials has an assigned cost shown in the "Materials" list on the bottom left of this page. Keep track of the total cost of all the materials you use in your design. You may test materials during your design stage and return them without cost if they are still usable. (For example, you cannot return used tape.)

Plan and Construct 

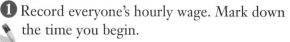

1. Record everyone's hourly wage. Mark down the time you begin.

2. With your team, plan different methods of making your pencil sharpener work.

3 Make a technical drawing of your design, using arrows to show how motion will change from one form to another. Include a list of the cost of materials next to your drawing.

4 Have your teacher approve your design. Then build and test your machine. Keep track of the time it takes to complete the project.

Evaluate

1. What was the most difficult part of building the machine?
2. Which team's machine worked best? Why?
3. Which team built the least expensive machine?
4. How did wages affect the cost of the product? What are some problems related to very high wages? What are some problems related to very low wages?
5. How might your team present its design to the company's managers to convince them that manufacturing your product is a good idea in terms of:
   - costs of materials and equipment needed to manufacture the product?
   - potential harmlessness of the product?

Extend Your Skills

6. You now work in the marketing department of the pencil sharpener manufacturer. You have been presented with the machine your team just built. How would you present this product to consumers as something that they need and want? Design an advertisement or poster for your company's new product. Select any media you like.
7. Write clear instructions for using the pencil sharpener. Be sure to include any safety warnings that you think might be necessary, especially for children.

Career CONNECT

Communication Skills in Manufacturing

As a former teacher and business writer, Marina Kovrig has had a great deal of experience reading, writing, and doing research. Language and communication skills are extremely important in her current job at a company that packages chemical products. Does this surprise you? Not everyone employed in manufacturing works with machines.

Consumers need to know if the mothballs or air fresheners or other chemical products they buy are poisonous, flammable, or potentially dangerous in any way. In Canada, laws specify that certain chemical products have to be correctly labelled to give consumers the information they need to use a product safely. Manufacturers follow Canada's *Hazardous Products Act*, which outlines what type of safe packaging must be used for each chemical product. This Act also indicates exactly what information must appear on the product's label.

As Vice President of Public and Government Affairs, Marina's responsibilities include finding the section in the *Hazardous Products Act* that applies to each product being packaged and making sure her company follows the rules laid out for that product. She also directs designers, who decide how each package will look. In addition, Marina supervises the printing of package labels and cartons to be sure all the required information appears on the product's package or container.

Marina Kovrig

Examine some chemical products in your classroom or at home. (Do this with an adult present.) What symbols or written warnings are printed on each product? Keep a class record of the warnings you and your classmates see and list on which product each warning appeared. Compare the information you find on two or more brands of the same product. Try to figure out what information the manufacturer was required by law to include on the label or container.

Old and Worn Out

No products last forever. Eventually old machines, vehicles, and equipment must be discarded. Have you ever heard someone say, "They just don't make things like they used to"? Major appliances (for example, television sets, stoves, and washing machines) and electronic items such as computers are not built to last a lifetime. Technology improves products quickly, and many consumers want only the latest model. For example, computer models become outdated in only one or two years. Why do you think manufacturers might not build appliances to last a lifetime, even if they could?

Find Out **ACTIVITY**

Is It User-Friendly?

Have you ever tried to set the timer on a video cassette recorder, only to be frustrated when you could not figure out the instructions? One factor to think about when buying any new product is how easy the item will be to assemble and use. In other words, how **user-friendly** is it? Write your own directions to see how challenging clear communication can be!

What to Do

Write a set of instructions to teach someone how to walk. Your teacher will select a few instructions and demonstrate them for the class. Alternatively, small groups could write instructions on cards, trade them with another group, and see whether they can follow the instructions.

What Did You Find Out?

1. Did your teacher or your classmates encounter any problems following the written directions? If so, what were the most common problems?
2. How could the directions be improved?
3. Would a labelled drawing or series of drawings make the directions easier to follow? Make drawings to accompany some written instructions and test them on a classmate.

Extensions

4. Write instructions telling someone how to perform a simple activity such as brushing teeth, buttoning a jacket, tying shoelaces, or making a peanut butter sandwich. Exchange your instructions with a partner and test the instructions. Recommend changes, if necessary, to make your partner's instructions more user-friendly.
5. Obtain a product manual for an item such as a telephone, a video game, a computer, a clock radio, a kitchen home appliance, or any item of your choice. Analyze the user-friendliness of the instructions in the product manual. Alternatively, analyze the user-friendliness of a web site of your choice.

Disposing of old equipment and other waste is becoming an increasingly challenging environmental problem. Part of the problem is that there is simply more waste material. With increasing urbanization, more people moved to cities and towns, farther from where their food was grown and raised. This meant farmers had to package their products to travel greater distances and to stay fresh longer. Factories began shipping goods to new markets — even in other countries. Over time, people developed many means of keeping products fresh and sanitary, for example, cans, jars, plastic bags, boxes, and bottles.

In the last 50 years or so, garbage disposal has become a crisis in some communities. More people, more products, and more packaging have meant more garbage. Landfills are filling up near many urban areas. Some cities now charge households by the bag to collect their garbage. A great deal of garbage is now made of synthetic material that does not decompose easily. Some of the material produces toxic residues. Harmful chemicals, radioactive materials, acids, and chemicals can seep into ground water and contaminate streams and rivers.

In some cases, there are no completely harmless ways to dispose of wastes. Currently, it is more expensive to remove the dirt and contaminants from used motor oil than it is to produce new oil. This means there are few disposal sites where people can take their used oil. Figure 15.15 shows one way to warn people not to pour used motor oil into the sewage system. Scientists and engineers are trying to develop more economical methods for cleaning used motor oil. Hydraulic fluid waste from industries causes much the same problem, especially because there is so much of it. Consider how much more fluid a large factory or airport or a fleet of trucks would use in its hydraulic systems, compared to a car. Figure 15.16 shows one community's response to the problem of toxic waste disposal.

Pause& Reflect

Would you be willing to pay more for your motor oil if it meant that you were protecting the environment? Why or why not? Write your responses in your Science Log.

Figure 15.15 The organization Ducks Unlimited paints yellow fish near storm sewers to remind people that if they dump used motor oil down the sewer, it will pollute streams and harm fish and other wildlife.

Figure 15.16 Some communities have a "toxic round-up" once or twice a year. Residents collect the remains of used paint, turpentine, cleaning agents, motor oil, or any chemicals that they do not know how to discard, and take them to a drop-off depot for safe disposal.

A Rubber Tire Dilemma

You might not think that worn-out, shredded tires like the ones shown in Figure 15.17 would cause a serious disposal problem. Why not use the material to make more tires? You may be surprised to discover that there is currently no method to recover rubber from old tires. The process for making tires is so complex that the final product cannot be separated back into its original ingredients. You could compare this to making gingerbread cookies and then trying to separate out all the ingredients to get them back into their original form.

Figure 15.17 Canadians dispose of about 26 million automobile tires every year.

What is wrong with dumping tires in landfills? For one thing, the decaying material in landfills often makes methane gas. Because tires have the shape of hollow donuts, the gas collects inside the tires and causes them to rise to the top of the pile. No matter what position a tire is in, it will hold water. Tires are black and therefore absorb heat from the Sun. These hot, water-filled tires become a breeding place for rats and mosquitoes that carry diseases. In fact, one type of mosquito thrives so well in these conditions that it is sometimes called the "tire pile mosquito." Figure 15.18 shows yet another risk posed by piling up old tires.

Figure 15.18 Tires do not catch fire easily, but when they do, the fire is extremely difficult to put out. Some tire fires can burn for months. Toxic fumes from smouldering tires escape into the air and oily products seep into the soil and contaminate nearby water supplies.

Figure 15.19 Shredded rubber tires make excellent surfaces for walking trails, playgrounds, indoor running tracks, and other athletic surfaces. Intact tires are also used in playgrounds, as shown here.

New Life for Old Products

Today, throughout the world, less than 5 percent of discarded tires are recycled. **Recycling** is the extraction and reprocessing of useful materials from waste for re-use. Increasingly, entrepreneurs are starting up companies that make products from old tires. Parking curbs and cushioned mats for truck beds, homes, and businesses are just some of the new uses for old tires. Figure 15.19 shows another use. Currently, more rubber from discarded tires is used in making asphalt for paving roads than for any other single product. If the ground-up rubber makes up only 3 percent of the ingredients in asphalt, 1 km of road surface would use about 7500 tires.

Recycling material can make a huge difference in the amount of waste deposited in landfills. An additional benefit is that recycled material often uses much less energy to process than new material. For example, recycled glass melts at a much lower temperature than raw glass material. About 75 percent of Canada's container glass is re-used or recycled. The energy saved by processing this recycled glass is enough to power the energy needs of a town the size of High River, Alberta, for an entire year. Glass containers such as bottles and jars are 100 percent recyclable and can be recycled over and over to make new glass containers.

Recycling products does not always mean destroying them. Many products can be re-used rather than thrown out or destroyed and recycled (see Figure 15.20). Think of some products that are good candidates for re-use.

What Do Consumers Want?

Manufacturers find out consumers' needs, attitudes, and values by doing market research, often in the form of **consumer product surveys**. As you probably know from your math studies, a survey is a sampling of information, often compiled by asking people questions or interviewing them. Surveys belong to a branch of mathematics called statistics. This branch of mathematics deals with the collection, interpretation, and presentation of numerical data.

Conduct your own consumer product survey in the next investigation.

Figure 15.20 Some people reject mass-produced fashion in favour of unique vintage and second-hand clothing. Thrift shops provide an interesting selection of inexpensive, recycled clothes.

CONDUCT AN

INVESTIGATION 15-D

A Consumer Product Survey

Many consumers are highly conscious of environmental issues. This consciousness has encouraged some manufacturers to adopt environmentally friendly methods of manufacturing and packaging. For example, aluminum pop cans are now 30 percent lighter than they were in 1960. Processing aluminum products from recycled material takes only 5 percent of the electricity required to process the original aluminum ore. Consumer concerns for the environment encouraged manufacturers to look for better ways to package pop.

Problem

How can responses to consumer product surveys help manufacturers make decisions about their products?

Procedure

Survey at least five people outside your classroom using the following questions.

Pop Quiz

1. Indicate how often you purchase pop in the following types of containers:

	Never	Sometimes	Frequently
(a) plastic container	☐	☐	☐
(b) aluminum can	☐	☐	☐
(c) glass container	☐	☐	☐

2. If you pay a deposit when you purchase the container, do you return the container for your deposit?

Never	Sometimes	Always
☐	☐	☐

3. If you did not pay a deposit when you purchased the container, do you recycle the container?

Never	Sometimes	Always
☐	☐	☐

4. How many servings of pop do you drink in a week? A serving consists of a glass or a can.

(a) 14 or more ☐

(b) 7–13 ☐

(c) 4–6 ☐

(d) 3 or fewer ☐

Sorting materials at a recycling depot

Skill POWER

To review how to make a stem-and-leaf plot, turn to page 546.

Analyze

1. Compile all the survey results in a class chart.
2. Organize the responses to question 4 of the quiz in a stem-and-leaf plot to find the average number of servings of pop consumed in a week by all the survey subjects.
3. If all the people surveyed recycled their pop containers, how many containers would be saved from landfills in one week?
4. Based on the class results, is a container deposit a good incentive to encourage consumers to recycle? Explain using information from the class results.
5. Do all people who drink seven or more servings of pop per week have the same preference about the containers they use? If you find a preference, hypothesize reasons for your results.

Conclude and Apply

6. If your survey was going to be used by a pop manufacturer to develop a package for a new soft drink, what would you recommend? Why?

Extend Your Knowledge

7. When professional market research firms conduct surveys, they often ask for the same information two or three times, each time worded differently, during the course of the survey. Why might they do this?

Extend Your Skills

8. Conduct a consumer survey for an item of your choice. You might survey people about the features they want most in a snowboard, blue jeans, sunglasses, or other product. Prepare questions on both price and performance. Show your questions to your teacher for approval. Remember that your results will be easier to analyze if all the questions require "closed" answers. For example, asking people "What kind of juice do you like?" will be harder to analyze than asking "Of the following juices, which do you prefer? Apple, orange, grapefruit, fruit punch."

   Survey at least ten people. Make a table to show your results and then write a set of recommendations for the "ideal" product to satisfy consumer demands.

A Complex Business

Pause& Reflect

It takes twice as much energy to make a 1 L glass bottle as a 1 L plastic bottle, but glass is recycled much more easily than plastic. In your Science Log, explain which alternative you think is better. Give reasons for your choice.

You have just learned about factors you might consider when buying new products, as well as some of the factors manufacturers might consider when making new products.

Some manufacturing decisions are easy to make. For example, many antiperspirants were once sold in boxes. To reduce waste, most are now sold without boxes. This helps the manufacturer save money and it helps everyone else by producing less waste for landfills. Figure 15.21 shows two types of take-out food containers. Which one do you think is more environmentally friendly?

Figure 15.21 The take-out food container on the left is made of paper, while the other is made of styrofoam.

However, not all decisions are that simple. Many businesses are discouraged from using recycled material in their products because it often costs more to buy recycled goods than to obtain new resources. This often results in higher price tags on goods made from recycled material. If consumers continue to buy products made with new material because they are less expensive, manufacturers will not support the use of recycled material. Blue boxes full of recyclable materials do nothing for the environment if the materials are never processed for re-use.

Check Your Understanding

1. List factors a consumer might consider when buying
   (a) a new winter coat
   (b) a new washing machine
   (c) a new cellular phone
   (d) a new bicycle helmet
2. List three problems that are caused by dumping old tires in a landfill.
3. Give three uses for worn-out tires.
4. (a) What is a consumer product survey?
   (b) What is its purpose?
5. (a) Give at least four reasons why a manufacturer might change the way a particular product is made.
   (b) On what evidence might a manufacturer decide to change the design or materials of a product?
6. **Design Your Own** Design your own experiment to test the effectiveness of two different brands of a household product, such as rust remover, scale remover (for bathtubs or sinks), or a consumer product of your choice. Select appropriate materials and show your experimental procedure to your teacher before you conduct your investigation.

CHAPTER at a glance

Now that you have completed this chapter, try to do the following. If you cannot, go back to the sections indicated.

Determine experimentally the effect of temperature on the volume of a fluid. (15.1)

Describe how an increase in temperature affects the pressure in a fluid. (15.1)

Using diagrams, explain how (a) a steam engine works, and (b) how an internal combustion engine works. (15.2)

Explain why a heat engine cannot be 100 percent efficient. (15.2)

Give examples of how the steam engine was a revolutionary invention. (15.2)

Explain how engines have improved since the first steam engine. (15.2)

Describe how to evaluate the user-friendliness of written instructions. (15.3)

Discuss the impact on the environment of a technological product such as rubber tires. (15.3)

Discuss connections among consumer choices, manufacturers' choices, and the environment. (15.3)

Consider new technological uses for discarded products that will help protect the environment. (15.3)

Prepare Your Own Summary

Summarize this chapter by doing one of the following. Use a graphic organizer (such as a concept map), produce a poster, or write the summary to include the key chapter ideas. Here are a few ideas to use as a guide:

- What effect does temperature have on the pressure and volume of a fluid? How can this effect be used to make machines run?

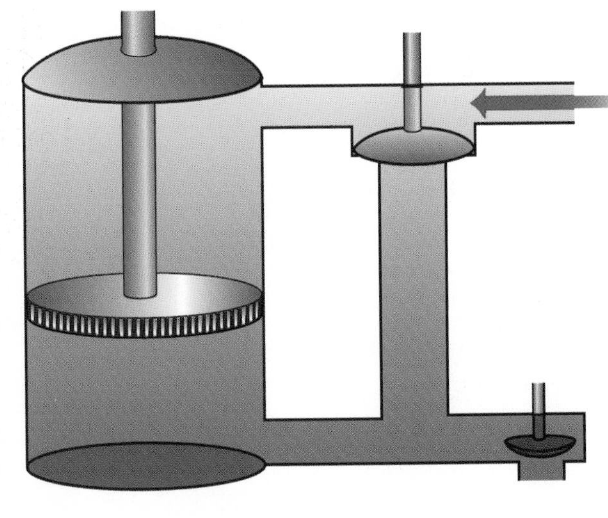

- What is the relationship between the temperature and the pressure of a gas that is confined within a certain volume?
- Sketch the diagram shown here and add labels and a detailed caption to describe how a steam engine works.
- How do turbines power ocean liners?
- Prepare a checklist for environmentally conscious consumers buying a new car or another product of your choice.
- What is the purpose of a "toxic round-up"?
- What are some ways in which new technology has helped the environment?
- Give one or two examples of a technological product or device that has caused a problem for the environment, and suggest an existing or potential solution for the problem.

CHAPTER

15 Review

Key Terms

steam
temperature
thermal energy
heat
steam engine
exhaust valves
mass production
turbine
internal combustion engine
crankshaft
heat engine
refrigerant
user-friendly
recycling
consumer product surveys

Reviewing Key Terms

If you need to review, the section numbers show you where these terms were introduced.

1. Write a sentence that correctly uses the terms "heat," "temperature," and "thermal energy." (15.1)
2. (a) What is the everyday meaning of the word "steam"? (15.1)
   (b) What is the scientific meaning of the word "steam"? (15.1)
3. How does a steam turbine differ from a steam engine? (15.2)
4. What makes a piston move? (15.2)
5. Where would you find pistons? (15.2)
6. Explain the meaning of the words "internal" and "combustion" as they are used in the term "internal combustion engine." (15.2)
7. Write a sentence that correctly includes the term "user-friendly." (15.3)
8. Give some examples of product recycling. (15.3)

Understanding Key Ideas

Section numbers are provided if you need to review.

9. How can you make the temperature of a fluid increase? (15.1)
10. When the temperature of a fluid increases, two different results could occur. (15.1)
    (a) What are these two possibilities?
    (b) What determines which result occurs?
11. When you say that water is "boiling," what is happening? (15.1)
12. Explain why all heat engines are inefficient. (15.2)
13. Steam engines and internal combustion engines both have pistons that go through an up-and-down cycle. State two ways in which the cycles for steam engines and internal combustion engines are different. (15.2)
14. Identify one feature that makes a diesel engine more efficient than a gasoline engine. (15.2)
15. Provide an example that demonstrates how the needs of consumers can influence the manufacturing of a product. (15.3)
16. How do manufacturers ensure reliable results in consumer product testing? (15.3)

Developing Skills

17. Complete the following spider map for heat engines.

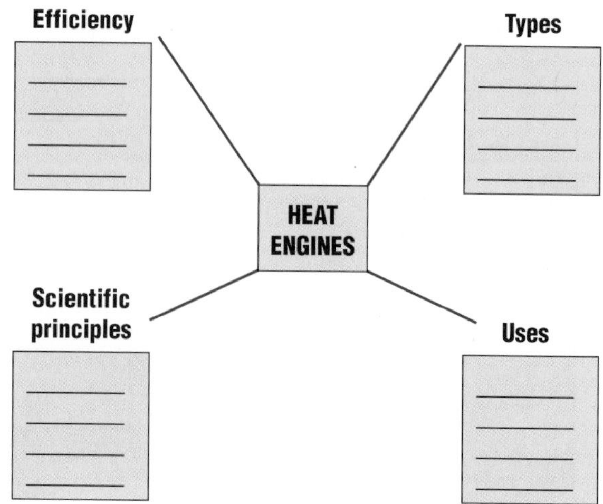

18. The following graph represents the total volume inside a piston cylinder over one cycle. Review Figure 15.10 on page 500. Sketch the graph below and label the times on this graph that the processes occurring in parts A, B, C, and D of Figure 15.10 are taking place.

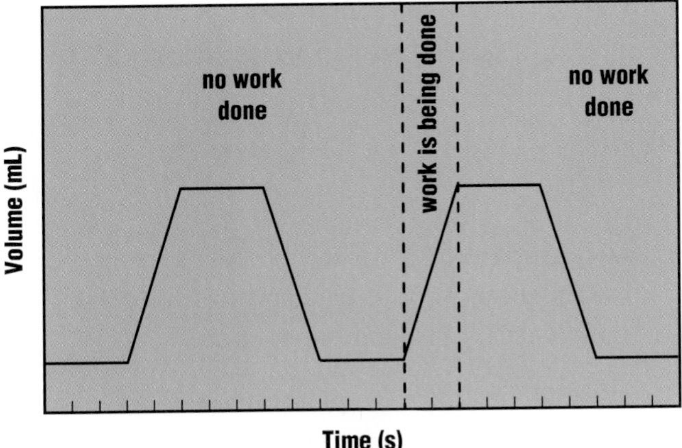

Problem Solving/Applying

19. In this chapter, you learned that adding heat to a gas and causing the temperature of the gas to increase will result in an increase in volume. However, the gas must be in a container that will allow it to expand. The reverse of this situation is also true. If you allow heat to escape from the gas and the temperature decreases, and if the pressure remains the same, the volume of the gas will decrease. Now consider another situation. If a gas suddenly expands, but no heat is added, what do you think will happen to the temperature? How might this effect be used in a practical way?

Critical Thinking

20. Why do you think it is important to continue looking for ways to make machines more efficient?

21. How do you think it was possible for Thomas Savery and James Watt to build steam engines when they did not know the scientific principles and theory of heat engines?

22. Diesel fuel is usually cheaper than gasoline and diesel engines waste less of the energy stored in the fuel. However, more consumers buy cars with gasoline engines. Hypothesize why consumers might prefer gasoline-powered cars to diesel-powered cars.

23. If you were to start your own company that made products from worn-out tires, what products would you manufacture, and why?

Pause& Reflect

1. Review the Starting Point Activity on page 487 and your answers to the questions. Now that you have learned the scientific principles relating heat, temperature, and pressure, rewrite your answers.
2. Review the Key Ideas on page 487. Write a paragraph that summarizes what you have learned about these ideas.
3. All thermo-electric plants produce heat to convert water to steam. This steam turns a turbine and generates electricity. However, hydro-electric generating plants do not follow this process. Hydro-electric power comes from generators in dams that hold back water in reservoirs. How do you think water turns the turbines to make electricity in hydro-electric generating plants? Write your ideas in your Science Log, and include a sketch if you like.
4. Go back to the beginning of this chapter on page 486 and check your original answers to the Getting Ready questions. How has your thinking changed? How would you answer those questions now that you have investigated the topics in this chapter?

UNIT 5 Ask an Expert

If you had to get someone out of an upside-down car that has been crushed in a collision, what tool would you use? Randy Segboer will tell you there is no simple answer to that question. As a firefighting instructor at the Alberta Fire Training School (AFTS), Randy trains firefighters and rescue workers in just about every skill they need to know. His specialty is rescue extrication — getting people out of dangerous situations.

Q How did you become a firefighting instructor?

A I was a mechanic for many years before I joined the fire service. I came to AFTS for specialized training as a firefighter, and later I became an instructor. I've been teaching here for three years now.

Q What do you teach students about rescue extrication?

A We teach them the proper use of tools at a rescue scene and give them hands-on experience in judging when to use each tool. We stage many kinds of rescue scenarios here at the school, and the students respond as though each scenario were the real thing.

Q Can you describe one of those scenarios?

A Let's say we have a car accident, a single-vehicle rollover, with an injured victim pinned inside the car. First, we have to stabilize the scene — make sure the car isn't going to roll farther, slide down a hill, or burst into flames. Next, we assess the victim's health and determine what's trapping the victim inside the car.

Q How do you decide which tools you will use to get the victim out?

A Unless a car door will open, we will have to force or cut some part of the car in order to get inside. It's important to use as little force as possible so that we don't make the situation worse, by causing the car to collapse, for example. So, we first try hand tools because they don't require any set-up time.

We may try to force open a jammed door using a leverage tool called a Halligan™. While one firefighter is prying with the Halligan™, others are setting up the next type of equipment in case the Halligan™ doesn't do the job. For many jammed doors, simple hand tools alone would take too long.

In this rescue simulation, Richelle Johnson, who is training to become an emergency services technician, uses a Halligan™ to pry a small opening in a jammed car door.

Randy shows Richelle how to use a heavy hydraulic spreader to force a car door open.

Q What type of equipment do the rescue workers try next?

A The next option is to use hand hydraulic tools. These are similar to a hydraulic car jack. We pump these tools by hand and the hydraulic action pulls or pushes apart two sections of the car. Hydraulic tools apply more force than a simple hand tool, but it takes two people to operate them. One person pumps to supply power while the other person manipulates the tool.

If the car door remains jammed, we move on to heavy hydraulic tools, which are powered by an engine or compressed air. Simple hand tools, hand hydraulic tools, and heavy hydraulic tools all do the same thing: push, pull, or cut. The advantage of the heavy hydraulic tools is their power and strength. In most situations, these tools get the job done.

Q If that is so, why not just use heavy hydraulic tools in every situation?

A Heavy hydraulic tools take time to set up. During that time, other workers might as well be trying the faster tools. Noise is another factor. The sound of the loud, heavy hydraulic tools can raise the victim's anxiety and blood pressure. Also, heavy hydraulic tools are extremely powerful. They apply a lot of force and when something finally gives, it gives in a big way. If you haven't correctly anticipated the point that will give, you may have made the situation worse.

Q Do you use any other kinds of rescue equipment?

A Rescue trucks and fire-pump trucks usually carry some pneumatic equipment as well, such as pneumatic wrenches, chisels, and jack-hammers. We also have air bags, which we place between the ground and a solid part of the vehicle. Then we inflate the air bag to raise the vehicle.

Our goal is to get the victim out as quickly as possible to improve chances of survival. Basically, we'll use anything that helps us achieve that goal.

EXPLORING Further

Tools of the Trade

Reread the information about the specific tools mentioned in this interview. For each tool, list as much information as you can, including

- the energy source
- any simple machine(s) involved
- advantages and disadvantages of using the tool in a rescue situation

Can you suggest other jobs that might use these same tools? List the jobs and compare your list with a classmate's.

An *Issue* to Analyze

A SIMULATION

Paper Making Goes Green

Think About It

Mistico Pulp and Paper Company in the town of Castlebrae has manufactured pulp, writing paper, and paperboard since 1922. Its pulp and paper mill is situated on the bank of the river on the east side of town. In 1992, a hugely popular bird sanctuary and the surrounding area west of Castlebrae were declared a national park. Since then, environmentalists and local residents alike have complained more and more frequently that Mistico's pollution-control practices are substandard. To improve pollution control at the mill, Mistico believes it needs to *expand* its existing facilities. The company will present its expansion bid at an upcoming town council meeting. Start preparing for the meeting by considering the Background Information.

Background Information

- For some time now, Canadian pulp and paper mills have taken pride in their effective treatment of waste to reduce water pollution. However, to satisfy the environmentalists' and citizens' concerns, Mistico would have to move toward more expensive pollution-prevention strategies. Mistico organized a task force to investigate exactly how pollution prevention might take place throughout the company.

- The task force quickly realized that the long-term effects of pollution-prevention measures would not only benefit the environment, but also improve the company's profits. Restructuring the mill to implement pollution prevention would actually begin to pay for itself. Why? Consider waste. A company loses money when incoming materials (which the company paid for) do not get turned into a finished product (selling the finished product is how the company makes money). In addition, in the pulp and paper industry, the waste material needs to be treated before it can be thrown away (either as solids dumped in landfills or as wastewater released back into rivers and lakes). The costs of treating waste are very high. If a company can reduce waste production, more of the processing materials will be transformed into finished products. Thus, waste-treatment costs can be greatly reduced.

Mistico's environmental restructuring resulted in reduced effluent (leftover chemicals or natural substances that are washed out during paper manufacturing), reduced energy use, reduced water use, better bleaching methods, and more efficient recycling. (In paper manufacturing, chemical pulp is prepared from wood chips boiled under pressure with chemicals to leave mostly cellulose fibre. The wood pulp is washed, bleached, blended, and then poured over a wire screen, leaving a thin layer of fibre.)

- Mistico's strategic restructuring involved "closing the loop" by recycling wood fibre. During paper production, machines can clog or malfunction, or paper can tear or crumple. Until the problem is caught or fixed, the product is ruined and considered wasted, without ever leaving the plant. Mistico developed a supplementary process to clean and re-pulp waste paper or paperboard to recover fibre. Not only did this save the company money on raw materials, but it also reduced the amount of solid waste that would end up in landfills. Encouraged by their effective recycled fibre methods, Mistico would like to expand its operations. It wants to build a state-of-the-art recycled fibre and newsprint facility, in the north end of Castlebrae, at a location just before the river divides.

Plan and Act

1. Plan to attend the town council meeting. Representatives from each of the following groups and organizations will be making a formal presentation at the meeting:

- National Parks and Recreation Association
  - **(a)** president of the Birdwatchers Society
  - **(b)** camp director
  - **(c)** hiker
- Mistico Pulp and Paper Company
  - **(a)** President
  - **(b)** Vice-President, Environment
  - **(c)** Vice-President, Research and Development
  - **(d)** Vice-President, Public Affairs
  - **(e)** Vice-President, Human Resources
- The Castlebrae Chamber of Commerce
  - **(a)** bank manager
  - **(b)** real estate developer
  - **(c)** merchant
- The Castlebrae and Vicinity Environmental Association (comprising private citizens)
  - **(a)** doctor
  - **(b)** teacher
  - **(c)** machinist
  - **(d)** student
- The Municipality of Castlebrae
  - **(a)** water-treatment department
  - **(b)** air-quality department
  - **(c)** hydro department

2. How do you think each person will react to the proposed expansion of the mill? What are some concerns or alternatives they might want to present and discuss at the meeting? If any facts or arguments might change their point of view, what do you think they might be?

3. Your teacher will give your group the role of one of these people, along with additional information to help you plan your presentation. As a group, research your role, and then present your position at the town council meeting. (Your teacher will provide you with a blackline master outlining the correct *Procedure for a Public Hearing*.)

4. Your task, as a class, will be to assess all the arguments presented at the meeting. You will weigh the arguments to try to determine the degree of risk to the environment and the existing community that you think might be posed by expanding the paper mill's operations. You will then devise a plan of action in response to your conclusions.

Analyze

1. Were all the arguments put forth at the meeting well researched and well presented? If not, how do you think they could be improved?

2. Were all participants given an equal opportunity to express their views and their concerns? If not, how might this be improved?

3. What plan of action did you decide upon as a class? How did your understanding of science and technology influence your plan of action?

UNIT 5 PROJECT

Adapting Tools

Have you ever broken your arm and had it set in a cast? If so, you probably had trouble doing simple, everyday tasks. Opening a jar of peanut butter or styling your hair would be awkward with your arm in a cast. Imagine what it must be like to experience the permanent loss of a hand or an arm.

Some people are born with conditions that make it hard to perform delicate hand and finger movements. Also, many older people do not have the strength to open a jar or a can. People with arthritis find it difficult to open bottles or jars with childproof lids. Common household utensils and tools are not usually designed for people with such physical challenges.

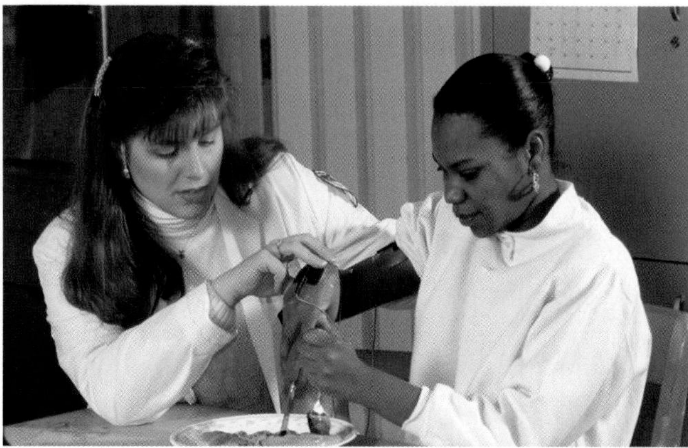

With the help of an occupational therapist, this woman is relearning how to use a knife after a hand injury.

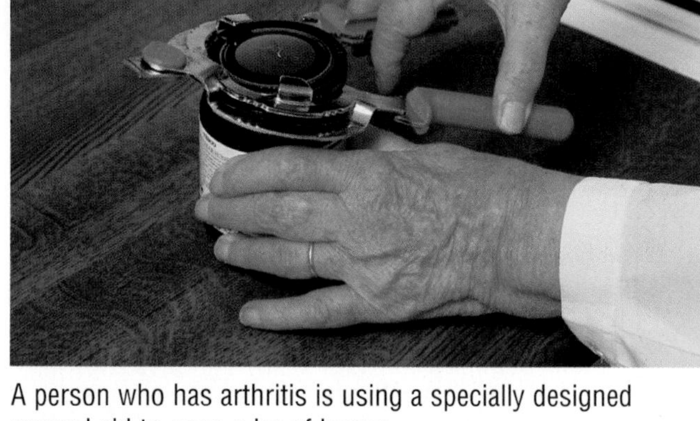

A person who has arthritis is using a specially designed manual aid to open a jar of honey.

One of the jobs of an occupational therapist is to find or adapt tools and gadgets for use by people who have been injured or disabled. The photograph on the right above shows a special tool designed for use by people who have conditions such as arthritis. (Arthritis is an inflammation of the joints characterized by pain, swelling, and stiffness.)

Challenge

Adapt or redesign tools, utensils, personal-care items, or craft or hobby items for use by an older person or a person with a physical injury or disability.

Materials

tools (e.g., pliers, wrench, hammer, screwdriver, putty knife)

utensils (e.g., table knife, fork, spoon, tongs, funnel, measuring spoons, spatula)

personal-care items (e.g., comb, hairbrush, toothbrush, soap, mirror, empty childproof prescription bottle), or other common household items of your choice

cardboard

scissors

dowels

tape

wood

Styrofoam™

glue or glue gun

Safety Precautions

- Be careful when using sharp objects such as scissors, knives, and screwdrivers.
- A glue gun is hot and the glue remains hot for several minutes.

Design Criteria

A. Choose two items from the tools, utensils, and personal-care items listed above. Adapt or redesign the devices for use by a person who is physically challenged in some way.

B. Each adaptation must include at least one type of simple machine that you have studied in this unit.

C. Your adaptation must be completed and ready to demonstrate to the class during the time allotted by your teacher.

D. Your adaptation may be either an actual device or a working model.

E. You must submit a summary of each team member's contribution to the design, development, and demonstration of your adapted device.

Plan and Construct

1. In your group, brainstorm a variety of tools or utensils that you could adapt to meet the Challenge outlined above. Decide what specific task each adapted device would allow the person using it to perform. Discuss why a person with a specific physical disability would be unable to perform the task without your device.

2. Of the items that you discussed, select two that the majority of the group members would like to adapt or redesign. If you wish, select one optional device in case one of your choices does not function as planned.

3. Make a list of the materials you will need for each device or model.

4. Assign tasks to each member of your group, such as the collection of materials, the assembly of the device or model, and the testing of the device or model. Set deadlines for each stage of the project.

5. As a group, draft an illustrated plan that clearly shows the materials and the design of the adapted device. Submit your plan to your teacher for approval.

6. Subdivide the group into two smaller groups and assign one of the two devices to each group. Assemble your device, testing it at each stage of assembly. When each group has accomplished as much as possible, the two groups will confer with each other. If one group has ideas that can help the other group improve its device, implement that idea. If either group finds that its device cannot be made to function properly, the group should start to work on the alternative.

7. When the devices or models are completed, the whole group will prepare a written and oral presentation describing and demonstrating the devices for the class.

Evaluate

1. As a group, discuss the effectiveness of your devices. Did they perform as well as you had intended? Why or why not?

2. Did you encounter problems in developing your devices? If so, how well did you solve the problems?

3. How practical would your devices be for use by a person who has a physical challenge?

4. What would you change about your design, if you were to begin again?

5. Write a summary of your group's evaluation of the two devices.

Working in a group, design an original tool or device for use by a person who has a physical disability. If time permits, construct a model of your new device and give it a name.

APPENDIX

Classifying Living Things

Over 2000 years ago, the Greek philosopher Aristotle developed a system of classification that grouped organisms according to whether they were plant or animal. Scientists used Aristotle's system for hundreds of years, but as they discovered more and more living things, the system did not work well because it did not show probable relationships between similar organisms.

In 1735, Carolus Linnaeus produced a new system that also classified all organisms as plant or animal, but this new system was very different in other ways from Aristotle's system.

Linnaeus' system gives a two-word name to each type of organism. This system of naming organisms is still in use today. The two-word name is called the organism's scientific name, and it is given in Latin, a language that is no longer spoken. The first word of the organism's name is its genus, and the second word is its specific name. A **genus** is a group of species that are related. A **species** is the smaller, more limiting classification grouping. A species name includes both the genus name and specific name. For example, the dog shown above on the left and the wolf shown on the right are members of the same genus, *Canis*. The dog, however, is the species *Canis familiaris*, while the wolf is the species *Canis lupus*.

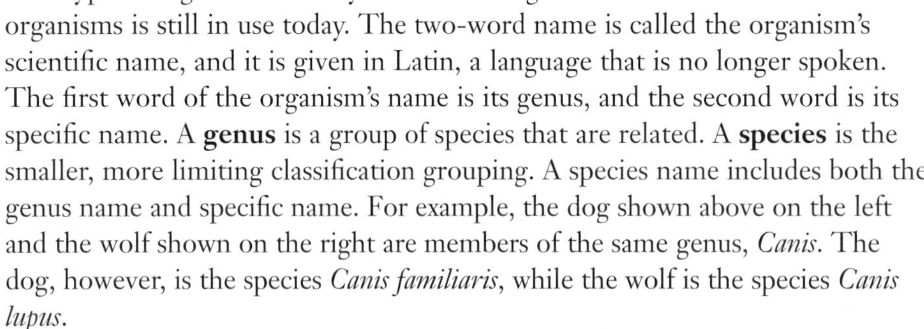

By the 1900s, scientists had discovered a great diversity of organisms on Earth. Separating organisms into only two main groups or **kingdoms**, plant and animal, began to seem inadequate. For example, bacteria are just too different from either plants or animals to be grouped with either. Similarly, fungi such as bread mould, yeast, and the many kinds of mushrooms are very different from plants and animals. In 1969, Robert Whittaker proposed a system that classifies organisms into five different kingdoms. The illustrated table on the next page shows the major groups of organisms and their kingdoms.

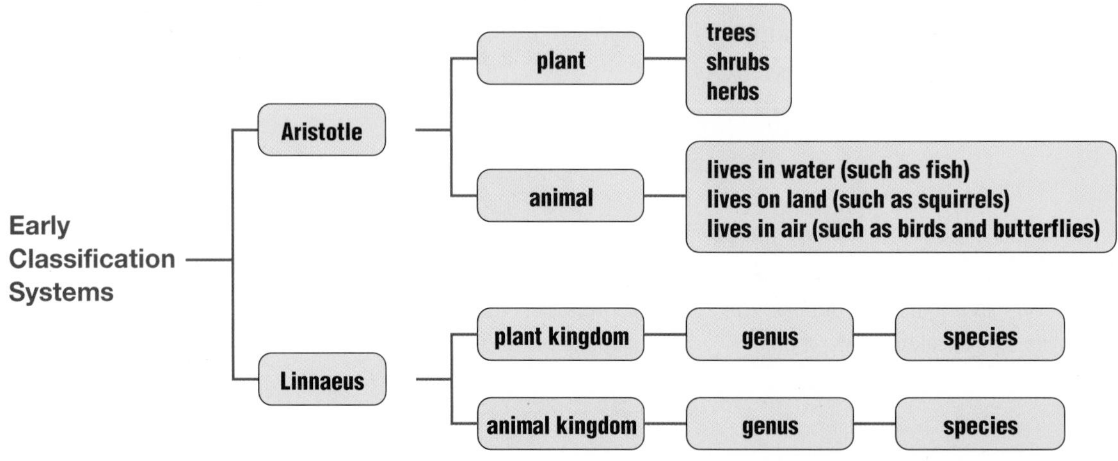

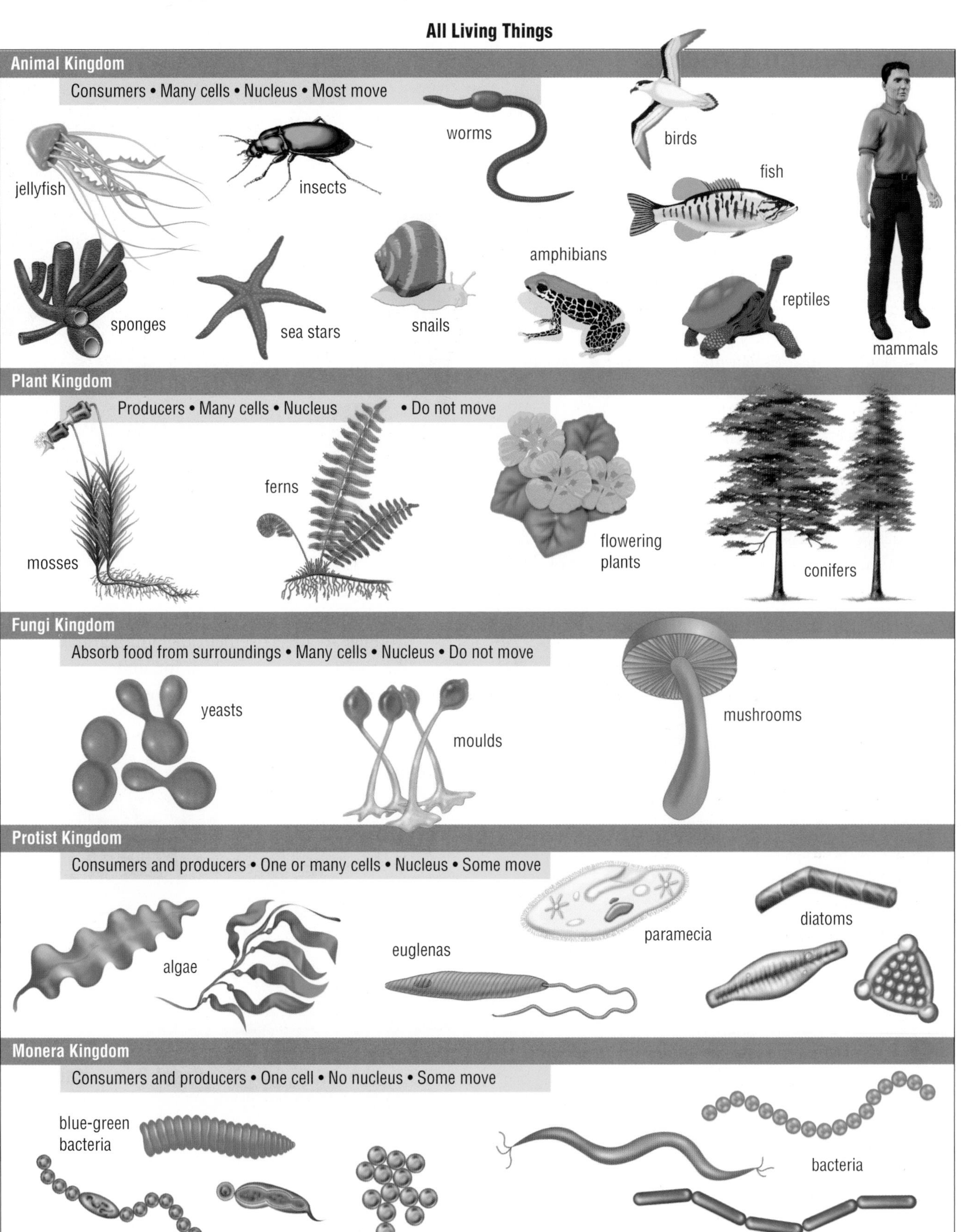
All Living Things
Animal Kingdom
Consumers • Many cells • Nucleus • Most move
jellyfish
insects
worms
birds
fish
sponges
sea stars
snails
amphibians
reptiles
mammals
Plant Kingdom
Producers • Many cells • Nucleus • Do not move
mosses
ferns
flowering plants
conifers
Fungi Kingdom
Absorb food from surroundings • Many cells • Nucleus • Do not move
yeasts
moulds
mushrooms
Protist Kingdom
Consumers and producers • One or many cells • Nucleus • Some move
algae
euglenas
paramecia
diatoms
Monera Kingdom
Consumers and producers • One cell • No nucleus • Some move
blue-green bacteria
bacteria

APPENDIX

B

The Design Process

The *Design & Do Investigations* in this book, as well as some of the end-of-unit *Projects*, give you an opportunity to design, plan, and construct your own devices, systems, or products. To help you develop and present your design concepts, you may find it useful to make two- or three-dimensional scale diagrams, a mock-up, or a prototype (model). In many projects, the construction of the end product is too costly or time-consuming. Diagrams and models, therefore, are simple, inexpensive ways of showing the final result of your ideas.

Start with Sketches

1. Sketch objects when a two-dimensional view is sufficient, such as a pair of scissors.

2. As you are getting started, use printed material provided by your teacher to copy shapes you would like to use. Tracing is a great way to copy shapes from printed material.
3. Use plastic templates as guides for producing smooth curves.
4. If the end product is to be hand-held, cut out the full-size templates (patterns) to get the feel of it.

5. Apply drawing and communication skills throughout the project. Create graphics to accompany your design. You might want to produce an advertisement including graphics like the one shown. In fact, for some projects, the graphics may be your end product.

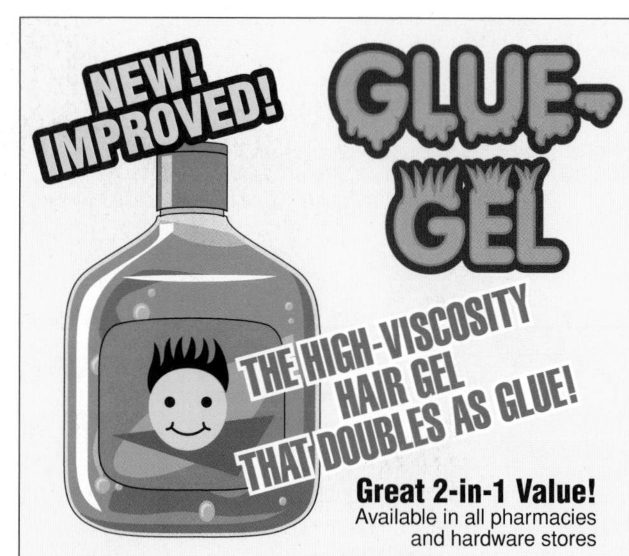

Design on Graph Paper

Use graph paper to produce your design. The grid will help you draw to scale. As well, use graph paper to produce clear flow diagrams.

6 Isometric grid paper is an excellent aid in making pictorial sketches. Use it to help plan your design.

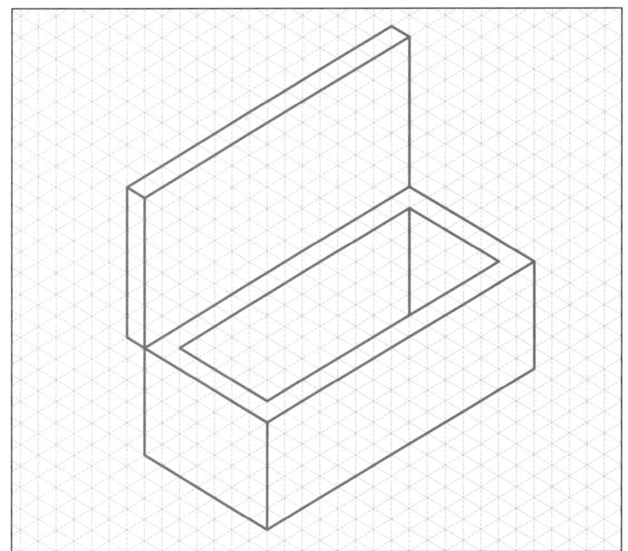

7 Sometimes showing exploded views of your planned project is useful. You can draw these on graph paper for better accuracy.

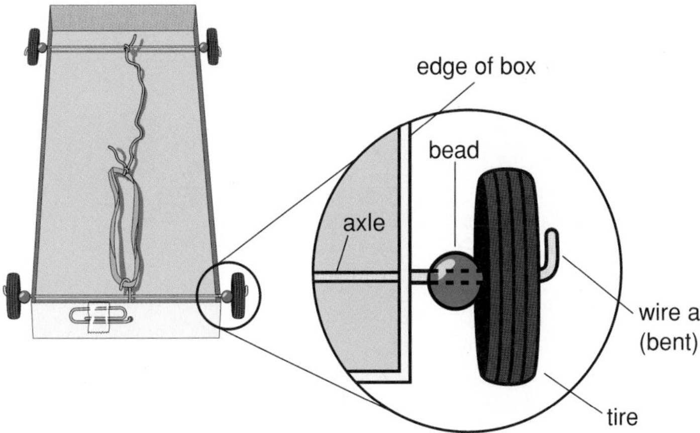

Plan, Construct, and Evaluate

8 Use a technological problem-solving approach like the one explained on pages IS-14–IS-17 to carry out your design plan.

9 As you design, plan, and construct more projects, you will find that your sketching and drafting skills will improve. You can even use computer programs to develop parts of your projects if you have access to the necessary equipment and software.

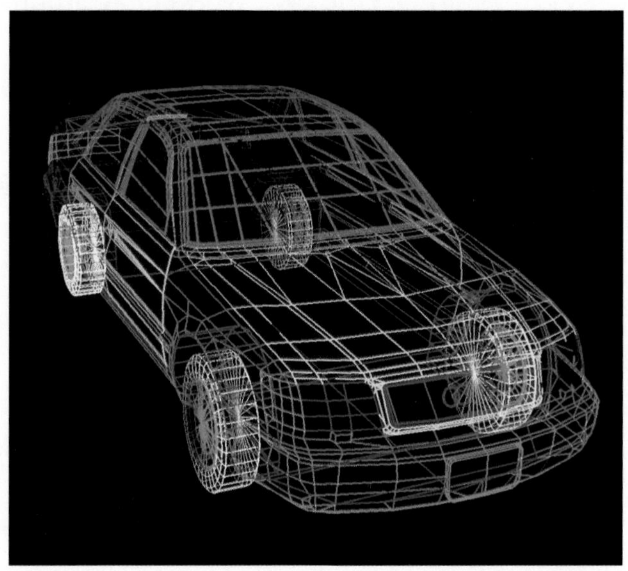

10 As you design, plan, and construct each project, be sure to continue thinking of how you can solve any problems you encounter in its development. The evaluation process at the end may take you in new directions, or you may decide to make only minor changes to your design.

Science and Technology Skills Guide

USING YOUR TEXTBOOK AS A STUDY TOOL

SCIENCEPOWER™ 8 contains a great deal of useful information. How can you read your textbook effectively in order to add information to your existing store of knowledge, and to identify areas of inquiry that you might like to pursue? This *SkillPower* will give you some ideas for remembering what you read.

Organizing the Information in Your Textbook

Look at all of the suggestions presented here. Use the learning strategies that work for you, but try others as well. Doing something in a different way often helps you see ideas more clearly and understand them better.

1. When you are starting a new unit, read the *Chapter Title*, the *Key Ideas*, and the *Key Skills* in each chapter. Try to predict some ideas you might learn about in each chapter. Write some questions of your own about the chapter material.
2. Try rewriting the chapter section headings and subheadings as questions. Then look for the answers to each question in the section.
3. Think about what you are reading, and write brief notes to help you remember the information in each paragraph.

Using Your Textbook Visuals

As you read each page, look at any photographs, illustrations, or graphs that appear on the page. Read the captions and labels that accompany the photographs, as well as the titles of graphs. Think about the information each visual provides, and note how it helps you to understand the ideas presented in the text. For example, look closely at the illustration on this page. What information does it convey to you?

Look, as well, at any terms that are in bold (dark, heavy) type. These terms will provide important definitions that you will need in order to understand and write about the information in the chapter. Make sure that you understand these terms and how they are used. Each boldfaced term appears in the *Glossary* at the back of this book.

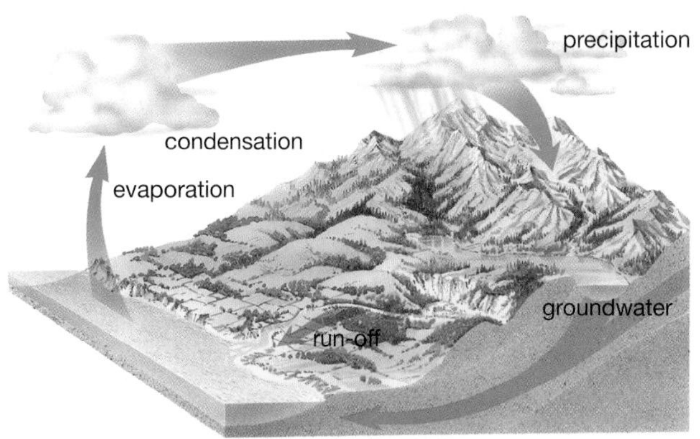

Water on Earth moves in an endless water cycle.

Making Sure You Understand

At the end of every section and every chapter, you will find review questions. If you are unable to answer them, reread the material to find the answers. If you feel that you do understand the chapter well, try using the questions provided at the end of the chapter under *Prepare Your Own Summary* to summarize the key ideas in the chapter. Preparing your summary is a good way to see if you have understood the ideas.

Instant Practice

1. Go to the unit your teacher has told you that you will be studying, and try strategy number 1 (under "Organizing the Information in Your Textbook").
2. In the first chapter of the unit, try out strategy number 2.
3. Find any terms that are in bold in the introduction and in the first section of the first chapter of the unit. Record the terms and their meanings.

Graphic Organizers

A good way to organize information you are learning is to use a **graphic organizer**. One kind of graphic organizer you will find useful is a **concept map**.

A concept map is a diagram that represents visually how ideas are related. Because the concept map shows the relationships among concepts, it can clarify the meaning of the ideas and terms and help you to understand what you are studying.

Study the construction of the concept map below called a **network tree**. Notice how some words are enclosed while others are written on connecting lines. The enclosed words are ideas or terms called concepts. The lines in the map show related concepts, and the words written on them describe relationships between the concepts.

As you learn more about the topic, your concept map will grow and change. Concept maps are just another tool for you to use. There is no single "correct" concept map, only the connections that make sense to you. Make your map as neat and clear as possible and make sure you have good reasons for suggesting the connections between its parts.

When you have completed the concept map, you may have dozens of interesting ideas. Your map is a record of your thinking. Although it may contain many of the same concepts as other students' maps, your ideas may be recorded and linked differently. You can use your map for study and review. You can refer to it to help you recall concepts and relationships. At a later date, you can use your map to see what you have learned and how your ideas have changed.

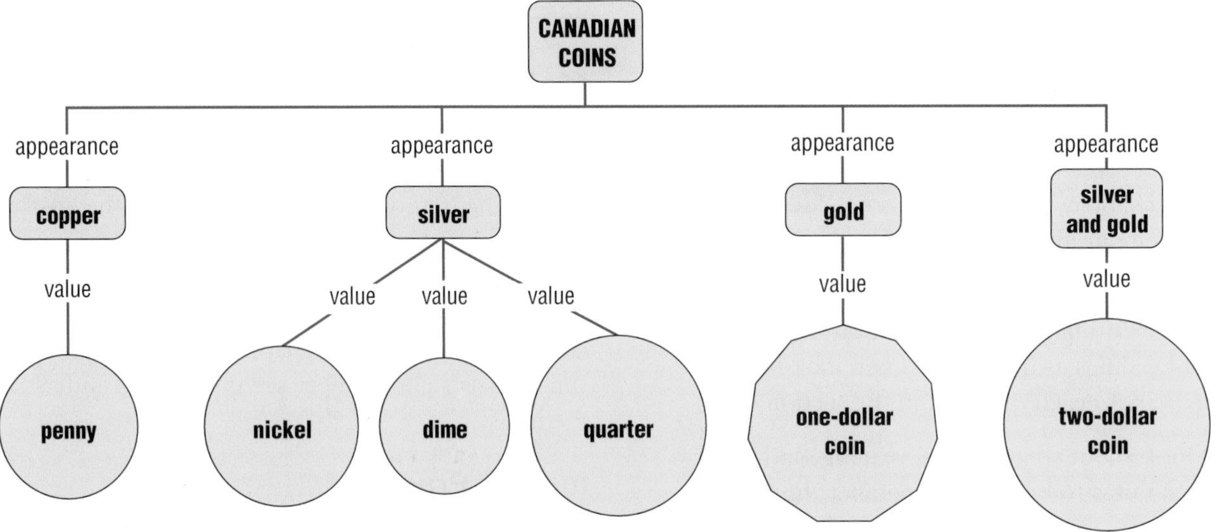

An **events chain map** describes ideas in order. In science, an events chain can be used to describe a sequence of events, the steps in a procedure, or the stages of a process. When making an events chain, you must first find out the one event that starts the chain. This event is called the initiating event. You then find the next event in the chain and continue until you reach an outcome. Here is an events chain concept map showing how an animal fossil may be formed.

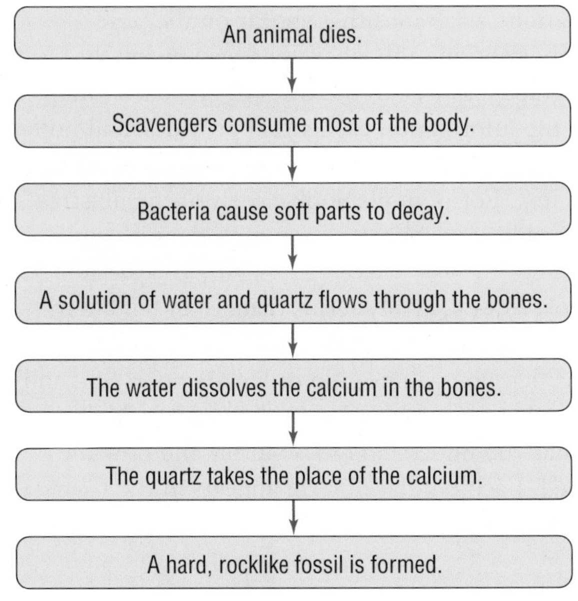

A **cycle concept map** is a special type of events chain map. In a cycle concept map, the series of events do not produce a final outcome. This type of concept map has no beginning and no end.

To construct a cycle concept map, you first decide on a starting point and then list each important event in order. Since there is no outcome and the last event relates back to the first event, the cycle repeats itself. Look at the cycle concept map below showing changes in the state of water.

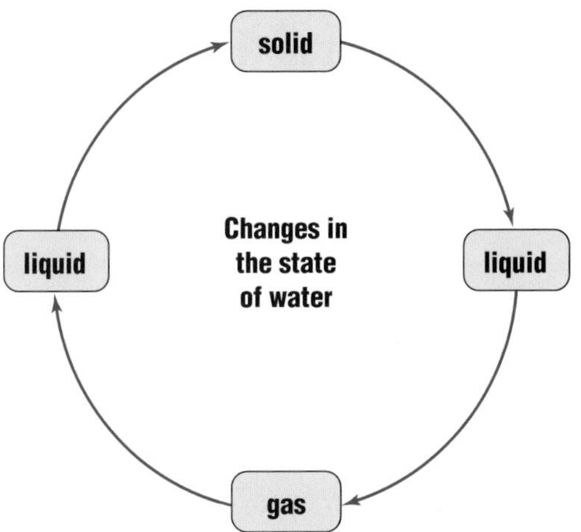

A **spider map** is a concept map that you may find useful for brainstorming. You may, for example, have a central idea and a jumble of associated concepts, but they may not necessarily be related to each other. By placing these associated ideas outside the main concept, you can begin to group these ideas so that their relationships become easier to understand. Examine the following spider map of the geological time scale to see how various concepts related to this time scale may be grouped to provide clearer understanding.

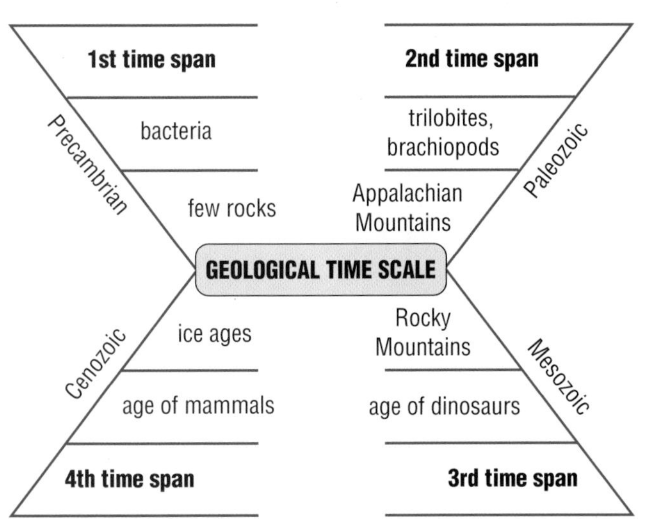

Comparing and contrasting is another way to help solidify your learning. When you compare, you look for similarities between two concepts or objects. When you contrast, you look for differences. You can do this by listing ways in which two things are similar and ways in which they are different. You can also use a graphic organizer called a **Venn diagram** to do this, using two circles.

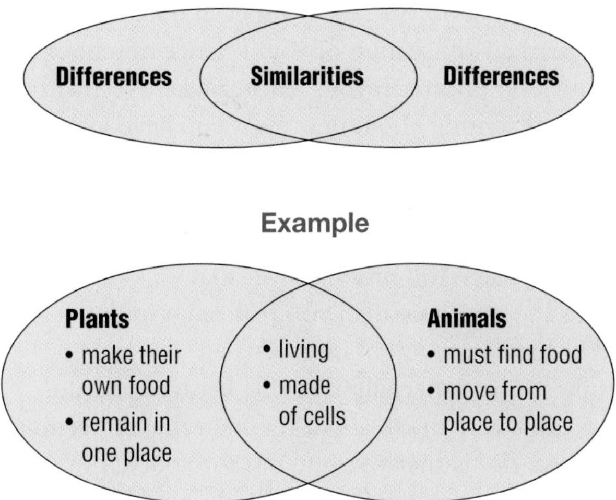

Testing Your Skills

Another way to make sure you understand what you have learned is to practise and use the special skills you learned at the beginning of this textbook and all through it. These are the skills of inquiry, problem solving, and decision making.

Instant Practice

1. Use the following words to produce a network tree concept map: music for listening, music for dancing, rap, classical, rhythm and blues, favourite CDs, rock, folk.
2. Make a Venn diagram to compare and contrast chocolate chip cookies with oatmeal cookies.
3. Produce an events chain concept map that starts with the buzzing of your clock radio and ends with your sitting at your desk as school begins.
4. Design a spider map to represent different means of transportation.

HOW TO USE A SCIENCE LOG

Scientists keep logs — detailed records — of their observations, new data, and new ideas. You can keep a *Science Log* (or *Science Journal*) to help you organize your thinking.

In your *Science Log*, you can record what you already know about a topic and add to that information as you learn more. Your teacher might ask you to keep a *Science Log* as a special booklet or as a marked-off section of your science notebook. Whichever approach your teacher takes, you will find that writing about new ideas you have learned will help you to solidify them in your own mind.

Stopping to reflect and then writing about what you already know may also help you to understand a topic better. You may discover that you know more about a topic than you realized. On the other hand, you might discover that you need to study a topic especially carefully because you do not know very much about it. The value in keeping a *Science Log* is that you find out for yourself how clear your understanding is. You do not have to wait until your teacher assesses your knowledge through a formal test or examination to know how well you understand a topic.

SCIENCEPOWER™ 8 has some special features to make sure you can add to your *Science Log* effectively. Each chapter begins with a set of questions called "Getting Ready...".

Getting Ready...

- What do you think a cell looks like?
- A wooden baseball bat is made up of dead tree cells. What cell part gives the bat its hardness?
- Why is it not possible for a single-celled organism to grow to be as large as you?

Are you able to answer any of these questions from your previous studies? You can write the answers, draw a sketch, or use whatever means you find best to explain what you know. Also, feel free to record that you know very little about a particular topic if that is the case.

Pause& Reflect

Here are some situations in which diffusion occurs: A sugar cube is left in a beaker of water for a while. Fumes of perfume rise from the bottle when the top is removed. Give some other examples of diffusion. Can solids diffuse? Why or why not? Write your responses in your Science Log.

Throughout each chapter, *Pause & Reflect* features help you keep thinking about what you already know. These features are designed to help you make connections between ideas and organize your thoughts. Your teacher may have you use these features all the time to record your new learnings, or your teacher may ask you to decide for yourself how often to answer the *Pause & Reflect* questions.

As the final question in each Chapter Review, a *Pause & Reflect* item asks you to look back at your original answers to the *Getting Ready* questions at the beginning of the chapter, and to compose new answers to those questions. You may be amazed at how much your answers have changed, based on the new knowledge you have gained by studying the chapter.

Other information you might want to add to your *Science Log* could include:

- questions that occur to you that you would like to be able to answer
- sketches and notes about models and processes in science
- graphic organizers (see examples in *SkillPower 1*)
- thoughts on what is difficult for you and ideas on how you might overcome the barriers to learning a new topic
- notes about interesting items in the news that involve a chapter topic and that spark more questions or answers to some existing questions
- profiles of leading Canadian scientists or technologists that you learn about in the media, plus profiles of careers related to science and technology that you find interesting
- connections between science and other subject areas that occur to you in the course of your learning

Your *Science Log* will help you become a better learner, so take the time to make entries on a regular basis.

Instant Practice

1. What do you know about science? What do you know about technology? What do you know about how both science and technology can help answer questions about societal issues? Formulate your own questions about science, technology, and societal issues to produce a set of *Getting Ready* questions. Then exchange your questions with a classmate so each of you can start your own *Science Log* using these questions.
2. Think of something that you consider to be an example of a simple technology. It can be any tool or device that you may have used. Do you necessarily have to be a scientist to produce such technological devices? What is an example of a complex technology? How do you think scientific knowledge might help to produce such technology? Write responses to these questions in your *Science Log*.
3. After you have finished reading pages xxviii – IS-17, produce a set of *Pause & Reflect* questions comparing and contrasting science and technology. (How are they similar and how are they different?) Again, exchange your set of questions with a classmate to continue the development of your *Science Log*.

WORKING IN GROUPS

In this program, you are often asked to work in a group to complete a task. Working in a group may help you to improve your ability to think critically, solve problems, and remember ideas and concepts you have learned. However, you must follow some important rules for effective learning in groups.

In a co-operative group, each of you will have one or more assigned tasks for which you are responsible. This means that a group can make the best use of each member's special skills, and help develop new skills. Teams often develop a special ability to work together, so that each presentation or project completed together is an improvement over the last.

Working well in a group is not always easy. The best way to develop your skills is to consider carefully what makes a group succeed at a task.

Assessing Group Performance

These are some behaviours you should aim toward when doing group work in this course and others:

- Share your ideas with others in the group.
- Show others respect, even if you disagree with them.
- Listen to another group member when that person is speaking.
- Encourage others to speak.
- Stay focussed on the group's task.
- Help the group to stay focussed on task.
- Do not allow yourself to be distracted.
- Keep your voice low enough that it will not distract other groups.
- Allow others to present their ideas, even when you think you know the answers.

Instant Practice

1. The first time you are asked to work in a group in this course, assess yourself on how well you did. Your teacher will provide a *Performance Task Self-Assessment* form for you to use. Before you begin your assessment, read over the form to note how you will be assessed.
2. Use the *Performance Task Group Assessment* form provided by your teacher to assess your whole group.

UNITS OF MEASUREMENT AND SCIENTIFIC NOTATION

Throughout history, people have developed systems of numbering and measurement. In time, when different groups of people began to communicate with each other, they discovered that their systems and units of measurement were different. Groups within societies created their own unique systems of measurement. Did you know, for example, that the height of a horse is still measured in "hands"?

Today, however, scientists around the world have agreed to use the metric system of numbers and units. The metric system is the official system of measurement in Canada.

The Metric System

The word "metric" comes from the Greek word, *metron*, which means "measure." The **metric system** is based on multiples of ten. For example, the basic unit of length is the metre. All larger units of length are expressed in units based on metres multiplied by 10, 100, 1000, or more. Smaller units of length are expressed in units based on metres divided by 10, 100, 1000, or more. Each multiple of ten has its own prefix (a word joined to the beginning of another word). For example, *kilo-* means multiplied by 1000. Thus, one kilometre is one thousand metres.

$$1 \text{ km} = 1000 \text{ m}$$

The prefix *milli-* means divided by one thousand. Thus, one millimetre is one one-thousandth of a metre.

$$1 \text{ mm} = \frac{1}{1000} \text{ m}$$

In the metric system, the same prefixes are used for nearly all types of measure, such as mass, weight, area, and energy. A table of the most commonly used metric prefixes is given on page 539.

Example 1

The distance from Halifax to Winnipeg is 3538 km. Express this distance in metres.

Solution

$$3538 \text{ km} = ? \text{ m}$$
$$1 \text{ km} = 1000 \text{ m}$$
$$3538 \text{ km} = 3538 \times 1000 \text{ m}$$
$$= 3\ 538\ 000 \text{ m}$$

Example 2

There are 250 g of cereal in a package. Express this mass in kilograms.

Solution

$$1000 \text{ g} = 1 \text{ kg}$$
$$250 \text{ g} \times 4 = 1000 \text{ g}$$
$$\frac{1000}{4} \text{ g} = 250 \text{ g}$$
$$\frac{1}{4} \text{ kg} = 0.25 \text{ kg}$$

The following table lists most of the frequently used metric quantities you will encounter in your science classes. You will learn more about units of electricity, the ampere and the coulomb, in your grade 9 studies.

Frequently Used Scientific Quantities, Units, and Symbols

Quantity	Unit	Symbol
length	nanometre micrometre millimetre centimetre metre kilometre	nm μm mm cm m km
mass	gram kilogram tonne	g kg t
area	square centimetre square metre hectare	cm^2 m^2 ha
volume	cubic centimetre cubic metre millilitre litre	cm^3 m^3 mL L
time	second	s
temperature	degree Celsius	°C
force	newton	N
energy	joule kilojoule	J kJ
pressure	pascal kilopascal	Pa kPa
electric current	ampere	A
quantity of electric charge	coulomb	C
frequency	hertz	Hz
power	watt	W

Instant Practice

1. A box is 35 cm wide. Express the width in metres.
2. You ride your bicycle 1.4 km to school. Express the distance in metres.
3. A teaspoon of water has a mass of 5.0 g. Express the mass in milligrams.
4. There are 600 mL of soft drink in a bottle. Express the volume in litres.
5. A glass of water contains 32 μg of sulfur. Express the mass in grams.
6. A student added 0.0055 L of cleaning solution to some water. Express the volume in mL.

SI Units

In science classes, you will often be instructed to report your measurements and answers in **SI** units. The term SI is taken from the French name *Le Système international d'unités*. In SI, the unit of mass is the kilogram, the unit of length is the metre, the unit of time is the second, the unit of temperature is the Kelvin (see "SkillPower 5: Estimating and Measuring," for an explanation of the Kelvin), and the unit of electric current is the ampere. Nearly all other units are defined as combinations of these units.

Example 1

Convert 527 cm to SI units.

Solution

The SI unit of length is the metre.

1 m = 100 cm

$$527\ \cancel{\text{cm}} \times \frac{1\ \text{m}}{100\ \cancel{\text{cm}}} = 5.27\ \text{m}$$

Example 2

Convert 3.2 h to SI units.

Solution

The SI unit of time is the second.

1 min = 60 s; 1 h = 60 min

$$\frac{3.2\ \cancel{\text{h}} \times 60\ \cancel{\text{min}}}{\cancel{\text{h}}} \times \frac{60\ \text{s}}{1\ \cancel{\text{min}}} = 11\ 520\ \text{s}$$

Instant Practice

Convert the following quantities to SI units.

1. 52 km
2. 43 min
3. 8.63 g
4. 45 973 mm
5. 537 891 cm
6. 1.75 h
7. 16 Mg (megagrams)
8. 100 km/h

Scientific Notation

Some common values in SI units can be very large or very small. The distance from Earth to the Sun is 149 600 000 000 m. The mass of an electron is 0.000 000 000 000 000 000 000 000 000 000 91 kg. The speed of light is 299 792 500 m/s. These numbers are difficult to read and tedious to write.

To solve the problem, scientists use **scientific notation** to express a number in exponents of ten. An **exponent**, or power, tells you how many times the number is multiplied by itself. For example, 10^3 means 10 × 10 × 10 or 1000. Notice that the exponent of ten is the number of zeros after the 1 in 1000. The following table shows some powers of 10 written in standard form (no exponents) and in exponential form.

In scientific notation, all numbers are written in the form, $X.Y \times 10^n$, where X and Y are numbers and n is the exponent of 10. The decimal point is always placed after the first digit. Examine the following examples to learn how to convert numbers from standard form to scientific notation.

	Standard form	Exponential form
ten thousands	10 000	10^4
thousands	1000	10^3
hundreds	100	10^2
tens	10	10^1
ones	1	10^0
tenths	0.1	$\frac{1}{10^1} = 10^{-1}$
hundredths	0.01	$\frac{1}{10^2} = 10^{-2}$
thousandths	0.001	$\frac{1}{10^3} = 10^{-3}$
ten thousandths	0.0001	$\frac{1}{10^4} = 10^{-4}$

Example 1

The circumference of Earth at the equator is about 127 500 000 000 000 m. Write this distance in scientific notation.

Solution

First, count the number of places you have to move the decimal point to leave only one digit to the left of it.

127 500 000 000 000. m ← The decimal point starts here. Move the decimal 14 places to the left.

$= 1.275 \times 100\ 000\ 000\ 000\ 000$ m
$= 1.275 \times 10^{14}$ m

Remember that the number of zeros after the 1 is the exponent of 10. This number is also the number of places that you moved the decimal point.

Example 2

The length of a bacterium is about 0.000 001 25 m. Write this distance in scientific notation.

Solution

First, count the number of places you have to move the decimal point to leave only one digit to the left of it.

The decimal point starts here. Move the decimal 6 places to the right. → 0.000 001 25 m

$= 1.25 \times 0.000\ 001$ m
$= 1.25 \times 10^{-6}$ m

Because you moved the decimal point 6 places, you might think that the exponent of 10 is 6. However, this time you moved the decimal point to the right. Therefore, the exponent of ten is negative, in this case –6. Use the following table to determine the sign of exponent of 10.

Direction decimal moved	Sign of exponent	Size of quantity
left	positive (+)	large
right	negative (–)	small

Instant Practice

1. Express the following measurements in scientific notation.
   (a) The distance from the north pole to the south pole, directly through Earth, is about 12 700 000 m.
   (b) Light travels at a speed of approximately 300 000 km/s.
   (c) The thickness of a cell membrane is about 0.000 000 008 m.
   (d) The mass of a blue whale is approximately 140 000 kg.
2. Convert the following measurements to standard form.
   (a) 5.82×10^{9} m
   (b) 1.773×10^{-5} kg
   (c) 3.15×10^{7} s
   (d) 8.3×10^{-4} L

Commonly Used Metric Prefixes

Prefixes	Symbol	Relationship to the base unit
giga-	G	$10^9 = 1\ 000\ 000\ 000$
mega-	M	$10^6 = 1\ 000\ 000$
kilo-	k	$10^3 = 1\ 000$
hecto-	h	$10^2 = 100$
deca-	da	$10^1 = 10$
–	–	$10^0 = 1$
deci-	d	$10^{-1} = 0.1$
centi-	c	$10^{-2} = 0.01$
milli-	m	$10^{-3} = 0.001$
micro-	μ	$10^{-6} = 0.000\ 001$
nano-	n	$10^{-9} = 0.000\ 000\ 001$

ESTIMATING AND MEASURING

Estimating

How long will it take you to read this page? How heavy is this textbook? What is the height of your desk? You could probably answer all of these questions fairly quickly by estimating — making an informed judgment about a measurement. You recognize that the estimate gives you an idea of the measure but an estimate is not totally accurate.

Scientists often make estimates, as well, when exact numbers are not essential. You will find it useful to be able to estimate as accurately as possible, too. For example, suppose you wanted to know how many ants live in a local park. Counting every ant would be very time-consuming — and the ants would be most unlikely to stay in one spot for your convenience! What you can do is count the number of ants in a typical square-metre area. Multiply the number of ants by the number of square metres in the total area you are investigating. This will give you an estimate of the total population of ants in that area.

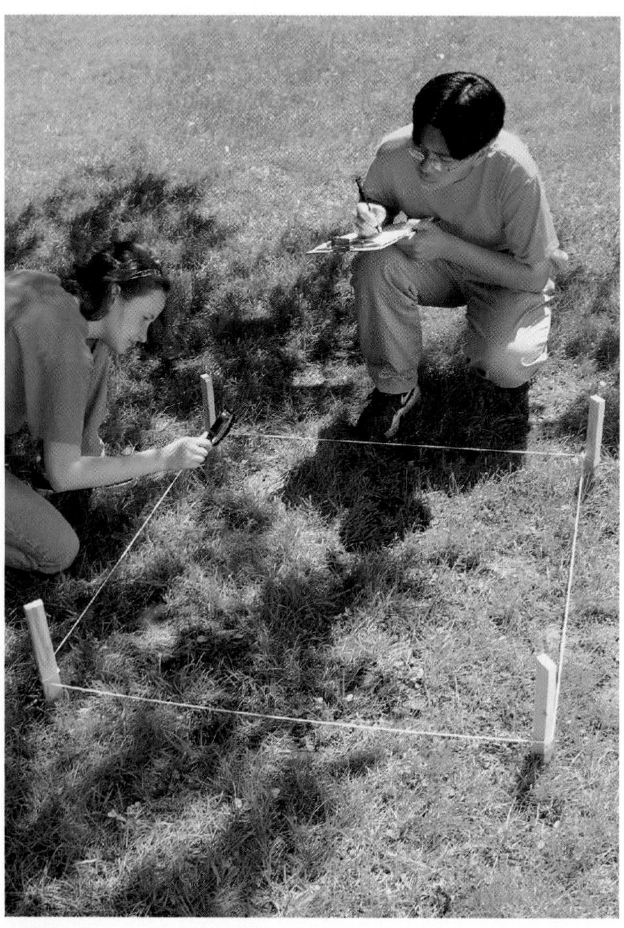

Instant Practice

1. A blue whale has a mass of about 140 000 kg. Students in your class have an average mass of about 60 kg. Estimate the number of students it would take to equal the whale in mass. Calculate the exact number to see how close your estimate was.
2. The arctic tern is a bird that nests in the Canadian Arctic and migrates 17 500 km to the Antarctic to spend the winter. It makes the trip in about 16 weeks. Estimate how far the tern flies in one week. Calculate the exact distance to see how close your estimate was.
3. A 1 L (1000 mL) jar is filled with popcorn kernels. How can you make a good estimate of the number of popcorn kernels in the jar?
   - **(a)** Decide how you can use a 100 mL container and a small number of popcorn kernels to estimate the number in the 1 L jar.
   - **(b)** Carry out your plan.
   - **(c)** Compare your results with those of two or three classmates.
   - **(d)** About how many popcorn kernels will a 1 L jar hold?

Measuring Length and Area

You can use a metre stick or a ruler to measure short distances. These are usually marked off in centimetres and/or millimetres. Use a ruler to measure the length in millimetres between points A and F, C and E, F and B, and A and D. Convert your measurements to centimetres and then to metres.

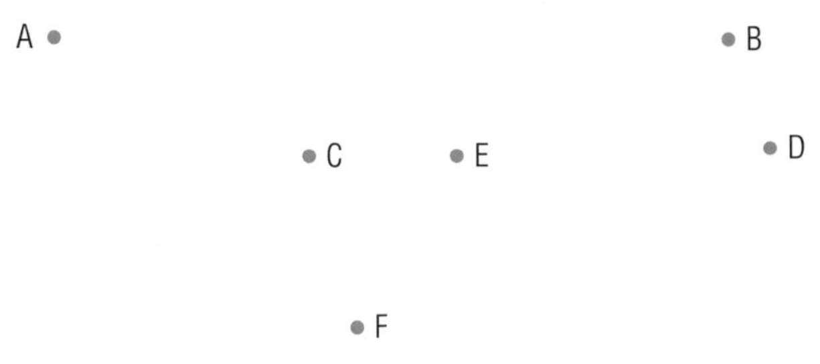

To calculate an area, you can use length measurements. For example, for a square or a rectangle, you can find the area by multiplying the length by the width.

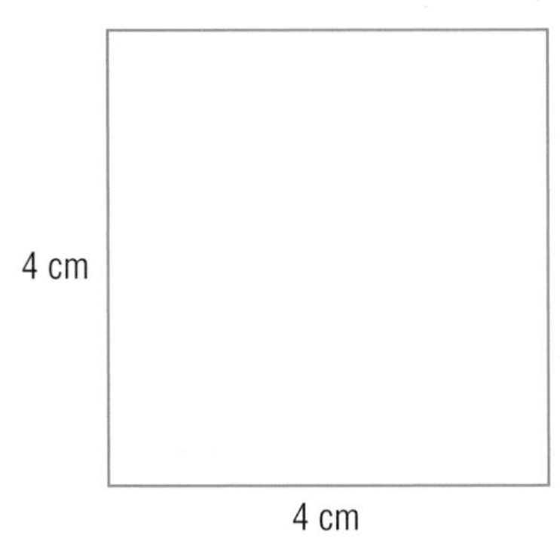

Area of square is 4 cm × 4 cm = 16 cm^2.

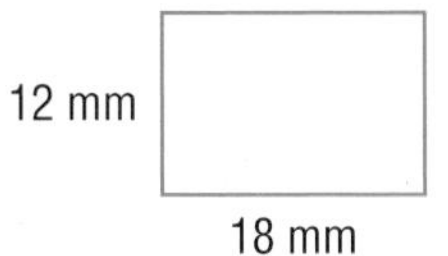

Area of rectangle is 18 mm × 12 mm = 216 mm^2.

Instant Practice

1. What is the area of a rectangle with a width of 3 cm and a length of 4.5 cm?
   - **(a)** Convert your measurements to millimetres.
   - **(b)** Convert your measurements to metres.
2. Imagine you are in charge of tiling the floor of your classroom. How many 10 cm × 10 cm tiles would you need to cover it?
   - **(a)** First, decide what unit would be most practical for the floor area — mm^2, cm^2, or m^2.
   - **(b)** Measure the length and width of your classroom.
   - **(c)** Calculate the floor area in the unit you have selected.
   - **(d)** Calculate how many tiles you would need to fill 1 unit of area.
   - **(e)** Multiply that number by the number of these units in the floor area.

Make sure you always use the same units — if you mix up centimetres and millimetres, your calculations will be wrong. Remember to ask yourself if your answer is reasonable (you could make an estimate to consider this).

Measuring Volume

The **volume** of an object is the amount of space that the object occupies. There are several ways of measuring volume, depending on the kind of object you want to measure. A cubic metre is the space occupied by a 1 m × 1 m × 1 m cube. This unit of volume is used to measure large quantities, such as the volume of concrete in a building. In this course, you are more likely to use cubic centimetres (cm^3) or cubic millimetres (mm^3) to record the volume of a an object. You can calculate the volume of a cube by multiplying its sides. For example, volume = 1 cm × 1 cm × 1 cm = 1 cm^3.

You can calculate the volume of a rectangular solid if you know its length, width, and height.

Volume = Length × Width × Height

If all the sides are measured in millimetres (mm), the volume will be in cubic millimetres (mm^3). If all the sides are measured in centimetres (cm), the volume will be in cubic centimetres (cm^3). The units for measuring the volume of a solid are called **cubic units**.

The units used to measure the volume of liquids are called **capacity units**. The basic unit of volume for liquids is the litre (L). In this course, you also measure volume using millilitres (mL). Recall that 1 L = 1000 mL. You have probably seen capacity in litres and millilitres printed on juice, milk, and soft drink containers.

Cubic units and capacity units are interchangeable, for example:

1 cm^3 = 1 mL
1 dm^3 = 1 L
1 m^3 = 1 kL

As you can see in Diagram A, the volume of a regularly shaped solid object can be measured directly.

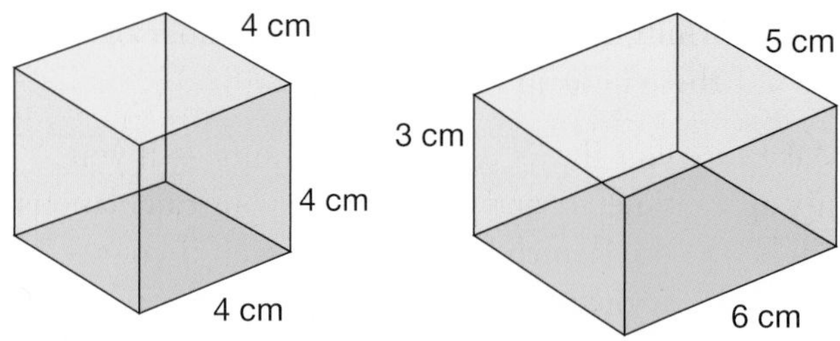

Measuring the volume of a regularly shaped solid

Similarly, the volume of a liquid can be measured directly, as shown in Diagram B. Make sure you measure to the bottom of the **meniscus**, the slight curve where the liquid touches the sides of the container. To measure accurately, make sure your eye is at the same level as the bottom of the meniscus.

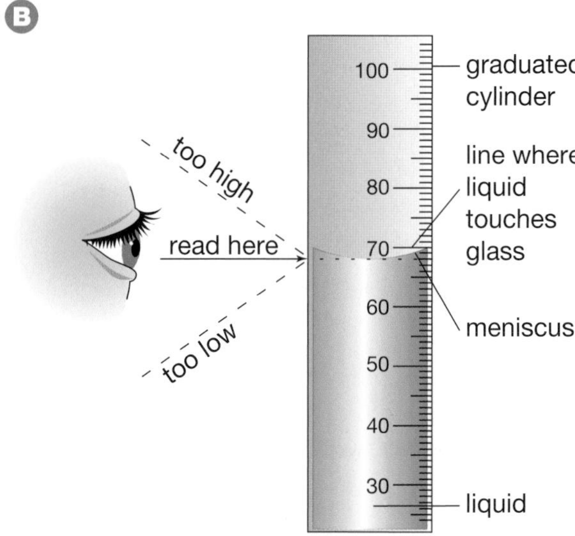

Measuring the volume of a liquid

The volume of an irregularly shaped solid object, however, must be measured indirectly. This is done by determining the volume of a liquid it will displace, as shown in the photographs below.

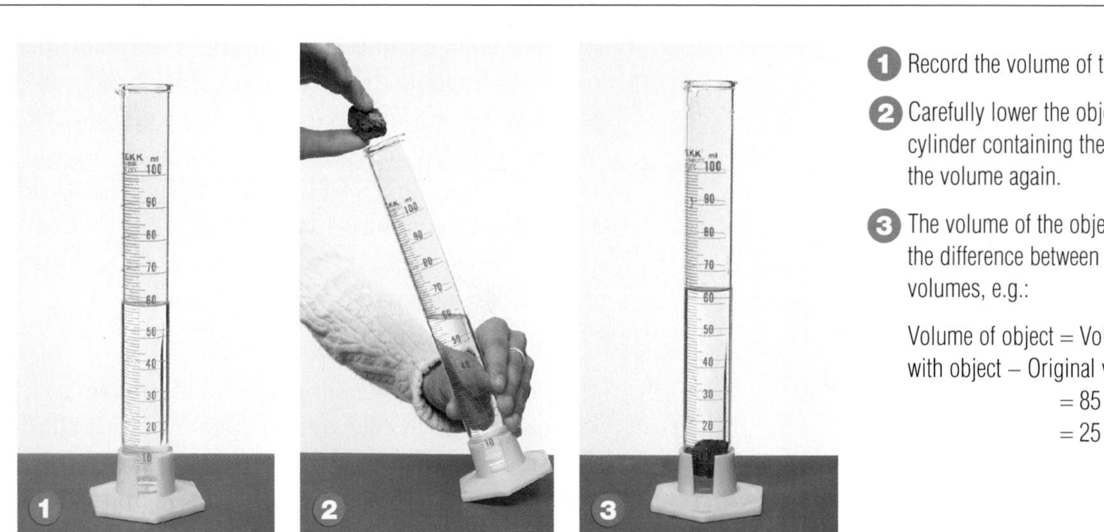

1. Record the volume of the liquid.
2. Carefully lower the object into the cylinder containing the liquid. Record the volume again.
3. The volume of the object is equal to the difference between the two volumes, e.g.:

Volume of object = Volume of water with object – Original volume of water
= 85 mL – 60 mL
= 25 mL

Measuring the volume of an irregularly shaped solid

Instant Practice

1. Imagine you are cooking potatoes for dinner. You decide to measure the volume of a potato so that you can figure out how much water to use.

What You Need

large potato
500 mL beaker
water

   (a) Add water to the beaker (your "saucepan"). How much do you think you can add and still leave room for the potatoes?
   (b) Carefully add the potato. What is its volume? (If your beaker does not have millilitre markings, use a wax pencil to mark the water levels before and after the potato is added. Pour out the water and use a graduated cylinder to measure and record the number of millilitres between the two wax pencil marks.)
   (c) Could you add a second potato of the same size without spilling water, or would you need to pour off some of the water?

2. Now that you know the volume of a potato in millilitres, what is its volume in cubic centimetres?

Measuring Mass

Is your backpack heavier than your friend's backpack? It can be difficult to check by holding a backpack in each hand. At such times, you need a way to measure mass accurately. The **mass** of an object is the measure of the amount of material that makes up the object. Mass is measured in milligrams, grams, kilograms, and tonnes. You need a balance for measuring mass. A triple beam balance is one commonly used type.

Following are the steps involved in measuring the mass of a solid object.

1. Set the balance to zero. Do this by sliding all three riders back to their zero points. Using the adjusting screw, make sure the pointer swings an equal amount above and below the zero point at the far end of the balance.
2. Place the object on the pan. Observe what happens to the pointer.
3. Slide the largest rider along until the pointer is just below zero. Then move it back one notch.
4. Repeat with the middle rider and then with the smallest rider. Adjust the last rider until the pointer swings equally above and below zero again.
5. Add the readings on the three scales to find the mass.

How can you find the mass of a certain quantity of a substance, such as table salt, that you have added to a beaker? First, find the mass of the beaker. Next, pour the salt into the beaker and find the mass of the beaker and salt together. To find the mass of the salt, simply subtract the beaker's mass from the combined mass of the beaker and salt.

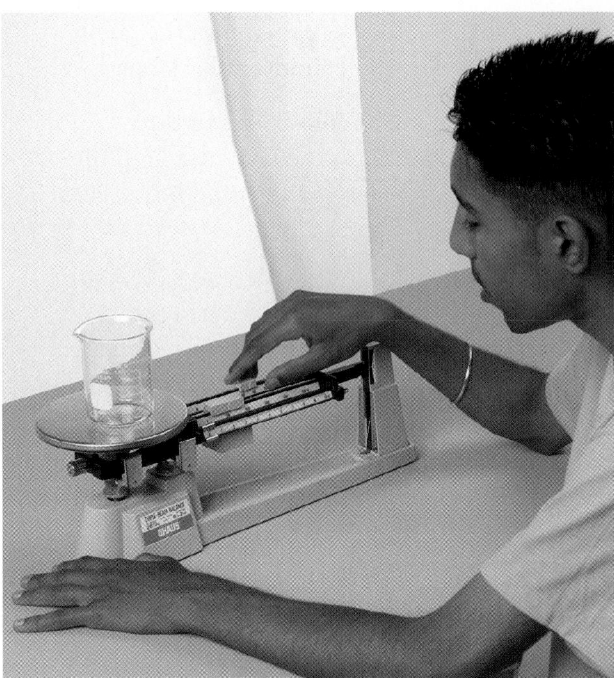

The mass of the beaker is 160 g.

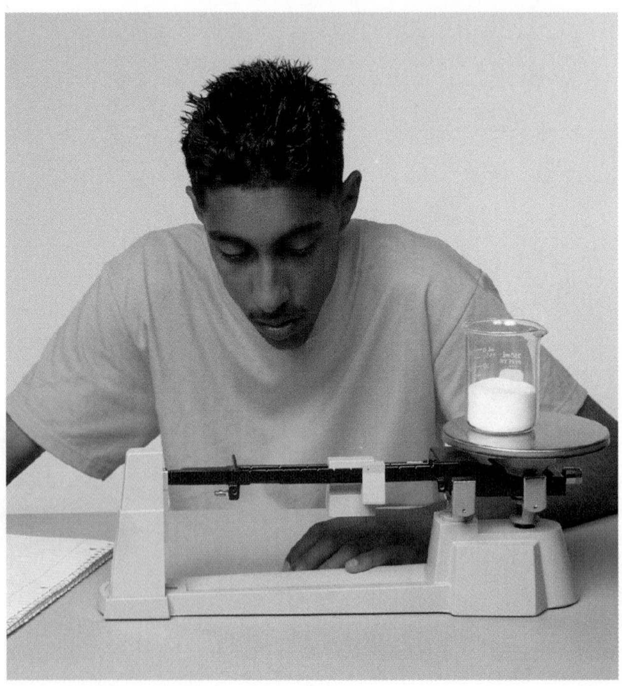

The mass of the table salt and beaker together is 230 g. Therefore, the mass of the salt is 70 g.

Instant Practice

1. Which takes more "muscle" to carry: your favourite paperback book or your favourite portable electronic game? Find out by using a balance to compare their masses.
2. Write the steps you would take to find the mass of the contents of a container of juice.

Measuring Angles

In Unit 3, Light and Optical Instruments, you need to be able to measure angles using a protractor. Protractors usually have an inner scale and an outer scale. The scale you use depends on how you place the protractor on an angle (symbol = ∠). Look at the following examples to learn how to use a protractor.

Example 1

What is the measure of ∠XYZ?

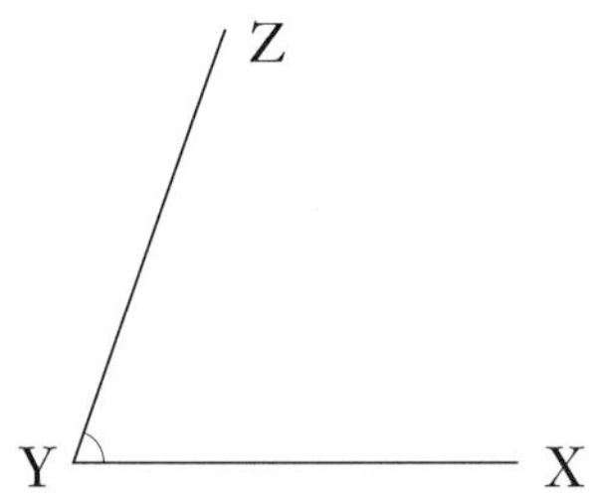

Solution

Place the centre of the protractor on point Y. YX crosses 0° on the outer scale. YZ crosses 70° on the outer scale. So ∠XYZ = 70°.

Example 2

Draw ∠ABC = 155°.

Solution

First, draw a straight line, AB. Place the centre of the protractor on B and line up AB with 0° on the inner scale. Mark C at 155° on the inner scale. Join BC. The angle you have drawn, ∠ABC, is equal to 155°.

Instant Practice

1. State the measure of each of the following angles using the following diagram.

   (a) DAE (b) HAG (c) EAH
   (d) DAG (e) GAE (f) EAF
   (g) IAH (h) DAH (i) FAI

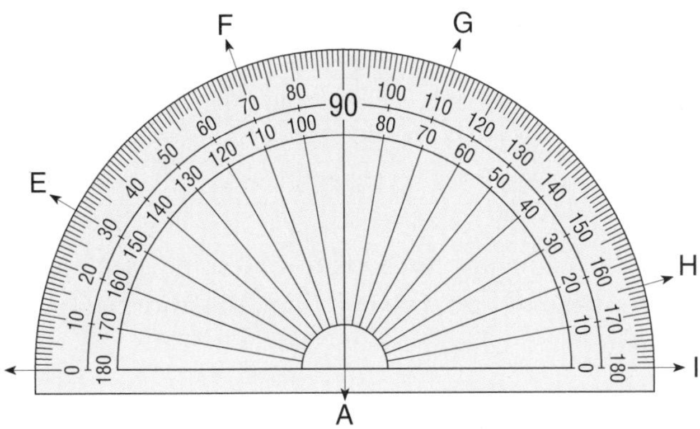

2. Use a protractor to draw angles with the following measures. Label each angle.

   (a) 60° (b) 10° (c) 45° (d) 90° (e) 32°

Measuring Temperature

"Temperature" is a measure of the thermal energy of the particles of a substance. In the very simplest terms, you can think of temperature as a measure of how hot or how cold something is. The temperature of a material is measured with a thermometer.

For most scientific work, temperature is measured on the Celsius scale. On this scale, the freezing point of water is zero degrees (0°C), and the boiling point of water is one hundred degrees (100°C). Between these points, the scale is divided into 100 equal divisions. Each division represents 1 degree Celsius. On the Celsius scale, average human body temperature is 37°C, and a typical room temperature may be between 20°C and 25°C.

The SI unit of temperature is the Kelvin (K). Zero on the Kelvin scale (0 K) is the coldest possible temperature. This temperature is also known as absolute zero. It is equivalent to –273°C, which is 273 degrees below the freezing point of water. Notice that degree symbols are not used with the Kelvin scale.

Most laboratory thermometers are marked only with the Celsius scale. Because the divisions on the two scales are the same size, the Kelvin temperature can be found by adding 273 to the Celsius reading. Thus, on the Kelvin scale, water freezes at 273 K and boils at 373 K.

Tips for Using a Thermometer

When using a thermometer to measure the temperature of a substance, here are three important tips to remember:

- Handle the thermometer extremely carefully. It is made of glass and can break easily.
- Do not use the thermometer as a stirring rod.
- Do not let the bulb of the thermometer touch the walls of the container.

ORGANIZING AND COMMUNICATING SCIENTIFIC RESULTS

In your investigations, you will collect information, often in numerical form. To analyze and report the information, you will need a clear, concise way to organize the data. You may choose to organize the data in the form of a table, stem-and-leaf plot, line graph, bar graph, or circle graph.

Making a Table

A data table is usually a good way to start organizing information. The table may be the final form in which you present the data or you may use it to help construct a graph or a diagram.

Example

Read the newspaper article below, then examine Table 1. What information do the article and table present? Which one makes the data easier to analyze?

Climatologist Dr. David Dillon studied the effect of a large neighbouring body of water on the climate of four nearby towns. He observed that the town of Ajax, located on the lakeshore, has an average of 205 frost-free days a year. The average monthly temperature range is 7.6°C and the annual precipitation is 73.6 cm. In comparison, Baxter, located 1.6 km from the lake, has 194 frost-free days a year with an annual precipitation of 81.4 cm. Baxter's average monthly temperature range is 8.8°C. The town of Carter, 48.3 km from the same lake, has an annual precipitation of 94 cm, 162 frost-free days, and an average monthly temperature range of 10.8°C. Finally, Delbert, 80.5 km from the lake, has an average monthly temperature range of 11.9°C, 154 frost-free days a year, and an annual precipitation of 97.5 cm.

Table 1 Distance from Lake and Climate

City	Distance from lake (km)	Average monthly temperature range (°C)	Number of frost-free days per year	Annual precipitation (cm)
Ajax	0.0	7.6	205	73.6
Baxter	1.6	8.8	194	81.4
Carter	48.3	10.8	162	94.0
Delbert	80.5	11.9	154	97.5

Look through your textbook to find some examples of data tables. Note why you think the information is presented in a table.

Instant Practice

In some areas of science, researchers use probability, or the likelihood that certain things will happen, to help analyze their data. In this activity, you will make a data table of the outcomes of three different types of experiments in probability.

1. Prepare a table with two main headings: "Experiment" and "Outcomes." Divide the Outcomes column into four columns with the following headings: (A = 3, B = 0), (A = 2, B = 1), (A = 1, B = 2), (A = 0, B = 3) as shown here.

Table 2 Probability Data

Experiment	Outcomes			
	(A = 3, B = 0)	(A = 2, B = 1)	(A = 1, B = 2)	(A = 0, B = 3)
Coin tossing A - heads B - tails				
Drawing cards A - black B - red				
Paper picking				

2. Obtain three identical coins. Toss all three coins and write down the number of heads (A) and tails (B). Repeat the process nine more times. Count the number of times you tossed three heads and no tails and record that number in the column (A = 3, B = 0), in the row for "Coin tossing." Under the heading (A = 2, B = 1), record the number of times you tossed two heads and one tail. In the third column, record the number of times you tossed one head and two tails; and in the fourth column, record the number of times you tossed three tails.

3. Obtain a deck of playing cards and shuffle the cards. Draw three cards and write down the number of black cards and the number of red cards. Replace the three cards, shuffle the deck, and draw three more cards. Record the outcome. Repeat the process until you have drawn three cards, a total of ten times. Record the card data in your table in the same way that you recorded the coin-toss data. Note that "A" represents the number of black cards and "B" represents the number of red cards.

4. Cut paper or cardboard into twelve 1 cm squares. On six of the squares, write "A." On the other six, write "B." Put the squares of paper in a bag and shake them up. Ask a friend to reach in and, without looking, draw three squares of paper. Record the number of A's and B's your friend drew. Replace the pieces of paper in the bag and shake it. Ask another friend to draw three squares of paper. Record the outcome. Repeat the process until friends have drawn three squares of paper, ten times. Record your outcomes in your table.

5. Analyze the outcomes in your table. Compare the data from the different experiments. Describe any pattern you observe. How does the table help you to analyze your data?

Making a Stem-and-Leaf Plot

A stem-and-leaf plot can sometimes give you a more visual representation of your data than a table can. Values on the stem represent a range of data, while the leaves include the data points that lie in that range.

Example

During one year, 16 Canadian airports reported the following peak wind speeds in kilometres per hour: 109, 129, 127, 161, 177, 146, 148, 132, 153, 129, 124, 135, 177, 193, 106, 106

To learn how to make a stem-and-leaf plot to display these data, examine the diagram below while reading the following steps.

Peak Wind Speeds at Canadian Airports

Stem	Leaf			
10	6	6	9	
11				
12	4	7	9	9
13	2	5		
14	6	8		
15	3			
16	1			
17	7	7		
18				
19	3			

1. Make a table with two main headings: "Stem" and "Leaf."

2. Look at the data above. The lowest number is 106, so your stem-and-leaf plot should start at 100. The highest number is 193, so your stem-and-leaf plot should go up to 200. In the "Stem" column, write the numbers 10, 11, 12, up to 19, where 10 represents all numbers having the first two digits 10_, and 11 represents all the numbers having the first two digits 11_. The pattern continues up to all numbers having the first two digits 19_.

3. Look for the lowest readings in the data table. You will find two numbers between 100 and 109. Next to the 10 "stem," write the last digits of each of these numbers in the "leaf" columns, as shown in the sample stem-and-leaf plot above.

CONTINUED

4. Notice that there are no data points between 110 and 119. The 11 "stem" has no "leaves."

5. Verify that the "leaves" on the remainder of the "stem" numbers are the last number in the data points that start with the numbers in the "stem."

A stem-and-leaf plot, turned on its side, looks similar to a bar graph (see page 550). It allows you to see, at a glance, where the largest and smallest numbers are in the range.

Instant Practice

Make a stem-and-leaf plot of students' scores on a science exam. Remember to give your plot a suitable title. The following numbers are the students' scores: 75, 82, 55, 69, 91, 52, 77, 63, 85, 81, 73, 62, 58, 76, 44, 57, 65, 68, 84, 69, 53, 67, 73, 59, 82, 79, 72, 67, 71, 55.

Graphing

A graph is the most visual way to present data. A graph can help you to see patterns and relationships among the data. The type of graph you choose depends on the type of data you have and how you want to present it. Throughout the year, you will be using line graphs, bar graphs, and circle graphs (pie charts).

Drawing a Line Graph

A line graph is used to show the relationship between two variables. The following example will demonstrate how to draw a line graph from a data table.

Example

Suppose you have conducted a survey to find out how many students in your school are recycling drink containers. Out of 65 students that you surveyed, 28 are recycling. To find out if more recycling bins would encourage students to recycle cans and bottles, you place temporary recycling bins at three other locations in the school. Assume that, in a follow-up survey, you obtained the data shown in the table below. Compare the steps in the procedure with the graph on page 549 to learn how to make a line graph to display your findings.

Table 3 Students Using Recycling Bins

Number of bins	Number of students using recycling bins
1	28
2	36
3	48
4	62

1. With a ruler, draw an x-axis and a y-axis on a piece of graph paper. (The horizontal line is the x-axis, and the vertical line is the y-axis.)

2. To label the axes, write "Number of recycling bins" along the x-axis and "Number of students using recycling bins" along the y-axis.

3. Now you have to decide what scale to use. You are working with two numbers (number of students, and number of bins). You need to show how many students use the existing bin, and how many would recycle if there were a second, a third, and a fourth bin. The scale on the x-axis will go from 0 to 4. There are 65 students, so you might want to use intervals of 5 for the y-axis. That means that every space on your y-axis represents 5 students.

4. On the x-axis, you want to make sure you will be able to read your graph when it is complete, so make sure your intervals are large enough.

5. To plot your graph, gently move a pencil up the y-axis until you reach a point just below 30 (you are representing 28 students). Now move along the line on the graph paper until you reach the vertical line that represents the first recycling bin. Place a dot at this point (1 bin, 28 students). Repeat this process until you have plotted all of the data for the four bins. Now, draw a line from one dot to the next.

6. Give your graph a title. Based on these data, what is the relationship between the number of students using recycling bins and the number of recycling bins?

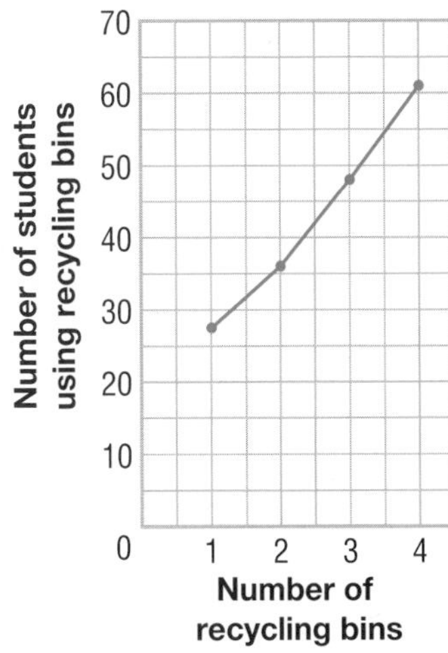

Instant Practice

Make a line graph using the following data on the development of a fetus. The first column represents the time since conception, in months. Plot these values along the x-axis. The second column is the average length of a fetus at that stage of development. Plot these values along the y-axis. Be sure to include units. Give your graph a title.

Table 4 Development of a Fetus

Time since conception (months)	Average length (cm)
1	0.6
2	3.0
3	7.5
4	18.0
5	27.0
6	31.0
7	37.0
8	43.0
9	50.0

Constructing a Bar Graph

Bar graphs help you to compare a numerical quantity with some other category, at a glance. The second category may or may not be a numerical quantity. It could be places, items, organisms, or groups, for example.

Example

To learn how to make a bar graph to display the data in Table 5, examine the corresponding graph on page 550 as you read the steps below. The data show the number of days of fog recorded during one year, at one weather station in each of the provinces and territories.

Table 5 Average Number of Days of Fog per Year in Canadian Provinces and Territories (prior to April 1, 1999)

Province	Number of days of fog
Newfoundland	206
Prince Edward Island	47
New Brunswick	106
Nova Scotia	127
Québec	85
Ontario	76
Manitoba	48
Saskatchewan	37
Alberta	39
British Columbia	226
Yukon Territory	61
Northwest Territories	196

1. Draw your x-axis and y-axis on a sheet of graph paper. Label the x-axis with the names of the provinces and the y-axis with the average number of days of fog.
2. Look at the data carefully in order to select an appropriate scale. Write the scale of your y-axis on the lines.
3. Decide on a width for the bars that will be large enough to make the graph easy to read. Leave the same amount of space between each bar.
4. Using Newfoundland and 206 as the first pair of data, move along the x-axis the width of your first bar, then go up the y-axis to 206. Use a pencil and ruler to draw in the first bar lightly. Repeat this process for the other pairs of data.

5. When you have drawn all of the bars, you might want to colour them so that each one stands out. If you use different colours, make a legend or key to explain the meaning of the colours. Write a title for your graph.

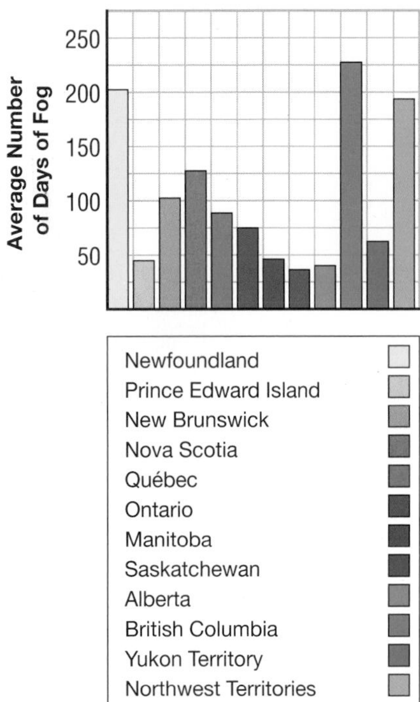

Instant Practice

Construct a bar graph to display the data in Table 6 showing the average heart rates of adult animals in several different species.

Table 6 Heart Rate and Species

Species	Heart rate (beats per min)
codfish (in water at 18°C)	30
iguana (in hot sun)	90
duck (resting)	240
dog (resting)	100
human (resting)	70
elephant (resting)	30
white rat (resting)	350

Constructing a Histogram

How does a histogram differ from a bar graph? The *x*-axis on a histogram represents continuous data such as time, mass, distance, or height. The histogram shown below displays the age distribution of 30 students in a grade 8 classroom. The *x*-axis is divided into eight height ranges. The size of each bar represents the number of students in that height range.

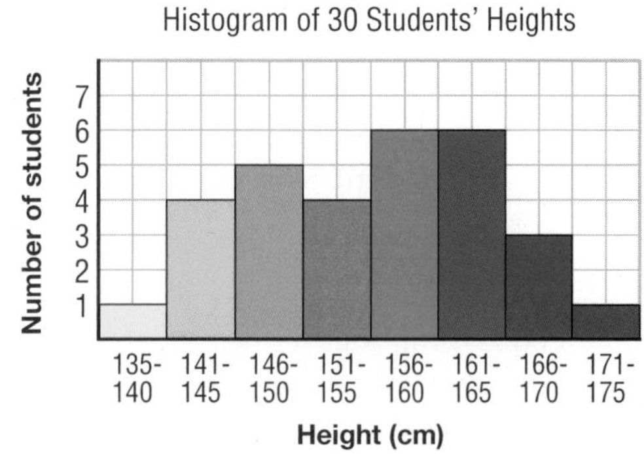

To learn how to make a histogram, examine the raw data given below, as well as the frequency table, as you read the procedure.

Students' heights (in cm): 148, 165, 152, 156, 149, 160, 164, 169, 157, 170, 150, 159, 141, 173, 144, 150, 152, 158, 161, 137, 142, 145, 162, 165, 170, 147, 154, 157, 164, 157

1. Make a frequency table like Table 7, to tally the data.

Table 7 Frequency Table

Height range (in cm)	Tally	Frequency
135–140	/	1
141–145	////	4
146–150	~~////~~	5
151–155	////	4
156–160	~~////~~ /	6
161–165	~~////~~ /	6
166–170	///	3
171–175	/	1

2. Start with the first number, 148, in the raw data. Because 148 is in the range 146-150, put a mark in the tally column beside this range. Go to the next number, find its range, and put a mark in the tally column. Continue until you have tallied all of the data.

3. Write the number of marks in the tally column in the "Frequency" column.

4. Make a histogram with height ranges on the x-axis and "Number of students" (frequency) on the y-axis. The size of the bar in each height range is the number of students that fell in that range, as shown in the frequency table.

Instant Practice

A battery manufacturer routinely tests batteries for the number of times the batteries can be recharged. The quality control officer for the company selected 30 batteries of a certain model, at random, and tested them. The following data represent the number of times the batteries could be recharged and still perform according to company standards. Make a histogram to display the results of the test. Use a range of 10 recharges for the bars along the x-axis. Make a frequency table to organize the data. Then choose an appropriate scale on the y-axis for the number of times the batteries could be recharged.
723, 756, 771, 758, 749, 754, 738, 766, 755, 747, 762, 751, 748, 761, 735, 755, 736, 754, 743, 751, 762, 742, 751, 778, 741, 756, 732, 769, 743, 768

Constructing a Circle Graph

A circle graph (sometimes called a pie chart) uses a circle divided into sections (pieces of pie) to show the data. Each section represents a percent of the whole. All sections together represent all (100%) of the data.

Example

To learn how to make a circle graph from the data in Table 8, study the corresponding circle graph on page 552 as you read the following steps.

Table 8 Birds Breeding in Canada

Type of bird	Number of species	Percent of total	Degrees in "piece of pie"
ducks	36	9.0	32
birds of prey	19	4.8	17
shorebirds	71	17.7	64
owls	14	3.5	13
perching birds	180	45.0	162
other	80	20.0	72

1. Use a mathematical compass to make a large circle on a piece of paper. Make a dot in the centre of the circle.

2. Determine the percent of the total number of species that each type of bird represents by using the following formula.

$$\text{Percent of total} = \frac{\text{Number of species within the type}}{\text{Total number of species}} \times 100\%$$

For example, the percent of all species of birds that are ducks is:

$$\text{Percent that are ducks} = \frac{\text{36 species of ducks}}{\text{400 species}} \times 100\% = 9.0\%$$

3. To determine the degrees in the "piece of pie" that represents each type of bird, use the following formula.

$$\text{Degrees in "piece of pie"} = \frac{\text{Percent for a type of bird}}{100\%} \times 360^\circ$$

Round your answer to the nearest whole number. For example, the "piece of pie" for ducks is:

$$\text{Degrees for ducks} = \frac{9.0\%}{100\%} \times 360^\circ = 32.4^\circ \text{ or } 32^\circ$$

CONTINUED ▶

4. Draw a straight line from the centre to the edge of the circle. Place a protractor on this line and use it to mark a point on the edge of the circle at 32°. Connect the point to the centre of the circle. This is the "piece" that represents the fraction of all bird species breeding in Canada that are ducks.

5. Repeat steps 2 to 4 for the remaining types of birds.

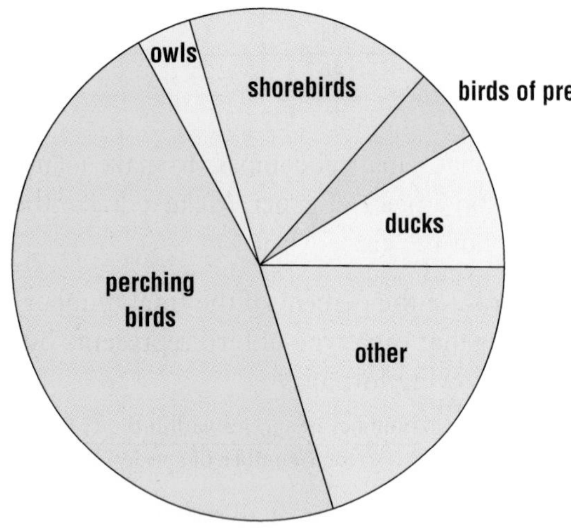

Instant Practice

Make a circle graph using the following data on the elements in Earth's crust. Notice that the data are given in percent.

Table 9 Percent of Elements in Earth's Crust

Element	Percent of Earth's crust (%)
aluminum	8.0
calcium	2.4
iron	6.0
magnesium	4.0
oxygen	46.0
potassium	2.3
silicon	28.0
sodium	2.1
other	1.0

SAFETY SYMBOLS

The following safety symbols are used in the *SCIENCEPOWER™ 8* program to alert you to possible dangers. Be sure you understand each symbol used in an activity or investigation before you begin.

	Disposal Alert This symbol appears when care must be taken to dispose of materials properly.
	Thermal Safety This symbol appears as a reminder to use caution when handling hot objects.
	Sharp Object Safety This symbol appears when a danger of cuts or punctures caused by the use of sharp objects exists.
	Electrical Safety This symbol appears when care should be taken when using electrical equipment.
	Skin Protection Safety This symbol appears when use of caustic chemicals might irritate the skin or when contact with micro-organisms might transmit infection.
	Clothing Protection Safety A lab apron should be worn when this symbol appears.
	Fire Safety This symbol appears when care should be taken around open flames.
	Eye Safety This symbol appears when a danger to the eyes exists. Safety goggles should be worn when this symbol appears.

Instant Practice

Find four of the *SCIENCEPOWER™ 8* safety symbols in activities or investigations in this textbook. Record the page number and the title of the investigation or activity in which you found the symbol. What are the possible dangers in the activity or investigation you have identified that relate to each symbol?

WHMIS Symbols

Look carefully at the WHMIS (Workplace Hazardous Materials Information System) safety symbols shown here. The WHMIS symbols are used throughout Canada to identify dangerous materials used in all workplaces, including schools. Make certain you understand what these symbols mean. When you see these symbols on containers in your classroom, at home, or in a workplace, use safety precautions.

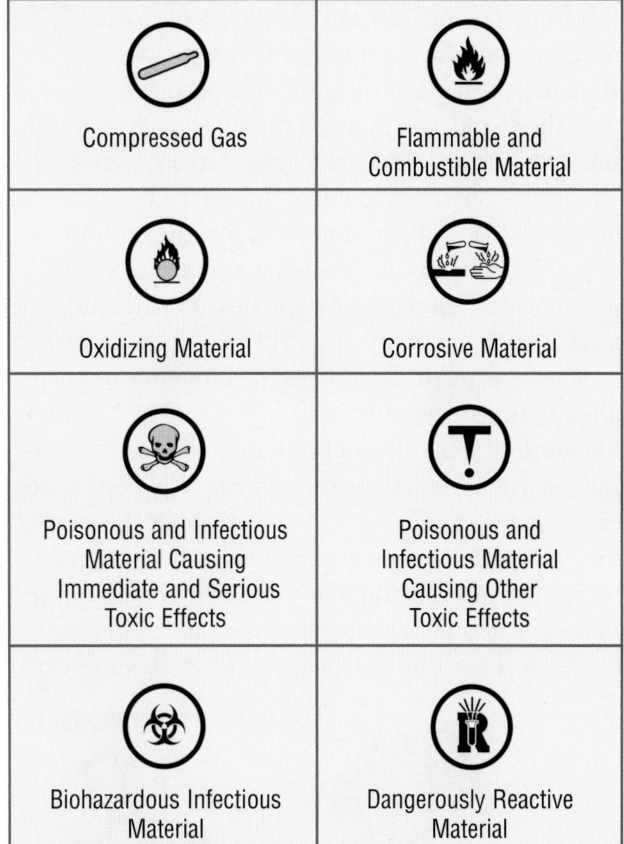

Compressed Gas	Flammable and Combustible Material
Oxidizing Material	Corrosive Material
Poisonous and Infectious Material Causing Immediate and Serious Toxic Effects	Poisonous and Infectious Material Causing Other Toxic Effects
Biohazardous Infectious Material	Dangerously Reactive Material

Instant Practice

Find any two WHMIS symbols on containers in your school, or ask your parent or guardian to look for WHMIS symbols in a workplace. Record the name of the substance on which the symbols are used, and where you or your parent or guardian saw the containers stored. What dangers are associated with the substance in each container?

USING MODELS IN SCIENCE

When you think of a model, you probably think of a toy such as a model airplane. Is a model airplane similar to a scientific model? If building a model airplane helps you learn about flight, then you can also say it is a scientific model.

In science, a model is anything that helps you better understand a scientific concept. A model can be a picture, a mental image, a structure, or even a mathematical formula. Sometimes, you need a model because the objects you are studying are too small to see with the unaided eye. In previous studies, you learned about the particle theory of matter, for example, which is a model that suggests that all matter is made of tiny, invisible particles. On the other hand, sometimes a model is useful because some objects are extremely large — the planets in our solar system, for example. In other cases, the object may be hidden from view, like the interior of Earth or the inside of a living organism. A mathematical model shows you how to perform a calculation.

Scientists often use models to communicate their ideas to other scientists or to students. They also use models to test an idea and to find out if a hypothesis is supported. Models assist scientists in planning new experiments in order to learn more about the subject they are studying. Sometimes, scientists discover so much new information that they have to modify their models. Examine the models in the illustrations on this page. How can they help you learn about science?

You can learn about day and night by using a globe and a flashlight to model Earth and the Sun.

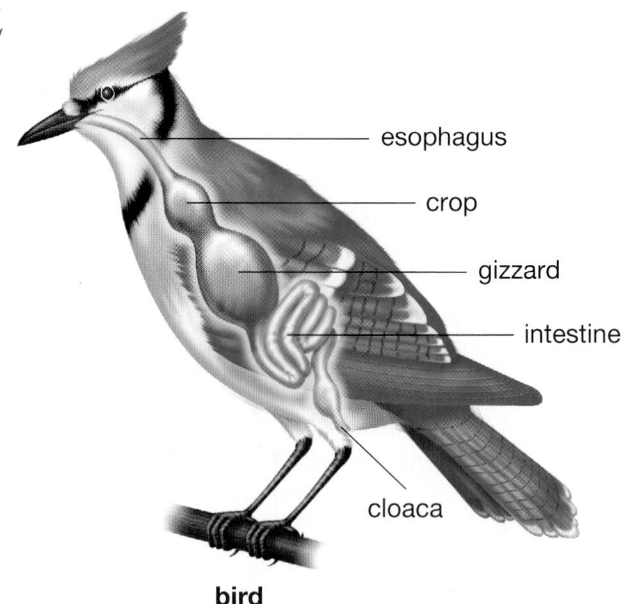

This model shows the digestive tract of a bird. You do not have a crop or a gizzard. Why do birds need these organs in their digestive systems?

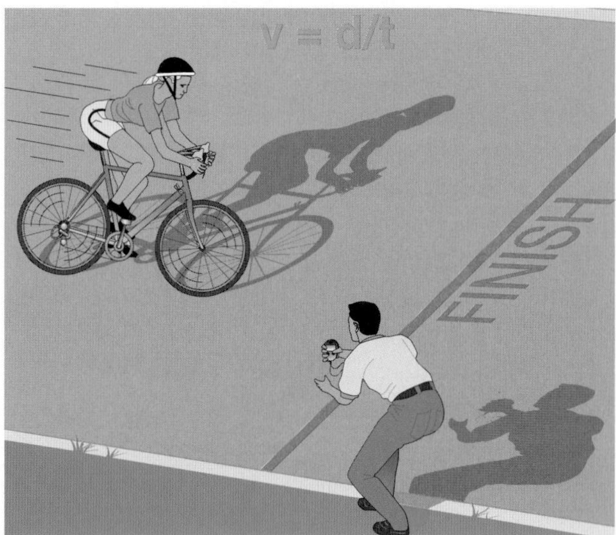

The formula shown here is a mathematical model for speed. The symbol *v* represents speed (or velocity), *d* is distance, and *t* is time. The model shows you how to calculate speed if you know the time it took for a person or object to travel a certain distance. If the cyclist in the illustration travelled 25 m in 10 s, what was her speed?

Instant Practice

Build a model of the layers of Earth's crust as it would appear after millions of years of movements far below the surface.

You will need at least four colours of modelling clay, a pencil, a kitchen knife, and a plastic bag or plastic wrap. The following directions will help you build the model.

Safety Precaution

Be careful when using sharp objects such as a knife.

1. Flatten each of the four colours of clay into layers about 1 cm thick. Stack three of the layers of clay on a piece of plastic bag or wrap, as shown.

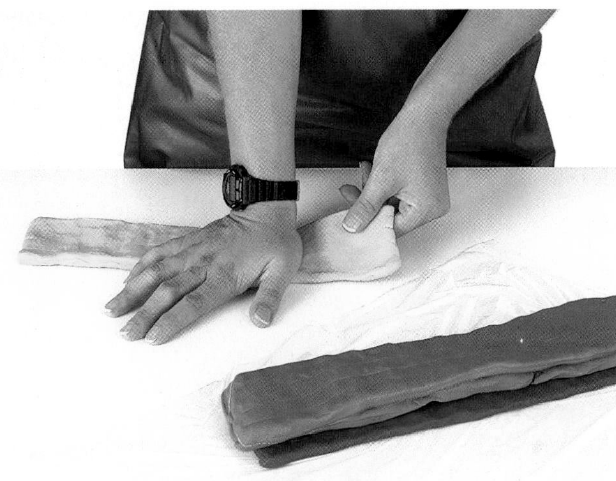

The clay represents layers of three different types of sediment deposited on an ocean floor many thousands of years ago.

2. Fold one side of the clay up and then lay the edge back down. Push the edges together gently.

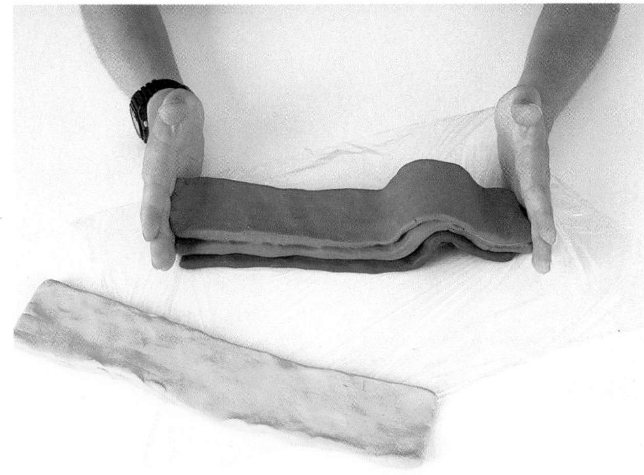

Movements of the tectonic plates beneath the sea caused the sea floor to rise and become dry land. Further movements of the tectonic plates far beneath the surface caused the crust to fold.

3. About 2 or 3 cm from the fold, use the kitchen knife to cut the layers diagonally downwards. Lift one side of the cut slightly upward and push the pieces of modelling clay together gently.

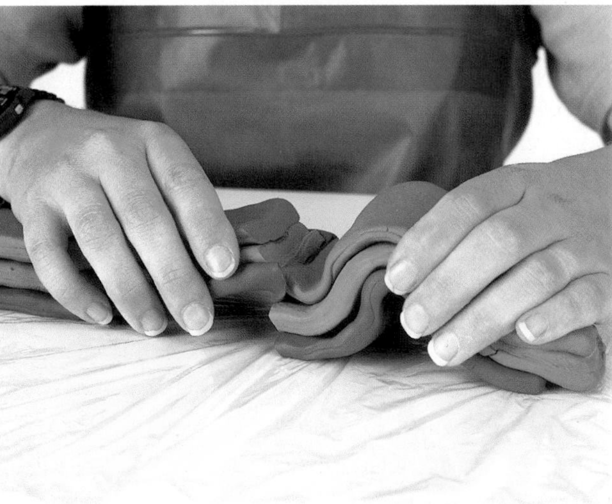

Faults form when pressure and movement deep underground cause the layers on the upper crust of Earth to break and slide up or down on each other.

4. Using the kitchen knife, carefully cut off the top of the model.

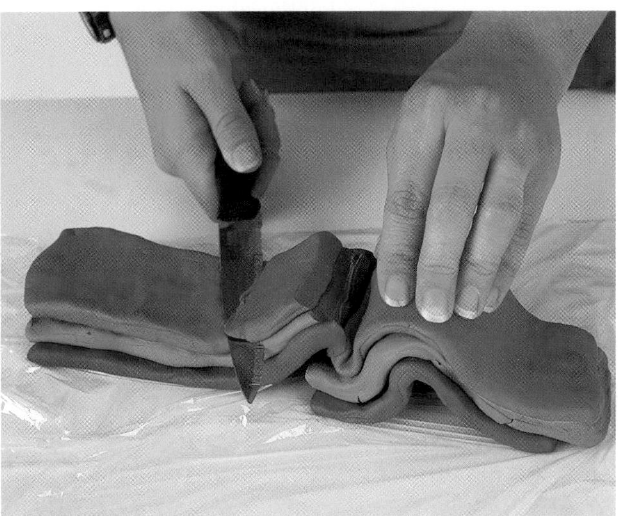

Thousands of years of wind and water, or erosion, wear down hills and mountains that were formed from folds and faults.

CONTINUED

5. Add the last slab of clay to the top of the stack, as shown.

Movements of Earth's crust, over thousands of years, can cause dry land to sink below sea level, causing it to become flooded. If the land remains below the ocean for thousands of years, another layer of sediment will form. Then this section of Earth's crust may rise up out of the sea and become dry land once again.

6. Make a hole directly downward through the clay with a pencil. Fill the hole with one colour of clay. Add more clay of the same colour above the hole and shape it like a volcano.

Deep cracks form in the layers of rock and allow molten rock to push upward, forming an intrusion. Heat and pressure force the molten rock upward through the crack, and a volcano erupts.

7. Using the knife, carefully cut the model in half. Cut perpendicular to both the fold and the fault.

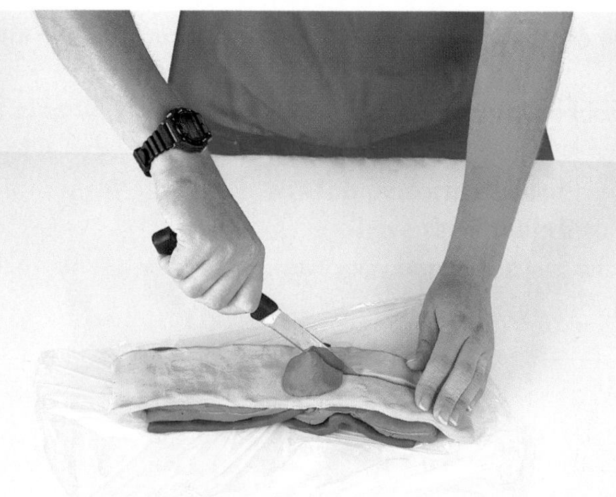

Flowing water wears away land and even solid rock. Over thousands of years, a river may cut a deep canyon.

8. Move the halves apart. Examine your model. Can you see how a geologist might learn about the history of Earth's crust by studying the walls of a deep canyon? Write a summary paragraph to describe the movements of Earth's crust that might have created the formations you see in your model.

SCIENTIFIC AND TECHNOLOGICAL DRAWING

Have you ever used a drawing to explain something that was too difficult to explain in words? A clear drawing can often assist or replace words in a scientific explanation.

In science, drawings are especially important when you are trying to explain difficult concepts or describe something that contains a lot of detail. It is important to make scientific drawings clear, neat, and accurate.

Example

Examine the drawing shown below. It is taken from a Grade 8 student's lab report on an experiment to test the expansion of air in a balloon. The student's verbal description of results included an explanation of how the particle theory can explain what happens to the balloon when the bottle is placed in hot water and in cold water. As you can see, the clear diagrams of the results can support or even replace many words of explanation. While your drawing itself is important, it is also important to label it clearly. If you are comparing and contrasting two objects, label each object and use labels to indicate the point of comparisons between them.

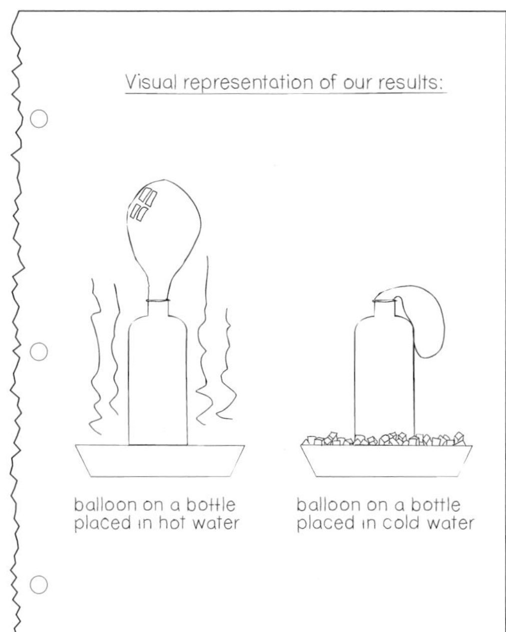

Making a Scientific Drawing

Follow these steps to make a good scientific drawing.

1. Use unlined paper and a sharp pencil with an eraser.
2. Give yourself plenty of space on the paper. You need to make sure that your drawing will be large enough to show all necessary details. You also need to allow space for labels. Labels identify parts of the object you are drawing. Place all of your labels to the right of your drawing, unless there are so many labels that your drawing looks cluttered.
3. Carefully study the object that you will be drawing. Make sure you know what you need to include.
4. Draw only what you see, and keep your drawing simple. Do not try to indicate parts of the object that are not visible from the angle you observed. If you think it is important to show another part of the object, do a second drawing, and indicate the angle from which each drawing is viewed.

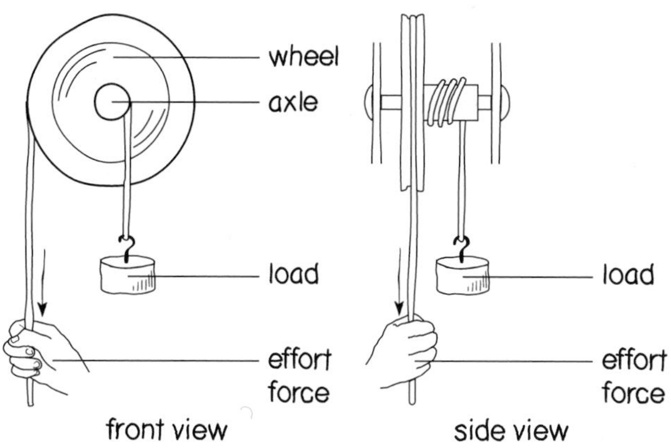

5. Shading or colouring is not usually used in scientific drawings. If you want to indicate a darker area, you can use stippling (a series of dots). You can use double lines to indicate thick parts of the object.
6. If you do use colour, try to be as accurate as you can and choose colours that are as close as possible to the colours in the object you are observing.
7. Label your drawing carefully and completely, using lower-case (small) letters. Pretend you know nothing about the object you have just observed, and think about what you would need to know if you were looking at it for the first time. Remember to place all your labels to the right of the drawing, if

possible. Use a ruler to draw a horizontal line from the label to the part you are identifying. Make sure that none of your label lines cross.

8. Give your drawing a title. **Note:** The drawing of a human skin cell shown here is from a grade 8 student's notebook. This student used stippling to show darker areas, horizontal label lines for the cell parts viewed, and a title — all elements of an excellent final drawing.

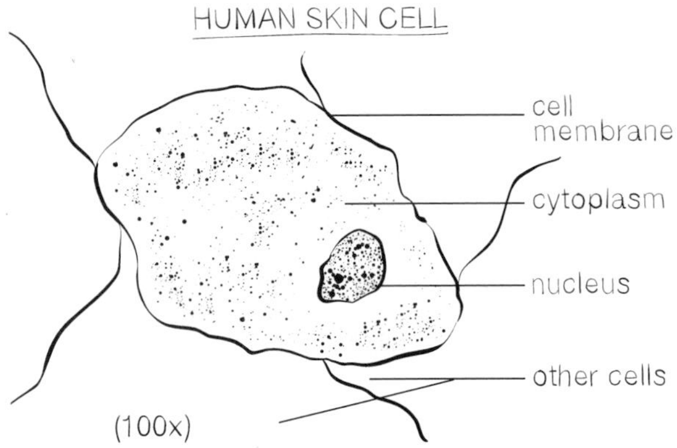

The stippling on this drawing of a human skin cell shows that some areas are darker than others.

Drawing to Scale

In Unit 1, you will be making drawings of objects that have been magnified using a microscope. When you draw objects seen through a microscope, the size of your drawing is important. Your drawing should be in proportion to the size of the object as the object appears when viewed through the microscope. This type of drawing is called a **scale drawing**. A scale drawing allows you to compare the sizes of different objects and to estimate the actual size of the object being viewed. Here are some steps to follow when making scale drawings.

1. Use a mathematical compass to draw an accurate circle in your notebook. The size of the circle does not matter. The circle represents the microscope's field of view.

2. Imagine the circle is divided into four equal sections (see the diagram). Use a pencil and a ruler to draw these sections in your circle, as shown here.

3. Using low or medium power, locate an object under the microscope. Imagine that the field of view is also divided into four equal sections.

4. Observe how much of the field of view is taken up by the object. Also note the location of the object in the field of view.

5. Draw the object in the circle. Position the object in about the same part of the circle as it appears in the field of view. Also, draw the object so that it takes up about the same amount of space within the circle as it takes up in the field of view, as shown in the diagram.

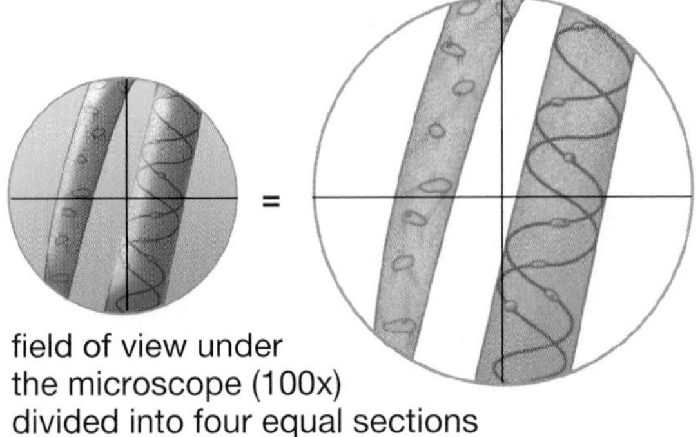

Tips on Technological Drawing

You will find that well laid-out drawings are a valuable learning tool. Please refer to "Appendix B: The Design Process" for ideas. Also, ask the advice of specialist teachers or engineering or technology experts.

Instant Practice

1. How might you improve the student's drawing on page 557 to show the results of the balloon experiments visually?

2. **(a)** Choose an object in your classroom and use stippling as a way of giving it a three-dimensional appearance.
   (b) Exchange your drawing with that of a classmate to see if each of you can identify the other's "object." As well, give each other feedback on how you think the drawing could be improved for greater clarity.

3. Select any mechanical system in your classroom or at home; for example, a wall clock (with "hands") or bicycle gears. Show two different views of the system that would help someone else understand how the system "works."

4. Use your new knowledge to make a scale drawing of a unicellular organism that you observed in Conduct an Investigation 1-C on page 22.

USING RESOURCES AND THE INTERNET EFFECTIVELY

Using Resources Effectively

You probably have some books and periodicals in your classroom that you can use to find out more about certain topics. However, for much of the information you need, you will want to use the library in your school or a branch of your local public library.

Searching for information in a library can be overwhelming, but if you approach your task in an organized way, you can quickly and efficiently find what you need.

First, be aware of the huge amount of material you can access through the library: general and specialized encyclopedias, directories (telephone, city, postal code, business, and associations), almanacs, atlases, handbooks, periodicals, newspapers, government publications, pamphlets, tapes, videos, CD-ROMs, databases, and the ever-changing store of information on the Internet.

To make the best use of your time and the resources available to you, ask yourself these questions before you start your research:

- *What* information do I need? In how much detail?
- *When* is the assignment due? (That might help you decide how much detail you need.)
- *Why* do I need the information? Am I preparing something for an audience (my teacher, another group, or another class)?
- *How* will I be presenting the information (as a written report, a poster, an oral presentation, a multimedia presentation)?

Next, identify which kinds of resources will give you what you need. Librarians are extremely knowledgeable and helpful. Consider them an initiating resource by asking them questions about library use when you cannot find out what you need to know.

What Is the Internet?

The Internet is an extensive network of interlinked yet independent computers. In less than two decades, the Internet went from being a highly specialized communications network used mostly by the military and universities to a massive electronic bazaar. Today, the network includes

- educational and government computers
- computers from research institutions
- computerized library catalogues
- businesses
- home computers
- community-based computers (called *freenets*)
- a diverse range of local computer bulletin boards.

Anyone who has an account on one of these computers can send electronic mail (e-mail) throughout the network and access resources from hundreds of other computers on the network. Here are some of the ways you will find the Internet most useful as a learning tool.

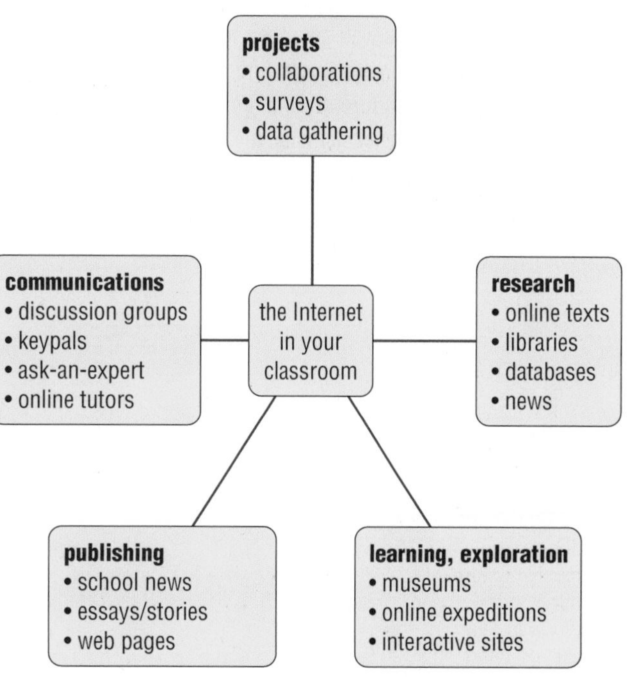

Using the Internet Effectively

The web site address for the publisher of this textbook is: http://www.mcgrawhill.ca
This URL (Universal Resource Locator, or web site address) will take you to the headquarters of McGraw-Hill Ryerson Limited in Whitby, Ontario. You can use the site to obtain information about specific topics in this textbook. When "surfing" the Internet yourself, remember that anyone, anywhere can develop a web site or post information on the Internet. They can use it to "publish" their own opinions. Sometimes, it is difficult to distinguish accurate scientific information from these opinions. Always check the source of the information. Be wary of an individual publishing alone. Government sites and educational association sites tend to have reliable information. Follow your own school's guidelines for "surfing" the Internet to do your research, and make good use of McGraw-Hill Ryerson's School Division web site at www.school.mcgrawhill.ca/resources/ to streamline your search.

INTERNET CONNECT

www.school.mcgrawhill.ca/resources/

Do you want to take an imaginary journey through a cell? Go to the above web site, then to **Science Resources**, and on to **SCIENCEPOWER 8** to find out where to go next. You can zoom in, turn around, and check out different organelles inside a virtual cell.

Answers to Think & Link Investigation 2-D, page 61

A bone in leg
B muscle in arm
C nerve in toe
D blood in heart
E skin on head

Answers to Find Out Activity, page 70

A skeletal muscle tissue
B bone tissue
C nerve tissue

Instant Practice

1. Assume your teacher assigns a research project asking you to do research on a famous Canadian scientist. You are to write a 300-word report. How would you decide on a scientist for your report? How would you begin to find the information you need?
2. Think of a question in science that interests you. For example, why is the sky blue? How far is it to Jupiter, and have any spacecraft travelled to this planet? Write down your question. Now explain how you would do research to find the answer(s) to your question. Show a step-by-step plan. (You could use an events chain concept map to do this.)

Glossary

How to Use This Glossary

This Glossary provides the definitions of the key terms that are shown in **boldface** type in the text. Other terms that are not critical to your understanding, but that you may wish to know, are also included in the glossary. A pronunciation key, in square brackets, will help you pronounce difficult words.

a = mask, back	i = simple, this
ae = same, day	oh = home, loan
ah = car, farther	oo = food, boot
aw = dawn, hot	u = wonder, sun
e = met, less	uh = taken, focus
ee = leaf, clean	uhr = insert, turn
ih = idea, life	

absorption in biology, the process by which dissolved food particles pass from the small intestine to the capillaries

abyssal plains on the deep ocean floor, wide, open, flat plains between the high mountain ranges at the centre and the deep trenches at the edges of the sea floor

accommodation in vision, the process of changing the shape of the lens (in the eye) to adjust for different distances from an object being viewed

acid rain rain that contains higher than normal levels of acid; caused by waste gases released into the atmosphere by industries and automobiles; damaging to the environment

active transport the process by which a protein attaches to a substance, uses energy to move it through a cell membrane, and releases it on the opposite side of the membrane

additive primary colours red, green, and blue; described as additive because adding all three colours together in the proper amounts makes white light

algal bloom a problematic population explosion of algae; usually caused by excessive nutrients (e.g., from fertilizers)

alveolus [ahl-vee-OH-luhs] in the lungs, a tiny air sac at the end of a bronchiole (plural alveoli)

amplitude in a wave, the height of the crest or the depth of the trough

aneroid barometer a type of barometer that contains no liquid

angle of incidence (*i*) in optics, the angle between the incident ray and the normal

angle of reflection (*r*) in optics, the angle between the normal and the reflected ray

angle of refraction (*R*) in optics, the angle between the normal and the refracted ray

annual precipitation the total amount of precipitation in a year

aperture [AP-uhr-chuhr] the opening in the diaphragm of a camera; can let more light reach the film by opening wider

aquifer [AK-wi-fuhr] an area of porous rock with a water system flowing through it

Archimedes' principle a scientific principle stating that the buoyant force acting on a submerged object equals the weight (force of gravity) of the fluid displaced by the object

area the amount of surface; measured in square units such as cm^2

artificial light source a human-made source of light

attitude in optics, the position (upright or upside down) of an image in relation to the object it reflects

average density the total mass of an object divided by the total volume

average monthly temperature range the average difference between the maximum and minimum temperature for a month

bar graph a diagram consisting of horizontal or vertical bars that represent (often numerical) data

barometer [buh-RAWM-uh-tuhr] the most common device for measuring air pressure

bay a part of a body of water that reaches into the land

bioluminescence [BIH-oh-LOO-min-E-sens] the emission of light produced by chemical reactions inside the bodies of living creatures

bioluminescent source [BIH-oh-LOO-min-E-sent] an organism that relies on chemical reactions inside its body to produce light (e.g., a firefly)

blind spot the point where the optic nerve enters the retina; has no light-sensing cells

block and tackle a combination of fixed and movable pulleys; may be used to lift very heavy or awkward loads

blood the fluid that transports substances to and from all parts of the body; consists of plasma, red blood cells, and white blood cells

boiling rapid vaporization occurring at a specific temperature called the boiling point

boiling point the temperature at which a liquid vaporizes rapidly

breaker in a body of water, a wave that breaks or collapses into foam when it reaches shallow water or a beach

bronchiole [BRAWN-kee-ohl] in the lungs, a small, narrow tube branching off from the bronchus

bronchus [BRAWN-kuhs] a tube connecting the trachea with the lungs

buoyancy the tendency to rise or float in a fluid

buoyant force the upward force exerted on objects submerged in or floating on a fluid

C

capacity the largest amount that can be held by a container (usually measured in litres or millilitres)

capacity unit unit used to measure the volume of liquids, e.g., the litre (L)

capillary [ca-PIL-uh-ree] the smallest type of blood vessel; connects arteries and veins

carbohydrates the main source of quick energy for cells; sugar and starch are carbohydrates

carpal tunnel syndrome a common workplace disorder causing numbness and pain in the thumb and first three fingers; often results from repetitive finger movements such as those involved in working at a computer keyboard

carrier protein in a cell membrane, a protein that controls the passage of substances into or out of a cell

cell the smallest unit that can perform the functions of life

cell membrane the selectively permeable structure enclosing the contents of a cell or organelle; regulates the passage of substances into and out of a cell or organelle

cell theory a major theory of living things, formulated by scientists Matthias Schleiden, Theodore Schwann, and Rudolf Virchow in the nineteenth century; the theory states that all living things are composed of one or more cells, that the cell is the basic unit of life, and that all cells come from already existing cells

cell wall a rigid structure surrounding the cell membrane of plants, fungi, and some unicellular organisms; protects and supports the cell

cellular respiration in cells, the process that releases food energy

cellulose the tough, carbohydrate material that in large part makes up the cell walls of many plants and fungi

change of state the transformation of a substance's physical state (whether solid, liquid, or gas) into another state

chemiluminescence [KE-mee-LOO-min-E-sens] the emission of light resulting from chemical action and not involving heat

chemiluminescent source [KE-mee-LOO-min-E-sent] chemical reaction that produces particles that give off visible light energy

chemosynthesis [KEE-mo-sinth-isis] the process of making food using energy from chemical reactions; carried out by bacteria around sea-floor vents

chloroplast an organelle (cell part) that contains chlorophyll and enables plants to make carbohydrates through the process of photosynthesis; found only in plants and in some unicellular organisms

chromosome [KROH-muh-sohm] in a cell nucleus, a threadlike structure that carries genetic material (instructions for producing new cells with the same characteristics as the parent cell)

ciliary muscle [SIL-yuh-ree] the muscle in the eye that controls the eye's accommodation for viewing objects at varying distances

circle graph a circle divided into sections (like pieces of a pie) to represent data; also called a pie chart

circulatory system the system that transports food and oxygen throughout the body; includes the heart, blood, and blood vessels

Class 1 lever a lever in which the fulcrum is between the effort and the load (e.g., a teeter-totter)

Class 2 lever a lever in which the load is between the fulcrum and the effort (e.g., a bottle-opener)

Class 3 lever a lever in which the effort is applied between the fulcrum and the load (e.g., tweezers)

classify to group ideas, information, or objects based on their similarities

climate patterns of weather in a particular place over a period of years; e.g., condition of temperature or precipitation

closed system a system having a boundary that separates it from its surroundings

coherent referring to light, having only one wavelength, with all waves lined up in a similar pattern (e.g., laser light)

colour blindness a condition in which people have difficulty distinguishing between some colours

complementary colours pairs of colours that, together, form white light (e.g., magenta and green)

compound pulley a combination of several pulleys working together

compressibility the ability to be squeezed into a smaller volume; a property of gases

compressible capable of being squeezed into a smaller volume

compressor an electrical device that compresses air

computer spreadsheet software that uses rows and columns to help organize information

concave curving inwards

concave lens a lens that is thinner and flatter in the middle than around the edges; causes refracting light rays to diverge (spread out)

concept map a diagram comprised of words or phrases in circles or boxes and connecting lines; used to show various relationships among concepts; can also contain references to events, objects, laws, themes, classroom activities, or other items or patterns related to the concepts

conclusion an explanation of the results of an experiment as it applies to the hypothesis being tested

condensation the process of changing from gas to liquid

cone in the retina of the eye, a cone-shaped cell that detects colour

consumer product survey a sampling of information about products, gained by asking consumers questions and recording and analyzing the responses

Continental Divide in North America, the continuous ridge of mountain summits dividing the continent into two main drainage areas. On one side, rivers and streams flow west to the Pacific Ocean; on the other side, rivers and streams flow northeast to Hudson Bay or southeast to the Gulf of Mexico.

continental glacier a glacier, or ice sheet, covering all or a significant part of a continent

continental shelf the gradually sloping area between a seacoast and the edge of an ocean basin

continental slope a steep slope dividing a continental shelf from an ocean basin

control in a scientific experiment, a standard to which the results are compared; often necessary in order to draw a valid conclusion; ensures a fair test

convex curving outwards

convex lens a lens that is thicker in the middle than around the edges; causes refracting light rays to converge (come together)

co-ordinate graph a grid that has data points named as ordered pairs of numbers, e.g., (4,3)

Coriolis effect a deflection (bending) of moving air and water currents; produced by Earth's rotation

crank shaft a shaft that turns or is turned by a crank. A crank shaft turns the wheels of an automobile.

crest the high part of a wave

cubic units the units used to report the volume of a substance (e.g., cm^3)

current a broad, continuous movement of water in an ocean or other body of water

cycle concept map an events chain map in which a series of events does not produce a final outcome; this type of concept map has no beginning and no end

cylinder in an engine, a hollow, tube-shaped chamber in which a gas or a liquid causes a piston to move

cytoplasm the gel-like substance within the cell membrane that contains and supports the structures of the cell

daughter cell either of the two new cells formed by the process of mitosis and cell division

decompress to release from pressure

density the amount of mass in a certain unit volume of a substance (density equals mass divided by volume)

density current a mass of cold water flowing beneath the ocean surface; the water is often more saline (salty) than surrounding water and often contains a significant amount of sediment

dependent (or responding) variable in an experiment, the factor that changes in response to a change in the independent variable

desalination plant [dee-salin-AE-shun] the buildings and machinery used in producing fresh water by removing salts from seawater

diaphragm [DIH-uh-fram] in a camera or microscope, a device that controls the amount of light that enters

diffuse reflection the type of reflection that occurs off a rough surface, resulting in no clear image

diffusion the movement of particles in liquids and gases from an area of higher concentration to an area of lower concentration

digestive system a group of organs that work together to break down food and eliminate wastes; includes the stomach and intestines

displace to move something out of the way (e.g., a solid object can displace water out of a container)

downstream the direction in which a river flows

drainage basin the total area from which precipitation drains into a single river or system of rivers

drainage divide the boundary between two drainage basins; usually the crest of a hill or mountain

driven gear a gear whose movement is caused by a driving gear; also called follower

driver *see* driving gear

driving gear a main gear that causes other gears to move; also called driver

efficiency the ratio of the useful work or energy provided by a machine or system with the actual work or energy supplied to the machine or system

effort arm in a lever, the distance between the fulcrum and the effort force

effort force the force supplied to any machine to produce an action

El Niño [EL NEE-nyoh] a change in water currents in the Pacific Ocean that produces weather changes over about two thirds of Earth; usually occurs every few years

electromagnetic spectrum the arrangement by wavelength of the different forms of electromagnetic radiation, including visible light energy

endocrine system [END-oh-krin] a set of glands that produce chemical messengers, called hormones, which are released into the bloodstream; includes the pituitary and thyroid glands, and many more

endoplasmic reticulum [END-oh-plasmik ret-IK-yoo-luhm] within a cell's cytoplasm, a folded membrane that forms a system of canal-like structures that transport materials to different parts of the cell

epidermal tissue the outermost layer of cells, which protects the outer surface of the plant or animal; also known as skin

epithelial tissue [epi-THEEL-ee-ul] in animals, the tissue that covers the body and internal structures such as the intestine

ergonomics the science of designing home or work environments that best suit the human body in its various dimensions

evaporation vaporization that occurs slowly over a wide range of temperatures

events chain map a concept map used to describe a sequence of events, the steps in a procedure, or the stages of a process

excretory system the system that regulates blood composition and gets rid of waste fluids; includes the kidneys

exhaust valve in an engine, a movable part that controls the amount of steam escaping from a cylinder

experiment an activity or procedure designed to test a hypothesis

exponent in science or mathematics, a number, or power, that tells you how many times the number is multiplied by itself; e.g. 10^3 means $10 \times 10 \times 10$ or 1000

eyepiece lens in a telescope or microscope, the lens that works as a magnifying glass to enlarge the image

fair test an investigation (experiment) carried out under strictly controlled conditions to ensure accuracy and reliability of results. In a fair test, all variables are controlled except the one variable under investigation.

far point (of the eye) the greatest distance at which an object is in focus

far-sighted unable to see close objects clearly

fat any of several yellow or white oily substances formed in the bodies of animals and also in some plant seeds; stores energy

feedback the return of information from the output of a system to the input

field of view the area seen through the eyepiece of a microscope or other optical instrument

filament in a light bulb, the metal strip that glows to produce light

film a sheet or strip of thin, flexible material coated with a light-sensitive material and used to make photographs

fixed pulley a pulley supported by attachment to something that does not move, such as a ceiling, wall, or tree

floating remaining suspended in a fluid; for example, not falling in air or sinking in water

flood plain bordering a river or stream, a generally flat area of land that is naturally subject to flooding; made up largely of soil deposited by floods

flow pressure pressure that is caused by a moving fluid

flow rate the volume of fluid that passes a point in a pipe or tube in a certain amount of time

fluid any substance that flows; includes liquids and gases

fluid friction the friction a fluid experiences when it flows past an object such as the solid wall of a pipe or tube

fluorescence [fluhr-E-sens] the process in which high-energy, invisible ultraviolet light is absorbed by the particles of an object, which then emits some of this energy as visible light, causing the object to glow

fluorescent source [fluhr-E-sent] a source that produces light when exposed to light of a particular wavelength

focus to bring (rays of light) to a point; for example, a concave mirror or a convex lens focusses light rays

follower *see* driven gear

force a push or a pull, or anything that causes a change in the motion of an object

freezing the process of changing from liquid to solid

freezing point the temperature at which a liquid freezes

frequency the number of entire cycles completed by a vibrating object in a unit of time; usually given in cycles per second, or hertz (Hz)

fulcrum the point of a lever that does not move

gamma rays the rays having the shortest frequency and highest energy of all radiant waves in the electromagnetic spectrum; gamma rays come from nuclear reactions

gas the state of matter in which a substance has neither a definite shape nor a definite volume (e.g., water vapour)

gear a rotating wheel-like device with teeth around its rim

gear train a group of two or more gears that are meshed together

genus a group of species that are related

graphic organizer a visual learning tool that helps clarify the relationship between a central concept and related ideas or terms

gravity the attractive force between masses; causes objects to be attracted to Earth

groundwater the water that has seeped down under Earth's surface to a depth of about 100 m

guard cells in a plant leaf, cells that surround openings called stomata; can expand to close off the stomata

headland an area of land that sticks out into a body of water

heat thermal energy transferred from one object or substance to another because of a temperature difference

heat capacity the thermal energy needed to raise the temperature of 1 kg of a substance, such as water, by 1°C

heat engine a device that continuously converts thermal energy to mechanical energy

hemoglobin [HEE-moh-gloh-bin] in red blood cells, an iron-rich chemical that binds oxygen, allowing the blood to carry more oxygen than it could otherwise

hertz (Hz) a unit of frequency equal to one cycle per second

histogram a type of bar graph in which each bar represents a range of values and in which the data are continuous

homeostasis [HOHM-ee-oh-stae-sis] the process by which body systems make constant adjustments to maintain a stable internal environment for the body's cells

hormones substances released from specific glands to control particular body activities. The hormone insulin, for example, regulates the body's burning and storage of sugar.

humour any of a number of fluids in the body, including those in the eye

hydraulic lift a mechanical system that uses a liquid under pressure in a closed system to raise heavy objects

hydraulic power power that comes from the pressure of a liquid in a hydraulic system

hydraulic system a device that transmits an applied force through a liquid to move something else by means of pressure

hydraulics the study of pressure in liquids

hydrometer an instrument designed to measure the density of a liquid

hypothalamus [hih-poh-THAL-amuhs] the part of the brain that regulates many body functions, such as temperature, hunger, thirst, sleep, and growth

hypothesis a testable proposal used to explain an observation or to predict the outcome of an experiment; often expressed in the form of an "If ..., then ..." statement.

ice age any of the major periods when glaciers covered much of Earth. The most recent ice age ended about 11 000 years ago.

image the likeness of an object

impermeable allowing no materials to pass through

incandescence [in-cand-E-sens] the emission of visible light by a hot object

incandescent source [in-cand-E-sent] an object that can be heated to such a high temperature that it emits visible light

incident ray the light that strikes a reflecting or refracting material

inclined plane a ramp or slope for reducing the force needed to lift something

incoherent referring to light, having many different wavelengths or one wavelength but the waves are not lined up (e.g., incandescent light)

incompressible incapable of being squeezed into a smaller volume

independent (or manipulated) variable in an experiment, a factor that is selected or adjusted to see what effect the change will have on the dependent variable

infer to conclude or decide by reasoning

inference a conclusion or opinion formed by inferring

infrared radiation heat radiation; anything that is warmer than its surroundings emits infrared rays

intensity brightness (of light); describes how much energy a surface will receive

internal combustion engine a type of engine in which fuel is burned internally, that is, inside the engine

internal friction the motion-resisting force between the surfaces of the particles making up a substance

iris in the eye, the coloured ring; works like the diaphragm of a camera

iris reflex the natural adjustment in the eye's pupil size in response to varying light levels

issue a problem with two or more possible resolutions of interest to members of society

joule (J) a unit used to measure energy or work; 1 J = a force of 1 N moving through 1 m

kilopascal (kPa) a unit of pressure equal to 1000 pascals

kilowatt hour (kW·h) a unit of electrical energy; the amount of energy transmitted by one thousand watts of power over a period of one hour; 1 kW•h = 1000 W of power used for 1 h

kingdom one of five main groupings for classifying living things on Earth; the five kingdoms are: animal, plant, fungus, protist, and monera

lake a large area of water surrounded by land

land bridge a narrow strip of land connecting larger land masses

laser a device for amplifying light to produce an intense, narrow beam; used in computer printers, surgical procedures, and other applications. The term stands for **l**ight **a**mplification by the **s**timulated **e**mission of **r**adiation.

law in science, a statement of a pattern, action, or condition that has been observed so consistently that scientists are convinced it will always happen

laws of reflection the two main predictable behaviours of reflected light: 1. The angle of reflection (*r*) is equal to the angle of incidence (*i*). 2. The incident ray, the normal, and the reflected ray are always in the same plane.

lens a curved piece of transparent material, usually glass or plastic. Light rays bend as they pass through a lens.

levee an elevated bank built along the side of a river valley to keep the river from overflowing

levels of organization in organisms, the arrangement of structures from the simplest (i.e., cells) to more complex (i.e., tissues, organs, and organ systems)

lever a machine consisting of a bar that is free to rotate around a fixed point, changing the amount of force that must be exerted to move an object

light the form of energy we can see

line graph a diagram that shows how one value depends on or changes according to another value; produced by drawing a line that connects data points plotted in relation to a y-axis (vertical axis) and an x-axis (horizontal axis)

liquid the state of matter in which a substance has a definite volume, but no definite shape (e.g., water)

load the weight of an object that is moved or lifted by a machine, or the resistance to movement that a machine must overcome

load arm in a lever, the distance from the load to the fulcrum

luminous giving off its own light

magnification the apparent amount of enlargement produced by a microscope or similar magnifying instrument

mass the amount of matter in a substance; often measured with a balance

mass production the manufacturing of large quantities of a standardized item by standardized mechanical processes

mass-to-volume ratio the density of an object; calculated by dividing the mass by the volume

mean discharge the volume of water flowing from a river into an ocean; measured in cubic metres per second (m^3/s)

mechanical advantage the ratio of the force produced by a machine or system (the load) to the force applied to the machine or system (the effort force)

melting the process of changing from solid to liquid

melting point the temperature at which a solid changes to a liquid

meltwater the run-off from melting snow

meniscus the slight curve at the top of a liquid where the liquid meets the sides of a container

metric system a system of measurement based on multiples of ten and in which the basic unit of length is the metre

micrometre a unit often used for measuring the size of cells; equal to one millionth of a metre; expressed by the symbol μm

microscope an instrument that makes objects appear larger by bending light through a lens

microwave a radio wave having a short wavelength and high frequency; used in microwave ovens, telecommunications satellites, and other applications

mid-ocean ridges long undersea mountain chains that run along the centres of some oceans

mitochondrion [mih-toh-KAWN-dree-awn] in a cell, an oval-shaped organelle that transforms energy for the cell (plural mitochondria)

mitosis [mih-TOH-sis] the process by which a cell's nucleus divides to form two identical copies of genetic material

mixed layer near the ocean surface, a layer of water in which winds and waves evenly mix heat from the Sun and other sources

model a verbal, mathematical, or visual representation of a scientific structure or process, which allows scientists to construct and test inferences and theories (e.g., the particle theory of matter)

movable pulley a pulley attached to something movable, such as a construction crane or oil derrick

multicellular having many cells

nanometre one billionth of a metre; a unit used to measure radiation wavelength; abbreviated nm

natural light source a non-human-made source of light, such as the Sun

neap tides the smallest tidal movements; occur when lines pointing from Earth to the Sun and the Moon are perpendicular to each other

near point (of the eye) the shortest distance at which an object is in focus

near-sighted unable to see distant objects clearly

nervous system the body system that senses internal and external changes, and controls and co-ordinates body activities; includes the brain and nerves

network tree a concept map in which some terms are circled while other terms are written on connecting lines

neutral buoyancy the condition in which the amount of force pulling down on an object immersed in a fluid (i.e., gravity) equals the amount of force pushing up (buoyancy)

non-luminous a substance that does not give off its own light

normal a reference line drawn perpendicular to a reflecting surface at the point where an incident ray strikes the surface

nuclear membrane [NOO-klee-uhr] the thin, outer membrane that surrounds the cell nucleus; separates the contents of the nucleus from the cytoplasm

nucleus in a cell, an organelle that controls all the cell's activities

nutrients the substances in foods that provide energy and materials for cell development, growth, and repair

objective lens the convex lens in a refracting telescope or microscope

objective mirror the mirror in a reflecting telescope; also called the primary mirror

observation the use of the senses to gather information; in science, often aided by instruments such as telescopes, thermometers, and balances

ocean the large body of salt water that covers almost three quarters of Earth's surface; any of its five main sections (the Pacific, Atlantic, Indian, Southern, and Arctic Oceans)

ocean basins deep, wide depressions in Earth's surface that contain the oceans

opaque [oh-PAEK] not allowing any light to pass through

optic nerve the nerve that connects the eye to the brain

organ a group of different tissues that work together to perform a specific function (e.g., the heart)

organ system a group of organs that work together to perform a major function (e.g., the respiratory system)

organelle a structure within a cell that has a specific function (e.g., mitochondrion)

osmosis [oz-MOH-sis] the diffusion of a solvent, usually water, through a selectively permeable membrane

ozone layer a "blanket" of ozone (a form of oxygen) surrounding Earth about 20–25 km above the ground; acts like a filter, absorbing much of the Sun's ultraviolet light

P

palisade cells in a plant leaf, a layer of cells that are filled with chloroplasts

particle theory of matter a scientific model of the structure of matter; one part of this theory states that all matter is made up of extremely small particles

pascal (Pa) a unit for pressure; newtons per square metre (N/m^2)

Pascal's law a law stating that when pressure is exerted on one part of a fluid, the same pressure is transmitted unchanged to all parts of the fluid, no matter what the shape of the container holding the fluid

permeable allowing materials to pass through

phloem tissue [FLOH-EM] in plants, the tissue that transports sugars manufactured in the leaves to the rest of the plant

phosphorescence [faws-fohr-E-sens] the persistent emission of light following exposure to and removal of a source of radiation

phosphorescent source [faws-fohr-E-sent] a substance that gives off visible light released after the light energy has been absorbed by certain particles that have stored this energy for a while. The light continues for some time even after the substance is no longer exposed to the light.

phytoplankton [fih-toh-PLANK-tuhn] the plant organisms in plankton

piston in an engine, a disc-shaped part that fits tightly inside a cylinder in which some force, such as steam pressure, moves it back and forth or up and down

plain a generally flat or gently sloping area of land

plane a flat or level surface

plane mirror a mirror having a flat surface

plankton tiny, drifting organisms in the surface waters of the ocean

plasma the liquid portion of blood

pneumatic system [noo-MAT-ik] a system in which a gas, such as air, transmits a force exerted on the gas in an enclosed space

pneumatics [noo-MAT-iks] the study of pressure in gases

porous having pores or tiny holes; allowing fluid to pass through

potable safe to drink

precipitation the water (in its liquid or solid state) that falls to Earth; rain, snow, sleet, hail, etc.

pressure the force acting perpendicular to a certain surface area

prisms in binoculars, glass blocks serving as plane mirrors; allow binoculars to be made shorter than telescopes

protein a complex nitrogen-containing substance essential for growth and repair of body tissues; an essential part of animal and plant cells

pulley a wheel with a grooved rim to guide a rope or chain that runs along the groove; used to transmit or change the direction of force

pupil the opening in the eye; appears as the dark centre of the eye

Q

qualitative data information gathered by observations in which no measurement takes place

qualitative property a characteristic of a substance that can be described but not measured

quantitative data data that consist of numbers and/or units of measurement; obtained through measurement and through mathematical calculations

quantitative property a characteristic of a substance that can be measured

R

radar a device that uses the reflection of radio waves to determine the distance and location of unseen objects; the term stands for **ra**dio **d**etecting **a**nd **r**anging

radiant energy energy transferred or emitted as waves or rays in all directions

radiate to send out energy in the form of waves or rays

radiation the transfer of radiant energy, such as light

radiation therapy the use of gamma rays to destroy cancer cells

radio wave any electromagnetic wave that is useful for carrying sounds or pictures through the air from a transmitter to a receiver

radius the distance in a straight line from the centre of a circle to the circle's circumference (outer boundary)

ray a single line or narrow beam of light that originates from a light source and that may bounce off a surface that it strikes

ray diagram a representation of the path of light by using a straight line with an arrow

real image an image located where reflected or refracted rays actually meet

recycling the extraction and reprocessing of useful materials from waste for re-use

reflected ray the light that is cast back (reflected) from a reflecting surface

reflecting telescope a telescope having a concave mirror to collect rays of light from a distant object

reflection the casting back of light from a surface

refracting telescope a telescope having a convex lens to collect and focus light from a distant object, and an eyepiece lens to magnify the image

refraction the bending of light when it travels from one medium (material) to another

refrigerant the cooling liquid in refrigerators

regeneration in living things, the process of repairing injured cells or growing lost body parts

resistance to flow the condition in which particles in a substance can move around but cannot easily pass each other

respiratory system the system that moves air in and out of the body; includes the lungs, bronchi, trachea, and nasal passages

retina a light-sensitive area at the back of the eye

rod in the retina of the eye, a cylinder-like cell that detects the presence of light

root hair on a plant root, an extension of a single epidermal cell. Water enters a root hair by osmosis.

run-off rainwater that flows off a land surface

salinity the measure of the amount of salts dissolved in a liquid

scale drawing a drawing in which the objects appear in the same proportions as they are in reality

science a body of facts or knowledge about the natural world, but also a way of thinking and asking questions about nature and the universe

science inquiry the orderly process of asking concise and well-focussed questions and designing experiments that will give clear answers to those questions

scientific investigation an investigation that involves the systematic application of concepts and procedures (e.g., experimentation and research, observation and measurement, analysis and sharing of data)

scientific notation a short form for writing very large or very small numbers (e.g., 3×10^{11} means $3 \times 10 \times 10 \times 10 \times 10 \times 10 \times 10 \times 10 \times 10 \times 10 \times 10 \times 10$)

sea-floor vent a crack in the ocean floor that releases heat and minerals

seamount a large mountain peak under the surface of the ocean

secondary colours colours produced by the light of two additive primary colours. The three secondary colours are cyan, magenta, and yellow; also called subtractive colours.

selectively permeable allowing only certain materials to pass through

shutter in a camera, a device that opens the aperture to allow light to reach the film; can let more light reach the film by staying open longer

SI (from the French *Le Système international d'unités*) the international system of measurement units, including such terms as kilogram, metre, and second

society a group of people united by common goals and interests

solar spectrum the pattern of colours in sunlight: red, orange, yellow, green, blue, indigo, and violet

solid the state of matter in which a substance has a definite shape and a definite volume (e.g., ice)

solute a substance that dissolves in a solvent (e.g., salt is a solute that dissolves in water)

solution a homogeneous mixture of two or more pure substances

solvent a substance that dissolves a solute to form a solution (e.g., water is a solvent that dissolves salt)

specialized in living things, cells having different structures and appearances and performing different functions

species a narrow classification grouping for organisms; e.g., a wolf is the species *Canis lupus*, while a dog is the species *Canis familiaris*

spectrum the series of coloured bands produced when white light is separated into its component wavelengths

speed the rate of motion, or the rate at which an object changes its position

sphygmomanometer [sfig-moh-ma-NAHM-e-tuhr] a device used to measure blood pressure

spider map a concept map used to organize a central idea and a jumble of associated ideas that are not necessarily related to each other

spontaneous generation [spawn-TAEN-ee-uhs gen-e-RAE-shun] historically, the idea that some living organisms could be produced from non-living matter such as air or water

spring a small stream consisting of groundwater flowing naturally out onto the surface of Earth

spring tides the largest tidal movements; occur when Earth, Moon, and Sun are lined up

sprocket a gear with teeth that fit into the links of a chain

static pressure the force on an object exerted by a fluid when the fluid is not moving

steam the invisible gas into which water is changed by boiling

steam engine a steam-powered engine; often burns fuel such as coal or wood to change water into steam in a boiler outside the engine

stem-and-leaf plot an organization of data into categories based on place values; values on the stem represent a range of data, while the leaves indicate the data points that lie in that range; when turned on its side, resembles a bar graph

stethoscope [STETH-uh-skohp] a device for listening to sounds in the heart, lungs, and other body parts

STSE an abbreviation for the interrelationships among science, technology, society, and the environment

stomata the tiny openings on the underside of a plant leaf that let air into the leaf (singular stoma)

sublimation the process of changing from solid to gas without going through the liquid state

subsystem a secondary or subordinate system that is part of a larger system

subtractive primary colours cyan, magenta, and yellow; described as subtractive because some portion of white light has been removed in order to get each colour; also called secondary colours

surface area the area of surface of an object

surface area-to-volume ratio the relationship between an object's surface area and volume; changes as the object gets bigger or smaller

swell in the ocean, a long, smooth wave near the shore; caused by winds and storms far out at sea

system a collection of parts that work together in such a way that a change in one part can result in a change in another part

table an orderly arrangement of facts or numerical data set out for easy reference; for example, an arrangement of numerical values in vertical or horizontal columns

technology the application of scientific knowledge and everyday experience in solving practical problems by designing and developing devices, materials, systems and processes

temperature a measure of the thermal energy of the particles in a substance

theory an explanation of an event that has been supported by consistent, repeated experimental results and has therefore been accepted by a majority of scientists

thermal energy the total energy of the random motion of particles making up a substance

thermocline 200–1000 m below the ocean surface, a region in which temperatures decrease rapidly with increasing water depth

tidal range the difference in the water level between high tide and low tide; varies with the shape of the shoreline

tide the slow rise and fall of the ocean surface, caused by the gravitational pull of the Sun and the Moon

tissue a group of cells having the same structure and function (e.g., muscle, nerve, skin)

trachea [TRAE-kee-uh] the windpipe; part of the respiratory system

trade winds near the equator, winds that blow from the east to drive ocean currents westward

translucent allowing some light to pass through. The light, however, is scattered from its straight path.

transmitted passed along

transparent allowing light to pass through

transpiration the loss of water from a plant through evaporation

trenches narrow, steep-sided canyons running along some ocean-floor margins

trough the low part of a wave, between crests

turbine a device used to convert the motion of a fluid into mechanical energy; consists of fan blades attached to a central hub

ultraviolet (UV) radiation wavelengths of 200 nm beyond violet light in the electromagnetic spectrum (nm = nanometre); causes tanning

unicellular having a single cell

upstream the direction from which a river flows

upwelling a type of vertical current flowing upward from the ocean floor to the surface

user-friendly easy to assemble and use

vacuole [VA-kyoo-ohl] in a cell, a fluid-filled organelle that stores water, food, wastes, and other materials

valley a low stretch of land between higher areas of land. A river may flow through it.

valve a device used to regulate the flow of a liquid in a hydraulic system

vaporization the process of changing from liquid to gas

variable a factor that can influence the outcome of an experiment

vascular tissues in plants, tissues that connect the root system and the shoot system

velocity the speed of an object in a specified direction

velocity ratio in a set of gears, a comparison of how much farther (and faster) a smaller gear turns than a larger gear

Venn diagram a graphic organizer consisting of overlapping circles; used to compare and contrast two concepts or objects

vent an ocean-floor opening that releases hot water and gases from deep within Earth

villus a tiny, fingerlike projection on the inner lining of the small intestine (plural villi)

virtual image an image located where reflected rays only seem to originate

viscosity the measure of how fast a fluid will flow; the "thickness" or "thinness" of a fluid

volume the measurement of the amount of space occupied by a substance; measured in litres or cubic units such as cubic centimetres (cm^3)

water cycle the process in which nearly all water on Earth moves continuously between the oceans, land, and atmosphere

water table in the ground, the level beneath which porous rocks are saturated with water

watt (W) a unit of power equivalent to one joule per second; 1 W = 1 J/s

wave in a body of water, a large ripple set in motion by steady winds; also, the way in which light travels

wave model of light a way of explaining the behaviour of light; involves picturing it travelling as a wave

wavelength the distance from crest to crest, or from trough to trough, of a wave of light as it travels through space; also used to describe the distance from crest to crest, or from trough to trough, of a water wave

weather atmospheric conditions at a particular time and place, including temperature, moisture, cloudiness, and windiness

weight the force of gravity exerted on a mass

westerly winds in temperate zones (e.g., near the Canada–United States border), winds blowing from the west that drive ocean currents eastward

wet mount a type of sample preparation using a microscope slide, a cover slip, and water

wheel and axle a machine consisting of two turning objects attached to each other at their centres. One object causes the other to turn (e.g., a winch).

WHMIS an acronym that stands for Workplace Hazardous Materials Information System

winch a machine consisting of a small cylinder, a crank or handle, and a cable; used for lifting and pulling

work the transfer of energy through motion

X ray electromagnetic radiation having a very short wavelength; can penetrate substances such as skin and muscle

xylem tissue [ZIH-lem] in plants, the tissue that conducts water and minerals absorbed by the root cells to every cell in the plant

zone of saturation a layer of porous rock in which all pores are full of water

zooplankton [zoh-PLANK-tuhn] the animal organisms in plankton

Index

The page numbers in **boldface** type indicate the pages where the terms are defined.
Terms that occur in investigations (*inv.*) and activities (*act.*) are also indicated.

Photo Credits

iii Mark Tomalty/Masterfile; **vii top** Mark Cullen; **viii top** Michael Keller/First Light, **centre** Clive Webster/Ivy Images; **ix top** Dale Stokes/Mo Yung Productions/Norbert Wu, **centre** Dick Hemingway, **bottom** Richard Megna/Fundamental Photographs; **x top** John Cunningham/Visuals Unlimited, **centre** Philip Norton/Valan Photos, **bottom** Guy Motil/First Light; **xi top** New Brunswick Electric Power Commission, **bottom centre** Terje Rakke/Image Bank Canada; **xii top** Air Canada, **centre right** Dick Hemingway, **bottom** Tom W. Parkin/Valan Photos; **xiii top** Frank Siteman/Tony Stone Images; **xiv-xv** Artbase Inc.; **xviii left** John Edwards/Tony Stone Images, **right** Ivy Images; **IS-4 top left** Willie McElligott/National Research Council, **top right** Jeff Foott/Valan Photos, **bottom left** J. Pasachoff/Visuals Unlimited, **bottom right** Scott Audette/Associated Press/Canapress; **IS-5 top left** Rod Brindamour/Orangutan Foundation International, **top right** David Vaughan/Science Photo Library/Photo Researchers, **bottom main photo** WHOI/R. Catanach/Visuals Unlimited, **bottom inset** D. Foster/WHOI/Visuals Unlimited; **IS-7 top right** Richard Megna/Fundamental Photographs, **bottom left** B & D Productions/First Light, **bottom right** Dick Hemingway; **IS-9 top** Robert Mathena/Fundamental Photographs; **IS-14 left** Picture Collection/Metro Toronto Reference Library, **right** The Farm Museum; **IS-15** Ross Harrison Koty/Tony Stone Images; **IS-18 top** Ivy Images, **bottom** NASA; **IS-19 bottom right** Ontario Hydro; **IS-21** New Brunswick Electrical Power Commission; **2-3** Johnny Johnson/Valan Photos, **2 inset** Biology Media/Photo Researchers; **4** Stephen Dalton/Animals, Animals; **5** Harold V. Green/Valan Photos; **7** Graphics & Photography Services/University of Toronto Press; **8** Sinclair Stammers/Science Photo Library/Photo Researchers; **9 top left** Lawrence Migdale/Science Source/Photo Researchers, **top right** Doug Martin/Photo Researchers, **9 bottom (TEM)** Phil A. Harrington/Peter Arnold, **bottom (SEM)** David Burder/Tony Stone Images; **11** Mark Cullen; **18 both** From Hooke's *Micrographia*, 1665; **21 Diatoms** Manfred Kage/Peter Arnold, **Chlamydomonas and Paramecium** M.I. Walker/Photo Researchers, **Euglena** Biophoto Associates/Photo Researchers, **Volvox** Carolina Biological Supply Company/Phototake NYC/First Light, **Stentor** Eric Grave/Photo Researchers; **25, 26** Ed Reschke/Peter Arnold; **27 brick wall** Artbase Inc., **bottom (both)** Biophoto Associates/Photo Researchers; **28 left** Dr. Gopal Murti/Science Photo Library/Photo Researchers; **29 top left** Dr. Dennis Kunkel/Phototake NYC/First Light; **31 right** James R. Page/Valan Photos; **35 left** Mark Cullen; **38** First Light; **40 top** V. Wilkinson/Valan Photos; **46 left** Dick Hemingway, **right** Dennis Kunkel/CNRI/Phototake NYC/First Light; **49 top** Dick Hemingway; **53 top** Canadian Tourism Commission; **bottom** E.R. Degginger/Photo Researchers; **55 Tree and roots** Artbase Inc., **Photos A and B** Biology Media/Photo Researchers, **Photos C and D** Ed Reschke/Peter Arnold; **61 Photo A** Carolina Biological Supply Company/Phototake NYC/First Light; **Photo B** Michael Abbey/Science Source/Photo Researchers, **Photos C and D** Dennis Kunkel/Phototake NYC/First Light, **Photo E** Biophoto Associates/Photo Researchers; **65 both** David Phillips/Science Source/Photo Researchers; **66** John Colwell/Grant Heilman Photography; **70 bottom left** M. Abbey/Photo Researchers, **bottom centre** J. & L. Weber/Peter Arnold, **bottom right** Martin M. Rotker/Photo Researchers; **72** Zig Leszczynski/Animals, Animals; **76 both** Ray Simons/Photo Researchers; **78 top left** Stephen J. Krasemann/Valan Photos, **bottom left** Ontario Ministry of Natural Resources, **bottom centre** Bill Ivy/Ivy Images, **bottom right** Kjell B. Sandved/Visuals Unlimited; **79 top** Michael Keller/First Light; **82 bottom** Fred Bruemmer/Valan Photos; **87** University of Toronto/Department of Public Affairs; **92 bottom** Neil Hokan/Canada in Stock/Ivy Images; **93 bottom** Scott Winter/*Career Connections: Great Careers for People Interested in Sports & Fitness*, published by Trifolium Books Inc./Weigl Educational Publishers; **100 top, 101 left & right** The Hospital for Sick Children; **102 left** David York/The Stock Shop-Medichrome, **right** Bert Gildart/Peter Arnold; **106-107** 20th Century Fox, The Kobal Collection, **106 inset** Canadian Tourism Commission; **108** Clive Webster/Ivy Images; **112 top** John Cancalosi/Valan Photos; **113 top** First Light; **114 bottom** NASA/First Light; **115 top** Charles D. Winters/Photo Reseachers, **bottom** Bill Ivy/Ivy images; **116 top right** Fred Ward/First Light, **bottom right** Sears Canada; **117 bottom** Claus Andersen/Canadian Olympic Association/ Canadian Sports Images; **120 bottom left** Kristen Brochmann/Fundamental Photographs, **bottom right** Bill Ivy/Ivy Images; **121 left** Francis Lepine/Valan Photos, **right** Ottmar Bierwagen Photo Inc./Canada in Stock/Ivy Images; **125** UPI/CORBIS/Bettmann; **130** Fred Bavendam/Peter Arnold; **131 left** Nova Scotia Information Service, **right** Wayne Lankinen/Valan Photos; **133 bottom** Richard Sears/Valan Photos; **136 left** Richard Megna/Fundamental Photographs; **144 left** Dick Hemingway, **144 right** Artbase Inc.; **146 top left** Courtesy Petro Canada, **bottom** Steve Satushek/Image Bank Canada; **149 top** Goodyear Canada Inc.; **151 bottom** Valan Photos; **154 top left** Runk & Schoenberger/Grant Heilman Photography Inc., **top right** Paul Silverman/Fundamental Photographs, **bottom** Norbert Wu; **158** Dr. Verena Tunnicliffe; **162-163** Alvin Uptis/Image Bank Canada, **163 inset** Patti McConville/Image Bank Canada; **164 bottom left** From Perrault, *Les Hommes Illustres*, Paris, 1696-1700; **167 bottom** Will & Deni McIntyre/Photo Researchers; **170 both** Dick Hemingway; **171 top left** Carson Ganci/First Light, **171 top right** Artbase Inc., **171 bottom** Dick Hemingway; **173 bottom** Dale Stokes/Mo Yung Productions/Norbert Wu; **176 top** Will & Deni McIntyre/Photo Researchers Inc.; **177 top three photos** Dick Hemingway/equipment supplied courtesy of the Toronto Fire Department, **bottom** Howard M. Paul/Emergency! Stock; **179 left** Ontario Ministry of the Environment; **180 bottom** Canadian Space Agency 2000/www.space.gc.ca; **185 both (bottom)** Runk & Schoenberger/Grant Heilman Photography; **187** From *Career Connections: Great Careers for People Who Like Being Outdoors*, published by Trifolium Books Inc./Weigl Educational Publishers Limited; **188 top** David Nunuk/First Light; **190 bottom** Patti McConville/Image Bank Canada; **194** Covey Island Boatworks; **195 left** *Belle Marie* designed by Dave Gerr, Gerr Marine Inc./built by Covey Island Boatworks, Photo by Paul Bremer **right** Stark Jett; **196 left** Ford Motor Company of Canada, **right** Ivy Images; **200-201** Canadian Tourism Commission, **201 inset** Will & Deni McIntyre/Photo Researchers; **202** Dick Hemingway; **204** Bill Ross/West Light/First Light; **206 top** Joyce Photographics/Valan Photos; **207** David Parker/Science Photo Library/Photo Researchers; **208 top** Dick Hemingway; **210** Tom Lyle/The Stock Shop-Medichrome; **211 left** Chris Parks/Mo Yung Productions/Norbert Wu, **right** Tom Lyle/The Stock Shop-Medichrome; **215 top** Jeff Isaac Greenberg/Visuals Unlimited; **216** Artbase Inc.; **224 bottom** Dick Hemingway; **231** John M. Dunay IV/Fundamental Photographs; **236** NOAO/TSADO/Tom Stack & Associates/First Light; **237** M. Abbey/ Photo Researchers; **242 left** Monarch Marking Devices; **243** Dick Hemingway; **244** Richard Megna/Fundamental Photographs; **245 both** David Parker/Science Photo Library/Photo Researchers; **246** Geoff Tompkinson/Science Photo Library/Photo Researchers; **251 bottom** Joseph Pobereskin/Tony Stone Images; **252 bottom** TV Ontario; **256 top** Bachmann/Photo Researchers; **257 left** David Nunuk/First Light; **261 top** Ontario Ministry of Natural Resources; **266** E.R. Degginger/ Photo Researchers; **267** John Cunningham/Visuals Unlimited; **268** Ontario Ministry of Natural Resources; **270 top** Don Johnston/Ivy Images; **273 bottom** Vaughan Fleming/Science Photo Library/Photo Researchers; **275 both (top)** Diane Hirsch/Fundamental Photographs; **280** Steve Short/First Light; **288** James R. Page/Valan Photos; **291 top left** Phil Silverman/Fundamental Photographs; **bottom** From *Career Connections: Great Careers in Communications and Technology*, published by Trifolium Books Inc.,/Weigl Educational Publishers Limited; **292** University of Toronto/Department of Public Affairs; **294 left** Philip Norton/Valan Photos, **right** Alfred Pasieka/Science Photo Library/ Photo Researchers; **295 bottom left** Susan Leavines/Photo Researchers, **bottom right** Dick Hemingway; **296 top** Dick Hemingway; **297 top** Scott Camazine/Photo Researchers, **297 bottom** Dr. Li-Hong Xu; **298** Richard Megna/Fundamental Photographs; **302, 303** Judy Kitto/Solar Power Systems Inc.; **304 left** Francis Lepine/Valan Photos; **right** Dick Hemingway; **308-309** Guido Alberto Rossi/Image Bank Canada, **308 inset** Richard Hartimer/First Light; **310** NASA, **311** Dan Kozlovic; **312 top** B & D Productions/First Light, **bottom** Joyce and Frank Burek/Earth Scenes; **314** Jean Bruneau/Valan Photos; **315 bottom left** Michel Bourque/Valan Photos, **right** Aubrey Long/Valan Photos; **319 left** G. Kopelow/First Light, **inset** Michael S. Green/Associated Press/Canapress; **321** Canadian Tourism Commission; **322 top left** John Cunningham/Visuals Unlimited, **top centre** OHSC Archives, Negative # 83.0505 (Ontario Hydro), **top right** Harold V. Green/Valan Photos, **322 centre** Artbase Inc., **bottom left** Ontario Science Centre, **bottom centre** The Farm Museum, **bottom right** Dick Hemingway; **323** Bill Ivy/Ivy Images; **326 top** Brian Milne/First Light, **bottom** Dick Hemingway; **327 top** Dan Kozlovic, **bottom** Chris Jones/First Light; **330** Kim Patterson/Valan Photos; **335 bottom** Travel Alberta; **337 top** Philip Norton/Valan ɔs; , **337 bottom right** From *Career Connections: Great Careers for* *le Who Are Concerned About the Environment*, published by Trifolium ɔks Inc./Weigl Educational Publishers; **342** Steve Wilkings/First ght; **343** Artbase Inc.; **344** Guy Motil/First Light; **346 bottom** H. ιichard Johnston/Tony Stone Images; **347 left** Val Wilkinson/Valan Photos, **right** Ian Barrett/First Light; **350** Joseph R. Pierce/Valan Photos; **351 top** John Sohlden/Visuals Unlimited, **bottom** Francis Lepine/Valan Photos; **354 bottom** Images of New Brunswick/Images du Nouveau Brunswick; **363 top** David Sandwell/Scripps Institution of Oceanography/UCSD, **bottom** Institute of Oceanographic Sciences/NERC/Science Photo Library/Photo Researchers; **364 top** C.C. Lockwood/Earth Scenes; **366 left** Peter Parks/Mo Yung Productions/Norbert Wu, **right** Peter Parks/Norbert Wu; **368** Carl Purcell/Photo Researchers; **378** New Brunswick Electric Power Commission; **379** Artbase Inc.; **381 top** John Eastcott/YVA Momatiuk/Valan Photos; **382** Artbase Inc.; **383 right** Richard Reid/Earth Scenes; **384** Gordon Petersen/First Light; **387** John Sylvester/First Light; **388 jellyfish, mollusc, tube worms, sea star, corals, sea cucumber** Kozmik/Ivy Images, **sea slug, crab, fan worms** Norbert Wu; **390** Ivy Images; **391** WHOI/D. Foster/Visuals Unlimited; **392 top** J. Eastcott/Y. Momatiuk/ Valan Photos, **bottom** Ontario Ministry of the Environment; **393 top** Science Vu/Visuals Unlimited; **396 bottom** Metro Toronto Region Conservation Authority; **397 top** New Brunswick Electric Power Commission, **bottom** Frank McGurk from *Career Connections: Great Careers for People Who Like to Work with Their Hands* by Julie Czerneda, published by Trifolium Books Inc./Weigl Educational Publishers Limited; **399 top** John Eastcott/YVA Momatiuk/ Valan Photos, **bottom** Francis Lepine/Valan Photos; **400** Imperial Oil; **401 bottom** Reynolds Aluminum/Ontario Science Centre; **403 top** Gerald & Buff Corsi/Visuals Unlimited, **bottom** Tim Krochak/ Canapress; **404** Grant Black/First Light; **408** Jody Butler-Walker; **409** Water Resources/Department of Indian Affairs and Northern Development; **410** Pam Hickman/Valan Photos; **411 right** Kennon Cooke/Valan Photos; **413 left** Inga Spence/Visuals Unlimited, **right** V. Wilkinson/Valan Photos; **414-415** Kjell B. Sandved/Visuals Unlimited; **416** North Wind Picture Archives; **418 right** Bill Ivy/Ivy Images; **419 bottom left** Dick Hemingway, **bottom right** John P. Kelly/Image Bank Canada; **420 top centre** Terje Rakke/Image Bank Canada; **421 both** The Canadian Space Agency; **428 left** Geoff Tompkinson/Science Photo Library/Photo Researchers; **431 top left** Jerry Wachter/Photo Researchers, **bottom right** Blair Seitz/Photo Researchers; **432 left** NASA, **centre** Ken Stratton/First Light, **top right** Bill Ivy/Ivy Images, **bottom right** Ralf-Finn Hestoft/Ivy Images; **433 left** Dick Hemingway, **right** Action for Disability, Newcastle upon Tyne/Simon Fraser/Science Photo Library/Photo Researchers; **434 top** Artbase Inc.; **436 top right** Dick Hemingway, **centre** Ford Motor Company, **bottom right** CORBIS/Kevin R. Morris, **bottom left** Artbase Inc.; **438 top left** North Wind Picture Archives, **top right** John Cancalosi/Valan Photos; **441 top** Darwin Wiggett/First Light; **443 top** Debbie Cote/Valan Photos; **445 top** Philip Norton/Valan Photos; **448** The Calgary Herald; **449 both** Courtesy of West Island College/Class Afloat (TM), **453** Michael J. Johnson/Valan Photos; **454 both** Paul L. Ruben; **456 top** David Young Wolff/Tony Stone Images; **461 left** Sargent Welch Exclusive Cenco Physics; **466 top left** Robert W. Allen/First Light, **top right, bottom left** Dick Hemingway, **bottom right** Ford Motor Company/New Holland Division; **467** Roger Czerneda/*Career Connections: Great Careers for People Who Like to Work with Their Hands*, by Julie Czerneda, published by Trifolium Books Inc.,/Weigl Educational Publishers Limited; **468 top left** Air Canada; **470 top** Joyce Photographics/Valan Photos, **bottom** Jeff Wasserman/Globe and Mail/Canapress; **471** Dick Hemingway; **472 top right** A. Farquhar/Valan Photos, **top left** Stephen Saks/Photo Researchers, **bottom left and right** Nova Scotia Information Service; **473 top** Aircast Incorporated; **476 top** J. A. Wilkinson/Valan; **481 top left** Bill Ivy/Ivy Images, **top right** Dick Hemingway/equipment supplied courtesy of Toronto Fire Department, **bottom** University of Victoria Photo Services; **482 left** Dick Hemingway/equipment supplied courtesy Toronto Fire Department, **right** Paratech Inc./David Young Wolff/Tony Stone Images; **486** Tom W. Parkin/Valan Photos; **487 top** Alan Marsh/First Light; **493** Bachmann/Photo Researchers; **494 left** Artbase Inc.; **496 top** British Columbia Archives/ A-00010; **497 left** Library of Congress, **right** Picture Collection/Metro Toronto Reference Library; **498 top** Ron Watts/First Light; **502** Dick Hemingway; **506 top left** OHSC Archives, Negative # 80.1735 (Ontario Hydro), **bottom left** Andy Sacks/Tony Stone Images, **bottom right** IKEA © Inter IKEA Systems B.V.; **509** Heidy Lawrence Associates, from the Packaging Careers Council of Canada's free *Careers in a Package* classroom kit; **511 left** Dick Hemingway, **right** Philip Norton/Valan Photos; **512 top** Kennon Cooke/Valan Photos, **bottom** Darryl Dahmer/Toronto Sun/Canapress; **513 top** Philip Norton/Valan Photos, **bottom** John Curtis/First Light; **515 right** Dick Hemingway; **520 top & bottom** Alberta Fire Training School, Vermilion, Alberta; **522** Nova Scotia Pulp Limited/Nova Scotia Information Service; **524 left** John Hopkins Medical Center/Photo Researchers, **right** Chris Priest/Science Photo Library/Photo Researchers; **526 left** Frank Siteman/Tony Stone Images, **right** Stephen Krasemann/Tony Stone Images; **529 right** Alfred Pasieka/Science Photo Library/Photo Researchers. Syringes used on pp. 448, 459, 460, 474 & 475 provided courtesy of S17 Science Supplies and Services Co. Ltd, 57 Glen Cameron Road, Thornhill, Ont., L3T 1P3 (905-709-2033). Other scientific supplies provided courtesy of Exclusive Educational Products, 243 Saunders Road, Barrie, Ont., L4N 9A3 (705-725-1166), **543 stamp** © Canada Post Corporation 1992. Reproduced with permission, **all others** Artbase Inc.; **Back Cover Unit 1** Johnny Johnson/Valan Photos, **Unit 2** 20th Century Fox, The Kobal Collection, **Unit 3** Canadian Tourism Commission, **Unit 4** Guido Alberto Rossi/Image Bank Canada, **Unit 5** Kjell B. Sandved/Visuals Unlimited.

Text Credits

104-105 Adapted from Bottle Biology Projects, © 1991, Department of Plant Pathology, University of Wisconsin, 1630 Linden Dr., Madison, WI, 53706; **383 left** The Canadian Press; **509** Profile of Marina Kovrig adapted from the Packaging Careers Council of Canada's free *Careers in a Package* classroom kit. For further information: www.packagingcareers.org or info@packagingcareers.org; **528-529** Adapted from *By Design: Technology, Exploration, and Integration*, Metropolitan Toronto School Board, published by Trifolium Books Inc.

Illustration Credits

76 top Precision Graphics from *Science Interactions 1* by Bill Aldridge, © 1998 Glencoe/McGraw-Hill; **79 bottom, 82 top** Laurie O'Keefe from *Life Science* by Lucy Daniel © 1997 Glencoe/McGraw-Hill; **100 bottom** Rob Romanek; **182** Jim Jobst, from *Physical Science* by Charles McLaughlin © 1997 Glencoe/McGraw-Hill; **209** Morgan-Cain and Associates from *Physical Science* by Charles McLaughlin © 1997 Glencoe/McGraw-Hill; **257 right, 258 top** Jim Shough from *Physical Science* by Charles W. McLaughlin © 1997 Glencoe/McGraw-Hill; **296 bottom** Morgan-Cain and Associates from *Physical Science* by Charles W. McLaughlin © 1997 Glencoe/McGraw-Hill; **315 top** John Edwards from *Earth Science* by Ralph Feather Jr. © 1997 Glencoe/McGraw-Hill; **324** Leon Bishop from *Earth Science* by Ralph Feather Jr. © 1997 Glencoe/McGraw-Hill; **331** From *Earth Science* by Ralph Feather Jr. © 1997 Glencoe/McGraw-Hill; **345 top, 346** John Edwards from *Earth Science* by Ralph Feather Jr. © 1997 Glencoe/McGraw-Hill; **357 bottom** John Edwards from *Earth Science* by Ralph Feather Jr. © 1997 Glencoe/McGraw-Hill; **361** From *Earth Science* by Ralph Feather Jr. © 1997 Glencoe/McGraw-Hill; **364** Dartmouth Publishing Inc., from *Earth Science* by Ralph Feather Jr. © 1997 Glencoe/McGraw-Hill; **371 bottom** Edwin Huff from *Earth Science* by Ralph Feather Jr. © 1997 Glencoe/McGraw-Hill; **375 top** John Edwards from *Earth Science* by Ralph Feather Jr. © 1997 Glencoe/McGraw-Hill, **bottom** Edwin Huff from *Earth Science* by Ralph Feather Jr. © 1997 Glencoe/McGraw-Hill; **385** Tom Kennedy from *Earth Science* by Ralph Feather Jr. © 1997 Glencoe/McGraw-Hill; **389, 405 top** Felipe Passalacqua/Worldwide Biomedic Images, from *Science Interactions 4*, by Robert W. Avakian © 1996 Glencoe/McGraw Hill; **393 left** David Ashby from *Life Science* by Lucy Daniel © 1999 Glencoe/McGraw-Hill, **right** From Mader, *Inquiry into Life, 8th edition* © The McGraw-Hill Companies Inc.; **401 bottom** John Edwards from *Earth Science* by Ralph Feather Jr. © 1997 Glencoe/McGraw-Hill; **419, 443, 445 centre** Morgan-Cain and Associates from *Physical Science* by Charles W. McLaughlin © 1997 Glencoe/McGraw-Hill; **420 top, 444** Thomas Gagliano from *Physical Science* by Charles W. McLaughlin © 1997 Glencoe/McGraw-Hill; **461 bottom** Henry Hill/John Edwards from *Physical Science* by Charles W. McLaughlin © 1997 Glencoe/McGraw-Hill; **469** Adapted from Boeing manual; **531** From *Earth Science* by Ralph Feather Jr. © 1997 Glencoe/McGraw-Hill.